# LAKE ECOSYSTEM ECOLOGY: A GLOBAL PERSPECTIVE

A DERIVATIVE OF ENCYCLOPEDIA OF INLAND WATERS

# LAKE ECOSYSTEM ECOLOGY: A GLOBAL PERSPECTIVE

## A DERIVATIVE OF ENCYCLOPEDIA OF INLAND WATERS

EDITOR

PROFESSOR GENE E. LIKENS
Cary Institute of Ecosystem Studies
Millbrook, NY, USA

**ELSEVIER**

AMSTERDAM • BOSTON • HEIDELBERG • LONDON • NEW YORK • OXFORD
PARIS • SAN DIEGO • SAN FRANCISCO • SINGAPORE • SYDNEY • TOKYO
Academic Press is an imprint of Elsevier

ACADEMIC
PRESS

Academic Press is an imprint of Elsevier
525 B Street, Suite 1900, San Diego, CA 92101-4495, USA
30 Corporate Drive, Suite 400, Burlington, MA 01803, USA
32 Jamestown Road, London, NW1 7BY, UK
Radarweg 29, PO Box 211, 1000 AE Amsterdam, The Netherlands

**British Library Cataloguing in Publication Data**
A catalogue record for this book is available from the British Library

**Library of Congress Catalog Number: Applied**

ISBN: 9780123820020

For information on all Academic Press publications
visit our website at elsevierdirect.com

# EDITOR

Professor Gene E. Likens is an ecologist best known for his discovery, with colleagues, of acid rain in North America, for co-founding the internationally renowned Hubbard Brook Ecosystem Study, and for founding the Institute of Ecosystem Studies, a leading international ecological research and education center. Professor Likens is an educator and advisor at state, national, and international levels. He has been an advisor to two governors in New York State and one in New Hampshire, as well as one U.S. President. He holds faculty positions at Yale, Cornell, Rutgers Universities, State University of New York at Albany, and the University of Connecticut, and has been awarded nine Honorary doctoral Degrees. In addition to being elected a member of the prestigious National Academy of Sciences and the American Philosophical Society, Dr. Likens has been elected to membership in the American Academy of Arts and Sciences, the Royal Swedish Academy of Sciences, Royal Danish Academy of Sciences and Letters, Austrian Academy of Sciences, and an Honorary Member of the British Ecological Society. In June 2002, Professor Likens was awarded the 2001 National Medal of Science, presented at The White House by President G. W. Bush; and in 2003 he was awarded the Blue Planet Prize (with F. H. Bormann) from the Asahi Glass Foundation. Among other awards, in 1993 Professor Likens, with F. H. Bormann, was awarded the Tyler Prize, The World Prize for Environmental Achievement, and in 1994, he was the sole recipient of the Australia Prize for Science and Technology. In 2004, Professor Likens was honored to be in Melbourne, Australia with a Miegunyah Fellowship. He was awarded the first G. E. Hutchinson Medal for excellence in research from The American Society of Limnology and Oceanography in 1982, and the Naumann-Thienemann Medal from the Societas Internationalis Limnologiae, and the Ecological Society of America's Eminent Ecologist Award in 1995. Professor Likens recently stepped down as President of the International Association of Theoretical and Applied Limnology, and is also a past president of the American Institute of Biological Sciences, the Ecological Society of America, and the American Society of Limnology and Oceanography. He is the author, co-author or editor of 20 books and more than 500 scientific papers.

Professor Likens is currently in Australia on a Commonwealth Environment Research Facilities (CERF) Fellowship at the Australian National University.

# CONTRIBUTORS

**R Abell**
WWF-United States, Washington, DC, USA

**J P Antenucci**
University of Western Australia, Nedlands, WA, Australia

**S Blanch**
WWF-Australia, Darwin, NT, Australia

**L Boegman**
Queen's University, Kingston, ON, Canada

**B Boehrer**
UFZ – Helmholtz Centre for Environmental Research, Magdeburg, Germany

**D K Branstrator**
University of Minnesota Duluth, Duluth, MN, USA

**J D Brookes**
The University of Adelaide, SA, Australia

**C Brönmark**
Lund University, Lund, Sweden

**A C Cardoso**
Institute for Environment and Sustainability, Ispra, VA, Italy

**J A Downing**
Iowa State University, Ames, IA, USA

**C M Duarte**
IMEDEA (CSIC-UIB), Esporles, Islas Baleares, Spain

**N Elkhiati**
Hassan II University, Casablanca, Morocco

**R J Flower**
University College, London, UK

**C M Foreman**
Montana State University, Bozeman, MT, USA

**G Free**
Institute for Environment and Sustainability, Ispra, VA, Italy

**U Gaedke**
University of Potsdam, Potsdam, Germany

**D Ghosh**
Jawaharlal Nehru University, New Delhi, India

**B Gopal**
Jawaharlal Nehru University, New Delhi, India

**D P Hamilton**
University of Waikato, Hamilton, New Zealand

**L-A Hansson**
Lund University, Lund, Sweden

**L Håkanson**
Uppsala University, Uppsala, Sweden

**V Istvánovics**
Budapest University of Technology and Economics, Budapest, Hungary

**H S Jensen**
University of Southern Denmark, Odense, Denmark

**E Jeppesen**
University of Aarhus, Denmark

**Ø Kaste**
Norwegian Institute for Water Research, Grimstad, Norway

**R Kipfer**
Swiss Federal Institute of Environmental Science and Technology (Eawag), Swiss Federal Institute of Technology (ETH), Ueberlandstr, Duebendorf, Switzerland

**G W Kling**
University of Michigan, Ann Arbor, MI, USA

**W M Lewis**
University of Colorado, Boulder, CO, USA

**G E Likens**
Cary Institute of Ecosystem Studies, Millbrook, NY, USA

**K E Limburg**
State University of New York, Syracuse, NY, USA

**M E Llames**
Instituto Tecnológico de Chascomús (CONICET – UNSAM), Buenos Aires, Argentina

**D M Lodge**
University of Notre Dame, Notre Dame, IN, USA

**A Lorke**
University of Koblenz-Landan, Landan/Pfaly, Germany

**A Lyche Solheim**
Norwegian Institute for Water Research, Oslo, Norway

**S MacIntyre**
University of California, Santa Barbara, CA, USA

**M Meerhoff**
University of Aarhus, Denmark
Universidad de la República, Uruguay

**J M Melack**
University of California, Santa Barbara, CA, USA

**O V Msiska**
Mzuzu University, Luwinga, Mzuzu, Malawi

**C Nilsson**
Umeå University, Umeå, Sweden

**P Nõges**
Institute for Environment and Sustainability, Ispra, VA, Italy

**F Peeters**
Universität Konstanz, Mainaustrasse, Konstanz, Germany

**J A Peters**
University of Notre Dame, Notre Dame, IN, USA

**S Poikane**
Institute for Environment and Sustainability, Ispra, VA, Italy

**J C Priscu**
Montana State University, Bozeman, MT, USA

**R Psenner**
University of Innsbruck, Innsbruck, Austria

**M Ramdani**
Mohammed V University, Rabat, Morocco

**W H Renwick**
Miami University, Oxford, OH, USA

**V H Resh**
University of California, Berkeley, CA, USA

**C Revenga**
The Nature Conservancy, Arlington, VA, USA

**C S Reynolds**
Centre of Ecology and Hydrology and Freshwater Biological Association, Cumbria, UK

**R D Robarts**
UNEP GEMS/Water Program, Saskatoon, SK, Canada

**F J Rueda**
Universidad de Granada, Granada, Spain

**M Schultze**
UFZ – Helmholtz Centre for Environmental Research, Magdeburg, Germany

**J P Smol**
Queen's University, Kingston, ON, Canada

**M Søndergaard**
University of Aarhus, Denmark

**K M Stewart**
State University of New York, Buffalo, NY, USA

**D L Strayer**
Cary Institute of Ecosystem Studies, Millbrook, NY, USA

**M Thieme**
WWF-United States, Washington, DC, USA

**B Timms**
University of Newcastle, Callaghan, NSW, Australia

**L J Tranvik**
Uppsala University, Uppsala, Sweden

**A-M Ventälä**
Pyhäjärvi Institute, Kauttua, Finland

**J E Vermaat**
Institute for Environmental Studies, Vrije Universiteit Amsterdam, Amsterdam, theNetherlands

**J Vidal**
Universidad de Granada, Granada, Spain

**W F Vincent**
Laval University, Quebec City, QC, Canada

**M J Waiser**
Environment Canada, Saskatoon, SK, Canada

**K F Walker**
The University of Adelaide, SA, Australia

**G A Weyhenmeyer**
Swedish University of Agricultural Sciences,
Uppsala, Sweden

**A T Wolf**
Oregon State University, Corvallis, OR, USA

**A Wüest**
Eawag, Surface Waters – Research and
Management, Kastanienbaum, Switzerland

**H E Zagarese**
Instituto Tecnológico de Chascomús (CONICET –
UNSAM), Buenos Aires, Argentina

# CONTENTS

## LAKE ECOSYSTEMS: STRUCTURE, FUNCTION AND CHANGE

## HYDRODYNAMICS AND MIXING IN LAKES

# LAKES AND RESERVOIRS OF THE WORLD

# LAKES AND RESERVOIRS: POLLUTION, MANAGEMENT AND SERVICES

# INTRODUCTION TO LAKE ECOSYSTEM ECOLOGY: A GLOBAL PERSPECTIVE

The scientific discipline of limnology is focused on the study of inland waters, fresh and saline, large and small, and young and old. For example, the 5 Laurentian Great Lakes in North America contain some 20% of the Earth's surface fresh water, and Lake Baikal in Siberia contains another 18% (Likens, 2009). The recently constructed Three Gorges Dam Reservoir in China, to be completely filled by the end of 2009, will contain about 40 $km^3$, or about 8% of what Laurentian Lake Erie (484 $km^3$) contains. Lake Baikal is the deepest and one of the oldest freshwater basins (>20 Mya); Three Gorges Reservoir is one of the newest major basins.

Whether large or small, young or old, shallow or deep, these water bodies function as ecosystems with strong and complicated interactions among all of the living and nonliving components of the ecosystem. Ecosystems are the basic units of nature (Tansley, 1935). In terms of area of water (<1%) or volume of fresh water (<0.3%) on our planet, inland lakes and reservoirs of the world are tiny, but in terms of interest and use for drinking, sanitation services, food, recreation, transportation and aesthetics, they provide hugely important resources and ecosystem services to humans.

Limnology is a quintessential form of ecosystem science because comprehensive limnological understanding is based on information from biology, physics, chemistry, geology, hydrology, meteorology and so forth. The recent Encyclopedia of Inland Waters (Likens, 2009) incorporates all of these various disciplines and more in addressing and describing the inland-water ecosystems of the Earth.

Limnology is a relatively old scientific discipline with a professional international society (the International Society of Limnology (SIL)) founded in 1922. From the beginning, SIL has focused on both theoretical and applied issues in aquatic ecosystems as do the topics of this volume.

This volume focuses on 5 topics about the lentic, inland waters of our planet, that is, lake and reservoir ecosystems: 1. Introduction to Lake Ecosystem Ecology: A Global Perspective, 2. Lake Ecosystems: Structure, Function and Change, 3. Hydrodynamics and Mixing in Lakes, 4. Lakes and Reservoirs of the World, and 5. Lakes and Reservoirs: Pollution, Management and Services. The information and perspectives contained in this volume are highly relevant to the understanding and management of a variety of current environmental problems, such as eutrophication, acid rain and climate change.

The articles in this volume are reproduced from the Encyclopedia of Inland Waters (Likens, 2009). I would like to acknowledge and thank the authors of the articles in this volume for their excellent and up-to-date coverage of these important topics in limnology.

Gene E. Likens
Cary Institute of Ecosystem Studies
Millbrook, NY
*December 2009*

## References

Likens GE (2009) *Chapter 1 in Encyclopedia of Inland Waters.*
Likens GE (2009) *Encyclopedia of Inland Waters.* Elsevier.
Tansley AG (1935) The use and abuse of vegetational concepts and terms. *Ecology* 16: 284–307.

# LAKE ECOSYSTEMS: STRUCTURE, FUNCTION AND CHANGE

Contents

## Lakes as Ecosystems

**W M Lewis,** University of Colorado, Boulder, CO, USA

### Introduction of the Ecosystem Concept

Scientific studies of lakes began as early as the seventeenth century, but at first were descriptive rather than analytical. Toward the end of the nineteenth century, measurements and observations on lakes became more directed. For example, the thermal layering of lakes was attributed to specific physical causes, and such phenomena as the movement of plankton in the water column were the subject of hypotheses that were tested with specific kinds of data.

Comprehensive studies of lakes began with the work of Alphonse Forel (1841–1912) on Lac Léman (Lake Geneva), Switzerland, as well as other Swiss lakes. In a three-volume monograph (*Le Léman*: 1892, 1895, 1904), Forel presented data on a wide variety of subjects including sediments and bottom-dwelling organisms, fishes and fisheries, water movement, transparency and color, temperature, and others. Thus Forel, who introduced the term 'limnology' (originally the study of lakes, but later expanded to include other inland waters), demonstrated the holistic approach for understanding a lake as an environmental entity, but without application of an explicit ecosystem concept.

The conceptual basis for studying lakes as ecosystems was first clearly given by Stephen Forbes (1844–1930; **Figure 1**) through a short essay, *The Lake as a Microcosm* (1887). Forbes was professor of biology at the University of Illinois in Champaign, IL, USA, and director of the Illinois Natural History Laboratory (subsequently the Illinois Natural History Survey), which was charged with describing and analyzing the flora and fauna of Illinois. Forbes realized that it was not possible to achieve a full understanding of a lake or, by implication, of any other environmental system such as a stream or forest, simply from knowledge of the resident species. Forbes proposed that the species in a particular environment, when interacting with each other and with the nonliving components of the environment, show collective (system) properties. The microcosm that Forbes described today would be called an ecosystem, although this term did not come into use until 48 years later through the work of the British botanist A. G. Tansley. By current usage, an ecosystem is any portion of the Earth's surface that shows strong and constant interactions among its resident organisms and between these organisms and the abotic environment.

Forbes not only described accurately the modern ecosystem concept, but also identified ways in which critical properties of ecosystems could be measured and analyzed. He named four common properties of lakes, each of which provides a cornerstone for the study of lakes and other ecosystems, as shown in **Table 1**. Although Forbes's concepts have been renamed, they are easily visible in the modern study of lakes as ecosystems.

Although the essay by Forbes now is considered a classic in limnology and in ecology generally, it caused no immediate change in the practices of limnologists or ecologists. Like many important discoveries in science, it was a seed that required considerable time to germinate.

**Figure 1**   Stephen A. Forbes. Reproduced from the Illinois Natural History Survey website, with permission from the Illinois Natural History Survey.

In studies of Cedar Bog Lake, Minnesota, for his Ph.D. at the University of Minnesota, Raymond Lindeman (1915–1942; **Figure 2**) gave limnologists the clearest early modern example of the study of lakes as ecosystems. Rather than focusing on a particular type of organism or group of organisms, which would have been quite typical for his era, Lindeman decided that he would attempt to analyze all of the feeding relationships ('trophic' relationships) among organisms in Cedar Bog Lake. Thus, his Ph.D. work extended from algae and aquatic vascular plants to herbivorous invertebrates, and then to carnivores, and conceptually even to bacteria, although there were no methods for quantifying bacterial abundance at that time. Lindeman's descriptions of feeding relationships were voluminous but straightforward to write up and publish, but he sought some more general conclusions for which he needed a new concept.

Lindeman took a postdoctoral position with G. Evelyn Hutchinson at Yale University in 1942. Hutchinson had become a limnologist of note through his quantitatively oriented studies of plankton and biogeochemical processes in the small kettle lakes near New Haven. He would in subsequent

**Table 1**   Four key properties of ecosystems identified by S. A. Forbes, along with their modern nomenclatural counterparts and some examples

| Forbes concept | Modern nomenclature | Modern studies |
|---|---|---|
| Web of interactions | Food-web dynamics | Food-web complexity and efficiency |
| Building up and breaking down of organic matter | Ecosystem metabolism | Total ecosystem photosynthesis and respiration |
| Circulation of matter | Biogeochemistry | Dynamics and cycling of carbon, nitrogen, and phosphorus |
| Distribution of organisms along gradients | Community organization | Vertical and horizontal patterns in the distribution of fishes, invertebrates, and algae |

**Figure 2**   R. L. Lindeman and his famous study site, Cedar Bog Lake, MN. Reproduced from People of Cedar Creek website, with permission from Cedar Creek Natural History Area.

decades become the world's most influential limnologist, and part of his reputation grew out of his contributions to the field of biogeochemistry, an important tool of ecosystem science.

Hutchinson made suggestions that no doubt were critical to Lindeman's groundbreaking paper, *The Trophic Dynamic Aspect of Ecology* (1942, published in the journal *Ecology*), which now is recognized as a landmark in limnology and in ecology generally. Lindeman proposed a way of converting the tremendous mass of highly specific information for Cedar Bog Lake into a format that would allow comparisons with any other lake or even with other kinds of ecosystems. Building on the work of the German limnologist August Thienemann and the British ecologist Charles Elton, Lindeman organized the feeding relationships as a feeding hierarchy within which each kind of organism was assigned to a specific feeding level (trophic level). He then proposed that the feeding relationship represented by any given link in the food web be quantified as an energy flow. Thus, the total energy flow from level 1 (plants) to level 2 (herbivores) could be quantified as the summation of energy flows between all pairs of plants and herbivores; a similar estimate could be made for all other pairs of trophic levels. In this way, the flow of energy across the levels of the food web could be expressed in quantitative terms.

The first important conclusion from Lindeman's energy-based approach was that each transfer of energy between trophic levels is governed by the second law of thermodynamics, which requires that significant energy loss must occur each time energy is transferred. Thus, Lindeman demonstrated why food webs have relatively few trophic levels: progressive dissipation of energy as it passes through the food web from the bottom (plants) to the upper levels (upper-level carnivores) ultimately provides insufficient energy for expansion of the food web to further levels. Also, analysis of a food web in this way sets the stage for calculating efficiencies of energy transfer, comparison of efficiencies across different ecosystem types, and the identification and analysis of bottlenecks restricting the flow of energy within the food web.

The contributions of G. Evelyn Hutchinson in the 1940s on the biogeochemistry of carbon in lakes also must be counted as landmarks in the development of ecosystem science. Even so, ecosystem science was scarcely represented in the research agenda of ecologists or in the academic curriculum as late as 1950. Penetration of the ecosystem thinking into research, teaching, and public awareness occurred first through the publication of a textbook, *Fundamentals of Ecology* (1953), written by Eugene P. Odum, and especially through the second edition of the same book (1959), written by E. P. Odum and Howard T. Odum. The Odums visualized ecology as best viewed from the top down, with ecosystems as a point of departure and studies of ecosystem components as infrastructure for the understanding of ecosystems.

The ecosystem perspective has not displaced the more specialized branches of ecology that deal with particular kinds of organisms or specific kinds of physical phenomena, such as studies of water movement, optics, or heat exchange. Rather, ecosystem science has had a unifying effect on studies of ecosystem components (**Figure 3**). Study of a specific ecosystem component produces not only a better understanding of that component, but also a better understanding of the ecosystem, which is a final objective for the science of an ecosystem type, such as lakes.

## Metabolism in Lakes

The dominant anabolic component of metabolism in lakes is photosynthesis based on carbon dioxide and water plus solar irradiance as an energy source. The dominant catabolic component is aerobic respiration based on the oxidation of organic matter with oxygen as an electron acceptor. Conversion of solar energy to stored chemical energy in the form of biomass seldom reaches 1% efficiency in lakes because of losses inherent in the wavelengths that can be intercepted by photosynthetic pigments, inefficiency in the interception process, and thermodynamic losses in the conversions leading to the production of biomass. Respiration also involves thermodynamic losses. Therefore, the solar energy source greatly exceeds photosynthetic output, and cellular capture of the energy released from organic matter by respiration greatly exceeds the energy stored in the organic matter.

Aerobic photosynthesis is characteristic of the aquatic vascular plants, attached algae, and phytoplankton of lakes and is universal wherever light and oxygen are present. Aerobic respiration is characteristic of aerobic autotrophs as well as consumers and most bacteria; it occurs wherever oxygen is present (**Table 2**). Other categories of metabolism occur under either of two more restrictive conditions: (1) where light penetrates into an anaerobic zone, and (2) where there is an interface or mixing between oxidizing and reducing conditions, as is common near a sediment–water interface.

A study of whole-ecosystem metabolism would require consideration of all of the metabolic categories listed in **Table 2**. The dominance of aerobic photosynthesis and aerobic respiration, however, often allows ecosystem studies to focus on these two metabolic components. In open water, photosynthesis

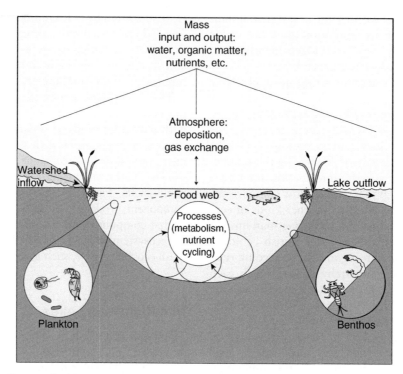

**Figure 3**   Ecosystem diagram of a lake.

**Table 2**   Summary of metabolic processes in lake ecosystems

| Metabolic process | Energy source | Capabilities | | | Occurrence conditions in lakes |
|---|---|---|---|---|---|
| | | Eukaryotic plants | Cyanobacteria | Other bacteria | |
| Aerobic photosynthesis | Solar | Yes | Yes | No | Universal, photic, oxic |
| Anaerobic photosynthesis | Solar | No | No | Specialists | Occasional, photic, anoxic |
| Aerobic chemosynthesis | Inorganic oxidation | No | No | Specialists | Common, oxic/anoxic interfaces |
| Aerobic respiration | Organic oxidation | Yes | Yes | Yes | Universal, oxic |
| Anaerobic respiration | Inorganic or organic reduction | No | No | Yes | Common, oxic/anoxic interfaces |

and respiration are often measured with a vertical series of paired transparent and darkened incubation bottles filled with lake water; rate of decline of oxygen in the dark bottles indicates respiration rate, and rate increase of oxygen in the transparent bottle indicates net photosynthesis. In unproductive lakes, uptake of $^{14}$C-labeled $CO_2$ can be used as an even more sensitive indicator of photosynthesis. Where macrophytes or attached algae are important, as is often the case in small lakes, separate measurements must be made of their photosynthesis, typically by the use of enclosures. The metabolic rates of microbes in deepwater sediments can be inferred from the rate of

oxygen loss from the hypolimnion of a stratified lake, or can be measured with enclosures.

Annual rate of photosynthesis and respiration per unit area (typically given as mg C m$^{-2}$ year$^{-1}$) is the most commonly used metabolic statistic for lakes. A complete metabolic accounting would also include processing of organic matter entering a lake from the surrounding terrestrial environment, primarily through stream flow.

Total annual net production of autotrophs (production in excess of respiration) is a measure of the capacity for a lake to generate biomass at higher trophic levels (**Figure 4**).

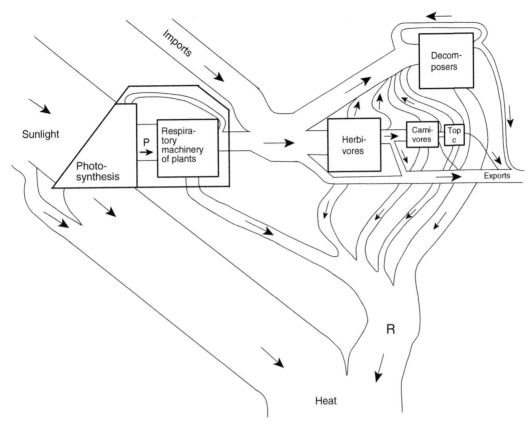

**Figure 4** Example of an early metabolic (energy-flow) diagram for an aquatic ecosystem. Reproduced from Odum H T (1956) *Limnology and Oceanography* 1: 102–118, with permission from the American Society of Limnology and Oceanography.

Also, the time-course of production, which shows seasonal and nonseasonal variation, is often a useful point of departure for the analysis of mechanisms that control biotic functions in a lake. Factors that may suppress production include deep mixing of the water column, exhaustion of key nutrients, grazing, and hydraulic removal of biomass.

Photosynthesis and respiration often respond differently to a specific physical or chemical change.

## Food-Web Analysis

Modern food-web analysis follows the example given by Lindeman. The sophistication of the analysis is much greater today than it was in 1942, however, and the uses of food-web analysis have diversified as well.

At the base of the food web is organic matter generated within a lake or coming to a lake from its watershed and atmosphere above (**Figure 5**). All of these sources of organic matter can be quantified, as explained in the preceding section, but require the application of several methods and must take into account both spatial and temporal variability in the synthesis or transport of organic matter.

Quantification of linkages in the food web begins with feeding relationships between autotrophs (primary producers) and herbivores (primary consumers). Analysis of gut contents is one method for establishing the linkage between a specific kind of consumer and one or more plant foods. A linkage drawn in this way has two disadvantages, however: (1) it is not quantitative because food items often cannot be identified fully, and (2) it is subject to errors of interpretation caused by the ingestion of foods that are not assimilated or only partially assimilated through the gut wall. The first problem can be overcome either by the use of feeding experiments or by the quantification of growth rate of the consumer, with some empirically-based assumptions about the growth efficiency of the consumer. The second problem can be resolved by experimental use of tissue labels (typically isotopes) or, more efficiently, by the use of stable isotopes as passive tracers (i.e., relying on the natural abundance of stable isotopes to infer food sources). Because primary producers of different categories may differ substantially in their concentration of the stable carbon isotope $^{13}C$, analysis of $^{13}C$ content of protoplasm from the consumer may allow a quantitative estimation of the relative

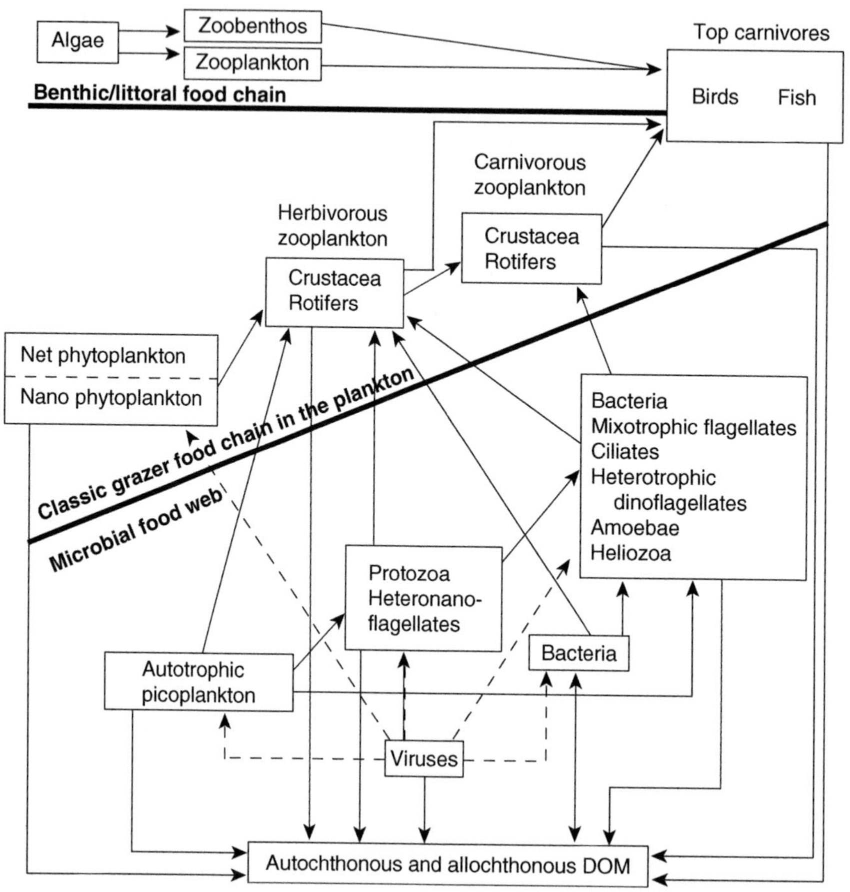

**Figure 5**  A modern diagrammatic view of a lacustrine food web, including both microbial and nonmicrobial components (from Weisse and Stockner, 1993 as modified by Kalff, 2002; with permission).

importance of several possible foods contributing to the synthesis of biomass by the consumer.

Above the level of primary consumers are carnivores, which may be secondary, tertiary, or even quaternary consumers, depending on their food source. Measurements of growth rate, along with gut-content analysis and use of passive or active tracers, can be applied to carnivores just as they are to herbivores. Within the carnivore trophic levels, however, the assignment of consumers to a specific trophic level may be difficult because carnivores often consume foods belonging to more than one trophic level. The stable nitrogen isotope $^{15}$N is useful in assigning a species to a fractional position on the food web because $^{15}$N shows increasing tissue enrichment from one trophic level to the next.

When completed, a food-web analysis shows the efficiency of energy transfer from one level to the next. Because of the thermodynamic limits on efficiency and the typical observed efficiency, which are a matter of record, a food-web analysis based on energy shows whether particular linkages are unusually weak or strong by comparison with expectations.

Such observations in turn lead to hypotheses about mechanisms of control for energy transfer within the food web. For example, modern ecosystem theory includes the concept of 'trophic cascades' involving 'top-down' and 'bottom-up' effects on trophic dynamics, which is easily applicable to lakes. Change in one trophic level may be visible in other trophic level, in the manner of a cascade (**Figure 6**). Top-down effects pass from any trophic level to the next trophic level below. For example, an unusual abundance of algal biomass in a lake could be traced to unusually efficient removal of herbivores through predation at such a rate as to leave algae mostly uneaten (a top-down effect).

Similarly, bottom-up control passes from any trophic level to the next level above. The inability of an oligotrophic (nutrient-poor) lake to grow substantial amounts of plant biomass, e.g., exerts a bottom-up effect on all higher trophic levels by restricting the amount of energy that is available at the base of the food chain.

Trophic-dynamic analysis also leads to other fundamental questions involving the structure of biotic communities. For example, food webs might or might

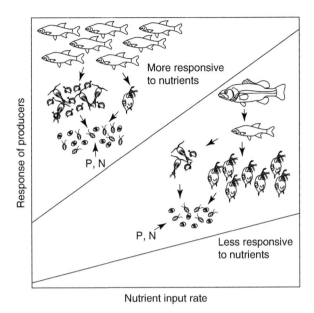

**Figure 6** Contrasts in nutrient response for lakes that have strong control by top carnivores (lower line) and lakes that have weak control by top carnivores (upper line). The top carnivore exercises top-down control by suppressing smaller fish that eat large grazing zooplankton. Survival of more large zooplankton holds phytoplankton in check, thus weakening response to plant nutrients (P, N). Reproduced from Carpenter S R (2003) *Regime Shifts in Lake Ecosystems*, with permission from the International Ecology Institute, Oldendorf/Luhe, Germany.

| | Inorganic | Organic |
|---|---|---|
| Dissolved | $CO_2$, $H_2CO_3$, $HCO_3^-$ | Humic and fulvic acids from soils<br><br>Protein, carbohydrate, and nucleic acid breakdown products from organisms |
| Particulate | $CaCO_3$ | Nonliving debris, organisms |

**Figure 7** Compartment matrix for analysis of carbon cycling in the water column of a lake.

not be more productive or more efficient with a high diversity of herbivores than with strong dominance from a very successful, specific type of herbivore. Such questions bear not only on the analysis of natural ecosystems, but also on ecosystem management.

## Biogeochemistry

Many of the functional attributes of lake ecosystems can be analyzed through biogeochemical studies. While metabolism and trophic dynamics are viewed in terms of energy flux, biogeochemistry typically is viewed in terms of mass flux. As with energy analysis, the foundation is basic physics: mass is neither created nor destroyed (under conditions that are compatible with the presence of life). The conservation of mass leads to the mass-balance equation, which can be given as follows for an ecosystem: $I - O = \Delta S$, where $I$ is input of mass of a particular type (carbon or phosphorus, for example), $O$ is output of mass, and $\Delta S$ is change in storage of mass. Input and output pathways for lakes are either hydrologic (flow) or atmospheric (deposition or gas exchange). Over the short term, change in storage may involve changes in concentration of mass in the water column, but over the long term, change in storage mostly reflects accumulation

of mass in sediments. Any element can be a target for biogeochemical studies, but the most frequently studied biogeochemical processes in aquatic ecosystems involve carbon, phosphorus, and nitrogen.

The mass-balance equation applies not only to the entire ecosystem, but also to compartments within the ecosystem (**Figure 7**). For example, carbon in the water column could be partitioned into compartments, each of which would have its own mass-balance equation. Some of these compartments would have a high turnover rate, while others would not. The dynamics of compartments explain restrictions on specific biological processes such as photosynthesis, and account for differences in individual lakes or categories of lakes, such as those that differ greatly in nutrient supply.

Studies of the carbon cycle are ideal complements to studies of lake metabolism and food webs. $CO_2$, which is the feedstock for photosynthesis, enters aquatic autotrophs either in the form of free $CO_2$ ($CO_2 + H_2CO_3$) or bicarbonate ($HCO_3^-$). $CO_2$ is converted to organic matter by the reduction process of photosynthesis. Either within an autotroph or through consumption of autotroph or other biomass by consumers or decomposers, reduced carbon is reconverted to its oxidized form as $CO_2$ through aerobic respiration.

Organic carbon resides not only within organisms, but also in water or sediments in dissolved or nonliving particulate form. Dissolved organic carbon derives from the watershed, atmosphere, or organisms within the lake. Watershed contributions to dissolved organic carbon in lakes are composed to a large extent of humic and fulvic acids, which are the byproduct of the degradation of organic matter within soils. Humic and fulvic acids are refractory (resistant to breakdown), although they are slowly decomposed by microbes and can be broken down by ultraviolet light in a water column. Humic and fulvic acids are generally present in concentrations of $1-10\,mg\,l^{-1}$ in lakes and, when present at concentrations above $5\,mg\,l^{-1}$, generally impart a brown or orange color to the water.

Organic compounds that are released to the water column of lakes by the resident organisms vary greatly in composition. All soluble organic compounds present in organisms can be found in the water column at measurable concentrations. Thus, a careful analysis of lake water would show a wide variety of amino acids, carbohydrates of varying complexity, and other metabolites of organisms. These compounds enter the water column through leakage (excretion), death, or the production of fecal matter. As might be expected, the turnover rate for metabolites is generally high because most of these compounds are labile (easily used), in contrast to humic and fulvic acids.

Particulate organic matter present in a water column may be either living or nonliving. Frequently, these two compartments are physically joined, as all nonliving particulate organic matter in water is colonized by bacteria and, under some conditions, fungi. Most of the carbon attributable to living organisms is accounted for by phytoplankton and zooplankton; fish and bacteria make smaller contributions in the sense of mass but have important ecosystem effects. Carbon is continually stored in sediments, which accumulate at rates often about 1 mm per year, much of which is organic. As sediments become buried by additional sediments, their decomposition slows because of the lack of oxygen and chemically hostile conditions for active metabolism of bacteria. Thus, while some of the organic matter in sediments is remobilized into the water column, other organic matter in the sediments becomes long-term storage.

A study of mass flux and storage, when viewed in terms of ecosystem compartments and subcompartments, tells a story about the mechanisms by which a lake functions. For example, lakes may differ greatly in the terrestrial contribution to carbon processing and carbon storage, and also may differ greatly in speed of carbon turnover in specific living and nonliving compartments.

The cycling of phosphorus in lakes has been studied intensively, because phosphorus is one of the two elements (nitrogen is the other) most likely to be critically depleted by aquatic autotrophs. Thus, phosphorus at the ecosystem level is commonly viewed as an ecosystem regulator; lakes in which it is abundant (eutrophic lakes) have potential to produce high amounts of plant biomass (phytoplankton, attached algae, or aquatic vascular plants) and have water-quality characteristics that are often viewed as undesirable (low transparency, green color, potential for odor production, and severe loss of oxygen in deep water). In contrast, lakes having low supplies of phosphorus (oligotrophic lakes) may show constant suppression of plant growth through phosphorus scarcity. Such lakes have much lower amounts of carbon in the water, higher transparency, often appear blue, and retain oxygen in deep water consistent with the requirements of fish and other eukaryotes.

Because the phosphorus requirement of plants is only approximately 1% of dry biomass, scarcity of phosphorus can be offset by relatively modest increases in the phosphorus additions to a lake. Addition of 1 kg of phosphorus per unit volume or area, e.g., could easily generate 100 kg of dry mass or 500 kg of wet mass of autotrophs. Thus, mobilization of phosphorus by humans has the potential to change lakes and has done so throughout the world wherever human populations liberate phosphorus through waste disposal, agriculture, and disturbance of soil. For this reason, the study of trophic state (nutrient status) has received more attention than any other ecosystem feature of lakes. The practical application of modeling or analysis of trophic status in lakes arises through the desire to prevent changes in trophic state or to reverse changes in trophic state that have already occurred, which requires ecosystem-level understanding of the lake and particularly of its nutrient budgets.

Lakes may also be limited by low concentrations of the forms of inorganic nitrogen that are readily available to aquatic autotrophs (ammonium, nitrate). In this case, nitrogen limitation rather than phosphorus limitation may control plant growth. In fact, the two types of nutrient limitation may occur sequentially across seasons or across years in a single lake.

Nitrogen cycling in lakes is much more complicated than is phosphorus cycling because nitrogen has a gaseous atmospheric component that phosphorus lacks, and because nitrogen exists in seven stable oxidation states within a lake, which sets the stage for the use of nitrogen as a substrate for oxidation reduction reactions supporting microbial metabolism (**Table 2**). Like phosphorus, nitrogen enters lakes through the watershed and, in small amounts, with precipitation. Unlike phosphorus, nitrogen also enters lakes as a gas by diffusion at the air–water interface in the form of $N_2$. $N_2$ is so inert chemically that it cannot be used as a nitrogen source by most organisms. Certain prokaryotes have the ability to use $N_2$ by converting it to ammonium, which then can be used in organic synthesis, provided that a substantial energy supply is available. This process is called nitrogen fixation, in that it converts the gaseous nitrogen to a solid that is soluble in water (ammonium). The dominant nitrogen fixers in lakes are the cyanobacteria. Not all cyanobacteria fix nitrogen, but certain taxa that have a specialized nitrogen-fixing cell (heterocyst) grow commonly in lakes that show nitrogen depletion.

Nitrogen fixers escape the limitation of growth associated with nitrogen depletion, thereby gaining an advantage over other autotrophs. Thus, lakes that are enriched with phosphorus but showing nitrogen depletion often have nuisance growths of nitrogen-fixing cyanobacteria, which produce all of the expected symptoms of nutrient enrichment, and sometimes also produce toxins. There is currently much interest in predicting and preventing the development of circumstances that lead to large and persistent blooms of nitrogen-fixing cyanobacteria.

Nitrogen is a multiplier element, just as phosphorus is. It constitutes approximately 5% of dry mass in plants. Therefore, when it is added to lakes, provided that phosphorus is added at the same time, it supports a 20-fold multiplication of dry plant biomass.

A detailed documentation of the carbon, nitrogen, and phosphorus cycles for a lake produces an understanding of factors that regulate the metabolic rates of lakes and the accumulation of biomass of various kinds in lakes. These ecosystem phenomena have numerous practical applications, ranging from interest in biomass production (e.g., fish) to interest in constraining water quality within boundaries that are either natural or that favor human purposes.

## Community Organization

As foretold by Forbes, ecosystems have a strong degree of spatial organization involving the arrangement of organisms according to abiotic constraints based on factors such as amount of irradiance, concentration of dissolved oxygen, or substrate characteristics. The requirement for irradiance is especially important because it dictates the distribution and growth potential for aquatic autotrophs. Phytoplankton grow strongly only within the euphotic zone, which corresponds approximately to water receiving at least 1% of the solar irradiance available at the water surface. Thus, vigorously growing phytoplanktons often are confined to the upper part of the water column (epilimnion and sometimes metaliminon). The same principle applies to periphyton and rooted aquatic plants, which can occupy substrates only over portions of the sediment that are exposed to at least 1% surface irradiance. The gradient of irradiance is controlled by the transparency of the water.

Spatial organization of animal communities and microbial communities is dictated partly by physical conditions and partly by the distribution of autotrophs.

Crustacean zooplankton, e.g., may migrate large distances vertically in the water, hiding in deep water during the daytime but rising to feed within surface waters at night. The zooplankton of the littoral zone differs from that of open water, in that the littoral zone offers food types (periphyton especially) that are not available in open water. Distribution of fishes may be dictated in part by the presence of structure associated with littoral zone. The effect of structure may even be related to life history in that immature or small fish may seek the structure of the littoral zone for shelter from predation.

Because organisms conduct the business of an ecosystem, an understanding of their habitat requirements, reflected as spatial organization, leads to explanations of the total abundances of certain categories of organisms, the failure of certain organisms to fare well in one lake but not in another, and the consequences of habitat disturbances of natural or human origin.

## Synthetic Analysis of Ecosystems

Synthetic analysis of ecosystems often involves the statistical study of empirical information, particularly if the objective is to test a particular hypothesis. A quantitative overview of multiple ecosystem functions or of the intricate detail for an ecosystem component typically involves modeling as a supplement to other types of quantitative analysis. When computers first became available, it was anticipated that ecosystem science would ultimately be driven almost entirely by mechanistic models that would predict ecosystem responses to natural or anthropogenic conditions. These expectations were not realized. Ecosystems, like other complex systems such as climate or economics, are moderately predictable from modeling that combines a modest number of well-quantified variables addressed to a specific question, but typically are unpredictable on the basis of large number of variables addressing more general questions. Even so, modeling of numerous variables in an ecosystem context can be useful in setting limits on expected outcomes or showing possible outcomes of multiple interactions that cannot be easily discerned from the study of individual variables. Therefore, modeling is useful in promoting the understanding of ecosystems.

## Conclusion

Lakes first inspired the ecosystem concept, and have been a constant source of ideas about ecosystem structure and function. Ecosystem science as applied to lakes, is supported by and is consistent with, other kinds of ecological studies that are directed toward specific organisms, groups of organisms, or specific categories of abiotic phenomena in lakes. The strength of ecosystem science lies in its relevance to the understanding of all subordinate components in ecosystems, and to its often direct connection to human concerns in understanding and managing lakes.

## Glossary

**Biogeochemistry** – Scientific study of the mass flux of any element, compound, or group of compounds within or across the boundaries of an ecosystem or any other spatial component of the environment. Mass flux within an ecosystem is often designated as nutrient cycling or element cycling.

**Cycling** – A biogeochemical term referring to the movement of elements or compounds within an ecosystem or any other bounded environmental system.

**Ecosystem** – Any portion of the earth's surface that shows strong and constant interaction among its resident organisms and between those organisms and the abiotic environment.

**Lake trophic state** – Fertility of a lake, as measured either by its concentrations of key plant nutrients (especially phosphorus and nitrogen) or the annual production of plant biomass (aquatic vascular plants and algae). Trophic-state categories include oligotrophic (weakly nourished), mesotrophic (nutrition of intermediate status), and eutrophic (richly nourished).

**Mass flux** – Movement of mass per unit time, often in ecosystem studies expressed as mass/area/time.

**Stable isotope** – Any isotope of an element that does not decay spontaneously.

**Trophic dynamics** – Fluxes of energy or mass caused by feeding relationships, including rates of grazing by herbivores on plant matter, rates of predation by carnivores on other animals, or rates of decomposition of organic matter by microbes.

*See also:* Modeling of Lake Ecosystems; Trophic Dynamics in Aquatic Ecosystems.

## Further Reading

Belgrano A, Scharler UM, Dunne J, and Ulanowicz RE (2006) *Aquatic Food Webs: An Ecosystem Approach.* New York: Oxford University Presss.

Carpenter SR (2003) *Regime Shifts in Lake Ecosystems: Pattern and Variation.* Oldendorf/Luhe, Germany: International Ecology Institute.

Golley FB (1993) *A History of the Ecosystem Concept in Ecology: More than the Sum of the Parts.* New Haven/London: Yale University Press.

Kalff J (2002) *Limnology: Inland Water Ecosystems.* Upper Saddle River, NJ: Prentice Hall.

# Ecological Zonation in Lakes

**W M Lewis,** University of Colorado, Boulder, CO, USA

## Introduction

Lakes show many kinds of spatial variation in both vertical and horizontal dimensions. Variation can be chemical, physical, or biotic, and is important to the understanding of ecosystem functions. Although some types of variation are unique to specific classes of lakes, others are common to most lakes, and correspond to an obvious spatial organization of the biota in lakes. The existence of certain common types of spatial organization in lakes has led to the naming of specific zones that have distinctive ecological characteristics.

A complete list of all zones that have been named by limnologists would be lengthy and complex (Wetzel, 2001). Complex nomenclatures have been abandoned by modern limnologists, however. Modern limnology focuses on simple zonation systems that are easily applied by limnologists and others interested in lakes. **Table 1** gives a summary of zonation systems that are currently in broad use.

## Horizontal Zonation: Littoral and Pelagic

The nearshore area of a lake (littoral zone) differs from the offshore shore area (pelagic zone). The only group of autotrophs in the pelagic zone is the phytoplankton, which consists of very small algae that are suspended in the water column. The littoral zone also has phytoplankton (which move freely between littoral zone and pelagic zone), but also has two other categories of autotrophs (**Figure 1**): aquatic vascular plants (aquatic macrophytes), and films of attached algae (periphyton). Periphyton grow on the leaves of macrophytes and on other solid surfaces such as mud, sand, rocks, or wood.

The outer margin of the littoral zone, beyond which is the pelagic zone, is the point at which significant growth of macrophytes and periphyton becomes impossible because of darkness. This boundary corresponds approximately to the location at which the amount of solar irradiance reaching the bottom of the lake is <1% of surface irradiance. At bottom irradiances <1%, there is little or no net photosynthesis, which prevents growth of the attached autotrophs (macrophytes and periphyton) that are typical of the littoral zone.

For a lake of a given size and shape, the width of the littoral zone depends on transparency of the water as well as shoreline slope (**Figure 2**). In oligotrophic lakes, which have low nutrient concentrations and therefore develop very small amounts of the phytoplankton biomass that could shade the lower water column of lakes, the littoral zone extends to depths of 4–20 m or even more, depending on transparency of the water. In the eutrophic category, the depth of 1% irradiance ranges between 0.1 and about 2 m, and the mesotrophic category spans ~2–4 m. In the most extreme cases of eutrophication (lakes highly enriched with nutrients), where the depth of 1% light corresponds to only a few centimeters, the littoral zone as defined by light is virtually absent, and the littoral zone may be defined instead by the zone of influence for traveling waves and corresponding disturbance of the bottom (0.5–1.5 m). In general, small lakes have a higher percentage of surface area in the littoral zone than do large lakes (**Figure 2**), although some large, shallow lakes have large littoral zones (e.g., Lake Okeechobee, Florida).

Although littoral zones are most easily defined on the basis of macrophytes and attached algae, a littoral zone can also be distinguished from a pelagic zone by its distinctive heterotrophic communities and by its food-web structure. Because littoral zones provide shelter, whereas pelagic zones do not, littoral zones often support dense populations of organisms that thrive when protected from predation. Larval and juvenile fish, for example, seek shelter within the littoral zone from predation by larger fish. Large invertebrates, such as dragonfly larvae or crayfish, typically are most abundant in littoral zones, where they are least likely to be consumed by fish. The littoral zone also has invertebrate communities that specialize in the consumption of attached algae by nipping or scraping the algal coatings on macrophytes or other solid surfaces. In the pelagic zone, there is no food source comparable to the periphyton of a littoral zone. In general, the communities of a littoral zone are more diverse than those of the pelagic zone, and the key species of the two zones differ.

In the pelagic zone of a lake, the autotroph community is composed of phytoplankton (**Figure 1**), which are adapted for life in an environment that is free of solid surfaces. Consumers, such as zooplankton, living and reproducing in the pelagic zone must escape predators by avoiding the upper, illuminated part of the water column during the day, or must be agile or so small as to be impractical as a food for many predators.

## Vertical Zonation: Water Column, Sediments, and the Benthic Interface

Lakes have a vertical zonation consisting of the water column, underlying lacustrine sediments (lake sediments), and the benthic zone, which occupies a few

**Table 1** Summary of the four major zonation systems for lakes

| Zonation | Temporal variability | Description |
|---|---|---|
| Horizontal | Stable | |
| Pelagic zone | | Offshore (bottom irradiance <1%) |
| Littoral zone | | Nearshore (bottom irradiance ≥1%) |
| Vertical | Stable | |
| Water column | | Water extending from lake surface to bottom |
| Lacustrine sediments | | Lake-generated solids below the water column |
| Benthic zone | | Interface of water column and lake bottom |
| Vertical | Seasonal | |
| Epilimnion (mixed layer) | | Uppermost density layer (warm) |
| Metalimnion | | Middle density layer (transition) |
| Hypolimnion | | Bottom density layer (cool) |
| Vertical | Dynamic | |
| Euphotic zone | | Portions of a lake with ≥1% light (photosynthesis) |
| Aphotic zone | | Portions of a lake with <1% light (no photosynthesis) |

centimeters above and below the sediment–water interface (**Figure 3**). The water column extends across both the pelagic and littoral zones. The water column of the pelagic zone is driven by wind-generated currents into the littoral zone where water is displaced from the littoral zone into the pelagic zone. Thus, water-column constituents such as dissolved gases, dissolved solids, suspended solids, and suspended organisms are constantly exchanged between the pelagic zone and the littoral zone whenever there are currents in the top few meters of a lake. Chemical differences between the top few meters of the pelagic zone and the littoral zone may develop under the influence of biological processes, however, when currents are weak.

Although the water column is shared by the pelagic zone and the littoral zone, lacustrine sediments always underlie the pelagic zone but may or may not cover all of the littoral zone. Sediments are produced by the settling of mineral and organic matter that is derived from the watershed of a lake, and from organic matter consisting of fecal pellets, organic debris (detritus), and skeletal fragments of organisms derived from the lake itself. A constant sedimentation of this fine mixed solid material occurs over the entire lake. Disturbance of sediments by moving water occurs primarily in shallow water, where most of the energy of wind-generated currents and traveling waves are expended against the bottom of the littoral

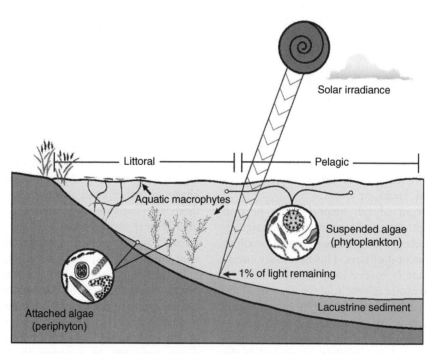

**Figure 1** Depiction of the littoral and pelagic zones of a lake. The littoral zone extends outward from the shoreline to approximately the location at which the solar irradiance at the bottom of the lake corresponds to about 1% of the solar irradiance at the top of the water column. Within the littoral zone, growth of aquatic macrophytes and attached algae (periphyton) is possible. The pelagic zone begins at the outer margin of the littoral zone. Phytoplankton are exchanged freely between the littoral and pelagic zones as well.

zone. Thus, energy near the shore may cause fine sediments, such as those that are characteristic of lakes, to be swept to deeper water. For this reason, lacustrine sediments may not accumulate in all parts of a littoral zone. Alternatively, in lakes that are small or strongly sheltered from wind, and thus not subjected to extremes of wind-generated disturbance, lacustrine sediments may occupy most or all of the littoral zone.

Lacustrine sediments are capable of supporting eukaryotic organisms (algae, protozoans, invertebrates, vertebrates) only when they are oxic. When the hypolimnion is oxic, the top few millimeters of sediment often (but not always) will be oxic. Below the top few millimeters, there typically is a decline in oxygen because microbial respiration supported by organic matter in the sediments leads to the depletion of oxygen, but some oxygen (e.g., 50%) may persist because invertebrates in the sediment pump oxygen through small tunnels into the sediment to as much as 10–20 cm within the sediments. Almost all of the deeper sediments (>10–20 cm) in lakes, which may be many meters thick, are anoxic and can support only microbes that are capable of anaerobic metabolism. When the water of the hypolimnion is anoxic, the entire sediment profile is anoxic, and can support only anaerobic microbes. There are many such microbes, and anoxic sediments show strong evidence of their metabolism, including accumulation of reduced substances such as ferrous iron, sulfide, and methane. At progressively greater depths in sediments, however, the metabolism of microbial anaerobes slows because the easily used portions of organic matter are exhausted or because oxidizing agents such as sulfate or nitrate may be depleted. Thus, there is a decline in microbial metabolic rate from the upper sediments to the deepest sediments, which are almost inert biologically.

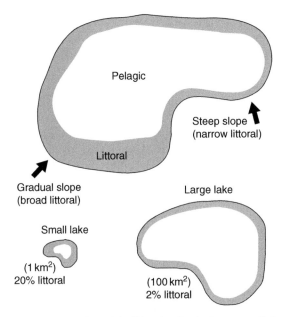

**Figure 2** A plan view of the littoral and pelagic zones of lakes, illustrating variation in the width of the littoral zone associated with changes in slope along the margin of any given lake, and the tendency of smaller lakes to have a higher percent areal coverage of littoral zone for the total lake area.

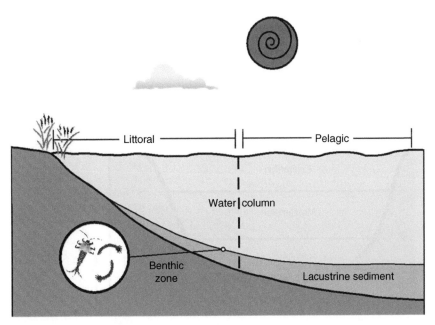

**Figure 3** Illustration of the division of a lake into three vertical components: water column, lacustrine sediments, and the boundary between the water column and sediments (benthic zone).

The interface between the water column and the lacustrine sediments carries its own name ('benthic zone') because it is exceptionally important from the ecosystem perspective, despite its narrow dimensions. Organisms that live on the sediment surface or just below it (down to about 20 cm) carry the name 'benthos'. In sediments below the pelagic zone, the benthos does not include autotrophs because there is no light reaching these sediments. The benthic zone is rich in invertebrates, provided that it is oxic at the surface, which is not always the case. Oxic benthic zones often support a number of important invertebrates, most of which are embedded within the sediment, as necessary to avoid predation. Examples include midge larvae and the larvae of other insects (**Figure 3**). Certain fish species (e.g., catfish) may be associated closely with the benthic zone, in that they are adapted to find and consume the embedded invertebrates by chemosensory means, without using vision. Oxic benthic zones also support protozoans and bacteria conducting oxic metabolism, including especially the oxic breakdown of organic matter. Anoxic benthic zones only support anaerobic bacteria and a few specialized protozoans.

The benthic zone extends not only across the bottom of the pelagic zone but also across the bottom of the littoral zone (**Figure 3**). The benthic component of the littoral zone includes not only the interface between lacustrine sediments and the water column but also between the water column and any parts of the littoral zone that happen to be swept free of lacustrine sediments. Thus, the entire solid surface at the bottom of a lake lies within the benthic zone.

## Seasonal Zonation: Vertical Layering Based on Density

Because the temperature of water affects its density, it is common for lakes to develop layers of different density corresponding to temperature differences across the layers. At temperate latitudes, all but the shallowest lakes develop a density stratification during spring that typically persists until late fall. Similar seasonal stratification is also common in subtropical and tropical lakes, but the duration of stratification is longer and the nonstratified period (mixing period) does not contain an interval of ice cover, as it often does at temperate latitudes.

Density stratification causes an ecologically important vertical zonation of lakes (**Figure 4**). An upper layer, which contains the air–water interface, is the epilimnion of a stratified lake; it may also be referred to as the 'mixed layer'. The epilimnion is the warmest and least dense of the three layers. Its thickness is strongly influenced by the size of the lake, in that larger lakes show a higher transfer of wind energy to water currents, which thickens the mixed layer during its period of formation. In sheltered water bodies, mixed layers may be as thin as two meters, but in larger ($>10 \, \text{km}^2$), windswept water bodies, they may be as thick as 15–20 m. In addition, the thickness of a mixed layer in a given lake increases over the fall cooling period, during which the bottom of the mixed layer erodes the layer below.

The mixed layer often shows sufficient irradiance throughout its full thickness to support photosynthesis. Even in lakes that have very low transparency,

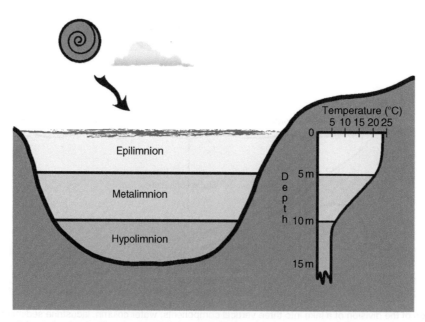

**Figure 4**  Illustration of the three density (temperature) layers determined by seasonal stratification of a water column.

the uppermost portion of the mixed layer is well illuminated.

The mixed layer extends continuously across the pelagic zone and littoral zone. Only in highly transparent lakes does the littoral zone extend below the mixed layer. Within the pelagic portion of the mixed layer, zooplankton herbivores feed vigorously on phytoplankton, but may move downward out of the mixed layer during the day in order to avoid predation. Other small invertebrates often migrate out of the mixed layer during daylight hours as well. Most of the energy of wind transferred to water is dissipated within the mixed layer; this energy contributes to the high degree of uniformity in the mixed layer at small to intermediate distance scales (up to 10 km or more), except during extended calm weather.

Below the mixed layer of a stratified lake is a thermal gradient, corresponding to a density gradient. The gradient makes a transition in temperature and density between the mixed layer and the coolest layer, which lies in contact with the bottom of the lake. The thermal transition is referred to as the thermocline, but the layer within which the thermocline lies is best referred to as the metalimnion (**Figure 4**). The metalimnion may or may not receive sufficient irradiance to support photosynthesis. In lakes that are quite transparent, phytoplankton often grows in the metalimnion, whereas lakes of low transparency seldom show growth of autotrophs in the metalimnion. In the pelagic zone, phytoplankton growing in the metalimnion of a transparent lake may be of different species composition than phytoplankton growing in the mixed layer. Some mass transfer occurs between the epilimnion and the metalimnion, the amount of which is dependent on the amount of turbulence at the interface of the two layers. The thickness of the metalimnion varies a great deal among lakes. It is seldom thinner than 2 m, and may be as thick as the mixed layer or even thicker.

The deepest seasonal layer in lakes is the hypolimnion. The temperature and density of the hypolimnion typically reflect conditions that occur when seasonal stratification becomes established. It is common for stratified lakes at temperate latitudes to have hypolimnetic waters that are near 4 °C, the temperature at which water is most dense, or slightly above 4 °C, reflecting the prevailing water temperature at the time of spring stratification. At lower latitudes, the minimum temperature increases, until it reaches approximately 24 °C within 10° latitude of the equator. Thus, a stratified lake might have a hypolimnion of 4 °C in Wisconsin and 24 °C in Venezuela.

The hypolimnion, in contrast to the mixed layer, has a very low degree of turbulence, is too dark for photosynthesis, and is isolated from the surface. Also, except at low latitudes, it is much cooler than the mixed layer (**Figure 4**).

Because of its physical isolation from photosynthesis and from atmospheric oxygen, the hypolimnion typically loses oxygen during the period of stratification. The loss of oxygen depends on the size of the hypolimnion, its temperature, the duration of stratification, and the amount of organic matter coming down to it from above, which is a byproduct of the trophic status of the lake. In lakes that are just deep enough to support stable stratification (e.g., 10–20 m), the hypolimnion may be absent, as the metalimnion reaches the bottom of the lake. In deeper lakes, the hypolimnion may equal the volume of the epilimnion, and in very deep lakes (e.g., >100 m), the hypolimnion may be much larger than the epilimnion. Lakes with a very large hypolimnion often maintain hypolimnetic oxygen throughout the stratification season, especially at temperate latitudes where the hypolimnion is cool. Lakes with a very small hypolimnion typically lose most or all of their oxygen, even if they have low productivity, because the sediments of a lake contain enough organic matter to demand most or all of the oxygen from a small hypolimnion. Oxygen concentration in lakes with a hypolimnion of intermediate size is quite sensitive to trophic state. Eutrophic lakes typically lose most or all of their hypolimnetic oxygen, thus producing an anoxic benthic zone, whereas lakes of lower productivity may retain an oxic hypolimnion overlying a benthic zone that has an oxidized surface.

Loss of oxygen is of great importance to the metabolism of a lake because eukaryotes (most protozoa, invertebrates, fish, algae) cannot live in anoxic waters. Thus, loss of hypolimnetic oxygen excludes nonmicrobial components of the biota from the deep waters of a lake.

## Zones with Dynamic Dimensions: Euphotic and Aphotic

Photosynthesis in the water column of lake is dependent on the availability of photosynthetically available radiation (PAR, wavelengths 350–700 nm). PAR, which corresponds closely to the spectrum of human vision, is removed exponentially as it travels through a water column. The availability of PAR is high during daylight hours at or near the surface of the water column. If nutrients are present, rates of photosynthesis are likely to be high near the surface. A progressive decline in PAR with depth is paralleled by a decline in rates of photosynthesis with depth. At a depth corresponding to ~1% of the surface irradiance, net photosynthesis reaches zero, which is a threshold beyond which accumulation of plant biomass is not possible. At depths below the 1% level, photosynthetic organisms (e.g., phytoplankton) lose mass and either die or become dormant unless they are returned to the surface by water currents, which commonly occurs in the mixed layer but not in the metalimnion or hypolimnion.

The depth between the water surface and the depth of 1% irradiance is referred to as the euphotic zone (**Figure 5**). Below the depth of 1% irradiance is the aphotic zone. The euphotic zone may not occupy the entire epilimnion of a lake, or may extend to the full thickness of the epilimnion. In some cases, the euphotic zone may extend into the metalimnion, but its extension into the hypolimnion of a lake is unlikely.

The thickness of the euphotic zone is dependent on transparency of the water, which in turn is influenced by dissolved color (colored organic acids from soils), inorganic suspended matter (silt and clay), and living organisms, and especially those that contain chlorophyll, an efficient absorber of light. Because the concentrations of each of these constituents can vary on relatively short time scales (e.g., weekly), the thickness of the euphotic and aphotic zones is dynamic; it is subject to both seasonal and irregular change over time.

The thickness of the euphotic zone may be small at times of high runoff, when suspended inorganic material and colored organic compounds enter lakes in the largest quantities. Other times when the euphotic zone may be thin it coincides with algal blooms, which can produce sufficient chlorophyll to reduce the transparency of the water column substantially. Conversely, the euphotic zone may thicken substantially when nutrients are exhausted because phytoplankton biomass is likely to decline, thus increasing transparency of the water. Also, strong grazing by zooplankton may thicken the euphotic zone by removing phytoplankton biomass.

The euphotic zone extends across an entire lake, including both pelagic and littoral zones. In fact, the mean thickness of the euphotic zone determines the outer boundary of the littoral zone because of its effect on the attached vegetation that is characteristic of littoral zones.

## Overview

The four sets of zones shown in **Table 1** define distinctive habitats within lakes that are associated with specific categories of organisms and biogeochemical or metabolic processes. The zones reflect some of the most important physical and chemical factors that control biotically driven processes and biotic community structure. Zonation, although generally a qualitative rather a quantitative concept, reflects accumulation of experience and measurements across lakes of many kinds. Therefore, knowledge of zonal

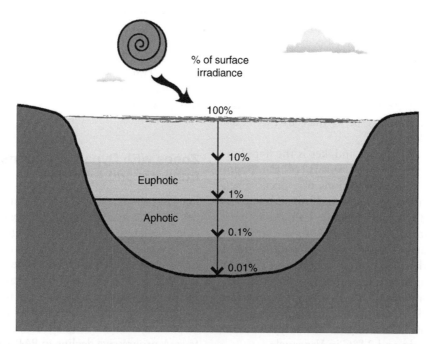

**Figure 5**    Illustration of the division of a lake into euphotic and aphotic zones. The euphotic zone, which is the lake volume within which positive net photosynthesis can occur, corresponds approximately to the depth at which ≥1% of incoming irradiance is found. Portions of the lake below this boundary have negative net photosynthesis or negligible total photosynthesis.

boundaries in a lake allows some general predictions to be made about the kinds of organisms and rates of biogeochemical processes that will occur in a given lake, and the spatial distribution of these organisms and processes.

## Glossary

**Aphotic** – Without light, generally interpreted limnologically as receiving less than 1% of solar irradiance reaching a lake surface.

**Epilimnion** – The uppermost and warmest layer (also called the mixed layer) of a lake that experiences density stratification induced by seasonal warming at the lake surface.

**Euphotic** – That portion of a lake receiving sufficient solar irradiance to support photosynthesis (typically more than 1% of full solar irradiance).

**Hypolimnion** – The most dense, deepest, and coolest layer of a thermally stratified lake. The hypolimnion does not support photosynthesis, because it lacks solar irradiance, and in many cases shows partial or complete depletion of dissolved oxygen.

**Littoral** – Near the shore of a waterbody, where irradiance reaching the bottom is above 1% of solar irradiance at the water surface.

**Metalimnion** – A layer of transitional density and temperature that connects the epilimnion to the hypolimnion.

**Pelagic** – Beyond the littoral zone of a lake.

**Benthic** – The zone of a lake extending a few centimeters above and below the bottom of the lake.

*See also:* Density Stratification and Stability; Geomorphology of Lake Basins.

## Further Reading

Hutchinson GE (1967) *A Treatise on Limnology, Volume II: Introduction to Lake Biology and the Limnoplankton.* New York: Wiley.

Hutchinson GE (1975) *A Treatise on Limnology, Volume III: Limnological Botany.* New York: Wiley.

Hutchinson GE (1993) *A Treatise on Limnology, Volume IV: The Zoobenthos.* New York: Wiley.

Kalff J (2002) *Limnology: Inland Water Ecosystems.* Upper Saddle River, NJ: Prentice Hall.

Wetzel RG (2001) *Limnology,* 3rd edn. New York: Academic Press.

# Littoral Zone

**J A Peters and D M Lodge,** University of Notre Dame, Notre Dame, IN, USA

## Introduction

The littoral zone of a lake is the nearshore interface between the terrestrial ecosystem and the deeper pelagic zone of the lake. It is the area where at least one percent of the photosynthetically active light (400–700 nm) entering the water reaches the sediment, allowing primary producers (macrophytes and algae) to flourish. The littoral zone is structurally and functionally an important part of most lakes for several reasons. First, most lakes on Earth are small and therefore, the littoral zone comprises a large proportion of total lake area (**Figure 1**). Second, as an interface, the littoral zone influences the movement and processing of material flowing into the lake from terrestrial runoff, groundwater, or stream connections, thus affecting the physical and biological processes in this zone and the rest of the lake ecosystem. Third, the littoral zone is generally the most productive area of the lake, especially in terms of aquatic plants and invertebrates. Finally, human uses of aquatic systems (swimming, fishing, boating, power generation, irrigation, etc.) often focus on the littoral zone.

In the first section, the physical structure and nutrient dynamics within the littoral zone are described. In the second section, interactions among organisms in the littoral zone are discussed, in addition to interactions between the littoral and terrestrial ecosystems, and between the littoral and pelagic zones. In the final section, anthropogenic effects on the littoral zone are described.

## Factors Influencing the Physical Structure and Nutrient Dynamics of the Littoral Zone

Many characteristics determine the percentage of the lake that is littoral zone and the type of lake-bottom substrates found there. Littoral zone area and substrate type, in turn, influence the processing of incoming nutrients, minerals, and organic matter and therefore the functioning of the entire lake.

### Physical Structure of the Littoral Zone

**Zonation** In general, the littoral zone can be divided into upper, middle, and lower zones, extending from the shoreline area sprayed by waves to the bottom of the littoral zone, below which light does not penetrate (**Figure 2**). Emergent vegetation is rooted in the upper littoral zone; floating vegetation is found in the middle littoral; and submergent vegetation often grows in the lower littoral. The littoriprofundal, which is inhabited by algae and autotrophic bacteria, is a transitional zone below the lower littoral zone. Below this transitional zone, fine particles permanently settle into the profundal zone because wind or convection current energy is not sufficient at these depths to resuspend the particles. The littoral zone depth commonly corresponds to the summer epilimnion depth in stratified lakes.

**Habitats** Compared with the homogeneous distribution of sediments in the profundal zone, the habitats and sediments of the littoral zone are distributed as heterogeneous patches (**Figure 3**).

Sizes of particles in the sediments range from very fine organic and inorganic particles (muck or silt) to large cobble and boulders. Macrophytes and fallen trees often provide vertical substrates within the littoral zone (refer to 'see also' section). The abundance and distribution of habitats within the littoral zone mediates the abundance, diversity, and interactions of biota. For example, cobble substrates provide a refuge for crayfish from fish predation; in contrast, fine organic substrates favor the growth of macrophytes that provide refuge for invertebrates, zooplankton, and juvenile fish. Invertebrate abundance and composition differ among different kinds of substrate. Overall benthic invertebrate diversity is greater in the heterogeneous littoral region compared with the homogeneous profundal region. As explained later, the types of habitats found in the littoral zone depend on lake morphometry, the surrounding landscape, wind patterns, and nutrient loads to the lake.

**Lake morphometry** The morphometric characteristics that influence the kinds of habitats within the littoral zone include lake area, depth, shoreline sinuosity, and underwater slope. The origin of a lake largely determines lake morphometry. For example, lakes formed through tectonic or volcanic activities are usually very large, steep-sided lakes with minimal littoral areas, whereas glacial lakes and reservoirs often have complex basin shapes and large littoral areas.

Lakes with greater lake area to depth ratios, more sinuous shorelines, more complex bathymetry, and shallow sloped basins will have a larger percent littoral zone compared with pelagic zone. For instance, shallow lakes with large surface areas have large littoral zones because the light is able to penetrate to the sediment in a high proportion of the lake area.

Lake morphometry characteristics also influence the types of substrates found within the littoral zone. Steep sloped littoral areas typically have rocky/cobble substrates, and areas with a gradual slope can be dominated by fine sediments with or without macrophytes (**Figure 3**). Lakes with a high shoreline sinuosity have more bays with macrophytes growing on sand or muck compared with circular shaped lakes, because wave action is reduced in protected bays, allowing the accumulation of fine

organic sediments, nutrients and minerals, and the establishment of macrophytes.

**Surrounding landscape**   The topography and geology of the land surrounding a lake influence the movement of water, associated nutrients, minerals, and organic matter into the littoral zone. The relative contribution of surface runoff and groundwater to a lake depends on water infiltration and transmission rates of surrounding soils, the productivity of terrestrial vegetation, and the slope and the drainage density of the watershed.

Elevation and hydrologic flow define the position of a lake in the landscape. High in the landscape, lakes tend to be small seepage lakes, which are fed primarily by precipitation and groundwater. Larger drainage lakes, which are fed by surface water, groundwater, and precipitation, tend to be lower in the landscape (**Figure 4**). Lakes lower in the landscape tend to have larger, more productive littoral areas because of greater watershed inputs of nutrients, minerals, and dissolved or particulate organic material, from both surface water and stream connections. This material input increases buffering capacity (ability to reduce affects of acidification) and the abundance and diversity of macrophytes and the invertebrates like snails that live on macrophytes. Also, lakes lower in the landscape usually have a more complex basin bathymetry, which also increases littoral area.

**Wind patterns**   Substrates found in the littoral zone are a function of wind patterns such as fetch and exposure. Fetch is the distance the wind blows across the lake. The windward and lee sides of the lake will have distinctly different substrate characteristics. The stronger the wave action caused by the wind, the more fine particles will be suspended and eventually deposited in the profundal zone of the lakes, and the more the littoral zone substrate will be characterized by rocks. Wave action will also be reduced by sinuous shorelines and macrophytes as described earlier.

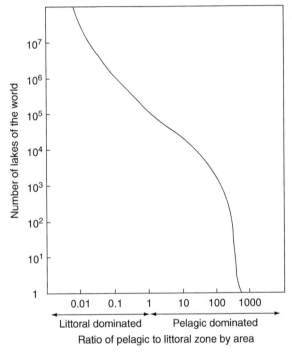

**Figure 1**   Number of lakes of the world dominated by littoral or pelagic zones. Modified from Wetzel RG (2001) *Limnology: Lake and River Ecosystems.* New York: Elsevier, Academic Press.

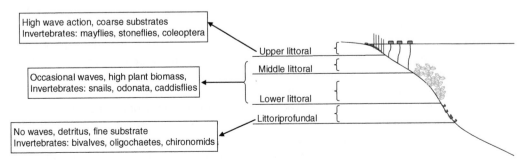

**Figure 2**   Zonation of the littoral zone. Associated wave conditions, substrate, and invertebrates are listed. The upper littoral zone can have emergent vegetation or cobble substrates depending on the type of lake and the wave action.

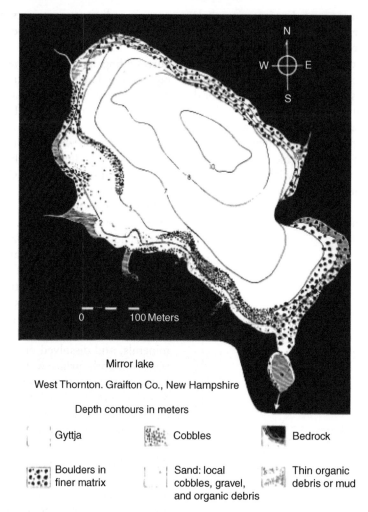

Mirror lake

West Thornton. Graifton Co., New Hampshire

Depth contours in meters

| | Gyttja | | Cobbles | | Bedrock |
| --- | --- | --- | --- | --- | --- |
| | Boulders in finer matrix | | Sand: local cobbles, gravel, and organic debris | | Thin organic debris or mud |

**Figure 3**　Example of habitat heterogeneity in the littoral zone and the influence of slope on substrate composition. From Moeller RE (1978) The hydrophytes of Mirror Lake: A study of vegetational structure and seasonal biomass dynamics, Ph.D. dissertation, 212 pp. Ithaca, NY: Cornell University.

## Nutrient Dynamics in the Littoral Zone

**Sources of nutrient inputs**　Detritus (dead organic matter) and associated nutrient inputs into the littoral zone are either allochthonous (derived from terrestrial sources) or autochthonous (aquatic sources). Allochthonous sources include groundwater, precipitation, fluvial inputs, terrestrial plant litter fall, and materials from soil erosion. Nutrients can also be transported into the littoral zone by animals moving between the terrestrial and the littoral zone for food resources (i.e., amphibians, waterfowl, or mammals such as beaver, etc.) or as food resources (i.e., small mammals) for aquatic organisms such as fish. Autochthonous sources of nutrients come from the death of aquatic organisms (plants and animals), and secretion, excretion, and egestion from living animals and plants.

The distribution of detritus influences the availability of dissolved organic matter and nutrients for biotic uptake. Detritus deposited in the profundal zone may become permanently lost from littoral food webs, whereas detritus deposited within the littoral zone can contribute to internal loading of dissolved organic matter and nutrients (i.e., phosphorus and nitrogen) for primary and secondary production. Much of the energy that drives ecosystem metabolism comes from allochthonous or autochthonous detritus, and shallower lakes with a greater percent of littoral area have a net deposition rate of detrital organic matter that is always greater than that of deeper lakes.

**Retention capacity of the littoral zone**　The retention time of water, nutrients, and detritus is influenced by the size and configuration of the littoral zone. Deep lakes have longer water retention times (up to hundreds or thousands of years) compared with shallow lakes (often less than a year) and the pelagic zone

High in landscape

Low in landscape

**Figure 4** Lakes in the landscape. Hydrologic connectedness ranges from isolated lakes to those connected by large rivers. Magnuson JJ, Kratz TK, Benson BJ (eds.) (2006) *Long-Term Dynamics of Lakes in the Landscape: Long-Term Ecological Research on North Temperate Lakes*, 51 pp. Oxford, UK: Oxford University Press.

has longer retention times compared with the littoral zone. The amount of time water is retained within the littoral zone influences the dynamics of nutrients within the lake. The longer it takes for water to pass through the littoral zone, the greater the amount of nutrients that will be used by plants and animals in the littoral zone.

Although deep lakes have a greater retention time of water, they usually have a small littoral zone that continuously loses detritus and nutrients to the profundal zone as detritus sinks through the metalimnion. In some stratified lakes, half of the total phosphorus can be lost to the hypolimnion (profundal zone) during the summer and only partially returned by the mixing of the lake in the spring and fall. Shallow lakes, on the other hand, do not have this constant nutrient loss because they have a greater proportion of the epilimnion volume in contact with the lake bottom. Thus in shallow lakes, nutrients are recycled within the littoral zone at a greater rate and less loss to the profundal zone occurs.

The littoral zone has therefore been described as a 'metabolic sieve' or 'trap' because of its ability to strain incoming water and nutrients before passing it on to the pelagic and profundal zone. In many cases, most of the dissolved organic matter and nutrients that are not used in the littoral zone will ultimately be lost to sedimentation and burial in the profundal zone.

**Major nutrients in the littoral zone** There are many nutrients and minerals (silica, calcium, iron, manganese, sulfur, etc.) that influence the type of chemical and biological processes that occur in the littoral zone. For example, high concentrations of ions such as calcium and magnesium increase the buffering capability of lakes. Iron and manganese bind to phosphorus (often the nutrient most limiting primary production) in aerobic conditions making it unavailable for biotic uptake. Calcium is used by snails and other invertebrates for shell or exoskeleton maintenance, while sponges and diatoms require silica for spicule and test construction.

Not only do littoral biota require nutrients and minerals, but in turn organisms such as bacteria, macrophytes, benthic invertebrates, and benthivorous fish alter the availability and composition of nutrients within the littoral zone. Autotrophic and heterotrophic bacteria can use and produce many different nutrients and gases including oxygen, carbon dioxide, iron, several nitrogen and sulfur products, and methane, depending on whether the conditions are aerobic or anaerobic. Macrophytes modify the chemical composition in the littoral zone by altering the oxygen and carbon dioxide concentrations and pH levels in the surrounding sediments and overlying water. Benthic invertebrates and fish increase nutrient release, such as phosphorus, through sediment resuspension.

## The Biota of the Littoral Zone

Biota of the littoral zone includes both permanent and transient species (**Figure 5**).

Transient species – those that move in and out of the littoral zone from the surrounding terrestrial ecosystem and pelagic zone create linkages between the littoral zone and surrounding environs. Both the biota and associated linkages are discussed in this section.

For ease of constructing the food web, organisms in **Figure 5** are grouped as either being terrestrial or aquatic. However, certain species within each group actually belong in both the terrestrial ecosystem and the littoral zone (i.e., amphibians and waterfowl) or in both the littoral and pelagic zones (i.e., zooplankton and fish). The important point is that many species, including some of those discussed below, use multiple food resources and zones within the lake, which can have cascading effects throughout terrestrial, littoral, and pelagic food webs.

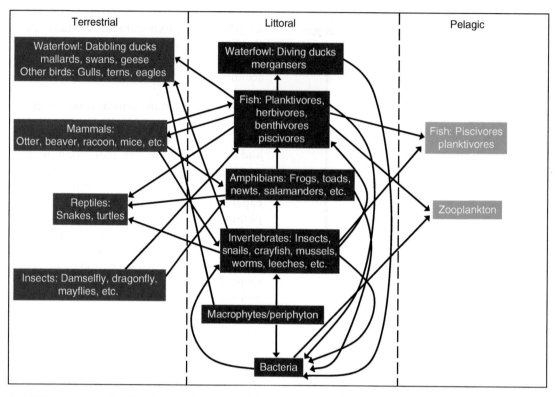

**Figure 5**   Linkages between the littoral zone with the terrestrial ecosystem as well as pelagic zone. Arrows represent energy flow. Only the interactions with the littoral zone are shown. There are interactions between biota on land and in the pelagic zone that are not depicted in this figure.

### Bacteria

Bacterial production is up to 120 times greater in the littoral zone than in the pelagic zone. Bacteria are one of the main biotic components that allow the littoral zone to act as a 'metabolic sieve.' The main function of bacteria in the littoral zone is to break down allochthonous and autochthonous organic material. As bacteria process detritus, different nutrients and gases, such as particulate and dissolved organic matter, nitrogen, phosphorus, methane and sulfur, etc., are produced and in many cases become available to other biota in the littoral zone. Lakes with large, shallow littoral zones will have increased bacteria metabolism and faster detrital processing.

### Periphyton

Periphyton is a mixture of autotrophic and heterotrophic microorganisms embedded in a matrix of organic detritus (refer to 'see also' section). Periphyton covers most submerged substrates, ranging from sand to macrophytes to rock. The metabolic importance of periphyton at the whole lake scale is constrained by the morphometry and substrate characteristics of the littoral zone. In oligotrophic lakes, even those

with few macrophytes for periphyton to grow on, periphyton can be an important component of whole-lake primary production. For example, in one oligotrophic lake, the littoral zone comprised only 15% of the lake, but the periphyton accounted for 70–85% of the lake primary production. In eutrophic lakes, however, phytoplankton is more abundant and shading by phytoplankton reduces periphyton and macrophyte abundance. Periphyton is a common food source for invertebrates and some amphibians.

### Macrophytes

Macrophytes require specific substrate types to thrive, and their growth provides a unique habitat for other organisms (refer to 'see also' section). Macrophytes grow best in a mixture of sand and muck, and are often found in areas with upwelling groundwater. Once macrophytes become established within the littoral zone they modify the microclimate through the reduction of wave energy and the creation of thermal gradients that prevents water from mixing. These conditions promote particle sedimentation. The degree of microclimate modification depends on the characteristics of the sediment structure, nutrient availability, and diffusion of oxygen through the sediment. Macrophytes are integral to nutrient cycling in the littoral zone as both sources and sinks of nutrients. Traditionally, limnologists have considered macrophytes a nutrient source, since they may incorporate nutrients from the anoxic sediment and then release them into the water column upon senescence. Others have found that nutrients removed from sediments or surrounding water column by plants are largely retained by plants until the plants decay.

In addition to their role in nutrient cycling, macrophytes provide important habitat for organisms such as bacteria, periphyton, zooplankton, invertebrates, amphibians, fish, and waterfowl. Invertebrates and small fish use macrophytes as a habitat refuge from predation by invertebrates (e.g., dragonfly or damselfly nymphs), fish (e.g., Esox), and amphibians, and as a place to reproduce. For many invertebrates (e.g., insects, crustaceans) and vertebrates (e.g., waterfowl, moose), macrophytes are a major food source.

### Invertebrates

Invertebrates are very diverse, and include: zooplankton, crayfish, insects, worms, and leeches. Invertebrates living on the bottom of lakes are referred to as zoobenthos, and are far more abundant and diverse in the littoral zone than in other lake zones. Therefore the ratio of the abundance of zoobenthos to zooplankton is inversely related to lake size. However, the absolute diversity and abundance of zoobenthos increases with lake size. Invertebrate diversity is also positively related to habitat complexity, macrophyte abundance, conductivity, and the presence of stream connections.

Habitat availability within the littoral zone influences the type of invertebrates that will colonize (**Figure 2**). For instance, ephemeroptera (mayflies) and plecoptera (stoneflies) generally prefer substrates that have higher wave action and coarser substrates, while lightly disturbed fine sediments are colonized by chironomids (midge larvae), bivalves (clams), and oligochaetes (worms). Substrate, macrophyte abundance, and detritus are the three main factors controlling the diversity and distribution of invertebrates, but water depth, wave exposure, and water clarity, (which influence the first three main factors) may consequently also affect invertebrate abundance and distribution.

### Fish

Like invertebrate diversity, fish diversity is positively related with lake size and habitat complexity. The use by fish of different littoral zone habitats also often varies seasonally and with the age or size of the fish. As mentioned above, macrophytes are both a refuge and a hunting ground for predatory fish. Some fish species (e.g., *Perca* spp.) also use macrophytes as substrate for egg-laying. Many other fish species lay eggs within cobble substrates, and some use cobble as a refuge (e.g., sculpin, darters, juvenile burbot).

Fish are often classified by their primary food source. Piscivores mainly eat other fish, planktivores consume plankton, benthivores feed on zoobenthos, and herbivores eat macrophytes and periphyton. Some fish species may change what they eat as they mature into adulthood. For instance, many fish species eat plankton as a juvenile and smaller fish as an adult. Even adult fish may have very broad diets as they move between littoral and pelagic zones (**Figure 5**).

### Terrestrial – Littoral Links

It is impossible to separate the processes that occur within a lake from the surrounding watershed and even the air above the watershed. Many organisms move resources and energy between the surrounding watershed and the littoral zone. Food resources from the littoral zone are an important source of energy for many terrestrial and semi-terrestrial organisms. Fish, for example, are eaten by many different terrestrial and amphibious species including waterfowl, hawks, herons, egrets, mammals, reptiles, and humans. Aquatic invertebrates found within the littoral zone provide an important source of protein for terrestrial and semi-terrestrial organisms. The reproductive success of ducks is closely related to the availability of chironomids and other insects emerging from their benthic larval form. Waterfowl, such as geese, feed on aquatic plants and can remove up to 50% of the standing stock of macrophytes in some areas.

Food resources from the terrestrial ecosystem are also important for littoral species. Depending on the time of the year and the species of fish, up to half the food consumed can be allochthonous insects and small mammals. Another type of resource that is moved from land to the littoral zone is large woody debris used by beavers to construct their lodges. The woody debris used by beavers also provides habitat for many fishes. Finally, transient organisms such as waterfowl, mink, otter, beaver, muskrat, snakes, and turtles, among others, move nutrients in and out of the littoral zone via feeding and excretion and egestion.

The riparian habitat is another resource that is important for species that use the littoral zone. For example, snakes and turtles sun themselves on logs and rocks found along the shoreline. Waterfowl and some mammals use the low lying shoreline habitat to make their nests, while eagles and some diving ducks use the trees surrounding lakes for their nesting sites. Hawks also use trees surrounding the lake as a perch to search for food. Riparian habitat is important for amphibians (e.g., newts and frogs) during different times within their lives.

Trophic cascades – food web interactions that strongly alter the abundance of three or more trophic levels – are well documented in the pelagic and littoral zone of lakes. They also cross the littoral-land interface (**Figure 6**). Plants near ponds with fish have more visits from pollinators than plants near ponds without fish. This is because in ponds with fish, larval dragonflies are reduced by fish predation, and thus the abundance of adult dragonflies is also decreased. Adult dragonflies have direct and indirect effects on insect pollinators. They directly prey upon the pollinators and indirectly reduce the number of pollinator visits just by being present.

### Littoral – Pelagic Links

The littoral and the pelagic zones are also strongly linked, especially by the diel horizontal migration of zooplankton, and by fish movements. Zooplankton sometimes move up to 30 m horizontally twice each day between zones. Zooplankton that normally reside in the pelagic zone will move into macrophyte habitats during the day to avoid pelagic predators

such as *Chaoborus* (phantom midge larvae) and visually feeding planktivores like small fishes. In some lakes, this movement can benefit zooplankton through reduced mortality from fish predation, food availability in the littoral zone (some zooplankton can become browsers in the littoral zone compared to being filter feeders in the pelagic zone), and enhanced growth. Predation by planktivores is often reduced by migration into the littoral zone, but in some lakes, littoral invertebrates (e.g., dragonfly larvae) pose a substantial risk of predation within the littoral zone. Thus, zooplankton movement depends on the complex interactions occurring in both the pelagic and littoral zones, which differ among lakes.

Fish movements also link the littoral and pelagic zones. The dependency of fish on littoral production differs by fish type, with planktivores, benthivores, and even piscivores relying on littoral food production to some degree (**Figure 7**). Fish that are often categorized as pelagic planktivores can derive up to 30% of their energy from the littoral zone, while fish categorized as piscivores sometimes derive almost all their energy, at least indirectly (e.g., from other fish that consume littoral-derived foods), from the littoral zone (**Figure 7**). Without the littoral zone, the production of many fish, including fish that may rarely venture into the littoral zone, would decline dramatically.

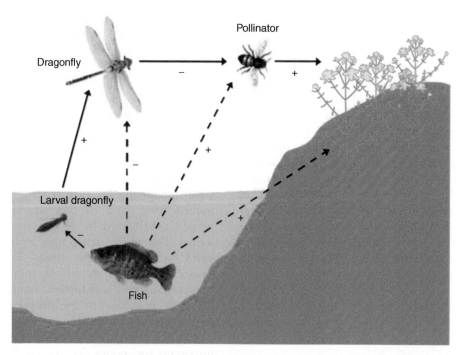

**Figure 6**  Example of a trophic cascade that links the terrestrial ecosystem with the littoral zone. Fish reduce the abundance of dragonflies, which leads to increased pollinators, and thereby facilitating the reproduction of terrestrial plants. Solid arrows indicate direct interactions; dashed arrows denote indirect interactions. + are positive interactions, − represents negative interactions. Modified from Knight *et al.* (2005) Trophic cascades across ecosystems. *Nature* 437: 880–883.

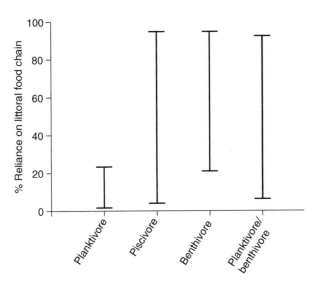

**Figure 7** The range of reliance different types of fishes have on littoral zone resources. Modified from Vadeboncoeur *et al.* (2005) Effects of multi-chain omnivory on the strength of trophic control in lakes. *Ecosystems* 8: 682–693.

## The Anthropogenic Influences on the Littoral Zone

Humans derive many ecosystem goods (e.g., harvested fish and waterfowl) and services (e.g., water purification, water supply) from the littoral zones of lakes. In turn, humans have immense impacts on the structure and function of the littoral zone. First, increased nutrient loading from activities such as logging, agriculture, and development causes eutrophication. Eutrophication leads to increased primary production in the littoral zone of many lakes, which can cause undesirable algal blooms, as well as increases in undesirable fish and cloudy water. Second, human-mediated spread of invasive species (e.g., zebra mussels, rusty crayfish, round gobies) alters nutrient cycling and food web composition in the littoral zone, causing changes that are generally undesirable to humans.

Third, fossil fuel combustion in industry and automobiles causes acid deposition and climate change. In many cases, acidification of lakes causes decreased abundance and diversity of macrophytes, invertebrates, and fish, while increasing filamentous green algal production, all of which has cascading effects through the food web. Acidification also causes the release of metals toxic to fish, e.g., aluminum and mercury. Some of these metals bioaccumulate in fish, which then makes the fish hazardous to humans. Climate change is expected to cause warmer lake waters, and in many parts of the world, will reduce runoff, increase water residence times, lower water levels, and increase evaporation. Concentrations of many ions will therefore increase, causing changes in nutrient and detritus availability as well as primary and secondary production within the littoral zone. Warming could also lead to poleward range expansion of many littoral species, further changing food web dynamics.

Finally, fluctuations in water level are often increased by irrigation and dams. At high water levels, flooding and erosion of riparian habitat occurs. This increases the dissolved organic matter input and turbidity in the littoral zone. On the other hand, water drawdown has both positive and negative effects on the littoral zone, depending on the lake basin morphometry. Steep sided littoral zones are not as affected as shallow sloping ones. At low water levels macrophytes are reduced, the percent of sandy/fine grained habitat increases, benthic invertebrate diversity and abundance decreases and fish refuges and spawning habitat can be reduced.

The response of the littoral zone to all these anthropogenic impacts is influenced by the structure and function of the littoral zone as well as the interaction between the littoral zone and the terrestrial ecosystem and the pelagic and profundal zones. The degree to which a lake responds or the length of time a lake can resist being effected by one of the human-mediated stressors described above depends on the size of the littoral zone, the position of the lake within the landscape, the abundance and distribution of different habitats within the littoral zone, and different biota present within that zone. Therefore, the littoral zone is important for whole lake functioning as well as the response of the whole lake to human beings.

*See also:* Benthic Invertebrate Fauna, Lakes and Reservoirs; Effects of Climate Change on Lakes; Eutrophication of Lakes and Reservoirs; Shallow Lakes and Ponds.

## Further Reading

Burks R. L, Lodge D. M, Jeppesen E, and Lauridsen T. L (2002) Diel horizontal migration of zooplankton: Costs and benefits of inhabiting the littoral. *Freshwater Biology* 47: 343–365.

Gasith A and Gafny S (1990) Effects of water level fluctuation on the sturucture and function of the littoral zone. In: Tilzer M. M and Serruya C (eds.) *Large Lakes: Ecological Structure and Function*, pp. 156–171. New York: Srpinger.

Jeppesen E, Sondergaard M, Sondergaard M, and Christoffersen K (eds.) (1998) *The Structuring Role of Submerged Macrophytes in Lakes*. New York: Springer.

Kalff J (2002) *Limnology: Inland Water Ecosystems*. Upper Saddle River, NJ: Prentice-Hall.

Knight T. M, McCoy M. W, Chase J. M, et al. (2005) Trophic cascades across ecosystems. *Nature* 437: 880–883.

Likens G. E (ed.) (1985) *An Ecosystem Approach to Aquatic Ecology.* New York: Springer.

Lodge D. M, Barko J. W, Strayer D, et al. (1988) Spatial heterogeneity and habitat interactions in lake communities. In: Carpenter S. R (ed.) *Complex Interactions in Lake Communities.* New York: Springer.

Magnuson J. J, Kratz T. K, and Benson B. J (eds.) (2006) *Long-Term Dynamics of Lakes in the Landscape: Long-Term Ecological Research on North Temperate Lakes.* Oxford, UK: Oxford University Press.

O'Sullivan P. E and Reynolds C. S (eds.) (2004) *The Lakes Handbook.* Oxford, UK: Blackwell Science.

Pieczynska E (1993) Detritus and nutrient dynamics in the shore zone of lakes: A review. *Hydrobiologia* 251: 49–58.

Scheffer M (1998) *Ecology of Shallow Lakes.* New York: Chapman and Hall.

Vadeboncoeur Y, McCann K. S, VanderZanden M. J, and Rasmussen J. B (2005) Effects of multi-chain omnivory on the strength of trophic control in lakes. *Ecosystems* 8: 682–693.

Weatherhead M. A and James M. R (2001) Distribution of macro-invertebrates in relation to physical and biological variables in the littoral zone of nine New Zealand lakes. *Hydrobiologia* 462: 115–129.

Wetzel R. G (1989) Land-water interfaces: Metabolic and limnological regulators. *International Association of Theoretical and Applied Limnology Proceedings* 24: 6–24.

Wetzel R. G (1989) *Limnology: Lake and River Ecosystems.* New York: Academic Press.

# Benthic Invertebrate Fauna, Lakes and Reservoirs

**D L Strayer,** Cary Institute of Ecosystem Studies, Millbrook, NY, USA

## Introduction

Benthos includes all animals that live in association with surfaces in lakes and reservoirs. This includes animals that live in and on sediments of all kinds (mud, sand, stones), as well as animals that live in, on, or around aquatic plants or debris. Animals large enough to be retained on a coarse sieve (usually 0.5-mm mesh) are called macrobenthos, those that pass through a coarse sieve but are retained on a fine sieve (usually $\sim$0.05 mm) are called meiobenthos, and those that pass through even fine sieves sometimes have been called microbenthos, although this last term is rarely used. It is customary to ignore the small benthic animals, but studies have shown that animals too small to be caught on a 0.5-mm mesh may constitute >95% of the individuals, $\sim$25% of the biomass, $\sim$50% of the metabolic activity, and >50% of the species in a zoobenthic community.

## Composition and Biological Traits of the Lacustrine Zoobenthos

The density of zoobenthos depends strongly on the mesh size of the sieve used for the study (**Figure 1**). Densities of macrobenthos (i.e., animals large enough to be caught on a 0.5-mm mesh) in lakes typically are $1000-10\,000\,m^{-2}$, while the total density of all benthic metazoans probably usually is $\sim$$1\,000\,000\,m^{-2}$. There is a great deal of variation in densities within and among lakes, so that densities reported for a given mesh size range over $\sim$100-fold within and among lakes.

Estimates of zoobenthic biomass usually include only the macrobenthos and exclude large bivalves. Because small benthic animals usually constitute a small part of zoobenthic biomass, estimates of the biomass of the macrobenthos probably are nearly equivalent to the entire zoobenthos. However, it appears that the meiobenthos may be especially important in unproductive habitats (**Figure 2**). Macrobenthic biomass without large bivalves ranges from $\sim$0.2 to 100 g dry mass $m^{-2}$, and dense populations of large bivalves can increase these values by >10 g DM $m^{-2}$. Zoobenthic biomass rises up to a point with increasing phytoplankton production, then asymptotes or even declines with further increase in phytoplankton production. Zoobenthic biomass also tends to be highest in shallow lakes, where rooted plants are abundant, where the lake bottom is flat or gently sloping, in warm lakes or in the epilimnion of stratified lakes, and in lakes of low color.

There are relatively few data on the production rates of entire macrobenthic assemblages and almost no estimates that include the small benthic animals. One would expect to see as much variation in rates of production as in biomass, and based on the expected ratio of annual production to biomass of $\sim$5 for animals of macrobenthic size, production of the lacustrine macrobenthos might range from $\sim$1 to 500 g DM $m^{-2}$ year$^{-1}$. Production of smaller benthic animals in lakes might be about the same order of magnitude, based on the sparse data now available. Production of the macrobenthos increases with increasing primary production (**Figure 3**), with no hint of an asymptote or downturn at very high primary production, as has been seen for macrobenthic biomass.

The lacustrine zoobenthos is enormously diverse; a typical lake contains hundreds of species from 12 to 15 animal phyla (**Table 1**, **Figure 4**). The macrobenthos often is numerically dominated by oligochaetes and chironomid and chaoborid midges, although large-bodied mollusks may dominate the biomass of the community. Aquatic insects other than dipterans (such as mayflies) may be abundant, especially in shallow lakes. Nematodes usually are by far the most numerous of the meiobenthos (and of the zoobenthos as a whole), although gastrotrichs, rotifers, and microcrustaceans often are abundant as well. Many important families are ecologically important in lakes around most parts of the world, including the Chironomidae and Chaoboridae in the Diptera; the Ephemeridae in the Ephemeroptera; the Tubificidae (including the 'Naididae') in the Oligochaeta; the Unionidae and Sphaeriidae in the Bivalvia; the Lymnaeidae and Planorbidae in the Gastropoda; the Chydoridae, Canthocamptidae, and Cyclopidae among the microcrustaceans; and the Chaetonotidae in the Gastrotricha. In contrast, several important groups are restricted to particular biogeographic regions (mysid shrimps and gammarid amphipods chiefly in the Northern Hemisphere, hyriid bivalves in the Southern Hemisphere) or habitats (ephydrid flies or brine shrimp in fishless or saline inland waters). Human introductions have vastly increased the ranges

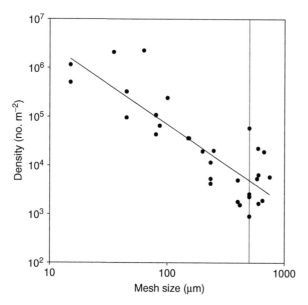

**Figure 1** Reported density of zoobenthos as a function of sieve mesh size in a series of lakes from the northern temperate zone ($r^2 = 0.72$, $p < 0.000001$). The vertical gray line marks the 500-μm mesh commonly used for macrobenthos.

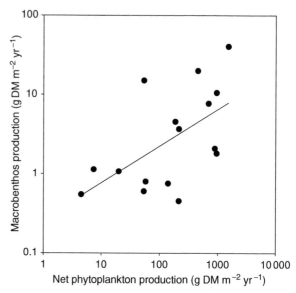

**Figure 3** Relationship between production of macrobenthos and production of phytoplankton in a series of lakes ($r^2 = 0.34$, $p = 0.02$). Adapted from data of Kajak et al. (1980).

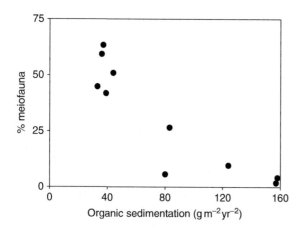

**Figure 2** Percentage of zoobenthic biomass belonging to the meiofauna, as a function of organic sedimentation at various sites in the profundal zone of Lake Paijanne, Finland ($r^2 = 0.79$, $p = 0.0006$). Adapted from Hakenkamp et al. (2002), after data of Särkkä (1995).

**Table 1** Estimates of species richness of the zoobenthos (excluding parasitic forms) in Mirror Lake, a small, unproductive lake in New Hampshire (USA), and typical densities of major phyla in lakes

| Phylum | Estimated number of species in Mirror Lake | Typical density in lakes (no. $m^{-2}$) |
| --- | --- | --- |
| Porifera (sponges) | 4 | NA |
| Cnidaria (hydras and jellyfish) | 2 | 100–1000 |
| Platyhelminthes (flatworms) | 40 | 1000–50 000 |
| Nemertea (ribbon worms) | 1 | <100 |
| Gastrotricha | 30 | 100 000–1 000 000 |
| Rotifera | 210 | 10 000–250 000 |
| Nematoda (roundworms) | 35 | 100 000–1 000 000 |
| Annelida (earthworms, leeches) | 30 | 5000–50 000 |
| Mollusca (snails, clams) | 6 | 100–1000 |
| Ectoprocta (moss animicules) | 2 | <100 |
| Crustacea (water-fleas, seed shrimp, copepods, and relatives) | 70 | 20 000–200 000 |
| Chelicerata (mites) | 50 | 1000–10 000 |
| Tardigrada (water-bears) | 5 | 1000–50 000 |
| Uniramia (insects) | 120 | 1000–50 000 |

Many lakes around the world contain the same phyla and a comparable numbers of species as Mirror Lake, but have not been so well studied.

of several ecologically important species of the lacustrine zoobenthos that once had small geographic ranges, most notably *Dreissena* spp. (zebra mussels, Bivalvia), *Mysis* (opossum shrimp, Crustacea), and several crayfish species.

Because the lacustrine zoobenthos is so diverse, it is difficult to generalize about the biological traits of its members. The largest animals in the community (bivalves and decapods, $>10^1$ g dry mass) are more

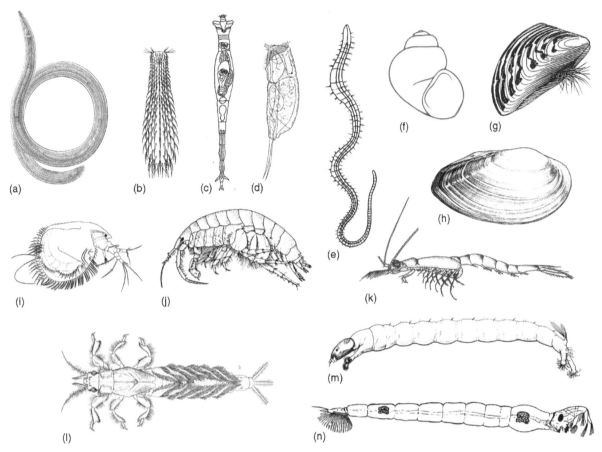

**Figure 4**  Some important members of the lacustrine zoobenthos. Typical adult body lengths are given in parentheses. (a) a nematode (2 mm); (b) a gastrotrich (0.2 mm); (c) a bdelloid rotifer (0.5 mm); (d) a ploimate rotifer (0.1 mm); (e) a tubificid oligochaete (50 mm); (f) a hydrobiid snail (3 mm); (g) the bivalve *Dreissena* (20 mm); (h) a unionid bivalve (75 mm); (i) a cladoceran (1 mm); (j) an amphipod (10 mm); (k) a mysid shrimp (20 mm); (l) an ephemerid mayfly (20 mm); (m) a chironomid (10 mm); (n) the phantom midge *Chaoborus* (10 mm).

than 10 billion times larger than the smallest (rotifers and gastrotrichs, $\sim 10^{-9}$ g DM), so this community spans an enormous range of body sizes. Life spans of zoobenthic animals range from less than a week (rotifers, gastrotrichs) to decades (bivalves). Some species have tough long-lived resting stages (sponge gemmules, ectoproct statoblasts, cladoceran ephippia) that allow populations to reestablish themselves after long unfavorable periods, and assist in passive dispersal between lakes. Various species burrow into the sediments (bivalves, tubificid oligochaetes, *Chaoborus*), glide at the sediment–water interface (gastrotrichs, ploimate rotifers, flatworms), attach (sessile rotifers) or mine (some chironomids in the genus *Cricotopus*) in aquatic plants, or crawl on or attach to solid object such as stones (*Dreissena*, many gastropods).

The zoobenthos includes species that suspension-feed on phytoplankton (many bivalves and some chironomids) or interstitial bacteria (*Pisidium*), graze on benthic algae (gastropods and many

rotifers), deposit-feed on sedimented detritus and bacteria (tubificid oligochaetes), are predators on other benthic animals (odonates, tanypodine chironomids, dicranophorid rotifers) or zooplankton (*Chaoborus*), feed on either the leaves (crayfish, chrysomelid beetles) or roots (dorylaimid nematodes, larvae of the beetle *Donacia*) of rooted plants, or shred leaves of terrestrial plants that fall into the water (isopods, caddisflies). Not surprisingly, passive suspension- feeders such as hydropsychid caddisflies and black flies are much rarer in lakes than in flowing waters, presumably because currents in lakes are too slow or undependable to provide them with food.

Benthic animals have several adaptations for dealing with low oxygen concentrations that occur in many benthic environments. Some burrowing animals (chironomids and mayflies) produce currents to bring oxygenated water into the otherwise anoxic sediments in which they burrow. Species in several taxonomic groups (e.g., cladocerans, chironomids, gastropods) produce hemoglobin, which improves oxygen transport under hypoxic conditions. A few species of nematodes, bdelloid rotifers, gastrotrichs, tubificid oligochaetes, copepods, ostracods, and chironomid and chaoborid midges are commonly found in the anoxic profundal sediments of productive lakes. These animals apparently survive extended anoxia either by using anaerobic metabolism of substances like glycogen or by going into extended diapause. Finally, many of the benthic animals (e.g., many insects and pulmonate snails) avoid the problem of low dissolved oxygen altogether by obtaining their oxygen from the air.

## Methods of Study

A few members of the lacustrine zoobenthos are large enough to be directly observed in situ (e.g., by SCUBA or mask-and-snorkel), but most species must be collected and brought into the laboratory for study. Scientists have invented a wide range of gear to collect benthic animals (**Figure 5**). Most often, scientists use grabs, corers, sweep nets, closing bags or boxes, or traps to collect animals or sediments. Often, animals need to be separated from the sediments or plants in which they live. This is usually accomplished by washing the sample through a sieve, most often of 0.1–1-mm mesh, sometimes followed by staining the sample with a dye such as Rose Bengal or examining the sample under a low-power microscope to aid in finding the animals. More or less experimental methods such as density-gradient centrifugation using silica sols (e.g., Ludox®) or application of ice or gentle heat are sometimes used as well.

## Variation in the Lacustrine Zoobenthos

One of the chief characteristics of the zoobenthos is its extreme patchiness. The abundance and species composition of the zoobenthos varies within lakes, at scales ranging from centimeters (replicate samples at a single site) to kilometers, as well as among lakes. A number of factors are known to affect the abundance and species composition of the zoobenthos.

### Variation within Lakes

The composition of the zoobenthos almost always varies greatly with water depth within a lake. The abundance of nearly every species in the zoobenthos varies with water depth (**Figure 6**), and species richness often is much lower in the deep waters of a lake than near the shore (**Figure 7**). Many studies have also reported variation in the total numbers or biomass of benthic animals with water depth in individual lakes. No single pattern of variation of abundance or biomass with water depth applies to all lakes, because the many mechanisms that link water depth to the zoobenthic community vary in strength across lakes and produce a wide range of patterns in different lakes.

The most important factors that cause zoobenthic composition and abundance to vary with water depth include dissolved oxygen, quantity and quality of organic matter inputs, temperature, sediment grain size and compaction, disturbance, and depth-dependent biotic interactions. The relative contributions of each of these factors and their interactions in driving zoobenthic community structure have not yet been disentangled. In stratified lakes, concentrations of hypolimnetic dissolved oxygen fall through the stratified period, and often reach zero at the sediment surface by the end of the summer. Only a few species of animals can tolerate hypoxic ($<2\,\text{mg}\,l^{-1}$) or anoxic conditions for any period of time. The hydrogen sulfide that often accompanies anoxia also is toxic to most animals. Thus, inadequate dissolved oxygen at the sediment surface probably is a major cause of the vertical zonation of the zoobenthos and low species richness in profundal sediments. Nevertheless, declines in richness often occur well above the depth of oxygen depletion (**Figure 7**), so factors other than oxygen must be important as well. Although few zoobenthic species can tolerate anoxia, these species may be abundant, so low dissolved oxygen does not necessarily reduce zoobenthic density or biomass. Even in unstratified lakes, periods of warm, windless weather may reduce mixing enough to cause short-term depletion of oxygen at the sediment surface and catastrophic losses of benthic animals. Such ephemeral stratification and oxygen depletion killed nearly the entire population of ~2500 metric tons (dry mass) of the mayfly *Hexagenia* in the shallow western basin of Lake Erie in late summer 1953.

Temperatures in the hypolimnion of a stratified lake are much lower and steadier than in the epilimnion. These low temperatures slow metabolic rates of benthic animals, may not meet thresholds for growth and reproduction of many species, and presumably exclude many species from profundal sediments.

The food base for the zoobenthos changes from the shoreline to the profundal zone. Aquatic plants and attached algae grow only in relatively shallow or clear water. In deep water, freshly settled phytoplankton supplements older detritus washed in from the watershed and littoral zone, and the amount and quality of organic matter reaching the profundal sediments of deep lakes probably declines with increasing water depth. The deposition of this sinking organic matter on the lake bed is highly uneven,

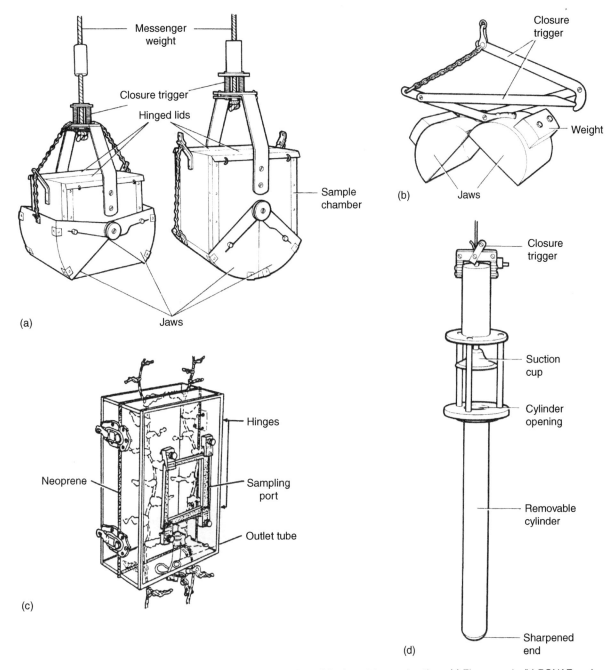

**Figure 5** Four samplers commonly used for quantitative studies of the lacustrine zoobenthos. (a) Ekman grab; (b) PONAR grab; (c) Downing box sampler for vegetation-dwelling invertebrates; (d) Kajak-Brinkhurst corer. Adapted from Downing (1984, 1986).

depending on wave energy and the slope of the bottom. It seems reasonable to suppose that the quantity and quality of food is usually highest in the littoral zone and in depositional areas, and leads to variation in zoobenthic biomass (**Figure 8**).

Sediment grain size and heterogeneity also vary with water depth, typically changing from a highly patchy mosaic of coarse sediments in shallow water to a monotonous plain of fine-grained sediments in the profundal zone. Because of the many mechanisms that link benthic animals to their sediments, this variation in sediments must have a large effect on the kinds of benthic animals that can live at different depths in a lake. High heterogeneity in shallow-water sediments probably is a major cause for the high species richness in the littoral zoobenthos.

Disturbance from wave-wash, ice-push, or fluctuating water levels often causes a zone of markedly

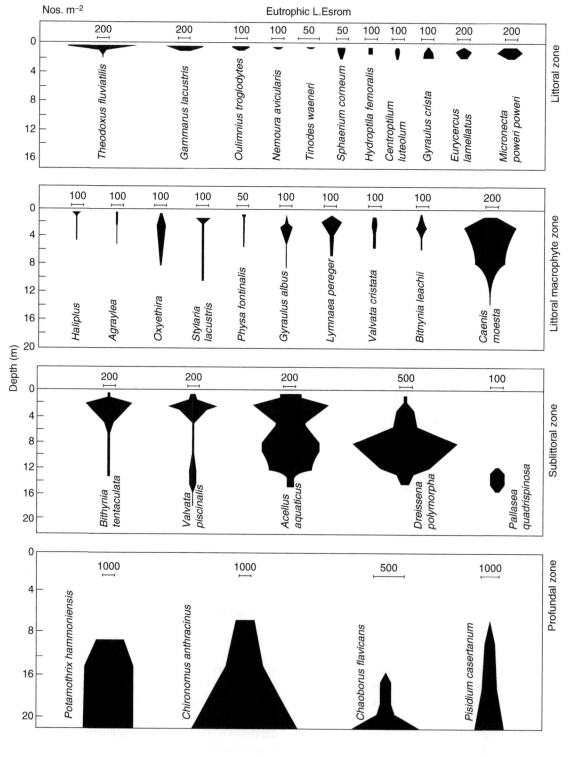

**Figure 6** Distribution of benthic animals with water depth in eutrophic Lake Esrom, Denmark. Adapted from Jonasson (2004).

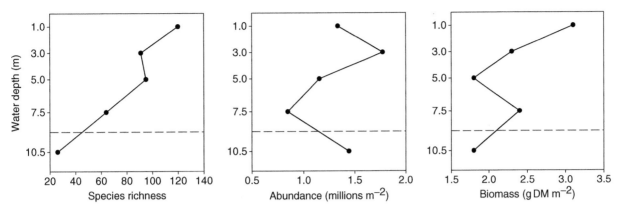

**Figure 7**  Species richness, abundance, and biomass of benthic animals (including all metazoans) as a function of water depth in Mirror Lake, a small unproductive lake in New Hampshire. The dashed horizontal line shows the depth at which oxygen at the sediment surface falls below 1 mg $l^{-1}$ by late summer. Adapted from Strayer (1985).

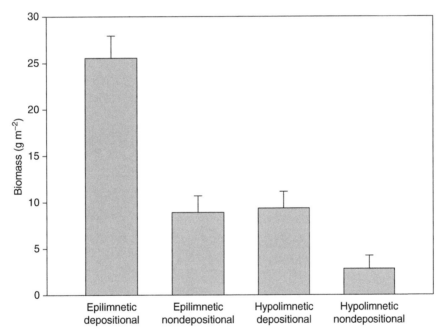

**Figure 8**  Mean (±1SE) biomass (wet mass) of macrobenthic animals at various sites in Lake Memphremagog, Quebec, as a function of thermal and depositional regime. Adapted from Rasmussen and Rowan (1997).

reduced zoobenthic density and diversity near the water's edge, particularly in large lakes and reservoirs.

Finally, many of the biotic interactions that affect the zoobenthos are depth-dependent. Fish predation can have very large effects on the number, size, and species composition of benthic animals (**Figure 9**). The numbers and kinds of fish change from the littoral to the profundal zone; indeed, anoxic profundal sediments may be free from fish predation. Likewise, the density and kind of macrophytes, which are important in providing surfaces for attachment, shelter from predation, and food for benthic animals, change markedly with water depth, and can drive major changes in the zoobenthos. Thus, biotic interactions underlie much of the within-lake variation in the zoobenthos.

Most benthic animals are found near the sediment surface (**Figure 10**), presumably because food (benthic algae and sinking phytoplankton) and oxygen are most available there. Nevertheless, animals can be found deeper in lake sediments, sometimes reaching depths of more than 50 cm. Tubificid oligochaetes feed head down in the sediments, so it is easy to understand why they reach deep below the sediment–water interface, but the activities of other benthic animals that live deep within-lake sediments (e.g., some candonid ostracods, midges and bdelloid rotifers) are less well understood.

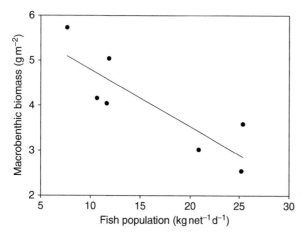

**Figure 9** Dependence of macrobenthic biomass (wet mass) on fish predation in the Finnish Lake Pohjalampi. Each point is one year during a long-term experimental manipulation of fish populations. Adapted from Leppä *et al.* (2003).

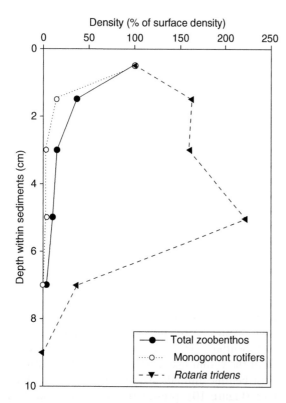

**Figure 10** Vertical distribution of benthic animals within the sediments of Mirror Lake, New Hampshire. The three lines show the distributions of all benthic animals, a surface-dwelling taxon (monogonont rotifers), and a taxon that lives deeper in the sediments (the bdelloid rotifer *Rotaria tridens*). Modified from Strayer (1985).

## Variation across Lakes

Organic matter inputs, especially phytoplankton production, affect the numbers and kinds of benthic animals in lakes (**Figure 11**). Typically, lakes with higher organic matter inputs support more benthic animals (**Figures 3** and **11**), although there is some evidence that very high inputs of organic matter may actually reduce numbers of benthic animals, possibly by increasing the area of anoxic sediments. The kinds of benthic animals change with lake productivity as well. Indeed, one of the earliest systems for classifying the productivity of lakes was based on the composition of the profundal zoobenthos.

Lake morphometry has a strong influence on zoobenthic communities. Small, shallow lakes tend to support higher densities of benthic animals than do

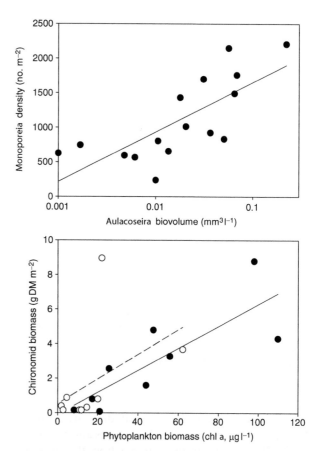

**Figure 11** Examples showing the dependence of the zoobenthos on phytoplankton. The upper panel shows the numbers of the amphipod *Monoporeia affinis* in Lake Erken, Sweden, as a function of the amount of the important planktonic diatom *Aulacoseira* in the previous year ($r^2 = 0.54$, $p = 0.0008$). The lower panel shows the biomass of chironomids in April–May at two sites in Lake Balaton, Hungary, as a function of phytoplankton biomass in the previous summer (for black circles, $r^2 = 0.66$, $p = 0.008$; for white circles, $r^2 = 0.20$, $p = 0.23$). Adapted from Johnson and Wiederholm (1992) and Specziár and Vörös (2001).

large, deep lakes. This pattern probably has at least three causes: (1) macrophytes, which support dense and diverse zoobenthic communities, are most abundant in small, shallow lakes; (2) shallow lakes are less likely to stratify than deep lakes, resulting in higher concentrations of dissolved oxygen at the sediment surface; (3) sinking phytoplankton (and other food particles) are less likely to be degraded before reaching the sediment surface in shallow lakes than in deep lakes.

Temperature affects both the number and kinds of benthic animals in lakes. Tropical lakes support different species than do temperate lakes (or arctic lakes), although groups such as chironomids, *Chaoborus*, and oligochaetes are abundant over wide latitudinal ranges. Warm lakes tend to support higher zoobenthic densities than do cold lakes.

Several aspects of water chemistry have strong effects on the zoobenthos. Saline lakes contain distinctive communities of benthic animals. Very salty lakes may contain just a few species of benthic animals, even though overall abundance or biomass of the zoobenthos may not be reduced. Soft water lakes of low pH and low calcium typically contain different species than do hard water lakes, and often are poor in shell-bearing animals (mollusks, ostracods) that need calcium to build their shells. Consequently, the recent acidification from atmospheric deposition has had important effects on the zoobenthos. High concentration of dissolved organic matter in lake water may reduce abundance of benthic animals.

Finally, differences across lakes in biotic interactions can have strong effects on the number and kinds of benthic animals. Only a few examples are well known. Fishless lakes usually contain large, active invertebrates (e.g., swimming or crawling insects) that are quickly eliminated if fish are introduced. Crayfish and large bivalves can have strong effects on other benthic animals, as has been shown when invasions of lakes by alien crayfish and *Dreissena* have caused very large changes in benthic animal communities. It seems likely that other unstudied biotic interactions are responsible for variation in the zoobenthos across lakes.

## Roles of the Lacustrine Zoobenthos

Secondary production by the zoobenthos in most lakes is modest (**Table 2**), and energy flow through the zoobenthos certainly is much smaller than that passing through microbial communities. Nevertheless, benthic animals play important roles in lake ecosystems, and even in human affairs. Broadly speaking, the ecological roles of the zoobenthos can be defined as food web or nontrophic roles.

As participants in lacustrine food webs, benthic animals consume phytoplankton, aquatic plants, animals, bacteria, and detritus. Consumption rates can be high enough to control the amount and composition of these food resources. The best-known examples involve the consumption of plankton by benthic animals. Suspension-feeders such as *Dreissena*, sponges, and cladocerans may be abundant enough to reduce phytoplankton biomass, or change its composition (**Figure 12**). Likewise, benthic animals that eat zooplankton (*Chaoborus*, mysids) can exert strong control on zooplankton communities,

**Table 2** Secondary production by the macrozoobenthos and zooplankton in several lakes, expressed as a percentage of net organic inputs to the lake

| Lake | Mean depth (m) | Net organic inputs $(g\ C\ m^{-2}\ yr^{-1})$ | Production of zoobenthos (%) | Production of zooplankton (%) |
|---|---|---|---|---|
| Tundra pond, AK | <1 | 26 | 7 | 0.8 |
| Marion, BC | 2 | 110 | 3 | 0.9 |
| Myvatn, Iceland | 2 | 330 | 6 | 1.3 |
| Wingra, WI | 3 | 610 | 0.4 | 4 |
| Kiev Reservoir, Ukraine | 4 | 280 | 6 | 7 |
| Rybinsk Reservoir, Russia | 6 | 93 | 0.3 | 2 |
| Mirror, NH | 6 | 49 | 12 | 5 |
| Red, Russia | 7 | 140 | 1 | 7 |
| Findley, WA | 8 | 12 | 6 | 2 |
| Naroch, Belarus | 9 | 160 | 0.8 | 5 |
| Mikołajskie, Poland | 11 | 260 | 2 | 21 |
| Esrom, Denmark | 12 | 160 | 6 | 7 |
| Pääjärvi, Finland | 14 | 60 | 3 | 12 |
| Dalnee, Russia | 32 | 260 | 1 | 22 |

Most benthic data exclude the meiofauna, and are therefore underestimates. Because of methodological differences among studies, data are only approximately comparable. Compiled from many sources.

with effects that ramify through the ecosystem (**Figure 16**). Although relatively few benthic animals (crayfish, plant–parasitic nematodes, and some aquatic insects) eat or destroy aquatic plants, they can strongly affect macrophyte biomass and species composition; indeed, benthic animals have been used for biological control of nuisance weeds such as milfoil. Much less is known about the influence of the zoobenthos on attached algae, bacteria, and detritus, although these are the primary foods of most benthic animals. Further, some benthic animals have highly specialized diets and might therefore have selective effects on food resources. Benthic animals probably often control the amount and kind of attached algae in lakes, and it seems likely that most particles of detritus pass through at least one benthic animal before being buried in the lake bottom. Thus,

there is ample evidence that benthic animals can control the amount and character of food resources in lakes, although we do not yet know how often and under what conditions such strong control occurs.

Benthic animals release nutrients such as inorganic nitrogen and phosphorus when they consume food. Such nutrient regeneration may be important to phytoplankton and attached periphyton in nutrient-limited lakes.

The predominant fate of zoobenthic production is to be eaten, whether by other members of the zoobenthos, fish, or terrestrial predators (**Table 3**). Consequently, the zoobenthos serves as an important link to higher trophic levels. Almost all lake-dwelling fish depend to some extent on zoobenthos, and benthic animals are the primary food of many fish species (**Figure 13**). Large populations of birds, bats, spiders, and other predators may be drawn to lakes or lakeshores to feed on benthic animals, including emerging insects.

Although it has been customary to treat the plankton and the benthos as belonging to separate food webs, many species of the zoobenthos depend either directly or indirectly on the plankton for food, and there are strong reciprocal links between the plankton and the zoobenthos. The pelagic and benthic zones of lake are linked by many strong connections (**Figure 14**), and function as an integrated system.

The major nontrophic role of the zoobenthos is sediment mixing (bioturbation). Three activities are important: (1) feeding, especially by conveyor-belt feeders such as tubificid oligochaetes, which feed in deep layers of the sediments and leave fecal pellets at the sediment surface; (2) burrow construction and ventilation, which are done by ephemerid mayflies and some chironomids; (3) ordinary locomotion by large, active animals such as *Chaoborus* and unionid bivalves. Bioturbation mixes sediments and increases

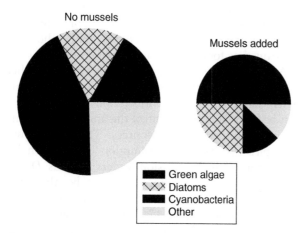

**Figure 12** Changes in the amount and composition of phytoplankton in two small Dutch lakes following the experimental addition of *Dreissena polymorpha* to one of the lakes. The area of each circle is proportional to the biovolume of the phytoplankton in each lake. Adapted from data of Reeder *et al.* (1993).

**Table 3** Estimated fate of zoobenthic production in several lakes

| Lake | Production (gDM m$^{-2}$ yr$^{-1}$) | Fate (% of production) | | | |
| --- | --- | --- | --- | --- | --- |
| | | Invertebrate predation | Fish predation | Bird predation | Emergence |
| Myvatn, Iceland | 42 | 5–52[a] | 43 | 9 | 48 |
| Mirror, NH | 14 | 80 | 15 | nd | 25 |
| Paajarvi, Finland | 3.8 | 50 | 40 | nd | 6 |
| Batorin, Belarus | 2.6[b] | 46 | 194 | nd | nd |
| Naroch, Belarus | 2.6[b] | 22 | 42 | nd | nd |
| Ovre Heimdalsvatn, Norway | 2.4[c] | 25–28[a] | 70 | 2 | 3 |
| Myastro, Belarus | 0.8[b] | 72 | 272 | nd | nd |

Data are approximate, so percentages do not sum to 100%; nd = not determined. Data are from after various sources.
[a]Estimated by difference.
[b]Growing season only; macrobenthos.
[c]Macrobenthos.

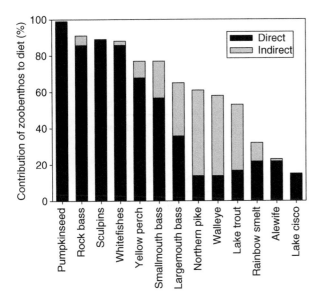

**Figure 13** Contribution of zoobenthos to the diets of common species of fish in northeastern North American lakes. 'Indirect' consumption of zoobenthos is consumption of fish that were supported by zoobenthos. Adapted from data compiled by Vadeboncoeur *et al.* (2002).

exchange of dissolved substances (e.g., oxygen, ammonium) between the overlying water and the sediment pore-water. These activities can have large effects on nutrient regeneration and cycling and burial of toxins. In addition, shell-building benthic animals (chiefly large bivalves) can alter the character of benthic habitats through accumulations of their living and dead shells. Such shell accumulations, which can probably reach masses $>1\,kg\,m^{-2}$, serve as habitat or spawning sites for other animals, as well as altering biogeochemical transformations at the sediment–water interface.

## Applied Issues

Some of the most important diseases of humans and wildlife are carried by benthic animals. Chief among these are diseases caused by trematodes (or flukes), in which freshwater snails, and occasionally decapods, serve as intermediate hosts. In a typical life cycle (**Figure 15**), eggs shed from the definitive host hatch into swimming larvae (miracidia) that seek out and penetrate an aquatic snail. Many species of freshwater snails serve as hosts for the various trematodes that affect humans, livestock, and wildlife. After undergoing development in the snail, a second free-swimming larval stage (the cercaria) may either enter the definitive host directly when the host is in the water, enter a second intermediate host (a fish or decapod), or encyst on an aquatic plant. The

trematode passes from the second intermediate host or aquatic plant into the definitive host when uncooked fish, decapods, or aquatic plants are eaten. Adult trematodes may live in the intestines, liver, blood vessels, or lungs of humans or other vertebrates, and often cause chronic, debilitating diseases. The most important of the snail-borne diseases is schistosomiasis, a debilitating disease caused by the genus *Schistosoma*, which affects ~200 million humans throughout the tropical world. Other significant snail-borne diseases include liver flukes (especially *Fasciola hepatica* and *Opisthorchis sinensis*), which cause large damage to sheep and cattle, as well as affecting millions of people, and various intestinal or lung flukes, which affect many people around the world. Trematodes using freshwater snails as intermediate hosts also affect many vertebrate species other than humans. In fact, cercaria of trematode species that use birds as definitive hosts sometimes burrow into humans. Although such cercariae do not develop normally in humans, they can cause a skin irritation known as 'swimmer's itch' and thereby limit recreational use of lakes. The acanthocephalans are another important group of parasites carried by freshwater benthic animals. Adult acanthocephalans are intestinal parasites of many fishes and other aquatic vertebrates, and benthic crustaceans usually serve as their intermediate hosts.

Human-caused changes to natural habitats sometimes increase disease problems (e.g., impoundments have increased prevalence of schistosomiasis), and ecological interventions (e.g., habitat management, introductions of predators or other biological controls) may be implemented as part of integrated programs of disease control.

Other benthic species are pests. Fouling species such as *Dreissena*, *Corbicula*, and occasionally sponges and ectoprocts block water intakes and may force plant operators to use mechanical cleaning, biocidal chemicals, or pipe coatings to keep water flowing. It appears that large emergences of chironomid midges may be a major cause of 'hay fever' and asthma in some parts of the world. Mass emergences of lake-dwelling insects (chiefly ephemerid mayflies or chironomid and chaoborid midges) may be so large that they cause traffic hazards, produce windrows of dead insects that need to be cleaned up with heavy equipment, and even short-circuit power plants.

Although there has been some interest in biomanipulation to increase the positive impacts of benthic animals, or reduce their negative impacts, such efforts have not proceeded as far as in the pelagic zone, where biomanipulation is now widely practiced. At least three types of biomanipulation have been considered involving benthic animals. Because benthic

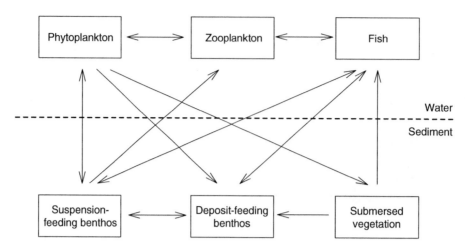

**Figure 14**    Major interactions between lacustrine communities (Strayer, 2006). Arrows show the hypothesized direction of control; note that many interaction arrows cross between the sediments and the pelagic zone.

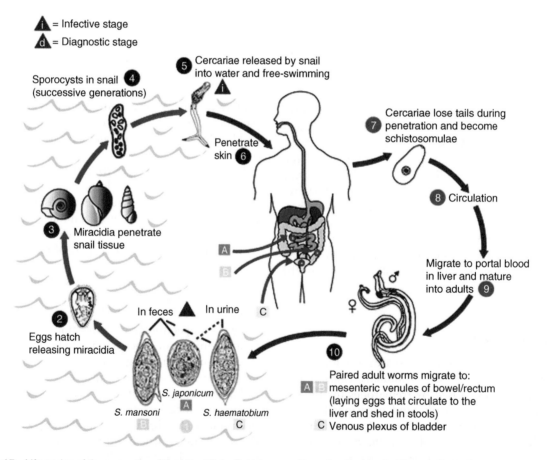

**Figure 15**    Life cycles of three species of the blood fluke *Schistosoma*. From the Centers for Disease Control.

suspension-feeders may control phytoplankton biomass and composition, some studies have sought to manage nuisance phytoplankton by increasing populations of benthic suspension-feeders (**Figure 12**). Benthic animals are valuable food for fish, so fisheries managers often have introduced large benthic animals

(crayfish and the opossum shrimp *Mysis*) to lakes to increase growth rates or biomass of fish populations. Such introductions rarely have been supported by a careful analysis of likely impacts, and introductions of forage invertebrates sometimes have led to undesirable and unforeseen impacts (**Figure 16**). Finally,

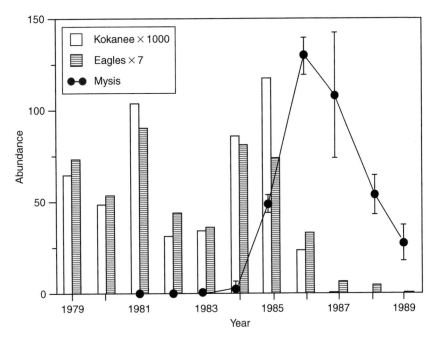

**Figure 16** Undesirable effects following the introduction of the opossum shrimp *Mysis* to Flathead Lake, Montana. Data on kokanee salmon and eagles are peak counts from a tributary stream used by the salmon for spawning. Adapted from Spencer *et al.* (1991).

there have been a few attempts to increase populations of benthic animals to control populations of animals that carry diseases. Probably, the most prominent examples have been the additions of predators or competitors of the snails that carry schistosomiasis. Clearly, biomanipulation of lakes using benthic animals is still in its infancy.

Like other freshwater plants and animals, human activities have harmed some species of the lacustrine zoobenthos, which are now extinct or imperiled. Several activities have been especially harmful. Large water diversions have caused some lakes to become excessively salty or even dry up altogether, endangering or eliminating the benthic animals that formerly lived there. Many lakes have been badly polluted by toxins from industry and other sources or sediment from poor land-use practices. Nutrient loading from fertilizers or domestic wastes has increased the productivity of many lakes, leading to anoxic sediments, with obvious consequences for benthic animals. Many of the alien species (sport fish, aquatic plants, and invertebrates) that have been introduced into lakes around the world have had strong effects on benthic animals. An accounting of the summed effects of these harmful activities on benthic animals does not exist, but surely many populations in individual lakes have been imperiled or eliminated. In the case of ancient lakes that support endemic species of benthic animals, local extirpations would translate into global extinctions for the species. Extinctions among the endemic fish species of ancient lakes

are well known, and presumably, such extinctions have occurred among the zoobenthos as well.

A few species of benthic animals are harvested from lakes or reservoirs for human use. Important fisheries for wild or cultured populations of various freshwater decapods (crayfishes, prawns, crabs) are scattered in lakes, ponds, and wetlands around the world. Several of the large bivalves have been harvested for millennia for food, pearls, and mother-of-pearl, although the largest fisheries for these animals are in rivers rather than in lakes. Particularly in the nineteenth century, large numbers (>10 million animals/year) of the medicinal leech (*Hirudo medicinalis*) were taken from the wild. Other members of the lacustrine zoobenthos (adult chaoborid midges, eggs and adults of corixid bugs, snails) are harvested in large numbers for human food locally.

## Glossary

**Benthos** – The community of organisms living around surfaces (e.g., sediments, plants) in aquatic ecosystems.

**Bioturbation** – Sediment mixing caused by the activities of organisms.

**Macrobenthos** – Benthic animals large enough to be retained on a coarse (usually ~0.5-mm mesh) sieve.

**Meiobenthos** – Benthic animals too small to be retained on a coarse (usually ~0.5-mm mesh) sieve, but large enough to be retained on a fine (usually ~0.05-mm mesh) sieve.

**Zoobenthos** – The community of animals living around surfaces (e.g., sediments, plants) in aquatic ecosystems.

*See also:* Biomanipulation of Aquatic Ecosystems; Littoral Zone; Trophic Dynamics in Aquatic Ecosystems.

## Further Reading

Brinkhurst RO (1974) *The Benthos of Lakes.* New York: St. Martin's Press.

Downing JA (1984) Sampling the benthos of running waters. In: Downing JA and Rigler FH (eds.) *A Manual on Methods for the Assessment of Secondary Productivity in Fresh Waters,* 2nd edn., pp. 87–130. Oxford: Blackwell.

Downing J (1986) A regression technique for the estimation of epiphytic invertebrate populations. *Freshwater Biology* 16: 161–173.

Hakenkamp CC, Morin A, and Strayer DL (2002) The functional importance of freshwater meiofauna. In: Rundle SD, Robertson AL, and Schmid-Araya JM (eds.) *Freshwater Meiofauna: Biology and Ecology,* pp. 321–335. Backhuys.

Hutchinson GE (1993) *A Treatise on Limnology, vol. IV. The Zoobenthos.* New York: Wiley.

Johnson RK and Wiederholm T (1992) Pelagic-benthic coupling – the importance of diatom interannual variability for population oscillations of *Monoporeia affinis. Limnology and Oceanography* 37: 1596–1607.

Jónasson PM (2004) Benthic invertebrates. In: O'Sullivan PE and Reynolds CS (eds.) *The Lakes Handbook, vol. 1. Limnology and Limnetic Ecology,* pp. 341–416. Malden: Blackwell.

Kajak Z, Bretschko G, Schiemer F, and Leveque C (1980) Zoobenthos. In: Lecren ED and Lowe-McConnell RH (eds.) *The Functioning of Freshwater Ecosystems,* pp. 285–307. Cambridge: Cambridge University Press.

Leppä M, Hamalainen H, and Karjalainen J (2003) The response of benthic invertebrates to whole-lake biomanipulation. *Hydrobiologia* 498: 97–105.

Merritt RW, Cummins KW, and Berg MB (eds.) (2007) *An Introduction to the Aquatic Insects of North America,* 4th edn. Dubuque: Kendall/Hunt.

Rasmussen JB and Kalff J (1987) Empirical models for zoobenthic biomass in lakes. *Canadian Journal of Fisheries and Aquatic Sciences* 44: 990–1001.

Rasmussen JB and Rowan DJ (1997) Wave velocity thresholds for fine sediment accumulation in lakes, and their effect on zoobenthic biomass and composition. *Journal of the North American Benthological Society* 16: 449–465.

Reeders H, bij de Vaate A, and Noordhuis R (1993) Potential of the zebra mussel (*Dreissena polymorpha*) for water quality management. In: Nalepa TE and Schloesser DW (eds.) *Zebra Mussels: Biology, Impacts, and Control,* pp. 439–451. Boca Raton, FL: Lewis Publishers.

Rundle SD, Robertson AL, and Schmid-Araya JM (eds.) (2002) *Freshwater Meiofauna: Biology and Ecology.* Leiden: Backhuys Publishers.

Särkkä J (1995) Profundal meiofauna in two large lakes: influence of pollution and bathymetric differences. *Archiv für Hydrobiologie* 132: 453–493.

Speczíar A and Vörös L (2001) Long-term dynamics of Lake Balaton's chironomid fauna and its dependence on the phytoplankton production. *Archiv für Hydrobiologie* 152: 119–142.

Spencer CN, McClelland BR, and Stanford JA (1991) Shrimp stocking, salmon collapse, and eagle displacement. *Bioscience* 41: 14–21.

Strayer D (1985) The benthic micrometazoans of Mirror Lake, New Hampshire. *Archiv für Hydrobiologie Supplementband* 72: 287–426.

Strayer D (1991) Perspectives on the size structure of the lacustrine zoobenthos, its causes, and its consequences. *Journal of the North American Benthological Society* 10: 210–221.

Strayer DL (2006) The benthic animal communities of the tidal-freshwater Hudson River estuary. In: Levinton JS and Waldman JR (eds.) *The Hudson River Estuary,* pp. 266–278. Cambridge University Press.

Thorp JH and Covich AP (eds.) (2001) *Ecology and Classification of North American Freshwater Invertebrates,* 2nd edn. San Diego: Academic Press.

Vadeboncoeur Y, Vander Zanden MJ, and Lodge DM (2002) Putting the lake back together: reintegrating benthic pathways into lake food web models. *BioScience* 52: 44–54.

# Trophic Dynamics in Aquatic Ecosystems

**U Gaedke,** University of Potsdam, Potsdam, Germany

## Introduction

An understanding of the role of species populations within an ecosystem is possible only when information on multiple species is systematically joined in a way that reflects mutual interactions. A synthesis of this type produces a view of ecosystem processes that are the by-product of many simultaneous interactions among populations.

Feeding relationships have proven to be an effective means of bridging the gap between populations and ecosystems. Studies of feeding at the ecosystem level may be referred to as 'food-web analysis' or 'trophic dynamics.' Early studies of this type involved diagrammatic portrayal of feeding levels (trophic levels) in an ecosystem (**Figure 1**). Such a diagram begins with photoautotrophs (level 1), which live by using inorganic substances and solar energy to synthesize biomass. Photoautotrophs are consumed by herbivores, which are the first level of consumers. Herbivores in turn are consumed by carnivores of the first level, which are consumed by carnivores of the second level. When the organisms are arranged by trophic level, they can be shown as a vertical stack of boxes that comprise a feeding pyramid (or trophic pyramid), or they can be shown in more detail as a diagram with spatially separated boxes connected by lines indicating the feeding pathways (**Figure 2**). Because there are numerous kinds of organisms at any given trophic level, a feeding diagram takes the form of a web, which gives rise to the term 'food web.' A diagrammatic analysis of a food web may be relatively simple if it focuses only on the dominant members of each level, or it may be much more detailed if it includes the subdominant members as well.

While a qualitative diagram is informative if it is accurate, a quantification of the amount of energy or materials passing through the food web produces more insight (**Figure 2**). Because quantification of a food web involves the linkage of compartments by dynamic pathways representing the flow of materials or energy, the quantitative study of food webs often is conducted through the construction of computer models.

Feeding relationships as portrayed in a food web may be quantified by any of several means, according to the interest of the investigator. For example, the amount of biomass passing from one level to another may be quantified. Because biomass has an energy equivalent per unit mass, quantification of energy flow also is possible, and is especially useful in connecting the flow of energy back to the solar source and to the metabolism of individual organisms or groups of organisms. In addition, it is possible to study food webs quantitatively in terms of the passage of an element or even a class of compounds through the food web. Carbon, a surrogate for biomass, is often traced through food webs, as are phosphorus and nitrogen, the two most critical inorganic nutrients in aquatic ecosystems. Also it would be possible to demonstrate or predict the flow of fatty acids or carbohydrates through food webs. Thus, food-web modeling is applicable to a wide range of interests that involve feeding interactions.

## Trophic Levels and Trophic Positions

Modern food-web analysis grew from the concept of the trophic pyramid (**Figure 1**), which was used by the British ecologist Charles Elton (1900–1991) and other founders of modern ecology to summarize information on the roles of organisms in ecosystems. Early portrayals of food webs typically were based on the number of organisms or on biomass. For many environments, such a diagram takes a pyramidal shape. The laws of nature do not require the pyramidal shape, however. For example, a higher trophic level may have more biomass than a lower trophic level adjacent to it.

The concept of the trophic pyramid was advanced significantly toward modern food-web analysis by Raymond Lindeman (1915–1942) who, working with Evelyn Hutchinson (1933–1991), introduced the concept of trophic dynamics. According to Lindeman, the linkages between trophic levels, when portrayed as energy flow, provide a view of the ecosystem that is linked to the solar energy source and to the metabolism of individual organisms. Furthermore, an analysis of trophic relationships based on energy leads to the calculation of energy-transfer efficiencies and other related phenomena that are readily associated with an energy-based analysis.

The early concept of trophic levels contained several important flaws, the solution of which has produced a number of innovations. The first and most obvious flaw is that individual kinds of organisms often cannot be assigned to a single trophic level. For example, an omnivorous organism, such as some types of crayfish, may be able to derive nourishment from both plant and animal matter, and therefore cannot be assigned either to the level of herbivore or to the level of carnivore. It is especially common for high-level carnivores to feed at multiple lower levels, including herbivores and lesser carnivores, at the same time.

The modern view is that species must be assigned fractional trophic levels that reflect their diet. For example, an organism feeding 50% at level 2 and 50% at level 3 would be assigned a level of 2.5. Thus, models can take into account proportionate differences in trophic level.

A second problem arises from organisms that use particulate organic matter (POM, detritus) or dissolved organic matter (DOM; also designated dissolved organic carbon, DOC) as food. Modern food-web analysis views such feeding relationships as a 'detrital food chain,' which is an important complement to the more traditionally recognized

feeding relationships, the 'grazer food chain,' based on photoautotrophs, which pass organic matter to herbivores. The detrital food chain, like the grazer food chain, contains multiple levels (**Figure 3**).

Inclusion of the detrital food chain leads to a more realistic view of bacteria in ecosystems. Bacteria are often the direct consumers of detritus (which they dissolve prior to consuming) and dissolved organic matter. Even though they create biomass at the base of the detrital food chain, they cannot do so through use of sunlight, as is the case of production of photoautotrophs, such as algae and aquatic vascular plants at the base of the grazer food chain. Because detritus and dissolved organic matter are abundant in aquatic ecosystems, the realities of the detrital food chain must be incorporated into quantitative models.

The increasingly realistic view of food webs leads away from the traditional pyramid or four-step energy diagram (**Figure 1**) and more toward diagrams that show various groups of organisms occupying noninteger positions in the grazing food chain, and connected to the detrital food chain, which incorporates bacteria as first-level producers of biomass (**Figure 2**).

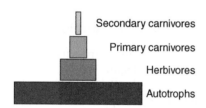

Secondary carnivores

Primary carnivores

Herbivores

Autotrophs

**Figure 1**  An example of a pyramid of biomass of organisms arranged by trophic level. Such pyramids led to modern trophic diagrams and food-web analysis.

## Use of Trophic Guilds to Analyze the Trophic Structure of Food Webs

A trophic guild consists of all organisms within a food web that have similar food sources and predators (e.g., boxes shown in **Figure 2**). In most cases, trophic guilds defined in this way would consist of multiple species, all of which would occupy similar positions

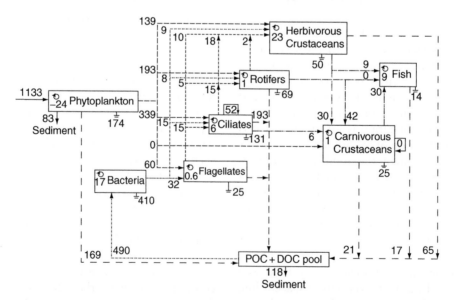

**Figure 2**  A modern food-web diagram for the pelagic zone (open water) of Lake Constance, adjacent to Germany, Switzerland, and Austria. Fluxes are given as mg C m$^{-2}$ d$^{-1}$. Source: U. Gaedke.

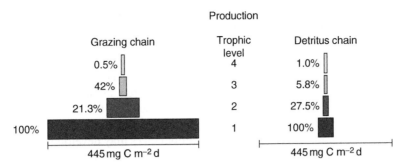

**Figure 3** Trophic pyramids for the grazing chain and the detritus chain of Lake Contance, shown in terms of annual production. Source: U. Gaedke.

in the trophic diagram or food-web model. Thus, application of the concept of guilds simplifies analysis and modeling of food webs.

One complication of the use of guilds is change in feeding habits for a species during its growth and development. For example, a carnivorous fish species may feed on zooplankton when it is small but feed on other fish when it is large. In such cases, it would be justified to treat the younger stages of a certain species as belonging to a different guild than the older individuals of the same species.

**Figure 2** shows the structure of a food-web model of the open waters of Lake Constance, which adjoins Germany, Switzerland, and Austria. The model is based on boxes corresponding to guilds of organisms as well as a pool of DOM and POM. Carbon comes into the food web through this DOM and POM pool or through a photosynthetic guild (in this case, phytoplankton) and leaves the food web either as sedimentation of POM to the bottom of the lake or release of $CO_2$ gas to the water as a by-product of respiration. Circles within each of the boxes indicate net growth. Arrows passing from one box toward another indicate consumption, and arrows passing out of each box indicate grazing, predation, fecal output, or other losses. The numbers indicate fluxes (biomass per unit time for a given unit area of the lake, in this case expressed as mg C m$^{-2}$ d).

A quantitative model such as the one shown in **Figure 2** is either an average for a considerable period of time or snapshot of an instant in time. The flow of mass from one compartment to another can be expected to change substantially from season to season and even week to week in a lake. Thus, modeling that encompasses changes occurring over any substantial period of time (e.g., a year) can require a great deal of information.

Although the progressive change in dynamics of food webs over time is seldom quantified because of the large amount of data that would be required, averages or snapshot views of food webs can reveal the main fluxes that account for ecosystem metabolism and the efficiency of transfer of mass or energy through the food web.

## Trophic Transfer Efficiency

The trophic structure of a food web is affected greatly by loss of energy that occurs between trophic levels. It is not unusual for loss of energy between any two adjacent trophic levels to reach 90% (**Figure 3**) because in **Figure 3** losses are between ca. 70 and 80% throughout. As recognized by Lindeman when he created the concept of trophic dynamics based on energy transfer, the second law of thermodynamics, when applied to trophic transfers, requires that dissipation of energy as heat will be a substantial loss even for the most efficient transfers. For example, the growth efficiency of individual cells seldom exceeds 40%, even when nonmetabolic losses are disregarded. Ecological factors introduce nonmetabolic inefficiencies related to nongrazing mortality, export of organic matter from the system (outflow or sedimentation), and elimination (fecal output).

Nongrazing mortality, which can occur when organisms are exposed to lethal physical or chemical conditions, physiological death related to age, or inadequate nutrition, divert biomass and energy from the transfer process that connects one trophic level to the next in the grazer food chain. Nongrazing mortality passes energy and mass to the detrital food chain. Elimination (fecal output) also is drain on efficiency in that mass or energy passes from the grazer food chain to the detrital food chain. Losses to elimination may be especially high for organisms feeding on organic sources that are difficult to digest. This is particularly the case for herbivores consuming vascular plant tissues, which are rich in complex carbohydrates that are difficult to digest.

All organisms must respire in order to live. Respiration involves the release of chemical energy from organic matter. The released energy is used in part for synthesis of new biomass, and is also used to maintain the integrity of basic metabolic functions that sustain life. Therefore, a considerable amount of organic matter is lost (converted to $CO_2 + H_2O$ and energy) for energy production purposes without passing to a higher trophic level.

The various modes by which energy and mass can be lost are the basis for quantifying several kinds of ecological efficiencies that are used in food-web analysis. The ratio of biomass synthesized (either growth or production of reproductive biomass such as eggs) to biomass ingested as food is symbolized $K_1$, the gross growth efficiency. $K_1$ rarely exceeds 0.30–0.35 under natural conditions, and often is considerably lower. The net growth efficiency of an organism ($K_2$) is defined as the ratio of new biomass produced to the amount of biomass that is assimilated (i.e., passing through the gut wall into the body of the organism). This efficiency has a maximum value of 50% for small organisms but is lower for large organisms (e.g., fish).

Trophic transfer efficiency is the ratio of biomass production at one trophic level to the biomass production of the next lower level. In plankton food webs, trophic transfer efficiencies may be high (0.15–0.30) (**Figure 3**) when compared with webs dominated by a transfer from vascular plants to herbivores. Temporal variability of the trophic-transfer efficiency is high, even within a given ecosystem, because of the instability of the numerous factors that can affect the trophic transfer efficiency. For example, the trophic transfer efficiency is strongly reduced when inedible algae (such as cyanobacteria) dominate or when animals with high respiratory needs (such as vertebrates and especially birds and mammals) prevail.

## Food Quality and Quantity

Heterotrophs (consumers, including bacteria) live by consumption of biomass or nonliving organic matter that is derived from biomass. Because the chemical composition of protoplasm in biomass (disregarding skeletal material or support structures) across all heterotrophs falls within a relatively narrow range, heterotrophs that feed on biomass are assimilating approximately the same mixture of elements that they will need in order to synthesize their own biomass (skeletal material and support structure typically pass through the gut, unassimilated). Detritivores also benefit from this carryover of elemental mixtures from one kind of organism to another, although detritus is more likely to show some selective loss of elements that would alter the balance typical of living biomass.

Unlike heterotrophs, photoautotrophs assimilate elements separately from water or, if they are rooted vascular plants, from sediments. For example, carbon is derived from $H_2CO_3$ and related inorganic carbon forms dissolved in water, and phosphorus is taken up separately as phosphoric acid that is dissolved in water. Because the inorganic substances required to synthesize biomass are taken up separately, large imbalances may develop when some essential components are much more abundant than others. Thus, autotrophs face greater challenges than heterotrophs in assembling the necessary ratios of elements to synthesize biomass, but even heterotrophs can experience imbalances of elements.

The approximate ratios of elements that are characteristic of autotrophic biomass have been extensively studied. Characteristic ratios of carbon to nitrogen and phosphorus are often the greatest focus of analysis because carbon is the feedstock for photosynthesis and phosphorus and nitrogen are the two additional elements that are often in short supply for conversion of photosynthetic products (carbohydrate) to other molecule types that are needed for the synthesis of protoplasm (e.g., amino acids). The importance of C:N:P ratios in aquatic organisms was first brought out by Alfred Redfield (1890–1983), who discovered that healthy oceanic phytoplankton show a characteristic atomic C:N:P ratio of about 106:16:1. Thus, the nutrient status of a phytoplankton community can be judged to some degree from the elemental ratios. For example, a phytoplankton community suffering phosphorus deficiency may show a C:P ratio of 500:1 rather than 106:1, as predicted by the Redfield Ratio for well-nourished phytoplankton. The analysis of elemental ratios for diagnosis of elemental imbalances is termed 'ecological stoichiometry.'

Imbalances in elemental ratios in one trophic level can create imbalances or inefficiencies at the next trophic level. This is particularly true between primary producers and herbivores. For example, phytoplankton suffering phosphorus scarcity may pass biomass with a high C:P ratio to grazers that consume phytoplankton. Because of an imbalance of elements in the food, the grazers must consume extra food in order to obtain the correct balance for the synthesis of their own biomass. Similarly, an especially low C:P ratio (e.g., 50:1) will provide an oversupply of phosphorus (typically this happens when bacteria are consumed), some of which would be released to the environment without generating any biomass.

One strategy that herbivores may employ in improving the elemental balance of food intake is to consume heterotrophs in addition to autotrophs (omnivory, which is feeding at multiple trophic levels). Thus, consumption of a phosphorus-rich food could be offset by consumption of a carbon-rich food, and the combination would provide more efficient use of ingested mass than a single food type.

## Trophic Structure of Food Webs in Open Water and Near Shore

In the open water of a lake at some distance from the shore (i.e., in the pelagic zone), the dominant autotrophs are phytoplankton, which live as individual cells or small colonies of cells suspended in the water. Some groups of phytoplankton (e.g., filamentous cyanobacteria) are rejected by most grazers because they are difficult to ingest or difficult to digest. Algae of low palatability may become quite abundant when grazing pressure is high in the pelagic zone. Overall, however, phytoplankton often is composed of a high proportion of edible and digestible species. Vascular plants (macrophytes), which typically grow near the shores of a lake (in the littoral zone), offer a less useful food supply because the digestible component of biomass is embedded in complex carbohydrates that maintain the shape of the plant. For this reason, much of the macrophyte biomass that develops during a growing season in the littoral zone of a lake dies and decomposes rather than being eaten while alive. The detrital food chain benefits from the biomass that was left uneaten by the grazer food chain.

Macrophytes may account for most of the autotroph biomass in a littoral zone, but attached algae colonize macrophytes and all other surfaces that are illuminated in the littoral zone. These attached autotrophs (periphyton) are an important food for herbivores in the littoral zone, even though the macrophytes themselves often are not.

The trophic structure in the pelagic zone often is dominated by four significant trophic levels. The top predator guild, which consists of multiple fish species, is not exposed to substantial predation pressure because of its size and mobility. The top predator guild therefore builds up biomass until it encounters a limitation caused by scarcity of appropriate food. This type of limitation is called a 'bottom up' limitation because the trophic level below is restricting increase in biomass of the trophic level above. While the top predator guild is controlled from the bottom up, its prey are controlled from the top down. 'Top down' control in this situation involves suppression of population biomass of the trophic level that serves as food for the abundant top predator.

In a four-level food web, the third level down from the top consists of herbivores. When top predators are abundant, the intermediate predators (e.g., fish that eat zooplankton) are suppressed, which relieves grazing pressure on herbivores. Therefore, herbivores, which consist mostly of zooplankton, may become abundant because of weak top-down control. This, in turn, will impose a strong grazing pressure on phytoplankton, which can cause a decline in phytoplankton biomass, yielding a higher water transparency. Thus, water clarity may be enhanced by stocking large predatory fish. This technique is an example of biomanipulation.

Physical factors may affect the structure of food webs. For example, the absence of any solid attachment points in the pelagic zone requires that all autotrophs be small. Thus, the food web must be based on grazers capable of harvesting large numbers of small autotrophs. The grazers that can do this work are also small, which creates a niche for small carnivores that eat the small herbivores. This leaves the opportunity for a top carnivore level that feeds on the small carnivores. Thus, predator and prey are related in body size through feeding behavior.

Food webs of the littoral zone in lakes are more similar to those of terrestrial environments or wetlands than to those of pelagic food webs of lakes if their production is dominated by vascular plants. Grazers (herbivores) may then be larger because larger units of food are available. Thus, top carnivores may feed directly on herbivores rather than on intermediate carnivores and body size and trophic level are not as strongly correlated as in pelagic food webs. In this case, the herbivores come under top-down control, and the efficiency of energy transfer from plants to herbivores is reduced because there are not enough herbivores to consume the total plant production. Thus, the length of the food chain in different portions of the ecosystem, or even in different ecosystems, affects the amount of biomass that can be produced by each trophic level.

## Transfer Efficiency along the Size Gradient in Pelagic Food Webs

In pelagic food webs, all autotrophs are small and predators exceed the size of their prey. Thus, the flow of matter and energy in pelagic food webs is from small to progressively larger organisms. The entire food web's trophic transfer efficiency along the size gradient of organisms reflects each of the trophic linkages and the number of trophic levels in the web. For example, efficiency of fish production in oligotrophic lakes is lower than the efficiency of fish production in eutrophic lakes when expressed as a proportion of the total amount of photosynthesis.

In unproductive lakes, food is more strongly dispersed, and is gathered by consumers that search for their prey. Consumers that locate their prey by searching usually are closer in size to their prey than consumers that feed without searching (by filtration of water, for example). When consumers are very close in size to their food source, the transfer of energy and materials along the size gradient up to large fish is less efficient because it requires more transfers to reach a given size. Thus, there is a theoretical explanation for the lower transfer efficiency along the size gradient in oligotrophic lakes. This is an example of the use of food-web analysis to explain observations about productivity and efficiency in lakes.

## Conclusion

Quantitative studies of food webs greatly magnify the value of information on individual species populations. Analysis of linkages in the food web provide explanations for the efficiency of energy or mass transfer to various points in the food web and the controlling influences that either enhance or suppress production at a given level in the food web. Thus, quantitative food-web studies support the understanding of mechanisms that govern the functioning of aquatic ecosystems.

## Glossary

**Detrital food chain** – Food-web components that begin with particulate organic matter (detritus) or dissolved organic matter, and pass through bacteria to other consumers.

**Ecological stoichiometry** – Study of element ratios within biomass in relation to element ratios within food or in the surrounding environment.

**Food web** – A diagram or model that shows the feeding connections between all major groups of organisms in an ecosystem.

**Grazer food chain** – Components of a food web that begin with photoautotrophs (algae, aquatic vascular plants) and pass through herbivores to carnivores.

**Trophic guild** – A group of organisms that share common food sources and common predators.

**Trophic level** – A group of organisms that obtain their food from sources of equal distance from the original source. Autotrophs equal level 1, herbivores equal level 2, primary carnivores equal level 3, and secondary carnivores equal level 4. Species feeding from more than one level may be assigned a fractional trophic level.

**Trophic transfer efficiency** – The fraction of total production at a given trophic level that is converted to production at the next trophic level.

*See also:* Lakes as Ecosystems.

## Further Reading

Begon M, Townsend CR, and Harper JL (2006) *Ecology-From Individuals to Ecosystems*, 4th edn., p. 738. Oxford, UK: Blackwell.

Gaedke U and Straile D (1994) Seasonal changes of trophic transfer efficiencies in a plankton food web derived from biomass size distributions and network analysis. *Ecological Modelling 77/76*: 435–445.

Gaedke U, Straile D, and Pahl-Wostl C (1996) Trophic structure and carbon flow dynamics in the pelagic community of a large lake. In: Polis G and Winemiller K (eds.) *Food Webs: Integration of Patterns and Dynamics*, pp. 60–71, Chapter 5. New York: Chapman and Hall.

Gaedke U, Hochstädter S, and Straile D (2002) Interplay between energy limitation and nutritional deficiency: Empirical data and food web models. *Ecological Monographs* 72: 251–270.

Lampert W and Sommer U (1997) *Limnoecology*, 398 pp. New York: Oxford University Press.

Morin PJ (1999) *Community Ecology*, 424 pp. Oxford, UK: Blackwell.

Sterner RW and Elser JJ (2002) *Ecological Stoichiometry*, 439 pp. Princeton, NJ: Princeton University Press.

Straile D (1997) Gross growth efficiencies of protozoan and metazoan zooplankton and their dependence on food concentration, predator–prey weight ratio, and taxonomic group. *Limnology & Oceanography* 42: 1375–1385.

# Eutrophication of Lakes and Reservoirs

**V Istvánovics,** Budapest University of Technology and Economics, Budapest, Hungary

## Definition

Eutrophication is the process of enrichment of waters with excess plant nutrients, primarily phosphorus and nitrogen, which leads to enhanced growth of algae, periphyton, or macrophytes. Abundant plant growth produces an undesirable disturbance to the balance of organisms (structural and functional changes, decrease in biodiversity, higher chance for invasions, fish kills, etc.) and to the quality of water (cyanobacterial blooms, depletion of oxygen, liberation of corrosive gases, and toxins, etc.).

## Introduction

Trophic status of a water body can roughly be assessed by using information about the concentration of the limiting nutrient (phosphorus), chlorophyll (an indicator of phytoplankton biomass), and transparency (dependent on both algal biomass and sediment resuspension, expressed as Secchi depth). The most widely accepted limits are those suggested by the Organization for Economic Cooperation and Development (OECD):

| Trophic category | Mean total, P ($\mu g\ l^{-1}$) | Mean ($\mu g$ chl-a $l^{-1}$) | Max. ($\mu g$ chl-a $l^{-1}$) | Mean Secchi depth (m) |
|---|---|---|---|---|
| Oligotrophic | <10 | <2.5 | <8 | >6 |
| Mesotrophic | 10–35 | 2.5–8 | 8–25 | 6–3 |
| Eutrophic | >35 | >8 | >25 | <3 |

During the last four decades, eutrophication has undoubtedly been the most challenging global threat to the quality of our freshwater resources. Survey of the International Lake Environmental Committee has indicated in the early 1990s that some 40–50% of lakes and reservoirs are eutrophicated. Many of these water bodies are extremely important for drinking water supply, recreation, fishery, and other economic purposes. Developing countries in hot and dry climate face particularly serious and rapidly increasing eutrophication-related ecological and economic challenges. Unlike this paper, eutrophication is not restricted to lakes and reservoirs. Large rivers, estuaries, coastal zones, and inland seas are also subject to this undesirable process.

Although nutrient cycling is far from being closed even in the most developed societies, eutrophication has been successfully reversed in several lakes by managing human nutrient emission (low-P detergents, P precipitation at sewage treatment plants, decreased fertilizer application, erosion control, etc.) and by cutting off nutrient loads to recipients (sewage diversion, buffer zones, etc.). Thus, eutrophication seems to be reversible. Nutrient management, however, does not result in an immediate oligotrophication (i.e., reversal of eutrophication) because of the resilience of the aquatic ecosystem. Various methods of in-lake physical, chemical, and biological interventions have been developed to facilitate the efficiency of load reduction, shorten the delay in recovery, and accelerate the rate of reversal. Early recognition of eutrophication, understanding the nutrient load–trophic response relationship in individual lakes, and planning the most suitable combination of management measures require an in-depth knowledge of functioning of aquatic ecosystems.

## Basics of Eutrophication

Water bodies are open systems, in which autotrophic organisms, mainly algae and macrophytes, convert inorganic carbon into organic matter using the energy of solar radiation. In addition to C, other nutrients (oxygen, hydrogen, nitrogen, phosphorus, sulfur, silica, trace metals) are universally or taxon-specifically needed for primary production. The German chemist, von Liebig recognized in the nineteenth century that the yield of a plant is proportional to the amount of nutrient, which is available in the lowest supply relative to the plant's demand. This observation is known as the Law of the Minimum. Studying the elemental composition of phytoplankton in the English Channel, Redfield found that healthy algae contained C, N, and P atoms in an average ratio of 106:16:1 (40:7:1 by weight). This means that 1 g of P available in the water allows the algae to assimilate 40 g of C and thus increase fresh weight by about 400 g.

Algae, like any other plants, require nutrients in ratios, which are radically different from the elemental composition of their environment. In freshwater, the mid-summer average demand to supply ratio of P is up to 800 000; that of N is 300 000; that of C is 6000; and that of all other elements is below 1000. Consequently, P and N are the nutrients, which most often determine the carrying capacity of lakes and reservoirs. This has been verified experimentally during the 1970s by

Schindler and his team. In the Canadian Experimental Lake Area, untreated lakes were compared with nutrient enriched ones. Standing crop of algae increased with the supply of P. Nitrogen addition alone failed to similarly trigger phytoplankton growth.

Another large-scale study of the OECD (1982) also confirmed the biomass-limiting role of P by the means of statistical modeling. This study considered over 120, mostly temperate lakes and reservoirs. Mean in-lake concentration of total nutrients was calculated from the annual load corrected for flushing. Annual mean and maximum concentration of chlorophyll was regressed against the estimated nutrient concentration. The relations were highly significant for P. In most lakes, biomass was determined by the amount of P, while N was the controlling factor in only a few cases.

Chemical nature of major N and P ions explain why P determines the carrying capacity of lakes more often than N. Both nitrate and ammonium are much more mobile in soils and sediments than orthophosphate, since the latter is chemisorbed by clay minerals, iron(III) oxy-hydroxides, carbonates, etc. As a consequence, nitrogen compounds are preferentially washed out from the soils and phytoplankton in pristine temperate lakes tends to be P limited. For the same reason, terrestrial plants are most often N limited. In productive lowland wetlands dominated by dense stands of emergent plants that release $O_2$ to the atmosphere, anaerobic conditions may develop in the water and denitrification speeds up. In such systems N deficiency prevails as frequently indicated by the presence of insectivorous plants. Carrying capacity of many subtropical and tropical lakes are also N determined because of the intense denitrification characteristic under warm climates.

However, widespread nutrient limitation is not a universal phenomenon in freshwaters. Besides nutrients, algal growth requires sufficient time and light. In several water bodies, biomass is determined by the availability of these latter factors. Such conditions prevail in running waters in which inorganic nutrients are usually in excess. To support abundant plankton, a river must either be sufficiently long or flow sufficiently slowly to permit seven to eight cell divisions required to raise an inoculum of several cells per milliliter to the order of a few thousand cells per milliliter. When the hydraulic residence time is increased by reservoir construction, the time available for 'undisturbed' algal growth and thus, the biomass may increase without any further nutrient enrichment. In addition to this, enhanced sedimentation within mainstream reservoirs improves light availability both in and downstream of the reservoir. Construction of a series of reservoirs on the upper Danube was

recognized as a key factor in downstream eutrophication of this large river.

Eutrophication is not merely an increase in the biomass of various organisms. Structural and functional changes accompany, and – as careful long term observations repeatedly testify – even precede quantitative changes at each trophic level.

## Changes in Primary Producers

One of the most conspicuous and the best-known changes associated with eutrophication is the mass development of cyanobacteria, be it $N_2$-fixing (*Anabaena*, *Aphanizomenon*, *Cylindrospermopsis*, etc.) or nonfixing (*Planktothrix*, *Microcystis*, etc.). Many species may develop toxic strains (e.g., *M. aeruginosa*, *Cylindrospermopsis raciborskii*) and thus, large blooms may directly harm both other aquatic organisms and humans. An enormous scientific literature discusses the ecological traits leading to mass development of Cyanobacteria as well as their manifold influence on the functioning of the aquatic ecosystem. Without going into details, one can recognize three basic lines of adaptations in the background of cyanobacterial success. Each bloom-forming species possesses a certain combination of the following traits:

1. *Nutrients.* Bloom-forming Cyanobacteria are capable of exploiting nutrient reserves that are unavailable, or not readily available, for most other algae. This is the reason why annual mean concentration of chlorophyll may show an abrupt rise upon the establishment of Cyanobacteria at high but 'steady' external nutrient load. Since the inorganic carbon concentrating mechanisms are highly efficient in the case of both $CO_2$ and $HCO_3^-$, Cyanobacteria continue to assimilate at high pH. Many species have high affinity for $NH_4^+$ uptake besides the ability to fix $N_2$. Either fast maximum rate of P uptake, or exploitation of vertical nutrient gradients, facilitates P acquisition, thanks to buoyancy regulation. The relatively large size allows the storage of considerable amounts of excess C, N, and P beyond the actual needs of growth, thereby providing independence from the fluctuating supply.

2. *Light.* The lowest light saturation values of both photosynthesis and growth have been observed among bloom-forming cyanobacteria (e.g., *Planktothrix rubescens*, *Cylindrospermopsis raciborskii*). In general, their light requirement tends to be lower than that of the eukaryotic algae. Under calm conditions, buoyancy regulation allows optimal positioning of Cyanobacteria in the light gradient. Simple prokaryotic structure results in relatively low maintenance costs that both decreases the light demand and leaves more energy to acquire the limiting nutrient.

3. *Low biomass loss.* Buoyancy regulation prevents sinking loss of healthy Cyanobacteria even under calm conditions. Large size and morphology (large colonies, filaments) reduce zooplankton grazing to negligible levels. The decreased loss of Cyanobacteria is equivalent to the slowdown of nutrient regeneration and diminished internal supply of the limiting nutrient.

Summer blooms of Cyanobacteria cause a major shift in the seasonal pattern of phytoplankton biomass in temperate lakes. In oligo- and mesotrophic lakes, the biomass maximum occurs during the spring when temperature and light increase rapidly, and relatively large amounts of nutrients are delivered into the water by spring floods as well as the spring overturn in deep lakes. In comparison, the summer biomass of phytoplankton is lower as a result of the diminishing external and internal nutrient supply. In productive lakes, increased nutrient availability differentially enhances summer production and leads to a virtually monomodal temporal distribution of biomass with a summer maximum.

The most important functional changes associated with the dominance of bloom-forming Cyanobacteria are (1) the involvement of formerly unavailable resources into aquatic production, (2) the decrease in the rate of turnover of nutrients, most importantly in that of the limiting ones, and (3) the decrease in the efficiency of energy transfer from primary producers to higher trophic levels. Because of these self-stabilizing functional changes, the shift from the noncyanobacteria dominated to Cyanobacteria dominated late summer phytoplankton assemblages can be seen as alternative stable states of the aquatic ecosystem. An important manifestation of the alternative stable states is the hysteresis that occurs when the system is forced from the one state to the other.

Establishment of bloom-forming Cyanobacteria is the last step in the course of restructuring of phytoplankton during eutrophication. Case Study 1 summarizes the history of compositional changes observed in a large, shallow temperate lake.

Submerged macrophytes may cover a significant portion of the area of shallow lakes that are protected from strong wave action or the small surface area of which restricts wind fetch. During eutrophication, such lakes may abruptly turn from a macrophyte-dominated clear water state into a phytoplankton-dominated turbid state. Similar to the case of Cyanobacteria, both states may prevail in a broad range of external nutrient loads since a number of feedback mechanisms stabilize the actual state. Thus, in the clear water state abundant macrophyte stands prevent sediment resuspension by

dampening wind work and provide shelter for zooplankton. Visual predatory fish keep efficient control on planktivorous fish, thereby promoting the growth of zooplankton. In turn, grazing may significantly influence growth, biomass, and succession of phytoplankton. In the turbid state, intense resuspension and shading by phytoplankton inhibits macrophyte growth and suppresses foraging of predatory fish. In the lack of sufficient refuge, zooplankton fall victim to planktivorous fish and phytoplankton are released from the top-down control.

### Changes in Consumers

Increased primary production supports an elevated production and biomass of consumers. Structural changes in the phytoplankton, first of all those in the size distribution, exert a strong influence on pelagic herbivores by differentially favoring or suppressing one or the other group of zooplankton. Since detritus and associated bacteria make up the main food of most benthic invertebrates, the zoobenthos is less sensitive to eutrophication-related changes in algal composition. At the same time, presence of aquatic macrophytes is beneficial for both pelagic and benthic consumers because of either the mere maintenance of habitat patchiness or to more specific biotic interactions. The retreat of macrophytes during eutrophication may result in a drastic reduction in the species diversity of consumers.

The four main groups of freshwater zooplankton – protists, rotifers, cladocerans and copepods – partition food primarily on the basis of size. Most aquatic grazers consume any particles in the appropriate size range, be it algae, bacteria, another grazer, detritus, or inorganic particle. Size-selective predation by planktivorous fish inserts a top-down control on grazer populations, the larger zooplankton being more vulnerable to predation. In addition to the direct biological interactions, deterrent environmental effects of eutrophication, including elevated turbidity, magnified daily and seasonal fluctuations in the oxygen concentration and pH may adversely affect the sensitive groups of zooplankton. Intense grazing alters species composition of phytoplankton by both selective removal of edible algae and nutrient regeneration. Because of this intricate net of interactions, eutrophication-related changes in the biomass, composition, size distribution, and seasonal pattern of various groups of zooplankton are highly lake-specific. The forthcoming discussion is restricted to a few general trends.

The most abundant freshwater zooplankters are crustaceans. Most cladocerans, such as *Daphnia*, are filter feeders whereas most cyclopoid and calanoid copepods are selective, raptorial grazers. In oligo-trophic lakes, the maximum community grazing rate is below 15% while in eutrophic lakes the entire volume of water can be filtered up to 4–5 times in a day (400–500%). *Daphnia* and other cladocerans usually account for up to 80% of the community grazing rate.

Filter feeders collect particulate matter from the water. Anatomy of feeding appendages prevents collecting particles smaller than 0.8 μm. The upper limit to ingestible particle size increase with the size of the animal up to about 45 μm in the case of spherical particles and larger in the case of nonspherical ones. Edible algae are small, naked green algae, nanoflagellates, cryptomonads, and certain diatoms. Algae that cannot be ingested are large, have biologically resistant cell walls and spines, or form colonies. Large, unicellular desmids and dinoflagellates, chain-forming diatoms, and colonial Cyanobacteria respresent this group. The copepods are inefficient at retaining small particles but can process much larger algae than cladocerans. Some copepods, for example, can break up chains of diatoms.

In oligo- and mesotrophic lakes, the spring bloom of phytoplankton is made up by small, fast growing, edible algae such as small diatoms, nanoflagellates, and small greens. This favors the growth of nonselective cladocerans that – similar to their prey – are specialists at fast reproduction by parthenogenesis. Seasonal succession proceeds towards summer and autumn associations of large, slowly growing species such as dinoflagellates, gelatinous greens, colonial or filamentous Cyanobacteria. With the increasing abundance of large or noxious algae, slower growing but more selective copepods become more common. Collapse of algal blooms is associated with an increased availability of detritus and bacteria that constitute an appropriate food for cladocerans. Smaller blooms of diatoms and nanoplankton (<60 μm) during the autumn also promote the growth of filter feeders. As a general trend, the aquatic food chain gradually shifts during the succession from one based on living algae to one based on bacteria and detritus.

Compared with that in unfertile lakes, the shift from the dominance of nanoplankton to that of larger 'net' plankton takes place early in the season in eutrophic lakes. Moreover, filamentous Cyanobacteria that take over the dominance in the summer assemblages inhibit filtration of nonselective grazers by clogging up the feeding appendages. Thus, the share of grazing-resistant algae increases during eutrophication, accelerating the shift toward a bacteria- and detritus-based pelagic food chain that may result in an increasing dominance of cladocerans.

The tendency of an eutrophication-related increase in the cladoceran to copepod ratio is best seen in deep, stratified lakes. In naturally turbid reservoirs and shallow lakes zooplankton inherently rely to a greater extent on detritus and associated bacteria than in deep ones of the same trophic state. With increasing productivity, however, enhanced detritus availability disrupts the natural balance between pelagic and benthic food webs and the share of the latter increases in the total energy budget of both shallow and deep lakes.

The zoobenthos is an extremely diverse group comprising nearly all phyla from protists through large macroinvertebrates to vertebrates. Similar to the zooplankton, eutrophication-related changes in the composition, biomass, and seasonal dynamics of benthic assemblages are determined by food availability, predation, and indirect environmental effects, particularly oxygen concentration. Oxygen conditions in the hypolimnion or in the uppermost sediment layer strongly depend on the downward flux of detritus, and the tolerance to low oxygen is extremely variable among benthic organisms. One of the pioneering discoveries of limnology was in the early twentieth century that the profundal benthic fauna is an excellent indicator of nutrient richness in deep lakes. Naumann postulated a direct relationship between phytoplankton and nutrient conditions in lakes and contrasted the extreme ends as 'eutrophic' (well-nourished) and 'oligotrophic' (poorly nourished). Thienemann recognized two lake types based on hypolimnetic oxygen concentrations and on correlated differences in the benthic chironomid fauna. The oligotrophic–eutrophic paradigm emerged originally from the crossing of these two lines of research. The oligotrophic water was deep with low nutrient supply, low algal production in the epilimnion, and low flux of detritus to the hypolimnion. Because of the small oxygen consumption, the hypolimnion had an orthograde oxygen profile and the corresponding stenoxybiont benthic fauna exploited mainly by whitefish (*Coregonus*). The eutrophic type was relatively shallow and nutrient-rich, water blooms appeared in the summer. The high flux of detritus to the shallow hypolimnion resulted in a fast depletion of oxygen. The oxygen profile was clinograde, and a euryoxybiont *Chironomus* fauna dominated the benthos. This paradigm had been rephrased during the 1950s and 1960s when eutrophication became a recognized environmental threat in the developed countries.

The two dominant groups of zoobenthos are the oligochaete worms and the dipteran chironomids. Although some oligochaetes are restricted to oligotrophic waters, the tubificids can be extremely abundant in highly eutrophicated as well as in organically polluted lakes if some oxygen is available from time to time and toxic products of anaerobic metabolism do not accumulate in large quantities. In such systems, they benefit from both the rich nutrient supply and the lack of competition with other benthic animals that cannot tolerate the poor oxygen conditions. In contrast to the tubificids, relative abundance of the more oxygen demanding chironomid larvae decreases during eutrophication.

In deep, mesotrophic lakes the biomass of zoobenthos exhibits two maxima: (1) a diverse fauna with high oxygen demand inhabits the littoral sediments and (2) less rich assemblages of species that tolerate low oxygen characterize the lower profundal zone. With increasing productivity, the littoral zone may loose its heterogeneity that results in a single profundal biomass maximum of the zoobenthos. A further increase in fertility and the associated hypolimnetic oxygen deficit may then lead to the decline in the biomass of the benthic fauna in the profundal zone, too.

Restructuring of the zooplankton and zoobenthos during eutrophication should not be perceived as a chain of smooth transitions. On the contrary, abrupt compositional shifts may be associated with stepping over the threshold values of critical environmental variables, including habitat patchiness, food availability, or simple physical and chemical factors. Case Study 2 demonstrates the threshold effect of food availability in the compositional change of the chironomid fauna in a large, shallow lake.

The shift from the pelagic to the benthic food web implies basic alterations in the functioning of the freshwater ecosystem. On the one hand, the consumer control of phytoplankton diminishes for two reasons. First, the diet of zooplankton relies to a much greater extent on living algae than that of benthic animals. Second, a major change occurs in the seasonal pattern and magnitude of nutrient regeneration. The zooplankton regenerate nutrients directly into the trophogenic zone, even though the sedimentation of fecal pellets represents a net outward flux of nutrients from the epilimnion to the sediments. In contrast to this, benthic animals release nutrients to the sediments from where the flux to the water is primarily regulated by abiotic factors. When, however, large oligochaetes or chironomids are present in high densities, their burrowing activity considerably enhances the sediment-water exchange rates of various nutrients (O, N, P, etc.). In shallow lakes and in the littoral zone of deep ones, bioturbation may result in a significant increase in the internal load of nutrients. On the other hand, the annual production to biomass ratios are higher among planktivorous fishes than among bentivorous ones. Therefore, the overall efficiency of

energy utilization decreases in the lake with the increasing proportion of the benthic food web.

Although the total biomass of fish tends to increase during eutrophication, the species composition shifts in undesirable directions. Most conspicuously, the relative abundance of visual predators drops drastically with increasing turbidity. This change was repeatedly shown to cascade down along the food chain to the phytoplankton. Unbalanced food availability may lead to enhanced mortality of the fry. Widely fluctuating oxygen conditions may result in mass killing of fish. Although intense fish production requires highly productive lakes, maintenance of a diverse native fish fauna conflicts with eutrophication.

## Oligotrophication

Comprehensive studies of lake recovery from eutrophication have shown that the trophic status of lakes is not a linear function of the external nutrient load. The same external load supports a higher biomass during oligotrophication than during eutrophication, that is a hysteresis can be observed (**Figure 1**). Sas and his team recognized four stages of recovery by examining 18 eutrophicated deep and shallow Western European lakes during the 1990s (**Figure 1**).

During Stage 1, the excess of available P is flushed out from the lake without any reduction in the standing crop of algae. This stage can be observed in lakes that received sufficiently high P loads for sufficiently long periods, and thus, phytoplankton biomass has no more been P-dependent. In such lakes, a considerable time may pass before P regains its biomass-limiting role. In deep lakes, the delay is essentially a function of the

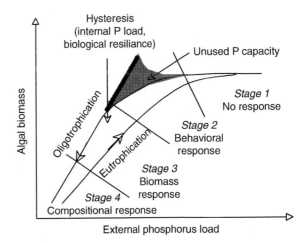

**Figure 1** Schematic representation of phytoplankton biomass as a function of the external nutrient load during eutrophication and oligotrophication with the four stages of lake recovery.

hydraulic residence time. In shallow lakes, where the excess of available P accumulates in the sediments, diagenetic processes and burial of P in deeper sediments are as, or even more important than flushing. A net annual internal P load is common among shallow lakes during the first years following the reduction of external load, whereas it is rare in the prerestoration period and does not occur in deep lakes.

During Stage 2, increasingly P limited algae disperse to greater depths in order to exploit vertical nutrient gradients characteristic of stratified lakes. As a result, transparency increases rapidly in spite of a negligible decrease in algal biomass. Shallow lakes do not provide such refuge options and therefore Stage 2 is lacking.

Stage 3 is the period of significant biomass decrease that occurs in proportion to load reduction. This phase is usually faster and more pronounced in stratified than in shallow lakes. The reason is the elevated internal load from the surface sediments of shallow lakes keep the 'memory' of past loading conditions.

In Stage 4, species composition of phytoplankton changes and a new steady state is established. Both absolute and relative abundance of Cyanobacteria decreases. Blooms disappear; species diversity and stability of the system increase. The delay in the retreat of Cyanobacteria may depend on the life history of species, gaining dominance during eutrophication in various types of lakes.

The data of Sas (1989) allowed investigating the reaction of perennial Cyanobacteria, like *Planktothrix*. Suppression of periodically planktonic species may take a longer time than that of the perennial ones. Akinetes and colonies of many common bloom-forming genera (e.g., *Aphanizomenon*, *Anabaena*, *Microcystis*, *Gloeotrichia*) overwinter in the sediments. In most species, a small portion (up to 5–6%) of the benthic forms is required to initiate the growth of the planktonic population next year. Thus, the benthic reserves of Cyanobacteria that might increase orders of magnitude during eutrophication may inoculate the water for many years after substantial reduction in the external load. Case Study 1 also describes the behavior of periodically planktonic Cyanobacteria in a recovering shallow lake.

One can complete the earlier-mentioned list by adding Stage 5, which covers compositional changes at higher trophic levels. Similar to the restructuring of phytoplankton, delayed responses and hysteretic effects are characteristic of this process, too (cf. Case Study 2). For this reason, reduction of the external nutrient loads is only the first and most crucial step during eutrophication management in lakes, particularly in shallow ones. A series of measures have been developed that aim at speeding up recovery after load reduction by manipulating biotic interactions within the lake. If, however, load reduction is analogous to stopping a tooth, biomanipulation is certainly related to brain surgery.

## Case Study 1

Three circumstances make large (596 km$^2$), shallow ($z_{mean} = 3.1$ m) Lake Balaton an excellent case study when studying eutrophication-related changes in phytoplankton. First, floristic studies date back to the end of the nineteenth century, while Sebestyén initiated quantitative phytoplankton research in the 1930s. Second, due to the specific morphological features (elongated shape, the main tributary enters at the southwestern end, the only outflow starts at the opposite end, relatively closed large-scale circulation patterns develop in the four basins under the influence of the dominant winds; **Figure 2**), the western areas become hypertrophic during the 1970s while the eastern ones remained mesotrophic. Third, after reducing the external P load by about 50% from 1.3 to 0.7 mg P m$^{-2}$ day$^{-1}$ during the late 1980s and early 1990s, the lake recovered surprisingly fast.

Although the development in the methods of phytoplankton counting somewhat biases the comparability of long-term data, neither eutrophication nor oligotrophication has substantially affected the eukaryotic algal flora. In the same time, appearance, disappearance, and reoccurrences of Cyanobacteria, especially Nostocales, have been detected during rapid changes in trophic conditions. Prior to eutrophication, three *Aphanizomenon* spp. (*A. flos-aquae*, *A. klebahnii*, *A. gracile*) and four or less *Anabaena* spp. represented the order in Lake Balaton. During the period of rapid eutrophication in the 1970s, many new species appeared abruptly in the flora, including *A. aphanizomenoides*, *A. issatschenkoi*, *Anabaenopsis elenkinii*, *Cylindrospermopsis raciborskii*, and *Rhaphidiopsis mediterranea*. Most cyanobacteria present in the pre-eutrophication period maintained at least modest populations but *A. gracile* became virtually extinct. In the 1980s, when the ecosystem 'stabilized' at a high trophic level, the only newcomer was *Anabaena contorta*. Fast oligotrophication from the mid-1990s was coupled with frequent appearance of new species (*Anabaena compacta*, *Anabena circinalis*, *Aphanizomenon hungaricus*, *Anabaenopsis cunningtonii*, *Komvophoron constrictum*), disappearance of some 'eutrophic' ones (*Anabaenopsis elenkinii*, *Rhaphidiopsis mediterranea*), and reappearance of *Aphanizomenon gracile*.

A definite increase had already been observed in the biomass of the dominant summer alga, *Ceratium*

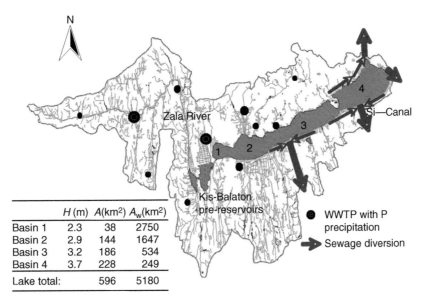

|  | H (m) | A(km²) | A_w(km²) |
|---|---|---|---|
| Basin 1 | 2.3 | 38 | 2750 |
| Basin 2 | 2.9 | 144 | 1647 |
| Basin 3 | 3.2 | 186 | 534 |
| Basin 4 | 3.7 | 228 | 249 |
| Lake total: |  | 596 | 5180 |

**Figure 2**  Lake Balaton and its catchment. The most important management measures are also indicated. $H$, average water depth; $A$, surface area; $A_w$, corresponding subwatershed area; WWTP, waste water treatment plant.

*hirundinella* between the 1930s and the 1950s. The annual mean biomass of phytoplankton increased from about 0.3 mg fresh weight per liter by a factor of 4. Up to the mid 1970, *Cyclotella bodanica* and *C. ocellata* dominated the spring plankton. Thereafter the abundance of pennate diatoms (*Synedra acus* and *Nitzschia acicularis*) increased conspicuously.

In the eastern areas of the lake, the biomass of summer phytoplankton showed a moderate further increase during the period of eutrophication compared with the 1950s, but annual maxima did not exceed 5 mg l$^{-1}$. In the same time, maxima reached 40–60 mg l$^{-1}$ in the western areas during the early 1980. Interannual differences in biomass among successive years increased substantially in both areas. Prior to eutrophication, late summer assemblages were dominated by *Ceratium hirundinella, Aphanizomenon klebahnii, and Snowella lacustris*. Depending on the wind regime, various meroplanctonic diatoms (*Aulacoseira granulata, Cyclotella radiosa, C. ocellata*, small *Navicula* and *Nitzschia* spp.) contributed significantly; in very windy years they could dominate. These species, however, have been gradually replaced by Cyanobacteria during eutrophication. The key cyanobacterium, *Cylindrospermopsis raciborskii* is a subtropical species the akinetes of which require exceptionally high and narrow temperature range (22–24 °C) for germination when compared with other Cyanobacteria present in Lake Balaton. Warm conditions are incidental in this temperate lake, and this is certainly one of the main factors leading to the irregularity of *C. raciborskii* blooms.

*C. raciborskii* was first detected in Lake Balaton in 1978, and it produced its first large bloom in 1982.

This bloom was exceptional compared with cyanobacterial blooms during the 1970s in two respects: (1) the maximum biomass exceeded those of the previous blooms by a factor of about 2, and (2) with a delay of about 3 weeks, the blooms spread to the mesotrophic areas. The external nutrient load was known in the westernmost basin from daily measurements since 1975. These data evidenced that *C. raciborskii* achieved a much higher biomass than other N$_2$-fixing Cyanobacteria under relatively constant external load conditions. This is clearly indicative of a superior exploitation of the available resources by *C. raciborskii*. Simple mass balance models indicated that in summers of *C. raciborskii* dominance, the internal P load was much higher than in other years. Indirect evidence suggests that it is not the enhanced internal P load that induces the blooms, but presence of *C. raciborskii* leads to an elevated internal load. Eastward extension of *C. raciborskii* blooms has repeatedly been observed in subsequent years including 1992 and 1994, when the external loads have already been reduced close to the present levels. It has been shown that dispersion along the longitudinal axis may result in an inoculation of the mesotrophic eastern areas from the eutrophic western ones.

### Case Study 2

The water of shallow, wind-exposed and highly calcareous Lake Balaton is always turbid, the mean vertical light attenuation coefficient varies between 2 and 4 m$^{-1}$. Gut content analysis of zooplankton revealed that grazers collect huge amounts of inorganic ballast and they starve even if the table is set in a near-optimal way. Because of this geomorphological drawback, community grazing rate of zooplankton is low (usually <5%), and phytoplankton production is channeled to fish primarily by zoobenthos. Investigation of the benthic macrofauna was rather sporadic in Lake Balaton before the mid-1990s. This sporadic information had to be supplemented with paleolimnological data, as well as with comparative data obtained along the west–east trophic gradient in order to reconstruct composition and biomass of the benthic macrofauna during the preeutrophication period.

Gastropods (*Potamopyrgus jenkinsi, Lithoglyphus naticoides*) comprised a quantitatively important group of the macrobenthos until about the middle of the twentieth century. In the late 1990, however, no living gastropods could be collected during a detailed, long-term zoobenthos survey. Disappearance of gastropods started shortly after the invasion of *Dreissena polymorpha* in the early 1940s, most probably independent of eutrophication. At the present time, chironomids and oligochaetes are the dominant groups of the benthos. Tubificidae, however, make up less than 20% of the total benthic biomass.

Of the more than 50 chironomid taxa known from Lake Balaton, only seven can be found in the profundal sediments. Three of the latter species, *Procladius* cf. *choreus, Tanypus punctipennis*, and *Chironomus balatonicus* give 80–90% of the biomass and annual production. These three species responded specifically to eutrophication and oligotrophication, and the reaction could sufficiently be explained by differences in their feeding habits.

*Procladius choreus* is a predator. It is able to utilize a wide range of food, including macrobenthic animals, zooplankton, microphytobenthos, and detritus, while the most important food is meiozoobenthos. As a consequence, the biomass of *P. choreus* was more or less independent of the changes in trophic conditions both in time (eutrophication and oligotrophication) and in space (the west-east trophic gradient).

*Tanypus punctipennis* preys other chironomid larvae, but it may also ingests a great deal of plant material and detritus. Its biomass increases towards the mesotrophic areas of Lake Balaton. Prior to eutrophication and in the mesotrophic eastern areas, a *Procladius–Tanypus* chironomid community was characteristic of Lake Balaton. The biomass of this community showed only a weak positive correlation with the biomass of phytoplankton.

*Chironomus* larvae are filter- and surface deposit feeders that take the advantage of the large amount of fresh detritus sedimenting after major algal blooms. In this way, their biomass and production strongly depends on the biomass of phytoplankton. In Lake Balaton, there was a strong positive correlation between the spring biomass of *Chironomus*-dominated benthic community and the late summer phytoplankton biomass in the previous year. The relationship, however, was not linear. *Chironomus* larvae showed a presence–absence type response. When the late summer algal biomass exceeds 20–30 µg chl-a $l^{-1}$, that is the annual primary production is higher than 220–250 g C $m^{-2}$ $year^{-1}$, *Chironomus* dominates the zoobenthos in the next year and the chironomid biomass attains high values (0.6–3.4 g $m^{-2}$). Below this level of primary production, *Chironomus* is absent, Tanypodinae dominate the zoobenthos, and the total benthic biomass varies from 0.3 to 0.5 g $m^{-2}$.

One of the reasons why *Chironomus* depends so strongly on the last year's primary production in Lake Balaton is the low organic content (2–4%) of the sediments. This is due to the frequent resuspension that keeps the water and surface sediments permanently aerobic, thus enhancing bacterial decomposition. It was estimated that only 1.1–3.2% of the primary production was assimilated by benthic chironomids. Fast aerobic decomposition may also explain that chironomid densities are rather low in Lake Balaton when compared with other European lakes of similar trophic state.

The threshold-like shift in the dominance of *Tanypus* to *Chironomus* with increasing trophy is accompanied by a 2.5-fold increase in energy transfer efficiency from the phytoplankton to chironomid fauna. The reason is that phytodetritiphagous *Chironomus* directly harvests primary producers, while a trophic loop of one or two steps length connects predatory *Tanypus* to algae. Variability of the energy transfer efficiency is of vital importance for benthivorous fish. For example, stock density and growth rate of the dominant fish species, the common bream (*Abramis brama*) is basically influenced by chironomid production.

*See also:* Effects of Climate Change on Lakes; Lake Management, Criteria.

## Further Reading

Istvánovics V and Somlyódy L (2001) Factors influencing lake recovery from eutrophication—The case of basin 1 of Lake Balaton. *Water Research* 35: 729–735.

Istvánovics V, Clement A, Somlyódy L, Specziár A, G-Tóth L, and Padisák J (2007) Updating water quality targets for shallow Lake Balaton (Hungary), recovering from eutrophication. *Hydrobiologia* 581: 305–318.

Padisák J and Reynolds CS (1998) Selection of phytoplankton associations in Lake Balaton, Hungary, in response to eutrophication and restoration measures, with special reference to the Cyanoprokaryotes. *Hydrobiologia* 384: 41–53.

Porter KG (1977) The plant–animal interface in freshwater ecosystems. *American Scientist* 65: 159–170.

Reynolds CS, Dokulil M, and Padisák J (eds.) (2000) *The Trophic Spectrum Revisited.* Springer.

Sas H (1989) *Lake Restoration and Reduction of Nutrient Loading: Expectations, Experiences, Extrapolations.* St Augustin: Academia Verlag Richarz.

Scheffer M, Hosper SH, Meijer M-L, Moss B, and Jeppesen E (1993) Alternative equilibria in shallow lakes. *Trends in Ecology and Evolution* 8: 275–279.

Schindler DW (1974) Eutrophication and recovery in experimental lakes: Implications for lake management. *Science* 184: 897–899.

Specziár A and Vörös L (2001) Long-term dynamics of Lake Balaton's chironomid fauna and its dependence on the phytoplankton production. *Archive für Hydrobiologie* 152: 119–142.

Vollenweider RA and Kerekes JJ (1982) Background and summary results of the OECD cooperative programme on eutrophication. OECD report, Paris.

Wetzel RG (2001) *Limnology. Lake and River Ecosystems.* San Diego: Academic Press.

## Relevant Websites

http://www.ceep-phosphates.org (European Union Eutrophication Guidance Document information. (2006). SCOPE Newsletter No. 64).

http://www.ilec.or.jp/eg/

http://www.umanitoba.ca/institutes/fisheries/

# Paleolimnology

**J P Smol,** Queen's University, Kingston, ON, Canada

## Introduction

The ecological and environmental conditions in lakes and rivers change on a variety of time scales. However, one of the greatest challenges faced by limnologists (as well as other environmental and ecological scientists) is the general lack of long-term data. Without such data, it is difficult to determine if changes are occurring in an aquatic system, and if so, to identify the likely drivers of these ecosystem changes. Such long-term data are critical in, for example, contrasting the effects of human activity versus natural environmental change on ecosystems.

Although direct instrumental data on limnological changes are often lacking, an important archive of past environmental and ecological conditions is preserved in the sediments of lakes and some river systems. Although incomplete, relatively comprehensive records of past limnological changes, as well as changes occurring in the surrounding terrestrial catchment and airshed, can often be reconstructed from information preserved in sediment profiles using paleolimnological techniques.

Paleolimnology can be broadly defined as the multidisciplinary science that uses the physical, chemical, and biological information preserved in aquatic sediments to track past changes in ecosystems. Paleolimnology has enjoyed considerable progress over the last three decades, with important advances in the ways paleolimnologists can retrieve and section sediment cores, new approaches to provide geochronological control for these sediment profiles, major advances in the quantity and quality of proxy information that paleolimnologists have learned to retrieve from sediments, as well as major improvements in the ways that paleolimnologists can interpret these stratigraphic changes in a statistically robust and defendable manner.

Examples using lakes will be emphasized throughout this article; however, similar approaches are being used in river and marine ecosystems. The overall focus is on paleolimnological approaches rather than applications.

## Sediments

Lakes slowly fill up with sediments. Typically, 24 hours a day, every day of the year, sediments are slowly accumulating at the bottom of lakes. The overriding principle in paleolimnology is that lake sediments accumulate in an organized fashion, with older material occurring deeper in the sediment core, and the most recent material at the surface of the core (i.e., the Law of Superposition). Of course, not all sediment profiles are ideal for historical analyses, as some mixing processes, such as bioturbation (i.e., the mixing of the sediments by benthic organisms), may occur in some profiles. Nonetheless, many of these potential problems can be assessed (e.g., using dating techniques) and it is now clear that the vast majority of lakes archive valuable records of past environmental change in a stratigraphically intact manner.

Incorporated in sediments is a tremendously rich library of information about the processes and the biological communities from within or external to the lake. Paleolimnologists typically divide the sources of lake sediments into two broad categories: (1) allochthonous sources, which refer to material originating from outside the lake basin (e.g., soil particles, pollen grains from trees, pollution from industries); and (2) autochthonous sources, which refer to material originating from within the water body itself (e.g., dead algae or invertebrates, chemical precipitates). The amount of sediment that has accumulated can vary tremendously between lakes, and even within a single lake basin. A typical glaciated temperate lake may contain 4 or 5 m (or even 10 m or more) of sediment, representing the lake's history since the time of the lake's formation following the retreat of the ice sheets at the end of the last ice age (c. 12 000 years ago). In contrast, a lake of similar age from a polar region may have far less sediment accumulated, as sedimentation rates are typically very low. In some regions, such as the rift valley lakes of Africa, records spanning several hundred thousand years are contained in 100s of meters of sediment.

A major advantage of paleolimnological approaches is that researchers can, to a large extent, set the time scale. For example, by taking longer sediment cores, a longer time frame can be studied. Similarly, the researcher can set the temporal resolution of the study. For example, by slicing the sediment core at finer intervals, more detailed records of past environmental changes can be inferred. Of course, there are also potentially serious limitations. For example, the sediment accumulation rate may be so low that even fine-resolution sampling will still not provide sufficiently fine temporal resolution to answer certain questions.

From the information preserved in sediments, paleolimnologists can reconstruct the history and development of aquatic ecosystems. There are two key steps involved in any paleolimnological investigation: (1) retrieval of the sedimentary material required for study, and (2) establishment of the depth–time scale of the profile using geochronological techniques. These first two critical steps are described in the following section.

## Retrieving Sediment Cores

A large number of ingenious sediment samplers have been developed to collect sediment cores in an undisturbed manner. Each sampler has various advantages and drawbacks, and so it is first important to determine the scientific questions that researchers hope to address. For example, what time frame will be required for the paleolimnological study? Will the program include a detailed analysis of the more recent sediments, which may be used to study, for example, the effects of recent human interventions, such as the effects of agriculture or acid rain, on the lake system? Or is the main focus on long-term changes, such as broad climatic shifts over millennial time scales? What temporal detail (e.g., annual, decadal) will be needed to answer these questions? The answers to these questions will determine which lake(s) will be chosen for study, as well as what type of equipment will be required to retrieve and section the most suitable sediment samples.

The earliest corers were developed to answer questions related to long-term (millennial) scale changes in lake ontogeny. Many of the original designs were inspired by the work of M. Jull Hvorslev (1895–1989), whose seminal work on sampling soil profiles for the U.S. Army Corps of Engineers contains many of the basic principles that paleolimnologists still use today to collect sedimentary profiles. Using some of these concepts, Daniel Livingstone published a piston coring design in 1955 that is still widely used today (often with some minor modifications). The typical piston corer consists of three components: the piston and cable assembly, the core tube, and the drive head and drive rods. Many variations of the original design are available, including those that can be used on a rope. Piston corers can remove, sequentially, sediment profiles of about 1 m in length, until the required depth of sample penetration is reached (**Figure 1**).

Since the mid-1980s, there has been an increased emphasis on using lake sediments to study more recent changes in lake histories, such as the effects

**Figure 1** Lake sediment coring on Lake Chala (Kenya), using a Livingstone-type piston corer (modified by UWITEC). Photograph courtesy of B. Cumming (Queen's University).

of anthropogenic impacts on lake ecosystems. This requires the use of a corer that will carefully sample the most recent, watery sediment records, which may be disturbed or lost if collected using standard piston corers. Two major types of samplers are used: gravity corers and freeze-crust corers.

A variety of gravity corers have been developed to specifically sample the most recent sediments. Many of these follow the general design of a benthic invertebrate sampler that Z. Kajak (1929–2002) developed; however, significant modifications have been made to the original design, such as those developed and modified by John R. Glew (**Figure 2**). In its simplest form, a surface sediment gravity corer is a hollow coring tube that is carefully lowered into the recent sediments on a rope from the surface, using its own weight for penetration. However, when sampling stiffer sediments, additional weight may be needed to achieve adequate penetration. A brass messenger is then delivered down the coring line to trigger a closing mechanism that seals off the top of the coring tube. The gravity corer, with the tube of collected sediment, is then retrieved to the surface for subsequent sectioning and subsampling.

Freeze-crust coring is another sampling method used for retrieving relatively undisturbed recent sediments. This process is also quite simple. A freeze-crust corer can take many designs, but most are either a metal box or tube that is filled with a coolant, such as frozen carbon dioxide (i.e., dry ice), as well as a fluid, such as ethanol. The filled freeze-crust corer, which would now be very cold, is then slowly lowered on a rope into the sediments where *in situ* freezing of the sediments onto the corer will occur. During this time, the sampling rope must be anchored on a solid

**Figure 2** Retrieving surface sediment cores with a Glew gravity corer from Lake Opinicon (Canada). Photograph courtesy of B. Cumming (Queen's University).

platform at the lake surface, such as the ice cover, to prevent it from sinking further into the sediments. After a certain period of time (the time that the sampler is left in the sediments depends on the design used, but typically from 10 to 15 min), a crust of sediment has been frozen onto the surface of the sampler. The sampler can then be retrieved, and the crust of frozen sediment can be removed and sectioned.

A large number of other coring designs are available, including those specifically developed to take long sediment cores in deep waters, such as the Kullenberg sampler. Other designs include corers designed to penetrate compact sediment profiles, such as hammer corers and vibracorers. Suffice to say that almost any lake system can now be sampled for its sediment history.

Just as there are many types of sediment samplers, there are also many ways to section the sediments. However, most researchers use an extrusion system

for recent sediments, whereby discrete intervals of sediment can be sliced off the core for further analyses (**Figure 3**). For some paleolimnological analyses, researchers may want to keep the core intact until some preliminary analyses are completed, such as X-raying the sediment profile.

## Dating Sediment Cores

A critical factor in deciphering lake sediment histories is to establish an accurate depth–time profile (i.e., geochronology) of the sediment core. Without this information, it would not be possible to place past environmental changes in a proper temporal perspective. A variety of approaches are available depending on the type of sediments, the location of the study lake, and the approximate age of the sediments themselves.

In some lakes, such as certain meromictic lakes, sediments may be annually laminated (i.e., varved), and superficially may resemble annual tree rings. A typical biogenic varve is composed of a couplet of laminae representing one year of sediment accumulation. The reason why couplets form in some sediment profiles vary from lake to lake, but varves represent seasonal differences in the materials (biological, chemical, or mineral) reaching the sediment surface. If varved sediments can be identified, the paleolimnologist can often attain a very high degree of temporal resolution (e.g., even subannual changes in some cases by splitting varves into seasons). However, the annual nature of varves must first be confirmed using other dating techniques (see later), as false laminae and missing laminae may obscure the record.

In most cases, lake sediments are dated using radiometric techniques. For older sediments (e.g., on millennial scales), radiocarbon-14 ($^{14}C$) dating is often used. As $^{14}C$ has a radioactive half-life of $5730 \pm 40$ years, it is useful for dating sediments on millennial time scales up to approximately 45 000 years of age. In the older paleolimnological literature, many $^{14}C$ dates were described as 'bulk dates,' as the $^{14}C$ content of bulk sediment was often used for dating. However, over the last 20 years, advances in accelerator mass spectrometry (AMS) technology have allowed paleolimnologists to date much smaller organic samples, such as a single seed or pine needle. As with all techniques, care must be taken to use the most appropriate material for dating, because a variety of potential error sources exist. For example, one of the most common problems with $^{14}C$ dating is the so-called 'hard water effect,' whereby 'old carbon' from the local bedrock can dilute the $^{14}C$ content of a bulk sediment sample. Ideally, paleolimnologists try to radiocarbon date material such as terrestrial

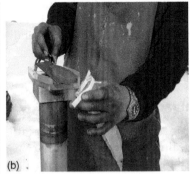

**Figure 3** An example of a vertical extrusion system used for sectioning recent sediment at close intervals. (a) A 1 cm slice of sediment has been extruded from the core tube. (b) The sediment section is sliced off and removed. Photographs courtesy of N. Michelutti (Queen's University).

macrofossils (e.g., seeds, twigs), which are not affected by some of these problems.

As noted previously, there has been a shift in some areas of paleolimnology to focus on the more recent histories of lakes. Because of the relatively long radioactive half-life of $^{14}$C, it is not appropriate for dating recent sediments. Instead lead-210 ($^{210}$Pb), a naturally occurring isotope, is often the dating method of choice. $^{210}$Pb has a half-life of about 22.3 years, and so it is ideal for dating sediments over the last century or so. In addition, other isotopes, such as cesium-137 ($^{137}$Cs), which is a by-product of nuclear explosions, can be used in some sedimentary sequences to further refine depth–time profiles. As stratospheric nuclear bomb testing became more common in the mid-1950s, a rise in $^{137}$Cs can often be seen in sediment profiles at this time. Bomb testing by the former Soviet Union and the United States peaked in 1963, and this is often seen as a peak in the $^{137}$Cs profile, followed by a marked decline in concentrations, with the signing of the nuclear Test Ban Treaty in that year. However, in some parts of Europe and elsewhere, a second peak in $^{137}$Cs can be found, marking the 1986 Chernobyl nuclear accident in the Ukraine.

Other dating methods include identifying ash layers (tephras) from known volcanic eruptions, as well as other episodic events, such as sedimentary charcoal from known forest fires. Accurate dating of sediment profiles remains a major challenge for many studies, and so, when possible, several dating methods should be used simultaneously in order to increase the confidence of the geochronology.

## The 'Top–Bottom' or 'Before and After' Approach

As noted earlier, paleolimnologists often have the important advantage that they can set the time scales for analysis. Typically, many sediment sections in a core are analyzed for a suite of indicators. However, this is also quite time consuming, as many paleolimnological analyses are time intensive (e.g., taxonomy, microscopy). For some research questions, though, it may be more important to study a larger number of lakes to obtain a regional assessment of environmental change. In such cases, detailed paleolimnological analyses are not practical because of time and resource restraints. Instead, paleolimnologists have developed a simple, regional assessment tool, which is often referred to as the 'top–bottom approach.'

The top–bottom approach is straightforward, and was originally developed to study the effects of recent human influences (e.g., acidification, eutrophication) on lake ecosystems. In this approach, surface sediment cores spanning the last several hundred years, such as those collected with gravity corers, are taken from the lakes included in the study. However, in contrast to a standard paleolimnological assessment, where many sediment samples would be analyzed from each core, typically only two sediment samples are analyzed for indicators: the surface (or top) sediment sample, representing recent environmental conditions; and a sediment sample that predates the period of marked human impacts (or bottom sample). For example, if one was attempting to assess the differences in lakewater pH as a result of acidic precipitation for a region, the bottom sample should predate $c$. 1850 AD. By comparing ecosystem changes, as inferred from the paleolimnological indicators present in the surface sediment (i.e., recent) and deeper sediments (i.e., preimpact), a regional assessment of environmental change can be attained. A major advantage of the top–bottom approach is that only two samples (not including samples to establish reproducibility) are analyzed for each core, as opposed to perhaps 20 or more sediment sections of a typical paleolimnological assessment of recent environmental change. The major disadvantage is that the top–bottom approach does not provide information on the trajectory or timing of environmental changes—it is simply a 'snapshot' approach.

## Paleolimnological Indicators

Lake sediments contain a wide array of physical, chemical, and biological indicators of past environmental change. Only the most commonly used indicators are discussed below.

Important information is contained in the physical structure and composition of the sediments themselves. For example, a simple visual inspection of the sedimentary sequence can provide information on color changes, texture, presence of laminae, etc. A variety of sediment logging and recording techniques are now available, especially with the advent of digital technology. In addition, radiography and X-ray techniques can often reveal important bedding features that may not be visible with the naked eye. Other procedures that are commonly used include particle size analyses of the sediments, which provide important insights to help determine the process, and, to some extent, the source of detrital sedimentation. Lake sediments also archive a magnetic signal, which can be used in a number of applications, including deciphering past erosion events.

Typically, one of the first analyses a paleolimnologist will undertake is to determine the relative proportions of water, organic matter, carbonate, and siliciclastic material (i.e., clastic sediment composed of silicon-bearing minerals such as quartz, feldspars, clay minerals) in the sediment sections. This is often done using weight loss techniques by successively exposing the sediment to higher temperatures, and then using the differences in sample weights to estimate the percentage of the various fractions. For example, wet sediment can be dried in a drying oven set at approximately 80–100 °C, and the percentage of water can be estimated by weight loss. Combusting the same sediment in a muffle furnace at about 450 °C for a few hours will provide an estimate of the organic matter content. A further ashing to about 950 °C can be used to determine the carbonate content. The remaining material represents mainly siliciclastic sediments. These types of data have many applications, although they also have to be interpreted cautiously. For example, intuitively, one might expect that the percentage of organic matter in a sediment profile would be directly related to the amount of organic production occurring in the lake basin. Although to some extent this relationship holds, it is important to consider that these variables are typically expressed as

percentages, and if, for example, human activities in a catchment have resulted in erosion as well as nutrient export, the siliciclastic material deposited in the lake as a result of the erosion may well obfuscate any signal of increased autochthonous production.

Lake sediments also archive a wide spectrum of geochemical information that can be used to interpret ecosystem changes. Typical approaches include analyses of various elements that can be used to track past biogeochemical changes in the lake and its catchment. Other geochemical data can be used to reconstruct past anthropogenic influences, such as mercury and lead, as well as other metals. Persistent organic pollutants, such as DDT or PCBs, also provide important records of contaminant trajectories to lake sediments. These pollution profiles can be strengthened by also studying the past deposition of fly-ash particles, such as spheroidal carbonaceous particles and inorganic ash spheres, resulting from various industrial activities.

The isotopic composition of sediments provides additional paleoenvironmental information. These include isotopes of carbon that have been used in, for example, lake eutrophication work, sulfur isotopes to track the effects of anthropogenic acidification, oxygen isotopes in climatic and hydrologic change research, and nitrogen isotopes to reconstruct past changes in sources and cycling of nitrogen (and even sockeye salmon populations). More recent approaches include determining the isotopic composition of specific indicators found in lake sediments, such as the carapaces of some invertebrates.

One of the most active areas of paleolimnological research, however, deals with biological indicators. A central theme in ecology has always been the attempt to link the distributions of organisms to environmental conditions. A diverse array of biological indicators is preserved in lake sediments, either as morphological remains (e.g., diatom valves) or as biogeochemical fossils (e.g., pigments). It is therefore not surprising that biological paleolimnology has provided important insights into past ecosystem changes. Below, I summarize some of the dominant indicators.

Pollen grains and spores are among the most common morphological indicators used in paleoenvironmental studies. The study of pollen grains and spores is a large scientific discipline, called palynology, which has many applications besides paleolimnology, such as in archeology and forensic science. Plants can produce large numbers of pollen grains, which are often well preserved in sediments. Similarly, the so-called lower plants, such as mosses, produce spores that are also well represented in sediment profiles. Although most pollen grains and spores are from terrestrial vegetation, aquatic macrophytes also produce pollen. As different species of plants produce morphologically distinct pollen grains (often identifiable to the genus level, but sometimes even to the species level), it is possible to reconstruct, at least in a general way, the composition of past forests and other vegetation. This provides important information on past climate and soil development, as well as the succession of different plant species. Some pollen grains can also assist in dating sediment cores. For example, in northeastern North America, the arrival of European settlers, and their related activities of clearing forests and initiating European style agriculture, resulted in the increase in pollen grains from ragweed (*Ambrosia*).

Plants are also represented in sedimentary deposits by macroscopic remains, typically referred to as macrofossils, which include seeds, twigs, and pieces of bark. Although pollen grains can be transported long distances via wind and other vectors, plant macrofossils are less easily transported, and so they more accurately reflect local vegetation changes. In addition, as noted earlier, they are also important sources for material used in $^{14}$C dating.

Algae and cyanobacteria (i.e., blue-green algae) represent major primary producers in most aquatic systems, and fortunately almost all of them leave reliable morphological or biogeochemical fossils. Foremost among these are the siliceous cell walls of diatoms (Bacillariophyceae, **Figure 4**). A diatom cell wall is composed of two similar halves (or valves) joined together; two valves forming a full diatom cell are called a frustule. Diatoms, which include many thousands of species, often dominate the plankton and periphyton of most lake systems. They have many characteristics that make them ideal paleoindicators. For example, they are abundant

and diverse, and their taxonomy is based on the size and sculpturing of the intricate glass cell walls that characterize individual species. As their cell walls are siliceous, they are very well preserved in most lake environments. The distributions of many diatom species are closely linked to specific limnological conditions, such as lakewater pH, salinity, nutrient levels, and so forth, and so are excellent paleoindicators.

Chrysophytes (classes Chrysophyceae and Synurophyceae) similarly can be tracked using siliceous microfossils. About 15% of chrysophyte taxa (including important genera such as *Mallomonas* and *Synura*) are characterized by an armor of overlapping siliceous scales. Similar to the diatom valves described earlier, these scales are species specific and can be used by paleolimnologists to track past populations of some groups. In addition to the scales, which are only produced by some taxa, all chrysophytes are characterized by the endogenous formation of siliceous resting stages, called stomatocysts (or statospores in the older literature). These cysts also appear to be species specific and are well preserved in sediments, However, unlike chrysophyte scales, the taxonomic identity of many cyst morphotypes have not yet been linked to the species that produced them, and instead cyst morphotypes are often simply referred to by temporary number designations.

Other algae also leave morphological microfossils, but are less widely used. This includes vegetative or reproductive cells of certain green (Chlorophyta) algae (e.g., *Pediastrum* colonies), or the akinetes and heterocysts of some blue-green algae. However, all algal and cyanobacterial groups are also represented in lake sediments by a variety of biogeochemical indicators. Fossil pigments (e.g., chlorophylls, carotenes, xanthophylls), which characterize certain groups and can be identified using high performance

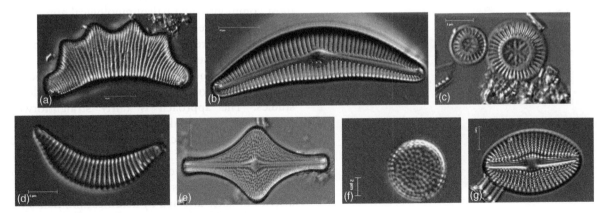

**Figure 4** Freshwater diatom valves. (a) *Eunotia*, (b) *Cymbella*, (c) *Cyclotella*, (d) *Semiorbis*, (e) *Brachysira*, (f) *Aulacoseira*, and (g) *Diploneis*. Photographs courtesy of B. Ginn (Queen's University).

liquid chromatography (HPLC) or other techniques from sediments, are most commonly used.

In addition to some of the primary producers described earlier, a large suite of zoological indicators is identifiable in sediments. Cladocera (**Figure 5**) are represented in sediments primarily by their chitinized body parts: head-shield, shell or carapace, and post-abdominal claws. Trained taxonomists can determine the species affinities of many of these body parts. In addition, the sexual resting stages (i.e., ephippia) of Cladocera are commonly encountered.

Lakes are also the habitat for many insect larvae, with midges or chironomids (Chironomidae, Diptera) dominating many assemblages. Fossil chironomids can be identified via their chitinized head capsules (**Figure 6**), which can often be identified to the generic level, or even species level. Chironomid head capsules are most often used in climate reconstructions and in tracking past deepwater oxygen levels. Other insect

remains include the chitinized mandibles of *Chaoborus* (phantom midge) larvae. Because certain taxa cannot coexist with fish predators, these mandibles have been used to document collapses of past fish populations (such as from lake acidification).

The Ostracoda (ostracods, also spelled ostracodes) is another widely used group of invertebrates in paleolimnological studies. Ostracods are characterized by two calcitic valves or shells that can be used to track a wide spectrum of environmental variables. Recent work has also focused not only on changes in species but also on the trace-element chemistry and stable isotope signatures of individual valves, as a memory of past limnological conditions is preserved in the chemistry and isotopic composition of ostracod valves.

In contrast to algal and invertebrate remains, vertebrates tend to be poorly represented in lake sediments. However, in certain environments, fish scales are common, and the inner ear bones of fish (i.e., otoliths) are increasingly being used in paleolimnological reconstructions.

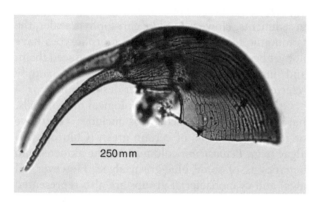

**Figure 5** Part of the chitinous exoskeleton of the cladoceran *Eubosmina longispina*. Photograph courtesy of J. Sweetman (Queen's University).

## Determining the Environmental Optima of Indicators

The previous section summarized some of the key indicators used in paleolimnology. However, in order to use, for example, diatom or chironomid species assemblages to infer environmental conditions, the environmental optima of the indicator taxa must first be determined. Some information on the ecological optima and tolerances of indicators is available in the scientific literature from previous biological surveys or similar studies. However, given the large number of taxa present, as well as the myriad of environmental variables that can influence species distributions, it can be a daunting task to link species distributions to environmental variables. The most commonly used paleolimnological approach to provide estimates of the environmental optima of indicator taxa are to use surface sediment calibration sets (or sometimes referred to as training sets).

Surface sediment calibration sets have become common components of many paleolimnological studies. The approach is fairly straightforward. A set of calibration lakes is chosen (perhaps 80 or so in number) that reflect the gradients of limnological conditions that might be expected in the paleolimnological study being undertaken. For example, suppose the goal of the study is to reconstruct lakewater pH in a series of lakes affected by acidification. Lakes with present-day pH values ranging from 4.5 to 8.0 would represent a broad gradient in pH values. The first

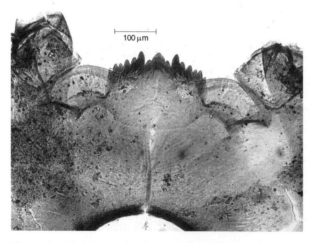

**Figure 6** A head capsule of the chironomid *Chironomus*. Photograph courtesy of J. Sweetman (Queen's University).

component of the calibration set would be a matrix of limnological and other environmental data available for the calibration lakes. The second step is to determine the distribution of indicator taxa in the 80 calibration lakes. This is done by taking surface sediment cores from each lake and identifying and enumerating the indicators in questions (e.g., diatoms) from the surface 0.5 or 1.0 cm of sediments (i.e., representing recently deposited biota). This represents the second matrix of data required for developing the calibration set, namely the species matrix. The next step is to use a variety of statistical techniques to link the species distributions in the surface sediments to the measured environmental variables for the calibration lakes. Multivariate techniques, such as canonical correspondence analysis (CCA), are often used to determine which environmental variables exert the greatest influence on species distributions. Transfer functions can then be constructed relating species distributions (e.g., diatom percentages) to the variables of interest (e.g., lakewater pH). The latter is typically done using statistical approaches such as weighted averaging regression and calibration. The resulting transfer

functions linking species distributions to environmental variables are often quite robust.

Although quantitative approaches, such as those described earlier, have become commonplace in paleolimnological studies, many environmental inferences can still be accomplished using qualitative approaches.

## Paleolimnological Applications

Paleolimnological approaches have now been used to assess a wide spectrum of limnological changes. Early work focused on long-term studies of lake ontogeny; however, a considerable volume of more recent research has focused on determining the effects of human impacts on aquatic ecosystems. Much of this applied paleolimnology began in the 1980s with work on acidic precipitation, where paleolimnology was used to determine if lakes had acidified, and if so, when and by how much. For example, **Figure 7** shows a typical diatom stratigraphic profile for an acidified lake in Nova Scotia (Canada). The change in the

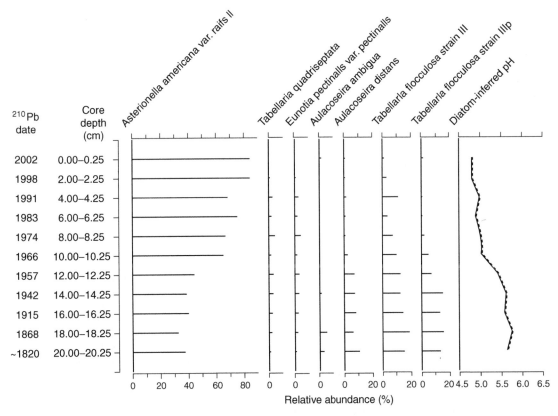

**Figure 7** A summary diagram of the dominant diatom changes (as percentages) in the recent sediments of Kejimkujik Lake (Nova Scotia, Canada). The dates of the sediment slices are estimated using $^{210}$Pb dating (left of figure). To the right is the diatom-inferred pH profile, which was generated using a transfer function developed from diatoms preserved in the surface sediments of a calibration set of east coast lakes. Reproduced from **Figure 3** from Ginn BK, Stewart LJ, Cumming BF, and Smol JP (2007) Surface-water acidification and reproducibility of sediment cores from Kejimkujik Lake, Nova Scotia, Canada. *Water, Air and Soil Pollution* 183: 15–24, with permission from Springer Science and Business Media.

dominant diatom assemblages (shown as percentages) clearly shows a shift in species assemblages (beginning *c.* 1940, as estimated from $^{210}$Pb dating) to taxa characteristic of more acidic waters. By using transfer functions developed from surface sediment calibration sets, the pH optima of the individual taxa can be used to provide quantitative estimates of past lakewater pH (as shown to the right of **Figure 7**). These types of analyses played a critical role in demonstrating the effects of acid precipitation on lake ecosystems. Other water quality issues can similarly be studied using these approaches.

Another major research focus is to use paleolimnological approaches to study long-term changes in climate, which has been receiving heightened attention with recent concerns about greenhouse gas induced warming. As climatic variables influence, to some extent, all biological distributions, it is not surprising that indicators such as chironomids, diatoms, and other proxies have been demonstrated to track climate variables, either directly (e.g., temperature reconstructions) or indirectly (e.g., via tracking past lakewater salinity changes which can be linked to past precipitation and evaporation ratios, past lake ice covers, etc.).

Many new paleolimnological studies are being initiated and are now often multidisciplinary, including a wide spectrum of techniques to provide more robust environmental inferences. In addition, important synergies between different groups of scientists (e.g., the study of fossil DNA in cladoceran ephippia) have heightened the interest in many paleolimnological approaches. As environmental problems continue to be identified, and as little monitoring data exist, it is clear that paleolimnological studies will continue to provide critical information on environmental change.

## Glossary

**Allochthonous material** – Material originating from outside the lake basin (e.g., soil particles, pollen grains from trees).

**Autochthonous material** – Material originating from within the water body (e.g., dead algal remains, chemical precipitates).

**Bioturbation** – The mixing of the sediments by benthic organisms.

**Paleolimnology** – The multidisciplinary science that uses the physical, chemical, and biological information preserved in aquatic sediments to track past changes in ecosystems.

**Palynology** – The study of pollen grains and spores.

**Siliciclastic sediment** – Clastic sediment composed of silicon-bearing minerals such as quartz, feldspars, and clay minerals.

*See also:* Benthic Invertebrate Fauna, Lakes and Reservoirs; Effects of Climate Change on Lakes; Eutrophication of Lakes and Reservoirs; Lake and Reservoir Management; Lakes as Ecosystems; Meromictic Lakes; Mixing Dynamics in Lakes Across Climatic Zones; Origins of Types of Lake Basins.

## Further Reading

Battarbee RW, Gasse F, and Stickley CE (eds.) (2004) *Past Climate Variability through Europe and Africa.* Dordrecht: Springer.

Birks HJB (1998) Numerical tools in palaeolimnology – Progress, potentialities, and problems. *Journal of Paleolimnology* 20: 307–332.

Cohen AS (2003) *Paleolimnology: The History and Evolution of Lake Systems.* Oxford: Oxford University Press.

Francus P (ed.) (2004) Image Analysis, Sediments and Paleoenvironments. Dordrecht: Springer.

Hvorslev MJ (1949) *Subsurface Exploration and Sampling of Soils for Civil Engineering Purposes.* Vicksburg, MS: American Society of Civil Engineers, Waterways Experiment Station, Corps of Engineers, U.S. Army.

Last WM and Smol JP (eds.) (2001) *Tracking Environmental Change Using Lake Sediments. Volume 1: Basin Analysis, Coring, and Chronological Techniques.* Dordrecht: Kluwer Academic.

Last WM and Smol JP (eds.) (2001) *Tracking Environmental Change Using Lake Sediments. Volume 2: Physical and Geochemical Methods.* Dordrecht: Kluwer Academic.

Leng MJ (ed.) (2006) *Isotopes in Palaeoenvironmental Research.* Dordrecht: Springer.

Maher BA and Thompson R (eds.) (1999) *Quaternary Climates, Environments and Magnetism.* Cambridge: Cambridge University Press.

Pienitz R, Douglas MSV, and Smol JP (eds.) (2004) *Long-Term Environmental Change in Arctic and Antarctic Lakes.* Dordrecht: Springer.

Smol JP (2008) *Pollution of Lakes and Rivers: A Paleoenvironmental Perspective*, 2nd edn. Oxford: Blackwell.

Smol JP, Birks HJB, and Last WM (eds.) (2001) *Tracking Environmental Change Using Lake Sediments. Volume 3: Terrestrial, Algal, and Siliceous Indicators.* Dordrecht: Kluwer Academic.

Smol JP, Birks HJB, and Last WM (eds.) (2001) *Tracking Environmental Change Using Lake Sediments. Volume 4: Zoological Indicators.* Dordrecht: Kluwer Academic.

Stoermer EF and Smol JP (eds.) (1999) *The Diatoms: Applications for the Environmental and Earth Sciences.* Cambridge: Cambridge University Press.

## Relevant Websites

http://www.paleolim.org/ – International Paleolimnology Association.

http://www.pages.unibe.ch/ – PAGES – Past Global Changes.

# Effects of Climate Change on Lakes

**W F Vincent,** Laval University, Quebec City, QC, Canada

## Introduction

Lake ecosystems are vital resources for aquatic wildlife and human needs, and any alteration of their environmental quality and water renewal rates has wide-ranging ecological and societal implications. The increasing accumulation of greenhouse gases in the atmosphere as a result of human activities has begun to affect the structure, functioning, and stability of lake ecosystems throughout the world, and much greater impacts are likely in the future. Current global circulation models predict an increase in air temperatures of several degrees by the end of the twenty-first century, combined with large changes in the regional distribution and intensity of rainfall. These changes will also be accompanied by massive disruption of the cryosphere, the ensemble of ice-containing environments on Earth. These shifts in climate forcing appear to have already begun, and the onset of changes in the physical, chemical, and biological attributes of lakes is affecting their ability to maintain the present-day communities of aquatic plants, animals, and microbes, and their capacity to provide ecosystem services such as safe drinking water and inland fisheries (**Figure 1**).

Lakes have always been subject to the impacts of climate change, and natural climate variations in the past have been one of the main reasons that lakes are ephemeral features of the landscape. Most of today's lakes are the result of climate amelioration and the retreat of the Pleistocene glaciers some 10 000 years ago, and so most present-day lakes are relatively young. A powerful approach toward understanding the potential impacts of future climate change on lakes is the application of paleolimnological methods in which lake sediment cores are dated and analyzed to infer climate impacts in the past. Most such analyses have been restricted to the time period of the last few 1000 years; however, detailed records of greater than 100 000 years are becoming available from studies of ancient lakes. Studies of present-day lakes of different ages (chronosequences) and across latitudinal gradients are also providing valuable insights into the consequences of climate change. Additional knowledge about climate impacts is coming from modeling and experiments, combined with multi-decade, regional analyses of lakes that are currently experiencing shifts in temperature and precipitation.

Some of the most striking examples of climate impacts to date are from limnological and paleolimnological studies in the polar regions. Global circulation models predict that the fastest and most pronounced warming will be at the highest latitudes because of a variety of feedback processes that amplify warming in these regions. These include the capacity of warm air to store more water vapor, itself a powerful greenhouse gas, and the reduced albedo (reflection of sunlight) as a result of the melting of snow and ice, leaving more solar energy to be available for heating. Some of the immediate impacts of climate change on high-latitude lakes include loss of perennial ice cover, increasing duration of open water conditions, increasing water temperatures, stronger water column stratification and shifts in water balance, in some cases leading to complete drainage or drying up of the waterbodies.

Changes in air temperature and precipitation have direct effects on the physical, chemical, and biological characteristics of lakes, and they also operate on lakes indirectly via modifications in the surrounding watershed, e.g., through shifts in hydrological flow pathways, landscape weathering, catchment erosion, soil properties, and vegetation. Of particular interest to limnologists (lake and river scientists) are the interactions between variables, the feedback effects that accelerate or dampen environmental change, and threshold effects by which lakes may abruptly shift from one environmental state to another.

## Physical Effects of Climate Change

### Basin Integrity

Changes in the precipitation regime that accompany climate change have the potential to cause shifts in the connectivity of lakes (with biological implications, e.g., for migratory fish species), as well as in erosion rates that could affect the inflow and outflow dynamics of lakes. The latter effects are particularly conspicuous in the tundra permafrost. Many thaw ponds (also called thermokarst ponds) on the Siberian and Alaska tundra are draining as a result of increased melting of the permafrost, while certain thaw ponds further south in discontinuous, subarctic permafrost have begun to expand as a result of the differential melting of such

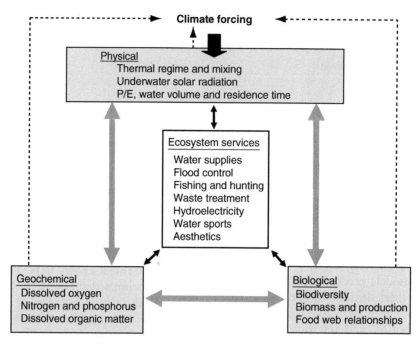

**Figure 1**  Changes in climate forcing affect the physical environment of lake ecosystems and thereby alter their chemical and biological properties. These changes affect the capacity of lakes to provide ecosystem services. P/E, precipitation to evaporation ratio. Dotted lines indicate positive feedback effects, e.g., via decreased ice cover or the release of greenhouse gases from lakes into the atmosphere.

soils. Ice-dammed lakes, e.g., those behind thick ice shelves (epishelf lakes), ice-jams on rivers, glacial dams, and behind coastal sea ice barriers (stamukhi lakes) are prone to complete loss because of climate-dependent ice fracturing and collapse.

### Water Balance

Shifts in precipitation relative to evaporation (the P/E ratio) cause changes in the water budget and hydraulic residence time of lakes, as well as in their depth and areal extent. Ponds and wetlands are especially vulnerable to changes in P/E because of their shallow depths and large surface to volume ratio. For example, some rock basin ponds in the Canadian High Arctic have been drying up as a result of climate warming in the region, perhaps for the first time in millennia. Even large, shallow lakes can undergo major climate-driven fluctuations and loss. Lake Chad in north-western Africa is less than 7-m deep and has responded greatly to changes in climate over the last few decades. From the 1960s onward it experienced a precipitous decline in rainfall, and this has led to a major contraction in lake area, from $25\,000\,km^2$ in the early 1960s to $1350\,km^2$ in the 1990s. This climate effect has been compounded by the increasing need to irrigate farmland during these drought conditions, and human water use appears to account for about half of the decrease in lake surface area.

Glaciers in many parts of the world are undergoing accelerated retreat, and glacial fed lakes and reservoirs are therefore particularly vulnerable to changes in their inflow regime. For example, the Chinese Academy of Sciences estimates that by 2050, up to 64% of China's glaciers may have disappeared, with serious consequences for the estimated 300 million people who live in China's arid west and who depend upon this vital water supply. At Lake Biwa, Japan, the reduced flow from snowpack in the surrounding mountains is thought to be a mechanism reducing cold underflows that play an important role in recharging the bottom waters of the lake with oxygen.

At the other extreme, global change may affect the frequency of extreme storm events and flooding in some regions. There is evidence that warming of the sea surface is resulting in an increasing intensity and duration of storms, and certain coastal lakes may therefore be exposed to increased flooding and wave erosion of their shorelines, as well as increased hydraulic flushing and storm-induced mixing.

### Ice Cover

Many lakes in the north temperate zone are covered by ice and snow through much of the year. Climate warming is already showing an effect on such lakes by causing earlier dates of ice break-up, and later dates of freeze-up. The 2007 Intergovernmental Panel on Climate Change report noted that for all available lake and river ice data over the last 150 years, the break-up date had become (mean ± 2SD) 6.5 ± 1.2 days earlier per century, while the freeze-up date had occurred 5.8 ± 1.6 days later per century. At polar latitudes, the earlier occurrence of rain instead of snow is likely to hasten the ice melting process, which is also affected by changes in the wind field. Loss of snow and ice cover results in a major shift in the underwater irradiance regime, with increased solar energy for radiative heating of the water column, increased availability of photosynthetically available radiation (PAR) for primary production, and increased exposure to ultraviolet radiation (UVR). The latter will accelerate photobiological and photochemical processes, with a variety of positive and negative effects on ecosystem structure and productivity. Modeling analyses show that loss of ice cover can have orders of magnitude larger effects on underwater UVR exposure than does moderate ozone depletion in the stratosphere. The prolonged loss of ice cover also allows a longer period of wind-induced mixing of the surface waters of lakes, and may change the transport pathways and fate of inflows that during ice conditions are retained immediately beneath the ice.

### Temperature and Stratification

Warmer air temperatures result in warmer surface waters via conduction, although this may be dampened in part by increased evaporation rates. For north temperate lakes, the increased duration of ice-free conditions will allow increased warming of the surface waters by sensible heat transfer (conduction) and by radiative transfer (penetration of solar radiation into the water column). These effects in turn cause changes in the density structure of lakes, with a layer of warmer, lighter water at the surface that prevents the transport of heat via turbulent mixing to deeper parts of the lake. This retention of heat in the upper waters acts as a positive feedback mechanism, and with increasing heating the thermal stratification becomes stronger and more resistant to wind-induced mixing. By trapping more heat in the surface-mixed layer, less is available for heating the lower column, and deep waters can become cooler as a result of climate warming. An example of the strong effects of climate change on water temperature is the current trend in Lake Superior, Canada–USA. The summer surface water temperatures of this lake increased by 2.5 °C over the period 1979–2006, well in excess of the air temperature increase over the same period. This has been primarily attributed to earlier ice break-up and the earlier onset of stratification and

warming each year. This differential effect is not apparent in tropical lakes that lack winter ice cover, and such waters appear to be more closely tracking the regional trend in air temperatures.

At high latitudes, increased warming may lead to a complete change in the stratification and mixing regime. Deep polar lakes are usually cold monomictic; that is, they remain stratified under the ice through most of the year, and then mix at temperatures less than 3.98 °C (the temperature of the maximum density of water) throughout their summer period of open water conditions. However, if the water temperatures warm above 3.98 °C in summer, the lake will become stratified, and therefore dimictic, with periods of mixing before and after this summer period. This shift from cold monomixis to dimixis is a critical threshold that allows the retention of heat in the surface waters and it may lead to a higher biological productivity. Paleolimnological analyses have suggested that such transitions occurred in certain Finnish lakes over the last 200 years, and that the resultant increased productivity caused biological shifts toward a cladoceran zooplankton community.

## Chemical Effects of Climate Change

### In-Lake Effects

The hydraulic residence time of a lake (the time required to completely replace all water in a lake by its river, groundwater and rainfall inputs) affects the chemical composition of lake waters by controlling the time available for biogeochemical and photochemical processes to operate, the extent of accumulation and loss of dissolved and particulate materials, and the duration of biogeochemical interactions with the lake sediments and littoral zone. In lakes that experience anoxic bottom water conditions and nutrient release from the sediments, a prolonged residence time caused by reduced precipitation and inflows can result in increased phosphorus accumulation (internal phosphorus loading) and eutrophication. Conversely, in regions that experience increased precipitation and water flow, the increased flushing of nutrients and phytoplankton may result in reduced algal production. This latter effect may be offset by increased catchment erosion and transport of nutrient-rich soils into the lake if storm events become more frequent and severe.

The stronger and more prolonged stratification of lake waters is also likely to lead to more pronounced gradients in chemical variables down through the water column. In enriched lakes, this may trigger the shift from oxygenated to anoxic conditions in the bottom waters, and crossing this threshold is typically accompanied by major upshifts in the concentrations of algal nutrients, including available phosphorus, and in various toxic compounds, notably hydrogen sulfide ($H_2S$). Anoxia can also lead to increased nitrogen losses from the ecosystem by denitrification, the bacterial process that converts nitrate to nitrogen gas.

### Catchment Effects

Climate affects many landscape properties, including vegetation and soils, which in turn exert a strong influence on water quality and quantity in the receiving lake basin. Increased temperatures and precipitation may lead to increased rock weathering, thereby affecting solute composition and concentrations in the runoff. Long-term studies at Toolik Lake Alaska have detected a doubling of alkalinity over time, and this may be the result of climate-dependent increases in active layer depth in the permafrost that have exposed new soil material to weathering. Changes in precipitation may affect the magnitude of overland versus subterranean flow, thereby changing the nature and quantity of nutrient, major ion, and dissolved organic carbon (DOC) inputs from different soil horizons to lakes. DOC loading is particularly responsive to climate change, through the leaching and erosion of soils (compounded by permafrost melting in the Polar Regions), changes in microbial transformation rates of particulate and dissolved organics, and through changes in catchment vegetation. The effects of the latter are apparent from high-latitude paleolimnological studies, which indicate an increase in DOC concentrations associated with colonization of the catchments by woody shrubs and trees during periods of climate warming. The increased transport of the colored fraction of this DOC (colored dissolved organic matter, CDOM) can have both positive and negative effects on aquatic photosynthesis, by shading cells against damaging UVR and by absorbing PAR that would otherwise be available for photosynthesis.

### Biological Effects of Climate Change

The biological responses to the physical and chemical changes noted here involve multiple interactions, feedbacks and complex nonlinear responses that are not possible to fully predict on the basis of current knowledge. However, a great variety of direct and indirect effects can be identified as some of the likely impact pathways of ongoing climate change. These operate at multiple scales, from changes in whole ecosystem structure and dynamics, down to physiological and molecular responses at the individual and cellular level that in turn feed back into overall ecosystem dynamics.

### Ecosystem Integrity

At the broadest scale, climate change has the potential to radically alter the physical structure of lake ecosystems, and thereby cause the extinction or alteration of aquatic biota. The melting of the Siberian permafrost is one of the most striking examples, where 11% of thaw ponds in a large study area have recently drained and been lost, thereby eliminating the habitat for aquatic biological communities. The contraction of large shallow lakes also results in major habitat loss, and even deep lakes may lose important ecological features as a result of relatively small fluctuations in water level. For example, the coastal wetlands of the North American Great Lakes are important for migratory birds, and the majority of species of fish in the region depend on these littoral and supralittoral environments for reproduction. Small changes in lake levels as a result of shifting water balance make these environments especially vulnerable to climate change. Changes in the connectivity between aquatic habitats as a result of rising or falling water levels will also influence the species composition of aquatic biota, especially fish communities.

### Vertical Habitat Structure

Changes in ice cover, stratification, and mixing, and thereby the vertical gradients in lake properties, will have far reaching effects on pelagic communities and production. The associated changes in light and nutrients will affect phytoplankton production, and therefore the availability of food to high trophic levels. For some high-latitude systems, decreased ice cover and stronger stratification may allow for increased irradiance supply and rates of primary production. For example, the strong climate warming at Signy Island in the maritime Antarctic has been accompanied by increased overwintering phytoplankton biomass. This was attributed to reductions in lake snow and ice cover and improved light conditions for summer growth, as well as increased nutrient inputs following deglaciation of the catchments. In contrast to this response, simulation analyses for the North American Great Lakes have shown negative effects of projected climate change. Earlier summer stratification results in a decreased period of mixing of nutrients from the sediments and from deeper waters into the euphotic zone, and therefore increased nutrient limitation of algal growth. These effects are compounded by projected increases in cloud cover, which would reduce the availability of light for photosynthesis.

In Lake Tanganyika, the present trend of climate warming, decreased wind speeds, and increased water column stability appears to have resulted in decreased nutrient supply to the euphotic zone by mixing, resulting in a 20% decrease in phytoplankton production. The consequences of this effect are serious given that a large portion of animal protein available to the surrounding human populations is derived from the pelagic fishery, and this decrease in algal photosynthetic rates is estimated to have caused a 30% drop in fish yield.

A prolonged ice-free period in northern, temperate lakes, and therefore a decreased duration of winter ice cover, may prove to be advantageous for some fish species. In certain locations (e.g., small, enriched lakes in southern Canada and northern United States), prolonged winter stratification leads to complete oxygen depletion of the hypolimnion, resulting in the "winter kill" of fish stocks, and such effects may lessen as the period of ice cover contracts.

### Photosynthetic Communities

Changes in light and nutrient availability that accompany climate change will also have a qualitative effect on species composition and diversity at the primary producer level, which in turn may impact on higher trophic levels. Paleolimnological studies have shown that the diatom composition of the periphyton and phytoplankton in ponds at many sites in the circumpolar Arctic has changed markedly over the last 150 years, consistent with a strong response to decreased ice cover and possibly increased nutrient supply. In ancient Lake Baikal, there is evidence that endemic, cold-dwelling diatom taxa that support the current food-web will be replaced by faster growing, nondiatom species, including picophytoplankton. Changes in lake level are likely to affect the species composition of aquatic macrophyte communities in the littoral zone of lakes, with implications for the birds, fish, and other biota that depend on these plants.

A major climate change concern at temperate latitudes is the prospect of a shift in phytoplankton species composition toward dominance by species of cyanobacteria that form noxious blooms. These organisms create many water quality problems in lakes and reservoirs, including the release of taste and odor compounds, the production of various toxins, and the overproduction of biomass that clogs water filters, disrupts zooplankton feeding, and causes oxygen depletion. Bloom-forming cyanobacteria are likely to be favored in a warming climate by several mechanisms. First, their temperature for

optimum growth tends to be high (25 °C or above), and warmer conditions will favor their more rapid accumulation and dominance. Second, these species can regulate their position in stable water columns by way of gas vacuoles that allow them to sink or rise to the optimal depths for photosynthesis or nutrient uptake. As noted above, future climate change scenarios predict increased stratification and water column stability that would favor this strategy. Third, a warmer climate may lead to increased anoxia and internal phosphorus loading (perhaps also in external loading in some circumstances), and bloom-forming cyanobacteria tend to become increasingly prevalent with increasing degree of phosphorus enrichment. Finally, a highly toxic species from tropical and subtropical regions, *Cylindrospermopsis raciborskii*, has been increasingly observed in temperate lakes. This apparent invasion has been ascribed to climate change, with warmer temperatures favoring the early germination of its resting spores (akinetes).

### Fish Communities and Migration

Many fish species are sensitive to even small changes in the temperature of their surroundings, and the warming trend associated with climate change is likely to cause a shift in the geographic distribution of many taxa. Cold water species such as walleye and trout may be driven to extinction in some lake waters because their thermal tolerance is exceeded, while other species may be able to colonize new habitats as a result of warmer conditions. In Toolik Lake, Alaska, a bioenergetics analysis showed that a 3 °C rise in summer lake water temperature could cause the young-of-the-year lake trout to consume eight times more food that is currently necessary to maintain their condition, which greatly exceeds the present food availability. In Ontario lakes, the northern range of smallmouth bass is currently limited by temperature because the length of the growing season determines the winter survival of the young-of-the-year. Under a climate change scenario, this species could invade some additional 25 000 northern lakes, and because of its strongly negative effects on other fish, such invasions would likely cause the extirpation from these lakes of four native cyprinid species.

Climate change may affect the migratory behavior of some fish species. Arctic char for example migrate from lakes to the sea where the higher food availability allows them to grow to a large size. If arctic lakes become more productive as a result of climate change, then char migrations are projected to decrease, and the resultant lake char would be of smaller size and less suitable as a food resource for northern aboriginal communities.

## Microbial Processes

Many lakes are net heterotrophic, with the respiration (R) of organic materials derived from autochthonous (within the lake) and allochthonous (outside the lake) sources exceeding photosynthesis (P). Heterotrophic microbial activity is likely to be enhanced in a warmer climate because of the direct physiological effects of temperature, and in some environments by the increased availability of organic carbon and nutrients, e.g., from increased catchment runoff. Photosynthesis on the other hand is strongly regulated by light availability and may be less responsive to the temperature increase, or even negatively affected via the mechanisms noted earlier. This combination of effects would exacerbate the negative P/R balance in lakes, and may result in lakes becoming even greater net sources of $CO_2$ to the atmosphere. Production of the more powerful greenhouse gas methane will be favored by prolonged anoxia in some environments, and in high northern latitudes by its release from permafrost surrounding thaw ponds. The production and release of greenhouse gases from arctic thaw lakes has been identified as a potentially important feedback mechanism that will accelerate climate warming.

*See also:* Littoral Zone; Reservoirs.

## Further Reading

Austin JA and Colman SM (2007) Lake Superior summer water temperatures are increasing more rapidly than regional air temperatures: A positive ice-albedo feedback. *Geophysical Research Letters* 34: L06604, doi:10.1029/2006GL029021.

Brooks AS and Zastrow JC (2002) The potential influence of climate change on offshore primary production in Lake Michigan. *Journal of Great Lakes Research* 28: 597–607.

Coe MT and Foley JA (2001) Human and natural impacts on the water resources of the Lake Chad basin. *Journal of Geophysical Research Atmospheres* 106(D4): 3349–3356.

Jackson DA and Mandrak NE (2002) Changing fish biodiversity: Predicting the loss of cyprinid biodiversity due to global climate change. In: McGinn NA (ed.) *Fisheries in a Changing Climate*, pp. 89–98. Bethesda, Maryland, USA: American Fisheries Society.

Korhola A, Sorvari S, Rautio M, *et al.* (2002) A multi-proxy analysis of climate impacts on the recent development of subarctic Lake Sannajaärvi in Finnish Lapland. *Journal of Paleolimnology* 28: 59–77.

Lemke P, Ren J, Alley RB, *et al.* (2007) Observations: Changes in snow, ice and frozen ground. In: Solomon S, Qin D, Manning M, *et al.* (eds.) *Climate Change 2007: The Physical Science Basis. Contribution of Working Group 1 to the Fourth Assessment Report of the Intergovernmental Panel on Climate Change.* Cambridge, UK: Cambridge University Press.

Mackay AW, Ryves DB, Morley DW, Jewson DH, and Rioual P (2006) Assessing the vulnerability of endemic diatom species in Lake Baikal to predicted future climate change: a multivariate approach. *Global Change Biology* 12: 2297–2315. doi: 10.1111/j.1365-2486.2006.01270.x.

McDonald ME, Herschey AE, and Miller MC (1996) Global warming impacts on lake trout in Arctic lakes. *Limnology and Oceanography* 41: 1102–1108.

McKnight D, Brakke DF, and Mulholland PJ (eds.) (1996) Freshwater ecosystems and climate change in North America. *Limnology and Oceanography* 41(special issue): 815–1149.

Mueller DR, Vincent WF, and Jeffries MO (2003) Break-up of the largest Arctic ice shelf and associated loss of an epishelf lake. *Geophysical Research Letters* 30: 2031 doi:10.1029/2003GL017931.

O'Reilly CM, Alin SR, Plisnier PD, Cohen AS, and McKee BA (2003) Climate change decreases aquatic ecosystem productivity of Lake Tanganyika, Africa. *Nature* 424: 766–768.

Prowse TD, Wrona FJ, Reist JD, Hobbie JE, Lévesque LMJ, and Vincent WF (2006) Climate change effects on hydroecology of arctic freshwater ecosystems. *Ambio* 35: 347–358.

Quayle WC, Peck LS, Peat H, Ellis-Evans JC, and Harrigan PR (2002) Extreme responses to climate change in Antarctic lakes. *Science* 295: 645.

Quesada A, Vincent WF, Kaup E, *et al.* (2006) Landscape control of high latitude lakes in a changing climate. In: Bergstrom D, Convey P, and Huiskes A (eds.) *Trends in Antarctic Terrestrial and Limnetic Ecosystems*, pp. 221–252. Dordrecht: Springer.

Schindler DW and Smol JP (2006) Cumulative effects of climate warming and other human activities on freshwaters of arctic and subarctic North America. *Ambio* 35: 160–168.

Smol JP and Douglas MSV (2007) Crossing the final ecological threshold in high arctic ponds. *Proceedings of the National Academy of Sciences USA* 104: 12395–12397.

Smith LC, Sheng Y, MacDonald GM, and Hinzman LD (2005) Disappearing Arctic lakes. *Science* 308: 1429. doi: 10.1126/science.1108142.

Vincent WF, Rautio M, and Pienitz R (2007) Climate control of underwater UV exposure in polar and alpine aquatic ecosystems. In: Orbaek JB, Kallenborn R, Tombre I, Hegseth E, Falk-Petersen A, and Hoel AH (eds.) *Arctic Alpine Ecosystems and People in a Changing Environment*, pp. 227–249. Berlin: Springer.

Walter KM, Zimov SA, Chanton JP, Verbyla D, and Chapin FS (2006) Methane bubbling from Siberian thaw lakes as a positive feedback to climate warming. *Nature* 443: 71–75.

Wiedner C, Rücker J, Brüggemann R, and Nixdorf B (2007) Climate change affects timing and size of populations of an invasive cyanobacterium in temperate regions. *Oecologia* 152: 473–484. doi: 10.1007/s00442-007-0683-5.

# Modeling of Lake Ecosystems

**L Håkanson,** Uppsala University, Uppsala, Sweden

## Introduction: Basic Concepts and Problems in Modeling of Lake Ecosystems

Every aquatic ecosystem is unique, but only a few are studied ecologically in great detail. For reasons related to the practical value of lakes, including fisheries, recreation, and water supply, there are demands for analytical information or predictions concerning lakes for which no detailed studies are available. Ecosystem modeling is the main tool by which predictions and analyses can be provided in such cases.

Ecosystem modeling of lakes and other aquatic ecosystems is important not only in practical matters related to lake management or environmental remediation, but also in demonstrating the ways in which multiple factors act simultaneously on ecosystem features such as community composition, biological productivity, or biogeochemical processes. Therefore, modeling is useful not only as a means of providing practical information for solving problems, but also in improving the basic understanding of lakes as ecosystems.

Lake ecosystem models are composed of linked predictive equations incorporating multiple environmental variables. The environmental variables may be physical, chemical, or biological, and they typically are expressed as quantities per unit or volume per unit area, or as fluxes (mass or energy per unit time). Relevant variables for lake ecosystem models include not only those applicable to the lake itself, but also to the watershed from which the lake derives water and dissolved or suspended substances that affect ecosystem processes (**Figure 1**).

Environmental variables that are used in the lake ecosystem models are of two types: those for which site-specific data must be available in support of modeling, and those for which typical or generic values can be used. Examples of site-specific variables include the dimensions of the lake (e.g., mean depth), hydrologic features (e.g., hydraulic residence time), and concentrations of dissolved or suspended substances in the water (e.g., nutrients, organic matter). Examples of variables for which typical or generic values can be used include rates of mass transfer from the water column to the sediment surface or from the sediment surface to water column. Variables that are specific to a given lake will be referred to here as 'lake-specific variables,' but

sometimes are also referred to as 'obligatory driving variables.' Variables for which typical or generic information is adequate will be referred to here as 'generic variables.'

Repeated testing of the predictive power of models, as judged by actual ecosystem characteristics, has shown which environmental variables must be site-specific and which can be generic. An important result of such experience with modeling over the last two decades is that a number of environmental variables that are very important to the predictive capability of models can be generic.

Mathematically speaking, it is possible to create ecosystem system models of great complexity. Experience has shown, however, that complexity degrades the predictive reliability of ecosystem models. Therefore, one important goal of modeling is to represent the relevant ecosystem processes in a manner that is both realistic and simple. A combination of the principle of simplicity with the principle of generic variables has greatly simplified ecosystem modeling for lakes.

Once prepared as a set of linked equations based on site-specific and generic variables, a model must be calibrated to make predictions for a specific lake. The calibration involves insertion of specific numeric values for site-specific variables. Often the first attempt at calibrating a model to make predictions for a specific lake shows that the model is making biased predictions. The modeler then attempts to find an error in calibration. If a suspected error is found, the calibration is adjusted until the model produces predictions showing low bias.

The calibration process involving adjustment of calibration to produce realistic results may be misleading. Because models contain coupled equations, many of which have mutual counteractive effects, it is almost always possible to make a model produce results of low bias by adjusting one or more of the calibrated variables. Thus, an improved fit does not necessarily indicate that the initial problem with the model has been solved; it only shows that the model has been forced to produce a better prediction by an adjustment that may or may not be correct.

The modeler's tool for finding errors in calibration is validation. Validation is the process by which the modeler uses a model that has been calibrated on one

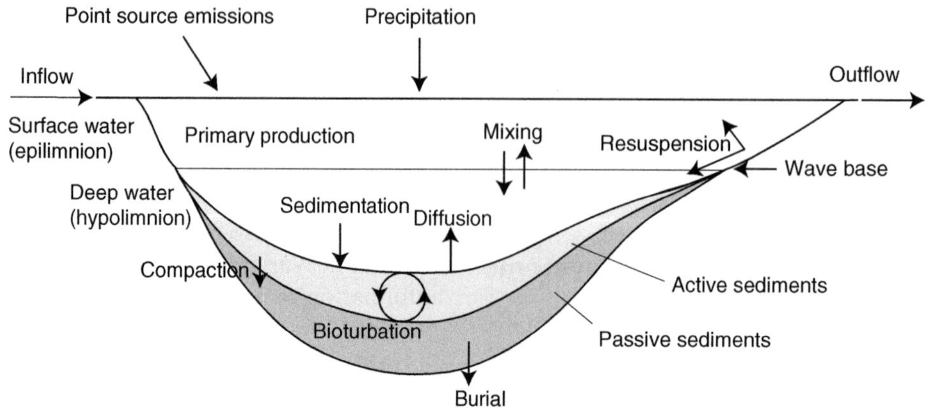

**Figure 1**  Schematic illustration of fundamental transport processes affecting lakes.

or a few ecosystems to make predictions on other ecosystems that were not involved in the calibration process. In other words, it is a test of the model outside the framework within which the model was first developed. If the model has been properly calibrated, it will perform well on systems that were not used in the calibration process, provided that these systems have a general similarity to the ones that were used in calibration.

The application of a lake ecosystem model produces several indicators of the value of the model. First, the accuracy and precision of the model predictions during validation is a quantitative index of the value of the model. A second indicator is the practical or fundamental importance of the model predictions; failure of a model to predict the most important types of outcomes is an indicator of weakness in the model. A third index of value is the range of conditions over which it can be applied. Some models perform well under conditions very close to those used in calibration, but fail to predict conditions in lakes that differ from those used in calibration. More robust models are more valuable. Finally, the most successful models are those that can be operated successfully on the basis of site-specific variables for which information is easily available. A demand for obscure information reduces the value of a model.

In emphasizing simplicity, as necessary in order to conserve the predictive power of models, the modeler must avoid any attempt to represent mathematically all of the possible connections of an ecosystem extending from a cellular level to whole organisms, populations, and communities. The modeler must search for connections between abiotic variables and biotic responses that are meaningful at the ecosystem level. One successful approach in this attempt at simplification is the development of quantitative relationships

between mass transport of environmentally important substances and biologically driven ecosystem responses. This concept may be called the 'effect-load-sensitivity' (ELS) approach to ecosystem modeling. It is well suited to modeling that is directly relevant to water management. It is based on the principle that lakes often have differing sensitivities to a given mass loading of a contaminant. For example, lakes showing low pH will sustain fish populations with higher mercury concentrations per unit mass than lakes of higher pH receiving the same mercury load per unit volume or per unit area.

## Classical Lake Modeling: The Vollenweider Approach

Richard Vollenweider first showed that phosphorus concentrations of lakes could be predicted from hydraulic residence time (water retention time) and the mean concentration of phosphorus for rivers or streams entering the lake using mass-balance modeling. He then showed that characteristic abundances of phytoplankton, as measured by concentrations of chlorophyll in the water column, could be predicted from the modeled concentrations of phosphorus. Calibration of the relationship between phosphorus and chlorophyll *a* was accomplished by simple regression analysis from field observations on numerous lakes. Vollenweider's approach was of great practical utility because it enabled lake managers for the first time to calculate how much reduction in the transport of phosphorus to a lake would be required in order to reduce the growth of algae in the lake to an acceptable level for management purposes.

The Vollenweider approach, although very influential, has shown some limitations. First, it does not deal with temporal variations in phytoplankton biomass, and therefore fails to account for the exceptional importance of peak algal biomass as contrasted with average biomass. In addition, the Vollenweider approach, as originally constructed, failed to account for the substantial escape of phosphorus from sediments in the most productive lakes ('internal loading of phosphorus'). Because such lakes can be essentially self-fertilizing, the predicted effects of controlling external sources of phosphorus may not be valid. Even so, the Vollenweider approach remains a keystone for lake modeling of eutrophication because it shows the two key features of successful models: simplicity and use of readily defined, lake-specific variables.

Lake models deal differently with abiotic and biotic variables. Mass fluxes of contaminants or nutrients, which carry units of gram per unit time, or the amounts or concentrations of any inorganic substance, are generally dealt with through the use of differential equations. Environmental variables that are under direct biological control (bioindicators) must be treated differently because they do not reflect the law of conservation of mass. For example,

nutrients present in a lake may not be completely incorporated into phytoplankton biomass because of the presence of biological removal agents (grazers) or hydraulic removal of biomass (washout). Therefore, bioindicators are related to abiotic variables by means of empirical calibrations, which often involve regressions of the type used by Vollenweider.

ELS models combine a mass-balance approach to the prediction of abiotic variables with an empirical approach such as regression analysis to relate bioindicators to an abiotic variables (**Figure 2**). In this way, such a model uses the principle of mass conservation to calculate the load or concentration of an important abiotic variable such as nutrient concentration. The model then converts the abiotic variable into a biotic signal, as shown by one or more bioindicators through an empirically established relationship between the abiotic variable and the bioindicator. Thus, the ELS modeling approach, as pioneered by Vollenweider, can be used in predicting a wide range of bioindicator responses through the consistent use of a few simple model development principles.

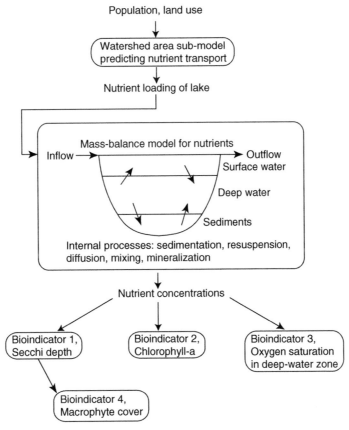

**Figure 2**  Basic elements in ELS modeling for aquatic eutrophication studies and management utilizing mass-balance modeling and regression analyses relating nutrient concentrations to bioindicators (e.g., Secchi depth, chlorophyll a concentrations, oxygen saturation in the deep-water zone and macrophyte cover).

## The Relevance of Scale to Modeling

Bioindicators, which typically are targets of high importance for modeling, can be quantified over wide ranges of temporal and spatial scales. Models designed to produce predictions at different scales may produce qualitatively different kinds of results. It is critical for the modeler to select scales of space and time that are appropriate to the application.

The flexibility in development of models of varying scale to some extent is restricted by the availability and degree of uncertainty for all types of empirical data that support modeling. As shown in **Figure 3**, uncertainty increases as temporal scale increases (the same would be true of spatial scale), but the confinement of modeling to short time scales produces an impossible requirement for empirical documentation. Therefore, optimal conditions for modeling involve intermediate scales of time and space. In addition, focus on the ecosystem as a natural entity imposes certain spatial constraints. Modeling of ecosystems cannot be built up easily from empirical data on specific organisms. Therefore, ecosystem modeling is facilitated by focus on bioindicators that have implications for the entire ecosystem. For example, ecosystem modeling of food chain phenomena such as bioconcentration of mercury can begin with a consideration of the abundance and type of top predators in a lake. The influence of top predators is spread through the lower links of the food chain, influencing trophic dynamics on an ecosystem basis.

## Recent Developments in Mass-Balance Modeling

The accidental release of large amounts of radiocesium to the atmosphere from the Chernobyl reactor site in the Ukraine during 1986 led to the transport of substantial quantities of radiocesium to Scandinavia and Europe. Although alarming from the viewpoint of environment and human health, the Chernobyl accident allowed unprecedented mass-balance tracking of an environmental constituent (cesium). Because radiocesium can be detected through its radioactive emissions, the transport of even small amounts from the atmosphere to soil surface, and through the soil surface to the drainage network and into food chains could be studied on a large spatial scale across many different types of aquatic ecosystems.

The radiocesium studies produced new understanding of rates and mechanisms for the transport not just of radiocesium, but also for many other substances, including contaminants and nutrients relevant to the modeling of lake ecosystems. This improved level of understanding has lead to increased sophistication of mass-balance prediction in the lake ecosystem models (**Figure 2**).

Sedimentation is the name for flux carrying mass from water to sediments. Return of mass from sediments to the water column can occur either through resuspension, which is generated by turbulence at the sediment–water interface, or by diffusion across concentration gradients at the sediment–water interface. In addition, some mass is buried through the

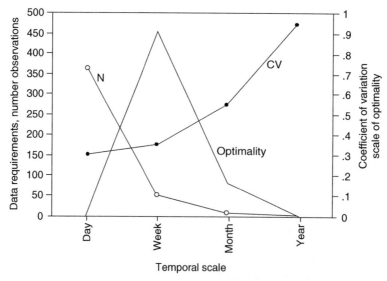

**Figure 3**   Illustration of factors controlling optimality of models in aquatic ecology. The curve marked CV shows that most lake variables show increasing coefficients of variation (CV) with increasing temporal spacing of samples. The curve marked N (number of data points needed to run or test a model) shows that data requirements are much higher for models designed to predict over short rather than long time scales. The optimality curve illustrates the combined effect of CV and N.

accumulation of mass on the bottom of the lake, and some leaves the lake through its outflow. The fluxes illustrated in **Figure 2** form the core of the mass-balance modeling of lakes.

Mass-balance modeling as illustrated in **Figure 2**, when improved through the new insights derived from the Chernobyl contamination, show very high predictive capabilities. For example, modeling of radiocesium concentrations from 23 very different European lakes accounted for 96% of variance and showed a slope relating protected to observe concentrations of 0.98. Such high predictability in mass-balance modeling is extremely useful.

**Figure 4** shows modeling of phosphorus in support of management through use of LakeMab, a lake

ecosystem model, as guided by insight gained through the Chernobyl disaster. The results shown in the model were achieved with no recalibration.

## Improvement in Modeling Practices

Abbreviations are used extensively in ecosystem modeling. Modern abbreviation systems follow rules that are intended to simplify the use of mathematical terminology. Contrary to custom in mathematics and physics, Greek letters are avoided in favor of mnemonic letter combinations. Consistency is used for like measures. For example, length measures are consistently designated as $L$, and subclassified with a

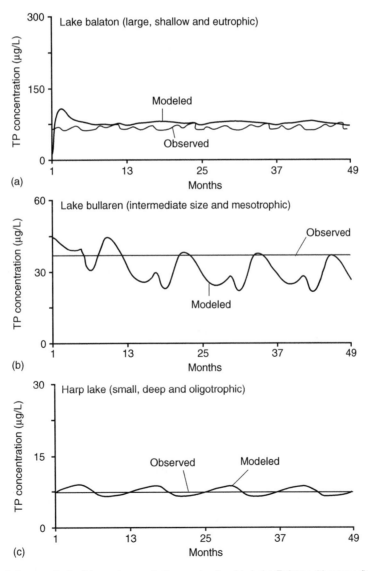

**Figure 4**   Example of modeling results for (a) very large, shallow and eutrophic Lake Balaton, Hungary, (b) Lake Bullaren, Sweden, which is one of moderate size (in this study) and mesotrophic and (c) Harp Lake, Canada, which is very small, deep and oligotrophic. The figures give modeled TP concentrations as well as observed long-term median values.

subscript. For example, $L_{max}$ for maximum length. Fluxes from one compartment to another of an ecosystem are designated by subscripts: $F_{ab}$ means flux from compartment A to compartment B.

Different forms of a particular substance that is being modeled may be distinguished by use of a distribution coefficient in a model. For example, a coefficient may differentiate between dissolved and particulate fractions in a water column. Such a distinction is important functionally because the particulate fraction can settle to the sediments, whereas the dissolved fraction cannot. Coefficients are also given for sedimentation into deep water as opposed to shallow water where resuspension is more likely. Overall, the use of distribution coefficients facilitates modeling of the important mass-flux processes within lakes: for example, sedimentation from water to sediments, resuspension from the sediments back to the water, diffusion from the sediment to the water, mixing of surface waters with deep waters, and conversion from organic to inorganic forms.

The model shown in **Figure 2** is typical in its use of spatial compartments: surface water, deep water, zones of erosion and transport, and zones of accumulation. Symbolic notation references these compartments. For example, $F_{SWDW}$ indicates flux from surface water to deep water.

## Modeling of Food Webs Based on Functional Groups

Management questions related to productivity, community composition, and concentrations of biomass commonly require modeling of food web interactions. **Figure 5** shows a typical application. For most lake ecosystem modeling, several functional groups are included in the model: predatory fish, prey fish, benthic animals, predatory zooplankton, herbivorous zooplankton, phytoplankton, bacterioplankton, benthic algae, and macrophytes. The groups are functionally linked. For example, predatory fish eat prey fish, which consume both herbivorous and predatory zooplankton as well as benthic animals. Changes in compartments through time are achieved by use of ordinary differential equations used at weekly or monthly time scales. Calibrated versions of such models produce predictions of typical patterns. Lakes responding to unusual influences such as contamination will show deviations from these patterns, which would indicate a need for data collection and empirical analysis.

Concepts that are fundamental to modeling of food webs include rates of consumption by predators, metabolic efficiencies, turnover rates for biomass in a compartment, selectivity of feeding, and migration rates (especially for fish).

## The Future of Lake Ecosystem Modeling

Recent improvements in modeling suggest that the lake ecosystem modeling will be more useful and more broadly applicable than in the past. Therefore, models dealing with issues important to management may be more extensively incorporated into university training programs and used in support of decision making. It will always be true, however, that each model has a specific domain of use, outside which its use will not be appropriate. Maybe the most crucial aspect for future model development, and hence also for our understanding of the structure and function of aquatic ecosystem, has to do with the access of reliable empirical data from different types of lakes. This is time-consuming and expensive work, which should have a high, and not a low, priority in spite of the fact that it may not always be regarded as very glamorous work.

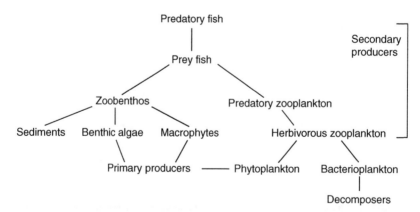

**Figure 5** Illustration of a typical lake food web including nine groups of organisms (phytoplankton, bacterioplankton, benthic algae, macrophytes, herbivorous zooplankton, predatory zooplankton, zoobenthos, prey fish and predatory fish).

## Glossary

**Calibration** – Adjustment of variables within a model in order to achieve minimum bias.

**Effect-load-sensitivity model** – A quantitative model that uses information on mass balance of nutrients or toxins to predict important bioindicators such as fish production or algal biomass.

**Food-web model** – A model whose main objective is to predict the interactions of primary producers, herbivores, and carnivores within a lake ecosystem.

**Generic variable** – A rate or quantity used in modeling that is typical of most ecosystems or of a group of ecosystems.

**Modeling scale** – Resolution of a model expressed in terms of dimensions of time or space.

**Site-specific variable** – A rate or quantity used in modeling that is unique to a particular ecosystem.

**Validation** – Use of information not used in a model calibration to determine whether the calibration has been successful.

*See also:* Ecological Zonation in Lakes; Eutrophication of Lakes and Reservoirs.

## Further Reading

Håkanson L and Peters RH (1995) *Predictive Limnology – Methods for Predictive Modelling.* Amsterdam: SPB Academic Publishers. p. 464.

Håkanson L (2000) *Modelling Radiocesium in Lakes and Coastal Areas – New Approaches for Ecosystem Modellers. A Textbook with Internet Support.* Dordrecht: Kluwer Academic. p. 215.

Håkanson L and Boulion V (2002) *The Lake Foodweb – Modelling Predation and Abiotic/Biotic Interactions.* Leiden: Backhuys Publishers. p. 344.

Monte L (1996) Collective models in environmental science. *Science of the Total Environment* 192: 41–47.

Odum EP (1971) *Fundamentals of Ecology,* 3rd edn. p. 574 Philadelphia: W.B. Saunders. p. 574.

OECD (1982) Eutrophication of waters. *Monitoring, assessment and control.* Paris: OECD. p. 154.

Vollenweider RA (1968) *The scientific basis of lake eutrophication, with particular reference to phosphorus and nitrogen as eutrophication factors.* Tech. Rep. DAS/DSI/68.27. Paris: OECD. p. 159.

# HYDRODYNAMICS AND MIXING IN LAKES

## Contents

## Biological-Physical Interactions

**C S Reynolds,** Centre of Ecology and Hydrology and Freshwater Biological Association, Cumbria, UK

### Introduction

As formal branches of science, limnology and ecology are each around a century in age. Both disciplines feature prominently in the evolving understanding of inland waters, where they are invoked to explain observable phenomena and their role in shaping the abundance, structure, and variability of the biotic communities. It is also true that, from the pioneer studies onwards, much of the scientific investigation has been focused on the mutual relationships among the biota and their chemical environments, most especially with regard to the important nutrient elements. Physical factors were not overlooked altogether: it is well known how the specific heat of water, its curiously variable coefficient of thermal expansion, and its transparency each influence the stability and duration of seasonal thermal stratification or to the underwater distribution of macrophytic plants and photosynthetic algae. The benefit to aquatic organisms of the mechanical support contingent upon the high density of water is also generally understood. On the other hand, the mathematics of fluid motion in lakes proved to be less amenable to solution, so its impact on the evolutionary ecology of the pelagic biota – those of the open water, mostly living independently of shores or the bottom oozes – for a long time remained wholly intuitive. Although broad patterns of wind- and gravity-generated currents could be described and modeled, the smaller scales of water movement and the quantitative description of turbulence were, until the last twenty years or so, relatively intransigent to solutions relevant to the function, selection, and evolutionary ecology of aquatic biota, or even to such matters as the transport and dispersal of organisms, particles, and solutes. This chapter reviews briefly the physical properties of fluid motion at the macro and microscale in inland water bodies, seeking to establish the spatial and temporal scales that impinge directly and indirectly on the organisms that live there, as well as on the functional adaptations that enable them to do so.

### Water Movements – The Concept of Scale

From the major oceanic circulations down to the (Brownian) behavior of the finest colloidal particles and the diffusion of solutes, water characteristically comprises molecules in motion. Curiously, the propensity of water molecules to polymerize into larger 'aquo complexes,' which is responsible for the relatively high density and viscosity of liquid water, makes them resistant to movement, so there is a constant battle between, on the one hand, the energy sources driving motion (the Earth's rotation, gravitational flow, convectional displacement due to thermal expansion and contraction and, especially, the work of wind forcing applied at the water surface of lakes; and, on the other, the resistance of internal viscous forces. Thus, the energy of external forcing is dissipated through a cascade of *turbulent eddies* of diminishing size and velocity, to the point that it is overwhelmed at the molecular level.

Turbulence in the upper water column affects pelagic organisms through several mechanisms operating at a range of temporal and spatial scales. Mixing influences the distribution of regeneration and recycling of dissolved nutrients, the dispersion of zooplankton, the location of rewarding feeding grounds for fish, and the resuspension of detrital particles, including biotic propagules. In relation to the depth of light penetration, the penetration of mixing may constrain the exposure of entrained photosynthetic algae and bacteria to light and to regulate their primary production. At the microscale, turbulence is relevant to individual organisms, conditioning their suspension in the mixed layer, the interaction with their own intrinsic motility and the fluxes of dissolved nutrients and gases to their cells. It is to the latter influences that this article is particularly addressed.

## Small-Scale Turbulence

The measurement of turbulence or its convenient components, such as the shear or friction velocity (symbolised as $u_*$), is not the concern of the present chapter. Turbulence in the upper water column is induced by wind, heat loss, and wave breaking. When wind is the predominant cause of turbulence, the turbulent velocity can be approximated from the shear stress at the air–water interface; if convective heat loss is the dominant driver, then a similar turbulent velocity scale, $w_*$, comes from the velocity of the resultant convectional plumes; if several processes are operating simultaneously, all are included in the calculation, and the resultant turbulent velocity scale is sometimes called the turbulent intensity $u$. Deeper in the water column, turbulence is often caused by breaking internal waves. Technically, the turbulent intensity is the root mean square velocity of the velocity fluctuations in a turbulent flow field. This measurement has only recently been applied in limnological studies. More commonly, turbulence is obtained from microstructure profiling as the rate of dissipation of the turbulent kinetic energy $\varepsilon$ with the assumption that turbulence production and dissipation are in balance. When turbulence is induced by wind in the surface layer, the turbulent velocity scale is roughly proportional to the square root of the quotient of the applied force per unit area ($\tau$, in kg $m^{-1}\,s^{-2}$) and the density of the water ($\rho_w$, in kg $m^{-3}$). Then

$$u_* = \left(\tau \rho_w^{-1}\right)^{1/2} \qquad [1]$$

The units are in $m\,s^{-1}$. The rate of dissipation of turbulent kinetic through the spectrum of eddy sizes ($\varepsilon$) is correlated to the dimensions of the largest eddies in the turbulence field ($l_e$) and their velocities ($u$) through eqn. [2]:

$$\epsilon = (u)^3 l_e^{-1} \qquad [2]$$

The units are thus $m^2\,s^{-3}$. The energy is lost, as heat, through progressively smaller eddies. If the dissipation rate is measured, and the root mean square sizes of turbulent eddies with microstructure profiling, then the turbulent velocity scales are obtained. Eventually, the spectrum collapses at the point where the driving energy is finally overcome by viscosity: the size of the smallest eddy ($l_m$) is predicted by:

$$l_m = \left[(\eta/\rho_w)^3 \epsilon\right]^{1/4} \qquad [3]$$

where $\eta$ is the absolute viscosity of the water (units: kg $m^{-1}\,s^{-1}$). These various equations have been used to calculate that the turbulence generated in the open waters of the unstratified Bodensee (Lake of Constance) by winds of $5$–$20\,m\,s^{-1}$ drive a spectrum of eddies penetrating to depths of between 45–180 m, dissipating at rates ($\varepsilon$) of between $1.4 \times 10^{-8}$ to $2.2 \times 10^{-7}\,m^2\,s^{-3}$ and culminating in eddy sizes ($l_m$) of between 2.9 and 1.5 mm. In stratified lakes, where the density gradient acts as a barrier to penetration by weak eddies, and in shallow lakes, where the water column is unable to accommodate the unrestricted propagation of turbulence, the same driving energy must be dissipated within a smaller spatial extent and, hence, at a faster rate and to a smaller spatial limit. The energy of a $20\,m\,s^{-1}$ wind applied to Lough Neagh (mean depth $< 9\,m$), is calculated to be dissipated at $\sim 4.3 \times 10^{-6}\,m^2\,s^{-3}$, i.e., at nearly twenty times the rate in Bodensee, under the same wind forcing, and culminating in an eddy size of $\sim 0.7\,mm$. In the most aggressively mixed estuaries and fluvial rapids, $\varepsilon$ may approach $5.5 \times 10^{-4}\,m^2\,s^{-3}$, with eddies as small as 0.2 mm across. The highest values observed in lakes are near the air–water interface and are of order $10^{-5}\,m^2\,s^{-3}$ but typical high values are of order $10^{-6}\,m^2\,s^{-3}$.

The capacity of turbulent motion to entrain particles, including living organisms, depends broadly upon the magnitude of the relation of the turbulent velocity scale to the intrinsic rates of gravitational settling of the particles in water ($w_s$, in $m\,s^{-1}$). Whereas a stone always drops almost unimpeded through water, particles of clay ($< 5\,\mu m$ in diameter but of similar density) persist in suspension for long periods and remain dispersed by the aggressive mixing evident in tidal estuaries. The sinking velocities of planktic diatoms such as *Asterionella* ($w_s \leq 10\,\mu m\,s^{-1}$), the cells of which form stellate colonies measuring 150–200 $\mu m$ across, are so trivial compared to the shear velocity of turbulence generated at the lake surface by a wind of (say) $8\,m\,s^{-1}$ ($\sim 1 \times 10^{-2}\,m\,s^{-1}$) that their entrainment is near complete ($u/w_s \sim 10^3$); the algae go more or less where the flow takes them. In metalimnia, or in the boundary layer adjacent to the lake bottom, wherein turbulent velocities are abruptly depressed to below $10^{-3}\,\mathrm{m\,s^{-1}}$, the same diatoms are readily disentrained, sinking almost unimpeded. For smaller and less dense algae and bacteria, the threshold of disentrainment may be an order of magnitude or two smaller. Equally, the rather larger cells of some dinoflagellates, such as *Ceratium*, and the colonies of the Cyanobacterium, *Microcystis*, that aspire to rising velocities of $\leq 300\,\mu\mathrm{m\,s^{-1}}$, have a correspondingly enhanced capacity to escape from weakly turbulent flows. The generally recognised criterion for the entrainment of particles is that $\psi = w_s/(15\ u) \leq 1$; based on the proportionality of the velocities of intrinsic settling and shear, however, entrainment can never be wholly complete. There is a persistent tendency for particles to be disentrained gradually into adjacent structures.

## Physical Influences on the Size and Structure of Organisms

The scaling of turbulence, observations as to orders of magnitude of turbulence, and understandings of boundary layer fluid dynamics provide insight into a number of biological problems. For instance, in the context of evolutionary adaptation to physical characteristics of the aquatic medium, it is plain that the foliage of submerged macrophytic plants and algae require tensile strength rather than rigidity; live biomass is supported by the density of the water but turbulent stresses are accommodated by their typically, flexible, flat, much-dissected or filamentous leaves. Self-evidently, macroinvertebrates and fish must be adequately robust to maintain their integrity in the face of turbulence, some of which is, of course, generated by their own movements. In contrast, the often delicate and unstrengthened substance of planktic protistans, algae and cyanobacteria, most of which are smaller or much smaller than 1 mm across, is normally protected within the viscous range of the eddy spectrum. While they have no power to prevent their entrainment by turbulence, they rarely experience its physical stresses. When larger algae are exposed to intense turbulence, however, their sizes relative to the smallest eddies expose them to shear; mostly they are insufficiently robust to resist fragmentation and disruption.

Generally, the adaptive problem for small organisms is rather to cope with viscosity – especially in the context of motility and the foraging requirement of phagotrophic protists to encounter food. The analogy has been made of a person collecting bananas suspended in a swimming pool filled with a fluid of the consistency of molasses or setting concrete.

Reciprocating swimming strokes are of less use than are boring or grapple lines, favouring slender flagella over flat paddles. In contrast, mesozooplanktic crustaceans are of a size (in the range 0.2 to 2 mm) that coincides with the transitional scale, in which swimming and the generation of shear currents are feasible; these currents may entrain small particles towards the feeding apparatus where it may be collected as food. In the filter-feeding cladocerans (including *Daphnia*), the flattened, paddle-like abdominal limbs (phyllopods) beat rhythmically within a feeding chamber formed by the abdomen and the carapace: an inhalant current is thus drawn into a chamber, where the particles are strained out by bristle-like setae on the phyllopods, and carried in viscous flow to the mouth. To pursue the analogy, another person immersed in the same fluid might gain more satisfaction through collecting kidney beans from a vessel closer to the size of a bath tub. It is only by being big enough and strong enough to escape the drag of viscosity and to be able to generate turbulent currents that the rate of particle encounter can be significantly increased.

Open turbulence is also constrained in the vicinity of solid surfaces, for instance, at the bottom and sides bounding the flow field, as well as the bodies of aquatic animals. Boundary layers are regions of reduced flow adjacent to the solid surfaces, characterized by a gradient of velocities perpendicular to the surface. Depending upon current velocity in the main direction of flow, boundary gradients may be compressed to fractions of a millimeter or extend over several meters. Their presence influences the structural and behavioral adaptations of organisms in a number of ways. A well-known example is the dorsoventral flattening of Ecdyonurid mayfly larvae that can apply themselves to the surfaces of stones in fast flowing streams. By presenting a protrusion of barely 2 mm above the stone surface, animals are able to exploit the boundary layer as a physical refuge from the flow, whilst they graze on the diatoms and other algae attached to the surface. These algae are of a small size (generally <0.3 mm) so they also experience minimal shear stress; if they are sufficiently dense, their presence augments the boundary effect by extending it further into the water. It should be noted that boundary layers are not necessarily free of turbulence: even weak eddies potentially increase the delivery of food particles to the vicinity of benthic filter feeders. In the case of stream-dwelling larvae of the black fly, *Simulium*, the firm attachment and attitude of the abdomen ensure the optimum presentation of the filtering mouth parts to the flow.

Subject to other conditions, broader boundary layers, variously adjacent to stony or silty river beds or within beds of submerged macrophytic plants, offer refuges of weakened turbulence exploitable by algae, fungi and macroinvertebrates that together comprise the *Aufwuchs* communities. There is no exact equivalent word in English – but the substratum-specific subdivisions of epilithic, epipelic, and epiphytic associations together convey the analogous concept; the term *periphyton* is also close but it should refer strictly to the microhabitats and species associated with aquatic macrophytes. Periphyton and the plants in shallow water with which it is associated together constitute some of the most productive habitats in the biosphere, in some instances turning over more than $1.5\,\mathrm{kg\,C\,m^{-2}}$ annually.

## Turbulent Extent and the Pelagic Habitat

Just as open-water turbulence dominates the environment of the planktic organisms in the near-surface layer, so the vertical extent of the entraining mixed layer has far-reaching consequences for the photosynthetic activity of phytoplankton. The vertical attenuation of light penetrating the surface, even in clear water, is such that significant net photosynthetic carbon fixation is severely restricted at depths greater than 60 m. At low angles of incidence and under cloud cover, this photic depth is further reduced. However, coloration of the water (due, for instance, to humic substances in solution) and the suspension of clay or other fine particles, as well as the presence of significant concentrations of phytoplankton, may restrict the photic depth to the order of a few meters or even centimeters only. In these instances, turbulent entrainment may well carry photosynthetic organisms to depths where net photosynthesis cannot be sustained. Under steady wind forcing, the probability is that the same chlorophyll-containing organisms will soon be carried back to the depth range wherein production may be resumed. Calculation of the proportionality of time during which net production is actually possible is complex, not least because, within limits, the photosynthetic apparatus itself is able to compensate to the average light exposure ('light adaptation'). However, the product of the day length and the ratio of the volumes of the photic and the wind-mixed layers (in large lakes, the ratio of their depths, $h_p/h_m$ is convenient) is adequately illustrative. Ignoring, for the present, the variables determining the evaluation of $h_p$ and focusing on the variability in wind forcing on the water columns of deep, non-stratified lakes, there is an approximate proportionality among wind velocity, $u_*$ and $h_m$, winds of $5-20\,\mathrm{m\,s^{-1}}$ being theoretically capable of driving turbulence to depths of about 44–177 m. For a mixed layer depth of 44 m and a dissipation rate of $10^{-7}\,\mathrm{m^2\,s^{-3}}$, as would be found on a windy, cloudy day, it would require, probabilistically, an average of 42 minutes for an alga entrained in open turbulence to be transported through the entire mixed layer. In mixed columns truncated by pycnoclines or the bottom of the lake, the mixing time is proportional to the (constrained) depth. In mixed layers of the order of one meter in depth, mixing times are in the order of one minute.

The interaction between $h_m$ and $h_p$ can be used to approximate the instantaneous phytoplankton-carrying capacity of the mixed layer to support a productive phytoplankton. Whereas an otherwise clear mixed layer of 10 m in depth might support a chlorophyll concentration $\leq 150\,\mathrm{mg\,m^{-3}}$, deepening it to 40 m reduces the maximum supportable concentration to $<20\,\mathrm{mg\,m^{-3}}$; mixing to 80 m takes the capacity to under $1\,\mathrm{mg\,m^{-3}}$. It follows that, in deep, temperate lakes like Bodensee, phytoplankton growth and net population increase in the winter months is weak, pending longer days, more surface heating and the onset of thermal stratification. Equally, as heat is lost in the autumnal months, so the stability of stratification and the resistance to wind mixing weaken, the mixed layer increases in depth and the downwelling light energy is increasingly diluted.

## Mixed Layer Depth and the Maintenance of Non-motile Plankton

Weak winds and a shrinking mixed layer depth are not, however, an unmitigated benefit to phytoplankton. For non-motile plankton, as well as fine particles, the thickness of the entraining mixed layer has an important relationship to the intrinsic rates of their settlement ($w_s$) and, thus, to the probability of their entry into the lower, less turbulent basal layer. Depletion from suspension is a first-order process, analogous to dilution; the rate of the decline in the suspended population is described by an exponent $(-r_S)$, equivalent to $-(w_s/h_m)$. Plainly, the greater is the particle-specific $w_s$, the greater is the dependence upon a deep mixed layer for its maintenance in suspension and the greater is its sensitivity to density stratification and the consequent contraction of $h_m$. It may seem paradoxical that planktic diatoms such as *Asterionella* ($w_s$ typically in the range $0.3-1.0\,\mathrm{m\,day^{-1}}$) require a shrinking mixed layer to promote their growth but then have to grow fast enough to balance sinking loss rates from a shrinking mixed layer! A pycnocline reaching to within 3 m of the surface will impose a sinking loss rate of $r_S \geq -0.1\,\mathrm{day^{-1}}$, which must be countered by a comparable rate of cell recruitment ($r'$) if a net collapse of the standing population is to be avoided. This is well within the capacity of the maximum performance of *Asterionella* but low temperature, shortage of nutrients and other losses (flow displacement, grazing by zooplankton) often make this hard to attain. Nevertheless, the onset of density stratification, especially in smaller lakes, is generally the principal correlative of the culmination of diatom abundance (as, for instance, in the so-called 'spring bloom') and settlement from suspension contributes prominently to their subsequent demise.

Not all plankton is heavier than water; diminution of the surface mixed layer promotes different strategies somewhat dependent upon the trophic status of the lake. In oligotrophic lakes, small, near-neutrally buoyant algae may proferate. In mesotrophic and eutrophic lakes, buoyant plankton may predominate. The most extreme case is also well known to freshwater ecologists: under calm conditions, physical entrainment may be insufficient to prevent suitably buoyant Cyanobacteria from accumulating at the water surface, where, if present in the water in significant amounts, they may constitute a striking surface scum. The exaggerated view of their abundance and the rapidity with which such 'water blooms' form have given them a notoriety among water users; that many of these are now known also to be toxic to livestock and to human consumers has increased the impetus to devise measures to better manage the frequency of their occurrence.

## Physical Influences on Nutrient Fluxes

A major problem for planktic photoautotroph is to accumulate from the suspending water a sufficient stock of the essential nutrients to be able not just to maintain itself but to accumulate the material requirements of the next generation (effectively harvesting its own mass again of carbon, nitrogen, phosphorus, etc. within each generation). There are at least twenty elements required in the production of living cells, several of which (C, N, P) are typically regarded as being in short supply relative to the needs and, hence, likely to limit the capacity to support biomass. For autotrophic phytoplankton, these elements are obtained essentially from relatively simple compounds and ions, dissolved in the water. In most lakes, the main proximal sources of these elements (even the gaseous ones, like carbon dioxide) are dissolved in rainfall and, especially, inflowing streams draining the catchment. There is thus a tendency for supplies to fluctuate seasonally, and often quite uncoupled from the demands of production. Moreover, while the growth of phytoplankton removes nutrients from the water in the surface circulation, they are unlikely to be released again only after cells have died, including through consumption by animals and through progressive settlement of

intact cells beyond the photic zone and the mixed layer. Other things being equal, the net flux of fecal pellets, detritus and disentrained phytoplankton is to greater depth. At the basin scale, there is an inevitable tendency for the resources of the lake to gravitate towards the bottom and for the surface waters to be depleted. Density stratification of the water masses only amplifies this segregation.

As a physical counter to vertical segregation, the mechanisms for transporting nutrient-rich water from depth may assume considerable biological importance. Increased wind action may expand the circulation of the surface mixed layer, depressing the pycnocline downwind, raising the shear stress of the return current and increasing the intensity of internal wave generation. Erosion and entrainment of deep, relatively nutrient-rich water into a deepened surface mixed layer can have a significant effect in refreshing nutrient availability and stimulating phytoplankton growth. Such macroscale events may pass, with the system returning to something approaching the earlier physical state. Alternatively, they may persist and, aided by surface cooling and seasonally more frequent wind episodes, lead eventually to the breakdown of stratification and substantial vertical mixing throughout most of the water column.

## Nutrient Fluxes at the Microscale

Biological-physical interactions influence the nutrient relationships at the microscale of individual phytoplankton cells. The resources present within healthy living cells are typically about one million times more concentrated than they are in the medium. This formidable gradient cannot be overcome by passive movement of molecules into the cell – they would move rapidly in the opposite direction – an elaborate apparatus and system of reactivity captures, retains and transports specific molecules from the exterior of the cell to the intracellular sites of deployment and assimilation. The operation involves kinases and the consumption of ATP; the cell expends significant energy in order to assimilate its essential resources.

Suspended in the relative vastness of the water mass, however, the phytoplankton cell can have no such influence on the external supply of nutrients or the frequency of the molecular encounters that it requires to satisfy its uptake demands. Within its immediate viscous environment, the cell is dependent upon diffusion of target molecules through the medium, in place of those taken up across the cell membrane. Delivery is governed by Fick's laws: the number of moles ($n$) of a solute that will diffuse across an area ($a$) per unit time, $t$, is a function of the gradient of solute concentration, C (i.e., dC/dx), and the coefficient of molecular diffusion of the substance ($m$). Then, introducing appropriate units, uptake per unit time is:

$$n = a\ m(dC/dx) \text{ mol m}^{-2}\text{ s}^{-1} \qquad [4]$$

Taking a small, spherical alga like *Chlorella* (diameter, $d \sim 4 \times 10^{-6}$ m; volume $\sim 33 \times 10^{-18}$ m$^3$; surface area $\sim 50 \times 10^{-12}$ m$^2$) and given that (i), for a small-sized molecule such as that of carbon dioxide, $m \sim 10^{-9}$ m$^2$ s$^{-1}$, (ii), the thickness of the water layer adjacent from which $CO_2$ molecules can be sequestered is equal to the cell radius, and (iii) that the carbon dioxide concentration in the water beyond is at air equilibrium (11 μmol l$^{-1}$, or $11 \times 10^{-3}$ mol m$^{-3}$), eqn. [4] may be solved to show diffusion should sustain a rate of acquisition by the cell equivalent to $275 \times 10^{-18}$ mol s$^{-1}$. Given a cell carbon content of $0.63 \times 10^{-12}$ mol carbon and that the doubling requirement is the uptake of a further $0.63 \times 10^{-12}$ mol carbon, it can be calculated that eqn. [4] reveals a theoretical capability of fulfilling the demand within $\sim$2300 s, or just over 38 min. This is rather less than the generation time ($\geq$9 h at 20 °C), so there is a considerable scope for $CO_2$ depletion before it impinges on the growth rate of the cell.

For comparison, a turbulent velocity of $10^{-2}$ m s$^{-1}$ would deliver $CO_2$ molecules at a the same concentration of $11 \times 10^{-3}$ mol m$^{-3}$ through an area equal to the area projected by the same *Chlorella* cell ($\sim$12.6 x $10^{-12}$ m$^2$) at a rate of $1386 \times 10^{-18}$ mol s$^{-1}$. The capacity of open turbulence to fulfill a minimum frequency of encounter with nutrient molecules becomes especially important at very low resource concentrations. Taking its requirement for phosphorus instead of carbon (say, one hundredth the number of carbon atoms), the *Chlorella* mother cell has to accumulate some $0.006 \times 10^{-12}$ mol if the cell is to sustain adequately the next doubling of biomass. Like most freshwater phytoplankters, *Chlorella* has a high capacity for the uptake of phosphate (maximum $\sim$ $13.5 \times 10^{-18}$ mol cell$^{-1}$ s$^{-1}$), with a sufficient affinity for phosphate molecules for the uptake to be half-saturated by an external concentration of soluble, reactive phosphorus of about $0.7 \times 10^{-3}$ mol P m$^{-3}$ (about 20 μg P l$^{-1}$). Given these conditions, the requirement could be met in less than 15 min). On the other hand, the external concentration required to sustain the maximum rate of growth (at 20 °C, 1.84 day$^{-1}$) may be sequestered from an external concentration of just $6.3 \times 10^{-6}$ mol P m$^{-3}$, about 0.2 μg P l$^{-1}$, before the rate of growth may be supposed to become phosphorus-limited.

Ultimately, the rate of renewal of the local water and the delivery of fresh nutrients to the vicinity of the algal cell affects the capacity of molecular diffusivity to meet growth demands. Being generally smaller than the turbulent eddies, most phytoplankton cells are eventually reliant upon movement of themselves and their viscous packets of water relative to the flow field to realize the environmental capacity to support their growth. The effect of advective transport of a given flow field, relative to diffusion, on a phytoplankton cell can be calculated from the ratio of the respective dimensionless Péclet numbers ($Pe$) of particles phytoplankton cells, of diameter $d$, either sinking passively in turbulence-free water or transported advectively through the water.

$$Pe = w_s\ d\ D^{-1} \qquad [5]$$

where $D$ is the appropriate diffusion coefficient in water. In the absence of turbulence, $D$ is substituted for $m$. The calculation with respect to advective transport is difficult but solutions have been presented by Riebesell and Wolf-Gladrow (see Further Reading). Their calculations showed $Pe$ values for smaller phytoplankton generally fall within the range 0.1–10, the larger values applying to larger algae moving more rapidly through the water. The Sherwood Number ($Sh$) expresses the ratio of the fluxes of nutrients arriving at the cell surface in the presence of motion with those of diffusion. For the small cells embedded deep within the eddy spectrum ($Pe < 1$), $Sh$ is always close to 1, signifying that any effect of motion is marginal. For larger organisms ($1 < Pe < 100$), the non-linear dependence of the nutrient supply on advection is approximated by

$$Sh = 1/2 + 1/2(1 + 2Pe)^{1/3} \qquad [6]$$

The effect of turbulence on the nutrient supply to cells of varying diameters ($d$) is also scaled to the Péclet numbers, calculated from the turbulent shear rate dissipation rate, as

$$Pe = (d/2)^2[\epsilon/(\eta/\rho_w)]^{1/2}m^{-1} \qquad [7]$$

Then, in the range $0.01 \leq Pe \leq 100$, the Sherwood scale conforms to:

$$Sh = 1.014 + 0.15Pe^{1/2} \qquad [8]$$

These derivations are interesting in the context of selection of evolutionary adaptations. It is at once apparent that the absolutely greater nutrient requirements of larger cells are more difficult to glean from low concentrations of specific nutrients in the medium, without the intervention of significant levels of turbulence. The further deduction is made that this dependence leaves larger phytoplankters less likely than smaller ones to be able to fulfill all their resource requirements in unit time, except under strongly turbulent conditions. With a less prevalent constraint on the development of picoplanktic and nanoplanktic than on the growth of large algae, their frequently-observed dominance over microplankton in chronically nutrient-deficient oligotrophic systems finds a compelling physical explanation. There are other alternative, probably additive, reasons for the dominance of picophytoplankton, invoking the inability of most filter-feeding mesozooplankters to harvest them adequately from low concentrations. Interestingly, it generally requires the intervention of a macroscale physical event, such as storm deepening, to raise nutrient levels for a period long enough for larger algae to flourish. This principle, that deep turbulent embedding and passive delivery of nutrients provides a superior strategy for the survival of photoautotrophs in an extensive, chronically nutrient-deficient medium, contrasts with what happens in the face of nutrient depletion in the epilimnion of a seasonally-stratified eutrophic lake. Here, the selective advantage passes to large-celled or large coenobial, motile, phytoplankton that are the most readily disentrained from weak turbulence and are capable of undertaking vertical migrations to scavenge the residual nutrient resources of the water column.

This regulation by nutrient fluxes of the size selection of planktic organisms is an especially satisfying instance of demonstrable physical–biological interaction in open waters. It is clear that the interactions between fundamental organismic processes and the physical motion of the fluid in which they may be suspended do not stop with transport and entrainment. Physical processes play a large part in sustaining nutrient fluxes in chemically rarefied pelagic environments and in underpinning their relative exploitability as habitats of autotrophic microorganisms.

*See also:* Currents in Stratified Water Bodies 1: Density-Driven Flows; Currents in Stratified Water Bodies 2: Internal Waves; Currents in Stratified Water Bodies 3: Effects of Rotation; Currents in the Upper Mixed Layer and in Unstratified Water Bodies; Small-Scale Turbulence and Mixing: Energy Fluxes in Stratified Lakes.

## Further Reading

Huisman J, van Oostveen P, and Weissing FJ (1999) Critical depth and critical turbulence: two different mechanisms for the development of phytoplankton blooms. *Limnology and Oceanography* 44: 1781–1787.

Hudson J, Schindler DW, and Taylor W (2000) Phosphate concentrations in lakes. *Nature* 406: 504–506.

Imberger J (1985) Thermal characteristics of standing waters: An illustration of dynamic processes. *Hydrobiologia* 125: 7–29.

Karp-Boss L, Boss E, and Jumars PA (1996) Nutrient fluxes to planktonic osmotrophs in the presence of fluid motion. *Oceanography and Marine Biology* 34: 71–107.

MacIntyre S (1993) Vertical mixing in a shallow eutrophic lake – Possible consequences for the light climate of phytoplankton. *Limnology and Oceanography* 38: 798–817.

Mann KH and Lazier JRN (1991) *Dynamics of Marine Ecosystems.* Oxford, UK: Blackwell.

O'Brien KR, Ivey GN, Hamilton DP, *et al.* (2003) Simple mixing criteria for the growth of negatively buoyant phytoplankton. *Limnology and Oceanography* 48: 1326–1337.

Reynolds CS (1998) Plants in motion: Physical–biological interaction in the plankton. *Coastal and Estuarine Studies* 54: 535–560.

Reynolds CS (2006) *Ecology of Phytoplankton.* Cambridge, UK: Cambridge University Press.

Riebesell U and Wolf-Gladrow DA (2002) Supply and uptake of inorganic nutrients. In: Wiliams J.leB, Thomas DN, and Reynolds CS (eds.) *Phytoplankton Productivity*, pp. 109–140. Oxford, UK: Blackwell Science.

Rothschild BJ and Osborn TR (1988) Small-scale turbulence and plankton contact rates. *Journal of Plankton Research* 10: 465–474.

Vogel S (1994) *Life in Mixing Fluids*, 2nd edn. Princeton, NJ: Princeton University Press.

# Mixing Dynamics in Lakes Across Climatic Zones

**S MacIntyre and J M Melack,** University of California, Santa Barbara, CA, USA

## Introduction

Interactions between climatic and geographic conditions lead to variations and gradients in hydrological, physical, chemical, and biological conditions in lakes. The annual cycle of solar irradiance with seasonal amplitude increasing at higher latitudes causes well-known gradients in photoperiod and temperature. Continental interiors differ from coastal regions that tend to be moister and to experience more moderate temperatures. Hence, because of the larger proportion of the Earth's land in the northern hemisphere, oceanic conditions tend to have a greater influence in the southern hemisphere. Globally, annual average precipitation is greater than evaporation in low and middle latitudes, but evaporation generally exceeds precipitation in the subtropics. However, hydrologically closed basins, where evaporation exceeds local precipitation, occur on all continents.

During the twentieth century, studies of lakes throughout the world have produced a wealth of information about individual lakes as well as regional syntheses. Based on these results, latitudinal correlations with limnological characteristics have revealed several patterns. Based on limited variations in solar irradiance in the tropics, maximum temperatures in the mixed layer of lakes are similar across the tropics but tend to decline through the temperate latitudes, and most of the geographical differences in near bottom temperatures are explained by latitude and elevation. Dissolved oxygen concentrations tend to be lower in deep waters of tropical lakes than higher latitude lakes. Though considerable monthly variation in photosynthetic rates of phytoplankton occurs in tropical lakes and corresponds to seasonality in rainfall, river inflows and vertical mixing, no latitudinal trend in primary productivity exists among tropical lakes. When extended to temperate and arctic lakes, maximum variability about doubles and a significant correlation with latitude emerges. In contrast to fish, species richness of phytoplankton, zooplankton, and zoobenthos does not tend to be higher in tropical lakes versus temperate lakes.

Recent investigations of physical processes in lakes have advanced our understanding and permit mechanistic explanations of climatic differences in stratification and mixing. Because these processes are fundamental to biological and chemical conditions, we focus our examination of lakes across climate zones on these processes.

## Patterns in Stratification and Mixing

### Background

Differences among lakes in stratification and mixing dynamics derive from momentum and energy exchanges with the atmosphere and inflows modulated by climatic conditions and basin morphometry. Schemes to classify lakes based on temporal differences in vertical mixing compiled in G.E. Hutchinson's *Treatise on Limnology* and modified by W.M. Lewis Jr. led to an empirical classification system. Eight types of mixing (amictic, cold monomictic, dimictic, warm monomictic, and four types of polymixis) were distributed in relation to water depth and latitude, adjusted for elevation. Meromictic lakes remain stratified for long periods of time due to processes that lead to accumulation of solutes at depth.

For lakes deep enough to seasonally stratify, the major mixing types defined by Lewis depend upon seasonal weather patterns. Cooling in autumn or during a monsoon combined with the shear due to wind forcing erodes seasonal stratification. Polymictic lakes are shallow and tend to mix frequently due to the combination of cooling and wind at night. Cold monomictic and polymictic lakes are found at high latitudes where water temperatures do not exceed 4 °C; these lakes mix in summer when ice free. Warm monomictic and polymictic lakes occur where water temperatures exceed 4 °C for some part of the year. Warm monomictic lakes mix in winter. Warm polymictic lakes are found over a wide range of latitudes including the Arctic.

As understanding of mixing processes in lakes has improved with applications of new instrumentation and numerical models, underlying mechanisms can be incorporated into analyses of the frequency, intensity, and depth of mixing. Detailed descriptions of the physical processes are provided in the Physical Limnology Section of the Encyclopedia. Computation of surface energy budgets using meteorological data and time series measurements of temperature is particularly important for understanding controls on mixed layer dynamics. With these, as will be illustrated later, we are beginning to develop a predictive understanding of how mixed layer dynamics vary in large and small lakes from the tropics to the poles and the implications for aquatic organisms. Dimensionless indices provide insights into the mechanisms inducing mixing in the metalimnion and hypolimnion.

Understanding how these vary with latitude will allow predictions of the connectivity between the upper and lower levels in the water column with implications for aquatic ecosystem function.

Via surface energy budgets, we learn how much heat is lost or gained via latent heat exchange (LE, evaporation), sensible heat exchange (SE, conduction), and long- and short-wave radiation and, with measurements of light attenuation, we can predict whether temperatures will be stably stratified to the surface, whether a mixed layer will form, and how deep it will be (**Figure 1**).

The heat fluxes at a lake's surface (**Figure 1**) are computed as $Q_{tot} = LW_{in} + LW_{out} + LE + SE$ where $Q_{tot}$ is the surface energy flux and $LW_{in}$ and $LW_{out}$ are incoming and outgoing long-wave radiation, respectively. $LW_{out}$ depends on surface water temperature in degrees Kelvin to the fourth power; $LW_{in}$ depends on cloud cover; LE depends on the vapor pressure gradient between the lake and overlying water, temperature, and wind speed; and SE depends upon the air–water temperature difference and wind speed. The effective heat flux into the actively mixing zone is computed as the sum of the surface energy fluxes plus the net short wave radiation into the actively mixing layer. When net solar radiation exceeds the heat losses at the air–water interface, the water column will stratify. When they are less,

as occurs at night, during cloudy periods, during fall cooling, and often during monsoon periods, the upper mixed layer will lose heat and deepen. The deepening is induced by thermals due to the cooling at the lake's surface. The turbulence produced by the heat loss can be quantified via the turbulent velocity scale for heat loss. This term, $w_*$, is computed from the buoyancy flux ($J_{b0}$) due to the effective heat loss ($\tilde{H}$) and the depth of the actively mixing layer ($h$). That is, $J_{b0} = \alpha g \tilde{H} / \rho c_p$ where $\alpha$ is the thermal coefficient of expansion, $g$ is gravity, $\rho$ is density, and $c_p$ is the heat capacity, and $w_* = (J_{b0} h)^{1/3}$; $w_*$ only exists when a lake is losing heat. Via comparisons of both the magnitude of this term across latitudes and the length of time it is positive during the day, we are positioned to understand the role of heat loss in mixed layer dynamics for lakes in different climate zones.

Wind-induced mixing is often assumed to dominate mixed layer deepening. However, whether this assumption is true can only be determined by comparing the turbulent velocity scale from wind with that from heat loss. The momentum flux from the wind depends upon the shear stress $\tau$ at the air–water interface and is generally computed based on the wind speed squared times a drag coefficient. That is, $\tau = \rho_a u_*^2 = \rho_w u_{w*}^2 = C_d \rho_a U^2$ where $u_*$ is friction velocity in the atmosphere, $u_{*w}$ is water friction velocity, $\rho_a$ and $\rho_w$ are density of air and water respectively, and $C_d$ is a drag coefficient. The shear stress is equivalent on both sides of the air–water interface. The turbulent velocity scale from wind can be characterized by the water friction velocity $u_{*w}$.

Wind forcing and heat loss can both generate turbulence in the upper water column. By computing the turbulent velocity scales for both, we learn whether heat loss contributes to the mixing and whether wind or heat loss is dominating the mixing in the upper waters (**Figure 2**). The turbulent kinetic energy flux (TKF) into the mixed layer depends upon the cube of both these terms. The magnitude of this term thus allows comparison of surface forcing during periods of calm and during storms. By computation of these different terms, we are poised for comparisons which enable us to understand the factors leading to temporal changes in mixed layer dynamics in one lake as well as between lake differences.

Wind forcing not only energizes currents and turbulence in the upper mixed layer, it also leads to upwelling and downwelling of the thermocline in stratified lakes and the ensuing production of internal waves and turbulence (**Figure 2**). By computing Wedderburn ($W$) or Lake numbers ($L_N$), we learn whether wind forcing at the surface is sufficient to tilt the thermocline (**Figure 3**) $W = g \Delta \rho h^2 / (\rho u_*^2 L)$, where $g$

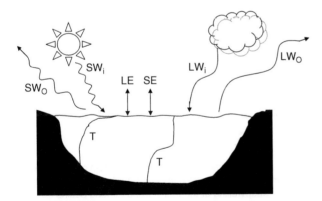

**Figure 1** Terms of a surface energy budget and examples of resulting temperature profiles. The heat losses or gains which occur at the surface of a lake include latent heat flux (LE) and sensible heat exchange (SE). Outgoing long wave radiation (LW₀) and incoming long wave radiation (LWᵢ) are added to the other terms to compute the surface energy fluxes (SEF). Incoming short wave radiation (SWᵢ) induces heating to a depth dependent on the diffuse attenuation coefficient. A portion of this heat is lost (SW₀) before entering the lake due to reflection with the amount lost dependent upon the albedo. The sum of the SEF and net shortwave is called the effective heat flux (EHF). Depending upon this latter term, plus mixing induced by wind, temperature profiles (T) will either be stratified to the surface (left profile) or a mixed layer will form with a thermocline below (right profile). Depending upon previous heating and cooling, multiple thermoclines may form.

Interactions leading to mixing

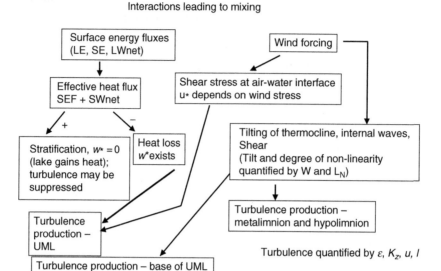

**Figure 2**   Schematic of processes leading to turbulence production in the upper mixed layer (UML), metalimnion, and hypolimnion of lakes. Turbulence is quantified in a number of ways, the most common being the rate of dissipation of turbulent kinetic energy ($\varepsilon$), the coefficient of eddy diffusivity ($K_z$), and the turbulent velocity ($u$) and length scales ($l$) where $u$ is computed from $u_{*w}$ and $w_*$ or from the relation $\varepsilon = u^3/l$ when $\varepsilon$ and $l$ have been measured directly.

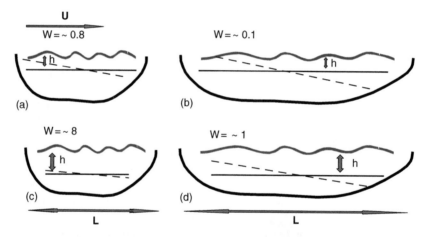

**Figure 3**   The degree of tilting of the thermocline depends on the wind speed and stratification as well as the depth of the mixed layer ($h$) and length of the lake ($L$) and can be computed from the Wedderburn and related Lake numbers. Four hypothetical cases with the same wind forcing and density stratification show the greater degree of upwelling of the thermocline for long basins relative to ones with shorter fetch (a vs b and c vs d) and the greater degree of upwelling for shallow relative to deep mixed layers (a vs c and b vs d) . For each case, a possible Wedderburn number is indicated, but values will shift with the assumed values of $h$ and $L$. The lower the value of the Wedderburn or Lake number, particularly as they drop below 10, the greater the degree of turbulence in the metalimnion and the hypolimnion below.

is gravity, $\rho$ is density, $\triangle\rho$ is density difference across the thermocline, $h$ is depth of the mixed layer, $L$ is length of the lake, and $u_{*w}$ is the water friction velocity derived from the shear stress and which can be approximated as $0.001U$ where $U$ is mean wind speed. $W$ is essentially the ratio of the buoyancy forces resisting mixing divided by the inertial forces that induce mixing times the aspect ratio of the lake.

Thus, for the same density difference across the thermocline and the same wind forcing, lakes with a shallower mixed layer or longer fetch will have lower values of $W$ (**Figure 3**). The Lake number is an integral form of the Wedderburn number. If these numbers are low ($\ll 1$), a lake mixes rapidly by wind forcing. If near 1, the thermocline upwells to the surface, if between $\sim$2 and 10, partial upwelling occurs, and if $>10$, no tilting occurs. If the thermocline tilts, increased shear occurs at the base of the mixed layer and can contribute to its deepening. Similarly, increased shear can occur within the metalimnion and hypolimnion leading to enhanced mixing and fluxes of biologically important solutes. In fact, the coefficient of eddy diffusivity ($K_z$), an index of turbulence used to compute fluxes in stratified flows, depends upon the Lake number. When $L_N > 10$, $K_z$ is $\sim 10^{-7} \, \mathrm{m}^2 \mathrm{s}^{-1}$, equivalent to the molecular conductivity of heat. As it decreases below 10, values of $K_z$ increase by up to three orders of magnitude. Comparison of these two dimensionless indices in lakes across latitude allows a predictive understanding of the influence of wind in generating fluxes below the surface mixed layer.

## Comparative Mixing Dynamics

Developing a predictive understanding of mixing processes in lakes at different latitudes requires consideration of the magnitude and phase of the forcing factors outlined in **Figure 1** and the resulting changes in terms of the surface energy budget, turbulent velocity scales, and dimensionless indices (**Figure 2**). As mentioned earlier, the large scale differences in forcing across latitudes are well known, and heat budgets based on monthly averages for a diversity of lakes have been published and led to the seasonal differentiation of lake mixing dynamics developed by Hutchinson and Lewis. What is not well known is the magnitude of parameters controlling and describing mixing dynamics on shorter time scales during the stratified period. To that end, we compare critical parameters based on a suite of lakes for which high frequency meteorological data are available (**Tables 1 and 2**).

The lakes for which we present data range in size from 1 ha to 670 km² and are found near the equator (Pilkington Bay, Uganda; Lake Calado, Brazil), in the Arctic (Toolik Lake, Alaska), and at a range of elevations in the temperate region. Three of the temperate lakes are only a few hectares in size and are dimictic; the three larger lakes, due to their larger size or salt content (Mono Lake, CA), are warm monomictic. Attenuation coefficients are variable, ranging from 0.09 in clear Lake Tahoe to 5 in highly strained Trout Bog. While mixed layer depths are anticipated to be shallower for higher attenuation, and deeper for longer fetches, the mixed layer depths noted here do not fully follow those patterns. In fact, the deeper mixed layer depths are found for the larger lakes in the temperate zone and in Pilkington Bay, and the smaller mixed layer depths occur in the sheltered dimictic lakes and in L. Calado, Brazil. Typically, the mixed layer depth in Toolik Lake is similar to the other small lakes, but when a cold front passes, it may deepen to a depth similar to those typically found in summer in larger lakes. Similar mixed layer depths occur in a nearby lake which is 10 times smaller than Toolik Lake. Thus because of differences in the surface energy budget and resulting stratification for Toolik Lake, it can have attributes similar to larger temperate lakes. Analysis of similar anomalies for other lakes will provide insights into how climate structures mixing dynamics.

## Comparative Mixing Dynamics: The Upper Mixed Layer

By comparing the surface energy budgets and turbulence parameters, we can begin to understand the factors controlling the differences in the hydrodynamics

**Table 1** Characteristics of lakes with high quality meteorological data in different climate zones arranged from small to large from the Arctic to the tropics

| Lake | Latitude, longitude | Surface area (km²) | $Z_{max}$ ($Z_{mean}$) (m) | $k_d$ (m⁻¹) | Depth UML (m) | Elevation (m) |
|---|---|---|---|---|---|---|
| Toolik Lake | 68°38′ N, 149°38′ W | 1.5 | 25 (7.1) | 0.5–0.9 | 0–10 | 760 |
| Emerald Lake | 36°35′ N, 118°40′ W | 0.027 | 10 (6) | 0.2–0.3 | 0–3 | 2800 |
| Trout Bog | 46°3′ N, 89°41′ W | 0.011 | 7.9 (5.6) | 2.5–5 | 0–1.5 | 494 |
| Lawrence Lake | 42°44′ N, 85°35′ W | 0.05 | 12.5, 5.9 | 0.4 | 0–4 | 270 |
| Mono Lake | 38° N, 119° W | 160 | 46 (18) | 0.3 | 0–10 | 1944 |
| Lake Tahoe | 39°1′ N, 20°1′ W | 490 | 501 (301) | 0.09 | 0–22 (near-shore) | 1899 |
| Lake Biwa | 35°11.5′ N, 135°58.8′ E | 670 | 103.8 (45.5 N basin); (3.5 m S basin) | 0.3–1.9 | 0–18 | 86 |
| Lake Calado | 3°15′ S, 60°34′ W | 2–8 | 1–12 m | 1.5 | 0–3 | 30 |
| Pilkington Bay | (00°17′ N, 33°20′ E | 40 | 16 (4.5) | 1–1.7 | 0 to bottom | 1240 |

Maximum depth, $Z_{max}$; mean depth, $Z_{mean}$; diffuse attenuation coefficient for photosynthetically available radiation ($k_d$); and depth of the upper mixed layer (UML) during stratification.

**Table 2** Comparison across latitudes of dominant terms in surface energy budgets, Wedderburn or Lake numbers, and coefficients of eddy diffusivity in the metalimnion when $W$ or $L_N$ are low using data from lakes in **Table 1**

|  | Tropical lakes | Temperate (small, sheltered) | Temperate (moderate to large) | Arctic |
|---|---|---|---|---|
| LE ($U = 4\,m\,s^{-1}$) | 250 | 180–300 | 150 | 50 |
| LE ($U = 8\,m\,s^{-1}$) | 400 | NA | 400 | <200 |
| Max $LW_{net}$ | ~100 | 150 | 200 | 130 |
| Largest term of SEF | LE | $LW_{net}$ | LE | LE |
| Relative size of $u_{*w}$ and $w_*$ | $u_{*w} <= w_*$ | $u_{*w} <= w_*$ | $u_{*w} >= w_*$ | $u_{*w} > w_*$ |
| Maximum of $u_{w*}$ and $w_*$ | 1, 1.3 | 0.75, 0.75 | 1.5, 1.2 | 1.5, 1 |
| Low values of $W$ or $L_N$ in seasonal thermocline | 1–3(Calado) | 100 | 1 | 0.1–5 |
| Metalimnetic $K_z$ | $10^{-6}$ | Molecular rates | $10^{-5}$–$10^{-4}$ | $10^{-6}$–$10^{-5}$ |

Units of LE and $LW_{net}$ are $W\,m^{-2}$, of $u_{*w}$ and $w_*$ are $m\,s^{-1}$, and of $K_z$ are $m^2\,s^{-1}$. Positive values of LE, SE and $LW_{net}$ represent heat losses.

of lakes. Cumulatively, the surface forcing and resulting surface energy budgets, in concert with the attenuation coefficient for light, determine lake temperatures. They contribute to the temperature difference across the thermocline, which is also set by air temperatures during the coldest time of year, either winter or the monsoon in tropical areas. The temperature difference is generally smallest for tropical lakes and largest for temperate lakes. Surface water temperatures in tropical lakes can be near 30 °C; they are between 20 and 30 °C in large temperate lakes; near 30 °C in small, low altitude temperate lakes and cooler as elevation or latitude increases. In arctic lakes in Alaska during summer, surface water temperatures range from 11 to 20 °C. These temperatures, in turn, lead to variations of different processes in the surface energy budget. For example, surface water temperatures determine the magnitude of evaporation and long wave outgoing radiation as both are higher at warmer temperatures. For many lakes, evaporation is the largest term in a surface energy budget, and thus the magnitude of evaporation is influential in determining how much the mixed layer will deepen due to heat loss (**Table 2**). For example, for low to moderate wind forcing, latent heat fluxes in tropical and temperate lakes in summer are 3–5 times larger than in arctic lakes. In addition, the rate of change of density with temperature is larger at warmer temperatures, thus cooling more quickly erodes stratification. Thus, given that sensible heat exchanges do not vary widely across latitudes once stratification has set up, and net long wave radiation rarely exceeds $150\,W\,m^{-2}$, the larger latent heat fluxes in warm water lakes contribute considerably to mixing the upper water column.

Wind is another major forcing factor. For the lakes in **Table 1**, wind forcing is somewhat dependent on lake size. For all the larger lakes, wind speeds in excess of $8\,m\,s^{-1}$ occur frequently. However, for the small, sheltered lakes, winds rarely exceed $5\,m\,s^{-1}$. Exceptions occur for L. Calado, where wind speeds are generally less than $6\,m\,s^{-1}$ and for Toolik Lake, where winds exceed $5\,m\,s^{-1}$ on a near daily basis and are frequently above $8\,m\,s^{-1}$. In consequence, maximum values of $u_{*w}$ are higher for the larger lakes and for exposed small lakes such as Toolik, indicating that the greater importance of wind forcing for turbulence production in the upper mixed layer in these lakes than in smaller lakes.

Based on the latitudinal differences in the magnitude of evaporation and the size, and in part latitudinal related differences in wind speed, patterns begin to emerge in the role of heat loss versus wind mixing in setting mixed layer depth in lakes in diverse locations. In general, the magnitude of $u_{*w}$ is lesser than $w_*$ in tropical lakes. In fact, in Pilkington Bay, $w_*$ is sometimes 6 times higher than $u_{*w}$. In contrast, in arctic lakes, $u_{*w}$ is generally larger than $w_*$. The relative magnitude also depends on the size of the lakes. $u_{*w}$ exceeds $w_*$ in the large temperate lakes; the converse is true for the small lakes. That is, heat loss contributes more toward turbulence production and mixed layer deepening in tropical lakes and in small temperate lakes than in lakes with cooler surface water temperatures. Because the magnitude of evaporation depends on wind speed, $w_*$ is lowest in the small sheltered temperate lakes; the combination of low $w_*$ and low $u_{*w}$ leads to the shallow mixed layer depths in such lakes.

The diel variations in $w_*$ are a major cause of the diurnal variations in mixed layer depth. That is, since $w_*$ goes to zero on sunny days, the upper water column may thermally stratify even if it is windy. It is through this mechanism that diurnal thermoclines form. With the higher evaporation rates in warm water bodies, the onset of cooling begins earlier in the day such that the combination of wind mixing and heat loss can lead to more rapid deepening and deeper mixed layers. For example, as a result of typical heat losses from sensible and latent heat exchanges and net long wave radiation of 200–300 W m$^{-2}$ at Pilkington Bay, the mixed layer is likely to deepen from about 0.5 to 6 m from late afternoon until early morning. For the arctic lake, similar heat losses occur with cold fronts with sustained cold temperatures and moderate winds.

Cloud cover or its absence can influence mixed layer dynamics in different ways at different latitudes. Cloud cover reduces net short- and long-wave radiation. In arctic and temperate regions, the decreased solar radiation is such that the effective heat loss drops below zero (heat loss from the lake) and the mixed layer deepens even during the day. In contrast, in tropical regions during the rainy season, warmer air temperatures, higher relative humidity, and greater cloud cover may lead to lower heat losses at the surface thus mitigating the lower insolation. Hence, stratification may be persistent even during the day. In contrast, for sheltered tropical lakes, cloud-free periods and the considerably enhanced heat loss, particularly at night, from long wave radiation may abet seasonal deepening of the mixed layer.

### Comparative Mixing Dynamics in the Upper Mixed Layer: Case Studies

In the following, we compare surface forcing and resulting surface energy budgets for specific lakes in the tropics, temperate zone, and Arctic (**Table 1**). We first compare the more weakly stratified tropical and arctic lakes to illustrate dominant mixing processes in these environments. We then contrast surface forcing in large and small temperate lakes found in regions that are dry and in regions that are more humid.

**Comparison of arctic and tropical lakes** – Pilkington Bay is a shallow embayment of Lake Victoria near the equator in Uganda. Due to warm air temperatures, high irradiance and fairly turbid water, it stratifies daily with temperature increases between 1 and 3 °C. For relatively low winds of 4 m s$^{-1}$, evaporative heat loss is 250 W m$^{-2}$; for moderate winds of 8 m s$^{-1}$, evaporative heat loss is 400 W m$^{-2}$. Once heat losses at the air–water interface exceed inputs from solar radiation, the mixed layer begins to deepen. For example, as a result of typical heat losses from sensible and latent heat exchanges and net long wave radiation of 200–300 W m$^{-2}$, the mixed layer is likely to deepen from about 0.5 m to 6 m from late afternoon until early morning. Turbulent kinetic energy fluxes with such heat losses and light winds are $(0.1–0.4) \times 10^{-6}$

m$^3$ s$^{-3}$. Passage of night-time squalls with winds up to 9 m s$^{-1}$ can increase surface heat losses to 600 W m$^{-2}$ and turbulent kinetic energy fluxes to $2.5 \times 10^{-6}$ m$^3$ s$^{-3}$. Even if wind speeds increase, the turbulent velocity scale for cooling exceeds that for wind mixing attesting to the role of convection due to heat loss for deepening the mixed layer.

Arctic lakes, such as Toolik Lake, have evaporative heat losses during summer for winds of 4 and 8 m s$^{-1}$ of 50 and 200 W m$^{-2}$, respectively (**Table 2**). Thus, with the cold temperatures, heat loss due to evaporation is much less than in tropical lakes. Consequently, net long-wave radiation and sensible heat exchange tend to be a larger proportion of the surface energy budget. Under stationary air masses with surface energy fluxes of 100–200 W m$^{-2}$, and turbulent kinetic energy fluxes less than $0.5 \times 10^{-6}$ m$^3$ s$^{-3}$, diurnal heating induces either linear stratification to the surface or diurnal thermoclines. Nocturnal cooling erodes these features to the depth of the seasonal thermocline. However when cold fronts with over a day of sustained cold temperatures and moderate winds occur, surface energy fluxes exceed 300 W m$^{-2}$, and fluxes of turbulent kinetic energy reach $2 \times 10^{-6}$ m$^3$ s$^{-3}$. The mixed layer deepens to depths that are reached within the tropical embayment in 36 h as opposed to 12 h. Due to the longer time scale for mixing, deepening is only possible during cloudy conditions; otherwise the heating from solar radiation would restratify the upper water column. In addition, the depth of mixing is reduced relative to tropical water bodies because of the higher temperature gradient across the thermocline.

Comparison of the turbulent velocity scale for cooling to that for wind mixing illustrates the different roles of heat loss and wind in the two environments. In the arctic lake, $u_{*w}$ and $w_*$ are similar or $u_{*w}$ is up to 2 times higher. Except during cold fronts, $w_*$ drops to zero in the day and a diurnal thermocline forms. During cold fronts, $w_*$ stays high and the mixed layer deepens due to the combination of heat loss and wind mixing. In contrast, in Pilkington Bay, the values of $u_{*w}$ and $w_*$ are similar or $w_*$ is up to 6 times higher. During cold fronts in the arctic lake, the Lake number drops to values between 1 and 3. This decrease indicates that the thermocline tilts and shear at the base of the mixed layer contributes to mixed layer deepening for much of the period. Although Lake numbers for Pilkington Bay are nearly always less than 1, a large fraction of the turbulent kinetic energy flux into the mixed layer at night is due to cooling. Thus, thermocline tilting does not appear to be essential for mixed layer deepening.

Overall, the picture emerges that deepening of the upper mixed layer in a tropical lake occurs readily by heat loss and is relatively rapid. In contrast, for similar turbulent kinetic energy fluxes in an arctic lake, wind shear plays a larger role in mixing the upper water column and shear near the base of the mixed layer further contributes. The typical nocturnal cooling and wind leads to a shallow upper mixed layer ($\sim$2–3 m) which overlays a seasonal thermocline. In contrast, in tropical lakes such as Pilkington Bay, the depth of nocturnal mixing is greater. The lake mixes to the bottom more frequently and a seasonal thermocline does not form or forms at deeper depths. Thus, the dominant processes causing mixing at these two latitudes differ.

**Mixing during the stratified period in moderate-sized temperate lakes** Mono Lake (California) is located in an arid region. Its winter climate is marked by the passage of frontal systems with high winds; in summer, diurnal winds predominate with magnitudes $\sim$5–8 m s$^{-1}$. Fronts are less frequent than in winter, but can induce winds in excess of 8–10 m s$^{-1}$ for periods over a day. With the absence of appreciable cloud cover in summer, heat losses due to net long-wave radiation range from 100 to 200 W m$^{-2}$ and are higher than in either the tropical or the arctic lake. Sensible heat exchanges are similar to those in the tropical and arctic lakes. Heat losses due to latent heat exchange are similar to those in the tropical and arctic lake when diurnal winds prevail. When winds increase to 10 m s$^{-1}$ and relative humidities drop to 25%, latent heat fluxes reach maxima of 500 W m$^{-2}$. The high winds which induce these large latent heat fluxes persist and latent heat fluxes can exceed 300 W m$^{-2}$ for over a day. During such conditions, $u_{*w}$ is only slightly larger than $w_*$. Turbulent kinetic energy fluxes during light winds are 0.2–0.4 $\times 10^{-6}$ m$^3$ s$^{-3}$; during high winds they increase to values between 2 and 3 $\times 10^{-6}$ m$^3$ s$^{-3}$. Similar to the tropical and arctic lakes, when winds were light, nocturnal mixing through the 10 m deep upper mixed layer is driven by heat loss and the surface layer restratifies in the day. However, when frontal systems pass, the combination of heat loss and wind shear induce mixing through much of the mixed layer even during the day. The Lake Number drops to values near 1; hence, mixing is induced near the base of the mixed layer by shear. Thus, we see that mixing during the passage of fronts at Mono Lake is similar to that during the passage of fronts in arctic lakes.

During periods with light to moderate winds, the depth of mixing at night in summer in moderate-sized lakes is greater for lakes in arid regions than where humidity is high. Lake Tahoe (California/Nevada) is similar to Mono Lake and experiences low humidity and cool nights. The upper mixed layer stratifies and

mixes on a daily basis, with nocturnal heat losses inducing turbulence to the top of the seasonal thermocline at 22 m. Diurnal heating restratifies the lake in the morning and dampens the turbulence induced at night. Light to moderate afternoon winds cause turbulence in only the upper 4 m. In contrast, Lake Biwa (Japan) has warm temperatures and high humidity. During periods with light winds, the upper mixed layer remains stratified even at night. For both lakes, the expectation is that as wind speeds increase, the mixing dynamics would be similar to those during fronts at Mono Lake.

**Mixing during the stratified period in small temperate lakes** Small, sheltered temperate lakes have received the majority of study by aquatic ecologists and experience conditions different from those in larger temperate lakes or small arctic lakes. Turbulent kinetic energy fluxes in high elevation Emerald Lake (California) and Trout Bog (northern Wisconsin) are less than $0.2 \times 10^{-6}$ m$^3$ s$^{-3}$. With rare exceptions, wind speeds are less than 5 m s$^{-1}$ at Emerald Lake, and less than 2 m s$^{-1}$ at Trout Bog. During spring in northern Wisconsin, cold fronts with cloudy skies and air temperatures 5–10 °C lower than surface water temperatures occur at $\sim$5 day intervals. A similar periodicity occurs in the fall, but in summer fronts pass through every 10–15 days. Surface water temperatures are almost always slightly warmer than air temperatures; typical values of SE are 20 W m$^{-2}$. Because of the low wind speeds, latent heat fluxes do not exceed 150 W m$^{-2}$. In consequence, net long-wave radiation is, on average, the largest heat loss term. Because the sum of the heat loss terms is at most 300 W m$^{-2}$, and, during daylight hours net solar radiation frequently exceeds this value, diurnal thermoclines form on most days. Thus, the pattern involves formation of a shallow diurnal thermocline each day followed by nocturnal cooling. $u_{*w}$ is always less than 0.5 m s$^{-1}$, and $w_*$ reaches 0.5 m s$^{-1}$ nearly every night. These low values attest to the reduced turbulence in these environments compared to larger or more exposed lakes. While the upper mixed layer mixes due to penetrative convection from heat loss at night; the reduced values of turbulent kinetic energy flux and the high density gradient at the top of the thermocline lead to a shallow mixing depth.

Cloudy periods with cool air temperatures occur infrequently at Emerald Lake in summer, but passage of fronts with air temperatures $\sim$10 °C colder than lake temperatures occur in autumn. During summer, relative humidity averages 50%. Similar to Trout Bog, sensible heat fluxes are low in summer although they increase to 50 W m$^{-2}$ during cold fronts in autumn. Latent heat fluxes increase steadily as surface temperatures warm with maxima in mid-September. They decrease as temperatures decrease in autumn. With a few exceptions, maximum values of latent heat exchanges are less than $150\,W\,m^{-2}$. Due to the low wind speeds and low latent heat fluxes, net long-wave radiation dominates the surface energy budget. $LW_{net}$ also varies with surface water temperatures. Consequently, surface energy fluxes increase steadily through summer. The upper water column heats diurnally and cools at night with $u_{*w}$ and $w_*$ similar in magnitude to values at Trout Bog. The magnitude of the heat losses at night sets the depth of nocturnal mixing and thus, with the increase in TKE due to warmer surface temperatures, the depth of the seasonal thermocline progressively deepens throughout the summer. The first cold front in autumn erodes the seasonal thermocline. Thus, the autumn cooling period is marked by diurnal heating and nocturnal mixing with successive cold fronts leading to overall decreases in water temperatures. Lake Numbers in summer at both Trout Bog and Emerald Lake always exceeded 100 and hence wind shear does not contribute to mixing at the base of the thermocline. Although Lake Numbers decrease due to the weakened stratification in fall, heat loss is the dominant source of turbulent kinetic energy due to the low winds. Hence, wind shear plays a minor role in the deepening.

**Summary** – For small, sheltered dimictic lakes, convective mixing due to heat loss sets the depth of the seasonal thermocline in summer with progressive deepening at night as surface temperatures increase with concomitant increases in evaporation and net long-wave radiation. Autumn cooling is induced by heat loss with a minor contribution from wind shear. This pattern occurs at Trout Bog where relative humidities vary from 30% to 90% and for Emerald Lake where they ranged from 15% to 85% in summer, and 5% to 90% in autumn. This dominance of convective cooling in mixed layer deepening is similar to that found in tropical lakes. For the moderate-sized to large temperate lakes, the mixed layer mixes by convection at night at low winds where relative humidities are low but not when relative humidities are higher. During periods with moderate to high winds, turbulent velocities from wind exceed those from heat loss and Lake Numbers decrease to values low enough such that mixing is induced at the base of the mixed layer by shear. Thus, for the moderate-sized lakes during the passage of fronts, wind shear plays a considerable role in energizing the mixed layer, and mixing dynamics are similar to those in the dimictic arctic lakes in summer.

## Mixing Dynamics: Thermocline and Hypolimnion

When Wedderburn and Lake numbers drop below threshold values (see Background section), vertical mixing is enhanced in the thermocline and hypolimnion by instabilities in the internal wave field. Lake numbers for Pilkington Bay rarely exceed 1. Because heat loss causes rapid deepening of the thermocline in tropical lakes and reduction of the density gradient across the thermocline, the low Lake numbers are not indicative of mixing induced by internal waves. In contrast, when Lake Calado, a dendritic lake on the Amazon floodplain, is thermally stratified, Lake numbers drop to values between 1 and 3 as winds increase diurnally. Eddy diffusivities increase to values 10 times the molecular diffusion of heat. Lake numbers during summer stratification in Emerald Lake and Trout Bog exceed 100 and indicate that the thermocline will not tilt and nonlinear internal waves will not form. Eddy diffusivities in the thermocline and hypolimnion of these lakes in summer are, with only a few exceptions in Emerald Lake, at molecular values. In Mono Lake, Toolik Lake, and Lake E5, a stratified lake near Toolik Lake but whose surface area is one-tenth larger, Lake numbers typically exceed 10 during periods with diurnal winds but drop to values near 1 when high winds occur in association with the passage of fronts. This greater wind forcing tilts the thermocline and induces instabilities in the internal wave field. In all three lakes, the coefficient of eddy diffusivity increases to values 10–100 times molecular in the thermocline and, in Toolik and E5, to values 10–1000 times molecular in the hypolimnion. These high values attest to the enhanced mixing associated with instabilities in the internal wave field.

In summary, depending on wind forcing relative to stratification, nonlinear waves form in the stratified waters of both small and large lakes. The instabilities are larger near the boundaries in nearly all cases examined. Thus fluxes will be larger near shore and other mechanisms will be required for transport to offshore waters. These include intrusions as well as the propagation of nonlinear waves. It is not surprising that nonlinear waves form in larger lakes with their greater fetch. Whether they are found in small lakes appears to depend upon the latitudinal gradient which sets the temperature difference across the thermocline and upon the degree with which wind forcing is reduced due to sheltering. Interestingly, the temperature gradient in small, stratified tropical lakes is small enough that Lake numbers drop to values indicative of instabilities in the metalimnion on a daily basis. In small temperate lakes, the temperature gradient is so high relative to the forcing that Lake numbers stay high. The temperature gradient in small arctic lakes is high enough that when diurnal winds predominate, Lake numbers exceed 10. However, the increase in wind speeds during the passage of fronts is sufficient to drop Lake numbers to values near 1 such that internal waves become unstable.

We thus see different patterns in lakes at the three latitudes. In tropical lakes, cooling events or events with low Lake numbers frequently occur allowing vertical exchange of solutes and particulates. In arctic lakes as small as 11 ha, exchanges occur in summer during the passage of fronts, and whether they involve full water column mixing or breaking internal waves depends on the depth of the basin and the extent of heat loss. Exchange is particularly effective during cold fronts when the mixed layer deepens and low Lake numbers occur. In moderate-sized temperate lakes in summer, exchanges occur when winds increase with the passage of fronts or major storms. In small temperate lakes during summer, the Lake and Wedderburn numbers stay high, exchange between the epilimnion and metalimnion and ultimately the hypolimnion occurs when entrainment occurs due to heat loss. The depth of this mixing depends upon the magnitude of heat loss relative to the stratification gradient. In summary, physically induced connectivity between the upper and lower water column occurs frequently in shallow, tropical lakes, occurs during the passage of fronts in stratified arctic and large temperate lakes, and rarely occurs in small, stratified temperate lakes during summer.

## Implications of Differences in Mixing Dynamics

Turbulent kinetic energy fluxes which cause mixing in lakes are similar during the stratified period in tropical, temperate, and arctic lakes. The highest values occur during windy periods; for sheltered lakes the values are always low. In tropical lakes, with their warm water and consequent high evaporation rates and the small temperature differences between the upper and lower water column, mixed layer deepening can occur rapidly at night. In temperate and arctic lakes, the overall depth of mixing in summer is reduced relative to tropical lakes for the same heat loss and flux of turbulent kinetic energy due to the greater work required to entrain the more stably stratified water below. Regardless, even with light winds, heat loss due to convection causes nocturnal mixing which can redistribute solutes and particles. The depth of nocturnal mixing is less where humidity is higher.

The phasing of wind relative to heat loss also influences mixed layer deepening. In particular, the implications of higher winds depend upon whether they occur during the day or night and whether it is cloudy. If they occur on days with low cloud cover, they distribute heat downwards in the water column and diurnal thermoclines may not form. If they occur on cloudy days or at night, heat loss occurs and the mixed layer deepens. In temperate and arctic regions, the passage of fronts induces increased wind speeds. At some locations, the winds are highest near mid-day, and little mixed layer deepening occurs. In arctic lakes, winds do stay high at night during fronts and considerable deepening occurs. In some summers the arctic lakes restratify, in others the combination of winds and cold temperatures is such that the lakes mix to the bottom and do not restratify.

When wind forcing is high relative to the density gradient across the thermocline, the thermocline tilts and nonlinear waves form. The ensuing turbulence induces connectivity between the hypolimnion and epilimnion. The latitudinal differences in mixed layer depth and density gradient across the thermocline determine the importance of this mechanism. Instabilities in the internal wave field occur in many small arctic lakes in summer. They are less frequent in small temperate lakes due to the stronger stratification. How frequently they occur in moderate-sized lakes depends upon the frequency of frontal systems with high winds. Their importance in tropical lakes depends upon the magnitude of the wind field and resulting latent heat fluxes. When latent heat fluxes are high, mixed layer deepening may be rapid and deep enough that connectivity between the upper and lower water columns is induced by heat loss. Thus, we see that comparisons of surface energy budgets and use of dimensionless indices provides a mechanistic understanding of latitudinal differences in lake hydrodynamics. This understanding will promote improved intuition as to differences in ecosystem level function in different latitudes as well as to ecosystem responses to climate change.

*See also:* Currents in Stratified Water Bodies 2: Internal Waves; Density Stratification and Stability; Small-Scale Turbulence and Mixing: Energy Fluxes in Stratified Lakes.

## Further Reading

Davies BR and Walmsley RD (eds.) (1985) *Perspectives in Southern Hemisphere Limnology*. Developments in Hydrobiology, vol. 28. Dr. W. Junk Publishers.

Hutchinson GE (1957) *A Treatise on Limnology*. John Wiley & Sons.

Lewis WM Jr. (1983) A revised classification of lakes based on mixing. *Canadian Journal of Fisheries and Aquatic Science* 40: 1779–1787.

Lewis WM Jr. (1987) Tropical limnology. *Annual Review of Ecology and Systematics* 18: 159–184.

Imberger J (1985) The diurnal mixed layer. *Limnology and Oceanography* 30: 737–770.

Melack JM (1979) Temporal variability of phytoplankton in tropical lakes. *Oecologia* 44: 1–7.

Melack JM (2006) Biodiversity in inland aquatic ecosystems: Natural gradients and human-caused impoverishment. In: Leybourne M and Gaynor A (eds.) *Water: Histories, Cultures, Ecologies*, pp. 182–190. ch. 14, University of Western Australia Press.

MacIntyre S (1997) Turbulent eddies and their implications for phytoplankton within the euphotic zone of Lake Biwa, Japan. *Japanese Journal of Limnology* 57: 395–410.

MacIntyre S, Romero JR, and Kling GW (2002) Spatial-temporal variability in mixed layer deepening and lateral advection in an embayment of Lake Victoria, East Africa. *Limnology and Oceanography* 47: 656–671.

MacIntyre S, Sickman JO, Goldthwait SA, and Kling GW (2006) Physical pathways of nutrient supply in a small, ultra-oligotrophic arctic lake during summer stratification. *Limnology and Oceanography*. 51: 1107–1124.

Robarts RD, Waiser M, Hadas O, Zohary T, and MacIntyre S (1998) Contrasting relaxation of phosphorus limitation due to typhoon-induced mixing in two morphologically distinct basins of Lake Biwa, Japan. *Limnology and Oceanography* 43: 1023–1036.

Talling JF and Lemoalle J (1998) *Ecological Dynamics of Tropical Inland Waters*. Cambridge University Press.

# Density Stratification and Stability

**B Boehrer** and **M Schultze**, UFZ – Helmholtz Centre for Environmental Research, Magdeburg, Germany

## Introduction

In most lakes, water properties change from the surface to greater depth, i.e., these lakes show a vertical stratification of their water masses at least for some extended time periods. Heat exchange with the atmosphere and the forming of gradients of dissolved substances controls internal waves and the vertical exchange of water within the lakes. This has decisive impact on the evolution of water quality and, as a consequence, on the community of organisms living in the lake. This article deals with processes contributing to the stratification of lakes and the forming of layers. The most common numerical approaches for the quantitative evaluation of stratification-relevant physical quantities, e.g., electrical conductivity, are included. The final section lists quantities for stability of density stratification and what conclusion can be drawn from them.

## Circulation Patterns

Surface temperatures of lakes show a pronounced temperature cycle over the year (**Figure 1**), in most latitudes. This is a consequence of heat exchange with the atmosphere and the seasonal variation of meteorological parameters, such as incoming solar radiation. The temperatures in the deep water follow the surface temperatures only for the time when the lake is homothermal, as in our example Lake Goitsche, Germany (**Figure 1**), during winter, from November until April.

Throughout summer, temperatures vary from the surface to the lake bed, and the lake remains stratified. Warmer and less dense water floats on top of colder, denser water (**Figure 1**). Thus, Lake Goitsche is called stably stratified, as overturning water parcels would require energy. On the contrary, during winter, no density differences obstruct the vertical transport. These seasons are commonly referred to as the stagnation period and the circulation period. Lakes that experience a complete overturn during the year are called holomictic.

During the circulation period, dissolved substances, such as oxygen or nutrients, get distributed over the entire water body (**Figure 2**). Hence, the circulation pattern is a decisive factor for the evolution of water quality and the biocenosis of the lake. In conclusion, the commonly used classification of lakes is according to their circulation patterns.

- *Holomictic lakes* overturn and homogenize at least once a year.
- In *meromictic lakes*, the deep recirculation does not reach the deepest point of the lake. As a consequence, a chemically different layer of bottom water is formed, the monimolimnion (see below), and remains there for at least 1 year.
- *Amictic lakes* do not experience a deep recirculation. Usually permanently ice-covered lakes are included in this class. Lakes, however, can circulate underneath an ice sheet by external forcing, such as solar radiation that penetrates to the lake bed and geothermal heat flux, or salinity gradients created when ice is forming on a salt lake.
- *Lakes with episodic partial deep water renewal* do not experience a complete overturn. The deep water however is partially replaced in episodic events.

Holomictic lakes are subdivided into classes indicating the frequency of complete overturn.

- *Polymictic* lakes are not deep enough to support a continuous stratification period throughout summer. The entire lake is mixed by sporadic strong wind events over the year or even on a daily basis in response to strong diurnal temperature variation.
- *Dimictic* lakes are handled as the prototype of lakes in moderate to cold climates. A closer look at the lakes, however, reveals that in most cases an ice cover or a great maximum depth is required to guarantee a stratification period during the cold season (see **Figure 2**). Between ice cover and summer stratification, the lake can be circulated completely in the vertical, the easiest when surface temperatures traverse the temperature of maximum density at 4 °C.
- *Monomictic* lakes possess one circulation period in addition to the stratification period. Many lakes in the temperate climate zone belong into this class, if they do not develop an ice cover during winter. Sometimes such lakes are also referred to as warm monomictic to distinguish them from cold monomictic lakes, which show an ice cover for most of the year and circulate during the short period without ice.
- *Oligomictic* lakes circulate less frequently than once a year, normally at irregular intervals, triggered by extreme weather conditions such as unusually cold winters for the respective location.

As a consequence of the natural variability of the weather conditions between years, the circulation patterns of the lakes also vary. A usually monomictic lake, for example, can show a dimictic circulation pattern when it freezes in an unusually cold winter. As another example, late during the twentieth century, Mono Lake turned meromictic for intermittent periods of 5 or 7 years, respectively, because of inflowing freshwater, but in other years showed a holomictic circulation.

## Density Differences and Formation of Layers

### Temperature Stratification

Although the surface water is exposed to solar radiation and thermal contact with the atmosphere, the

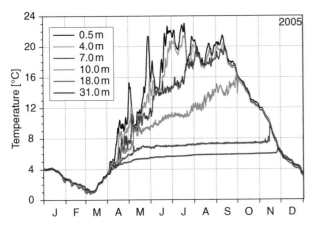

**Figure 1**  Temperatures (24 h mean) on several depths in Lake Goitsche near Bitterfeld, Germany during the year 2005. Reproduced from Boehrer B and Schultze M (2008) Stratification of lakes. *Reviews in Geophysics*, 46, RG2005, doi:10.1029/2006RG000210, with permission from American Geophysical Union.

deeper layers are shielded from major sources of heat. Diffusive heat transport on a molecular level is very slow and requires a month for the transport of heat over a vertical distance of 1 m. A much more efficient heat transport can be accomplished by turbulent transport. The energy for the turbulence is mainly supplied by wind stress at the lake surface and transferred via instabilities through friction at the side walls and internal current shear.

Heating a lake over 4 °C at the surface results in a stable stratification. As a consequence, transport of heat to greater depths requires energy. The limited budget of kinetic energy available for mixing limits the depth to which a certain amount of heat can be forwarded over the stratification period. In sufficiently deep lakes, the thermal stratification holds until cooler autumn and winter temperatures permit a deeper circulation. The warm surface water layer is called *epilimnion*, while the colder water layer beneath, which has not been mixed into the epilimnion is called *hypolimnion*. A sharp temperature gradient (*thermocline*) separates both layers (**Figure 3**).

Epilimnion and atmosphere are in thermal contact and exchange volatile substances with each other. In addition, the epilimnion is recirculated by wind events or periods of lower temperatures during the stratification period. During those periods, dissolved substances are distributed within the epilimnion. On the contrary, the hypolimnion is insulated from exchange with the atmosphere during the stratification period. Transport of dissolved matter across the vertical density gradient of the thermocline usually is small.

In general, wind determines the thickness of the epilimnion, with few exceptions, e.g., where light penetrates beyond the mixing depth because of wind, or where the stratification is determined by

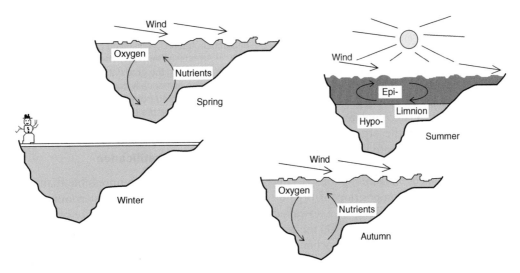

**Figure 2**  Annual cycle of a dimictic lake with ice cover during winter.

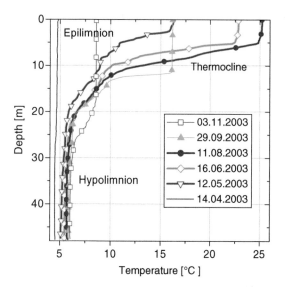

**Figure 3**   Temperature profiles of Lake Goitsche/Germany in station XN5 in 2003. Symbols are added for every sixteenth data point to distinguish between acquisition dates.

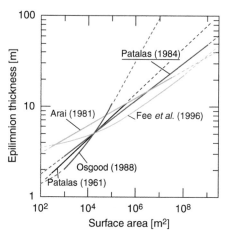

**Figure 4**   Graphical representation of several approximations of epilimnion thickness $z_{epi}$ versus surface area of the respective lakes. Adapted from Jöhnk KD (2000) 1D hydrodynamische Modelle in der Limnophysik – Turbulenz, Meromixis, Sauerstoff. Habilitationsschrift, Technical University of Darmstadt, Germany.

inflow and water withdrawal (reservoirs). As inferred from **Figure 3**, the thickness of the epilimnion is not constant over the stratification period. In spring, a thin layer is formed, which gradually thickens over the summer because of the cumulative input of wind energy and diurnal heating and cooling. It takes until autumn, when colder temperatures at the lake surface can erode the stratification. During this later period of thermal stratification, substances dissolved in hypolimnetic waters, such as nutrients, become available in the epilimnion again. Eventually epilimnion and hypolimnion are homogenized.

The epilimnion thickness is a crucial factor for living organisms. Hence limnologists have tried to correlate epilimnion thickness with lake morphometry to achieve an a priori estimate (**Figure 4**). The most central regression is $h_{epi} = 4.6 \times 10^{-4} A^{0.205}$, which includes the higher energy input from winds over lakes with larger surface area $A$.

Because of its high gradients, the thermocline forms a special habitat. Organisms controlling their density can position themselves in the strong density gradient. Also, inanimate particles can accumulate on their level of neutral buoyancy and motile organisms dwell in the thermocline to profit from both layers, epilimnion and hypolimnion. As a result, a layer of distinctive properties can form, called *metalimnion*. Especially in nutrient-rich lakes, the decomposition of organic material can deplete oxygen resulting in a so-called metalimnetic oxygen minimum (**Figure 5**). On the contrary, if light can penetrate to the thermocline and photosynthesis can overcome the oxygen consumption locally, a metalimnetic oxygen maximum occurs.

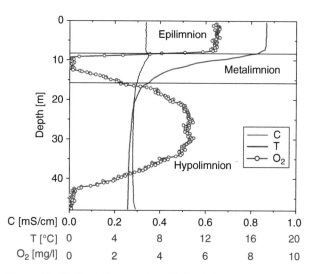

**Figure 5**   Profiles of temperature ($T$), (in situ) conductivity ($C$), and concentration of dissolved oxygen ($O_2$) from 6 September 2000 in Arendsee/Germany. The boundaries between layers were drawn along the gradients in the oxygen profiles. Oxygen concentration numerically corrected for response time of 7.5 s of the sensor. Adapted from Boehrer and Schultze, 2005, *Handbuch Angewandte Limnologie*, Landsberg: ecomed.

**Thermobaric Stratification**

Cold water is more compressible than warmer water, in the range of temperatures encountered in lakes. As a consequence, the temperature of maximum density $T_{md}$ decreases as pressure, i.e., depth, increases (by about 0.2 K over 100 m water depth). Hence in cold enough regions, very deep freshwater lakes can show temperature profiles during summer stratification

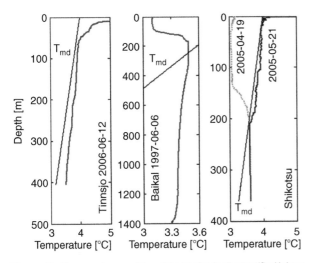

**Figure 6** Temperature profiles of thermobarically stratified lakes: Left panel: Tinnsjø, Norway, during (early) summer stratification; Central panel: Lake Baikal, Siberia, Russia, during (late) winter stratification with the vertical transition through $T_{md}$; Right panel: Lake Shikotsu, Hokkaido, Japan, with the nearly isothermal deep water body below the $T_{md}$ transition. Reproduced from Boehrer B and Schultze M (2008) Stratification of lakes. *Reviews in Geophysics*, 46, RG2005, doi:10.1029/2006RG000210, with permission from American Geophysical Union.

that extend below $4\,^{\circ}C$, though limited to the cold side by the $T_{md}$ profile (see **Figure 6**, left panel). During winter, surface temperature can be lower than $T_{md}$, while at greater depth the temperatures may be above $T_{md}$ (**Figure 6**, central panel). Stable density stratification is achieved in these profiles, if above the $T_{md}$ intersection, colder temperatures overlie warmer temperatures, while below this it is the opposite. At the intersection itself, the vertical temperature gradient must disappear. The water body below the $T_{md}$ profile is not directly affected by the annual temperature cycle (**Figure 6**, right panel).

Nevertheless, observations in such lakes (e.g., Lake Baikal, Russia; Crater Lake, USA; Lake Shikotsu, Japan) show deep waters well supplied with oxygen. This can in part be attributed to the fact that temperatures are low, and productivity and depletion of oxygen happen at a slow rate. However, it is also an indication for a considerable amount of mixing between mixolimnion and the deep water. As a consequence, chemical gradients do not appear in these lakes, and scientists have refrained from calling these lakes meromictic, although a complete overturn does not occur.

### Salinity Stratification

A considerable portion of lakes is salty. As there is no compelling boundary, lakes are called salt lakes, if the salt content lies above 3 g in a kilogram of lake water; i.e., $3\,g\,kg^{-1} = 3‰$. From this concentration, humans can clearly taste the salt, and ecological consequences become obvious. The salt content in lakes can be as high as $300\,g\,kg^{-1}$. However, salinities normally lie below $0.5\,g\,kg^{-1}$. Even smaller salinity gradients can determine the circulation of lakes.

Many large salt lakes, e.g., Caspian Sea, Issyk-Kul, Aral Sea, Lake Van, Great Salt Lake, and the Dead Sea, are located in endorheic basins, i.e., areas on the Earth without hydraulic connection to the world ocean at the surface. Salt lakes also occur outside these areas, as solar ponds or basins filled with sea water that lost the connection to the sea. In addition, some inland lakes are fed by saline groundwater.

Salt modifies the properties of lake water. As a quantitative expression, the familiar magnitude of salinity has been transferred from oceanography to limnetic water. One kilogram of ocean water contains about 35 g of salt. Brackish water, i.e., water mixed from sea water and freshwater, shows a similar mix of dissolved substances, while the composition of salts in lakes can greatly deviate from ocean conditions. Consequently, salinity is better replaced by total dissolved substances TDS in the limnetic environment. For quantitative investigations, usually the physical quantity of electrical conductance is suited much better (explained later).

In some lakes, dissolved substances raise the density of the deep waters enough that part of the water column is not recirculated at any time during the annual cycle. The remaining bottom layer, the monimolimnion, can show very different chemical conditions (**Figure 7**). Such lakes are termed meromictic. Well-known examples are the meromictic lakes of Carinthia, Austria, and Lake Tanganyika, or Lake Malawi. Some small and deep maar lakes as well as natural lakes in southern Norway and Finland are permanently stratified by small concentration differences between mixolimnion and monimolimnion. In addition, deep pit lakes tend to be meromictic.

The monimolimnion is excluded from the gas exchange with the atmosphere over long time periods. Diffusive and turbulent exchange across the chemocline usually is small. As a consequence, anoxia establishes in most cases after sufficient time. Under these chemical conditions, nitrates and sulphates serve as agents for the microbial oxidation of organic material and substances can be produced that would chemically not be stable in the mixolimnion. Permanently exposed to the hydrostatic pressure, gases ($CO_2$, $H_2S$, and others) can accumulate in monimolimnion in concentrations far beyond the concentrations encountered in mixolimnia (e.g., Lake Monoun in Cameroun, Africa, **Figure 8**).

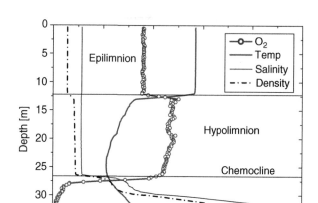

**Figure 7** Profiles of temperature, salinity, dissolved oxygen and density from Rassnitzer See in former mining area Merseburg-Ost on 7th October 2003. Oxygen concentrations are numerically corrected for a sensor response time of 7.5s. Adapted from Boehrer and Schultze, 2005, *Handbuch Angewandte Limnologie*, Landsberg: ecomed.

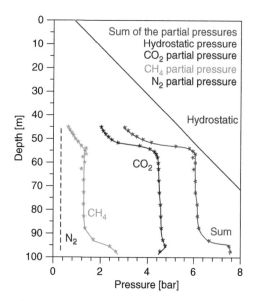

**Figure 8** Profiles of partial pressures of dissolved gases in deep water of Lake Monoun, Cameroun, in direct comparison with hydrostatic pressure (solid line). Reproduced from Halbwachs *et al.* (2004) *EOS* 85(30): 281–288, with permission of American Geophysical Union.

In many meromictic lakes, the recirculation of the mixolimnion erodes the monimolimnion leaving a sharp gradient at the end of the circulation period. The transition of all water properties happens within few decimeters from mixolimnetic to monimolimnetic

values (see **Figure 7**). This sharp gradient is called *halocline*, *chemocline*, or *pycnocline*, depending on whether the salinity, chemical, or density gradient is referred to. From observations, cases of intensive colonization with only few different species are known, where some plankton species obviously take advantage of such gradients (e.g., Lago Cadagno in the Swiss Alps, Lake Bolvod, Gek Gel, and Maral Gel).

### Processes Forming Gradients of Dissolved Substances

In general, diffusive and turbulent diffusive processes distribute dissolved substances more equally throughout a lake over time. On the contrary, inflows of different concentration of dissolved substances or internal processes can produce gradients within a lake. Freshwaters can flow onto a salt lake or enter the epilimnion during the stratification period and hence induce a difference between epilimnion and hypolimnion. These gradients contribute to the density gradients and if strong enough they may even control the circulation pattern of the lakes. In extreme cases, saline waters have been captured in deep layers of lakes for several thousands of years (e.g., Rorhopfjord, Norway; Salsvatn, Norway; Powell Lake, Canada). Such stable stratifications have been induced on purpose in mine lakes to confine heavy metals in the salty monimolimnion for further treatment (Island Copper Mine Lake, Vancouver Island, BC, Canada).

Also, the opposite case of lakes being exposed to high evaporation can encounter salinity gradients in the water column with the saltier layer above. In the case of Lake Svinsjøen (Norway), salty water from deicing roads has removed the previously present permanent stratification. In another case, when ice is forming on salt lakes the residual water can be highly loaded with salts, while after ice melt, relatively fresh water floats on more saline water (lakes in Antarctica). Also, groundwater inflows can form such gradients of dissolved substances (e.g., Lago Cardagno in Swiss Alps, Rassnitzer See; see **Figure 7**). Especially, lakes in volcanic areas are known for the continuous recharge of dissolved substances (e.g., Lake Nyos and Lake Kivu in East Africa, Lake Monoun see **Figure 8**). In particular, the latter case is interesting, as the dissolved gases contribute decisively to the stable density stratification, but they also supply the buoyancy for catastrophic limnic eruptions, in which poisonous gases escape abruptly from a lake with disastrous consequences for living organisms in the area.

Chemical reactions and biological activity in preferred layers can locally change the composition of dissolved substances and impact on the density structure. Although in most lakes at most depths physical transport mechanisms prevail, in some cases the density stratification is controlled by chemical and biological transformations of dissolved material and even control the circulation pattern of lakes. For example, photosynthetically active plankton uses incoming solar radiation for the production of organic material. In addition, allochthonous material is carried into the lake by surface inflows and wind. A portion of the organic material settles on the lake bed. Its decomposition is facilitated by the presence of oxygen or other oxidizing agents and the end products $CO_2$ and $HCO_3^-$ dissolve in the deep layers of the lake where they contribute to the density. Also iron (and manganese) cycling, calcite precipitation, and sodium sulfate precipitation have been documented to control density in lakes.

In some cases, gradients are strong enough to prevent overturns. These meromictic lakes are classified according to the dominant process sustaining the density difference between mixolimnion and monimolimnion in ectogenically (surface inflow; e.g., Rorhopfjord and Salsvatn in Norway; Powell Lake and Island Copper Mine Lake in Canada, Lower Mystic Lake in the United States), crenogenically (groundwater inflow; e.g., Lago Cardagno, Rassnitzer See, Lake Nyos), or biogenically (or endogenically) meromictic lakes (decomposition of organic material; e.g., Woerther See and Laengssee in Austria, iron cycle: Lakes in Norway, mine lakes, manganese: Lake Nordbytjernet, Norway; calcite: Lake La Cruz, Central Spain; sodium sulfate: Canadian prairie lakes). Also, basin depth and basin shape play an important role in the erosion of a monimolimnion and the formation of deep water renewal. Often monimolimnia are found in well-defined depressions in a lake bed, which are only marginally impacted by basin scale currents in the water body above.

### Episodic Partial Deep Water Renewal

A number of lakes do not experience a complete overturn. Episodic events replace parts of the deep water. For example, by cooling, water parcels of high density are formed within the mixolimnion and manage to proceed through the surrounding waters down into the deep water. This process is similar to the deep ocean circulation. Thus, it is not surprising that some of the largest lakes undergo this process. Issyk-Kul (central Asia) recharges its deep water by surface cooling and channelling the cold waters through submerged valleys to the abyss, while in Lake Baikal, wind forcing can push water below the compensation depth so that (temperature-dependent) compression

under high pressure raises the density enough compared with surrounding water that its buoyancy becomes negative; the water parcel continues to proceed deeper. Issyk-Kul and Lake Baikal do not show significant chemical gradients and hence are not termed meromictic. On the contrary, Lake Malawi (East Africa) shows an anoxic monimolimnion, and hence is called meromictic, although the deep water formation is similar to that of Issyk-Kul.

In parallel to heat, concentration gradients of dissolved substances also can force deep water renewal in perennially stratified water bodies. Salinity is increased when water is exposed to high evaporation in a side basin (Dead Sea before 1979), or salt accumulates underneath a forming ice cover (Deep Lake, Antarctica). As soon as water parcels become dense enough, they can proceed along slopes into deep parts of the lake.

## Quantifying Stability

### Temperature

Temperatures recorded in lakes are so-called in situ temperatures. Without any further annotation, temperature data will be understood as such. Nearly all calculations refer to this value, as it is the physically, chemically, and ecologically relevant magnitude. However, if detailed considerations on stability and vertical temperature gradients are envisaged, the reference to potential temperature may be useful. This latter quantity includes the effect of energy required for the expansion, when a water parcel is transferred to atmospheric pressure:

$$\left(\frac{dT}{dz}\right)_{ad} = \frac{g\alpha(T + 273.15)}{c_p} \qquad [1]$$

where $\alpha$ is the thermal expansion coefficient for a water parcel of temperature $T$ along the path from depth $z$ to the surface. In lakes where the deep water is close to temperatures of maximum density $T_{md}$, the thermal expansion coefficient is very small, $\alpha \approx 0$, and, as a consequence, the difference between in situ temperature and potential temperatures is small $\Theta \approx T$. In lakes with warmer deep waters, $\alpha$ can be considered constant. **Figure 9** shows a monotonous potential temperature profile in Lake Malawi, Africa, which indicates stable stratification by temperature only.

### Salinity, Electrical Conductivity and Electrical Conductance

Many substances in lake water are dissolved as ions. Hence electrical conductivity has been used to

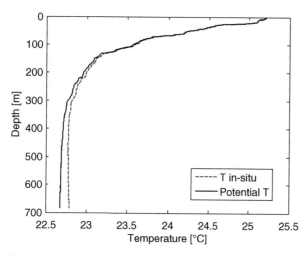

**Figure 9** Profiles of (in situ) temperature $T$ and potential temperature near the deepest location of Lake Malawi on 13 September 1997. Reproduced from Boehrer B and Schultze M (2008) Stratification of lakes. *Reviews in Geophysics*, 46, RG2005, doi:10.1029/2006RG000210, with permission from American Geophysical Union.

quantify dissolved substances. For compensation purposes, the temperature dependence of electrical conductivity of a water sample is recorded, while scanning the relevant temperature interval. In most cases, a linear regression $C(T) = aT + b$ is satisfactory to define the conductance, i.e., the electrical conductivity $\kappa_{ref} = C(T_{ref}) = aT_{ref} + b$ at a certain reference temperature $T_{ref}$. Most commonly, 25 °C is used for the reference:

$$\kappa_{25} = \frac{C(T)}{\alpha_{25}(T - 25°C) + 1} \text{ where } \alpha_{ref} = (T_{ref} + b/a)^{-1} \quad [2]$$

In most surface waters, a value close to $\alpha_{25} = 0.02 \, \text{K}^{-1}$ is appropriate. Electrical conductance is used for a bulk measurement of concentrations of ionically dissolved substances, quantifying transports from changes in the conductance profile, and to base density regression curves on.

Oceanography uses electrical conductivity and temperature to calculate salinity in practical salinity units (psu), which gives a good indication for dissolved salt in grams per kilogram for ocean water and brackish water (water mixed from ocean water and fresh water). In limnetic systems, however, the composition of dissolved substances differs from that of the ocean. Even within some lakes, there are pronounced vertical gradients. As a consequence, salinity can only be used with reservation in limnic systems.

## Density

As direct density measurements in the field are not accurate enough, indirect methods based on easy to

measure temperature and conductivity are implemented. As in most practical applications the difference between in situ and potential temperature is small; we use $T$ for temperature. For lakes of low salinities (<0.6 psu), density can be approximated:

$$\rho = \rho(S, T) = \sum_{i=0}^{6} a_i T^i + S \cdot \sum_{i=0}^{2} b_i T^i \quad [3]$$

using

$a_i = [999.8395; 6.7914 \times 10^{-2}; -9.0894 \times 10^{-3}; 1.0171 \times 10^{-4}; -1.2846 \times 10^{-6}; 1.1592 \times 10^{-8}; -5.0125 \times 10^{-11}]$

$b_i = [0.8181; -3.85 \times 10^{-3}; 4.96 \times 10^{-5}]$

In lakes of a composition of dissolved substances similar to the ocean, the so-called UNESCO formula may be applied (e.g., Rassnitzer See in **Figure 6**), which is applicable for salinities above 2 psu. In cases where salinity cannot be used, calculation of density may directly be based on measurements of temperature and conductivity. For Lake Constance – Obersee, the following formula was proposed, where 20 °C was used as reference temperature for conductance $\kappa_{20}$:

$$\rho = \rho_T + \Gamma = 999.8429 + 10^{-3} \\ \times (0.059385 \, T^3 - 8.56272 \, T^2 + 65.4891 \, T) + \Gamma \quad [4]$$

adding the conductivity contribution in separate

$$\Gamma = \gamma \kappa_{20} \text{ and } \gamma = 0.67 \times 10^{-3} \, \text{kgm}^{-3}\text{mS}^{-1}\text{cm} \quad [5]$$

Alternatively, if the dissolved substances are known, e.g., from chemical analysis, density can be calculated by adding the separate contributions:

$$\rho = \rho_T \left(1 + \sum_n \beta_n C_n\right) \quad [6]$$

where $C_n$ is the concentration of the substance $n$ (g kg$^{-1}$). A short table of coefficients

$$\beta_n = \frac{1}{\rho}\left(\frac{\partial \rho}{\partial C_n}\right)_{\Theta, p, C_m} \quad m \neq n$$

is given in **Table 1**. In most limnological applications, density is used for stability considerations. Hence, density refers to potential density, i.e., the density of a certain water parcel under normal atmospheric conditions (1013 hPa). As a consequence of the (small) adiabatic compressibility of water, in situ density increases with pressure, i.e., water depth, by about $5 \times 10^{-10}$ Pa. This means that at 200 m depth, (potential) density and in situ density differ by about $10^{-3}$.

## Stability

Stability of a water column derives from the density increase in the vertical. Hence, it is a measure for the

**Table 1** Contribution of dissolved or suspended substances to the density of water

| Substance | $\beta_n[(kg/kg)]$ |
|---|---|
| $Ca(HCO_3)_2$ | 0.807 |
| $Mg(HCO_3)_2$ | 0.861 |
| $Na(HCO_3)$ | 0.727 |
| $K(HCO_3)$ | 0.669 |
| $Fe(HCO_3)_2$ | 0.838 |
| $NH_4(HCO_3)$ | 0.462 |
| $CO_2$ | 0.273 |
| $CH_4$ | −1.250 |
| Air | −0.090 |

Modified from Imboden, DM and Wüest A (1995). *Physics and Chemistry of Lakes*, pp. 83–138. Berlin: Springer-Verlag.

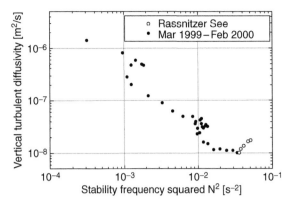

**Figure 10** Turbulent diffusive transport of an artificial tracer ($SF_6$) in the strongly stratified monimolimnion of Rassnitzer See, versus density gradient, $N^2 = -g/\rho \; d\rho/dz$. Adapted from von Rohden and Ilmberger (2001) *Aquatic Sciences* 63: 417–431.

potential energy required for vertical excursion and for overturning of water parcels. Stability considerations can be made for an interface in the water column or for the entire stratified water body as a whole. If an energy source is known, the ratio of required and supplied energy yields a nondimensional number.

**Differential Quantities** The stability of a density stratification is quantified by

$$N^2 = -\frac{g}{\rho}\frac{d\rho}{dz} \qquad [7]$$

where $g$ is the acceleration due to gravity, and $z$ is the vertical coordinate. The magnitude $N$ is also called stability frequency or Brunt–Väisälä frequency ($s^{-1}$), which indicates the maximum frequency ($\omega$) for internal waves that can propagate in the respective stratification. $N^2$ indicates how much energy is required to exchange water parcels in the vertical.

As a consequence, chemical gradients can only persist for longer time periods where density gradients limit the vertical transport of dissolved substances (see **Figure 10**). Lake basin Niemegk of Lake Goitsche (Germany) has been neutralized by introducing buffering river water to the epilimnion. During summer 2000, the vertical transport through the temperature stratification was limited and a chemical gradient in pH could be sustained (**Figure 11**). However, in winter the temperature stratification vanished, vertical transport was enhanced, and consequently, chemical gradients were removed. Gradients in the pH close to the lake bed were stabilized by increased density because of higher concentration of dissolved substances.

In a stratified water column, a current shear can supply kinetic energy for producing vertical

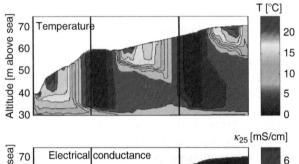

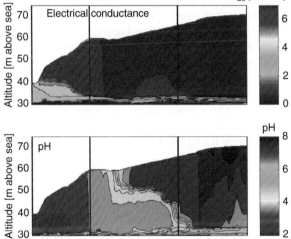

**Figure 11** Contour plot of temperature, electrical conductance and pH value versus time and depth in mining Lake Goitsche (station XN3 in Lake basin Niemegk); period of neutralization by flooding with river water; the rising water level is represented by the increasing colored area.

excursions and overturns. A comparison between density gradient and current shear yields the nondimensional gradient Richardson number:

$$Ri = \frac{N^2}{(du/dz)^2} \qquad [8]$$

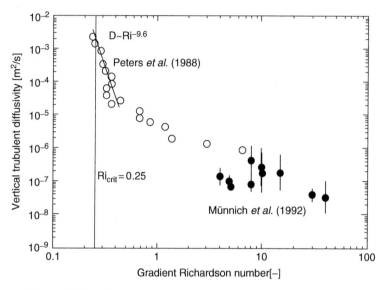

**Figure 12** Relation between diapycnal diffusivities and gradient Richardson number. Adapted from Imboden DM and Wüest A (1995) *Physics and Chemistry of Lakes*, pp. 83–138. Berlin: Springer-Verlag.

where $u = u(z)$ represents the horizontal current velocity profile. The critical value of $Ri = 1/4$, when the shear flow supplies enough energy to sustain overturning water parcels, is found by considering the energy balance in the centre of mass frame. As a consequence, diapycnal transports rapidly increase, if Richardson numbers get close to 0.25 or even fall below this critical value (see **Figure 12**). Although in zones of high shear, e.g., in the bottom boundary layer, supercritical Richardson numbers can be found, they appear only sporadically in the pelagic region of lakes, at least if measured on a vertical scale of meters. The measurements in **Figure 12** suggest a correlation between vertical transport coefficients and gradient Richardson number of

$$D = 3 \times 10^{-9} Ri^{-9.6} + 7 \times 10^{-6} Ri^{-1.3} + 1.4 \times 10^{-7} (\text{m}^2\text{s}^{-1})$$
[9]

if the value of molecular diffusivity of heat is included. On the basis of this, vertical diffusivities can be calculated from gradient Richardson number measurements, which are displayed in **Figure 13**, where gradient Richardson number was measured in pelagic waters over a depth resolution of 3–5 m.

**Bulk Quantities** For the bulk stability of a stratified water body, various quantities have been proposed, based on potential energy integrated from lake bottom $z_b$ to the surface $z_s$ (e.g., Birge work). We list the definition of Schmidt stability $S_t$ as the most often used reference for the work required for mixing a stratified lake:

$$S_t = \int_{z_b}^{z_s} (z - z_V)(\rho(z) - \bar{\rho})A(z)\mathrm{d}z$$
[10]

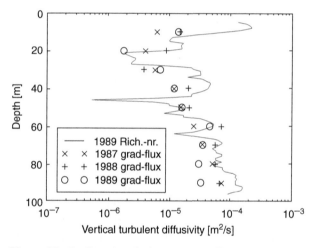

**Figure 13** Profiles of vertical transport coefficients in Lake Constance during the stratification period, based on gradient Richardson number measurements in the area of high shear at the Sill of Mainau (solid line), in comparison with results of the gradient flux method for the entire lake based on the evolution of temperature profiles in Überlinger See in years 1987, 1988 or 1989. Adapted from Boehrer et al. (2000) *Journal of Geophysical Research* 105(C12): 28,823.

where $z = z_V = \frac{1}{V}\int_{z_b}^{z_s} zA(z)\mathrm{d}z$ is the vertical position of the centre of lake volume $V$, and $\bar{\rho}$ is the density of the hypothetically homogenized lake.

In a two-layer system, like thermally stratified lakes, it is reasonable to compare the potential energy needed for vertical excursion with the wind stress applied to the surface as done by the Wedderburn number $W$:

$$W = \frac{g' h_{\text{epi}}^2}{u_*^2 L}$$
[11]

where $g' = g\frac{\Delta\rho}{\rho}$, with $\Delta\rho$ representing the density difference between epilimnion and hypolimnion, $u_*^2 = \tau/\rho$ is the friction velocity resulting from the surface stress $\tau$ implied by the wind, and $L$ stands for the length of the fetch.

Although the common use of the Wedderburn number is connected to its simplicity, the more sophisticated Lake number $L_N = M_{bc}/\left(z_V \int_A \tau dA\right)$ compares the wind stress applied to the lake surface with the angular momentum $M_{bc}$ needed for tilting the thermocline, and hence represents the integral counterpart of the Wedderburn number.

Small values of both, $W$ and $L_N$, indicate that wind stress can overcome restoring gravity forces because of density stratification. Under such conditions, upwelling of hypolimnion water is possible and intense mixing of hypolimnion water into the epilimnion can be expected. A typical consequence of ecological importance facilitated by this process is the recharging of nutrients in the epilimnion from the hypolimnion.

## Nomenclature

| | |
|---|---|
| $a, a_i, b, b_i$ | coefficients |
| $A$ | area, especially surface area of a lake (m$^2$) |
| $c_p$ | specific heat (J K$^{-1}$ kg$^{-1}$) |
| $C$ | electrical conductivity (mS cm$^{-1}$) |
| $C_n$ | concentration of substance (g kg$^{-1}$) |
| $D$ | vertical turbulent diffusivity (m$^2$ s$^{-1}$) |
| $h_{epi}$ | thickness of epilimnion (m) |
| $g$ | acceleration due to gravity (m$^2$/s) |
| $g'$ | reduced acceleration due to gravity (m$^2$/s) |
| $L$ | length of lake or wind fetch (m) |
| $L_N$ | lake number |
| $M_{bc}$ | angular momentum (N m) |
| $N$ | stability frequency (s$^{-1}$) |
| $Ri$ | gradient Richardson number |
| $Ri_{crit} = 0.25$ | critical gradient Richardson number |
| $[\ ]_{ref}, [\ ]_{25}$ | at reference temperature, mostly 25°C |
| $S$ | salinity for fresh water or ocean conditions (psu) |
| $S_t$ | Schmidt stability (kg m) |
| $T$ | (in situ) temperature (°C) |
| $T_{md}$ | temperature of maximum density (°C) |
| $T_{ref}$ | reference temperature (°C) |
| $u$ | horizontal current velocity (m s$^{-1}$) |
| $u_*$ | friction velocity (m s$^{-1}$) |
| $V$ | lake volume (m$^3$) |
| $W$ | Wedderburn number |
| $z$ | vertical coordinate (m) |
| $z_b, z_s, z_V$ | vertical coordinate of lake bed, surface, centre of volume (m) |

| | |
|---|---|
| $\alpha = \left(\frac{\partial\rho}{\partial T}\right)_{S,p}$ | thermal expansion (K$^{-1}$) |
| $\alpha_{ref}, \alpha_{25}$ | coefficient at reference temperature, at 25 °C (K$^{-1}$) |
| $\beta_n$ | coefficient for specific density contribution of salts |
| $\gamma$ | conductivity specific (potential) density contribution (kg m$^{-3}$ mS$^{-1}$ cm) |
| $\Gamma$ | (potential) density contribution by dissolved substances (kg m$^{-3}$) |
| $\kappa_{ref}, \kappa_{25}$ | electrical conductance at reference temperature, at 25 °C (mS cm$^{-1}$) |
| $\rho$ | (potential) density (kg m$^{-3}$) |
| $\rho_{in\ situ}$ | (potential) density (kg m$^{-3}$) |
| $\rho_T$ | (potential) density of pure water (kg m$^{-3}$) |
| $\bar{\rho}$ | reference density (kg m$^{-3}$) |
| $\Theta$ | potential temperature (K) |
| $\tau$ | surface stress (Pa) |
| $\omega$ | wave frequency (rad/s) |

*See also:* The Benthic Boundary Layer (in Rivers, Lakes, and Reservoirs); Currents in Stratified Water Bodies 1: Density-Driven Flows; Currents in Stratified Water Bodies 2: Internal Waves; Currents in Stratified Water Bodies 3: Effects of Rotation; Currents in the Upper Mixed Layer and in Unstratified Water Bodies; Effects of Climate Change on Lakes; Lakes as Ecosystems; Meromictic Lakes; Mixing Dynamics in Lakes Across Climatic Zones; Paleolimnology; Saline Inland Waters; Small-Scale Turbulence and Mixing: Energy Fluxes in Stratified Lakes.

## Further Reading

Boehrer B and Schultze M (2008) Stratification of lakes. *Reviews in Geophysics.* 46, RG2005, doi:10.1029/2006RG000210.

Boehrer B Ilmberger J, and Münnich K (2000) Vertical structure of currents in western Lake Constance. *Journal of Geophysical Research* 105(C12): 28,823–28,835.

Chen CTA and Millero FJ (1986) Precise thermodynamic properties for natural waters covering only the limnological range. *Limnology and Oceanography* 31: 657–662.

Gat JR (1995) Stable isotopes of fresh and saline lakes. In: Lerman A, Imboden D, and Gat J (eds.) *Physics and Chemistry of Lakes*, pp. 139–165. Berlin: Springer-Verlag.

Halbwachs M, Sabroux J-C, Grangeon J, Kayser G, Tochon-Danguy J-C, Felix A, Beard J-C, Vilevielle A, Vitter C, Richon P, Wüest A, and Hell J (2004) Degassing the 'Killer Lakes' Nyos and Monoun, Cameroon. *EOS* 85(30): 281–284.

Hongve D (2002) *Endogenic Meromixis: Studies of Nordbytjernet and Other Meromictic Lakes in the Upper Romerike Area*. PhD thesis, Norwegian Institute of Public Health, Oslo Norway.

Hutchinson GE (1957) *A Treatise on Limnology* vol. 1. New York: Wiley.

Imberger J and Patterson JC (1990) Physical limnology. *Advances in Applied Mechanics* 27: 303–475.

Imboden DM and Wüest A (1995) Mixing mechanisms in lakes. In: Lerman A, Imboden D, and Gat J (eds.) *Physics and Chemistry Lakes*, pp. 83–138. Berlin, Germany: Springer-Verlag.

ISO standard 7888 (1985) Water quality: Determination of electrical conductivity. International Organization for Standardization. www.iso.org.

Jöhnk KD (2000) 1D hydrodynamische Modelle in der Limnophysik – Turbulenz, Meromixis, Sauerstoff. Habilitationsschrift, Darmstadt, Germany, Technical University of Darmstadt.

Kalff J (2002) *Limnology*. Upper Saddle River, NJ: Prentice Hall.

Kjensmo J (1994) Internal energy, the work of the wind, and the thermal stability in Lake Tyrifjord, southeastern Norway. *Hydrobiologia* 286: 53–59.

Sorenson JA and Glass GE (1987) Ion and temperature dependence of electrical conductance for natural waters. *Analytical Chemistry* 59(13): 1594–1597.

Stevens CL and Lawrence GA (1997) Estimation of wind forced internal seiche amplitudes in lakes and reservoirs, with data from British Columbia, Canada. *Aquatic Science* 59: 115–134.

von Rohden C and Ilmberger J (2001) Tracer experiment with sulfur hexafluoride to quantify the vertical transport in a meromictic pit lake. *Aquatic Sciences* 63: 417–431.

## Relevant Websites

http://www.ilec.or.jp/ – International Lake Environment Committe; data on various lakes on Earth.

http://www.ioc.unesco.org/ – Intergovernmental Oceanographic Commission of UNESCO; provides on-line calculator for salinity following the so-called UNESCO formula.

http://www.cwr.uwa.edu.au/ – Centre for Water Research (CWR) at The University of Western Australia. Research Institution focussing of physical limnology.

http://www.eawag.ch/ – Swiss Federal Institute of Aquatic Science and Technology. Institution dealing with water related issues.

# Small-Scale Turbulence and Mixing: Energy Fluxes in Stratified Lakes

**A Wüest,** Eawag, Surface Waters – Research and Management, Kastanienbaum, Switzerland

**A Lorke,** University of Koblenz-Landau, Landau/Pfaly, Germany

## Introduction

### Density Stratification and Mixing – the Basin Scale

Nearly all lakes, reservoirs, and ponds that are deeper than a few meters, experience cycles of density stratification and destratification. Most important for this variation is the temperature-dependence of water density. During spring/summer – or the wet season in the tropics – the water is heated from above and a surface layer (SL: typically a few m thick) with warmer and hence lighter water develops on top of the cooler and heavier water below (**Figure 1**). In addition, although more important in saline lakes than freshwater ones, biological and hydrological processes may strengthen the density stratification by generating a vertical gradient in the concentration of dissolved substances (salinity). The resulting stratification is usually depicted by a strong density gradient (also called pycnocline), separating the SL from the deeper reaches of the water column (indicated as metalimnion and hypolimnion in **Figure 1**). Mixing of heavier water from greater depth with lighter water from the SL implies that water parcels of different densities are exchanged in the vertical direction (**Figure 2**). It is evident that mechanical energy is needed to move these water parcels against the prevailing density gradient, which forces lighter water up and heavier water down. The amount of energy needed to overcome vertical density stratifications is therefore determined by the potential energy $\Delta E_{pot}$ (**Figure 1**) stored in the stratification. $\Delta E_{pot}$ is calculated from the vertical separation of the centre of volume of the water body and its center of mass. Density stratification results in a lowering of the centre of mass by the vertical distance $\Delta h_M$ (**Figure 1**) and the energy needed to overcome the stratification and to mix the entire water column is $\Delta E_{pot} = H\rho g\Delta h_M$ (J m$^{-2}$), where $H$ is the average depth of the water body, $g$ is the gravitational acceleration, and $\rho$ is the density. Density stratification thus imposes stability on the water column and reduces – or even suppresses – vertical mixing.

Besides convective mixing in the SL – caused by seasonal or nocturnal surface cooling – in most lakes and reservoirs, the major source of energy for vertical mixing is the wind, whereas river inflows usually play a minor role (**Figure 1**). As water is 800 times denser than air and as momentum is conserved across the air–water interface, SLs receive only about 3.5% of the wind energy from the atmosphere above. Surface waves transport a portion of this energy to the shore where it is dissipated; the remaining energy causes large-scale currents, with surface water flows of 1.5–3% of the wind velocity. Moreover, surface currents cause a stratified water body to pivot with warm water piling up at the downwind end (causing downwelling) and deep-water accumulating at the upwind end (causing upwelling). After the wind ceases, the water displacement relaxes and various internal waves develop – including basin-scale seiches – inducing motion even in the deepest layers.

These deep-water currents are usually one order of magnitude less energetic than those in the SL. Typical deep flows of a few centimeters per second (or ~1 J m$^{-3}$) with energy dissipation of less than 1 mW m$^{-2}$ are able to reduce the potential energy of the stratification by only ~0.01–0.05 mW m$^{-2}$. Compared to the potential energy stored in the stratification (order of 1000 J m$^{-2}$; **Figure 1**) it would take much longer than one season to entirely mix a moderately deep lake. This implies that wind energy input (**Figure 1**) forms the vertical hypolimnion structure at times of weak stratification (beginning of the season), whereas the wind is not able to substantially change the vertical structure once the strong stratification is established. Therefore, in most regions on Earth, only very shallow waters (less than a few meters deep (such as Lake Balaton, Hungary) are found to be entirely nonstratified, even during the summer season. The majority of lakes and reservoirs deeper than a few meters are thus only 'partially' mixed to a limited depth, which is basically defining the SL. For those lakes that show a pronounced SL, its maintenance is mostly supported by night-time cooling. In this article, we focus on the 'limited' mixing below the SL, which occurs in the metalimnion and hypolimnion (**Figure 1**).

### Density Stratification and Mixing – the Small Scale

The same concepts of stability and mixing – described in the preceding section for the entire water body – also apply locally within the water column for small-scale vertical mixing of stratified layers. Local stability of the density stratification is quantified by the Brunt-Väisälä frequency (also buoyancy frequency) $N$ (s$^{-1}$), defined by:

$$N^2 = -\frac{g}{\rho}\frac{\partial\rho}{\partial z} \qquad [1]$$

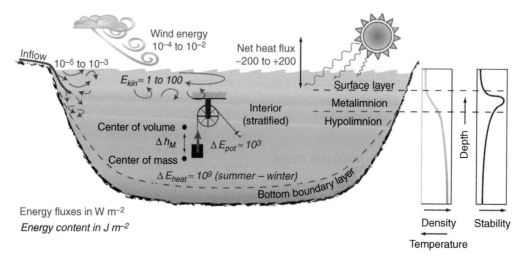

**Figure 1** Energy fluxes (heat, wind, and river inflow; in red) into the water ($W m^{-2}$) and energy content (heat, kinetic energy, potential energy; in blue) stored in the lake water body ($J m^{-2}$). Note that the energy fluxes and contents related to heat are many orders of magnitude larger than those of kinetic and potential energy. The effect of mixing by the river is only local and less effective than wind. The stratified part of the lake (below surface layer) has historically been divided into the metalimnion (see large stability, right) and the deep hypolimnion (weak stratification) below. The lower water column can also be differentiated into an interior region (away from the boundaries) which is quiescent except during storms and a bottom boundary layer where turbulence is enhanced. Adopted from Imboden DM and Wüest A (1995) Mixing mechanisms in lakes. In: Lerman A, Imboden D, and Gat JR (eds.) *Physics and Chemistry of Lakes*, vol. 2, pp. 83–138. Berlin: Springer-Verlag.

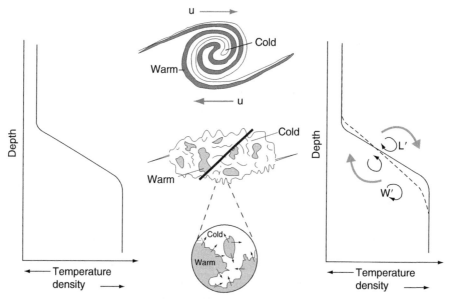

**Figure 2** The effect of turbulent mixing in a stable stratification: if the vertical gradient of horizontal currents (current shear $\partial u/\partial z$) is stronger then the stability of the water column (eqn. [1]), Kelvin Helmholtz instabilities can develop (top of middle panel) bringing warmer (lighter) and cooler (heavier) water in close proximity (bottom of middle panel). Finally, heat (or any other water constituent) is mixed by molecular diffusion across the manifold small-scale interfaces, which are generated by turbulence. The turbulent exchange of small water parcels leads to a fluctuating vertical heat flux (see example in **Figure 3**) which averages to a net downward heat flux. As a result, the original temperature profile (left) is modified (right): the gradient is weakened and expanded vertically with heat transported from top to bottom, and density vice versa, across the interface. Figure after the idea of Winters KB, Lombard PN, Riley JJ, and D'asaro EA (1995) Available potential-energy and mixing in density-stratified fluids. *Journal of Fluid Mechanics* 289: 115–128. Experiments were first performed by Thorpe SA (1973) Experiments on instability and turbulence in a stratified shear-flow. *Journal of Fluid Mechanics* 61: 731–751; and the phenomenon of sheared stratification in lakes was reported by Mortimer CH (1952) Water movements in lakes during summer stratification; evidence from the distribution of temperature in Windermere. *Philosophical Transactions of the Royal Society of London B: Biological Sciences* 236(635): 355–398; and by Thorpe SA (1977) Turbulence and Mixing in a Scottish loch. *Philosophical Transactions of the Royal Society of London A: Mathematical Physics and Engineering Sciences* 286(1334): 125–181.

$z$ is the depth (positive upward). As a result of wind-forced motions, a vertical gradient of the horizontal current $u$ (shear $\partial u/\partial z$) is superimposed on the vertical density gradient $\partial\rho/\partial z$. Depending on the relative strength of $N$ compared to the current shear $\partial u/\partial z$ such a stratified shear flow may eventually become unstable and develop into turbulence (**Figure 2**).

Although the large-scale (advective) motions are mainly horizontal, the turbulent eddies are associated with random velocity fluctuations in all three dimensions ($u'$, $v'$, $w'$). Turbulent kinetic energy (TKE) ($J\,kg^{-1}$) is defined as the energy per unit mass of water which is contained in these velocity fluctuations:

$$\mathrm{TKE} = \tfrac{1}{2}\left(\overline{u'^2} + \overline{v'^2} + \overline{w'^2}\right) \qquad [2]$$

In stratified turbulence, the vertical velocity fluctuations $w'$ are of particular importance as they transport water parcels and their contents in the vertical direction (**Figure 3**). The product of the vertical velocity fluctuations $w'$ and the associated density fluctuations $\rho'$ describes an instantaneous vertical flux of

density ($\overline{w'\rho'}$ ($kg\,m^{-2}\,s^{-1}$)). Resulting from many irregular and uncorrelated fluctuations (**Figure 3**) the averaged flux $\overline{w'\rho'}$ leads to a net upward mass flux, which is usually expressed as a buoyancy flux $J_b$:

$$J_b = -\frac{g}{\rho}\overline{w'\rho'} \qquad [3]$$

Therefore, we can interpret vertical mixing as an upward flux of mass, which causes a change of the potential energy of the stratification (**Figures 1** and **2**), expressed as a buoyancy flux (eqn. [3]). The required energy originates from the TKE, which is itself extracted from the mean (horizontal) flow. Approximately 90% of the TKE, however, does not contribute to the buoyancy flux (and hence to vertical mixing) but is instead dissipated into heat by viscous friction, without any further effect. By defining local rates of production $P$ ($W\,kg^{-1}$) and viscous dissipation $\varepsilon$ ($W\,kg^{-1}$) of TKE, the simplest form of TKE balance can be formulated as:

$$\frac{\partial}{\partial t}\mathrm{TKE} = P - \varepsilon - J_b \qquad [4]$$

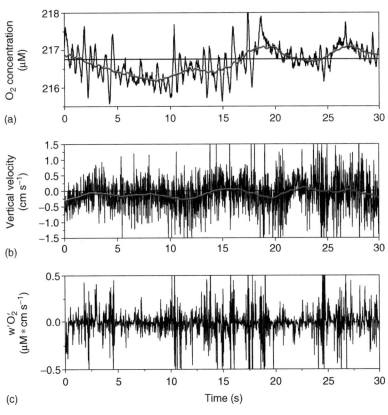

**Figure 3** Time series of $O_2$ concentration (thin line, a) and vertical velocity $w'$ (thin line, b; positive = upward), as measured 10 cm above the sediment in reservoir Wohlensee (Switzerland) at a frequency of 64 Hz. Red lines indicate the temporally varying averages, determined as running mean, whereas the black horizontal line marks the averages. Panel (c) shows the instantaneous eddy flux – covariance of $w'$ and $O_2'$: The average downward $O_2$ flux over the 30 s (~1900 data pairs) is $-6.4\,mmol\,m^{-2}\,day^{-1}$. Data source: Claudia Lorrai, Eawag.

As mentioned above, the dissipation rate is usually much larger than the buoyancy flux, and hence the mixing efficiency $\gamma_{mix}$, which is defined as the ratio

$$\gamma_{mix} = \frac{\mathcal{J}_b}{\varepsilon} = \frac{-g\rho^{-1}\overline{w'\rho'}}{\varepsilon} \qquad [5]$$

is much smaller than 1. A number of studies in stratified lakes and reservoirs have revealed typical mixing efficiencies in the range of 10–15%.

### Density Stratification and Mixing – the Turbulent Transport

The local flux of a water constituent is given by the product of the velocity times the concentration. In stratified waters, the time-averaged vertical velocity is often close to zero (negligible) and thus, the vertical fluxes stem only from the fluctuations of velocity and concentration, such as explained above for the vertical mass flux $\overline{w'\rho'}$ caused by the turbulence. This concept holds for any other water constituent, such as for oxygen, as exemplified in **Figure 3**, where the in situ measured $w'$, $O_2'$ and the product $w'O_2'$ is shown for a 30-s-long record. Although the momentary fluxes up and down are almost of equal variations and amounts, the averaging $\overline{w'O_2'}$ reveals slightly larger fluxes downwards to the sediment, where the oxygen is consumed.

Until recently, direct measurements of turbulent fluxes had not been possible and therefore turbulent fluxes in stratified waters are commonly expressed using the eddy diffusivity concept. Applied to the mass flux $\overline{w'\rho'}$ it implies assuming that (i) a well-defined local density gradient $\partial\rho/\partial z$ exists (due to the stratification) and (ii) the flux – in analogy to molecular diffusion – can be expressed by the eddy (or turbulent) diffusivity $K_z$ ($m^2\,s^{-1}$) multiplied by this local gradient:

$$\overline{w'\rho'} = -K_z\frac{\partial\rho}{\partial z} \qquad [6]$$

In this formulation, $K_z$ describes the vertical transport of density caused by turbulent velocity fluctuations $w'$ over a typical eddy distance $L'$ given by the level of turbulence and the strength of the stratification. Therefore, in contrast to the molecular diffusion process, eddy diffusivity is neither a function of medium (water) nor of the water constituents (particulate or dissolved), but rather a property of the turbulent flow within the stratified water itself. In particular, $K_z$ reflects the extent of the velocity fluctuations $w'$ and the eddy sizes $L'$: $K_z$ can be interpreted as the statistical average $\overline{w'L'}$ of a large number of eddies, which exchange small water parcels as a result of the turbulent flow (**Figures 2 and 3**).

In addition to density, all other water properties – such as temperature or substances – are transported and mixed in the same way via the turbulent exchange of small eddies or parcels of water (**Figure 3**). The eddy diffusivity concept can be applied to any dissolved or particulate substance and the associated vertical fluxes $F$ can be readily estimated in analogy to eqn. [6] by

$$F = -K_z\partial C/\partial z \qquad [7]$$

where $C$ is the appropriate concentration.

Assuming steady-state conditions, i.e., by neglecting the left-hand side of eqn. [4], and combining eqns. [1], [5], and [6] yields:

$$K_z = \gamma_{mix}\frac{\varepsilon}{N^2} \qquad [8]$$

This equation provides an expression to estimate $K_z$ from field measurements of $\varepsilon$ and $N^2$ and, moreover, it demonstrates the direct proportionality of $K_z$ on the level of turbulence ($\varepsilon$) and the inverse proportionality on the strength of stratification ($N^2$). In the last decades, two fundamentally different approaches have been used for the estimation of $K_z$: (i) the microstructure method and (ii) the tracer method. Method (i), is based on eqn. [8] where the dissipation of TKE or of temperature variations are measured by usually free-falling profilers which measure either temperature or velocity over small spatial scales. For example, spectral analysis of the temperature gradient signal provides estimates of $\varepsilon$ and the local buoyancy frequency is obtained from density computed from the temperature and salinity profiles. For the application of tracer method (ii), one has to measure the three-dimensional spreading of a tracer (artificial or natural) and then infer the diffusivities ($K_z$) from the observations. Heat is also used as a tracer and $K_z$ is obtained by computing a time series of the heat budget below the respective depth of a lake. Typical values in stratified natural waters are listed in **Table 1** and **Figures 4 and 5**.

Turbulence is caused by current shear, breaking surface waves, and instabilities in the internal wave field. Currents induce shear near boundaries regardless of whether the flow is stratified. Thus, the concept of eddy diffusivity is also applied to surface mixing layers and to nonstratified systems such as rivers.

## Turbulence and Mixing in Stratified Lakes and Reservoirs

### Turbulence Production in the Surface and Bottom Boundaries

There are fundamentally two mechanisms generating turbulence in the SL: (i) the action of wind causing

**Table 1** Typical values of dissipation, stability and vertical diffusivity in stratified waters

|  | Dissipation[a] $\varepsilon$ $(W\,kg^{-1})$ | Stability $N^2$ $(s^{-2})$ | Diffusivity* $K_z$ $(m^2\,s^{-1})$ |
| --- | --- | --- | --- |
| Ocean thermocline | $10^{-10}$–$10^{-8}$ | $\sim 10^{-4}$ | $(0.3$–$3) \times 10^{-5}$ |
| Surface layer | $10^{-6}$–$10^{-9}$ | $0$–$\sim 10^{-5}$ | $10^{-5}$–$10^{-2}$ |
| Lake interior only (without BBL) | $10^{-12}$–$10^{-10}$ | $10^{-8}$–$10^{-3}$ | $10^{-7}$–$10^{-5}$ |
| Metalimnion (basin scale) | $10^{-10}$–$10^{-8}$ | $\sim 10^{-3}$ | $(0.5$–$50) \times 10^{-7}$ |
| Near-shore metalimnion | $10^{-10}$–$10^{-6}$ | $\sim 10^{-3}$ | $(0.3$–$3) \times 10^{-4}$ |
| Deep hypolimnion (basin scale) | $10^{-12}$–$10^{-10}$ | $10^{-8}$–$10^{-6}$ | $(0.03$–$3) \times 10^{-4}$ |

[a]During storm events values are larger by orders of magnitudes for short.

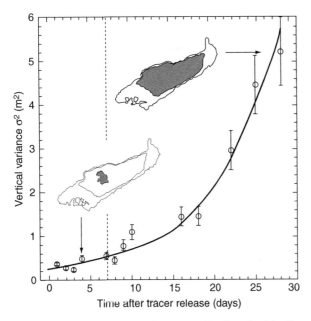

**Figure 4** Vertical spreading of the tracer Uranine after injection at 25 m depth in Lake Alpnach (Switzerland). The vertical line demarcates the initial period of 7 days, during which Uranine resided in the interior of the stratified deep water. The two insets show the lake area at the surface and at the depth of the Uranine injection, as well as the horizontal distribution of the Uranine cloud (shaded in gray) after 4 and 28 days. The slow growth of the spreading in the first 7 days illustrates the quietness in the interior. The fast growth of the vertical spreading after day 7 is due to the increasing contribution of BBL mixing after the tracer has reached the sediment at 25 m depth. Reproduced from Goudsmit GH, Peeters F, Gloor M, and Wüest A (1997) Boundary versus internal diapycnal mixing in stratified natural waters. *Journal of Geophysical Research* 102: 27903–27914, with permission from American Geophysical Union.

wave breaking and shear in the top few meters of the water column and (ii) surface cooling causing the sinking of heavier water parcels. Temperature-driven mixing (case (ii)) leads to homogenization of the SL and therefore to nonstratified conditions – at least for a few hours or days before heat fluxes from/to the atmosphere restratify the SL. This process is discussed in detail elsewhere in this encyclopedia. Only in shallow ponds or basins with relatively high through-flow will turbulence have other case-specific sources.

For wind-driven mixing (case (i)), the crucial parameter governing the dynamics of turbulence in the SL is the surface shear stress $\tau$ $(N\,m^{-2})$, the force per unit area exerted on the water by the wind. This stress is equal to the downward eddy-transport of horizontal momentum from the atmosphere. Part of $\tau$ is consumed in the acceleration and maintenance of waves $(\tau_{Wave})$, whereas the remaining momentum flux $\tau_{SL}$ generates currents and turbulence in the SL. By assuming a constant stress across the air–water interface, the two momentum fluxes on the water side equal the total wind stress $(\tau = \tau_{SL} + \tau_{Wave})$.

Immediately below the waves, the momentum flux, $\tau_{SL}$, drives the vertical profiles of horizontal velocity $u(z)$ in the SL. If the wind remains relatively constant for hours, quasi-steady-state conditions may develop in the SL: $u(z)$ then depicts the Law-of-the-Wall $\partial u/\partial z = u_*(\kappa z)^{-1} = (\tau_{SL}/\rho)^{1/2}(\kappa z)^{-1}$, where $u_* = (\tau_{SL}/\rho)^{1/2}$ is the frictional velocity and $\kappa$ $(= 0.41)$ is the von Karman constant. Because the buoyancy flux in the SL (defined in eqn. [3]) is not a large contribution in eqn. [4], we can assume a balance between the production of TKE and the rate of viscous dissipation $(\varepsilon)$ of TKE. This local balance between production and dissipation of turbulence determines the turbulence intensity as a function of depth throughout the SL. Under those assumptions, the dissipation

$$\varepsilon = (\tau_{SL}/\rho)\,\partial u/\partial z = u_*^3(\kappa z)^{-1} \qquad [9]$$

is only a function of the wind-induced stress $\tau_{SL}$ (here expressed as $u_*$) and of depth $z$. Several experiments have demonstrated that dissipation is indeed inversely proportional to depth (eqn. [9]), if averaged for long enough. However, one has to be critical about the validity of eqn. [9] for two reasons: First, at the very top of the water column, breaking waves, in addition to shear stress, produce a significant part of the turbulence in the SL. This additional TKE generation at the surface can be interpreted as an injection of TKE from above. Therefore, in the uppermost layer, the turbulence exceeds the level described by eqn. [9], depending on the intensity of the wave breaking. Second,

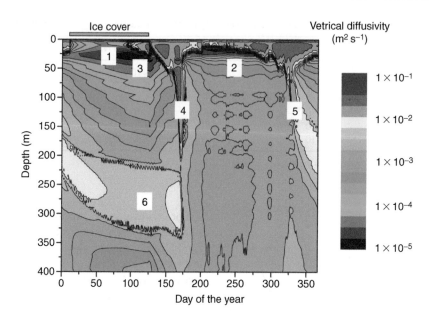

**Figure 5** Vertical diffusivities in Lake Baikal simulated with a k-epsilon model. The numbers (1–6) on the contour plot indicate the main features of the seasonal stratification and changes in diffusivity: the formation of thermal stratification with weak mixing (1) during winter under the ice and (2) during summer; (3) the formation of a convectively mixed layer in spring under the ice; the deep convective mixing in (4) June and (5) November; and (6) the formation of a mixed layer near the temperature of maximum density. Here the emphasis is on the temporal and vertical structure of the turbulent diffusivity and not on the absolute accuracy, which may be difficult to achieve with turbulence modeling better than a factor of 2–3. Reproduced from Schmid M *et al.* (2007) Sources and sinks of methane in Lake Baikal: A synthesis of measurements and modeling. *Limnology and Oceanography 52:* 1824–1837, with permission from American Society of Liminology and Oceanography.

eqn. [9] relies on quasi-steady-state conditions which may hold applicable for limited episodes only.

Despite these restrictions, eqn. [9] gives a good estimate of the diffusivity in the SL, if it is weakly stratified. Equations [8] and [9] reveal that the rate of mixing increases substantially within the SL as the surface is approached. The corresponding stability $N^2$ decreases at the surface and maintains rapid mixing. Therefore, gradients of temperature, nutrients, and particulates are usually smallest at the surface and increase with depth. During sunny days, diurnal thermoclines form with mixing reduced below them. On cloudy, windy days, the SL may mix fully and may even deepen depending upon the surface forcing. Factors that affect the depth of mixing are discussed elsewhere in this encyclopedia. It is typically a few m during the warm season and a few tens of meters during the cold season. Below, a strong density gradient (pycnocline) can develop leading to the separation between the SL and the metalimnion/hypolimnion. In the stratified interior (away from the BBL; see below), the effect of wind is shielded and the mixing regime is completely different.

As discussed in greater detail (see **The Benthic Boundary Layer (in Rivers, Lakes, and Reservoirs)**), turbulence generation and mixing along the bottom boundaries of water bodies can be described in analogy to the SL. Under steady-state conditions, the resulting bottom boundary layer (BBL) follows

a similar vertical structure of (i) current shear (see above), (ii) rate of TKE dissipation (eqn. [9]) and (iii) rate of vertical mixing. Although the original indirect driving force for turbulence in the BBL is also the wind, it is not the direct turbulent momentum flux from the atmosphere to the water which is the cause. Rather, the mechanism is indirectly induced by wind which causes large-scale currents and basin-wide internal waves (such as seiches) which act as intermediate energy reservoirs that generate TKE by bottom friction. Along sloping boundaries in particular, the breaking of propagating internal waves and convective processes – a secondary effect of bottom friction – can produce additional TKE, leading to dissipation and mixing in excess of that predicted by eqn. [9]. As with the SL, the BBL is also usually partly (and weakly) stratified. Again, mixing (eqn. [8]) increases substantially when approaching the sediment and often a completely homogenized layer a few m thick develops at the bottom.

### Internal Waves and Turbulence in the Stratified Interior

In the lake interior, away from surface and bottom boundaries (**Figure 1**), the water body is stratified and quiescent, and it does not feel the direct effects of the turbulence sources at the surface and above the sediment. This stratified interior consists of an upper region, the metalimnion where gradients in temperature and density are strongest, and a lower region, the hypolimnion, which is only weakly stratified and most water properties are homogeneous. Internal waves are prevalent.

The rate of mixing in the interior water body is low because (i) currents and shear are weak and the resulting turbulence production is reduced and (ii) stratification suppresses the turbulent mixing. The mechanical energy originates mainly from basin-scale internal currents and waves (see above), whereas the waves of smaller scale and higher frequencies – potentially generated at a few specific locations – are not contributing much to the energy budget of the deep-water. At the transition between small- and large-scale waves are the near-inertial currents, which can carry – especially in large lakes – a significant portion of the mechanical energy typically in the order of $\sim 1\,\mathrm{J\,m^{-3}}$. Given that observed energy residence time-scales are days (small lakes) to weeks (deepest lakes), the dissipation of the internal energy is $\sim 10^{-12}$–$\sim 10^{-10}\,\mathrm{W\,kg^{-1}}$ (**Table 1**). Considering typical values for stratification $N^2$ (eqn. [1]) of $10^{-8}$–$10^{-3}\,\mathrm{s^{-2}}$ and $\gamma_{\mathrm{mix}} \approx 0.1$ (eqn. [5]), interior diffusivities of $10^{-7}$–$10^{-5}\,\mathrm{m^2\,s^{-1}}$ can be expected (**Table 1**; **Figure 4**). The stratified interior – away from the SL and the BBL – is by far the most quite zone in lakes.

Important for the generation of small-scale mixing are local instabilities related to internal (baroclinic) motions, such as illustrated in **Figure 2**. Instabilities occur mostly where the usually weak background shear is enhanced by nonlinear steepening of internal waves or by superposition of the shear with small-scale propagating internal waves.

Direct observations of turbulence and mixing, using microstructure and tracer techniques, confirm that turbulence is indeed very weak in the stratified interior. Typically, only a few percent of the water column is found to be actively mixing. The occurrence of such turbulent patches is highly intermittent in space and time. During periods when the fluid is nonturbulent, we can expect laminar conditions and thus the dominance of molecular transport. The observable average diffusivity can be considered the superposition of a few turbulent events separated by molecular diffusion for most of the time. The resulting transport in the stratified interior will therefore be close to molecular. Tracer experiments and microstructure profiling conducted in small and medium-sized lakes confirm these quiet conditions in the interior and enhanced turbulence in the bottom boundary. In **Figure 4** the vertical spreading of a tracer, injected into the hypolimnion, is shown for the interior (first few days) and for a basin-wide

volume including the BBL (after a few days). From **Figure 4** it is evident that turbulent diffusivity in the interior is at least one order of magnitude lower than in the basin-wide deep-water volume, including the bottom boundary. In addition to these spatial differences, one has to be aware of the temporal variability. During storms, turbulence can be several orders of magnitude larger for short episodes. The transition from quiescent to actively mixing occurs rapidly once winds increase above a certain threshold relative to the stratification. The internal wave field is energized and turbulence can develop. But the greatest increases occur in the benthic BBL. It is during such storms that most of the vertical flux takes place.

The turbulent patches, where vertical fluxes are generated (as exemplified in **Figure 3**) vary in size depending in part upon the turbulence intensity $\varepsilon$ and the stratification $N^2$. Several length scales have been developed to characterize the sizes of turbulent eddies. One is the Ozmidov scale

$$L_{\mathrm{O}} = (\varepsilon / N^3)^{1/2} \qquad [10]$$

and the other is the Thorpe scale, $L_{\mathrm{T}}$, which is based on direct observations of the size of unstable regions. The ratio of the two numbers varies depending upon the strength of stratification and is useful for predicting the efficiency of mixing, $\gamma_{\mathrm{mix}}$ in eqn. [5] Typical values for $L_{\mathrm{O}}$ and $L_{\mathrm{T}}$ range from a few centimeters to a meter but for weak stratification eddies are larger and on scales of tens of meters to 100 m as found in weakly stratified Lake Baikal.

The spatial and temporal dynamics of mixing challenges not only the experimental estimation, but also the numerical simulation of its net effect, in terms of a turbulent diffusivity $K_z$. Local measurements of $K_z$ following eqn. [8] often neither resolve its spatial nor its temporal dynamics and the coarse grid sizes used in numerical simulations do not capture the small scales relevant for mixing processes in the interior.

## Turbulent Energy Flux through the Water Column – Synthesis

From the discussion above, we can draw the following overall scheme of the energy flux through the stratified waters of a lake. The origin of the energy for turbulent mixing is usually wind, which is imposing momentum onto the surface of the water. Approximately 3% of the wind energy from the atmosphere reaches the epilimnion in the form of horizontal currents and about 10% thereof is finally transferred to the stratified water body underneath. The major part of the energy is dissipated by bottom interaction, and the minor part is dissipated in the interior by shear instabilities and breaking of internal waves. Of this dissipated energy, only about 10% produce buoyancy flux (mixing efficiency $\gamma_{mix}$, eqn. [5]) increasing the potential energy of the stratification. Compared to the wind energy flux in the atmosphere, only a small fraction of ∼0.0003 actually causes the mixing against the stratification in the deep water, whereas the large fraction of ∼0.9997 is dissipated somewhere along the flux path. Although this partitioning depends on many factors, the overall scheme likely holds within a factor of 2–3 based on comparisons between different lakes.

The small amount of energy available for mixing – compared to the potential energy stored in the stratification – explains why lakes deeper than a few meters remain permanently stratified during the warm season. Consistent with this conclusion, the enhanced turbulence in the surface and bottom boundary layers cannot erode the stable – and partly very strong – stratification in the interior. Turbulent patches are intermittent and the eddies within them are small compared to the depth. As an example, the timescales to transport heat, solutes or particulates over a distance of 10 m would be $(10\,m)^2\,K_z^{-1}$; i.e., several years for a $K_z = 10^{-6}\,m^2\,s^{-1}$ in the metalimnion (**Table 1**). Therefore, two-dimensional processes, such as upwelling become important for vertical exchanges as well.

*See also:* The Benthic Boundary Layer (in Rivers, Lakes, and Reservoirs); Currents in Stratified Water Bodies 1: Density-Driven Flows; Currents in Stratified Water Bodies 2: Internal Waves; Currents in Stratified Water Bodies 3: Effects of Rotation; Currents in the Upper Mixed Layer and in Unstratified Water Bodies; Density Stratification and Stability; Mixing Dynamics in Lakes Across Climatic Zones.

## Further Reading

Goudsmit GH, Peeters F, Gloor M, and Wüest A (1997) Boundary versus internal diapycnal mixing in stratified natural waters. *Journal of Geophysical Research* 102(C13): 27903–27914.

Mortimer CH (2005) *Lake Michigan in Motion: Responses of an Inland Sea to Weather, Earth-Spin, and Human Activities.* University of Wisconsin Press. ISBN 978-0299178345.

Imberger J (1998) *Physical Processes in Lakes and Oceans. Coastal Estuarine Studies*, vol. 54. Washington, DC: American Geophysical Union.

Imberger J and Ivey GN (1991) On the nature of turbulence in a stratified fluid. Part II: Application to lakes. *Journal of Physical Oceanography* 21: 659–680.

Imberger J and Patterson JC (1990) Physical limnology. *Advances in Applied Mechanics* 27: 303–475.

Ivey GN, Winters KB, and Koseff JR (2008) Density stratification, turbulence, but how much mixing? *Annual Review of Fluid Mechanics* 40: 169–184.

Kantha LH and Clayson CA (2000) *Small Scale Processes in Geophysical Fluid Flows. International Geophysical Series*, vol. 67. London: Academic Press. ISBN-10: 0-12-434070-9.

Imboden DM and Wüest A (1995) Mixing mechanisms in lakes. In: Lerman A, Imboden D, and Gat JR (eds.) *Physics and Chemistry of Lakes*, pp. 83–138. Berlin: Springer-Verlag.

MacIntyre S, Flynn KM, Jellison R, and Romero JR (1999) Boundary mixing and nutrient flux in Mono Lake, CA. *Limnology and Oceanography* 44: 512–529.

Schmid M, et al. (2007) Sources and sinks of methane in Lake Baikal: A synthesis of measurements and modeling. *Limnology and Oceanography* 52: 1824–1837.

Thorpe SA (2007) *An Introduction to Ocean Turbulence.* Cambridge, UK: Cambridge University Press. ISBN: 978-0-521-85948-6.

Winters KB, Lombard PN, Riley JJ, and D'asaro EA (1995) Available potential-energy and mixing in density-stratified fluids. *Journal of Fluid Mechanics* 289: 115–128.

Wüest A and Lorke A (2003) Small-scale hydrodynamics in lakes. *Annual Review of Fluid Mechanics* 35: 373–412.

Wüest A and Lorke A (2005) Validation of microstructure-based diffusivity estimates using tracers in lakes and oceans. In: Baumert HA, Simpson J, and Sündermann J (eds.) *Marine Turbulence – Theories, Observations and Models.* Cambridge, UK: Cambridge University Press. ISBN: 0521837898.

# The Benthic Boundary Layer (in Rivers, Lakes, and Reservoirs)

**A Lorke,** University of Koblenz-Landau, Landau/Pfaly, Germany
**S MacIntyre,** University of California, Santa Barbara, CA, USA

## Introduction

### Definition and Relevance of the Benthic Boundary Layer

The benthic boundary layer (BBL) of lakes, reservoirs, and rivers constitutes that part of the water column that is directly influenced by the presence of the sediment–water interface. Similar to the surface mixed layer, it represents a hot spot not only of dissipation of kinetic energy, but also of biological activity and of geochemical transformation processes. These different processes are strongly coupled and interact with each other: While the hydrodynamic conditions are modified by biological activity, which changes the structure of the sediment surface, the release of dissolved solids from the sediment can modify the density stratification in the BBL. Moreover, the actual sediment surface cannot always be regarded as rigid since the BBL flow does not only modify the structure of the sediment surface, but it can also bring sediment particles into suspension, whereas at other sites or at other times, the particles resettle.

The BBL definition provided here and the more detailed discussions later explicitly refer to *direct* influences of the sediment surface. From a more general point of view, the BBL is of great importance for the entire water body, almost independent of the dimensions of the basin. Strong turbulence and mixing along the boundaries are known to be important for vertical mixing, and transport on a basin scale and biogeochemical processes at the sediment surface or within the sediment effect the distribution of relevant water constituents on scales much larger than the actual dimensions of the BBL. These larger-scale effects, however, require additional transport processes for energy and matter into or out of the BBL and are considered elsewhere.

A major characteristic of the BBL is the magnitude and the temporal dynamics of the physical forcing, i.e., the current velocity at the top of the BBL. Although in most rivers this forcing can be regarded as a steady-state unidirectional flow, its nature in deep and stratified lakes and reservoirs is more complex. In these, usually stratified, water bodies the major energy is provided by surface waves in the shallow littoral zone, by high-frequency internal waves at the depth of the thermocline and by basin-scale internal waves (seiches or Kelvin and Poincare

waves) in the hypolimnion. Hence, the magnitude and temporal dynamics of the different forcing mechanisms range from current velocities of some $10 \, \text{cm s}^{-1}$ and time scales of seconds for surface waves, to typical current velocities of a few centimeters per second and time scales of several hours to days for basin-scale internal waves (**Figure 1**).

### Structure of the BBL

The BBL can be structured vertically according to the physical processes governing the vertical transport of momentum and solutes (**Figure 2**). In an outer layer (turbulent BBL) up to several meters above the sediment surface, this transport is governed by turbulent eddies and the associated mixing rates are high. While approaching the sediment surface down to scales where viscous forces suppress overturning turbulent motions, the vertical transport of momentum is governed by molecular viscosity and a viscous sublayer with a typical height of $O(1 \, \text{cm})$ develops. The exchange of heat and dissolved solids and gases is eventually controlled by a diffusive sublayer with a height of $O(1 \, \text{mm})$ directly at the sediment–water interface.

## The Transport of Momentum

### The Turbulent BBL

The equation for total average shear stress $\tau$ in a turbulent boundary layer is

$$\tau = \mu \frac{\mathrm{d}u}{\mathrm{d}z} + \rho \overline{u'w'} \qquad [1]$$

where $\mu$ is dynamic viscosity, $\mathrm{d}u/\mathrm{d}z$ is the vertical velocity gradient, $u'$ and $w'$ are the fluctuating horizontal and vertical velocities, and the overbar denotes temporal averaging. While the first term on the right describes viscous shear, the second term is related to momentum transport by turbulent velocity fluctuations. In most aquatic systems, the Reynolds number associated with near-bottom flows is sufficiently high to sustain a turbulent boundary layer. Under such conditions, the first term on the right-hand side of eqn [1] may be negligible and turbulent shear stress is likely to dominate the shear stress computation.

On the basis of dimensional arguments, it can be assumed that the shear stress, $\tau$, on the sediment

115

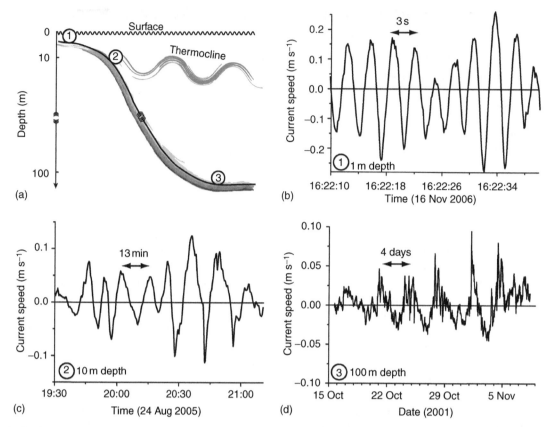

**Figure 1** Typical near-bottom current velocities measured at various locations (depths) in a large Lake (Lake Constance) emphasizing the different periods and magnitudes of BBL forcing. (a) A schematic cross-section of the lake with the three sampling sites indicated by numbers. Near-bottom current velocities induced by surface waves at 1-m depth are shown in panel (b). Typical periods of surface waves in lakes are in the order of seconds. (c) Near-bottom current velocities measured at the depth of the seasonal thermocline (10-m depth). The observed current velocities are driven by propagating internal waves with periods between 8 and 20 min. Near-bottom currents at 100-m depth (d) are mainly driven by basin-scale internal waves. The major period of about four days is associated with a Kelvin wave. Note that several other basin-scale modes of oscillation (e.g., 12 h) are superimposed on this four-day period.

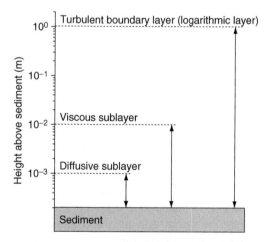

**Figure 2** Idealized structure of the BBL on a flat sediment surface. Note that the heights provided by the logarithmically scaled axis represent order of magnitude estimates for typical conditions found in inland water bodies.

surface is related to the current speed at a certain height above the sediment:

$$\tau = \rho C_D U_{1m}^2 \qquad [2]$$

where $\rho$ is the density, $C_D$ an empirical constant ($\approx 1.5 \times 10^{-3}$), the so-called drag coefficient, and $U_{1m}$ refers to the current speed measured at a height of 1 m above the sediment surface. Note that $C_D$ depends on the reference height where the current speed was measured and a standard height of 1 m is assumed from now on. The bed shear stress $\tau$ is assumed to be constant throughout the boundary layer (constant stress layer) and it is used to define a turbulent velocity scale $u_*$, the so-called friction velocity:

$$u_* = \sqrt{\frac{\tau}{\rho}} \qquad [3]$$

Dimensional analysis can then be used to deduce the velocity distribution $u(z)$ near the sediment surface, where $z$ is the distance from the sediment surface. For a layer far enough from the boundary so that the direct effect of molecular viscosity on the flow can be neglected (the outer layer), the analysis results in

$$u(z) = \frac{u_*}{\kappa} \ln \frac{z}{z_0} \qquad [4]$$

where $\kappa \approx 0.41$ is von Karman's constant and $z_0$ is the roughness length, which is related to the drag coefficient in eqn [2] by

$$C_{1m} = \left( \frac{\kappa}{\ln(1m) - \ln(z_0)} \right)^2 \qquad [5]$$

and will be discussed later.

Equation [4] is called 'law of the wall,' and by assuming a local steady-state equilibrium between production and dissipation of turbulent kinetic energy (TKE), it can be used to estimate the vertical distribution of the turbulence dissipation rate $\varepsilon$

$$\varepsilon = \frac{u_*^3}{\kappa z} \qquad [6]$$

Thus, in analogy to the wind-forced surface layer, the level of turbulence increases with decreasing distance from the boundary. This increasing turbulence leads, again in analogy to the surface mixed layer, to the development of a well-mixed boundary layer of up to several meters in height.

### The Viscous Sublayer

It should be noted that eqn [4] is strictly valid for turbulent flows only for which the vertical transport is governed by cascading turbulent eddies. The maximum (vertical) size of the turbulent eddies is determined by the distance from the sediment surface, and by approaching the sediment surface down to scales where overturning turbulent motions are suppressed by the effect of molecular viscosity, the momentum transport becomes governed by viscous forces (first term on the right-hand side of eqn [1]). Within this layer, which is called the viscous sublayer, current shear becomes constant and the resulting linear velocity profile can be described by

$$u(z) = \frac{u_*^2}{v} z \qquad [7]$$

On a smooth sediment surface, the viscous sublayer extends to a height $\delta_v$ of about $10v/u_*$, which is comparable to the Kolmogorov microscale describing the size of the smallest turbulent eddies (typically O (1 cm), cf. **Figure 2**).

Since viscosity is reduced to its molecular value, current shear within the viscous sublayer is greater than that in the turbulent layer above (cf. **Figure 3**),

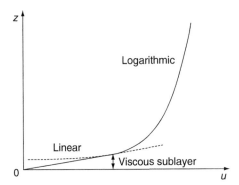

**Figure 3** Velocity distribution above a smooth and rigid bottom (solid line). Within the viscous sublayer the velocity $u$ increases linearly with distance from the sediment surface $z$. Above the viscous sublayer the velocity distribution follows the law of the wall (eqn [4]) and increases logarithmically with distance from the surface. Extrapolated continuations of the linear and logarithmic velocity distributions are shown as dashed lines.

a fact which has major consequences for organisms living within the viscous sublayer on the sediment surface because they have to withstand these strong shearing and overturning forces. It is further interesting to note that an appreciable amount of energy entering the BBL is dissipated within this layer (about 40%).

### Effects of Bottom Roughness

The roughness length $z_0$ in eqn [3] determines the effective height above the bottom $z$ at which the current velocity approaches zero. It is determined by the topographic structure of the sediment surface and hence by the typical height, width, and spacing of individual roughness elements on a stationary bed. When the scale of these roughness elements $z_S$ is on the order of the height of the viscous sublayer $\delta_v$ or less, $z_0$ is solely determined by $\delta_v$ and $z_0 \approx 0.1v/u_*$. This flow regime is called *smooth*. When the size of the roughness elements exceeds $\delta_v$, the flow regime is called *rough* and the corresponding roughness length is given by $z_0 \approx z_S/30$. Note that the drag coefficient $C_D \approx 1.5 \times 10^{-3}$ provided earlier (eqn [2]) corresponds to a roughness length $z_0 \approx 2.5 \times 10^{-5}$ m (eqn [5]) and hence is valid for smooth flows unless $u_*$ exceeds $0.4$ m s$^{-1}$ or $U_{1m}$ exceeds $10$ m s$^{-1}$.

In addition to the shear stress derived from viscous forces as described above (the so-called *skin friction*), larger-scale roughness elements can cause a *form drag*, which results from pressure gradients between the upstream and downstream side of particular roughness elements. Although skin friction is important for the lower part of the turbulent BBL and for the viscous sublayer, form drag resulting from, e.g., ripples, sand waves, or submerged vegetation is more important for the upper part of the turbulent BBL and for the total drag on flow. When form drag is significant, the turbulent BBL may consist of more than one logarithmic layer, described by different roughness lengths $z_0$, respectively.

## Oscillatory Boundary Layers

The turbulent BBL equations derived here are based on steady-state conditions, i.e., on a local balance between production and dissipation of TKE, which is in equilibrium with the applied forcing. As described later, however, many forcing mechanisms for near-bottom flows are related to surface or internal waves and are hence associated with oscillatory flows. Well above the viscous sublayer, such oscillatory BBL show, similar to the effect of form drag, deviations of the of the velocity distribution from its steady-state logarithmic pattern. One major characteristic of oscillatory BBL is a pronounced maximum of the current speed at some decimeters or meters above the bed. The analytical solution to this problem (Rayleigh flow or Stokes' second problem) is an exponentially damped vertical oscillation of the current profile with a vertical wave number of $\sqrt{\omega/2\nu}$, where $\omega$ is the forcing frequency and $\nu$ the turbulent viscosity, which, however, is a function of time and distance from the sediment surface. Depending on the overall energetics of the BBL flow, characteristic current speed maxima at 2–3 m above the bed could be observed in lakes where the internal wave forcing had a period as long as 24 h. Another major consequence of oscillatory BBL is that the maximum in turbulent intensity near the bed does not coincide with the maximum of the current speed, at the top of the BBL.

## Stratified BBL

### Effects of Density Stratification

Density stratification affects turbulent mixing in the outer layer by causing buoyancy forces that damp or even suppress overturning turbulent eddies. The vertical distribution of velocity and turbulence described earlier for unstratified BBL may hence change significantly under stratified conditions. In addition, the vertical structure of density stratification along with the presence of sloping boundaries can introduce additional mixing phenomena in BBL of enclosed basins.

Similar to the surface mixed layer, increased production of TKE along the boundaries of a water body often leads to the generation and maintenance of a well-mixed BBL of height $h_{mix}$. On a flat bottom (away from the slopes) and where a logarithmic boundary

layer occurs, $h_{mix}$ can be estimated by applying scaling laws from the wind-mixed surface layer

$$h_{mix} = 2^{3/4} \frac{u_*}{\sqrt{Nf}} \qquad [8]$$

where $u_*$ is the friction velocity in the BBL (eqn [2]), $N$ is the Brunt–Väisälä frequency, and $f$ is the Coriolis parameter. In small- to medium-sized water bodies, where the effect of Earth's rotation is unimportant, $f$ has to be replaced by the respective forcing frequency, e.g., the frequency of internal seiching.

In productive water bodies, in particular, the sediment can be a significant source of remineralized nutrients as a result of microbial degradation of organic matter at the sediment surface or within the sediment. The diffusion of solutes across the sediment–water interface (see Solute Transport and Sediment–Water Exchange section) has the potential to set up density stratification within the BBL, which could suppress turbulent mixing. Hence, whether or not a turbulent and mixed BBL can be developed and maintained depends not only on the amount of available TKE, which is typically extracted from basin-scale motions, but also on the buoyancy flux from the sediment that has to be overcome by turbulent mixing. Geothermal heating, in contrast, can result in unstable stratification and convective mixing in the BBL. A mean geothermal heat flux of 46 mW m$^{-2}$ results in a mean vertical temperature gradient of about $-8 \times 10^{-2}$ K m$^{-1}$, which can be observed when the BBL is chemically stratified and turbulent mixing is suppressed.

## 2-Dimensional Mixing Processes in Enclosed Basins

The occurrence of mixed BBL (in terms of density) is a straight consequence of the application of a zero-flux boundary condition at the sediment surface, i.e., no exchange of heat and dissolved solids across the sediment–water interface. This boundary condition forces the isopycnals (or isotherms if density changes are mainly caused by temperature) to intersect the boundary at a right angle, leading to a mixed density or temperature profile in the vicinity of the boundary. Enhanced mixing along the boundaries is thus not a necessary requirement for the development of such mixed layers. Along the sloping boundaries of enclosed basins, these mixed BBL cause horizontal density gradients and hence drive horizontal currents – a process which is believed to have important consequences for basin-wide diapycnal transport.

Measurements, however, have revealed that $h_{mix}$ is not constant along the sloping boundaries, as demonstrated in **Figure 4**. There the upper limit of the mixed BBL can be defined by the depth of the 5 °C isotherm

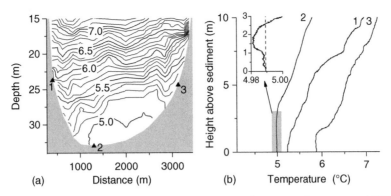

**Figure 4** (a) Isotherm depths along the main axis of Lake Alpnach (Switzerland) calculated from repeated CTD profiling. The increment between neighboring isotherms is 0.1 °C and the numbers refer to the respective temperature (in °C) of the isotherms plotted as thick lines. Note that the figure is not to scale and only the lower portion of the water column is shown. The numbered symbols show the position of temperature profiles shown in (b). (b) Near-bottom temperature profiles at selected positions along the transect shown in (a). Numbers at the top of the profiles refer to the positions indicated by symbols in (a). The inset emphasizes the inverse temperature stratification observed in the BBL of profile 2. Panel (a) is adopted from Lorke A, Wüest A, and Peeters F (2005) Shear-induced convective mixing in bottom boundary layers on slopes. *Limnology and Oceanography* 50: 1612–1619, with permission from American Society of Limnology and Oceanography.

(chemical stratification can be neglected in this particular lake), leading to $h_{mix} \simeq 2.3$ m in the central part of the lake, whereas $h_{mix} \lesssim 1$ m further up on the slopes (**Figure 4(b)**). The pronounced tilt of the isotherms in **Figure 4(a)** further emphasizes the importance of basin-scale internal waves for local estimates of $h_{mix}$ on the slopes because the associated currents push the well-mixed BBL from the central part of the lake up and down the respective slopes during the course of the seiching.

The flow within the BBL remains parallel to the sediment surface on the sloping boundaries because the vertical velocity component must vanish at a rigid surface. Hence the flow is no longer in parallel to the isopycnals (or isotherms in **Figure 4(a)**) and, in combination with the fact that the flow velocity is increasing with increasing distance from the sediment surface (eqn [4]), convective instabilities can occur on the slopes when heavier water is moved on top of lighter water. Unstable stratification, as shown in the inset of **Figure 4(b)**, can only be generated when the current is directed up-slope, a down-slope current leads to a stabilization of the BBL by the same principle. Thus, in a periodic, internal wave-driven flow, stratified and convectively mixing BBL occur alternately on the two opposing slopes of the water body. This shear-induced convection provides an additional source for mixing in BBL on slopes. Its general importance for BBL turbulence, however, is not yet fully understood.

### Turbulence Induced by Internal Wave Interactions with Bottom Boundaries

Turbulence production in stratified regions of lakes is linked with instabilities in the internal wave field.

A considerable portion of the turbulence occurs within or near the BBL and can be due to internal wave breaking at critical frequencies or internal wave steepening. Recent studies have shown that the form and degree of nonlinearity of internal waves in the thermocline can be predicted from the Wedderburn and Lake numbers, two dimensionless indices which indicate the balance between buoyancy forces and shear forces and which further take into account basin morphometry. As illustrated in **Figure 5**, the hypolimnion is also an internal wave field with similar wave forms to those observed in the thermocline. When Lake numbers, $L_N$, drop below 3, turbulence associated with the internal wave field increases (**Figure 5**). Thus, not only is turbulence induced in the thermocline when nonlinear waves form, but also in the hypolimnion with the greatest increases in the BBL. Values of the coefficient of eddy diffusivity increase 1–3 orders of magnitude above molecular. In small lakes, flow speeds in the BBL increase from a few millimeters per second to $\sim 2$ cm s$^{-1}$ with the decreases in Lake number.

## Solute Transport and Sediment–Water Exchange

### The Diffusive Sublayer

In the immediate vicinity of the sediment surface, the vertical transports of momentum and solutes (dissolved gases, solids, and heat) are reduced to their respective molecular levels, as described above. In analogy to the viscous sublayer, where momentum transport is governed by molecular viscosity, a diffusive sublayer exists, where the vertical transport of

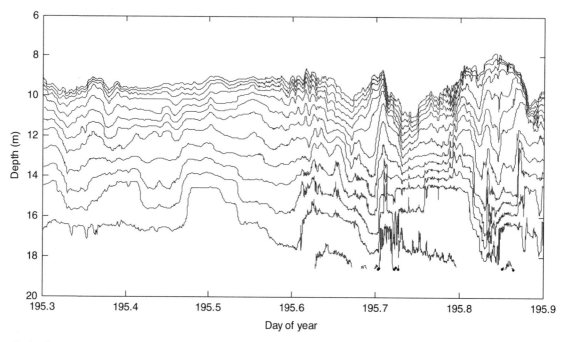

**Figure 5** Isotherms illustrating the internal wave field in the hypolimnion of Toolik Lake, Alaska, prior to and after wind forcing increased to $9 \, m \, s^{-1}$ (day 195.6) and the Lake number decreased to 1.5. Isotherms are at 0.1 °C intervals with uppermost isotherm 6 °C. Thermistors were 80 cm apart between 10.2 and 12.6 m depth and 2 m apart deeper in the water column. Deepest thermistor was within 50 cm of the lake bottom. Increased temperature fluctuations after $L_N$ decreases below 1.5 are indicative of turbulence either beginning or increasing in the lower water column (unpublished data, S. MacIntyre).

heat and solutes is governed by molecular diffusion. Since the molecular diffusivity of solutes $D$ is about 3 orders of magnitude smaller than the molecular viscosity $v$ (the Schmidt number $Sc$, defined as $Sc = v/D$, is about 1000), the height of the diffusive sublayer $\delta_D$ is with $\delta_D = O(1 \, mm)$ considerably smaller than the height of the viscous sublayer $\delta_v$ (**Figure 2**). A typical profile of dissolved oxygen concentrations measured through the sediment–water interface is shown in **Figure 6**. Although turbulent transport is already suppressed within the viscous sublayer, straining of concentration gradients by viscous shear results in an efficient vertical transport of solutes and to typically well-mixed solute distributions within most of the viscous sublayer. Concentration gradients are hence compressed to the diffusive sublayer overlaying the sediment surface and the constancy of the molecular diffusivity results in a linear concentration gradient $C$ $(z)$ (**Figure 6**). At the top of the diffusive boundary layer, the concentration gradient decreases gradually to zero and the solute concentration reaches its constant bulk value $C_\infty$.

Within the sediment, molecular diffusivity is reduced by the porosity (reduction of surface area) and by turtosity (increase of diffusion path length), and the concentration profile is additionally determined by chemical and microbial production and loss processes. In the case of oxygen (**Figure 6**),

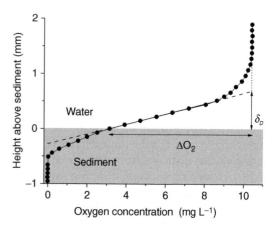

**Figure 6** Oxygen concentration profile across the sediment–water interface measured in Lake Alpnach (Switzerland). The diffusive sublayer is characterized by the linear concentration gradient above the sediment surface where transport is governed by molecular diffusion. The effective height of the diffusive sublayer $\delta_D$ is defined by the intersection of the extrapolated linear concentration gradient (dashed line) with the constant oxygen concentration above the diffusive sublayer.

a reaction–diffusion model with rather simple zero-order kinetics of oxygen consumption resulting in a parabolic oxygen profile provides surprisingly good agreement with measured oxygen distributions.

The fluxes $F$ of solutes through the diffusive boundary layer can be derived from Fick's first law:

$$F = \frac{D}{\delta_D}(C_\infty - C_0) \qquad [9]$$

where $C_0$ is the solute concentration at the sediment surface. Hence, for a given concentration gradient $(C_\infty - C_0)$ and by ignoring the temperature dependence of the molecular diffusivity, the magnitude of the flux is determined by the thickness of the diffusive sublayer $\delta_D$. It has been demonstrated in numerous laboratory and field measurements that $\delta_D$ depends strongly on the flow regime in the turbulent BBL above. With increasing levels of turbulence (e.g., with increasing $u_*$) the diffusive boundary layer becomes more compressed, and according to eqn [9], the fluxes increase. The thickness of the diffusive sublayer can be related empirically to $u_*$ or to the thickness if the viscous sublayer, which in turn is related to $u_*$, as described earlier:

$$\delta_D = \delta_\nu Sc^a \qquad [10]$$

where the Schmidt number $Sc$ accounts for the different kind of 'solutes' (e.g., heat or dissolved oxygen) and observed Schmidt number exponents $a$ range between 0.33 and 0.5. As $u_*$ is not always an appropriate parameter for describing BBL turbulence (e.g., in oscillatory BBL or in the presence of form drag or density stratification), $\delta_D$ can also be described in terms of the Batchelor length scale, which describes the size of the smallest fluctuations of a scalar tracer in turbulent flows as a function of the turbulence dissipation rate.

It is most interesting to note that the sediment–water exchange in productive water bodies is often flux-limited by the 'bottleneck' of the diffusive sublayer

and that it is actually the wind acting at the water surface that provides energy for turbulence within the BBL and hence effects the magnitude of the sediment–water fluxes by controlling the thickness of the diffusive sublayer.

### Effects of Small-Scale Sediment Topography

Increased roughness of the sediment surface (e.g., due to biological activity) affects the sediment–water exchange by increasing the mass and momentum transfer as well as by increasing the surface area of the sediment–water interface. Detailed observations have demonstrated that the diffusive sublayer tends to smooth out topographic structures that are smaller than the average height of the sublayer, but smoothly follows larger roughness elements (**Figure 7**). The detailed structure of the oxygen distribution within the diffusive sublayer is then not only determined by diffusion (normal to the local sediment surface) but also by advection with the flow (in parallel to the local sediment surface) and it can be expected that the degree of smoothing increases with decreasing flow velocities. Detailed comparisons of measured 3-dimensional fluxes over rough topography with the fluxes calculated from the respective average diffusive sublayer heights and concentration gradients are enhanced by factors up to 49%. It must be noted, however, that fluxes estimated from concentration profiles measured at one particular location on a rough sediment surface can severely overestimate or underestimate the average flux, as demonstrated in **Figure 7**.

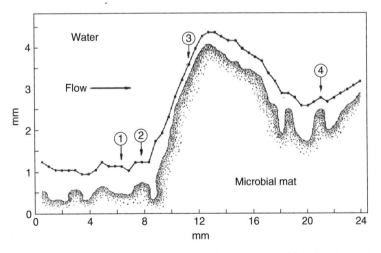

**Figure 7** Horizontal transect along the direction of flow showing how the upper limit of the diffusive sublayer (solid line with data points) follows the surface topography of a microbial mat. The diffusive sublayer limit was defined by the isopleth of 90% air saturation of oxygen. Notice different vertical and horizontal scales. Flow velocity at a height of 1 cm was 4 cm s$^{-1}$. Numbers indicate specific measuring positions discussed in the original publication. Reproduced from Jørgensen BB and Des Marais DJ (1990) The diffusive boundary layer of sediments: Oxygen microgradients over a microbial mat. *Limnology and Oceanography* 35:1343–1355, with permission from American Society of Limnology and Oceanography.

## Nondiffusive Fluxes

Besides the diffusive fluxes, there exist several additional pathways for the exchange of solutes across the sediment–water interface. Convectively driven transport of pore water through the interface can occur in shallow waters where shortwave solar radiation penetrates the water column to the sediment surface and heats the sediment. Similarly, changes in temperature of the water overlaying the sediment surface, e.g., due to internal waves in stratified water bodies, have been observed to drive convective transport across the sediment–water interface. The existence of larger roughness elements, such as ripples, on permeable sediments can further result in advective pore water exchange driven by pressure differences. Higher dynamic pressure at the upstream side of such topographic structures give rise to the transport of water into the sediment, whereas the lower pressure at the downstream side sucks pore water out of the sediment.

*Bioturbation* and *bioirrigation* are processes by which benthic fauna or flora enhances the sediment–water exchange. Whereas bioturbation mainly refers to the displacement and mixing of sediment particles by, e.g., worms, bivalves, or fish, bioirrigation refers to the flushing and active ventilation of burrows with water from above the sediment surface. These processes are particularly important in oligotrophic water bodies where the sediment surface remains oxic and provides suitable conditions for a diverse benthic fauna. In more eutrophic systems the emanation of gas bubbles (mainly methane or carbon dioxide) which are formed by biogenic production and a resulting supersaturation of pore water with these gases may have similar effects.

## In Situ Flux Measurements

The sediment oxygen demand or the release of nutrients from the sediment can be of major importance for the overall productivity and for the geochemical composition of a particular water body and quantification of sediment–water fluxes is often essential for understanding biogeochemical cycles within the water column. As these fluxes depend strongly on the hydrodynamic conditions in the BBL and as these conditions have a strong spatial and temporal dynamics, in situ measurements are often desirable. The measurement of concentration profiles through the sediment–water interface capable of resolving the diffusive sublayer are one way for estimating the fluxes. From a measured profile of dissolved oxygen, as shown in **Figure 6**, the sediment–water flux can be readily estimated by applying eqn [9]. Such measurements are carried out using microelectrodes, which are available for a large number of solutes, mounted on a benthic lander system. However, there are two major problems associated with this method: First, although these microelectrodes have tiny tip diameters (down to $10\,\mu m$ or less for oxygen sensors), they were demonstrated to disturb the concentration distribution within the diffusive boundary layer while profiling. The second and more severe problem results from the complexity of the spatial distribution of the fluxes resulting not only from the small-scale sediment topography (cf. **Figure 7**) but also from the strongly localized effects of advective pore water exchange and bioturbation. To overcome these problems, the flux can be measured within the turbulent BBL at some distance from the actual sediment surface. By neglecting any sources or sinks within the water column between the sampling volume and the sediment surface, this flux represents an areal average of the sediment–water flux including all nondiffusive flux contributions. The turbulent flux $F_{turb}$ is determined by the cross-correlation of turbulent vertical velocity ($w'$) and turbulent concentration ($C'$) fluctuations:

$$F_{turb} = \overline{w'C'} \qquad [11]$$

where the overbar denotes temporal averaging.

## Particle Dynamics

BBLs are often characterized by enhanced concentrations of suspended particles as compared to the water column above. Such nepheloid layers are generated by resuspension of particles from the sediment surface and subsequent upward transport. The potential of a turbulent flow to entrain sediment particles of size $D$ is often described in a Shields diagram where empirical thresholds of sediment motion are provided as a function of a nondimensional shear stress $\theta = \rho u_*^2/((\rho_p - \rho)gD)$ and a particle Reynolds number $Re_* = u_*D/\nu$, in which $\rho_p$ is the particle density and $g$ the gravitational constant. The $\theta$ can be interpreted as the ratio between the lift force provided by the turbulent shear stress defined in eqn [3] and the gravitational force acting on the particle. While in suspension, the fate of the particle is determined by the balance between upward transport by turbulent diffusion and Stokes settling.

The quantitative characterization of resuspension and particle transport, however, is often complicated by cohesive properties of the particles. Cohesive particles require greater shear stresses to become resuspended; moreover, they tend to form aggregates when in suspension, which alters their settling velocities.

The resuspension–settling cycles increase the contact area between particle surfaces and water and thus enhance the fluxes from and to the particles. In addition, suspended particles contribute to water density and locally enhanced resuspension, generated, e.g., by high near-bottom current velocities in the littoral zone or at the depth of the thermocline (cf. **Figure 1**), may lead to the formation of turbidity currents.

## Glossary

**Dissipation rate of TKE** – Rate of dissipation of TKE per unit volume of water and per unit time. This energy is dissipated into heat by internal friction among fluid elements described by viscosity.

**Nepheloid layer** – Particle-rich layer above the sediment. This layer is sustained by a balance between gravitational settling and turbulent vertical diffusion counteracting it.

**Pore water** – Water that fills the interstitial space between sediment grains.

**Reynolds number** – The dimensionless Reynolds number $Re$ is the ratio of inertial to viscous forces acting on a fluid element, obstacle in the flow, or submerged organism and describes the transition from laminar to turbulent flow regimes.

**Shear stress** – Force per unit area acting in parallel (tangential) to the surface of a fluid element or interface (e.g., bed shear stress).

**Turbulent eddies** – Turbulence is composed of eddies: patches of zigzagging, often swirling fluid, moving randomly around and about the overall direction of motion. Technically, the chaotic state of fluid motion arises when the speed of the fluid exceeds a specific threshold, below which viscous forces damp out the chaotic behaviour (see also *Reynolds number*).

**Turbulent kinetic energy** – Kinetic energy per unit volume of water, which is contained in the random fluctuations of turbulent motions. Turbulent velocity fluctuations $u'$ can be separated from the mean current velocity $\bar{u}$ by Reynolds decomposition of the actual velocity $u$ ($u = \bar{u} + u'$). Turbulent kinetic energy (TKE) is then defined by, $TKE = 1/2\rho u'^2$, where $\rho$ denotes density of water.

*See also:* Biological-Physical Interactions; Currents in Stratified Water Bodies 1: Density-Driven Flows; Currents in Stratified Water Bodies 2: Internal Waves; Currents in Stratified Water Bodies 3: Effects of Rotation; Currents in the Upper Mixed Layer and in Unstratified Water Bodies; Density Stratification and Stability; Small-Scale Turbulence and Mixing: Energy Fluxes in Stratified Lakes.

## Further Reading

Ackerman JD, Loewen MR, and Hamblin PF (2001) Benthic–Pelagic coupling over a zebra mussel reef in western Lake Erie. *Limnology and Oceanography* 46: 892–904.

Berg P, *et al.* (2003) Oxygen uptake by aquatic sediments measured with a novel non-invasive eddy-correlation technique. *Marine Ecology Progress Series* 261: 75–83.

Boudreau BP and Jørgensen BB (2001) *The Benthic Boundary Layer.* New York: Oxford University Press.

Caldwell DR and Chriss TM (1979) The viscous sublayer at the sea floor. *Science* 205: 1131–1132.

Gloor M, Wüest A, and Münnich M (1994) Benthic boundary mixing and resuspension induced by internal seiches. *Hydrobiology* 284: 59–68.

Gundersen JK and Jørgensen BB (1990) Microstructure of diffusive boundary layers and the oxygen uptake of the sea floor. *Nature* 345: 604–607.

Lorke A, Müller B, Maerki M, and Wüest A (2003) Breathing sediments: The control of diffusive transport across the sediment-water interface by periodic boundary-layer turbulence. *Limnology and Oceanography* 48: 2077–2085.

Lorke A, Umlauf L, Jonas T, and Wüest A (2002) Dynamics of turbulence in low-speed oscillating bottom-boundary layers of stratified basins. *Environmental Fluid Mechanics* 2: 291–313.

Lorke A, Wüest A, and Peeters F (2005) Shear-induced convective mixing in bottom boundary layers on slopes. *Limnology and Oceanography* 50: 1612–1619.

Mellor GL (2002) Oscillatory bottom boundary layers. *Journal of Physical Oceanography* 32: 3075–3088.

Miller MC, Mccave IN, and Komar PD (1977) Threshold of sediment motion under unidirectional currents. *Sedimentology* 24: 507–527.

Wüest A and Gloor M (1998) Bottom boundary mixing: The role of near-sediment density stratification. In: Imberger J (ed.) *Physical Processes in Lakes and Oceans. Coastal and Estuarine Studies*, pp. 485–502. American Geophysical Union.

# Currents in Stratified Water Bodies 1: Density-Driven Flows

**F Peeters,** Universität Konstanz, Mainaustrasse, Konstanz, Germany
**R Kipfer,** Swiss Federal Institute of Environmental Science and Technology (Eawag), Swiss Federal Institute of Technology (ETH), Ueberlandstr, Duebendorf, Switzerland

## Introduction

Vertical transport of dissolved substances and heat in lakes mainly results from two different mechanisms: (a) mixing by turbulence that is usually described as a diffusive transport and (b) density-driven exchange that can be considered as an advective transport. A typical example of the latter is convection owing to surface cooling in fall, which often leads to isothermal conditions in shallow lakes of the temperate zone. Because entrainment of ambient water limits the depth of convective plumes, density-driven transport to large depth in deep lakes usually occurs along the lake boundaries and is often the result of specific and localized processes, which are discussed later.

The important role of density-driven transport for vertical exchange in lakes becomes evident if one considers that temperature stratification typical for most lakes is characterized by a decrease in water temperature with increasing water depth. Turbulent diffusion causes heat to flow from high to low temperatures and hence typically leads to a gradual continuous warming of cold deep-water regions. Thus, on a long-term average, advective processes transporting cold surface water downwards must be sufficient to compensate for the heat flux due to turbulent diffusion. The low temperatures in the deep water are usually either the remnant of isothermal conditions generated by buoyancy-driven overturn during the cold season or originate from cold density currents propagating to largest depth. Because vertical transport due to density currents plays an important role in overall deep-water renewal and heat exchange, density driven exchange processes are central to the understanding of oxygenation and nutrient transport especially in deep lakes.

In the world's largest and deepest water bodies several processes have been identified that lead to advective deep-water renewal by density currents: river inflow, e.g., in Lake Constance, Lake Geneva, and Lake Baikal; inter-basin exchange, e.g., in Lake Lucerne, Lake Baikal, or even in the Caspian Sea; differential cooling, e.g., in Lake Geneva, Lake Constance, Lake Issyk-Kul, and Lake Malawi; thermal-bar mixing, e.g., in the Lake Ontario, Lake Ladoga, Lake Michigan, and Lake Baikal; and transport due to thermobaric instabilities, e.g., in Lake Baikal and possibly in Crater Lake. All these processes have been shown to significantly contribute to deep-water renewal in lakes, although advective transport to the lake bottom was not conclusively demonstrated in all cases. More details on the different processes are given below.

In the following, we first describe the principal characteristics of density currents and the associated signals of intrusions in vertical profiles of water constituents and temperature. Then, we present several mechanisms that lead to the generation of density plumes in deep freshwater lakes and discuss which of these processes can also be responsible for deep-water renewal in tropical and saline lakes. Finally, we discuss the potential impact of changes in the catchments of lakes and in the meteorological conditions on deep-water renewal by density currents.

## Characteristics of Density Currents

Density currents are driven by differences in water density, which can result from gradients in water temperature, salinity, dissolved uncharged substances, or suspended particles and are also affected by pressure. If a water mass with higher density is situated above a water mass with lower density, the stratification is unstable and buoyancy causes the upper water mass to sink. In the sinking process ambient water is mixed into the sinking water mass and thus alters its density (**Figure 1**), thereby reducing the density difference between the density plume and the ambient water.

Furthermore, the properties of the ambient water change along the path of the sinking water mass. Hence the buoyancy of the sinking plume changes continuously as it sinks into deeper depth. Eventually, a depth is reached where the density of the density plume and the density of the surrounding water become equal. At this depth the sinking process ceases and the plume water spreads out laterally into the ambient water forming an intrusion (**Figure 1**). In many cases the sinking plume meets the lake boundary and then continues to sink along the lake boundary (**Figure 1**). In this case, entrainment of ambient water is limited to the upper side of the density plume and less ambient water is entrained per unit sinking depth. Hence the characteristic properties of the water within the density plume (e.g., temperature, salinity, suspended particles, dissolved oxygen) change more

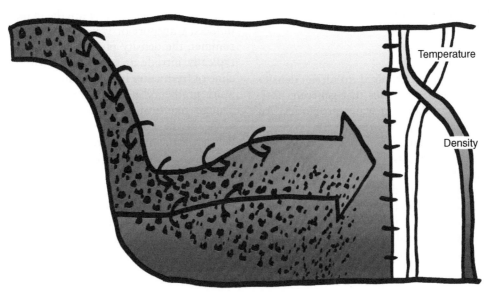

**Figure 1** Schematic illustration of density currents. The shading on the left-hand side of the figure indicates an increase in density with increasing depth.

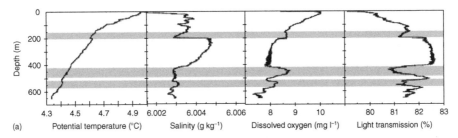

**Figure 2** Intrusions as indicators of density currents. Vertical profiles of temperature, salinity, dissolved oxygen, and light transmission measured in Lake Issyk-Kul. The distinct features in these profiles suggest intrusions from density currents. Grey bars mark depth regions with high concentrations of dissolved oxygen and low light transmission, suggesting water originating from shallower depth regions. Redrawn from **Figure 2** in Peeters FD, Finger M, Hofer M, Brennwald DM, and Livingstone R Kipfer (2003). Deep-water renewal in Lake Issyk-Kul driven by differential cooling. *Limnology & Oceanography* 48(4): 1419–1431.

slowly and the density plumes typically can propagate to larger depths than would be possible in the open water.

The occurrence of density plumes propagating from shallow to deep water are indicated by intrusions that can be identified in CTD-profiles (conductivity as measure of salinity, temperature, and depth) and in profiles of dissolved substances and suspended particles. **Figure 2** presents an example from Lake Issyk-Kul (Kyrghystan), where intrusions are characterized by a higher dissolved oxygen concentration and a lower light transmission than is observed in the ambient water.

High oxygen levels in the intrusions indicate oxygen-rich surface water that must have been transported recently because oxygen levels have not yet been significantly reduced by degradation processes.

The low light transmission in the intrusions indicates water with a high load of suspended particles suggesting either, that the water that generated the density current was enriched in suspended particles and thus may have originated from river inflow, or, that the density plume responsible for the intrusions has propagated along the lake boundary and caused resuspension of sediments during the sinking process. Temperature is usually not a good indicator of intrusions because it is a key parameter determining plume density. Thus, at the depth of the intrusion, plume water and surrounding water often have about the same temperature. However, in cases where the density plume propagates down to the largest depths, as it is sometimes the case e.g., in Lake Baikal, temperature anomalies at the lake bottom can be used to identify density plumes.

## Density Plumes Generated by External Inputs

### River Inflows

The density differences required to drive density plumes originate from processes that generate horizontal or vertical gradients in water properties. An obvious example is river inflow (**Figure 3**).

River water usually contains an increased load of suspended particles and has a different temperature and salinity than the lake water. Hence, river inflow is commonly associated with density plumes propagating from the river mouth to larger depths. The kinetic energy associated with the inflow of the river water is usually rapidly dissipated and the horizontal density gradients resulting from the different densities of

river water and lake water are the main cause of river induced transport of water masses in lakes. During summer, the density plumes induced by river inflow typically intrude at some depth within the thermocline of freshwater lakes. Because of the large temperature gradients in the thermocline, water densities change significantly within a rather narrow depth range in the lake. Hence, the probability that the density of the plume water agrees with the density in the water column of the lake is especially large within the thermocline. This fact explains that river water typically intrudes in this depth range. The depth reached by the density plumes varies during the course of the year since water properties and hence the density of lake and river water changes seasonally (**Figure 3**). Density currents containing a high load of

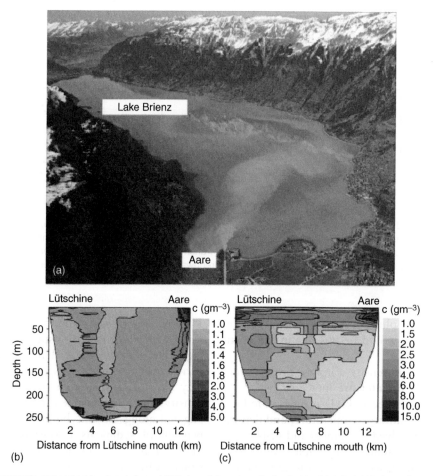

**Figure 3** Density currents generated by river inflow. (a) Water of river Aare indicated by high turbidity intruding near the surface of Lake Brienz. The sharp boundaries of this surface plume indicate plunging of river water to larger depth. (b, c) Suspended particle distribution inferred from light transmission measurements in a longitudinal cross-section of Lake Brienz measured in February (b) and October (c). The particle distributions suggest that, in February water introduced by the river Aare (inflow on the right-hand side) sinks as density plume along the lake bottom towards largest depth (b). In October river Aare and river Lütschine both intrude at intermediate depth (c). (**Figure 3(a)** was provided by Ueli Ochsenbein; **Figure 3(b)** and (c) are redrawn from **Figure 7(a)** and 7(d) in Finger D, Schmid M, and Wüest A (2006). Effects of upstream hydropower operation on riverine particle transport and turbidity in downstream lakes. *Water Resources Research* 42, W08429, doi:10.1029/2005WR004751. Reproduced/modified by permission of American Geophysical Union.

suspended particles are often called turbidity currents. Sedimentation of particles out of intrusions resulting from turbidity currents can reduce the density of the intruding water sufficiently that the depth of the intrusions becomes shallower over time. Density currents induced by river inflows usually propagate along the sloping bottom boundary before the water intrudes laterally. If the concentrations of solutes and particles are very different between the river and the lake water, the plumes can propagate down to the deepest parts of the lake. In cases where the sinking water is confined to underwater channels, e.g., density plumes propagating down the Kukui Canyon of the Selenga delta in Lake Baikal, entrainment of ambient water is reduced, and the density plume can propagate over a depth range of more than 1000 m down to 1640 m (see also **Figure 4(a)**).

### Interbasin Exchange

River inflows not only result in localized density plumes propagating from the river mouth to larger depths but also can generate subtle large scale gradients in water properties that affect the horizontal density distribution on a basin scale. These small density gradients on a large spatial scale may contribute to the generation of density currents far from the river inflow that occur especially in the vicinity of sills separating sub-basins of the lake. This mechanism is exemplified in **Figure 4** for two lakes of different size, Lake Baikal and Lake Lucerne, that both are structured into several sub-basins separated by sills.

In both lakes large scale horizontal salinity gradients are generated by river inflows introducing water with different ion concentrations into the different sub-basins. In the case of Lake Baikal, the River Selenga introduces more saline water into the Central Basin than the Upper Angara River introduces into the Northern Basin. In case of Lake Lucerne, the River Sarner Aa introduces more saline water into the sub-basin Lake Alpnach than the River Reuss introduces into the sub-basin Lake Uri. Because of the salinity gradients, the density of the water in the different sub-basins differs if temperature is the same. In Lake Lucerne the densest water can be found in sub-basin Lake Alpnach when winter cooling reduces surface water temperature to ~4 °C. Horizontal transport of the dense water from Lake Alpnach to the sub-basin Lake Vitznau across the sill separating the two sub-basins induces a density current renewing the deep-water of sub-basin Lake Vitznau (**Figure 4(b)**). The density plume causes upwelling of cold dense water within Lake Vitznau. Horizontal transport of water across the sills between sub-basin results into a cascading of density driven transport within all

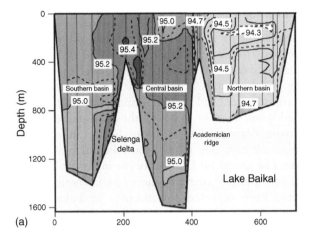

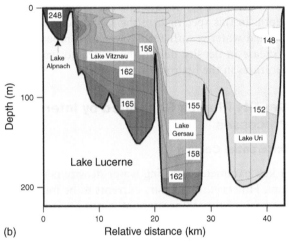

**Figure 4** Density currents generated at sills between sub-basins. Vertical transects of salinity in Lake Baikal (a) and in Lake Lucerne (b). In both lakes the horizontal gradients in salinity are generated by river inflows introducing water with different ion concentration. The salinity distributions suggest that density currents are not only generated directly by river inflow as is the case in Lake Baikal at the Selenga delta (see panel a) but that density currents also occur in both lakes at the sills between sub-basins most likely driven by horizontal transport across the sill. Contours depict salinity in mg kg$^{-1}$. **Figure 4(a)**: redrawn from Kipfer R and Peeters F (2000). Speculation on consequences of changes in the deep water renewal in Lake Baikal, in K Minoura (ed.) *Lake Baikal a mirror in time and space for understanding global change processes*, pp. 273–280. Amsterdam, Netherlands: Elsevier. **Figure 4(b)**: drawn using data from Aeschbach-Hertig W, Kipfer R, Hofer M, Imboden DM, and Baur H (1996) Density-driven exchange between the basins of Lake Lucerne (Switzerland) traced with the $^3$H-$^3$He method. *Limnology and Oceanography* 41: 707–721.

sub-basins (**Figure 4(b)**). A similar process is operating in the different basins of Lake Baikal (**Figure 4(a)**). Prerequisite of these density currents generated at sills between sub-basins is (1) the structuring of the lake basin into sub-basin that prevents homogenization of water properties by horizontal mixing and (2) a heterogeneous input of water properties, as e.g., the salinity by the river inflows in **Figure 4**.

### Subsurface Inflows

Besides the input from rivers, external water sources derived from groundwater inflows and from hydro-thermal vents can generate density currents depending on the depths at which the inflows are located. Groundwater and hydrothermal water is usually highly enriched in ions and thus can cause salinity driven density plumes. Groundwater inflows into lakes are common in artificial lakes such as mining lakes or gravel-pit lakes or occur in karstic environments. Density currents due to hydrothermal vents have been reported for instance in Lake Baikal, where hydrothermal water is introduced in Frohliha Bay at a depth of 200–400 m and propagates as a bottom following density current down to 1400 m depth. In this specific case the salinity of the hydro-thermal water is sufficiently large to compensate the decrease in density due to the increased water temperatures in the hydrothermal water.

## Density Plumes Generated by Internal Processes

### Differential Cooling

A key parameter affecting water density is temperature. However, for density currents to be induced on the basis of the temperature of water, horizontal gradients in temperature are required. Horizontal temperature gradients are generated by external surface and subsurface inflows (see earlier text) but also by internal processes. In most lakes the heat flux at the lake surface (expressed per unit area) can be considered as horizontally homogeneous because meteorological parameters and radiation do not vary significantly at the length scale of the lake basin. Nevertheless, differential cooling can generate significant temperature differences within lakes and thus generate density currents (e.g., Wellington Reservoir, Lake Geneva, Lake Constance, Lake Banyoles). Heat loss at the lake surface causes vertical convection and thus mixing of surface water with water from layers below. This process continuously mixes the cooled surface water with water from deeper layers containing heat stored during the warm season. In shallow-water regions, the reservoir of warmer deep water is exhausted earlier than in regions with large water depth. Hence, in shallow-water regions heat loss at the lake surface leads to a faster cooling of the water column than in deep-water regions. Because the cold water in the shallow-water regions has a larger density than the warmer water in the pelagic, deep-water regions, the cold water propagates downwards as a density current. Such density plumes often occur only sporadically. They are typically generated during

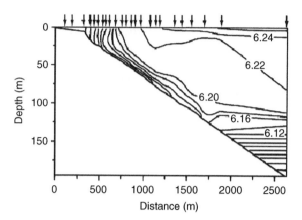

**Figure 5** Density current generated by differential cooling. Contour lines represent isotherms in °C indicating a density plume generated by differential cooling in Lake Geneva. Note that the temperatures are well above 4 °C. The isotherms are constructed from CTD-data collected at the locations indicated by the arrows. Redrawn from **Figure 2(b)** in Fer I and Lemmin U. Winter cascading of cold water in Lake Geneva. *Journal of Geophysical Research* 107, NO. C6, 10.1029/2001JC000828, 2002. Reproduced/modified by permission of American Geophysical Union.

night-time cooling in fall as has been demonstrated in e.g., Lake Constance and Lake Geneva (**Figure 5**) or during events that also induce cooling such as cold fronts or monsoons.

Differential cooling can result in density driven currents from any shallow region in a lake basin. Therefore it may affect a large volume of water and thus may significantly contribute to overall vertical transport. The process is particularly effective if large shelf regions are located around a deep basin. The density currents induced by differential cooling propagate along the lake bottom and can reach large depths especially if channels exist along which the density current can propagate without significant entrainment of ambient water, as is the case e.g., in Lake Issyk-Kul. Note, that in freshwater lakes differential cooling can only generate density plumes if water temperatures are above the temperature of maximum density ($T_{md}$) which is about 4 °C at the lake surface. Cooling below $T_{md}$ implies a decrease in water density and thus prohibits density plume development by differential cooling. Hence, in freshwater lakes the temperature of density plumes associated with differential cooling is at least 4 °C or higher.

### Thermal Bar

Temperature-driven density currents can also result from horizontal mixing of two adjacent surface water masses, one having a temperatures above and the other below $T_{md}$. Because of the non-linear temperature dependence of the equation of state, mixing of water masses with different temperatures always results in an increase in the mean density of the water. This process is called cabbeling. If the temperature of the mixed water is closer to the $T_{md}$ than the water below, it sinks as a density plume. A so-called thermal bar develops which is characterized by a vertically isotherm water column with a temperature close to $T_{md}$ separating an open water region with temperatures below $T_{md}$

from a warmer shore region with temperatures above $T_{md}$ (**Figure 6**).

The downward flow of dense surface water at the thermal bar can cause significant renewal and oxygenation of deep water. As time progresses the thermal bar moves further away from the shore to the open water. The dynamics of this process depends on the morphometry of the lake and on the atmospheric forcing. Mixing associated with the thermal bar has been

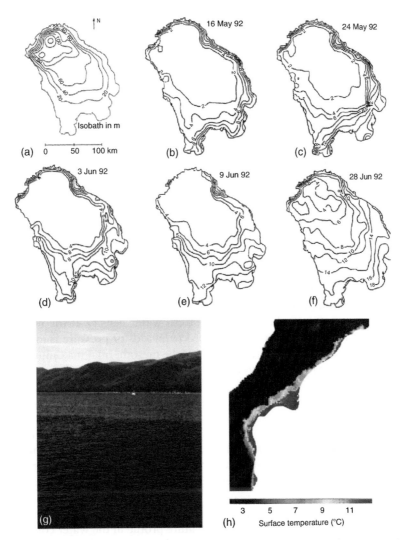

**Figure 6** The thermal bar. The development of a thermal-bar in Lake Ladoga indicated by surface temperatures measured with satellites (a–f) and the observation of a thermal bar near Selenga delta in Lake Baikal (g and h). Panel (a) provides the morphometry of Lake Ladoga with depth contours given in m. Panels (b–f) depict surface temperatures with isotherms in °C. Panels (b–f) show how the thermal bar, which is located at the 4 °C isotherm, moves towards deeper water as the season progresses. The position of the thermal bar changes more rapidly in the gently sloping shallower south-eastern part of the lake than in the steep and deep northern part. In (g), the sharp boundary between near shore water and open water indicates the position of the thermal bar located near the Selenga delta in Lake Baikal. The color differences result from differences in the load of suspended particles. The water trapped near shore by the thermal bar has an increased load of suspended particles owing to the nearby inflow of the Selenga River. Panel (h) shows an image of the surface temperatures near Selenga delta derived from satellite data. Density currents are generated at the sharp transition between warm shore water and cold open water characterized by a temperature close to 4 °C. White areas are land. **Figure 6(a–f)** are redrawn from **Figures 1** and **3** in Malm J, Mironov D, Terzhevikl A, and Jiinsson L (1994) Investigation of the spring thermal regime in Lake Ladoga using field and satellite data. *Limnology and Oceanography* 39(6): 1333–1348. Copyright 2000 by the American Society of Limnology and Oceanography, Inc.

reported for Lake Ontario, Lake Ladoga, Lake Michigan, and Lake Baikal. Because the thermal bar requires open-water temperatures below $T_{md}$, it occurrs mainly in lakes which become ice covered in winter.

The horizontal temperature gradients required for thermal bar development are generated by differential heating during spring warming, a process similar to differential cooling. Initially in spring, when surface-water temperatures are below $T_{md}$, an increase of water temperature during spring warming leads to an increase in the density of the surface waters and thus to convection. Because convection mixes the warmer surface water with colder water from below and the reservoir of the cold water from below is smaller in shallow near-shore than in deeper open-water regions the temperature in the shallow-water regions increases faster than in the deep off-shore regions. When the temperature in the shallow regions exceeds $4\,°C$ the water column is stratified and further influx of heat is not connected to convection, but leads to an even faster increase in the water temperature. Thus, a horizontal temperature gradient typical for a thermal bar situation develops with temperatures below $T_{md}$ at the surface of the open-water region and temperatures above $T_{md}$ at the surface in the shallow near-shore regions. A reverse thermal bar situation with temperatures below $T_{md}$ in the shallow-shore region and temperatures above $T_{md}$ in the open-water region could develop in fall as a consequence of differential cooling if cooling in the shore region progresses to temperatures below $T_{md}$. However, in the case of the reverse thermal bar exchange processes will be dominated by horizontal density gradients below the surface (see the section on 'Differential cooling').

**Thermal Baricity**

Another process that can generate density plumes as a consequence of the nonlinearity of the equation of state of freshwater is the thermobaric effect. The thermobaric effect results from the fact that $T_{md}$ decreases with increasing pressure. The generation of density currents by the thermobaric effect requires a very specific temperature stratification that occurs only in few stably stratified deep freshwater-lakes, e.g., Lake Baikal or Crater Lake. To generate density currents by the thermobaric effect, water temperatures must be below $4\,°C$ throughout the water column. The temperature in the surface layer must increase with increasing depth, whereas the temperature in the deep-water must decrease with increasing depth. Then, the temperature profile has a maximum at intermediate depth, the so-called mesothermal maximum (**Figure 7**).

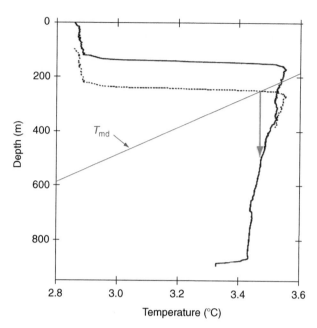

**Figure 7** Schematic on the generation of density currents by the thermobaric effect. The temperature profile presented has been measured in the northern basin of Lake Baikal. The temperature of maximum density as function of depth shown for comparison is labelled with $T_{md}$. Vertical displacement of the temperature profile (indicated by the dashed line) leads to a density driven vertical transport that is self supporting over the depth range indicated by the arrow.

A water column with such a temperature profile is stably stratified because of the effect of pressure on fresh water density. If the water column is displaced downwards or pressure is increased, the temperature at the mesothermal maximum is higher than the local $T_{md}$ and the water column becomes unstable. Cold water from above the mesothermal maximum can sink downwards as a density plume (**Figure 7**). Similarly, a water mass from the cold upper layer can be pushed downwards below the mesothermal maximum to a depth where its temperature is closer to $T_{md}$ than the temperature of the ambient water. Then, it will continue to sink driven by its buoyancy until the surrounding water has the same temperature as the sinking water mass. Besides the specific temperature profile in the water column the exchange due to the thermobaric effect also requires a mechanism by which the water pressure is altered substantially and/or the water is locally pushed downwards across the depth of the mesothermal maximum. Hence density plume generation by the thermobaric effect is not very common. Several investigations have claimed that the thermobaric effect may be important for deep-water oxygenation in Lake Baikal. However, the mechanism that could cause the required downward displacement in the open water column remained unclear. Recently, wind-driven Ekman transport near the coast of the

Southern Basin of Lake Baikal has been suggested to cause thermobaric instabilities.

### Turbidity Currents Generated by Waves

Internal processes not only can generate temperature gradients but also can cause gradients in suspended particles that are sufficient to drive density currents. Shear stress at the lake bottom associated with surface waves and high-frequency internal waves can induce sediment resuspension and thus increase the load of suspended particles in the water column. If the density increase induced by the change in particle load is sufficiently large to compensate vertical density stratification, turbidity currents are generated. Because the turbidity gradient is generated by the interaction of waves with the sediment, the resulting turbidity currents usually originate either near the shore at the lake surface (surface waves) or at the depth of the thermocline (high-frequency internal waves). The turbidity currents usually propagate along the lake bottom to larger depth until they intrude into the open water.

### Horizontal Density Currents Generated under Ice Cover

Temperature-driven density currents can also occur under ice cover due to differential heating by solar radiation. This process is invoked if the optical properties of the ice and snow cover vary horizontally either due to a variation in the ice structure (e.g., white ice and black ice) or in the thickness of snow cover. Then, penetration of solar radiation through the ice and snow cover varies horizontally and thus induces differential heating in the surface water below the ice. This process causes convection below the ice and results in horizontal density gradients that can drive horizontal density currents. Such under-ice currents are believed to be essential for the development of algal blooms in early spring in Lake Baikal.

## Density Currents in Tropical and Saline Lakes

In tropical lakes and also in saline lakes several of the processes mentioned above do not occur. Because the development of a therma bar requires a lateral transition of water temperature from above to below $T_{md}$ at the lake surface, a thermal bar never occurs in tropical lakes where water temperatures are above $4\,°C$ all year round and thus always exceed $T_{md}$. Because $T_{md}$ decreases with increasing salinity reaching freezing temperature at about $25\,g\,kg^{-1}$, thermal bar development and the associated density

currents also do not play an important role for the vertical exchange in saline lakes. The same arguments exclude the thermobaric effect as a significant cause of density currents in tropical and saline lakes.

In tropical lakes temperature gradients play the dominant role in the generation of density currents because at high water temperatures, slight temperature gradients imply large differences in density. Therefore, salinity gradients play a smaller role for density plume generation in tropical lakes than for lakes of temperate regions during the cold season. Convection due to night-time cooling and density currents due to differential cooling and river inflows can be expected to be the most important processes for advective deep-water renewal in tropical lakes.

In saline lakes, however, river inflows usually cannot drive density currents because rivers typically introduce freshwater that in most cases has a much lower density than the saline lake water, even if the inflow has a low temperature. For example water with a salinity of $4\,g\,kg^{-1}$ and a temperature of $24\,°C$ has a greater density than freshwater with $4\,°C$. Hence, river inflows can only directly drive density currents in saline lakes, if the riverine water carries a substantial load of suspended particles. Nevertheless, in saline lakes river inflows can generate large scale differences in salinity and thus may indirectly lead to density currents at sills induced by inter basin exchange as is probably the case in the Caspian Sea. Another process that is likely to cause density currents in saline lakes is differential cooling. In contrast to freshwater lakes this process can trigger density currents with temperatures well below $4\,°C$ and even down to freezing temperature, because $T_{md}$ can be substantially reduced or even does not exist depending on the salinity of the water.

## Impact of Changes in the Environmental Conditions on Density Currents and Deep-Water Renewal

Changes in climatic conditions and human activities in catchments may affect density currents and thus vertical mixing in lakes. An increase in precipitation and in the percentage of the land made impervious by human development typically leads to a higher discharge of rivers. Enhanced river discharge is usually associated with increased erosion and a higher load of suspended particles in the river water such that density currents will propagate to larger depth. Enhanced deep-water renewal is thus anticipated in freshwater lakes. In saline lakes, however, the increase in freshwater input associated with increased precipitation reduces the density of the surface waters and thus can significantly limit the generation of density currents. Consequently, deep-water renewal may be reduced or suppressed. Deep-water exchange decreased drastically as a consequence of increased riverine discharge in the Caspian Sea and in Mono Lake, CA.

Water storage dams in the catchments of lakes have the opposite effect on density currents than soil sealing. Retention by water storage dams reduces peak discharge and the load of suspended particles in downstream rivers. This results in a decrease in the intensity of density currents and in the depth of intrusions in lakes located downstream of dams.

Climate warming can lead to a reduction of deep-water renewal in lakes because additional input of heat at the lake surface may result in an increase in density stratification of the water column and in an extension of the stratified period. Persistence of the increased stratification over many years likely depends on the lake's latitude and depth and whether warming is intensified in winter or summer or, for tropical lakes, during the monsoon period or during less windy periods.

In a warmer climate, however, mixing in freshwater lakes due to density currents associated with the thermal bar and/or the thermobaric effect may cease, if climate warming leads to an increase in surface water temperature to values above 4 °C all year round, i.e., to values above $T_{md}$. Because density is a nonlinear function of temperature, warming of surface water may also shift the relative importance of turbidity and salinity gradients towards temperature gradients as agent to drive density plumes.

The potential consequences of environmental change on density currents are exemplified for Lake Baikal, the deepest lake on earth. Because of the peculiar temperature profile with the mesothermal temperature maximum (see **Figure 7**), deep-water renewal in Lake Baikal is predominantly driven by salinity differences between river and lake water and between the basins of the lake. The salinity differences result in density plumes associated with riverine inflows and inter-basin exchange. Hence, changes in the catchments leading to an increase in the concentration of dissolved ions and suspended particles in river inflow will intensify deep-water mixing by density plumes. Climate warming on the other hand will not severely affect density plumes and thus deep-water renewal in Lake Baikal, as long as the lake has an annual ice cover. Higher air temperatures most likely result in a shift of ice break-up to earlier times in the year, but will not have an affect on the thermal conditions immediately after ice break-up. Hence, the conditions required to generate density plumes will not change, but may occur earlier in the season.

In summary, density currents significantly contribute to deep renewal, especially in deep and very deep lakes. The density currents can result from a variety of processes. Which of these processes are relevant in a specific lake depends on the temperature regime of the lake, its salinity and also its morphometry. The environmental conditions, e.g., precipitation in the catchments and heat flux at the lake surface affect the occurrence of density currents and the depth reached by the density plumes. Hence, changes in the environmental conditions will have consequences for deep-water renewal and oxygenation of deep lakes not only because of a change in turbulence levels, but also by their effect on the intensity of density currents.

*See also:* The Benthic Boundary Layer (in Rivers, Lakes, and Reservoirs); Density Stratification and Stability.

## Further Reading

Aeschbach-Hertig W, Kipfer R, Hofer M, Imboden DM, and Baur H (1996) Density-driven exchange between the basins of Lake Lucerne (Switzerland) traced with the $^3$H-$^3$He method. *Limnology and Oceanography* 41: 707–721.

Fer I and Lemmin U (2002) Winter cascading of cold water in Lake Geneva. *Journal of Geophysical Research* 107: NO. C6, 10.1029/2001JC000828.

Finger D, Schmid M, and Wüest A (2006) Effects of upstream hydropower operation on riverine particle transport and turbidity in downstream lakes. *Water Resources Research* 42: W08429, doi:10.1029/2005WR004751.

Fischer HB, List EJ, Koh RCY, Imberger J, and Brooks NH (1979) *Mixing in Inland and Coastal Waters.* San Diego, CA: Academic Press.

Hamblin PF and Carmack EC (1978) River induced currents in a fjord lake. *Journal of Geophysical Research* 83: 885–899.

Hohmann R, Kipfer R, Peeters F, Piepke G, Imboden DM, and Shimaraev MN (1997) Processes of deep water renewal in Lake Baikal. *Limnology and Oceanography* 42: 841–855.

Malm J, Mironov D, Terzhevikl A, and Jiinsson L (1994) Investigation of the spring thermal regime in Lake Ladoga using field and satellite data. *Limnology and Oceanography* 39: 1333–1348.

Monismith SG, Imberger J, and Morison ML (1990) Convective motions in the sidearm of a small reservoir. *Limnology and Oceanography* 35: 1676–1702.

Peeters F, Finger D, Hofer M, Brennwald M, Livingstone DM, and Kipfer R (2003) Deep-water renewal in Lake Issyk-Kul driven by differential cooling. *Limnology and Oceanography* 48: 1419–1431.

Weiss RF, Carmack EC, and Koropalov VM (1991) Deep-water renewal and biological production in Lake Baikal. *Nature* 349: 665–669.

Zilitinkevich SS, Kreiman KD, and Terzevik AY (1992) The thermal bar. *Journal of Fluid Mechanics* 236: 27–42.

# Currents in Stratified Water Bodies 2: Internal Waves

**L Boegman,** Queen's University, Kingston, ON, Canada

## Introduction

In the surface layer of lakes, exchange with the atmosphere and energetic mixing from wind and convection provides oxygen and light, thus enabling growth of plankton and other aquatic life. The nutrients required to sustain surface layer ecology are primarily found in the benthos, where sediment resuspension, nutrient release from the sediments, and oxygen consumption occur. In the interior of lakes, seasonal stratification of the water column suppresses vertical mixing, effectively isolating the surface layer from the sediments. However, the stratification simultaneously provides an ideal environment for internal waves, whose oscillatory currents energize a quasi-steady turbulent benthic boundary layer (TBBL) that drives vertical biogeochemical flux.

The wave motions in lakes are initiated by the surface wind stress. Waves will occur on both the free surface and internal stratifying layers (e.g., the thermocline) and these are referred to as surface or barotropic and internal or baroclinic motions, respectively. The waves are categorized according to their length scale. Basin-scale waves have wavelengths that are of the same order as the lake diameter and are manifest as standing wave modes – or seiches. Sub-basin-scale waves have wavelengths of 10–1000 m. These waves are progressive in nature and will break where they shoal on sloping topography at the depth of the thermocline.

## Characteristic Geometry and Water-Column Stratification

The internal waves described in this chapter are in lakes that are not affected by the Coriolis force due to the Earth's rotation: for example, small lakes in the arctic, mid-sized lake (i.e., diameter $> \sim 5$ km) in the mid latitudes, and large lakes near the equator. These lakes have a Burger number $>1$. As we shall see later, the wave modes that are supported in such lakes depend upon the nature of the water column stratification.

We limit the analysis to several characteristic types of stratification that are commonly observed to occur. The simplest case is that of a homogenous lake of length $L$ and depth $H$, as may be typical for a shallow system or one that has recently experienced a turnover event (**Figure 1(a)**). During the summer months, solar heating causes a lake to become stratified with a layered structure consisting of an epilimnion, metalimnion, and hypolimnion. If the vertical density gradient is abrupt through the metalimnion, the lake may be approximated as a simple two-layer system of thickness $h_1$ and density $\rho_1$ over thickness $h_2$ and density $\rho_2$, where $H = h_1 + h_2$ is the total depth (**Figure 1(b)**). In lakes where a strong diurnal thermocline is present or the metalimnion is thick, the vertical density structure may be approximated with three contiguous fluid layers of density $\rho_1$, $\rho_2$, and $\rho_3$ with thicknesses $H = h_1 + h_2 + h_3$ (**Figure 1(c,d)**). The layered model for the stratification is inappropriate for shallow lakes ($H < \sim 15$ m), where the entire water column may be composed of weakly stratified water (e.g., western Lake Erie). In these lakes a transient diurnal thermocline may still occur. Shallow weakly stratified lakes are best characterized as having a continuous stratification (**Figure 1(e)**). Very deep lakes (with a thick laminar region between the metalimnion and TBBL) and those with a significant chemical (saline) component will also have a continuous stratification beneath the metalimnion (**Figure 1(f)**). In general, the strength of the stratification is measured according to the Brunt-Väisälä or buoyancy frequency $N = \sqrt{-(g/\rho_o)d\rho/dz}$, where $z$ is the vertical coordinate direction, $g$ is the gravitational constant, and $\rho_o = 1000 \, \text{kg m}^{-3}$ is the characteristic water density; in the thermocline of lakes the maximum $N \sim 10^{-2}$ Hz.

## Surface Momentum Transfer and Wind Set-Up

### Wind Set-Up of the Free Surface

The action of the wind across the lake surface results in frictional momentum transfer from the wind to the water. This transfer occurs in the form of a stress ($\text{N m}^{-2}$) applied at the free surface. The stress may be parameterized as

$$\tau = C_D \rho_a U_{10}^2$$

where $C_D$ is the drag coefficient, $\rho_a = 1.2 \, \text{kg m}^{-3}$ is the air density, and $U_{10}$ the wind speed measured at 10 m above the water surface. Typically $C_D = 1.3 \times 10^{-3}$, but this value may vary by $\pm 40\%$ depending upon the wind speed, water depth, and relative temperature

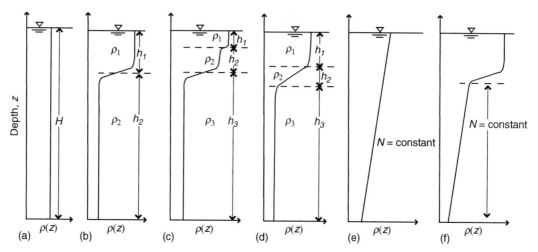

**Figure 1**  Characteristic continuous water-column stratifications as found in lakes and typical layer approximations. (a) Homogeneous water-column of constant density. (b) Two-layer approximation of the continuous stratification, where the layer separation occurs at the thermocline. (c) Three-layer approximation of the continuous stratification, where the layer separation occurs at the diurnal and seasonal thermoclines. (d) Three-layer approximation of the continuous stratification, where the layer separation occurs at the upper and lower surfaces of the metalimnion. (e) Continuous stratification throughout the water column with constant buoyancy frequency. (f) Continuous stratification where the hypolimnion is characterized by a constant buoyancy frequency.

difference between water surface and adjacent air column.

The momentum transfer associated with steady winds will push the surface water to the leeward shore, causing a displacement of the free surface due to the presence of the solid boundary (**Figure 2a**); for long and shallow lakes this may be as large as several meters (e.g., ~2 m in Lake Erie) (see **Currents in the Upper Mixed Layer and in Unstratified Water Bodies**).

This displacement is called wind set-up. If the wind stress is applied for sufficient time (one quarter of the fundamental seiche period as defined below), a steady-state tilt of the free surface will occur where there is a balance between the applied wind force ($\tau \times$ surface area) and the hydrostatic pressure force due to the desire of the free surface to return to gravitational equilibrium. Balancing these forces at steady state, given the equation for the slope of the free surface

$$\frac{\partial \eta_s}{\partial x} = \frac{u_*^2}{gH}$$

where $u_* = \sqrt{\tau/\rho_o}$ is the surface wind shear velocity, $\eta_s(x, t)$ is the interfacial (surface) displacement from the equilibrium position, and $x$ is the longitudinal coordinate. The equation for the free-surface slope may be integrated to give the maximum interfacial displacement, as measured along the vertical boundary

$$\eta_s(t = 0, x = 0, L) = \pm \frac{u_*^2}{gH}\frac{L}{2}$$

## Wind Set-Up of the Internal Stratification

In a manner analogous to the set-up at the free surface, wind induced displacements can also occur along the thermocline. Consider a simple two-layered lake. Water piled-up at the leeward shore by windward drift simultaneously pushes down the thermocline while pushing up the free surface (**Figure 2(b)**, **Figure 3**). The free surface remains nearly horizontal owing to a return flow that develops in the hypolomnion, leading to vertical velocity shear through the metalimnion. A corresponding upwelling occurs at the windward shore (**Figure 3**). The steady-state slope of the free surface is given by a balance between the baroclinic gravitational pressure force from the tilted thermocline and the force due to the wind-stress acting through the epilimnion

$$\frac{\partial \eta_i}{\partial x} = \frac{u_*^2}{g'h_1}$$

where $g' = g(\rho_2 - \rho_1)/\rho_2$ is the reduced gravity across the interface (thermocline) (see **Currents in Stratified Water Bodies 1: Density-Driven Flows**). The equation for the thermocline slope may be obtained through integration over the basin length

$$\eta_i(t = 0, x = 0, L) = \pm \frac{u_*^2}{g'h_1}\frac{L}{2}$$

The effect of buoyancy can be seen by decreasing the density difference between the two layers resulting in a decrease in $g'$ and corresponding increase in $\eta_i$.

Surface seiche (V0H1)

Internal seiche (V1H1)

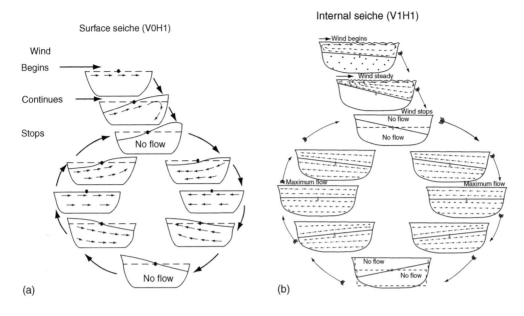

(a)  (b)

Internal seiche (V2H1)

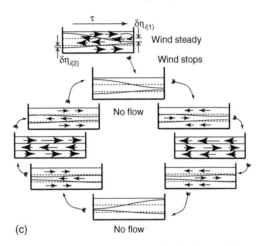

(c)  No flow

**Figure 2** Movement caused by steady moderate wind stress on a hypothetical layered lake and subsequent internal seiche motion neglecting damping. (a) Horizontal mode one surface seiche in a homogeneous one-layered system, (b) horizontal mode one vertical mode on internal seiche in a two-layered system, both adapted from Mortimer CH (1952) Water movements in lakes during summer stratification: Evidence from the distribution of temperature in Windermere. *Proceedings of the Royal Society of London Series B*. 236: 355–404 and (c) horizontal mode one vertical mode two internal seiche in a three-layered system. Arrows denote distribution and magnitude of water particle velocities. At $t = 0$, $(1/2)T_1$, $T_1$, $(3/2)T_1$, etc. the wave energy is purely in the potential form, isotherms are at their maximum tilt and there is no seiche induced flow, while at $t = (1/4)T_1$, $(3/4)T_1$, $(5/4)T_1$, etc. the energy is purely kinetic, giving rise to strong horizontal currents within the lake-basin and horizontal isotherms.

A simple comparison $\eta_s/\eta_i \sim (\Delta\rho/\rho_o)(h_1/H)$ shows that for weakly stratified deep systems, internal displacements ($\sim1$–$100$ m) may be more than an order of magnitude greater than their surface counterparts ($\sim0.01$–$1$ m); for example in Lake Baikal $\eta_s/\eta_i \sim 0.11\,\text{m}/75\,\text{m} \sim 10^{-3}$.

The available potential energy (APE) embodied in the tilted interface is readily calculated for a two-layer system by integrating the interfacial displacement over the basin length

$$\text{APE} = g\frac{\rho_2 - \rho_1}{2}\int_0^L \eta_i^2(x,t)dx$$

which may be integrated for the initial condition of a uniformly tilted thermocline

$$\text{APE} = \frac{1}{6}gL(\rho_2 - \rho_1)\eta_i^2$$

After the thermocline tilt has reached steady state, work done by continued winds is either dissipated

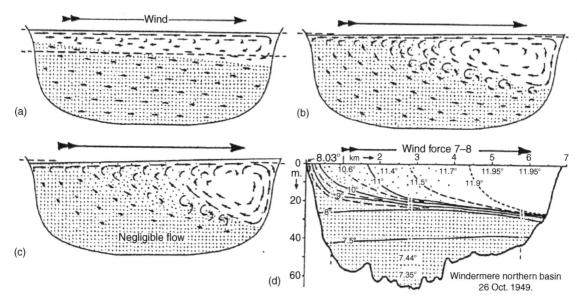

**Figure 3** Schematic showing the response of a stratified lake to a surface wind stress. (a–c) show the stages of development of a steady state thermocline tilt. The hypolomnion is shaded and the arrows show the relative speed and direction of the flow. (d) Isotherm distribution and temperatures in Lake Windermere, northern basin, after a steady wind for 12 h. Reprinted from Mortimer CH (1954) Models of the flow-pattern in lakes. *Weather* 9: 177–184.

internally as heat or acts to mix the water column by further deepening the surface layer. Surface layer deepening has been characterized into four distinct regimes based on the strength of the stratification and winds (**Figure 4**). For strong stratification and weak winds (Regime A) the thermocline set-up is small, internal seiches persist for long times, mixing is weak and the thermocline remains sharp. If the stratification is weaker or the winds stronger (Regimes B and C), seiche amplitudes increase and become a predominant feature, shear instabilities (e.g., Kelvin–Helmholtz billows) form leading to entrainment of the metalimnion into the epilimnion, enhanced mixing and causing rapid damping of the internal seiches. For weak stratification and strong winds (Regime D), shear instabilities are strong; the thermocline becomes diffuse with a steep slope and rapidly deepens toward the lake bed. This creates a sharp downwind interface and a broad upwelling at the upwind shore. The upwelled fluid creates a longitudinal temperature gradient, which subsequently mixes the lake horizontally. The colder upwelled water is nutrient rich and as it mixes rapid fluctuations in temperature and biogeochemistry result. In deeper lakes, stratification can be strong during summer and upwelling of metalimnetic (partial upwelling) or hypolimnetic (full upwelling) water is unlikely. In these lakes upwelling is favored just after spring turnover or prior to fall turnover when the thermal stratification is weak or near the surface.

## Wedderburn and Lake Numbers

The degree of tilt of the base of the surface layer resulting from an applied wind stress may be quantified using the dimensionless Wedderburn number $W$, which as the ratio of the wind disturbance force to the gravitational baroclinic restoring force is given by

$$W \sim \frac{g' h_1^2}{L u_*^2}$$

Here, $g'$ is evaluated across the base of the surface layer. The cases of $W \gg 1$, $W \sim 1$ and $W \ll 1$ correspond to Regimes A, B/C, and D, respectively.

For idealized laboratory studies and back-of-the-envelope calculations, substitution of $\eta_i$ into the equation for $W$ leads to a leads to two-layer form

$$W \sim \frac{h_1}{\delta \eta_i}$$

where $\delta \eta_i$ is the steady wind induced vertical displacement of the seasonal/diurnal thermocline measured at the boundary. Due to the order of magnitude scaling, the factor of 2 has been dropped as is commonly found in the scientific literature. Moreover, the somewhat counterintuitive nature of $W \to \infty$ as $\delta \eta_i \to 0$, leads to frequent use of the inverse form of the Wedderburn number ($W^{-1} \sim \delta \eta_i / h_1$).

For lakes which are not well approximated using a two-layer stratification, $W$ has been generalized into the Lake Number, $L_N$ (see **Density Stratification and Stability**). This accounts for the depth

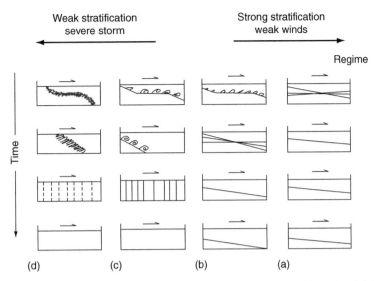

**Figure 4**   Schematic showing the mixed layer deepening response of a lake to wind stress. (a) Regime A: internal waves; (b) Regime B: internal waves and slight billowing; (c) Regime C: strong billowing and partial upwelling; (d) Regime D: intense billowing and full upwelling. Adapted From Fischer HB, List EJ, Koh RCY, Imberger J, and Brooks NH (1979). *Mixing in Inland and Coastal Waters*. San Diego, CA: Academic Press. After Spigel RH, and Imberger J (1980) The classification of mixed layer dynamics in lakes of small to medium size. *Journal of Physical Oceanography* 10: 1104–1121.

dependence of stratification and horizontal area. For a constant wind stress over a lake with an arbitrary basin shape and stratification

$$L_N = \frac{S_t(H - h_2)}{u_*^2 A_o^{3/2}(H - h_v)}$$

Here $A_o$ is the surface area of the lake and $h_v$ is the height from the lake-bed to the centre of volume of the lake. The stability of the lake

$$S_t = \frac{g}{\rho_o} \int_0^H (z - h_v)A(z)\rho(z)dz$$

incorporates the variable stratification $\rho(z)$ and irregular bathymetric area $A(z)$. For large Lake numbers, the stratification will be severe and dominate the forces introduced by the wind stress. The isotherms will be horizontal, with little or no seiching and associated turbulent mixing in the benthic boundary layer and interior. Changes in $S_t$ with latitude and season cause $L_N$ to vary spatially and temporally around the globe (see **Mixing Dynamics in Lakes Across Climatic Zones**). Under comparable wind conditions $L_N$ is maximal in the mid-latitudes during summer.

## Basin-Scale Standing Wave Motions (Seiches)

### Interfacial Waves in a Layered Stratification

**Horizontal modes**   When steady winds cease and the surface stress condition is relaxed, the gravitational restoring force associated with the tilted interface (water surface or thermocline) becomes unbalanced. The available potential energy embodied in the tilt is released under the action of gravity and converted to kinetic energy as the interface oscillates in the form of a sinusoidal standing waves or seiche. Antinodes are found at the basin end walls and nodal points in the basin interior (**Figures 2 and 5**). Seiches are commonly called linear waves because the evolving wave-field is well described in space and time by the linear wave equation

$$\frac{\partial^2 \eta}{\partial t^2} = c_o^2 \frac{\partial^2 \eta}{\partial x^2}$$

where $\eta(x,t)$ is the interfacial displacement and $c_o$ the linear shallow water phase speed (speed at which the crests/troughs propagate). This equation is equally applicable to interfacial waves travelling on the free-surface or thermocline by applying the appropriate form of $c_o = \sqrt{gH}$ or $c_o = \sqrt{g'h_1h_2/H}$, for the cases of surface and internal seiches, respectively. Due to the reduced effect of gravity across the thermocline relative to the free surface ($g' \ll g$), surface waves travel at ~50 times the speed of internal waves.

The familiar standing wave patterns associated with seiches forms as symmetric progressive waves of equal amplitude and wavelength, but opposite sign, propagate from the upwelled and downwelled fluid volumes at the opposite ends of the basin (**Figure 2**). These waves are most commonly represented with cosine functions (**Figure 5**), which have central node(s) and antinodes at the basin walls. Summing cosine

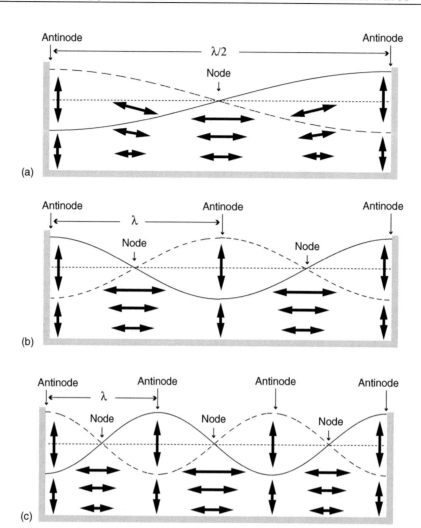

**Figure 5**  Schematic diagram showing the first three horizontal interfacial seiche modes: horizontal mode one ($n = 1$), mode two ($n = 2$), and mode three ($n = 3$). Arrows denote direction of water particle velocities. Solid and dashed lines denote the interfacial displacement at one-half period intervals. Upper layer velocities for the baroclinic case are not shown and can be inferred from symmetry.

equations for waves propagating in opposite directions, gives the equation for the horizontal mode one (H1) standing wave pattern (**Figure 5(a)**)

$$\eta(x,t) = a\cos(kx + \omega t) + a\cos(kx - \omega t) = 2a\cos kx \cos \omega t$$

The component wave amplitude $a = \delta\eta_i/2$ or $\delta\eta_s/2$ depending on the interface under consideration, the angular frequency $\omega = c_o k$, where $k = 2\pi/\lambda$ is the wavenumber and $\lambda$ the wavelength and $T = 2\pi/\omega$ the wave period. For an enclosed basin, there is one half wavelength of an H1 seiche in a lake, giving $\lambda = 2L$ and a period of

$$T_n = \frac{2L}{nc_o}$$

where $n = 1$ for a H1 seiche is the number of nodal points or half wavelengths in the horizontal direction.

The layer-averaged horizontal velocities associated with the H1 seiche are maximum the centre of the basin and are given by

$$U_1 = g'\frac{h_2}{H}\frac{\delta\eta_i}{L/2}t$$

$$U_2 = g'\frac{h_1}{H}\frac{\delta\eta_i}{L/2}t$$

These velocities are zero at the vertical boundaries, where the motion is purely vertical (**Figure 5**). Similarly for the surface seiche, the mid-lake depth-averaged velocity is

$$U = g\frac{\delta\eta_s}{L/2}t$$

The oscillatory seiche currents are low-period and quasi-steady. Observations from a variety of lakes and reservoirs show the surface and internal seiche

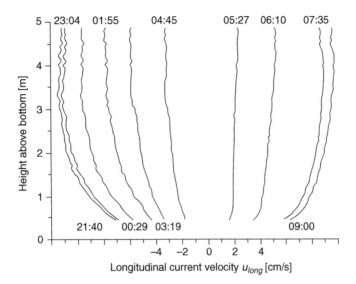

**Figure 6** Near-bed velocity profiles in a small lake showing the oscillatory nature and no-slip boundary associated with seiche currents. Observations are over one-half of the seiche period taken at times as indicated. The profiles are offset and are all plotted with the given velocity scale. From Lorke A, Umlauf L, Jonas T, and Wüest A (2002) Dynamics of turbulence in low-speed oscillating bottom boundary layers of stratified basins. *Environmental Fluid Mechanics* 2: 291–313.

currents to have a typical range of $0.02$–$0.20\,\mathrm{m\,s^{-1}}$, with a maximum $\sim 0.2\,\mathrm{m\,s^{-1}}$ during storms and to approach zero at the no-slip sediment boundary where the flow is impeded by friction (**Figure 6**).

Higher horizontal mode seiches ($n > 1$) are also observed in lakes. These are described by a more general solution of the linear wave equation, where the initial condition is that of a uniformly tilted interface, made up of the superposition of all higher horizontal modes. The general solution is

$$\eta(x,t) = \sum_{n=1}^{\infty} \left( \frac{-8\delta\eta}{n^2\pi^2} \right) \cos\left( \frac{n\pi}{L}x \right) \cos\left( \frac{c_0 n\pi}{L}t \right) \quad n = 1, 3, 5, \ldots$$

$n = 1$ for the horizontal mode-one (H1) seiche, $n = 2$ for the horizontal mode-two (H2) seiche, $n = 3$ for the horizontal mode-three (H3) seiche (**Figure 5**). An infinite number of modes are possible, each with decreasing amplitude and energy as the modal number increases. The fundamental solution is composed only of odd modes (i.e., $\eta(x,t) = 0$ when $n$ is even) as is intuitively expected because only odd odes have a nodal point at the mid-basin location where there is zero displacement associated with an initial uniform initial tilt (**Figures 2** and **3**). By calculating the APE associated with each mode, it can be shown that that more than 98% of the wave energy is contained in the H1 mode, but the energy distribution between modes may be significantly affected by resonant forcing and basin shape.

Examples of surface and internal seiche periods for various horizontal modes are given in **Table 1**. Energy will pass between potential and kinetic forms as the wave periodically oscillates with time (**Figure 2**). At

$t = 0$, $(1/2)T_1$, $T_1$, $(3/2)T_1$, etc. the wave energy is purely in the potential form, while at $t = (1/4)T_1$, $(3/4)T_1$, $(5/4)T_1$, etc. the energy is purely kinetic, giving rise to horizontal currents within the lake-basin (**Figure 6**). For a non dissipative system, the modal energy distributions represent the sum of kinetic and potential energy and are independent of time. Dissipative processes will lead to a decrease in wave amplitude, but not period, with time (**Table 1**); unless there is sufficient mixing across the thermocline to cause a change in the stratification and hence $c_0$. For surface seiches the decay in amplitude with each successive period can range from 3% (Lake Geneva) to 32% (Lake Erie).

**Vertical modes** When the vertical density structure may be approximated with three or more fluid layers (**Figure 1c,d**), in addition to vertical mode-one, horizontal mode-one (V1H1) seiches (**Figure 2b**), higher vertical mode seiches are supported; for example V2H1, etc. (**Figure 2c**). For a three-layer system $c_0$ becomes

$$c_0 = \frac{1}{2H}\left( \gamma - \sqrt{\gamma^2 - 4\alpha H} \right)$$

where

$$\gamma = (1 - \rho_1/\rho_2)h_1 h_2 + (1 - \rho_1/\rho_3)h_1 h_3 + (1 - \rho_2/\rho_3)h_2 h_3$$

and

$$\alpha = h_1 h_2 h_3 (1 - \rho_1/\rho_2)(1 - \rho_2/\rho_3).$$

Substitution into the equation for the wave period $T_n = 2L/nc_0$ gives the period of a vertical mode-two

**Table 1** Observed surface and internal seiche periods from a variety of lakes and the associated amplitude decay

| Lake and location | $T_1$ (h) | $T_2$ (h) | $T_3$ (h) | $T_4$ (h) | $T_5$ (h) | Fractional decrease in amplitude with each successive period |
|---|---|---|---|---|---|---|
| Surface seiches | | | | | | |
| Tanganyika (Africa)[1] | 0.075 | 0.038 | 0.028 | | | |
| Loch Earn (Scotland)[1] | 0.24 | 0.14 | 0.10 | 0.07 | 0.06 | |
| Yamanaka (Japan)[1,2] | 0.26 | 0.18 | 0.09 | | | 0.099 |
| Garda (Italy)[1,2] | 0.72 | 0.48 | 0.37 | 0.25 | 0.20 | 0.045 |
| Geneva (Switzerland–France)[1,2] | 1.2 | 0.59 | | | | 0.030 |
| Vättern (Sweden)[1,2] | 3.0 | 1.6 | 1.3 | 0.97 | 0.80 | 0.113 |
| Baikal (Russia)[1] | 4.6 | | | | | |
| Michigan (Canada–USA)[1] | 9.1 | 5.2 | 3.7 | 3.1 | 2.5 | |
| Erie (Canada–USA)[1,2] | 14.3 | 9.0 | 5.9 | 4.2 | | 0.322 |
| Internal seiches | | | | | | |
| Baldegg (Switzerland)[3] | 9.3 | 4.6 | 3.1 | 2.1 | | |
| Lugano (Switzerland–Italy)[4] | 24 | 12 | 8.0 | 6.2 | | |
| Windermere (England)[3] | 24 | 13 | 9 | | | |
| Zurich (Switzerland)[5] | 45 | 24 | 17 | | | |
| Loch Ness (Scotland)[3] | 57 | 27 | 18 | 14 | 11 | |
| Geneva (Switzerland–France)[3] | 74 | 46 | 30 | 22 | 18 | |

[1]Wilson W (1972) Seiches. In Chow VT (ed.) *Advances in Hydroscience* 8: 1–94.

[2]Wüest AJ and Farmer DM (2003) Seiches. In *McGraw-Hill Encyclopedia of Science and Technology*, 9th Edition. New York: McGraw-Hill.

[3]Lemmin U and Mortimer CH (1986) Tests of an extension to internal seiches of Defant's procedure for determination of surface seiche characteristics in real lakes. *Limnology and Oceanography* 31: 1207–1231.

[4]Hutter K, Salvadè G, and Schwab DJ (1983) On internal wave dynamics in the northern basin of the Lake of Lugano. *Geophysical and Astrophysical Fluid Dynamics* 27: 299–336.

[5]Horn W, Mortimer CH and Schwab DJ (1986) Wind-induced internal seiches in Lake Zurich observed and modeled. *Limnology and Oceanography* 31: 1232–1254.

wave, where the horizontal modal structure is defined by $n$.

Vertical mode-two seiches can be generated when there is an asymmetry in the tilting of upper and lower interfaces of the stratifying layers (the diurnal and seasonal thermoclines or the upper and lower boundaries of the metalimnion). Laboratory and limited field data shows that such asymmetries are introduced from the compression and expansion of the metalimnion that occurs along the downwind and upwind shores, respectively, under upwelling conditions (e.g., **Figure 3(d)**). The strength of the vertical mode two response has been hypothesized to depend upon the relative values of the Wedderburn and Lake Numbers. A V2 response occurs for small $W$ and large $L_N$ (strong tilting of the upper interface, large shear across the base of the surface layer and a relatively undisturbed lower interface); whereas a V1 response occurs for small $W$ and large $L_N$ (comparable tilts of both interfaces and a strong velocity in the hypolimnion).

Higher vertical mode basin-scale internal waves have been observed in several lakes, generally after a sudden wind pulse has excited an initial V1H1 response, which then evolves into a V2 seiche (e.g., Wood Lake, Upper Mystic Lake, Lake Constance). Resonance with the wind forcing (e.g., Lake Alpnach), sloping basin topography and unequal

density differences between stratifying layers can cause this preferential excitation of higher vertical modes.

## Internal Modes in a Continuous Stratification

The two-layer assumption for the stratification in lakes is inappropriate for the many shallow lakes ($H < \sim 15$ m; e.g., Frains Lake) and in the hypolomnion of lakes that are very deep (e.g., Lake Baikal) or where the stratification has a significant chemical (saline) component (e.g., Mono Lake). These systems are better modeled using a continuous stratification where the tilting of the isopycnals due to wind set-up is captured by $L_N$. Upon relaxation of a wind stress, a continuous stratification will support a spectrum of vertical basin-scale modes. The frequency associated with each mode is given by

$$\omega = \frac{N}{\left(1 + \frac{n^2}{m^2}\frac{L^2}{H^2}\right)^{1/2}}$$

which is dependent upon the basin geometry and may be used to calculate the wave period $T = 2\pi/\omega$. The structure of each vertical mode $m$ is described by the wavefunction $\psi_m$, which is obtained from the linear long-wave equation with constant $N$

$$\psi_m(z) = \sin\left(\frac{m\pi}{H}z\right) \text{ and } c_0 = \pm\frac{NH}{m\pi} m = 1, 2, 3, \ldots.$$

and is a measure of the wave-induced vertical displacement of the internal strata at a particular depth. For example, the V1 wave has a maximum internal displacement at mid-depth, whereas the V2 wave has a positive displacement at $1/4H$ and a negative displacement at $3/4H$ (**Figure 7a**); showing the characteristic opening of the strata.

The wavefunction contains no knowledge of the actual wave amplitude and consequently is often normalized $-1 \leq \psi_m \leq 1$. The utility of $\psi_m$ is that it may be numerically calculated from the Taylor–Goldstein equation for an arbitrary $N(z)$ profile, thus determining the wave modes that are supported by a particular water column stratification. The basin-scale vertical modal structure, either a single mode or combination of modes, may be calculated when the wave amplitude is known $\eta(z, t) \approx \psi_m(z)a(t)$. The vertical modal structure can then be projected in space and time throughout the basin by assuming a wave profile $\eta(x, z, t) \approx \psi_m(z)\eta(x, t)$; this is typically a cosine for seiches or $\mathrm{sech}^2$ function for solitary waves (described later).

The horizontal velocity profile induced by the wave motion scales with the gradient of the wavefunction $c_0 d\psi_m/dz$ (**Figure 7(b)**). Analytical expressions for the velocity field can be found in stratified flow texts. The continuous velocity profiles in **Figure 7** are consistent with the depth averaged currents presented in **Figure 2**. In both models, although technically incorrect (**Figure 6**), a free-slip bottom boundary has been assumed. Accordingly the velocities from these models are not representative of flow near the bottom boundary.

## Degeneration of Basin-Scale Internal Waves in Lakes

Understanding the factors leading to wave degeneration has been a major goal of limnologists. This is because internal waves ultimately lose their energy (degenerate) to dissipation (viscous frictional heating of the fluid at mm scales) and diapycnal mixing (mixing of fluid perpendicular to isopycnals or surfaces of constant density) in regions where the flow is turbulent. In turn, mixing drives biogeochemical fluxes. Turbulence is produced directly from the seiche induced currents through fluid straining in the lake interior and TBBL, and from processes that are uncoupled from seiche generation, such as surface wave breaking and inflows, which also act to disrupt the seiche motion. As a general rule, for deep lakes, the period of internal seiche decay is about 1 day per 40 m of water-column

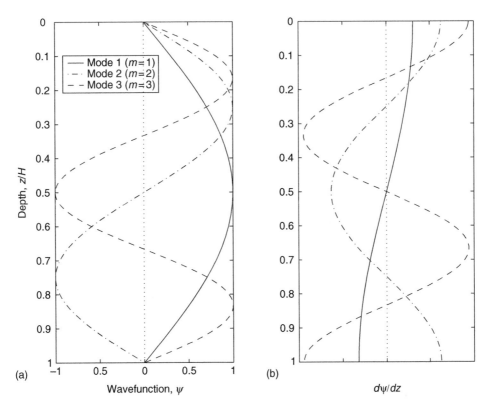

**Figure 7** (a) Wavefunction profiles for vertical modes one ($m = 1$), two ($m = 2$), and three ($m = 3$) supported by a constant $N$ stratification. (b) Characteristic velocity profiles $c_0 d\psi/dz$ for wavefunctions shown in panel (a).

depth. For very deep lakes, e.g., Lake Baikal where $H = 1637$ m, this equates to a decay period of more than one month.

Field observations show the degeneration of basin scale internal waves to occur primarily as a result of turbulence production in the TBBL rather than the interior; observations of dissipation and mixing are more than ten times greater in the TBBL than the interior (see **The Benthic Boundary Layer (in Rivers, Lakes, and Reservoirs)**). The degeneration can occur through four possible mechanisms: (1) viscous damping of seiche currents in the TBBL, (2) the formation of shear instabilities in the interior, (3) the production of nonlinear internal waves that will break on sloping topography; and (4) the formation of internal hydraulic jumps. By calculating the timescales over which each mechanism will occur, regimes have been delineated in which a particular mechanism will dominate (**Figure 8**). The regimes are defined according to the inverse Wedderburn number $W^{-1} = \delta\eta_i/h_1$ and the depth of the seasonal thermocline ($h_1/H$). Although strictly applicable to long narrow lakes that match the rectangular system used in the analysis, the regime diagram has been shown to suitably predict field observations from a variety of lakes (**Table 2**). The degeneration regimes are described below.

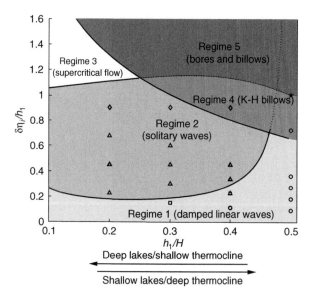

**Figure 8** Analytical regime diagram showing the degeneration mechanisms of seiches in long rectangular lakes. The regimes are characterized in terms of the normalized initial forcing scale $W^{-1} = \delta\eta_i/h_1$ and the depth of the seasonal thermocline ($h_1/H$). Laboratory observations are also plotted (*, Kelvin–Helmholtz (K-H) billows and bore; ◇, broken undular bore; △, solitary waves; □, steepening; ○, damped linear waves). From Horn DA, Imberger J, and Ivey GN (2001) The degeneration of large-scale interfacial gravity waves in lakes. *Journal of Fluid Mechanics* 434: 181–207.

### Regime 1: Damped Linear Waves

Under relatively calm conditions ($W^{-1} < \sim 0.2$) weak seiches develop, which are damped by viscosity in the TBBL. Seiche amplitudes and currents are not sufficient for solitary wave production, shear instability and/or supercritical flow. This regime corresponds to regime A in **Figure 4**. The time scale associated with viscous damping of a basin-scale seiche $T_D$ is estimated from the ratio of the seiche energy to the rate of energy dissipation in the benthic boundary layer $\epsilon_{TBBL}$:

$$T_D \sim \frac{\text{Seiche energy}}{\epsilon_{TBBL}}$$
$$\times \frac{\text{Lake volume}}{\text{TBBL volume}} \sim 1 \text{ to } 10 \text{ d (for moderately sized lakes)}$$

The energy dissipation in the lake interior $\epsilon_{Interior}$ is neglected because observational studies show that $\epsilon_{TBBL} > 10\epsilon_{Interior}$. More complex models for viscous seiche decay may be found in references below.

### Regime 4: Kelvin–Helmholtz Billows

Kelvin–Helmholtz instabilities can form due to a number of mechanisms. Under strong forcing conditions and as the thermocline approaches mid-depth ($> \sim 0.8$ and $> \sim 0.3$), the seiche-induced currents will be sufficiently strong to overcome the stabilizing effects of stratification. Shear instabilities will occur before solitary waves can be produced and/or the seiche is damped by viscosity. Shear from other processes, such as the surface wind stress, can augment that from seiche induced currents and induce instabilities under weaker forcing conditions.

Shear instabilities manifest themselves as high-frequency internal waves that are common features through the metalimnion of lakes and oceans. The waves are the early stages of growth of the instabilities and have sinusoidal profiles, frequencies $\sim 10^{-2}$ Hz, wavelengths $\sim 10$–50 m and amplitudes $\sim 1$–2 m. Depending on the particulars of the stratification and velocity shear profiles, the shear instabilities may be classified according to their profile (e.g., Kelvin–Helmholtz billows, Holmboe waves or combinations thereof). Instabilities grow exponentially from small random perturbations in the flow, leading to rapid degeneration as patches of localized turbulent mixing or billowing (see **Small-Scale Turbulence and Mixing: Energy Fluxes in Stratified Lakes**). This process is shown schematically in **Figure 9**.

In a continuously stratified flow, the stability behavior is governed by the Taylor–Goldstein equation, from which it can be shown that a gradient Richardson number

**Table 2**   Comparison between predicted regime and observations in several lakes

| Lake | Dates | Observations | $W\,(W^{-1})$ | Regime | Source |
|------|-------|--------------|-----------|--------|--------|
| Loch Ness | Oct. 2–3, 1971 | 'Pronounced front or surge' | 3 (0.3) | 2 | 1 |
| Lake of Zurich | Sept. 11–14, 1978 | 'Steep fronted solitary wave' | 3 (0.3) | 2 | 2 |
| Windermere | Aug. 14–20, 1951 | 'Damped harmonic oscillations' | 5 (0.2) | 1–2 | 3 |
|  | Sept. 13–17, 1951 | 'Oscillatory waves' with some steepening | 3 (0.3) | 1–2 | |
| Babine Lake | July 5–10, 1973 | 'Surges' | 2 (0.5) | 2 | 4 |
|  | Aug. 12–15, 1973 | 'Steep shock front' | 3 (0.3) | 2 | |
|  | Oct. 2–7, 1973 | 'Surges' and 'solitary waves' | 2 (0.5) | 2 | |
| Seneca Lake | Oct. 14–21, 1968 | 'Surges' consisting of trains of 'solitons' | 2 (0.5) | 2 | 5 |
| Kootenay Lake | July 13–Aug. 17, 1976 | 'Surges' consisting of waves resembling 'solitons' | 2 (0.5) | 2 | 6 |
| Balderggersee | Nov. 1–15, 1978 | 'Asymmetrical waves' | 9 (0.1) | 2 | 7 |
|  | Nov. 16–22, 1978 | 'Steepened wave front…described as an internal surge' | 2 (0.5) | 2 | |
| Lake Biwa | Sept. 4–13, 1993 | 'Undular bores and solitary waves' | 1 (1) | 2 | 8 |

Most observations fall in regime 2 ($W^{-1} > \sim 0.3$) and the internal seiche will degenerate into nonlinear internal waves. For Windermere, the baroclinic tilting is weaker ($W^{-1} < \sim 0.3$) and the predominant response is a damped seiche. After Horn DA, Imberger J, and Ivey GN (2001) The degeneration of large-scale interfacial gravity waves in lakes. *Journal of Fluid Mechanics* 434: 181–207.

Sources

1. Thorpe SA, Hall A, and Crofts I (1972) The internal surge in Loch Ness. *Nature* 237: 96–98.
2. Mortimer CH and Horn W (1982) Internal wave dynamics and their implications for plankton biology in the Lake of Zurich. *Vierteljahresschr. Naturforsch. Ges. Zurich* 127(4): 299–318.
3. Heaps NS and Ramsbottom AE (1966) Wind effects on the water in a narrow two-layered lake. *Philosophical Transactions of the Royal Society of London A* 259: 391–430.
4. Farmer DM (1978) Observations of long nonlinear internal waves in a lake. *Journal of Physical Oceanography* 8: 63–73.
5. Hunkins K and Fliegel M (1973) Internal undular surges in Seneca Lake: A natural occurrence of solitons. *Journal of Geophysical Research* 78: 539–548.
6. Wiegand RC and Carmack E (1986) The climatology of internal waves in a deep temperate lake. *Journal of Geophysical Research* 91: 3951–3958.
7. Lemmin U (1987) The structure and dynamics of internal waves in Baldeggersee. *Limnology and Oceanography* 32: 43–61.
8. Saggio A and Imberger J (1998) Internal wave weather in a stratified lake. *Limnology and Oceanography* 43: 1780–1795.

$$Ri_g = \frac{N(z)^2}{(du/dz)^2} < \frac{1}{4}$$

is a necessary but not sufficient condition for instability (Miles–Howard criterion). Billowing occurs along thin layers in the flow ($\sim 10$ cm thick) where $Ri_g$ is low and there are interfaces with sharp density gradients. Through billowing, the interfaces become more diffuse and are replaced by shear layers of thickness $\delta \sim 0.3(\Delta U)^2/g'$, where $\Delta U$ is the velocity jump over the interface. As a result of billowing, the flow becomes stable unless $\Delta U$ increases (e.g., due to increasing wind stress) or $\delta$ decreases (e.g., due to mixed layer deepening). A spectrum of growing instabilities are theoretically possible, but the most unstable mode wave will have a wavelength $\lambda = 2\pi/k \sim 7\delta$.

Application of the Taylor–Goldstein equation to field observations shows that the frequency of the most unstable mode is just below the maximum buoyancy frequency through the metalimnion. This is because a fluid parcel displaced vertically from its equilibrium density position, as occurs during the growth of an instability, will be subjected to buoyancy forces arising from the sudden density anomaly with respect to its surroundings. The fluid parcel will oscillate as a wave at frequency $N$ until the motion is frictionally damped by viscosity or evolves into a billow and collapses into turbulence. Fluid parcels will not naturally oscillate at frequencies greater than $N$, and so waves will not propagate at these frequencies; $N$ is thus the limiting high-frequency cut-off for internal wave motions.

Internal seiches generate substantial vertical shear due to the baroclinic flow reversals across layer interfaces (**Figures 2–5**). The magnitude of this shear is periodic and has a maximum value when the interfaces are horizontal (e.g., $T_1/4$) and all wave energy is in the kinetic form. A bulk $Ri$ may be applied over the interfaces separating discrete layers

$$Ri = \frac{g'\delta}{(\Delta U)^2} < \frac{1}{4}$$

to predict the formation of instabilities. If the background flow is time-variable, either the condition $Ri = 1/4$ must be maintained for longer than the growth and billowing period of the instability $T_b \sim 20(\Delta U)/g'$ or $Ri$ must be $\ll 1/4$. Values of $T_b$ in lakes are of the order of minutes or less.

Shear instabilities can occur at the nodal locations in lakes, where the vertical shear is greatest, at the base of the surface layer under strong wind conditions, on the upper and lower surfaces of thermocline

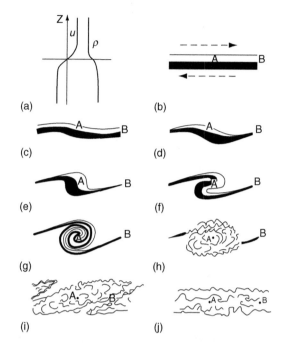

**Figure 9** Schematic showing the growth and turbulent degeneration of a Kelvin–Helmholtz shear instability leading to diapycnal mixing of the stratified fluid. The condition is shown as velocity ($u$) and density ($\rho$) profiles in (a) and as isopycnal surfaces in (b), where the arrows denote the flow direction. In (b–j), A and B are fixed points in the flow and the lines represent surfaces of constant density (isopycnals). Adapted from Mortimer CH (1974). Lake hydrodynamics. *Mitteilungen Int Ver Limnol* 20: 124–197, after Thorpe SA (1987) Transitional phenomena and the development of turbulence in stratified fluids: A review. *Journal of Geophysical Research* 92: 5231–5248.

jets that result from vertical mode-two compression of the metalimnion, near river influents and reservoir withdrawal layers, and in regions where there is flow over rough topography.

## Regime 3: Supercritical Flow

Internal hydraulic jumps occur in stratified flows at the transition from supercritical ($Fr_1^2 + Fr_2^2 > 1$) to subcritical ($Fr_1^2 + Fr_2^2 < 1$) flow conditions, where the upper and lower layer Froude numbers are defined as $Fr_1^2 = U_1^2/g'h_1$ and $Fr_2^2 = U_2^2/g'h_2$. Although they are more commonly observed in the ocean as a result of tidal flow over topographic features (e.g., Knight Inlet sill); internal hydraulic jumps can occur in lakes.

Progressive jumps form when supercritical flow (resulting from a gravity current, inflow, wind event or thermocline jet) propagates into an undisturbed region. Flow over a topographic feature can lead to a stationary jump in the lee of the obstacle. Localized energy dissipation and mixing occur near jumps and waves radiate from the critical point where $Fr = 1$.

The impacts and distribution of hydraulic phenomena in lakes are not well understood.

## Regime 2: Solitary Waves

In small- to medium-sized lakes subjected to moderate forcing ($0.3 < W^{-1} < 1.0$), nonlinearities become significant and the linear wave equation no longer completely describes wave evolution. In addition to the linear standing wave (composed of symmetric cosine components combined in a standing wave pattern), asymmetrical nonlinear wave components are generated from the wind induced thermocline tilt. The asymmetric components combine into a progressive internal wave pattern. The downwelled fluid becomes a dispersive packet of sub-basin scale internal waves of depression (**Figure 10**) called nonlinear internal waves (NLIWs), while upwelled fluid evolves into a progressive nonlinear basin-scale wave, referred to as a rarefaction or internal surge.

The weakly nonlinear Korteweg-de-Vries (KdV) equation mathematically describes the generation and unidirectional progression of NLIW from the wind-induced thermocline setup

$$\frac{\partial \eta}{\partial t} + c_0 \frac{\partial \eta}{\partial x} + \alpha \eta \frac{\partial \eta}{\partial x} + \beta \frac{\partial^3 \eta}{\partial x^3} = 0$$

where the nonlinear coefficient $\alpha = (3c_0/2)(h_1 - h_2)/h_1 h_2$ and the dispersive coefficient $\beta = c_0 h_1 h_2/6$. Initially, the internal surge propagates under a balance between the unsteady ($\partial \eta/\partial t$) and nonlinear $\alpha \eta(\partial \eta/\partial x)$ terms. As nonlinearities become more apparent as the waveform steepens and the wavefront approaches vertical (**Figure 10c**). This occurs at the steepening time scale

$$T_s = \frac{L}{\alpha \delta \eta}$$

Wave steepening causes the dispersive term $\beta \eta(\partial^3 \eta/\partial x^3)$ to become significant, eventually balancing nonlinear steepening at $t = T_s$, leading to the production of high-frequency NLIWs (**Figure 10d**).

In many lakes, NLIWs have a wave profile that matches a particular solution to the KdV equation

$$\eta(x, t) = a\,\text{sech}^2\left(\frac{x - ct}{\lambda}\right)$$

These are called solitary waves. The maximum amplitude of the solitary wave $a$, the solitary wave speed $c$, and characteristic horizontal length-scale $\lambda$ are given by

$$c = c_0 + \frac{1}{3}\alpha a \text{ and } \lambda^2 = 12\frac{\beta}{a\alpha}$$

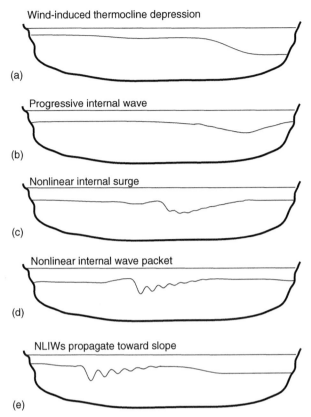

**Figure 10** Schematic showing the evolution of a NLIW packet. (a) Initial wind induced thermocline depression, (b–c) formation of a progressive surge through nonlinear steepening, (d–e) evolution of a dispersive NLIW packet at $t = T_s$. For the case shown, the wind had been blowing for less than $T_1/4$ and a steady state tilt of the entire thermocline was not achieved (i.e., upwelling did not occur). This is a common occurrence in long (~100 km) narrow lakes (e.g., Seneca Lake, Babine Lake, etc.).

The KdV equation reveals some interesting characteristics of solitary waves. In a two-layer system they will always protrude into the thicker layer and so are generally observed as waves of depression upon the thermocline. If the interface occurs at mid-depth, $\alpha \rightarrow 0$; thus preventing nonlinear steepening and subsequent solitary wave generation. Moreover, the dependence of $\alpha$ on $h_1 - h_2$ demonstrates that the degree of nonlinearity depends not only on the magnitude of the interfacial displacement, but also on the relative heights of the stratifying layers.

The dispersive nature of the wave packet is evident from the relationship between wave amplitude $a$ and wave speed $c$; a spectrum of waves in a particular packet will be rank ordered according to amplitude (**Figure 10e**) and will disperse with time as they propagate. An estimate of the number of solitary waves and their amplitudes, while beyond the scope of this article, may be obtained from the Schrödinger wave equation.

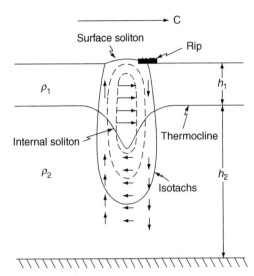

**Figure 11** Schematic showing the passage of an internal solitary wave in a two-layer stratified fluid. Dashed lines are contours of water particle speed (isotachs) and arrows indicate the magnitude and direction of the flow. A small surface solitary wave of amplitude $\sim (\rho_2 - \rho_1)a$ accompanies the solitary wave and causes the rip or surface slick. From Osborne AR and Burch TL (1980) Internal solitons in the Andaman Sea. *Science* 208 (4443): 451–460.

Laboratory experiments show that NLIWs can contain as much as 25% of the energy (APE) introduced to the internal wave field by the winds. In large lakes they may have amplitudes and wavelengths as large as $\sim 20$ m and $\sim 50$–$1000$ m, respectively, and travel at $c \sim 0.5$–$0.75$ ms$^{-1}$. The velocity field associated with large amplitude NLIWs, commonly found in the ocean, will form a slick (rip) on the water surface (**Figure 11**) allowing them to be located and tracked using shore-based, aerial or satellite imagery.

## Shoaling of Nonlinear Internal Waves

Progressive nonlinear internal waves travelling along the thermocline will break when they shoal upon the sloping lake bottom (**Figure 12**). Wave breaking irreversibly converts wave energy to mixing and dissipation, thus contributing to localized intense turbulence in the TBBL. The internal breaking process is similar to surface wave breaking at a beach and may be interpreted in the same manner using an internal form of the Iribarren number form of the Iribarren number $\xi$, which is the ratio of the boundary slope(s) to the offshore wave slope ($a/\lambda$)

$$\xi = \frac{s}{(a/\lambda)^{1/2}}$$

The relationship between $\xi$ and the breaking mechanism is an ongoing area of active research. Laboratory experiments show that for small $\xi$ (**Figure 13(a)**),

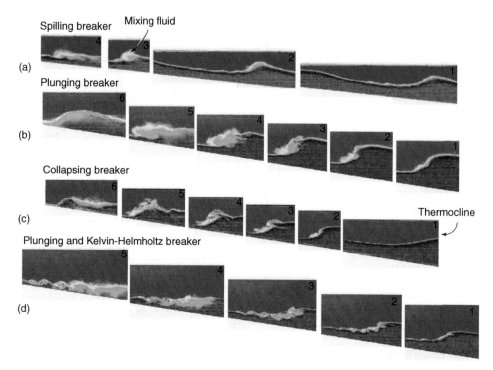

**Figure 12**   False color images showing types of internal wave breakers: (a) spilling breakers, (b) plunging and (c) collapsing breakers, (d) mixed plunging and Kelvin–Helmholtz breaking. Note the steepening of the rear face of the incident wave of depression, the transformation to a wave of elevation and subsequent breaking. Small numbered figures denote different stages of the breaking process. Adapted from Boegman L, Ivey GN, and Imberger J (2005) The degeneration of internal waves in lakes with sloping topography. *Limnology and Oceanography* 50: 1620–1637.

spilling breakers occur when small-scale shear instabilities form on the wave crests prior to breaking, causing mixing to be suppressed by viscosity. The mixing efficiency $R_f = b/(b + \epsilon)$, where $b$ is the energy lost irreversibly to mixing and $\varepsilon$ the energy dissipated by viscosity as heat, is less than 15%. As $\xi$ increases, plunging (**Figure 12(b)**) then collapsing (**Figure 12(c)**) breakers occur. In the plunging region, wave inertia dominates and the most energetic overturns approach the Ozmidov scale (the largest scale where inertia can overcome buoyancy); the potential energy available for mixing is maximized and $R_f > 15\%$. For collapsing breakers, the wave breaking processes is not sufficiently energetic to overcome the stratification, mixing is suppressed by buoyancy and $<R_f \ 15\%$. For $\xi \rightarrow 1$ surging breakers occur, $R_f \rightarrow 0$ and the wave energy is reflected from the slope. Mixed-mode convective and shear-driven breaking is also possible when the wave shoals through a strong background flow field (**Figure 12(d)**).

## Progressive Internal Wave Rays in a Continuous Stratification

We have conveniently described standing waves in a continuous stratification in terms of wave modes. A continuous stratification will also support

progressive sub-basin scale internal waves; however, it is more insightful to describe these waves in terms of rays (both methods of analysis can be shown to be equivalent). Unlike NLIWs, which require a thermocline waveguide, progressive waves in a continuous stratification are described by linear equations and occur in regions of the water column where $N > 0$ and slowly varying (e.g., **Figure 1(e,f)**). A disturbance in the flow (e.g., flow over rough topography) with a particular excitation frequency will generate a range of wavelengths that will radiate from the source at the same frequency. The wave rays will propagate through the fluid at a fixed angle to the horizontal $\beta$ given by the dispersion relation

$$\omega = N \sin \beta = \frac{kN}{K}$$

where $\omega$ is the wave frequency and the wavenumber vector $K = \sqrt{k^2 + m^2}$ has horizontal $k$ and vertical $m$ components (**Figure 13(b)**). The angle at which the rays propagate is chosen such that the vertical component of their frequency matches $N$, leading to a four-ray St. Andrew's cross pattern (**Figure 13(a)**). From the dispersion relation, the wave frequency is independent of the magnitude of the wavelength and only depends upon $\beta$. This property is quite different than for interfacial waves, where the wave frequency and period depend only on the magnitude of the wavelength.

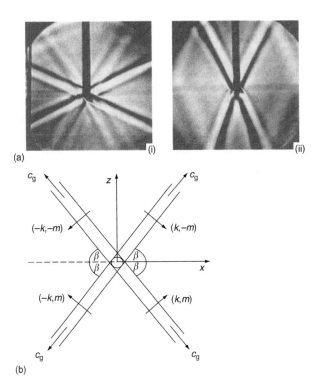

(a)

(b)

**Figure 13** Propagation of internal wave rays in a stratified fluid with constant $N$. (a) Laboratory images showing internal wave rays propagation from an oscillating cylinder. The light and dark bands are lines of constant phase (wave crests and troughs). In (i) and (ii) $\omega = 0.4N$ and $0.9N$ giving $\beta = 25°$ and $64°$, respectively. From Mowbray DE and Rarity BSH (1967) A theoretical and experimental investigation of the phase configuration of internal waves of small amplitude in a density stratified liquid. *Journal of Fluid Mechanics* 28: 1–16. (b) Schematic showing the direction of the group velocity $c_g$, wavenumber vectors (showing the direction of phase propagation), and angle of inclination of the rays relative to the horizontal $\beta$, for the experiments in (a). Adapted from Thorpe SA (2005) *The Turbulent Ocean*. Cambridge, UK: Cambridge University Press.

Equations for the velocity and density perturbations induced by the wave passage are beyond the scope of this review and can be found in physical oceanography texts.

Another property of the dispersion relation is that the wave frequency must lie in the range $0 < \omega < N$, mathematically showing that $N$ is indeed the cut-off frequency for internal waves. Excitation at frequencies $\omega > N$ generates motions that are exponentially damped. Wave energy will propagate from an excitation region at the group velocity $c_g$ of the wave envelope, which is perpendicular to the phase velocity $c$; the wave rays carry the energy at right angles to the motion of the crests and troughs! These waves are difficult to visualize, internal waves generated by a local source do not have the concentric circle pattern of crests and troughs familiar to those who observe

a stone thrown into a pond, but are composed of crests that stretch outward as spokes (rays) carrying energy radially from the source. The wave crests and troughs slide perpendicularly across the rays, seeming to appear from and disappear to nowhere.

From the discussion above, it is not surprising that internal wave rays do interesting things as they reflect from sloping topography and propagate into regions of variable $N(z)$. The intrinsic frequency $\omega$ is always conserved causing rays propagating into depths of diminishing $N(z)$ to refract towards the vertical and be totally reflected at the turning depth where $\omega = N(z)$. If generated in the seasonal thermocline these waves can be trapped between the upper and lower surfaces of the metalimnion where $\omega < N(z)$. Waves propagating into depths of increasing $N(z)$, such as toward the seasonal thermocline, will refract toward the horizontal.

Upon reflection from the lake surface or a sloping bottom with angle $\alpha$, the intrinsic wave frequency is conserved and, by the dispersion relation, the reflected ray must propagate at the same angle $\beta$ as the incident ray (**Figure 14**). The wavelength and group velocity will change as is evident in the change in concentration of wave rays upon refection. If $\alpha = \beta$, the rays are reflected parallel to the slope and have zero wavelength and group velocity. A turbulent bore will from and propagate along the slope and wave energy is rapidly converted into local dissipation and mixing. In this case, both the slope angle and wave frequency are considered critical.

Subcritical waves $\beta < \alpha$ will be reflected back in the direction from which they came and the may escape to deeper water (**Figure 14(c,d)**). However, supercritical waves $\beta > \alpha$ will continue in the same direction (**Figure 14(a)**) and if propagating towards shallower water in the littoral zone may thus be trapped, repeatedly reflecting off the surface, lakebed and turning depths (**Figure 15**). Eventually the rays will break when on a critical slope $\alpha = \beta$, where they have a critical frequency $\omega = N\sin\alpha$.

Progressive internal waves are produced by small localized disturbances such as flow over rough topography, patches of shear and turbulence and wave-wave/wave-flow interactions, when the excitation frequency $\omega < N$. Progressive waves are found ubiquitously in lakes and those in the $10^{-5}$ to $10^{-3}$ Hz bandwidth generally have critical frequencies relative to the sloping boundaries found where the metalimnion intersects the lake bed.

## Resonant and Forced Internal Waves

The periodicity inherent in weather patterns creates over-lake wind fields that occur at regular frequencies. For example, the winds over Lake Erie have

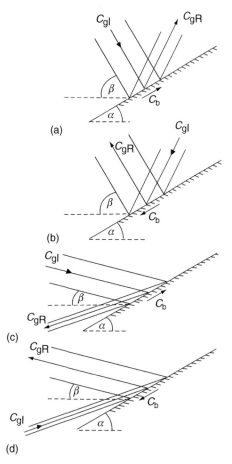

(a)

(b)

(c)

(d)

**Figure 14** Schematic showing the reflection of internal wave rays from a uniform slope. The lines indicate crests and troughs (lines of constant phase) and $c_{gI}$ and $c_{gR}$ the direction of the incident and reflected group velocity, respectively; $c_b$ is the direction of phase propagation on the slope. (a–b) show the supercritical case where $\alpha < \beta$ and rays continue upslope or downslope. (c–d) show the subcritical case $\alpha > \beta$ where rays are reflected out to deeper water. The distance between rays is proportional to the wavelength and can increase or decrease upon reflection. As $\beta \rightarrow \alpha$, the reflected rays become parallel to the slope and their wavelength goes to zero. Adapted from Thorpe SA and Umlauf L (2002) Internal gravity wave frequencies and wavenumbers from single point measurements over a slope. *Journal of Marine Research* 60: 690–723.

periodicities of 10 d and 24 h, associated with frontal weather systems and diurnal land/sea breeze phenomena, respectively. When the forcing frequency matches one of the natural frequencies of the basin-scale wave modes, resonant amplification will occur. Seasonal heating and deepening of the thermocline will adjust the natural frequencies relative to the forcing frequencies, thus tuning the system into and out of resonance.

Laboratory experiments show that the relative frequencies of the wind forcing $f_w$ and H1 internal seiche $f_{H1}$ can be used to model internal wave response to

periodic forcing conditions (**Figure 16**). When the forcing frequency of the wind stress is less than the natural frequency of the H1 internal seiche ($f_w < \sim 2/3 f_{H1}$), the phase of the basin-scale oscillations are reset with each wind event and a forced internal seiche is generated at the same frequency as the wind forcing. A resonant H1 seiche will occur when the frequency of the wind forcing is near the frequency of the H1 seiche ($2/3 f_{H1} < f_w < 2 f_{H1}$) and when the forcing frequency is greater than the natural frequency of the H1 seiche ($f_w > \sim 2 f_{H1}$), higher-mode horizontal seiches are generated. Resonance appears to be particularly effective in amplifying the response of the second vertical mode (e.g., Lake Alpnach) and the even horizontal modes that are not naturally energized by a wind-induced interfacial setup.

Resonance between the wind and the H1 internal mode leads to increased seiche amplitudes under relatively weak wind forcing conditions. For example, when $W^{-1} > 0.03$, nonlinearities become significant (compared to $W^{-1} > 0.3$ under non-resonant conditions), favoring the formation of steepened internal wave fronts (e.g., Loch Ness) and progressive NLIWs. If the wind induced tilting increases, $W^{-1} > 0.2$ will cause Kelvin–Helmholtz instabilities to form within the progressive NLIWs, leading to significant diapycnal mixing within the basin interior. The energy content of the H1 seiche, NLIWs and shear instability modes thus appear to be capped and continued energy input via resonant amplification is transferred between these discrete modes, ultimately being lost to dissipation and mixing at turbulent scales.

## Analysis of Timeseries Data

The various wave, instability and turbulence processes described in this chapter are shown schematically in **Figure 17**. These processes occur beneath the lake surface and so practicing limnologists do not have the luxury of being able to directly observe them in the field. Some insight regarding their spatial structure is gained from idealized laboratory and computational models but for the most part, limnologists must resort to deciphering timeseries data from thermistor chains, which are the most useful tools in their arsenal. **Figure 18** shows many of the processes from **Figure 17** as they would appear on thermistor data, which has been contoured to show isotherm displacement timeseries with depth.

During the strong wind event, shear instabilities form at the base of the surface layer (**Figure 18(d)**). Shortly thereafter, a thermocline jet occurs within a compressed region of the metalimnion and causes a rapid expansion of the strata and localized mixing

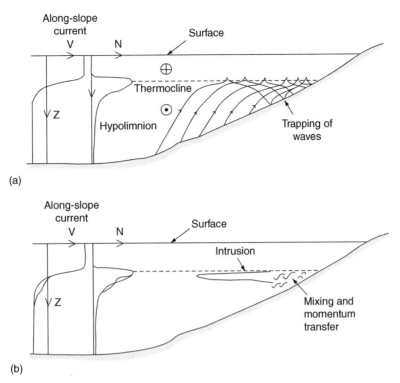

**Figure 15** Schematic showing wave rays radiating from along-slope flow over bottom topography and propagating toward the littoral zone. (a) The waves are trapped between the mixed surface layer and the lake bed and will eventually break where the bed slope $\alpha$ is critical and the wave frequency $\omega = N\sin\alpha$. (b) The mixed fluid created during wave breaking collapses and intrudes into the lake interior, carrying sediment and nutrients from the littoral zone. The local strength of the stratification $N'$ is reduced and the along-slope flow is modified. Adapted From Thorpe SA (1998). Some dynamical effects of internal waves and the sloping sides of lakes. Coastal and estuarine studies: Physical processes in lakes and oceans. *American Geophysical Union* 54: 441–460.

(**Figure 18(e)**). The basin-scale internal wave energized by the wind event has steepened into an internal surge supporting large amplitude NLIWs of depression and series of step like features resembling internal hydraulic jumps (**Figure 18(c)**). The surge wave interacts with the lake bed leading to significant mixing over the bottom 20 m of the water column.

The motions described above all result from a single intense wind event. Considering the periodic nature of surface winds, it is not surprising that periodic wave motions occur in lakes over a range of length scales and frequencies. Spectral frequency analysis is used to conveniently analyze timeseries data and determine the relative amount of energy found at each frequency.

Observations from many lakes suggest the existence of a universal frequency spectrum model for lakes. The main features of the internal wave spectrum are shown for several lakes in **Figure 19**. These lakes range in diameter from 20 km (Lake Biwa) to 10 km (Lake Kinneret) to 1 km (Lake Pusiano). Unlike the spectral energy cascade occurring in turbulent flows, internal waves are generated at discrete frequencies throughout the spectrum. Motions are

bounded at the low frequency end of the spectrum by H1 seiches that contain the most energy with frequencies between zero and $10^{-4}$ Hz. At the high-frequency end of the spectrum, motions are bounded by the high-frequency cut-off $N$. Shear instabilities cause a sub-$N$ spectral peak with frequency $\sim 10^{-2}$ Hz and five orders of magnitude less energy than the basin-scale seiches. The middle portion of the spectrum contains freely propagating linear and nonlinear internal waves. The NLIWs are generated under moderate forcing conditions with $\mathrm{sech}^2$ or solitary wave profiles and frequencies $\sim 10^{-3}$ Hz. These waves are short lived because they break upon shoaling topography at the depth of the thermocline. The portion of the spectrum between the basin-scale seiches and NLIWs (i.e., $\sim 10^{-4}$ Hz) appears to consist of freely propagating gravity waves that have linear or sinusoidal profiles; similar to the broadband background internal wave field associated with the Garrett-Munk spectrum in the ocean. These waves are generated by disturbances within the flow field (radiation from flow over rough topography, wave–wave interactions, mixing regions, intrusions, nonlinear surges and internal hydraulic jumps) where gravity acts as

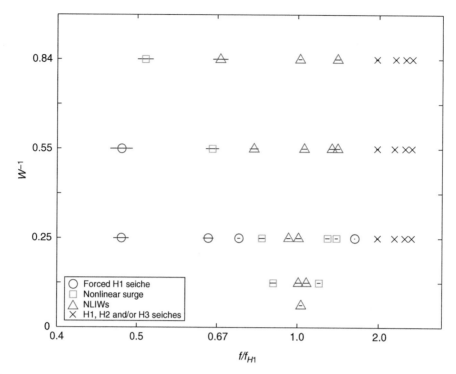

**Figure 16** Regime diagram showing the dominant internal wave response under periodic forcing conditions in a long rectangular laboratory tank. Error bars denote the variation in forcing during an experiment (mean ± standard deviation). After Boegman L and Ivey GN (2007) Experiments on internal wave resonance in periodically forced lakes. In *Proceedings of the 5th International Symposium on Environmental Hydraulics*, 4–7 Dec. 2007, Tempe, Arizona.

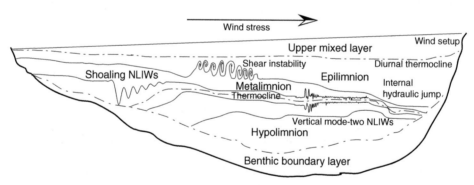

**Figure 17** Pictorial representation of the various regions of a lake and some of the wave-like physical processes that occur. See also **Figure 6.13** in Fischer HB, List, EJ, Koh RCY, Imberger J, and Brooks NH (1979) *Mixing in Inland and Coastal Waters*. San Diego, CA: Academic Press; **Figure 15** in Imberger J (1985) Thermal characteristics of standing waters: An illustration of dynamic processes. *Hydrobiologia* 125: 7–29; and **Figure 7** in Imboden DM and Wüest A (1995). Mixing mechanisms in lakes. In Lerman A, Imboden DM, and Gat J (eds.) *Physics and Chemistry of Lakes*. pp. 83–138. Berlin: Springer.

restoring force on fluid parcels displaced from their equilibrium position. The waves in this frequency bandwidth are interacting with one another making it difficult to identify their source. The bandwidth limits on the spectrum depend only upon the stratification and basin size and are independent of the strength of the wind forcing. Stronger winds lead to sharper energy peaks with higher energy content (larger amplitude seiches and more shear instabilities).

## Summary

The dynamical interplay between stratification, waves, and wind can is best summed up in using the conceptual model proposed by J. Imberger in 1990 and supported by the more recent measurements by A. Wüest and others. A lake behaves like an engine that is powered by the wind and does work against the potential energy gradient embodied in the stratification. Approximately 2% of the wind energy flux

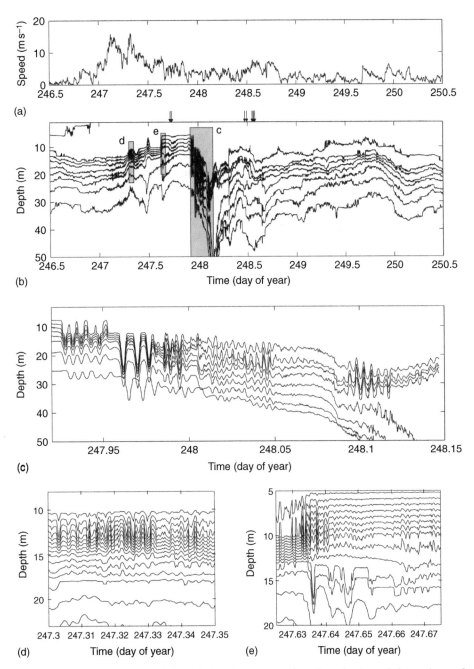

**Figure 18** Observations from Lake Biwa (Japan) in 1993 showing steepened nonlinear basin-scale internal wave front and associated nonlinear internal waves. (a) Wind speed collected at 10-min intervals showing the passage of a storm event; (b) nonlinear response of basin-scale internal wave field showing steepened wave front (2 °C isotherms); (c) magnified view of shaded region in panel b showing details of NLIWs (2 °C isotherms). (d) Magnified view of shaded region d in panel b showing shear instabilities resulting from enhanced shear at the base of the surface layer during the strong wind event (1 °C isotherms); (e) Magnified view of shaded region e in panel b showing a V2 expansion of the metalimnion resulting from a thermocline jet that forms after a period of metalimnion compression (1 °C isotherms). The bottom isotherm in panels b and c is 10 °C. Adapted from Boegman L, Imberger J, Ivey GN, and Antenucci JP (2003). High-frequency internal waves in large stratified lakes. *Limnology and Oceanography* 48: 895–919.

enters the lake; of this ∼80% is dissipated in the surface layer and ∼20% is transferred to the basin-scale internal wave field. The basin-scale seiches energized by the wind are frictionally damped as they swash along the lake bed and by degeneration into progressive high-frequency internal waves, shear instabilities, and eventually turbulence. Approximately 90% of the seiche energy is lost energizing turbulent dissipation and mixing in the TBBL; <1/4 of which first passes through the nonlinear internal

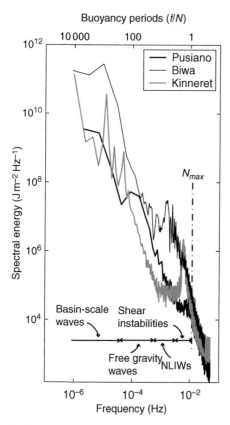

**Figure 19** Spectra of the vertically integrated potential energy signals from Lakes Pusiano, Kinneret and Biwa showing basin-scale seiches ($0-10^{-4}$ Hz), freely propagating nonlinear wave groups with sinusoidal profiles ($\sim 10^{-4}$ Hz) and solitary wave profiles ($\sim 10^{-3}$ Hz), and shear instabilities ($\sim 10^{-2}$ Hz). $N_{max}$ denotes the maximum buoyancy frequency. Adapted from Boegman L, Ivey GN, and Imberger J (2005). The degeneration of internal waves in lakes with sloping topography. *Limnology and Oceanography* 50: 1620–1637.

wave field prior to wave breaking and $>3/4$ of which is lost to frictional swashing in the TBBL. The remaining $\sim 10\%$ of the seiche energy results in intermittent shear instability in the basin interior. The overall mixing efficiency of the turbulence is $\sim 15\%$ leading to an upwards buoyancy flux that works to weaken the stratification and raise the centre of gravity of the lake. The lake engine is extremely inefficient with only $\sim 0.06\%$ of the wind work acting to irreversibly mix the stratification; the bulk of the wind work is lost to frictional viscous dissipation, which due to the large heat capacity of water has no significant effect on the lake temperature. From this model, it is clear that while internal waves control biogeochemical mixing and transport within a stratified waterbody, they are not able to significantly influence the stratification upon which they propagate. Consequently, their existence depends entirely upon the seasonal stratification cycle set up by the surface thermodynamics.

*See also:* The Benthic Boundary Layer (in Rivers, Lakes, and Reservoirs); Currents in Stratified Water Bodies 1: Density-Driven Flows; Currents in Stratified Water Bodies 3: Effects of Rotation; Currents in the Upper Mixed Layer and in Unstratified Water Bodies; Density Stratification and Stability; Mixing Dynamics in Lakes Across Climatic Zones; Small-Scale Turbulence and Mixing: Energy Fluxes in Stratified Lakes.

## Further Reading

Boegman L, Imberger J, Ivey GN, and Antenucci JP (2003) High-frequency internal waves in large stratified lakes. *Limnology and Oceanography* 48: 895–919.

Boegman L, Ivey GN, and Imberger J (2005) The degeneration of internal waves in lakes with sloping topography. *Limnology and Oceanography* 50: 1620–1637.

Fischer HB, List EJ, Koh RCY, Imberger J, and Brooks NH (1979) *Mixing in Inland and Coastal Waters*. San Diego, CA: Academic Press.

Helfrich KR and Melville WK (2006) Long nonlinear internal waves. *Annual Review of Fluid Mechanics* 38: 395–425.

Horn DA, Imberger J, and Ivey GN (2001) The degeneration of large-scale interfacial gravity waves in lakes. *Journal of Fluid Mechanics* 434: 181–207.

Imberger J (1998) Flux paths in a stratified lake: A review. Coastal and estuarine studies: Physical processes in lakes and oceans. *American Geophysical Union* 54: 1–18.

Imberger J and Patterson JC (1990) Physical limnology. *Advances in Applied Mechanics* 27: 303–475.

Imboden DM and Wüest A (1995) Mixing mechanisms in lakes. In: Lerman A, Imboden DM, and Gat J (eds.) *Physics and Chemistry of Lakes*, pp. 83–138. Berlin: Springer.

Lazerte BD (1980) The dominating higher order vertical modes of the internal seiche in a small lake. *Limnology and Oceanography* 25: 846–854.

MacIntyre S, Flynn KM, Jellison R, and Romero JR (1999) Boundary mixing and nutrient fluxes in Mono Lake, California. *Limnology and Oceanography* 44: 128–156.

Mortimer CH (1974) Lake hydrodynamics. *Mitteilungen Internationale Vereinigung Limnologie* 20: 124–197.

Mortimer CH (2004) *Lake Michigan in Motion*. 310 pp. Madison: The University of Wisconsin Press.

Münnich M, Wüest A, and Imboden DM (1992) Observations of the second vertical mode of the internal seiche in an alpine lake. *Limnology and Oceanography* 37: 1705–1719.

Thorpe SA (1998) Some dynamical effects of internal waves and the sloping sides of lakes. Coastal and estuarine studies: Physical processes in lakes and oceans. *American Geophysical Union* 54: 441–460.

Thorpe SA (2005) *The Turbulent Ocean*. Cambridge, UK: Cambridge University Press.

Turner JS (1973) *Buoyancy Effects in Fluids*. Cambridge, UK: Cambridge University Press.

Wüest AJ and Farmer DM (2003) Seiches. *McGraw-Hill Encyclopedia of Science and Technology*, 9th ed. New York: McGraw-Hill.

Wüest A and Lorke A (2003) Small-scale hydrodynamics in lakes. *Annual Review of Fluid Mechanics* 35: 373–412.

Wüest A, Piepke G, and Van Senden DC (2000) Turbulent kinetic energy balance as a tool for estimating vertical diffusivity in wind-forced stratified waters. *Limnology and Oceanography* 45: 1388–1400.

# Currents in Stratified Water Bodies 3: Effects of Rotation

**J P Antenucci,** University of Western Australia, Nedlands, WA, Australia

## Introduction

In this article, we outline the role of the Earth's rotation in modifying currents in inland waters. The first investigation into these dynamics was conducted by Lord Kelvin in the nineteenth century, and the analytical model he developed is still useful in describing some of these dynamics today. More recently, theoretical developments, laboratory experimentation, and field measurements have allowed for the development of a relatively complete picture of the role of the earth's rotation in inland waters.

It is important to note that the effect of the earth's rotation on currents in stratified lakes is predominantly through periodic, oscillatory motions such as gravity waves and vorticity waves, which will be defined later. It is not possible to discuss one without the other, and so in this article we discuss both waves and the currents they induce simultaneously, remembering that waves can be identified through fluctuations in potential energy (typically measured as thermocline or isotherm oscillations) or kinetic energy (typically measured as currents). It is therefore essential that the reader has a good understanding of the material presented in the preceding article. We follow on from this material by investigating how the wind induces motion in large lakes, where the rotation of the earth cannot be ignored.

We begin by defining several parameters that will assist in describing the impacts of the earth's rotation. The most important parameter is the Coriolis parameter (or the inertial frequency)

$$f = \frac{4\pi}{T}\sin\theta \qquad [1]$$

where $T$ is the period of rotation of the earth (1 day or 86 400 s), $\theta$ is the latitude, and the units of $f$ are radians per second. This parameter is zero at the equator (meaning that the effects of the earth's rotation on internal waves and currents can be ignored at the equator), and reaches a maximum value at the poles. The inertial period is defined as

$$T_I = \frac{2\pi}{f} \qquad [2]$$

which is infinite at the equator, and has a minimum value of 12 h at the poles. We also define the Rossby radius of deformation

$$R = \frac{c}{f} \qquad [3]$$

where $c$ is the celerity (speed) of long gravity waves in the water body of interest, where we define long waves as those whose wavelength is far greater than the water depth. For surface ('barotropic') waves $c = \sqrt{gH}$, where $g$ is the gravitational constant (9.8 m s$^{-2}$) and $H$ is the water depth, as described in the preceding article.

Note that it is possible to represent a stratified system as an equivalent depth of homogenous fluid so that the internal ('baroclinic') dynamics can be represented by the same equations. For example, for a two-layer stratification, we can define the equivalent depth as

$$H_e = \frac{\rho_2 - \rho_1}{\rho_2}\frac{h_1 h_2}{h_1 + h_2} \qquad [4]$$

where $\rho$ is the density and $h$ the depth and the subscripts refer to the upper and lower layer. This allows for simple definition of the barotropic phase speed as $c = \sqrt{gH}$, and the baroclinic phase speed as $c_i = \sqrt{gH_e}$, and also allows us to define an internal Rossby radius of deformation that applies to baroclinic processes (those due to stratification) as

$$R_i = \frac{c_i}{f} \qquad [5]$$

The equivalent depth can also be defined for a continuous stratification, using $c_m$ in eqn. [17] discussed later in this chapter.

We also define the Burger number

$$S = \frac{R}{L} \qquad [6]$$

where $L$ is a length characterizing the basin length and/or width and $R$ can represent either the Rossby radius or the internal Rossby radius. This dimensionless number is used to help classify both vorticity and gravity waves, and has been called by various names in the literature, such as the stratification parameter or nondimensional channel width. It is simply the ratio of the length scales at which rotation effects become important to the length scale of the lake in question, so for $S \to 0$ rotation is very important for the dynamics, and for $S \rightarrow \infty$ rotation can be ignored as the lake is physically small. Note that there is no abrupt transition, where the effects of rotation are suddenly felt at $S = 1$, but a gradual transition – this will be discussed later in this article.

## Governing Equations

The dynamics described herein are based solely on the linear inviscid equations of motion for a homogenous fluid. The $x$-momentum, $y$-momentum, and conservation of mass equations are

$$\frac{\partial u}{\partial t} - fv = -g\frac{\partial \eta}{\partial x} \qquad [7]$$

$$\frac{\partial v}{\partial t} + fu = -g\frac{\partial \eta}{\partial y} \qquad [8]$$

$$\frac{\partial \eta}{\partial t} + \frac{\partial}{\partial x}(Hu) + \frac{\partial}{\partial y}(Hv) = 0 \qquad [9]$$

where $\eta$ is the height of the water surface above equilibrium, $H$ is the water depth, and $f$ is the Coriolis parameter. The momentum equations listed here are nothing more than the application of the Newton's famous equation $F = ma$, where the terms on the left-hand side represent acceleration terms (unsteady and Coriolis, respectively), and the term on the right-hand side represents the restoring force due to gravity. The same equations can be applied for barotropic and baroclinic motions, where for the baroclinic case, we replace the actual water depth $H$ by the equivalent depth $H_e$ described earlier.

In describing the effects of the earth's rotation on currents in inland waters, we consider two classes of motion based on the above equations. As we are interested in rotational effects, we first assume $f \neq 0$, such that we are sufficiently far away from the equator. For the first class of motions, which we will term 'gravity waves,' we also assume that the body of water under consideration is sufficiently small such that $f$ can be considered constant (i.e., the lake is at a constant latitude) and that the bottom is flat, and therefore the restoring force is due to gravity only. For the second class of motions, which we will term 'vorticity waves,' we assume $f$ is constant as for gravity waves, but we allow for variable water depth. This variable water depth allows for waves that arise due to the conservation of angular momentum. Dynamically, the second class of motions have similar characteristics to planetary Rossby waves in the ocean and atmosphere (i.e., where the latitude is not considered constant). In most cases, gravity waves dominate the dynamics of lakes and hence are explained later in detail. Only a brief summary

description is given of the dynamics of vorticity waves, and for additional information the reader is referred to the references in Further Reading.

## Gravity Waves

Of the two classes of periodic motions outlined earlier, gravity waves are the most well-studied and best understood in inland waters. We will consider only linear waves, that is, motions where the amplitude of the oscillations of the thermocline is small compared with the depth of the surface and bottom layer. This is not a major restriction on the analysis, as the inclusion of nonlinear effects has been shown in most cases to require only a minor correction to the linear approximation. In this article, we focus on cases where the Burger number is ~1, such that rotational effects can be expected. For surface (barotropic) waves, this would typically require lakes of more than 300 km width, of which there are very few. For baroclinic motions, where the phase speed $c$ is far less than for barotropic motions, there are many lakes in which the Burger number is ~1. For typical values of the baroclinic internal wave phase speed ($0.05-0.4 \text{ m s}^{-1}$), the internal Rossby radius is ~1–5 km (**Figure 1**), indicating that internal gravity waves in lakes of this scale (or larger) should experience the rotational effects of the earth. Note also that unlike the barotropic phase speed, which depends on water depth alone, the baroclinic phase speed varies as a function of stratification and so changes through the year. Rotation may therefore play a more important role in the internal wave dynamics during the strongly stratified period when the internal Rossby radius (and therefore the Burger number) are minimal than at other times of the year.

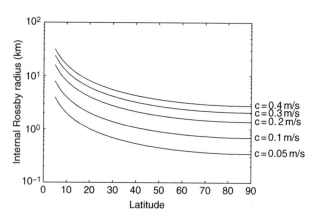

**Figure 1**   Internal Rossby radius as a function of latitude for several internal wave speeds. For horizontal length scales similar to or greater than the internal Rossby radius, rotational effects should be observed.

To understand the form of these motions in inland waters, it is instructive to build up our knowledge from simpler systems. We begin in a rotating system without boundaries, such as in the middle of the ocean far from the coast, where the classic gravity wave solutions are called plane progressive Poincaré waves. The amplitude ($\eta$) and velocity structure ($u,v$) associated with these waves can be described by

$$\eta = \eta_0 \cos(kx - \omega t) \quad [10]$$

$$u = (\omega\eta_0/kH)\cos(kx - \omega t) \quad [11]$$

$$v = (f\eta_0/kH)\sin(kx - \omega t) \quad [12]$$

where $u$ is the velocity in the direction of propagation of the wave, $v$ is the velocity in the transverse direction, $\eta_0$ is the maximum amplitude, $k$ is the wave number ($=2\pi/\lambda$, where $\lambda$ is the wavelength), $\omega$ is the wave frequency ($=2\pi/T$, where $T$ is the wave period), $H$ is the water depth (or equivalent water depth $H_e$ for an internal wave), and $f$ is the inertial frequency. The fluid particle trajectories (in plan view, as the vertical motion is small due to the linear wave assumption) are ellipses with major axes in the direction of propagation, with the ratio of the ellipse axes equal to $\omega/f$ and the direction of rotation anticyclonic (i.e., opposite to the direction of rotation of the earth). For short waves, which have high frequency, $\omega/f$ is large and so the trajectory ellipses are long and thin. Long waves are the opposite, with $\omega/f$ approaching one as the wave frequency is low and therefore the particle tracks are circular and trace out the well-known 'inertial circles' in the ocean (**Figure 2**). The radius of these circular tracks is $U/f$, which can be reformulated using eqns. [11] and [12] as $\eta_0/kH$.

An important aspect of the dynamics of internal waves influenced by the earth's rotation is that energy is generally not equally split between kinetic and potential forms. For the plane progressive Poincaré wave (**Figure 2**), the mean kinetic energy per unit area is

$$\text{KE} = \frac{1}{4}\left(\frac{\omega^2 + f^2}{\omega^2 - f^2}\right)\rho g\eta_0^2 \quad [13]$$

where $\rho$ is the water density, and the potential energy per unit area is

$$\text{PE} = \frac{1}{4}\rho g\eta_0^2 \quad [14]$$

such that the ratio of potential to kinetic energy is

$$\frac{\text{PE}}{\text{KE}} = \frac{\omega^2 - f^2}{\omega^2 + f^2} \quad [15]$$

This indicates that waves with a frequency much greater than the inertial frequency will have a potential

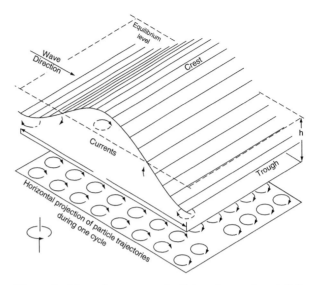

**Figure 2** A long plane progressive Poincaré wave, in an infinite ocean, where $\omega \sim f$. Note the rotation of the current vectors is opposite to the direction of the earth's rotation. Reproduced from Mortimer CH (1974) Lake hydrodynamics. *Mitt. Int. Ver. Theor. Angew. Limnol.* 20: 124–197, with permission from E. Schweizerbart (http://www.schweizerbart.de/).

to kinetic ratio approaching one (as in the nonrotating case where $f \to 0$), and that waves close to the inertial frequency will have close to zero potential energy signal (such as the wave shown in **Figure 2**). This has implications for measurement of these waves, as they will only generally be observed by current measurements (a measure of kinetic energy) and not by fluctuations in stratification (a measure of potential energy variation).

The introduction of a boundary allows for the existence of Kelvin waves. The classical Kelvin wave solution is one in which the velocity perpendicular to the shore is considered to be zero (**Figure 3**). These waves propagate parallel to the boundary with the maximum amplitude at the shore, where the waves crests to the right (in the Northern Hemisphere) when looking along the direction of propagation. The amplitude decreases exponentially offshore at a rate equal to the Rossby radius of deformation $R$,

$$\eta = \eta_0 e^{-y/R}\cos(kx - \omega t) \quad [16]$$

where $x$ is both the alongshore direction and the direction of propagation, and $y$ is the offshore direction (**Figure 3**). Note that the phase speed of the wave is $c = \sqrt{gH}$, the same as for a wave in a nonrotating system. Current vectors, by definition, are rectilinear and oscillate in the alongshore direction only. As with waves in a nonrotating frame, the ratio of potential to kinetic energy is unity. For internal Kelvin waves, the dynamics are the same, except that the baroclinic phase speed applies and the wave amplitude decreases

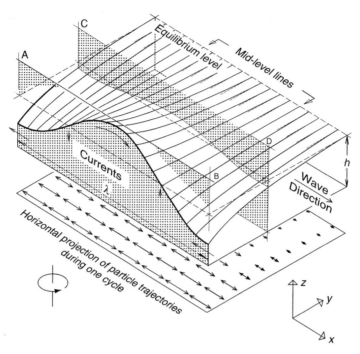

**Figure 3** A long Kelvin wave progressing in the x-positive direction, with the shore located at $y = 0$. Channel walls can be placed vertically at any point of constant $y$, for example indicated by the planes AB and CD. Reproduced from Mortimer CH (1974) Lake hydrodynamics. *Mitt. Int. Ver. Theor. Angew. Limnol.* 20: 124–197, with permission from ● ● ● (http://www.schweizerbart.de/).

exponentially offshore with the internal Rossby radius of deformation $R_i$.

The introduction of a second boundary significantly complicates the waves supported in a rotating system. A channel (defined as two parallel walls with open ends) is able to support progressive Poincaré waves, made up of an obliquely incident plane progressive Poincaré wave with its reflection, and standing Poincaré waves, consisting of two progressive Poincaré waves traveling in opposite directions. These waves consist of cells similar to those presented in **Figures 2** and **3**, however the velocity at the border of each cell approaches zero. Closing a basin, and therefore creating a 'lake,' significantly complicates the wave field. For a rectangular basin, due to the complexity of the corners, an incident plus a reflected Kelvin wave is required along with an infinite number of Poincaré waves of the same frequency to satisfy the boundary conditions. Far simpler solutions can be found by assuming lakes to be represented by circular or elliptical basins of uniform depth, which we will use in the following discussion. As outlined earlier, a key nondimensional parameter controlling this response is the Burger number $S$. Based on this parameter, it is possible to determine the wave frequency for both circular and elliptic basins, the ratio of potential to kinetic energy in the wave response, and the response of a basin to external forcing. We

will also rely on the simplified Kelvin and plane progressive Poincaré waves described earlier to assist in interpreting the results.

To understand the spatial structure of the waves (and the currents they induce) in a rotating system, it is helpful to consider the two end points: strong rotation ($S \rightarrow 0$) and no rotation ($S \rightarrow \infty$). For inland waters, these might also be considered the 'large lake' and 'small lake' case. For $S \rightarrow 0$, in the lake interior, we might expect plane progressive Poincaré waves to be present as outlined earlier (**Figure 2**), where the frequency approaches the inertial frequency, the current vectors rotate anticyclonically and the majority of energy is in the kinetic form. At the lake boundary, we might expect the classical Kelvin wave solution, where the offshore decay in amplitude is exponential at a rate $R$, the velocity at the boundary is parallel to the shore, the ratio of potential to kinetic energy is unity and the frequency approaches zero (**Figure 3**). Data collected from the North American Great Lakes support this conceptual model, with motion in the interior dominated by near-inertial frequencies and motion at the boundary appearing in the form of 'coastal jets,' which are the manifestation of the Kelvin wave solution. As the lake gets smaller (i.e., for $S \rightarrow \infty$), we should expect waves that are similar to the nonrotating case, where the ratio of potential to kinetic energy is unity, and the offshore decay of amplitude is no longer exponential, and the current vectors become rectilinear (i.e., the ellipses become long and thin).

How the characteristics of these waves vary as a function of Burger number and aspect ratio of the lake is graphically presented in **Figure 4**. The nondimensional frequency $\omega/f$ and the ratio of potential to kinetic energy are presented for analytical solutions to the circular and elliptical basin case for cyclonic (Kelvin-type) and anticyclonic (Poincaré-type) waves. Note that the potential to kinetic energy ratios are integrated over the entire lake, and do not represent the character at a particular point in space. We first consider waves in a circular basin (where the aspect

ratio by definition is 1:1). In the strong rotation case ($S \rightarrow 0$), the cyclonic wave frequency goes to zero and the energy ratio approaches unity. This is the Kelvin wave limit for a semi-infinite boundary outlined earlier (**Figure 3**). For the anticyclonic wave, the frequency approaches the inertial frequency and the energy is predominantly kinetic – the plane progressive Poincaré wave solution outlined earlier (**Figure 2**). As $S$ increases, the wave frequency for both types of waves increases and slowly converges as the lake gets smaller. The energy ratio for the anticyclonic wave also increases and asymptotically approaches unity, the nonrotation limit ($S \rightarrow \infty$). For the cyclonic wave, the energy ratio increases to a maximum of $\sim 1.5$, before asymptotically approaching unity as $S \rightarrow \infty$. Importantly as $S \rightarrow \infty$ these two solutions have the same characteristics (frequency, energy ratio, cross-basin structure), except that they rotate in opposite directions. They will thus manifest themselves at high $S$ as a standing wave.

The distribution of potential (i.e., thermocline oscillations) and kinetic energy (i.e., currents) in the basin also changes as the importance of rotation changes (**Figure 5**). For the strong rotation case ($S \rightarrow 0$), the cross-shore potential energy structure of the cyclonic waves has the exponential decay associated with Kelvin waves propagating along a shoreline, where we can rewrite eqn. [14] $\eta = \eta_0 e^{-y/S_L}$ so that for small $S$ the exponential decay is rapid relative to the lake width (**Figure 5(a)**). The kinetic energy is also predominantly located close to the shore (**Figure 5(c)**), hence the term 'coastal jet' being applied to these motions in the North American Great Lakes. As the importance of rotation decreases ($S$ increases), there is a stronger signal of the cyclonic waves present in the interior. It is important to note that for this case, the currents are not parallel to the boundary everywhere in the lake – next to the shoreline they remain parallel as in **Figure 3**; however, towards the interior, the current ellipses become more circular and actually rotate in a cyclonic direction. For the anticyclonic (Poincaré-type) waves, the structure changes very little as the importance of rotation changes (**Figure 5(b)** and **5(d)**). Note that as $S \rightarrow \infty$, the distribution of the potential and kinetic energy in the cyclonic wave approaches that of the anticyclonic wave.

We now consider the impact of changing the aspect ratio by moving towards elliptical basins from a circular basin shape. Note that the Rossby radius is defined in the elliptical basin based on the length of the major axis, not the minor axis, in **Figure 4**. The effect of decreasing the aspect ratio is that the system approaches the nonrotating case for lower values of $S$. Unlike the case of the circular basin, the frequencies of

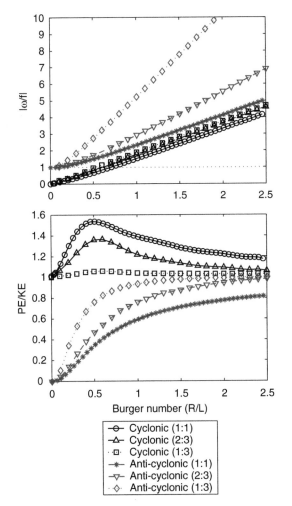

**Figure 4** Nondimensional frequency (upper panel) and ratio of potential to kinetic energy (lower panel) as a function of wave type and aspect ratio, where the numbers in parentheses refer to the aspect ratio. The absolute value of the nondimensional frequency is presented as the results are independent of hemisphere. Reproduced from Antenucci JP and Imberger J (2001) Energetics of long internal gravity waves in large lakes. *Limnology and Oceanography* 46: 1760–1773, with permission from the American Society for Limnology and Oceanography.

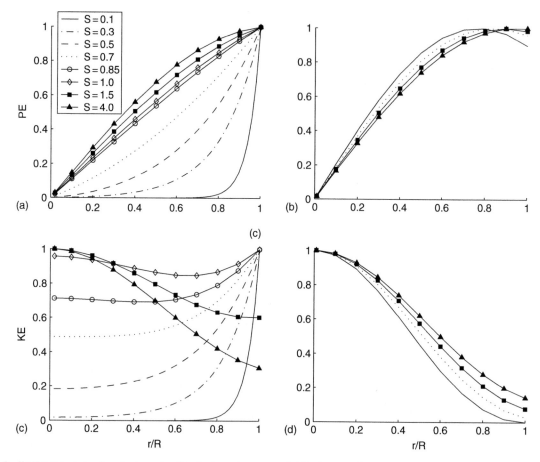

**Figure 5** Radial structure of cyclonic and anticyclonic wave energy distribution as a function of Burger number for a circular lake for the lowest frequency motion (fundamental mode). The cyclonic wave structure is shown in panels (a) and (c). The anticyclonic wave structure is shown in panels (b) and (d). The radial structure for each Burger number has been normalized by its maximum value. Note that not all Burger numbers are shown in panels (b) and (d). Note the exponential decay in (a) is the same as that represented in **Figure 3** and eqn. [16]. Reproduced from Antenucci JP and Imberger J (2001) Energetics of long internal gravity waves in large lakes. *Limnology and Oceanography* 46: 1760–1773, with permission from the American Society for Limnology and Oceanography.

each cyclonic and anticyclonic wave pair diverge rather than converge. In the limit of $S \to \infty$, the cyclonic waves transform into longitudinal seiches, whereas the anticyclonic waves become the transverse seiche solution. It is for this reason that wind-forcing in the transverse direction has been observed to more easily generate anticyclonic, Poincaré-type, waves.

## Current Structure and Measurement

As we have made the linear wave assumption, the vertical velocities induced by these motions are small; however, the horizontal current structure can show significant complexity both in the horizontal and vertical dimension. The complexity in the horizontal direction is due primarily to the presence of boundaries (and hence the horizontal structure of the waves), whereas the vertical complexity is due to both the stratification and the vertical mode.

It is important to tie the vertical position of the measurement location with the likely motion that dominates the flow. The simplest method to determine the likely points of maximum displacement and maximum current is to solve the long linear internal wave problem in a rotating system

$$\frac{d^2 w_m(z)}{dz^2} + \frac{N^2(z)}{c_m} w_m(z) = 0 \qquad [17]$$

where $w_m$ is the vertical velocity eigenfunction for waves of vertical mode $m$, $z$ the vertical dimension, $N^2(z)$ is the vertical profile of the square of the buoyancy frequency,

$$N(z)^2 = -\frac{g}{\rho_0} \frac{\partial \rho(z)}{\partial z} \qquad [18]$$

where $g$ is the gravitational acceleration, $\rho_0$ is a reference density (typically the maximum density), $\rho(z)$ is the vertical profile of density, and $c_m$ is the internal wave phase speed for the particular vertical mode in question. This is an eigenvalue problem, such that an infinite number of solutions exist for an infinite number of vertical modes $m$. For the case of a constant $N^2$, the equation has the sinusoidal solution

$$w_m(z) = \sin\left(\frac{m\pi}{H}z\right)$$ [19]

$$u_m(z) = u_m^0 \cos\left(\frac{m\pi}{H}z\right)$$ [20]

$$c_m = \frac{H}{m\pi}N$$ [21]

where $u_m$ is the horizontal velocity induced by the wave and $u_m^0$ is a constant. It is quite clear from the above that the position of maximum vertical displacement is offset from the position of maximum horizontal current, which the selection of measurement points needs to take into account. Note that the equivalent depth for each vertical mode in a continuously stratified system can be calculated as $H_{em} = c_m/\sqrt{g}$.

For nonconstant $N^2(z)$, eqn. [17] is relatively easily solved numerically as it takes the form of a Sturm–Liouville equation, so from a depth profile of temperature a vertical profile of $N^2$ can be computed and fed into the eigenvalue solver, which will return both the eigenvalues $c_m$ and the eigenvectors $w_m$. These eigenvectors will indicate which region of the water column will experience the maximum isotherm oscillations, and the derivate of this eigenvector with respect to $z$ will give the position of the maximum horizontal velocity fluctuations. It is at these locations that thermistors and current meters should be concentrated, respectively. For most vertical modes, concentrating instruments in and around the thermocline is sufficient, except for capturing the velocity signal of the first vertical mode in which current measurements are best made either near the surface or bottom.

A key method to link currents with the wave motion described earlier is the use of rotary spectra of currents. By analyzing the different direction of propagation of the currents at different depths and points in space, it is possible to understand not only the predominant frequency of oscillation but also the predominant direction of rotation. **Figure 6** shows data from Lake Kinneret in which the basin-scale wave field is dominated by a cyclonic vertical mode one Kelvin wave of period ~24 h, an anticyclonic vertical mode one Poincaré wave of period ~12 h, and an anticyclonic vertical mode two Poincaré wave of period ~22 h. Analysis of current data collected at this location indicates that the Kelvin wave effect on currents in the thermocline (where the 24 °C

isotherm is located) is weak at this station as the current in the 20–24 h bandwidth is dominated by anticyclonic rotation.

## Vorticity Waves

From the above equations, allowing either $f$ to vary as a function of $y$ (the $\beta$-plane) or allowing $H$ to vary as a function of $x$ and $y$ allows a similar class of waves to exist. In the ocean, where $f$ does vary, a class of waves called Rossby waves (or planetary Rossby waves) exist because of the conservation of angular momentum.

In inland waters such as lakes, these effects can be ignored as they are generally smaller than 500 km and $f$ can be assumed to be constant. However, variations in water depth $H$ result in a similar type of wave being possible, again due to the conservation of angular momentum. The structure of these motions is typically more complex than planetary Rossby waves as variations in water depth can occur in all directions, whereas variations in $f$ are limited to the north–south

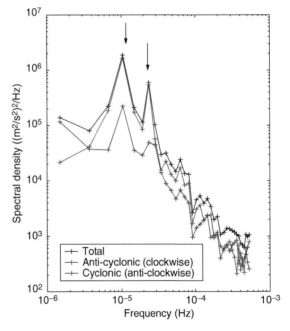

**Figure 6** Spectra of currents along the 24C isotherm in Lake Kinneret during summer 1998 at station T3 on the western margin, showing the total spectrum, the component due to anticyclonic motion ('Poincaré waves'), and the component due to cyclonic motion ('Kelvin waves'). The arrows denote periods of 24 and 12 h from left to right. Adapted from Antenucci JP, Imberger J, and Saggio A (2000) Seasonal evolution of the basin-scale internal wave field in a large stratified lake. *Limnology and Oceanography* 45: 1621–1638.

direction. In this section, we use the term 'vorticity waves' to describe these motions, though they are also called 'topographic waves,' 'vortical modes,' 'second class waves,' or 'quasi-geostrophic waves.' These waves have been observed in large lakes such as Lake Ontario, Lake Michigan, Lake Zurich, and Lake Lugano.

The frequency of these motions is always subinertial (i.e., less than the inertial frequency at that latitude), and the frequency depends primarily on the topography of the basin as it is the topography (through the variation in $H(x,y)$) that causes changes in angular momentum. Importantly, the frequency of these motions is not a function of stratification, and so does not vary on a seasonal basis. This simplifies the measurement of these waves as they are existing at the same frequency year-round.

These waves propagate their phase cyclonically (anticlockwise in the Northern Hemisphere); however, the currents measured rotate both cyclonically and anticyclonically dependent on the horizontal structure of the wave. The currents induced by these waves in the bottom layer consist of a barotropic component only, whereas in the surface layer both a barotropic and a baroclinic current component exists, provided the upper layer is relatively thin than the lower layer. The exact structure of these waves is difficult to determine as there are multiple solutions that have similar frequencies, though it is recognized that the fundamental modes are the most likely to be generated.

## Practical Guide to Measurement of these Waves

On the basis of the aforementioned disucssion, a step-by-step guide is provided on the method that should be applied when investigating these waves-
    Gravity Waves:

1. Compute inertial frequency for the latitude of the lake in question using eqn. [1].
2. Make a simple two-layer approximation to the stratification, and compute the equivalent water depth $H_e$ using eqn. [4].
3. Compute the internal wave speed using $c = \sqrt{gH_e}$. Typically this value will be between 0.1 and 0.3 m s$^{-1}$.
4. Compute the internal Rossby radius using eqn. [5].
5. Based on the dimension of the lake, compute the Burger number using eqn. [6]. If the lake is approximately circular, use the radius for the length-scale $L$. If the lake is approximately elliptical, use the major axes half-length for the length-scale $L$. If the Burger number is greater than 2, rotational effects will be minimal. If the Burger number is less than 1, rotational effects will be very important.
6. From **Figure 4(a)**, read off the nondimensional frequency $\omega/f$ for both the fundamental (lowest horizontal mode) cyclonic and anticyclonic wave for the aspect ratio of your lake.
7. Compute the angular frequency $\omega$ from the non-dimensional frequency $\omega/f$ and the inertial frequency $f$ for each of these two fundamental modes.
8. Compute the period $T$ of these two waves from $T = 2\pi/\omega$.
9. Install measuring equipment (thermistor chain and/or current meters) for a sufficient period to measure more than 10 cycles of each wave. For example, if $T = 2$ days, at least 20 days of measurement will be required to achieve significant confidence in the data analysis. The location of these instruments should be carefully selected and is nontrivial. Typically the best location is halfway between the lake center and the lake boundary. Multiple sampling points are generally necessary – if two stations are deployed they should *not* be placed 180° apart as the direction of propagation can not be determined. It is best to orient stations such that they are 45–135° offset.
10. Compute spectra of temperature signals, isotherm depths, or integrated potential energy to determine the dominant frequencies in the field.
11. Compute rotary spectra of currents to determine the dominant rotation direction.
12. Compute phase and coherence between stations to assist in determining rotation direction and spatial structure. This can be done graphically by simply overlaying signals from the two stations or by using spectral analysis techniques.

    Vorticity Waves:

1. Compute inertial frequency for the latitude of the lake in question using eqn. [1].
2. Deploy current meters for many periods longer than the inertial period, typically several months of record will be required. Two stations are required at the minimum, as with gravity waves they should *not* be placed 180° apart. Current meters should be placed in both the upper and lower layer, away from the thermocline.
3. Compute spectra of current signals to determine dominant frequencies, whether these frequencies change as a function of stratification, and whether the upper and lower layer current structure differs.

Armed with this information, it should be possible to determine which waves are dominating the temperature and current signals. Going into greater detail would typically require the reader to conduct further reading.

## Glossary

**Anticyclonic** – Rotating in the opposite direction as the earth's rotation (clockwise in the Northern Hemisphere, anticlockwise in the Southern Hemisphere).

**Baroclinic** – A flow in which lines of constant pressure are not parallel with lines of constant density, in the context of lake dynamics this applies to all motion that is dependent on stratification.

**Barotropic** – A flow in which lines of constant pressure are parallel with lines of constant density, in the context of lake dynamics this assumption can be made when considering surface motion only.

**Cyclonic** – Rotating in the same direction as the Earth's rotation (anticlockwise in the Northern Hemisphere, clockwise in the Southern Hemisphere).

**Equivalent depth** – Depth of homogeneous fluid where the long-wave barotropic phase speed $c = \sqrt{gH_e}$ is equal to the long-wave baroclinic internal wave phase speed. As internal waves travel slower than surface waves, the equivalent depth is always smaller than the actual depth.

**Gravity wave** – Oscillatory motion where the restoring force is due to gravity.

**Rectilinear motion** – Movement in a straight line.

**Vertical mode** – Number of maxima in the vertical velocity structure for a particular wave.

**Vorticity** – The curl of the velocity field, or circulation ('spin') per unit area about a local vertical axis.

**Vorticity wave** – Oscillatory motion that results from the conservation of angular momentum in flow over varying topography.

## Further Reading

Antenucci JP and Imberger J (2001) Energetics of long internal gravity waves in large lakes. *Limnology and Oceanography* 46: 1760–1773.

Antenucci JP and Imberger J (2003) The seasonal evolution of wind/internal wave resonance in Lake Kinneret. *Limnology and Oceanography* 48: 2055–2061.

Antenucci JP, Imberger J, and Saggio A (2000) Seasonal evolution of the basin-scale internal wave field in a large stratified lake. *Limnology and Oceanography* 45: 1621–1638.

Csanady GT (1967) Large-scale motion in the Great Lakes. *Journal of Geophysical Research* 72: 4151–4162.

Csanady GT (1972) Response of large stratified lakes to wind. *Journal of Physical Oceanography* 2: 3–13.

Gill AE (1982) *Atmosphere-Ocean Dynamics*. San Diego: Academic Press.

Huang JCK and Saylor JH (1982) Vorticity waves in a shallow basin. *Dynamics of Atmospheres and Oceans* 6: 177–186.

Hutter K (ed.) (1984) *Hydrodynamics of Lakes*, CISM Courses and Lectures, vol. 286. Wien: Springer-Verlag.

Mortimer CH (1974) Lake hydrodynamics. *Mitteilungen Internationale Vereinigung für Theoretische und Angewardte Limnologie* 20: 124–197.

Mysak LA, Salvade G, Hutter K, and Scheiwiller T (1985) Topographic waves in a stratified elliptical basin, with application to the Lake of Lugano. *Philosophical Transactions of the Royal Society of London A* 316: 1–55.

Simons TJ (1980) Circulation models of lakes and inland seas. *Canadian Bulletin of Fisheries and Aquatic Sciences* 203: 146.

Stocker R and Imberger J (2003) Energy partitioning and horizontal dispersion in a stratified rotating lake. *Journal of Physical Oceanography* 33: 512–529.

Stocker T and Hutter K (1987) *Topographic Waves in Channels and Lakes on the f-Plane*. Lecture Notes on Coastal and Estuarine Studies, vol. 21, 176 pp. Springer-Verlag.

Thomson SW (Lord Kelvin) (1879) On gravitational oscillations of rotating water. *Proceedings of the Royal Society of Edinburgh* 10: 92–100.

# Currents in the Upper Mixed Layer and in Unstratified Water Bodies

**F J Rueda and J Vidal,** Universidad de Granada, Granada, Spain

## Introduction

Water parcels in lakes and reservoirs change their position at a wide range of spatial and temporal scales, describing trajectories that are the result of random and periodic displacements superimposed on more orderly or coherent patterns. It is the coherent and orderly pattern of motions with long-(seasonal) time scales and large-(basin) spatial scales, what we refer to as circulation (or currents). Random motions, in turn, acting at subseasonal and sub-basin scales are responsible for mixing and diffusion processes. Our focus, in this chapter, is on the large-scale circulation in homogeneous or unstratified water bodies (here on HB) and in the upper mixed layer (here on SML) of stratified lakes. The behavior of HB and SML are considered together, since both are homogeneous layers directly forced at the free surface by wind. The exact circulation that develops in response to wind forcing depend on the specific spatial and temporal patterns characterizing the wind field over the lake, together with morphometric characteristics of the layers themselves. In HB, currents interact with the bottom and variations in topography help create gyres and other secondary flows. Under stratified conditions, in turn, currents in the SML are influenced by the time-varying topology of the thermocline (the lower limit of the SML), which is controlled by internal or density driven motions in complex interaction with the wind field and the Earth's rotation (Coriolis forces). Hence, in generating the currents in the SML, both internal wave motions and wind forcing are intricately linked. In this chapter we will focus on the response of homogenous layers to wind forcing, leaving aside any discussion on the time evolution of their lower boundaries, i.e., whether they remain constant (as is the case of the bottom of the lake in HB) or they change with time, as a consequence of internal wave motions, as is the case of the SML. The temporal evolution of the thermocline, or in general, that of isopycnal surfaces is discussed in depth in other chapters.

The goal of this chapter is to introduce concepts and tools that are needed to understand the mechanisms by which circulation in these homogenous layers is generated in response to external forcing, mainly wind. Wind forcing over lakes is episodic in nature, characterized by a sequence of events of varying intensity and duration interspersed with periods of calm. To describe the response of a lake to such forcing, we will assume that, to first order, lakes behave as linear systems. Under that assumption, the state of motion in a given lake and at any given instant can be described as the result of superimposing the responses to all individual wind events that have acted over its free surface in the past (i.e., a convolution exercise). Owing to frictional losses, water bodies have 'limited' memory, and only those wind events in the 'closest' past (within a frictional adjustment time scale) will effectively determine the circulation patterns exhibited at any given point in time. Here, our focus will be on describing the currents that, according to the linear theory, will develop in an initially quiescent lake in response to a suddenly imposed wind.

Our rationale for using the linearized equations of motion as the starting point for the description of circulation (and not the full non-linear Navier-Stokes equations, governing the motion of fluids in nature) is that linear theory can accurately predict the spatial scales of the large-scale motions, along with their build up or decay time. Furthermore, considerable insight can be gained into the mechanisms involved in the generation of currents by using a simplified set of equations. However, the reader should be aware that the description of the dynamics of circulation provided by the linear theory is, at most, approximate and 'other' features appear as a consequence of the nonlinearity of the fluid motion. The nonlinearities make the problem of studying circulation patterns intractable with analytical tools. The analysis of the nonlinear dynamics of motion needs to be approached with sophisticated numerical models that simulate the hydrodynamic behavior of lakes. Much of the research in the last few years in the study of lake circulation has been in this direction. Considerable advances, also, have been done in the identification of circulation patterns in lakes, with the help of new observational technologies not previously existing, as high-resolution remote sensing (**Figure 1**), autonomous satellite-tracked drogues, acoustic Doppler velocimetry, and others. We will review some of the studies conducted in several lakes located throughout the world in which the large-scale circulation has been described and studied.

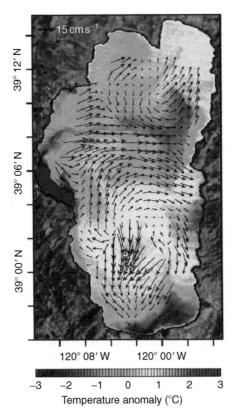

**Figure 1** ETM+ Band 6 (high gain) temperature anomaly, June 3, 2001 18:28 UTC, in Lake Tahoe. The ETM+ image was interpolated to a 90 m grid using bilinear interpolation. The satellite image-derived surface current vector field is overlaid on the image. Adapted from Steissberg TE, Hook SJ, and Schladow SG (2005). Measuring surface currents in lakes with high spatial resolution thermal infrared imagery. *Geophysical Research Letters* 32: L11402, doi:10.1029/2005GL022912, 2005. Copyright (2005) American Geophysical Union. Reproduced with permission from the American Geophysical Union.

## Fundamental Concepts

### Shallow Water Equations

The starting point in our analysis of currents in the SML or HB is the linear set of equations governing the motion for a shallow layer of homogeneous fluid in direct contact with the atmosphere, i.e.,

$$\frac{\partial u}{\partial t} = fv - g\frac{\partial \eta}{\partial x} + \frac{1}{\rho}\frac{\partial \tau_{xz}}{\partial z} \quad x\text{-momentum} \quad [1]$$

$$\frac{\partial v}{\partial t} = -fu - g\frac{\partial \eta}{\partial y} + \frac{1}{\rho}\frac{\partial \tau_{yz}}{\partial z} \quad y\text{-momentum} \quad [2]$$

$$\frac{\partial (\eta - \eta')}{\partial t} + \frac{\partial}{\partial x}(Hu) + \frac{\partial}{\partial y}(Hv) = 0 \text{ continuity} \quad [3]$$

Here, $t$ is time, $u$ and $v$ are the velocity components along the $x$- and $y$- Cartesian directions, $g$ is the acceleration of gravity, $f$ is the Coriolis parameter (or inertial frequency), $H(x, y)$ is the depth of the layer in equilibrium, $\eta$ and $\eta'$ denote the vertical

displacement of the free surface and the bottom boundary from the equilibrium level, and $h = H + \eta - \eta'$. The symbols $\tau_{yz}$ and $\tau_{xz}$ represent the $x$- and $y$-components of the shear stress applied to a horizontal surface at a vertical location $z$. The $z$- coordinate is here considered positive upwards. These symbols will be used consistently throughout this chapter. Equations [1] and [2] are expressions of the second Newton's Law (i.e., force equal mass times acceleration). The first term in their right-hand sides represents the Coriolis force, through which the Earth's rotation influences the dynamics of geophysical flows; the second term accounts for the effects of pressure gradients on fluid motion; and, finally, the third term represents the transfer of horizontal momentum in the vertical $z$ direction by turbulent diffusion. It is presumed, in the derivation of eqns [1]–[3], that the vertical dimension of the layers is much smaller than their horizontal length dimension $L$, i.e., $H << L$. Consequently the ratio of vertical to horizontal velocities is small (the motions are quasi-horizontal) and the distribution of pressure is hydrostatic in the layer. Furthermore, it will be presumed that $\eta << H$ and $\eta' << H$. Only the components $\tau_{xz}$ and $\tau_{yz}$ of the stress tensor are considered here, since they are probably the most important in determining the dynamics of the homogenous layers. It will be presumed that they can be modelled using a gradient transport relationship, i.e.,

$$\tau_{xz} = \rho K \frac{\partial u}{\partial z} \quad \text{and} \quad \tau_{yz} = \rho K \frac{\partial v}{\partial z} \quad [4]$$

where $\rho$ is water density and $K$ is an eddy momentum diffusivity of (or kinematic viscosity), which in the most general case is dependent on the vertical $z$ coordinate. Boundary conditions at the free surface and the bottom of the homogenous layers are

$$\tau_{xz}|_{z=0} = \tau_x^w \quad [5]$$

$$\tau_{yz}|_{z=0} = \tau_y^w \quad [6]$$

$$\tau_{xz}|_{z=-h} = \tau_x^b \quad [7]$$

$$\tau_{yz}|_{z=-h} = \tau_y^b \quad [8]$$

where $(\tau_x^b, \tau_y^b)$ are the $x$- and $y$- components of the stresses applied to the bottom surface of the layer $\boldsymbol{\tau}^b$ and $(\tau_x^w, \tau_y^w)$ represent the $x$- and $y$- components of the wind stress $\boldsymbol{\tau}^w$.

### Lateral and Vertical Circulations

Consistent with the linear approach and for the clarity of presentation, the circulation is considered as the result of superimposing a depth-integrated (or 'lateral') circulation and a vertically-varying horizontal motion (vertical or overturning circulation). The vertical distribution of horizontal velocities can be estimated by solving the momentum equations (eqns [1] and [2]), once a suitable approximation to a distribution of $K(z)$ has been adopted and using the gradients of the free-surface elevation at any $(x, y)$ location as external inputs. This problem is referred to as the 'local problem'. The remaining problem (referred to as the 'global' problem) is to calculate the basin-wide distribution of pressure or lake level (see wind setup). This, in turn, involves solving the set of equations that result from integrating in depth eqns [1]–[3]. The depth-integrated form of eqs. [1]–[3] are also referred to as the 'transport' equations, and they are posed in terms of the horizontal transports $U = \int_{-H}^0 u\,dz$ and $V = \int_{-H}^0 v\,dz$.

### Rotational Effects in Shallow Layers

Rotational effects need to be taken into account when analysing the circulation of lakes with horizontal dimensions larger than the Rossby radius of deformation $R_0$ and when considering processes with time scales which are on the order of $f^{-1}$ or larger. The Rossby radius of deformation is estimated as the ratio between $c$, a characteristic speed at which information is propagated, and $f$, i.e., $R_0 = c/f$. For example the speed at which long- surface waves presure perturbations in the free surface of shallow layers travel can be estimated as

$$c = \sqrt{gH} \qquad [9]$$

For typical values of $f = 10^{-4}$ s$^{-1}$ and $H = 10$ m, $R_0$ for long surface waves is approximately 100 km, larger than the dimensions of most lakes around the world. Hence, the rotational effects can be safely ignored in most lakes when analysing free surface dynamics. Internal waves (which modify $\eta'$ in eqn [3]) however, have periods which are on the order of the inertial period ($2\pi/f$) and their Rossby radius (referred to as the internal Rossby radius) is of order 1000 m. Moreover, surface currents are typically on the order of $10^{-1}$ ms$^{-1}$, and $R_0$ is also $\approx 1000$ m). Hence, it is likely that circulation is affected by rotational effects in most lakes. Only in very small or narrow lakes, with widths $b \ll R_0$ can the influence of the Earth rotation be safely neglected and the motion of water can be described using the same governing eqns. [1]–[3] but setting $f = 0$.

### Wind Setup

The response of a homogenous layer, initially at rest (i.e., with its free surface being horizontal), to a suddenly imposed wind of constant speed and direction consists of two parts: (1) a steady response, manifested by the piling up of water against a leeward coast (referred to as wind setup) and, as consequence, by tilted isopycnals and free surfaces, and (2) a transient or oscillatory response (seiching) of those same surfaces, as a consequence of the interplay between inertial and gravity forces. The period of the oscillatory responses will vary from lake to lake according to their geometry, and for stratified lakes, depending upon the vertical density distribution. The seiching motion of isopycnals (internal waves) in stratified waters and will not be discussed here. As for the surface seiches, their frequency $\sigma$ can be, in most cases estimated using Merian's formula

$$\sigma_n = \frac{n\pi c}{L} \qquad [10]$$

where $c$ is the speed of propagation of long surface waves in a basin of depth equal to the averaged depth of the basin $H$, $L$ is the length of the basin along the direction of the wind and $n$ is the mode of the oscillation. In lakes of complex geometry, such as those with multiple basins, the frequency of the surface seiches needs to be determined through more complicated procedures, but its magnitude is, in any case, on the same order as given by eq. 10. For example, in a lake with $H \approx 10$ m and $L \approx 10$ km, the frequency $\sigma_1$ of the gravest first mode is $O(10^{-3})$. Being their frequency much larger than $f$, surface seiches are seldom affected by rotational effects. They are subject, though, to large dissipative losses. Velocity observations collected in straights (for example connecting large lakes and embayments or several basins in multibasin lakes) are often contaminated and even dominated by large-amplitude oscillations affecting the whole water column, which are associated to the seiching motion of the free surface. The wind setup is characterized by a constant-slope free surface whose magnitude follows from a simple stress pressure-gradient balance (the two terms in the right hand side in eqns [1] and [2]). For a lake of constant depth $H$ acted upon a uniform wind stress in the $y$-direction, the free surface displacement can be estimated as

$$\eta = \frac{u_*^2}{gH} y \qquad [11]$$

where $y = 0$ at the center of the lake and $u_*$ is the shear velocity at the surface of the lake. Equation [11] is obtained by presuming that the lake is sufficiently narrow (i.e., $b/R_0 \ll 1$, where $b$ is the width of the basin and the Rossby radius is evaluated for long surface waves) so that both components of the transport $(U, V)$ vanish everywhere in the basin and Coriolis effects can be effectively neglected. Coriolis forces will only affect the wind setup solution in the limit of $b/R_0 \gg 1$.

## Lateral Circulation

### Generation of Circulation in Homogenous Layers

To characterize the spatial patterns of the horizontal currents use is customarily made of a variable called vorticity $\zeta$ defined as the curl (a mathematical operation, denoted by the symbol $\nabla x$, involving spatial derivatives) of the horizontal velocity vector **u**, i.e.,

$$\zeta = \nabla \times \mathbf{u} = \frac{\partial v}{\partial x} - \frac{\partial u}{\partial y} \qquad [12]$$

A flow field where a fluid particle describes counter-clockwise (or cyclonic) loops will have a positive vorticity; a flow field with negative vorticity, in turn, will make a fluid particle to rotate clockwise (or anti-cyclonically). The magnitude of $\zeta$ is equal to twice the angular velocity of a fluid parcel: the larger $\zeta$ becomes the higher will be the speed at which the particle circulates. Here, the equation governing the evolution of the vorticity $\zeta$ of the depth-averaged flow is horizontal velocity analyzed to understand the mechanisms generating circulation in HBs and SMLs. Ignoring advection and diffusion of vorticity, the governing equation for $\zeta$ in a homogenous layer of thickness $h(x, y, t)$ can be written as

$$\underbrace{\frac{\partial \zeta}{\partial t} = \left(\frac{f+\zeta}{h}\right)\left(\frac{\partial h}{\partial t} + u\frac{\partial h}{\partial x} + v\frac{\partial h}{\partial y}\right)}_{(1)} + \underbrace{\nabla \times \left(\frac{\tau^w}{Ph}\right)}_{(2)} - \underbrace{\nabla \times \left(\frac{\tau^b}{Ph}\right)}_{(3)} \quad [13]$$

The three terms on the right hand side of eq. 13 represent the sources and sinks of vorticity in a closed basin. The generation of cyclonic or anticyclonic vorticity depends on the balance of those terms, which represent:

1. generation of vorticity by temporal changes in the layer thickness, or by flow running into regions with larger or smaller thickness (e.g., a sloping bottom), which results in the stretching and/or squashing of the fluid columns (**Figure 2.1**);
2. flux of vorticity at the top boundary caused by spatial variations of the wind stress $\tau^w$ and/or the bottom bathymetry (**Figure 2.2**);
3. flux of vorticity at the bottom boundary caused by the curl of the ratio of the bottom stress ($\tau^b$) to the depth of the basin.

Term 3 is the only sink of vorticity. All other terms are sources. Term (1) is associated with spatial and/or temporal changes (usually of oscillatory nature) in the thickness $h$ of the homogenous layer. In a stratified water body, it is intricately linked to the existence of internal wave motions. Under the influence of oscillatory currents induced by internal waves, water parcels will describe open trajectories, with a small net displacement after each oscillation. The net basin-wide circulation created by Term (1), hence, is characterized by mean velocities (when averaged over several wave periods) that are much less than the instantaneous values. Such circulation is referred to as *residual circulation*, and it is cyclonic in nature. Residual circulations are defined in contrast with *direct circulations*, which are driven directly by spatially variable winds (Term 2). Water parcels in this type of circulations will move continuously in large gyral patterns, and the mean velocities of water parcels are similar in magnitude and direction to the instantaneous velocities. In contrast with the residual circulations, direct circulation can be either cyclonic or anticyclonic depending on the spatial variations of the wind over the lake surface (see Clear Lake in **Table 2**).

### Topographic Gyres in Homogenous Basins

The evolution of $\zeta$ in a HB acted upon by a suddenly imposed wind can be to first order, described as a result of the spatial variations of wind and the lake bathymetry (i.e. the second term in the right hand side of eqn [13]. This description is only valid for a period of time after the onset of the wind stress which must be short compared to $f^{-1}$, so that Coriolis forces (see Fundamental Concepts) and bottom friction can be safely neglected. The extent to which circulation is controlled by topographic effects or the spatial variability of wind can be assessed using simple scaling arguments, by breaking the second source term in eqn [13] as follows

$$\nabla \times \left(\frac{\tau^w}{h}\right) = \frac{1}{h}\nabla \times \tau^w + \tau^w \times \nabla\left(\frac{1}{h}\right) \approx O\left(\frac{\Delta\tau^w}{h}\right) + O\left(\frac{\tau^w}{\Delta h}\right) \quad [14]$$

Here, the symbol $\Delta$ applied to a variable refers to changes in that variable's magnitude across the basin. The first term represents the changes in magnitude and/or direction of the wind stress over the lake. The second is the result of wind acting in a direction perpendicular to depth variations, and it is referred to as the topological moment. The spatial variability of wind stress will dominate over the topological moment when

$$\left(\frac{\Delta\tau^w}{\tau^w}\right) > \left(\frac{\Delta H}{H}\right) \qquad [15]$$

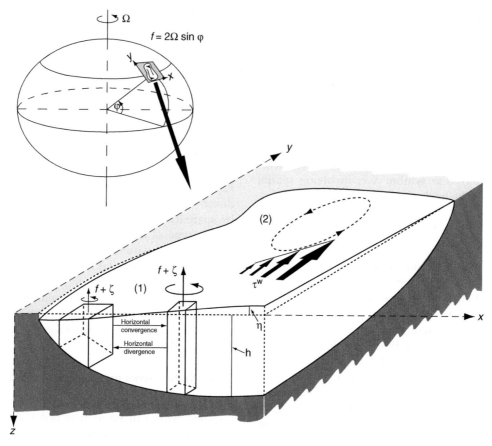

**Figure 2**   Three dimensional section of a lake. (1) Conservation of volume and circulation of a fluid parcel undergoing squeezing or stretching. Vorticity decreases as a fluid parcel travels to shallower regions, and vice versa. (2) Vorticity induced by the curl of the wind stress $\tau^w$.

In flat basins ($\Delta H \approx 0$) the main source of circulation is the curl of the wind stres. In lakes of more realistic varying bathymetry, though, both wind stress and depth variations control the lake circulation. For the case of nearly uniform winds acting along the main axis of a narrow and elongated lake with sloping sides, the only source of circulation is the topological moment and the resulting circulation is characterized by in a double gyre pattern: to the right of the axis in the direction of the wind the curl of $\tau^w/h$ is positive and cyclonic vorticity is produced, whereas to the left anti-cyclonic vorticity is produced (**Figure 3**). Bottom friction (not included, so far in our analysis of circulation) tends to cancel the vorticity input by the wind, but only after the topographic gyres have become established.

### Topographic Waves in Homogenous Bodies

As a consequence of the Coriolis forces acting on an initial double-gyre pattern, this will tend to rotate anti-clockwise (see **Figure 4**). This mode of motion is referred to as *topographic or vorticity wave*. It is free,

since it does not depend on the existence of external forcing (i.e., it only requires that the initial double gyre has been previously established), and has a characteristic frequency $\sigma$ which scales as

$$\sigma \approx \lambda f l k \qquad [16]$$

Here $\lambda$ is a constant of order unity, $l$ is a length scale characterizing the size of the sloping sides in a lake, and $k$ is the wave-number of the wave, which for the gravest first mode, is given by $k = 2\pi/P$ ($P$ being the perimeter). For typical values $f = 10^{-4}$ s$^{-1}$, $l = 10^4$ m and $k = 10^{-5}$ m$^{-1}$, $\sigma$ is of the order $10^{-5}$ s$^{-1}$, hence, $\sigma << f$.

Even though the possiblity of vorticity waves occurring in lakes has been recognized in classical hydrodynamics, evidence of their existence in real lakes, revealed by the cyclonic rotation of the velocity vector at a frequency given by eqn [16], is sparse and weak (see **Figure 4**). This is, in part, due to the fact that winds are almost never absent and will, most likely, interact with existing topographic waves. Moderate and strong winds with enough impulse can change existing circulation patterns, and establish a new basin-scale topographic gyre. Furthermore,

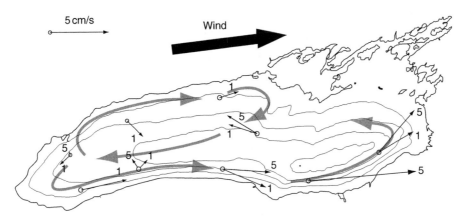

**Figure 3** Large-scale winter circulation in Lake Ontario, reconstructed from observations collected ad different stations (shown as black thin arrows). The numbers in the figure indicate the depth at which measurements were taken: 1 = 15 m and 5 = 75 m. The size of the arrows shows the magnitude of the currents. The gray thick and curved arrows represent the circulation interpreted from the data. Black thick arrow indicates the predominant wind direction. From Beletsky D, Saylor JH, and Schwab DJ (1999) Mean Circulation in the Great Lakes. *Journal of Great Lakes Research* 25: 78–93. Reproduced with permission from the International Association for Great Lakes Research.

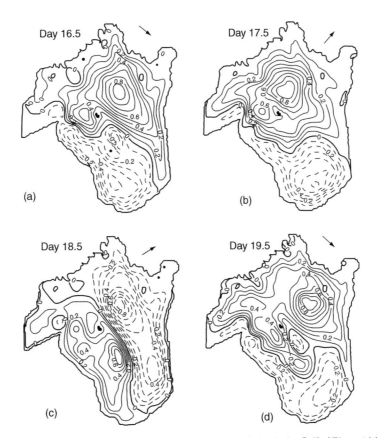

**Figure 4** Normalized stream function calculated with a 2D-model of lake circulation in the Gulf of Riga, at (a) 16.5, (b) 17.5, (c) 18.5 and (d) 19.5 days after the start of simulations. Currents will be parallel to lines of equal stream function. Continuous lines show the cyclonic gyre while dashed lines show the anticyclonic gyre of the double-gyre. Bold arrows show the direction of wind. Adapted from Raudsepp U, Beletsky D, and Schwab DJ (2003) Basin-scale topographic waves in the Gulf of Riga. *Journal of Physical Oceanography* 33: 1129–1140. Copyright 2003 American Meteorological Society. Reproduced with permission from the American Meteorological Society.

time-varying wind directions can also affect circulation patterns: cyclonically rotating winds have been shown in numerical experiments to reinforce the basin-scale topographic waves, while anticyclonically rotating winds tend to destroy the wave. In shallow areas, where the direct wind forcing and bottom friction dominate the vorticity balance, the topographic waves are not likely observed. Hence, only in the deepest points, evidence of topographic waves can be found.

## Spatial Variability of Wind Forcing

The specific wind-driven circulation patterns that develop in the SML or HB are tightly linked to the spatial and temporal variations of the wind stress over the lake (see eqn [13]). While the time variability of wind stress at a single point in space can be characterized with high-resolution wind sensors (e.g., sonic anemometers), characterizing its spatial variability, though, has proved to be a difficult task. Considerable effort in the recent years has been devoted to characterize the spatial variability of the wind stress field over lakes. This is done either by applying dynamic models of atmospheric circulation or by measuring wind speed and direction in arrays of wind sensors located on and around the lake (see, for example, **Figure 5** in Lake Kinneret). Bulk aerodynamic formulations are typically used to derive wind stress values from the wind speed. Technologies developed to characterize wind stress fields directly (scatterometry) have, so far, only applied to oceanic scales, given the very low resolution of existing sensors.

Studies conducted to characterize wind fields over lakes ranging in size from small to large demonstrate that a considerable degree of spatial variability exists both on synoptic and local scales. *Synoptic* scale variability of the wind field will only affect lakes of large dimensions (e.g., Great Lakes). On a *local* scale, factors such as spatial variations in the land surface thermal and/or moisture properties, surface roughness, or the topography can modify and even generate flows in the atmospheric boundary-layer. All of them are, most probably, at play over or in the immediate vicinity of all lakes. The most significant effect of the topography is the aerodynamic modification of ambient synoptic winds. The topographic features, existing around lakes will, among some other effects, cause the ambient wind to change direction (deflection effect) and will create areas of momentum

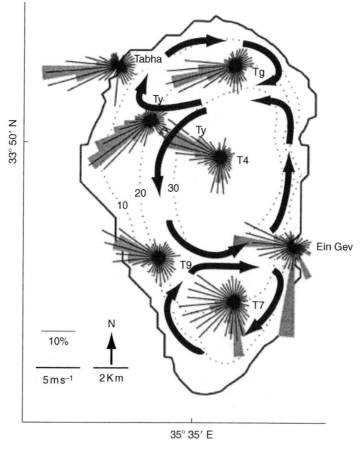

**Figure 5** Lake Kinneret, Israel with 10-, 20-, and 30-m depth contours; wind measurements over the lake and on the shore line. Average wind speed and frequency of occurrence in 108 direction bins, from which the wind is coming (meteorological convention), during days 170–183 are plotted as black lines and grey bars, respectively. Adapted from Laval B, Imberger J, and Hodges BR (2003) Modelling circulation in lakes: Spatial and temporal variations. *Limnology and Oceanography* 48(3): 983–994. Copyright 2003 by the American Society of Limnology and Oceanography, Inc. Reproduced with permission from the American Society of Limnology and Oceanography.

deficit or wakes (sheltering effect). These effects are particularly relevant for lake applications, as they induce spatially variable wind fields over the lower levels in the regional topography where lakes are usually located.

### Other Sources of Circulation

Shoreline irregularities (headlands, peninsulas, islands, bays, etc) or inflow/outflow features (i.e., open boundaries) can also act as sources of vorticity in the SML and HB (see **Figure 6**). The extent to which shoreline irregularities or inflow/outflow features can create vortex structures will depend on geometrical and hydraulic characteristics of the flow such as

- the length scale $L$ of the features; in the case of the peninsulas this length scale, compared with the width of the lake determines the extent to which the flow is blocked and the peninsula effectively seperates the basin in sub-basins;
- their sharpness (i.e., whether the changes in shoreline direction are abrupt or smooth), which can by characterized be a radius of curvature $R_w$;
- the inflow velocity or, more generally, the spatial gradients of horizontal velocity induced by shoreline features:
- the rate at which momentum is transferred horizontally; and
- the Coriolis force, which represents the rate at which flow tend to veer in response to the Earth's rotation, which is characterized by the inertial frequency $f$.

For smooth features and small velocities gradients, currents will be diverted but they will tend to follow the shoreline, and no gyral patterns will develop.

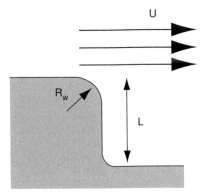

**Figure 6** Schematic representation of a shoreline feature where velocity gradients form. The magnitude of the undisturbed velocity is U; L represents a characteristic length scale for the feature and Rw characterizes its sharpness. This scheme could represent a portion of a river entering a lake, part of a peninsula, or a section of a bay.

For sharp features and large velocity gradients, gyral patterns will, most likely, develop. Dimensional analysis and laboratory experiments suggest that gyral patterns will develop near inflows or shoreline irregularities if the sharpness of the exit corner is larger than the inertial radius, i.e.,

$$\frac{u}{R_w f} \gg 1 \qquad [17]$$

The dividing line between gyre formation and shoreline attachment is not well established and may depend on details such as the actual current profile across in the main body or in the inflowing water. In any case, the behavior of currents near shoreline irregularities is controlled by the nonlinear nature of water motions.

The horizontal extent of the gyres generated near shoreline irregularities will most likely scale with the size of the irregularities, and in most cases, these gyres will have a only be local (not basin) scale (see **Figure 7**). In the case of peninsulas or islands that are large compared to the total width of the lake, the large-scale circulation can be modified to the extent that the basins on each side of the obstruction respond to wind as if they, effectively, were independent, with gyral structures similar to those described above for individual lakes. Vortices associated to river jets will only drive basin-scale gyral patterns for very large inflows. For the Upper Lake Constance, for example, it has been estimated that the observed basin-scale cyclonic gyres (12 000 m wide) can only be driven by extreme inflow events of about 1500 $m^3$ $s^{-1}$ from the Alpine Rhine.

### Vertical Circulation

The wind stress applied to the lake surface is directly transmitted through internal friction to a surface shear layer which, in general, is much thinner than the water column. In stratified water bodies, the thickness of that shear layer is determined by stratification: the vertical flux of momentum beyond the bottom of the SML, by viscous or turbulent processes, is considerably reduced by the existence of large density gradients. In homogeneous water bodies, those shear layers are referred to as Ekman layers and their thickness $D$ is calculated as $(2K/f)^{1/2}$. For turbulent flows, $D$ is empirically found to scale as

$$D \propto u_* f^{-1} \qquad [18]$$

where $u^*$ is the friction velocity at the free surface (defined as the square root of the result of dividing wind stress by water density), and the proportionality factor ranges from 0.1 to 0.4. For typical values of

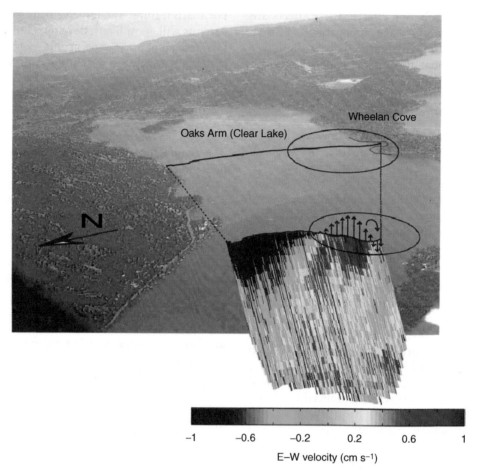

E–W velocity (cm s⁻¹)

**Figure 7**   ADCP velocity measurements taken in the Oaks Arm of Clear Lake, California in the course of a field experiment (Data taken and provided by S.G. Monismith, Stanford University). The ADCP was mounted upside-down on a boat and operated on bottom-tracking mode. The depth of the water column was 10 m. The transect starts on the N-end of the basin and ends in Wheelan Cove at its southern shore. The East–West (EW) velocity component of the velocity field is shown as a cross-section in the lower part of the figure. Color marks the magnitude of the EW velocity: blue color represents velocities towards the West; red color represents velocities directed towards the East. Black arrows represent superficial velocity. Note that at the southern end, within Wheelan cove the measurements reveal the existence of a vortex structure, which forms as a result of the shoreline irregularity. The photograph was provided by S.G. Schladow, University of California – Davis.

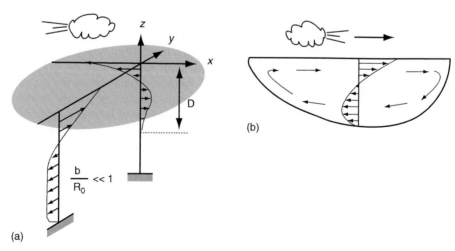

**Figure 8**   (a) Interior velocities caused by wind acting in the negative $x$-direction in a deep basin of limited horizontal extent. Symbols defined in the text. (b) Vertical circulation in a small narrow lake.

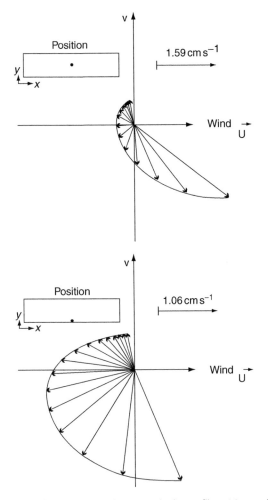

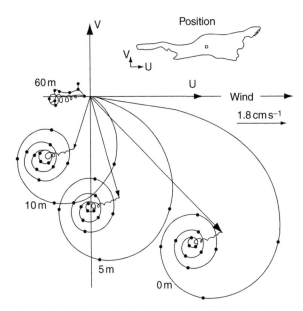

**Figure 10**   Hodographs, i.e., time series of the horizontal velocities in the middle of the Obersee in Lake Constance, at 0, 5, and 10 m depths. The motion is set up from rest. The small circles mark time intervals of approximately four hours. Adapted from Wang Y, Hutter K, and Bäuerle E (2001) Barotropic response in a lake to wind-forcing. *Annales Geophysicae* 19: 367–388. Reproduced with permission from European Geophysical Society.

**Figure 9**   Simulated steady state velocity profiles at two points in a 100 m deep rectangular flat bottom basin, with sides of length 65 km (in the $x$-direction) and 17 km (in the $y$-direction). The arrows correspond to velocity vectors and are shown at 5-m depth intervals from the free surface to the bottom. These are results of simulations conducted with three-dimensional hydrodynamic model, where $K(z)$ is set constant and equal to 0.02 $m^2\,s^{-1}$. The top frame corresponds to a point located near the center of the basin while the lower frame is for a point located near the southern end. Adapted from Hutter K, Bauer G, Wang Y, and Guting P (1998) Forced motion response in enclosed lakes. In Imberger J (ed.) *Physical Processes in Lakes and Oceans*. American Geophysical Union, Washington D.C. pp 137–166. Copyright (1998) American Geophysical Union. Reproduced with permission of American Geophysical Union.

$u_* = 10^{-2}\,ms^{-1}$ in mid-latitude lakes ($f = 10^{-4}\,s^{-1}$) the Ekman boundary layer depth is of the order 10 m. Within the Ekman layer, the stresses reduce to zero, and both velocity magnitude and direction undergo significant changes.

The simplest possible model of velocity change with depth for lakes, is that of the steady-state local solution of eqns [1]–[3] (with $K$ constant), for a flat bottom deep basin of limited horizontal dimensions

acted upon by a constant and uniform wind stress in the $y$-direction, is (see **Figure 8(a)**)

$$u = -\frac{u_*}{fb} + \frac{u_*}{fD}e^{z/D}\left[\cos\frac{z}{D} - \sin\frac{z}{D}\right] \qquad [19]$$

$$v = \frac{u_*}{fD}e^{z/D}\left[\cos\frac{z}{D} + \sin\frac{z}{D}\right] \qquad [20]$$

For $z \gg D$, i.e., outside the Ekman layer, the velocity is in geostrophic balance with the pressure gradient that develops in response to wind forcing (or wind setup, eqn [11]). Closer to the surface, the velocity is a sum of the geostrophic velocity and the Ekman layer velocity (the exponentially decaying terms in eqns [19] and [20]). In the derivation of eqns [19] and [20], bottom stresses have been ignored. For deep basins, this is a reasonable assumption since the velocity near the bottom from eqn [19] and [20] is, in this case, negligible $u_*/fb$.

In basins of intermediate depth, the velocity profiles will have both a surface (driven by wind stress) and bottom (driven by bottom stress or currents) Ekman layers. In both layers, the horizontal velocity vectors would tend to rotate clockwise (in the northern hemisphere) as one moves out of the boundary. The rate at which velocity vectors rotate with vertical

**Table 1**   Circulation patterns of HB in natural systems (lakes and reservoirs): observations and models

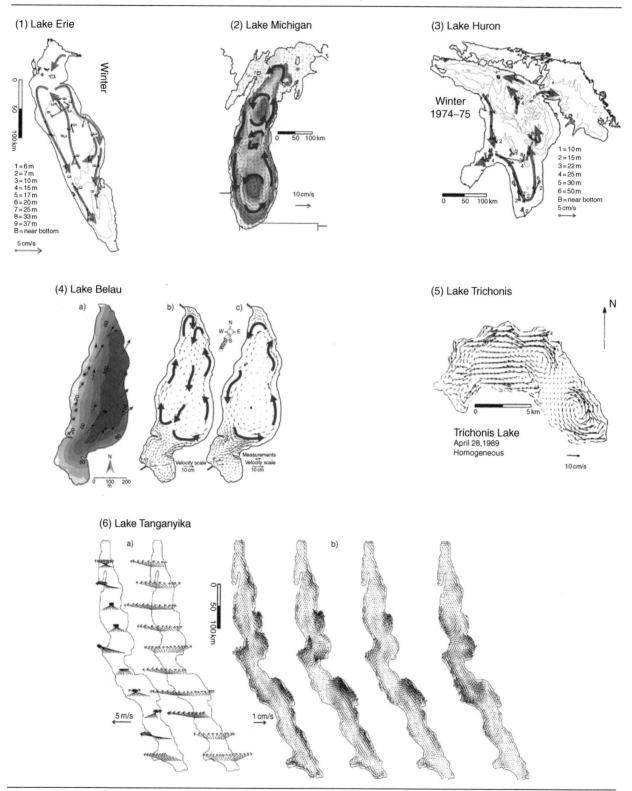

distance away from the boundaries (and, in general, the velocity profiles) is very sensitive to the vertical distribution of the eddy viscosity $K$. In deriving eqns [19] and [20], $K$ has been presumed constant, for simplicity. However, this assumption is not strictly correct since $K$ is known to vary rapidly in the vertical direction within 'wall layers' existing near the top and bottom boundaries and where the velocity varies logarithmically. Only outside these wall layers, but within an homogenous turbulent shear flow region, such as the top and bottom mixed shear layers, the eddy viscosity is approximately constant and proportional to the velocity and the virtual extent of the flow. In any case, the velocity profiles are very sensitive to the magnitude of $K$ and the location (near or far from lateral boundaries) where they are monitored (**Figure 9**). In the limit of very shallow layers ($U*/fh>> 1$) or, equivalently, for very large Ekman numbers ($K f^{-1} h^{-2} >> 1$) it can be assumed that the effects of Coriolis forces in the vertical distribution of velocities are negligible. Large Ekman numbers occur also in narrow or small lakes. In those cases, the balance between pressure and frictional forces (the last two terms in eqns. [1] and [2]) results in a velocity profile in the direction of the wind (no currents perpendicular to the wind) of the form shown in **Figure 8(b)**. For constant $K$ the following analytical expression can be derived that represents the vertical distribution of flow velocities

$$v(z) = \frac{u_*^2}{4Kh} z(3z - 2h)$$  [21]

Note that at the bottom (here $z = 0$) the velocity is zero. Observed velocity profiles in laboratory experiments are, however, closer to a double-logarithmic curve, which corresponds to a parabolic distribution of the eddy viscosity $K$.

In response to a suddenly imposed wind stress on its free surface, the vertical velocity distribution in a homogenous layers of thickness $h >> D$, consists of a steady-state Ekman spiral and inertial oscillations (of frequency $f$) that penetrate to gradually increasing depths. At time $t$ inertial oscillations exists up to a depth of order $(2Kt)^{1/2}$, which is the depth to which momentum diffusion penetrates in a nonrotating system. However, the magnitude of the inertial oscillations will decrease with time as

$$\frac{u_*^2}{f\sqrt{2Kt}}$$  [22]

Thus, after a long period, the inertial oscillations are imperceptibles, even if we do not consider frictional losses at the bottom (see **Figure 10**).

## Circulation Patterns in Natural Systems (Lakes and Reservoirs)

Large-scale circulation patterns for a suite of lakes of different sizes and geometries are illustrated in **Tables 1** (for HB) and **2** (for SML). The patterns were obtained from observations using drogues, velocimetry or other measurements, numerical model simulations, or both.

---

(1) Double-gyre wind-driven circulation in Lake Erie. Figure shows the main winter circulation from observed patterns.
   - Source: Beletsky D, Saylor JH, and Schwab DJ (1999) Mean circulation in the Great Lakes. *Journal of Great Lakes Research* 25(1), 78–93. Reproduced with permission from the International Association for Great Lakes Research.
(2) Cyclonic Gyres in Lake Michigan. Figure shows the winter averaged currents from numerical model simulations. Colored shading represents stream function values; yellow is positive (generally anticyclonic vorticity) and the green-blue-purple areas are negative stream function (generally cyclonic vorticity).
   - Source: Schwab DJ, and Beletsky D, Relative effects of wind stress curl, topography, and stratification on large-scale circulation in Lake Michigan. *Journal of Geophysical research* 108(C2): 3044, doi:10.1029/2001JC001066, 2003. Copyright (2003) American Geophysical Union. Reproduced with permission from the American Geophysical Union.
(3) Circulation in Lake Huron. Figure shows the main winter circulation from observed patterns.
   - Source: Beletsky D, Saylor JH, and Schwab DJ (1999) Mean circulation in the Great Lakes. *Journal of Great Lakes Research* 25(1): 78–93. Reproduced with permission from the International Association for Great Lakes Research.
(4) Cyclonic circulation induced by wind sheltering in Lake Belau. Figure (a) shows the sheltering effect of the wind in the lake. Arrows indicate the measured wind and direction. (b) Double-gyre wind-driven circulation obtained by numerical model using an uniform spatially wind field. (c) Cyclonic circulation obtained with the model and observations using the measured, spatially variable, wind of (a).
   - Source: Reprinted from Podsetchine V and Schernewski G (1999) The influence of spatial wind inhomogeneity on flow patterns in a small lake. *Water Research* 33(15): 3348–2256. Copyright (1999). Reproduced with permission from Elsevier.
(5) Gyres in the wind-driven circulation in Lake Trichonis. Figure shows the simulated water circulation.
   - Source: Reproduced from Zacharias I and Ferentinos G (1997) A numerical model for the winter circulation in Lake Trichonis, Greece. *Environmental Modelling & software* 12(4): 311–321. Copyright (1997). Reproduced with permission from Elsevier.
(6) Gyres induced by diurnally varying winds in Lake Tanganyika. Figure (a) shows the simulated near-surface winds over the lake at midnight (left) and midday (right). Figure (b) shows the depth-averaged flow field, from left to right, at 00:00, 06:00, 12:00, and 18:00, for an average July day.
   - Source: Podsetchine V, Huttula T, and Savijärvi H (1999) A three dimensional-circulation model of Lake Tanganyika. *Hydrobiologia* 407: 25–35. Copyright (1999). Reproduced with kind permission of Springer Science and Business Media.

**Table 2** Circulation patterns of the SML in natural systems (lakes and reservoirs): observations and models

(1) Lake Constance

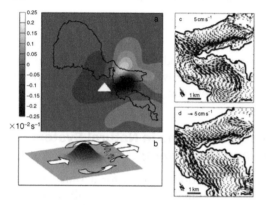

(2) Lake Kinneret

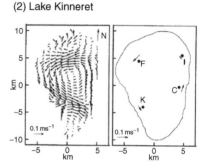

(3) Clear Lake

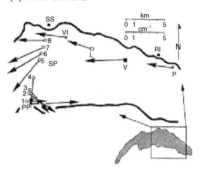

(4) Lake Geneva

(5) Lake Biwa

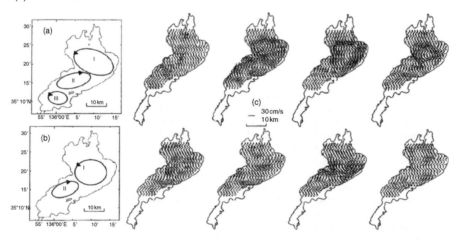

(1) Gyres induced by wind and inflow-driven currents in Upper Lake Constance. Figure shows a qualitative map of typical surface currents. The inflow may induce synoptic eddies; however only extreme discharges could induce the largest eddies of about 12 Km diameter.

  – Source: Hollan E. Large inflow-driven vortices in Lake Constance. In Imberger J (ed.) *Physical Processes in Lakes and Oceans*, pp 123–136. AGU, 1998. Copyright (1998) American Geophysical Union. Reproduced with permission of the American Geophysical Union.

(2) Gyres in Lake Kinneret. Figure shows daily mean currents simulated with a numerical model (left panel) and measured (right panel).

  – Source: Pan H, Avissar R, and Haidvogel DB (2002) Summer circulation and temperature structure of Lake Kinneret. *Journal of Physical Oceanography*, 32: 295–313. Reproduced with permission from the American Meteorological Society.

(3) Circulation induced by windfield vorticity in Clear Lake. Figure shows windfield vorticity generated by Mount Konocti (white triangle) in (a) and (b); and, model simulation of velocity field 2.5 m below free surface, with (c) spatially variable wind and (d) with a uniform wind field.

  – Source: Rueda FJ, Schladow SG, Monismith SG, and Stacey MT (2005) On the effects of topography on wind and generation of currents in a large multibasin lake. *Hydrobiologia* 532: 139–151. Copyright (2005). Reproduced with kind permission from Springer Science and Business Media.

## Conclusions and Future Research Needs

Currents in homogenous bodies and the upper mixed layers of stratified lakes are mainly the result of wind forcing applied at the free surface. The specific basin-scale patterns of motion that develop in response to wind depend on the specific spatial and temporal patterns characterizing the wind field (the frequency, intensity, duration, and location at the time of the events), interacting with the topography, internal waves, and Coriolis effects. By using a simplified and linear set of governing equations and boundary conditions, we have analyzed the mechanisms by which some specific patterns of motion, observed in the field, are generated. However, the governing equations of motion are not linear and the description presented here is, at most, approximate. In shallow lakes, stratification is weak and one cannot easily determine where the bottom surface of the upper mixed layer is. Even if one could decide where that surface is, upwelling events frequently occur in these lakes, when bottom water surfaces, and the upper mixed layer disappears from the upwind end of the lake. These are cases when the nonlinear effects become important. In those cases, an appropriate description of currents involves using the fully non-linear governing equations, which can only be solved with numerical methods. Even if the response of lakes were linear (or one had a 'perfect' numerical model), one major difficulty that needs to be faced in interpreting and analyzing quantitatively circulation and currents in HB and SML is to characterize the wind stress field (i.e., define realistic boundary conditions). Partly, because of this difficulty, there continues to be a considerable amount of uncertainty on what is the relative importance of wind stress curl, topography, and internal waves in generating the large-scale circulation in lakes. Another difficulty that we also face in analyzing the velocity field in lakes is the fact that any description (or observation) of such fields is, at most, incomplete. Many of the techniques used are based on the study of discrete points (Eulerian current meters, Lagrangian drifters, etc.), at single locations or at a series of locations along a transect.

One of the most promising techniques in describing and analyzing the currents in lakes consists in integrating the numerical solution of nonlinear equations and field observations (data-assimilation) in creating a coherent description of circulation. These techniques are now used in the ocean but very few studies have applied this technique to lakes.

*See also:* The Benthic Boundary Layer (in Rivers, Lakes, and Reservoirs); Currents in Stratified Water Bodies 1: Density-Driven Flows; Currents in Stratified Water Bodies 2: Internal Waves; Currents in Stratified Water Bodies 3: Effects of Rotation; Small-Scale Turbulence and Mixing: Energy Fluxes in Stratified Lakes.

## Further Reading

Beletsky D, Saylor JH, and Schwab DJ (1999) Mean circulation in the Great Lakes. *Journal of Great Lakes Research* 25(1): 78–93.

Bormans M and Garret C (1989) A simple criterion for gyre formation by the surface outflow from a straight, with application to the Alboran Sea. *Journal of Geophysical Research* 94(C9): 12637–12644.

Csanady GT (1982) *Circulation in the Coastal Ocean.* Dordrecht, Holland: D. Reidel Publishing Company.

Chubarenko BV, Wang Y, Chubarenko IP, and Hutter K (2000) Barotropic wind-driven circulation patterns in a closed rectangular basin of variable depth influenced by a peninsula or an island. *Annales Geophysicae* 18: 706–727.

Hunter JR and Hearn CJ (1987) Lateral and vertical variations in the wind-driven circulation in long, shallow lakes. *Journal of Geophysical Research* 92(C12): 13106–13114.

Laval B, Imberger J, and Hodges BR (2003) Modelling circulation in lakes: Spatial and temporal variations. *Limnology and Oceanography* 48(3): 983–994.

Lemmin U and D'Adamo N (1996) Summertime winds and direct cyclonic circulation: observations from Lake Geneva. *Annales Geophysicae* 14: 1207–1220.

Mathieu PP, Deleersnijder E, Cushman-Roisin B, Beckers JM, and Bolding K (2002) The role of topography in small well-mixed bays, with application to the lagoon of Mururoa. *Continental Shelf Research* 22: 1379–1395.

(4) Cyclonic circulation in Lake Geneva. Figure shows the observed mean current vectors recorded at around 10 m depth in the central basin of Lake Geneva.
- Source: Lemmin U and D'Adamo N (1996) Summertime winds and direct cyclonic circulation: Observations from Lake Geneva. Ann. Geophysicae, 14: 1207–1220. Reproduced with permission from European Geophysical Society

(5) Spatio-Temporal evolution of Gyres in Lake Biwa. Figures (a) and (b) shows the described gyres I, II, and III at different stages; while figure (c) from left to right and up to down shows the monthly evolution (April–Dec 1994) of the epilimniun velocities measured with ADCP transects. Gyre I was described to move south as temperature stratification became stronger and north again as the stratification weakened. Gyres II and III were not stable, varying position and size accordingly to the wind field.
- Source: Kumagai M, Asada Y, and Nakano S (1998) Gyres Measured by ADCP in Lake Biwa. In Imberger J (ed.) *Physical processes in lakes and oceans*, pp. 123–136. Washington, DC: AGU. Copyright (1998) American Geophysical Union. Reproduced with permission from the American Geophysical Union.

Raudsepp U, Beletsky D, and Schwab DJ (2003) Basin-scale topographic waves in the Gulf of Riga. *Journal of Physical Oceanography* 33: 1129–1140.

Rubbert S and Köngeter J (2005) Measurements and three-dimensional simulations of flow in a shallow reservoir subject to small-scale wind field inhomogeneities induced by sheltering. *Aquatic Sciences* 67: 104–121.

Rueda F, Schladow GS, Monismith SG, and Stacey MT (2005) On the effects of topography on wind and the generation of currents in a large multi-basin lake. *Hydrobiologia* 532: 139–151.

Schwab DJ and Beletsky D (2003) Relative effects of wind stress curl, topography, and stratification on large-scale circulation in Lake Michigan. *Journal of Geophysical Research* 108(C2): doi:10.1029/2001JC001066.

Strub PD and Powell TM (1986) Wind-driven surface transport in stratified closed basins: direct versus residual circulations. *Journal of Geophysical Research* 91(C7): 8497–8508.

Wang Y, Hutter K, and Bäuerle E (2001) Barotropic response in a lake to wind-forcing. *Annales Geophysicae* 19: 367–388.

# Meromictic Lakes

**K M Stewart,** State University of New York, Buffalo, NY, USA
**K F Walker,** The University of Adelaide, SA, Australia
**G E Likens,** Cary Institute of Ecosystem Studies, Millbrook, NY, USA

## Introduction

Meromictic lakes are unusual bodies of water that fascinate scientists from many disciplines. Most deep lakes follow a seasonal cycle of stratification and complete mixing, but meromixis is a condition in which a lake does not mix completely. The term was introduced in 1935 by an Austrian limnologist, Ingo Findenegg, to describe the circulation behavior of subalpine lakes in Corinthia. Since then, examples have been reported from many other parts of the world.

Lakes are warmed by the sun and mixed by the wind. In shallow lakes, wind-driven currents keep the water mixed all year round, but in deep lakes there comes a time, during the warming cycle, when the currents are no longer able to overcome the density differences that develop between top and bottom waters, and the lake becomes stratified into layers (an upper epilimnion, a lower hypolimnion, and an intervening boundary layer termed the metalimnion or thermocline). Later, as the air cools, the density difference is diminished, the currents penetrate deeper, and the lake eventually is restored to complete circulation (holomixis). In Temperate Zones the cycle of temperature is aligned with seasonal changes, but even in tropical and Polar Regions the interplay of sun and wind may impose short-lived periods of incomplete circulation. Meromixis, then, is a global phenomenon.

In meromictic lakes the bottom water usually contains dissolved salt, so that the density difference between top and bottom is reinforced by both salinity and temperature. The effect may be aided by the size, shape, and orientation of the lake basin. A deep hole in a shallow basin, or a lake sheltered from the prevailing wind by surrounding mountains or tall vegetation, would have a greater propensity to retain stratification. Depending on the shape and exposure of the basin, and the intensity of the stratification, meromictic conditions can last anywhere from one year to centuries.

A meromictic lake has an upper mixolimnion (a stratum where mixing by wind occurs), a lower monimolimnion (the more dense stratum that is not mixed by wind), and an intervening boundary layer or chemocline (usually the thinnest stratum, but contains the greatest chemical or density gradient). The mixolimnion may behave like a typical lake and become seasonally stratified (thus, an epilimnion and hypolimnion) and later mix completely. However, the monimolimnion remains isolated from the atmosphere and does not participate in these periods of mixing. The resistance of the total water column to mixing depends mainly on the density differences across the chemocline.

In a meromictic lake, the monimolimion is usually anoxic (without oxygen) and uninhabited by organisms other than bacteria, some protozoans, and possibly nematodes. Some of the dipteran 'fly' larvae of the Chironomidae and Chaoboridae families are able to survive for considerable periods of time without oxygen. The hypolimnion of a thermally stratified lake may also be seasonally depleted in oxygen, if the lake is nutrient rich or *eutrophic*, but oxygen is redistributed throughout the water column when the lake does undergo mixis.

## Meromixis as a Pattern

In limnology (the physics, chemistry, and biology of inland waters), the circulation patterns of lakes are classified by the extent to which the water mixes vertically, from top to bottom, during the course of a year (**Table 1**).

The scheme is merely a guide because lake circulation patterns, like most natural phenomena, do not fall easily into discrete categories. A more accurate, but certainly more arduous, approach would be to characterize lakes only after long term data are available.

Another term, oligomixis, is used sometimes to refer to lakes with limited episodes of complete mixing (perhaps 1–4 times in a decade), whereas lakes that do not mix over decades, centuries, and millennia are unequivocal examples of meromixis. Oligomictic lakes are not distinguished in **Table 1** because the distinction is arbitrary, but they underscore the point that circulation patterns form a continuum rather than a series of discrete types. The boundaries between types are also pliable in the sense that lakes may be meromictic (or some other condition) for many years and then, through some natural event, or human influence, the pattern of mixing may change.

**Table 1** Categories of annual vertical mixing in lakes

| | Category | Definition |
|---|---|---|
| A | Amixis | A (no) mixing |
| B | Meromixis | Mero (partial) mixing; incomplete mixing |
| C | Holomixis | Holo (complete) mixing at least once per year |
| | Monomixis | Mono (one) period of mixing |
| | Dimixis | Di (two) periods of mixing |
| | Polymixis | Poly (multiple) periods of mixing |

Adjectives are formed by appending '-ictic', thus amictic, meromictic etc.

'Spring meromixis' is another phenomenon that blurs the boundaries between discrete types. In some cold, Temperate Zone lakes, stratification formed under ice in winter persists into spring and summer. The surface water may warm rapidly after melting of the ice, reinforcing the winter stratification, or the winds in spring may be unusually mild, so that the stratification is not destroyed by mixing. As a consequence, there is no spring 'overturn,' as in a dimictic lake. 'Spring meromixis' effectively converts a dimictic lake to a monomictic lake for that year.

## Origins of Meromixis

The modes of origin of meromixis may also be classified, with the same kinds of caveats. (Note: To avoid creating new 'Type' designations in this chapter, and possibly adding confusion, we will use the same Roman numeral designations in the text and in **Table 2** as in the 1975 usage of Walker and Likens. However, we are not including the Type II category of Walker and Likens. See Further Reading.)

### Ectogenic Meromixis (Type I)

This type results when more dense, usually saline, water from an external source enters the waters of a less dense lake. The heavier inflowing water moves down into the deeper regions and forms a monimo-limnetic pool. This is how ectogenic meromixis is caused by salt water intrusions to low-lying coastal basins, as in Tessiarrsuk Lake in Labrador, Canada, and Lake Suigetsu, Japan. Ectogenic meromixis may also be initiated if fresher or less saline water flows over a more saline lake. A likely example of the latter is Soap Lake in Washington, USA. In either case, a chemocline develops as a barrier to complete vertical mixing and the deeper water quickly becomes anoxic. Of course, the chemical stratification may be eroded, even destroyed, by extensive inflows of fresh water or saline water, depending on the circumstances. Examples of water bodies with ectogenic meromixis are shown in **Table 2**.

### Crenogenic Meromixis (Type III)

This condition develops when saline water enters a lake from (1) discrete springs within the basin, (2) lateral seepage from intersecting geological strata, or (3) some combination of (1) and (2). Two examples are Lake Kivu, Africa, and Junius Ponds 5 in New York, USA. Three European lakes once considered as classic examples of crenogenic meromixis, namely Lakes Girrotte (France), Ritom (Switzerland), and Ulmener Maar (Germany), appear to have lost their meromictic condition. Lakes Girrotte and Ritom were modified owing to the development of hydroelectric power, and Ulmener Maar was affected by the addition of water through a permanent tunnel from the nearby Jungfernweiher Reservoir and by wells, drilled to substantially below mean lake level, to provide 'bank-filtered' water for the Town of Ulmen.

### Biogenic Meromixis (Type IV)

This type is biologically induced, and results from an accumulation in the bottom water of solutes from photosynthetic carbonate precipitation or from diffusion of chemicals from sediments. The density of the bottom water increases, and prevents the lake from mixing completely throughout the year. Two examples are Lake Mary, in northern Wisconsin, USA, and Devil's Bathtub, in New York, USA; others are listed in **Table 2**. Because the density differential in biogenic meromictic lakes is governed by 'home-grown' rather than 'imported' contributions of salts, the morphometric influence (depth relative to surface area) tends to be high. In Lake Mary, for example, the basin has a maximum depth of 21.5 m but a surface diameter of only about 100 m, and there is little opportunity for the wind to mix the deeper water. If Lake Mary were only 5 m deep, it is unlikely that biogenic meromixis would develop.

### Cryogenic Meromixis (Type V)

This less familiar mode involves the effects of 'freezing-out' of salts and gases as the ice cover develops on a lake during winter, and is likely to be the genesis of meromixis in some polar lakes. As ions are separated during ice formation, relatively dense 'salt fingers' form below the ice and descend. More salts are expelled as the ice thickens, reinforcing the density differences between top and bottom waters. If the ionic ratios of both the upper and lower waters were similar, as is the case for Lakes Garrow and Sophia in the Canadian High Arctic, then meromixis by cryogenic means would be more likely than through episodic inputs of waters with different chemistries.

**Table 2** List of worldwide water bodies exhibiting different types of meromixis

| Region | Lake | Lat. | Long. | If Known...? Area (ha) | Zm | Mean-m | L (km) | masl | Type | Source |
|---|---|---|---|---|---|---|---|---|---|---|
| NORTH AMERICA USA (States) | | | | | | | | | | |
| Alaska | Long | 62° 57' N | 141° 52' W | 3.2 | 3.2 | | | 515 | III, IV | Likens, 1967 |
| | Nuwuk | 71° 25' N | 156° 28' W | | 6 | | | | Ib, IV | Mohr et al., 1967 |
| | Pingo | 65° 40' N | 144° 20' W | 2.4 | 8.8 | | | 206 | III, Ia | Likens & Johnson, 1966 |
| | Redoubt | 56° 54' N | 135° 15' W | 1660 | 266 | | | | Ib, IV | McCoy, 1977 |
| | Rosetead | 57° 29' N | 152° 27' W | | | | | | | McCoy, 1978 |
| | Vee Pond | 62° 58' N | 141° 56' W | 4.5 | 1.6 | | | 515 | III, V | Likens, 1967 |
| Florida | Beville's Pond | 29° 35' N | 82° 20' W | | >6 | | | | IV | Shannon & Brezonik, 1972 |
| | Deep | 26° 03' N | 81° 23' W | 0.7 | 30 | | | | IV | Hund, 1958 |
| | Lake 27 | 29° 35' N | 82° 20' W | | 7 | | | | IV? | Shannon & Brezonik, 1974 |
| | Mize | 28° 40' N | 82° 11' W | 0.9 | 25.3 | | | | IV? | Shannon & Brezonik, 1975 |
| Hawaii | Kauhako (on Molokái) | 21° 11' N | 156° 57' W | 0.35 | 248 | | ~0.035 | 0 | | University of Hawaii, Dr. Alam web site |
| Maine | Basin Pond | 44° 27' N | 70° 03' W | 1.3 | 32.3 | | | 124 | | p.c., S. Cooper |
| | Wellman Pond | 44° 24' N | 69° 49' W | 0.4 | 20.1 | 7.9 | 0.29 | 82 | | p.c., S. Cooper |
| Massachusetts | Mystic (Lower) | 42° 25' N | 71° 08' W | 36 | 24 | 8.3 | | 0 | I | Duval & Ludlam, 2001 |
| | Oyster Pond | 41° 20' W | 70° 35' W | 91 | 14 | | | | | Walsh, in Carter, 1967 |
| Michigan | Canyon | 50° 00' N | 93° 35' W | 26 | 18 | 4.3 | 0.53 | 264 | IV | Smith, 1941 |
| | Forest | 42° 35' N | 83° 18' W | 1.0 | 90 | | | 277 | | p.c., P. Sawayko |
| Crystal falls | Fountain Pond Pit | 46° 06' N | 88° 23' W | 1.98 | 18.6 | 8.6 | 0.164 | 421 | | p.c., W. Ziegler, MI-DNR |
| | Hemlock | 45° 11' N | 84° 24' W | 2.3 | 17.7 | | | 271 | | Fast & Tyler, 1981 |
| | Sodon | 42° 19' N | 83° 17' W | | | | | 569 | IV | Newcombe & Slater, 1958 |
| | Turtle | 42° 36' N | 83° 18' W | | | | | 279 | | p.c., P. Sawayko |
| | Walnut | 42° 33' N | 83° 19' W | | | | | 268 | | p.c., P. Sawayko |
| Minnesota | Arco | 47° 09' N | 95° 10' W | 1.4 | 10.2 | | | | IV | Baker & Brook, 1971 |
| | Brownie | 44° 58' N | 93° 19' W | 7.1 | 15.2 | 6.8 | | | | p.c., Sara Aplikowski |
| | Budd | 47° 10' N | 95° 10' W | 2 | 10.8 | | | | IV | Baker & Brook, 1971 |
| | Deming | 47° 10' N | 95° 11' W | 5 | 17 | | | | IV | Baker & Brook, 1971 |
| | Josephine | 47° 09' N | 95° 10' W | | 10.3 | | | | | Baker & Brook, 1971 |
| | Lake of the Clouds | 48° 8.5' N | 91° 6.7' W | 12 | 31.0 | | | | IV? | Swain, 1973 |
| Ely, MN | Miner's Pit | 47° 54' N | 91° 51' W | 55.9 | 47.8 | | 2.54 | 388 | | Janzen, J. 2004. The diver's guide to Lake Wazee. Inland Sea Corp. Andover, MN 55304 |
| Crosby, MN | Pennington Pit | 46° 29' N | 93° 59' W | 23.1 | 78.9 | | 1.20 | 362 | | Janzen, J. 2004. The diver's guide to Lake Wazee. Inland Sea Corp. Andover, MN 55305 |

Continued

**Table 2** Continued

| Region | Lake | Lat. | Long. | If Known…? | | | | | Type | Source |
|---|---|---|---|---|---|---|---|---|---|---|
| | | | | Area (ha) | Zm | Mean-m | L (km) | masl | | |
| W. of Itasca Park | Spring | 45° 06′ N | 93° 13′ W | 1.2 | 8.5 | 3.0 | | | | p.c., Sara Aplikowski |
| | Squaw | 47° 14′ N | 95° 17′ W | 61.0 | 24.0 | | | | | Baker & Brezonik, 1971 |
| | Tin Cup | | | | | | | | IV | p.c., E. Gorham |
| Montana | Berkeley Pit | 46° 01′ N | 112° 30′ W | 276 | 260 | | 1.76 | | | p.c., T. Duaime |
| Nevada | Big Soda | 39° 31′ N | 118° 52′ W | 161.6 | 64.5 | 26.2 | 2.0 | 1217 | Ia | Kimmel et al., 1978, L&O |
| New Mexico | Cinder Cone | | | | | | | | Ia | Bradbury, 1971 |
| New York | Ballston | 42° 55′ N | 73° 52′ W | | 50–60 | | | 77 | ? | p.c., D. Rodbell, Geol. Dept., Union College, Schenectady, NY |
| | Devil's Bathtub | 43° 01′ N | 77° 34′ W | 0.7 | 14.3 | | 0.011 | | IV, III | Stewart, 1997 |
| | Green (Clarke Res.) | 42° 59′ N | 76° 05′ W | 3 | 15.5 | 5 | | | III | Brunskill et al., 1969 |
| | Green (Fayetteville) | 43° 03′ N | 78° 58′ W | 25.8 | 52.5 | 28 | 1.141 | 127 | IV, III | Brunskill et al., 1970 |
| | Junius Pond 5 | 42° 57′ N | 76° 57′ W | 6.5 | 17.6 | | 0.27 | 515 | | Pendl & Stewart, 1986 |
| | Round (Fayetteville) | 43° 01′ N | 76° 01′ W | 13.2 | 51 | 34 | 0.46 | 127 | IV, III | Brunskill & Ludlum, 1969 |
| Rhode Island | Basin (U), Pettaquamscutt R. | 41° 30′ N | 71° 26′ W | | 12.5 | | 0.71 | 2.1 | | Gaines & Pilson, 1972, L&O |
| | Basin (L), Pettaquamscutt R. | 41° 30′ N | 71° 27′ W | | >12 | | 2.19 | 2.1 | | Gaines & Pilson, 1972, L&O |
| South Dakota | Medicine | | | | | | | | | Hayden, 1972 |
| Washington | Blue | 41° 39′ N | 119° 46′ W | 44.3 | | 19.4 | 1.3 | 543 | Ia | Walker, 1974 |
| | Hall, no longer meromictic | | | | | | | | | p.c., Culver, 2006 |
| | Hot | 48° 58′ N | 119° 29′ W | 1.3 | 3.25 | 1.1 | | 594 | Ia, III | Walker, 1974 |
| | Soap (Grant Co.) | 47° 23′ N | 119° 30′ W | 339 | 27 | 7.5 | | 327 | Ia, III | Anderson, 1958, L&O |
| | Soap (Okanogan Co.) | 48° 14′ N | 119° 39′ W | 62.6 | 17.5 | 8.8 | 1.3 | 384 | Ia | Walker, 1974 |
| | Lenore, no longer meromictic | | | | | | | | | Walker, 1974 |
| | Lower Goose | 46° 57′ N | 119° 17′ W | 21.8 | 28 | 4.8 | 1.03 | 259 | Ia | Walker, 1974 |
| | Wannacut | 58° 52′ N | 119° 34′ W | 166.6 | 48.8 | 17.1 | 3.2 | 563 | Ia | Walker, 1974 |
| | Cottage | 47° 45′ N | 122° 53′ W | | 7.5 | | | | | T. Edmondson? |
| | Langlois | 47° 38′ N | 121° 53′ W | | 29 | | | | IV | T. Edmondson? |
| Wisconsin | Hell's Kitchen | 46° 11′ N | 89° 42′ W | 2.8 | 19.2 | | | 503 | | p.c., D. Robertson & T. Kratz |
| | Knaack | 44° 37′ N | 88° 54′ W | 1.1 | 22 | 7 | | 264 | IV | Parkin, Winfrey, & Brock, 1980 |
| | Mary | 46° 04′ N | 90° 09′ W | 1.2 | 21.5 | 7.7 | 0.125 | | IV | Stewart et al., 1966 |
| | Max 2 | 45° 16′ N | 91° 21′ W | | 6 | | | 329 | IV | Stewart et al., 1966 |
| | Max 3 | 45° 16′ N | 91° 21′ W | | 8 | | | 329 | IV | Stewart et al., 1966 |
| | Scaffold | 45° 57′ N | 89° 38′ W | 7.9 | 9.1 | 4.3 | | 497 | IV | Manning & Juday, 1941 |
| | Stewart's Dark | 46° 18′ N | 91° 27′ W | 0.7 | 8.8 | | | | | Likens & Hasler, 1960 |
| | Wausau Granite Pit | | | <0.5 | 20.6 | | | | Crenogenic | Stewart et al., 1966 |

| Region | Lake | Latitude | Longitude | | | | | | Type | Reference |
|---|---|---|---|---|---|---|---|---|---|---|
| | Wazee Mine | 44° 17' N | 90° 43' W | 59.1 | 104 | | | 297 | | p.c., D. Robertson |
| *Canada (provinces)* | | | | | | | | | | |
| British Columbia | Island Copper Mine | 50° 36' N | 127° 28' W | 171 | 350 | 4.8 | 2.1 | 8? | IV | Pieters et al., 2003 |
| | Lyons | 50° 57' N | 120° 07' W | 35.7 | 12 | 7.2 | 1.6 | 686 | IV | Northcote & Halsey, 1969 |
| | Mahoney | 49° 17' N | 119° 32' W | 21.6 | 18.9 | 99.6 | 0.9 | 472 | Ib | Northcote & Halsey, 1969 |
| | Nitinat | 48° 42' N | 124° 50' W | 2760 | 205 | | 23 | 0 | Ib | Richards et al., 1965 |
| | Powell | 49° 53' N | 124° 32' W | ~8 | 358 | 43 | 50 | 56 | Ib | Williams et al., 1961 |
| | Sakinaw | 49° 38' N | 124° 02' W | 99.6 | 140 | 4.6 | >8 | | | Northcote & Johnson, 1964 |
| | White | 51° 22' N | 121° 53' W | 32.3 | 15 | 20.5 | 2.4 | 1037 | IV, III | Northcote & Halsey, 1969 |
| | Yellow | 49° 10' N | 119° 45' W | | 40 | 24.5 | 1.9 | 750 | Ib | Northcote & Halsey, 1969 |
| Labrador | Tessiarsuk | 56° 30' N | 61° 57' W | 545 | 654 | | | 0 | Ib | Carter, 1965 |
| NW Territories | Baker | ~65° 30' N | 95° W | | | | | | | Johnson, 1964 |
| | Campbell | | | | | | | | | Richie et al., 1976 |
| | Sunday | | | | | | | | | Johnson, 1964 |
| | Winton Bay | 63° 24' N | 64° 39' W | | | | | | | Carter, 1967 |
| Nunavut | | | | | | | | | | |
| L. Cornwallis Is. | Garrow | 75° 24' N | 96° 47' W | 0.043 | 49 | 22.9 | 3.2 | 6.7 | | Stewart & Platford, 1986 |
| Ellesmere Is. | Lake A | | | 4.9 | 57 | | | 3.3 | Ectogenic | Hattersley-Smith et al., 1970 |
| Ellesmere Is. | Lake B | | | 0.9 | 40 | | | 23 | | Hattersley-Smith et al., 1970 |
| Ellesmere Is. | Lake C | | | 0.8 | 57 | | | 2.3 | Ectogenic | Hattersley-Smith et al., 1970 |
| Baffin Is. | Ogac | 65° 52' N | 67° 21' W | 1.481 | 60.5 | 23.2 | 2 | | Ib | McLaren, 1963 |
| Cornwallis Is. | Sophia | 75° 05' N | 93° 38' W | 0.034 | 50 | 22.4 | 4.8 | | | Stewart & Platford, 1986 |
| Ellesmere Is. | Tuberg | | | | | | | | | Hattersley-Smith et al., 1970 |
| Ontario | Crawford | 43° 28' N | 80° 57' W | 5 | 23.5 | | 0.272 | | IV | McAndrews et al., 1973 |
| | ELA 120 | 49° 39' N | 93° 50' W | 0.093 | 19 | 7.6 | | | IV | Brunskill & Schindler, 1971 |
| | ELA 241 | 49° 40' N | 93° 33' W | 0.018 | 12.5 | 5.5 | | | IV | Brunskill & Schindler, 1971 |
| Algonquin Park | Jake | 48° 33' N | 78° 33' W | 9 | 13 | | | | | Smol, 1979 |
| Near Picton | Lake of the Hills | 44° 45' N | 76° 33' W | | 18.5 | | | | | p.c., McNeely, 1979 |
| Near Picton | Lake on the Mountain | | Losing its meromixis | | | | | | | p.c., McNeely, 2007 |
| | Little Round | 44° 47' N | 76° 41' W | 7.39 | 16.8 | 8.3 | 0.425 | 201 | | McNeely, 1974 |
| Near Picton | Loon | 46° 09' N | 81° 40' W | 6.5 | 33 | | | | | p.c., Davis at Queen's University |
| Near Peterborough | McGinnis | | | | | | | | | p.c., M. Dickman, 1979 |
| In Ottawa | Mckay | | | 7.43 | 10.9 | 0.37 | | 150 | | p.c., McNeely, 2001 |
| Near Aurora | Simeon | | | | | | | | | p.c., M. Dickman, 1979 |
| W. of Waterloo | Sunfish | 43° 28' N | 80° 38' W | 8 | 20 | 10.4 | 0.551 | 365 | IV | Duthie & Carter, 1970 |
| Quebec | Bedard | 47° 16' N | 71° 07' W | 4 | 10 | | | | | Bernard & Lagueux, 1970 |
| | Green | 46° 11' N | 76° 19' W | 20 | 25 | | | | IV | Dickman et al., 1971 |
| | Pinks | 45° 30' N | 76° 00' W | 12 | 20 | | 0.78 | 163 | Ib, IV | Dickman et al., 1975 |

Continued

**Table 2**  Continued

| Region | Lake | Lat. | Long. | If Known...? Area (ha) | Zm | Mean-m | L (km) | masl | Type | Source |
|---|---|---|---|---|---|---|---|---|---|---|
| Saskatchewan | Waldsea | 52° 17′ N | 105° 12′ W | 464 | 14.3 | 8.1 | 3.71 | 532 | Ia | Hammer et al., 1978 |
| | Dead Moose | 52° 19′ N | 105° 10′ W | 1000 | 48.0 | 6 | 5.54 | 532 | Ia | Hammer et al., 1978 |
| | Sayer | 52° 24′ N | 105° 24′ W | 204 | 5.1 | ~2.7 | 1.95 | ~545 | Ia | Hammer et al., 1978 |
| | Arthur | 52° 34′ N | 105° 26′ W | 250 | 5.6 | ~3 | 3.90 | 540 | Ia | Hammer et al., 1978 |
| | Marie | 52° 31′ N | 105° 29′ W | 160 | 3.1 | ~1.7 | 1.70 | ~530 | Ia | Hammer et al., 1978 |
| Greenland | | | | | | | | | | |
| Mexico and Cuba | | | | | | | | | | |
| Mexico, Nayarit | Isabela Crater L, | 21° 52′ N | 105° 54′ W | ~5.7 | 17.5 | | 0.27 | 7 | | Alcocer et al., 1998. |
| Cuba | Valle-de-San-Juan | | | 0.1 | 25 | | | 9 | 1b | Romanenko et al.,1976; Straskraba et al., 1967 |
| Central America, Carribean, & Galapagos | | | | | | | | | | |
| Guatemala | Encantada, El Salto | | | | 14 | | | | | Brezonic & Fox, 1974 |
| | Eckixil, Peten | | | | 21 | | | | | Brezonic & Fox, 1974 |
| | Juleque, Peten | | | | 25.5 | | | | | Brezonic & Fox, 1974 |
| | Lago de Peten | | | | 32 | | | | | Brezonic & Fox, 1974 |
| | Macanche, Peten | | | | >40 | | | | | Brezonic & Fox, 1974 |
| | Paxcamen, Peten | | | | 30 | | | | | Brezonic & Fox, 1974 |
| Panama | Third Lock, Canal Zone, post Canal | | | | | | | | | Boznaik et al., 1969 |
| Bahamas | Devil's Hole, Abaco Is. | 26° 41′ N | 77° 18′ W | | 50 | | | | | Goehle & Storr, 1978 |
| Galapagos | | | | | | | | | | |
| Genovesa Is. | Arcturus | | | 20 | 30 | | 0.5 | | IV | Howmiller, 1969 |
| Isabela Is. | Pond 2 | 00° 41′ S | 91° 16′ W | 0.12 | 2.5 | | 0.06 | | Ia | Howmiller, 1969 |
| Isabela Is. | Pond 3 | 00° 41′ S | 91° 16′ W | 0.1 | | | 0.05 | | Ia | Howmiller, 1969 |
| SOUTH AMERICA & ANTARCTICA | | | | | | | | | | |
| South America (?) | | | | | | | | | | |
| Antarctica | | | | | | | | | | |
| Vestfold Hills | Ace | 68° 28′ S | 78° 11′ E | 13.2 | 23 | 0.35 | 0.026 | <23 | V | Hand & Burton, 1981 |
| Ross Is. | Algal | 77° 38′ S | 166° 24′ E | 0.045 | 0.8 | 7.1 | | | | Walker & Likens, 1975 |
| Taylor Valley | Bonney | 77° 43′ S | 162° 26′ E | 2408 | 3.2 | | | | Ia, V | Walker & Likens, 1975 |
| Vesfold Hills | Clear | 68° 37′ S | 78° 95′ E | 55 | 55 | | | | | Hand & Burton, 1981 |
| Taylor Valley | Fryxell | 77° 35′ S | 163° 35′ E | 706 | 18 | | | 18 | Ia, V | Walker & Likens, 1975 |
| Taylor Valley | Hoare | 77° 38′ S | 162° 53′ E | 180 | 34 | | | 58 | | Wharton et al., 1986 |
| Deception Is. | Irizar | | | | | | | | | Burton, 1981 |
| Victoria Land | Joyce | | | | | | | | | p.c., G. Likens |

| Location | Lake | Latitude | Longitude | | | >15 | | >23 | | Reference |
|---|---|---|---|---|---|---|---|---|---|---|
| Victoria Land | Miers | | | | | | | | | p.c., G. Likens |
| Lutzou-Holm Bay | Nurume | | | | | | | | | Burton, 1981 |
| Ross Is. | Skua | 81° 38′ S | 170° 24′ E | 0.017 | 0.22 | 0.75 | 0.07 | <23 | IV, V | Waker & Likens, 1975 |
| | Trough | | | | | | | | | p.c., G. Likens |
| Victoria Land / Taylor Valley | Vanda | 82° 38′ S | 171° 24′ E | 521 | | 66.1 | 5.0 | | Ia, V | Walker & Likens, 1975 |
| Victoria Land | Vida | | | | | | | | | p.c., G. Likens |
| | Vostoc & many other 'still liquid' lakes deep within the Antartic ice. Meromixis possible but not known | | | | | | | | | |
| **EUROPE** | | | | | | | | | | |
| Austria | Goggausee | 46° 30′ N | 14° 09′ E | 215 | 12.8 | 6 | 0.1 | 770 | IV | Loffler, 1975 |
| | Hallstattersee | 47° 35′ N | 13° 39′ E | 2 | 125 | 65 | | 508 | IV | Walker & Likens, 1975 |
| | Klopinersee | 46° 38′ N | 14° 35′ E | 130 | 26 | 30 | | 448 | IV | Walker & Likens, 1975 |
| | Krottensee | 47° 45′ N | 13° 35′ E | 82 | 47 | 26 | | 579 | IV | Anon. 1987 Austrian Report |
| | Langsee | 46° 30′ N | 14° 17′ E | 9 | 21 | 11 | | 543 | II, IV | Walker & Likens, 1975 |
| | Lahngangsee | 47° 46′ N | 13° 37′ E | 80 | 77 | 33 | | 1555 | III | Walker & Likens, 1975 |
| | Millstattersee | 46° 45′ N | 13° 37′ E | 34 | 140 | 86 | | 530 | IV | Walker & Likens, 1975 |
| | Piburger | | | 13.4 | 25 | | 0.8 | 915 | | Findenegg, 1968 |
| | Toplitzee | 47° 39′ N | 13° 56′ E | 660 | 106 | 62 | | 718 | III | Walker & Likens, 1975 |
| | Weissensee | 46° 43′ N | 13° 20′ E | 1940 | 99 | 36 | | 930 | IV | Walker & Likens, 1975 |
| | Worthersee | 46° 37′ N | 14° 10′ E | 640 | 84 | 43 | | 439 | IV | Walker & Likens, 1975 |
| | Zellersee | 47° 19′ N | 12° 52′ E | 6.4 | 68 | 39 | | | IV | Walker & Likens, 1975 |
| Finland | Hannisenlampi | 62° 05′ N | 30° 12′ E | 0.2 | 16 | 5.2 | | 84 | | Hakala, 2004 |
| | Laukunlampi | 62° 40′ N | 29° 10′ E | 8.8 | 27 | 6.3 | | 108 | | Hakala, 2004 |
| | Lovojarvi | 60° 05′ N | 25° 02′ E | ~5.1 | 17.5 | 7.7 | | 93 | | Hakala, 2004 |
| | Vaha-Pitkusta | 60° 54′ N | 23° 39′ E | 1.1 | 35 | 12 | 0.50 | 110 | III, IV | Hakala, 2004 |
| | Valkiajarvi | 61° 54′ N | 23° 53′ E | 7.8 | 25 | 8.4 | 0.58 | | | Hakala, 2004 |
| | Horkkajarvi | 61° 13′ N | 25° 10′ E | | 13 | 7 | 0.60 | | | Hakala, 2004 |
| Aland Is. | Vargsundet | | | 11 | 35 | | 5.00 | | | Hakala, 2005 |
| Aland Is. | Langsjon | | | 14.3 | 18 | | 4.50 | | | Hakala, 2006 |
| Aland Is. | Vastra Kyrksundet | | | 6 | 18 | | 2.50 | | | Hakala, 2007 |
| Aland Is. | Ostra Kyrksundet | | | 20 | 22 | | 4.00 | | | Hakala, 2008 |
| Nauvo Is. | Vasterholmarna | | | | 8 | | | | | Hakala, 2009 |
| France | Girotte, no longer meromictic | | | | | | | | | |
| | Pavin | 45° 34′ N | 2° 53′ E | 44 | 92.1 | | 0.76 | 1198 | III, IV | Stewart & Hollan, 1975 |
| Germany | Hemmelsdorfersee and Ulmener Maar, both no longer meromictic | | | | | | | | | |

Continued

**Table 2**   Continued

| Region | Lake | Lat. | Long. | Area (ha) | Zm | Mean-m | L (km) | masl | Type | Source |
|---|---|---|---|---|---|---|---|---|---|---|
| | Goitsche Mine | | | 13.3 | 47 | | | | Crenogenic | Boehrer & Schultze, 2006 |
| | Meresburg-Ost-1b Mine | | | 2.8 | 36 | | | | Crenogenic | Boehrer & Schultze, 2006 |
| Greenland | Saleso | 77° 03' N | 20° 35' W | | 112 | | | | Ectogenic | Hutchinson, 1957 |
| Italy Sicily Is. | Lugano | 45° 59' N | 8° 58' E | 2750 | 288 | | | | Ectogenic | Truper & Genovese, 1968, L&O, 13: 225–232 |
| | Faro | 38° 16' N | 15° 38' E | 21.2 | 21 | | 0.56 | 271 | Ectogenic | |
| Norway | Aurtjern | 60° 14' N | 11° 08' E | 9.5 | 16.5 | | | | III, IV | Walker & Likens, 1975 |
| | Bakketjern | 60° 12' N | 11° 09' E | 3 | 14.5 | | | | III, IV | Walker & Likens, 1975 |
| | Blankvatn | 67° N | | | 55 | | | | IV | Walker & Likens, 1975 |
| | Botnvatn | 60° 12' N | 11° 14' E | 10.00 | 113 | | | | Ib | Walker & Likens, 1975 |
| | Gravtjern | | | | 7 | | | | III, IV | Walker & Likens, 1975 |
| | Kyllaren | 61° 25' N | 5° 10' E | | 29 | | | | III, IV | Breugel, Y. et al., 2005 |
| Spitsbergen | Kongressvatn | 78° 01' N | 13° 59' E | 82.00 | 52 | | 1.0 | 94 | III, IV | Boyum et al., 1970 |
| | Ljogottjern | 60° 09' N | 11° 09' E | 20.00 | 17 | | | | III, IV | Walker & Likens, 1975 |
| | Maena | | | 34 | 39 | | | | IV | Walker & Likens, 1975 |
| | Nordbytjern | 60° 09' N | 11° 10' E | 1330 | 23.5 | | | | III, IV | Hongve, 1974 |
| | Rorhopvatn | | | 41 | 92 | | | | III, IV | Walker & Likens, 1975 |
| W. Norway | Saelenvannet | | | | | | | 3 | Ib | Boersheim, K.Y. et al., 1985, W. Norway |
| | Skjennungen | 60° 00' N | 10° 41' E | 3.4 | 17.8 | 8.3 | | 418 | IV | Walker & Likens, 1975 |
| | Skratjern | 60° 12' N | 11° 09' E | 2 | 12.5 | | | | III, IV | Walker & Likens, 1975 |
| | Store Aaklungen | 60° 00' N | 10° 44' E | 13.2 | 32.5 | 8.4 | | 293 | IV | Walker & Likens, 1975 |
| | Svinsjoen | | | 11 | 33 | | | 180 | III, IV | Walker & Likens, 1975 |
| | Tokke | 59° 35' N | 8° 00' E | 9.4 | 147.5 | | | 60 | Ib | Strom, 1957 |
| | Transjoen | 60° 13' N | 11° 08' E | 2560 | 22 | | | | IV | Walker & Likens, 1975 |
| | Traunsee | 47° 50' N | 13° 40' E | 0.3 | 197 | 90 | | 422 | III | Walker & Likens, 1975 |
| | Vesle Bakleetjern | 60° 12' N | 11° 09' E | 2.4 | 10.5 | | | | IV | Walker & Likens, 1975 |
| | Vibergtjern | 60° 11' N | 11° 10' E | | 17.1 | | | | IV | Walker & Likens, 1975 |
| Poland | Czarne (N. Poland) | | | 2.1 | 32 | | | | | Golebiowska et al., 2005 |
| | Zakrzwek Pit | | | 106 | 30 | | | | Ectogenic | Slusarczky, 2003 |
| Spain | Banyoles | 42° 7" N | 2° 45' E | 4.0 | 8.25 | 3.3 | | | Ectogenic | Garcia-Gil et al., 1996 |
| | Cibollar | | | 67.3 | | | | | | |
| | El Tobar | | | | 19.5 | 8.21 | 1.48 | 1250 | Crenogenic | Reviriego et al., 1993 |
| | La Cruz | | | 1.11 | 23.5 | 9 | ~0.13 | | Biogenic | Vicente et al., 1993 |
| | Vilar | | | 29 | 10 | | 0.17 | | | Rodrigo et al., 1993 |
| Sweden | Ravlidmyran Pit | | | | | | | | | Garcia-Gil & Figueras, 1993 |
| Switzerland | Ritom, no longer meromictic | | | | | | | | Ectogenic | Boehrer & Schultze, 2006 |

| Location | Lake | Latitude | Longitude | | | | | | Type | Reference |
|---|---|---|---|---|---|---|---|---|---|---|
| **AFRICA** | | | | | | | | | | |
| Cameroon | Monoun | 5° 34' N | 10° 35' E | | 95 | | 1.3 | | | Halbwachs et al., 2004 |
| | Nyos | | | | 210 | | | | | Halbwachs et al., 2004 |
| Dem.Rep. of Congo | Kivu | 1° 50' S | 29° 11' E | 237 000 | 480 | 240 | 97 | 445 | | Damas, 1972 |
| Ghana | Bossumpti | (Probably not meromictic. Lake does mix but stays anoxic) | | | | | | | | p.c., R. Hecky |
| Kenya | Bogonia | | | | | | | | | Robarts & Allanson, 1977 |
| South Africa | Swartvlei | 34° 00' S | 22° 46' E | 108.5 | 12.5 | | 4.5 | 0 | | Hutchinson, 1957 |
| Tanzania | Malawi | 12° 00' S | 34° 30' E | 3 080 000 | 706 | 273 | | | IV | Hutchinson, 1957 |
| | Tanganyika | 7° 00' S | 30° 00' E | 3 400 000 | 1470 | 572 | 1900 | | IV | |
| Uganda | Nkugute | 0° 19' S | 30° 06' E | | 58 | | 1.4 | 1500 | | Beadle, 1966 |
| | Bunyoni | 1° 10' S | 29° 50' E | 5700 | 45 | | | 1800 | IV | Beadle, 1966 |
| **RUSSIA** | | | | | | | | | | |
| Azerbaijan | Geck-Gel | | | | 90 | | | | Ia | Sorokin, 1969 |
| | Mara-Gel | | | | 60 | | | | Ia | |
| Somewhere in USSR | | | | | | | | | | |
| Vladimir Dist. | Belevod | | | | | | | | | Sorokin, 1968 |
| Middle Volga | Bolshoj Kucheyer | | | | 15 | | | | III | Steppe region of Khakasia |
| Middle Volga | Chornoe Kucheyer | | | | 9 | | | | III | Steppe region of Khakasia |
| Middle Volga | Kononjer | | | | 23 | | | | III | |
| Middle Volga | Kusnechiha | | | | 20 | | | | | |
| Middle Volga | Mogilnoe | | | | 16.3 | | | | | |
| Kildin Is. | Okha-Lampi | | | | | | | 0 | Ib | Sololova-Dubinina, et al. 1969 |
| Ukraine SSR | Repnoe | | | | 6.5 | | | | III | Chebotarev et al., 1975 |
| | Shira | | | | | | | | | Savvichev et al., 2005 |
| | Shunet | | | | | | | | | Savvichev et al., 2006 |
| | Wesijsowoe | | | | | | | | III | Chebotarev et al., 1975 |
| Petrozavodsk Dist. | Urozero | | | | 37 | | | | Ia | |
| | Veisovo | | | | | | | | | |
| **AUSTRALIA, TASMANIA, NEW GUINEA** | | | | | | | | | | |
| Australia, W. | Serpentine | ~32° S | 115° 40' E | | | | | | | Bunn & Edward, 1984 |
| Australia, W. | Government House | ~32° S | 115° 40' E | | | | | | | Bunn & Edward, 1984 |
| Australia, W. | Herschell | ~32° S | 115° 40' E | | | | | | | Bunn & Edward, 1984 |
| Australia, Victoria | West Basin | 38° 20' S | 143° 27' E | 15.8 | 13.5 | | | | | Timms & Brand, 1973; Last & DeDeckker, 1990 |

Continued

**Table 2**  Continued

| Region | Lake | Lat. | Long. | If Known...? Area (ha) | Zm | Mean-m | L (km) | masl | Type | Source |
|---|---|---|---|---|---|---|---|---|---|---|
| Australia, Queensland | Hidden | 25° 14' S | 153° 10' E | 9 | 8.2 | | c.0.4 | | | Longmore et al., 1983; Bayly et al., 1975 |
| Australia, Queensland | Barrine | 17° 15' S | 145° 38' E | 103.3 | 67 | | | | | Timms, 1976, 1979; Walker, 1999 |
| Tasmania | Fidler | 42° 30' S | 145° 40' E | 1.28 | 7.6 | | 0.186 | | | Hodgsson et al., 1996; Tyler et al., 2002 |
| Tasmania | Barrington | 41° 21' S | 146° 17' E | 780 | 7.2 | | 18.8 | | | Tyler & Buckney, 1974 |
| New Guinea, Papua | Nagada | | | | 10 | | | | | Vyverman & Tyler, 1994; DeMeester & Vyverman, 1997 |
| New Guinea, Papua | Siar | | | | | | | | | Vyverman & Tyler, 1994; DeMeester & Vyverman, 1997 |
| ASIA (parts only) | | | | | | | | | | |
| Japan | Abashiriko | 43° 16' N | 143° 07' E | 3404 | 17.6 | | 20.5 | 1 | Ib? | Yoshimura, 1938 |
| Shizuoka | Hamana | 34° 45' N | 137° 35' E | 7350 | 16.6 | 4.8 | | 0 | Ib, III | Motojima & Maki, 1958 |
| Hokkaido | Hangetsuko | 42° 51' N | 140° 45' E | 4.5 | 18.2 | 4.4 | 0.45 | 270 | IV? | Yoshimura, 1934 |
| Hokkaido | Harutoriko | 42° 58' N | 144° 23' E | 3800 | 8.5 | 3.4 | 2.00 | 5 | Ib | Kusuki, 1937 |
| Fukui | Hirugako | 36° 04' N | 136° 12' E | 95 | 33 | 22.2 | 1.40 | 0 | Ib | Yoshimura, 1932 |
| | Kai-ike | 31° 51' N | 129° 40' E | 16 | 11.5 | | | 1 | Ib? | Yoshimura, 1938 |
| Tokyo | Kisaratsu | 35° 30' N | 140° E | 0.2 | 10 | | 0.05 | 0 | Ib | Takahashi & Ichimura, 1968, 1970 |
| Kagoshima | Namako-ike | 31° 37' N | 130° 32' E | 43 | 21.3 | 9.3 | | 0 | Ib? | Kobe Mar, Observ. 1935, Yoshimura 1938 |
| | Onneto | 44° 16' N | 139° 30' E | 83 | 22.6 | | | 5 | I | Yoshimura, 1938 |
| Riu Kiu Is. | Simmiyo | 34° 03' N | | 6.2 | 34.7 | 24.2 | | 0 | III | Yoshimura, 1938 |
| Fukui | Suigetsuko | 35° 35' N | 135° 53' E | 510 | 34 | 14.3 | 3.5 | 0 | Ib | Yoshimura, 1934; Matsuyama, M. et al., 1978 |
| | Togo-ike | 35° 29' N | 134° E | 300 | 7.5 | | | 1 | Ib? | Yoshimura, 1938 |
| Aomori | Towadako | 40° 28' N | 140° 53' E | 5005 | 334.0 | 71.0 | 11.0 | 401 | III | Yoshimura, 1934 |
| Miyagi | Zao-okama | 38° 08' N | 140° 27' E | 8 | 27.1 | 15.0 | 0.4 | 1570 | III | Yoshimura, 1934 |
| Tibet | Zige Tangco | 32° 00' N | 90° 44' E | 18 700 | 38.9 | | 18.75 | 4560 | Crenogenic | Li & Li, 1999; Li et al., 2005 (NOTE: Highest meromictic lake in world) |
| Turkey | Eski Acigol | 37° 49' N | 29° 53' E | | | | 11.2 | | Crenogenic | Roberts et al., 2001 |
| | Koycegiz | 36° 55' N | 28° 39' E | | | | 13 | | Crenogenic | Bayari, 1995 |

## Other Meromixis?

Another mechanism whereby lakes could become meromictic is through the runoff from deicing salts applied to roads during winter. This effect may occur in urban areas or along major highways in cool Temperate Zones, but it is so far little studied. Finally, it should again be stressed that classifications are tools for study and not necessarily real. Some meromictic conditions may result from a combination of more than one mode of origin, just as meromixis itself is prone to vary in space and time. In all cases, it is important to remember that the main physical limits to vertical mixing are the differences in the upper and lower water densities.

## 'In the Pits'

Because of their morphometry, man-excavated surface depressions or 'pits' (usually created during the extraction of minerals) tend to have high relative depths. Following mineral extraction, unused pits usually fill with groundwater and runoff water. When a water body has a small surface area, and yet is relatively deep, its 'relative depth' tends to be high and such water bodies may have a 'predisposition' to meromixis. The 'relative depth' $Zr$, regarded as the ratio of the maximum depth ($Zm$) (in meters) to the square root of the mean diameter of the lake surface area ($Ao$) in square meters can be quantified by the formula: $Zr = (50\,Zm)(\sqrt{\pi})/(\sqrt{Ao})$ or $Zr = (Zm \times 88.6)/(\sqrt{Ao})$, which converts directly into percentages.

There are large numbers of water bodies around the world that are located in pits, probably many more than generally realized. We cannot possibly cover here all examples of meromixis in water-filled pits. However, the following examples illustrate some variations in depths and locations of just six different pits containing meromictic bodies of water: (1) Island Copper Mine (>400 m deep) on Vancouver Island, Canada; (2) Goitsche Lignite Mine (47 m deep) in Germany; (3) St. Louis Coal Mine (60 m deep) in France; (4) Ravlidmyran Zinc/Silver Mine (29 m deep) in Sweden; (5) Fountain Pond Iron Mine (~90 m deep) in Michigan, USA; (6) Wausau Granite Quarry (~21 m deep) in Wisconsin, USA.

## Selected Examples

**Table 2** provides an extensive list of examples of meromixis from many countries. The list is incomplete but gives a rich indication of this unusual lake type around the world. Other examples may be reported in the literature and, of course, many more probably remain unrecognized. Obviously, we need far more information and data but, based on current knowledge, it seems safe to say that less than 1 out of 1000 water bodies are known to have meromictic characteristics.

However, even among the hundreds of meromictic lakes known, there are still a few that are different or noteworthy (the unusual among the unusual?) enough to merit special attention. It may be something about their size, morphometry, chemical stratification, educational utility, or other features that makes them worthy of special interest. The following 9 examples, in no special order of importance, are meant to demonstrate the history, diversity, and distribution of meromictic lakes that have fascinated limnologists for many years.

1. *Lake Tanganyika* (>1450 m deep) and
2. *Lake Malawi/Nyasa* (>700 m deep), tectonic lakes in Africa, are the second- and third-deepest freshwater lakes in the world, after Lake Baikal, and the two deepest meromictic lakes in the world. Below roughly 150 m in Lake Tanganyika and 200 m in Malawi, the waters are continuously anoxic. Interestingly, huge numbers of fish species (including endemic ones) have evolved over time within the upper oxygenated 'mixolimnetic' waters. Indeed, Lake Malawi is regarded as having more species of fish than any other lake in the world! The fish in both these ancient lakes are under increasing pressure from a growing proximate human population and overfishing. The biotic resources of these huge lakes, like resources anywhere, are not endless and sustainability appears doubtful. The *Nyanza* program (sponsored by the US National Science Foundation and the Tanganyika Biodiversity Program, and carried out at the town of Kigoma on the northeastern shore of Lake Tanganyika in Tanzania) provides some limnological training and research opportunities for students from Africa and other parts of the world.
3. *Skua Lake*, on Ross Island, Antarctica, is the shallowest (0.75 m deep) known meromictic lake. Its very small area (170 m$^2$) is probably less significant for the maintenance of meromixis than its salinity gradient.
4. *Hot Lake*, in northern Washington State, USA, is also small (13 000 m$^2$) and shallow (3.2 m deep), but has a strong salinity gradient that fosters the 'trapping' of solar radiation into a saline lens where the summer water temperature may exceed 50 °C and, even under the winter ice, the water is warmer than 25 °C. These are so-called heliothermal lakes. Hot Lake (with Lake Vanda, below) has the

warmest known 'under ice' temperatures of any lake in the world. Solar Lake, a small coastal pond by the Gulf of Aqaba near Elat, Israel, also has a marked salinity gradient, and bottom-water temperatures exceeding 50 °C, but it is monomictic and does not develop an ice cover. Interestingly, nearly a century ago, limited studies were done on a small wind-protected water body (Medveto Lake) in Hungary and temperatures up to 56 °C, attributed exclusively to solar radiation, were found in a saline stratum not far below the water surface. Unfortunately, the 'mictic' status of this latter small lake, and information about ice cover, are unknown.

5. The volcanic crater lakes *Lake Monoun* (95 m deep) and *Lake Nyos* (210 m deep), in Cameroon, Africa, have a history of explosive degassing of carbon dioxide (Monoun in 1984; Nyos in 1986). Some 1800 people and hundreds of cattle and other animals, some 25 km away, were killed by gases resulting from these explosions. Efforts to artificially degas these 'killer lakes' are underway, by reducing the build-up of monimolimnetic gases.

6. *Lake Mary* (~21 m deep) and *Green Lake* (Fayettville) (~52 m deep), in Wisconsin and New York, respectively, USA, are of interest for their educational utility. For decades hundreds of students from several universities and colleges have learned about and visited these lakes on field trips. Consequently, these lakes have provided first hand knowledge about meromixis for generations of students.

7. *Lake Vanda* (66 m deep), in Victoria Land, Antarctica, appears to be a crenogenic meromictic lake. The lake surface is covered by ice year-round, and even in the Antarctic summers of 1962–1963 the ice was 3.5–4.3 m thick. The waters of Lake Vanda are exceptionally clear, and comparable to the clearest ocean waters. The lake also maintains a monimolimnetic temperature, beneath thick ice, of more than 25 °C. The latter lower water temperatures are thought to be caused by geothermal heating from below, rather than by solar heating from above.

8. *Lake Kauhako* is a small (~3500 m$^2$), deep (248 m) water body located in a formerly drained volcanic conduit, on the Kalaupapa Peninsula of the Island of Molokai, one of the Hawaiian islands, in the Pacific Ocean. Because of its small surface area and relatively great depth, Lake Kauhako has the greatest 'relative depth' (Zr = 371%) of any known natural (not human-made) water body in the world! Initial studies by some Hawaiian scientists suggest that only the top 4–4.5 m of water show

substantial stratification of salinity, temperature, and dissolved oxygen. Hydrogen sulfide becomes prominent within five meters of the surface and extends to the bottom. If so, most of Lake Kauhako resembles a giant anoxic (and stinky?) tube of sea water. Other than the top few stratified meters, most of the water column below about five meters of depth is relatively homogeneous in temperature and salinity. Kauhako (248 m) is the deepest natural meromictic lake in the USA and the sixth deepest lake in the United States. Some comparative depths for other deep, *but nonmeromictic*, lakes of the USA are Lakes Crater (608 m), Tahoe (501 m), Chelan (458 m), Superior (407 m, border shared with Canada), and Michigan (282 m).

9. *Soap Lake* (27 m deep) is located in the Lower Grand Coulee area of the central part of Washington State, USA. The lake probably originated from fresh water flowing over a saline lake that today remains as the monimolimnion (salinity 146 g L$^{-1}$). In winters when the surface freezes, the upper waters (mixolimnion) behave as a typical dimictic lake. However, in milder winters, mixing probably persists and the mixolimnion behaves as a monomictic lake. In that sense, Soap Lake may be reflective of a variety of meromitic lakes. Indeed, one way of viewing meromixis is to consider that there is an irregular seasonal stratification within the mixolimnion that effectively generates a holomictic lake on top of more dense or saline monimolimnetic waters.

## Acknowledgments

We thank Dr. Mike Dickman, Ontario, Canada, for the loan of research articles.

*See also:* Africa: South of Sahara; Antarctica; Asia; Australia and New Zealand; Europe; North America; South America.

## Further Reading

Alcocer J, Lugo A, Sanchez MR, and Escobar E (1998) Isabela Crater-lake: a Mexican insular saline lake. *Hydrobiologia* 381: 1–7.

Anati DA (1998) Dead Sea water trajectories in the T-S space. *Hydrobiologia* 381: 41–49. (Note: The Dead Sea seems to alternate between uneven periods of meromixis and holomixis).

Anderson RY, Dean WE, Bradbury JP, and Love D (1985) Meromictic lakes and varved lake sediments in North America. *U.S. Geological Survey Bulletin* 1607: 19.

Aplikowski S (Ecologist) Minneapolis Park & Recreation Board, Minneapolis, Minnesota (Personal Communication).

Baxter RM, Prosser MV, Talling JF, and Wood RB (1965) Stratification in tropical African lakes at moderate altitudes. *Limnol. Oceanogr.* 10: 510–520.

Bayari CS (1995) The determination of the origin of the waters of Koycegiz lake. *J. Hydrol.* 166(1–2): 171–191.

Beadle LC (1963) Prolonged stratification and deoxygenation in tropical lakes. I. Crater Lake Nkugute, Uganda, compared with Lakes Bunoni and Edward. *Limnol. Oceanogr.* 6: 152–163.

Binati NL (1967) Estudio limnologico del lago Irizar, Isla Decepcion, Shetland del Sur. *Contribuciones, Istituto Antartico Argentino* 111: 1–36. (ref. from Burton 1981, not seen).

Boehrer B and Schultze M (2006) On the relevance of meromixis in mine pit lakes. Poster Abstract: 7th Internat. Conf. Acid Rock Drainage (ICARD), 26–29 March 2006, St. Louis, MO.

Boyum A and Kjensmo J (1970) Kongressvatn. A crenogenic meromictic lake at Wester Spitsbergen. *Arch. Hydrobiol.* 67: 542–552.

Brunskill GJ, Ludlam SD, and Diment WH (1969) A comparative study of meromixis (Abstract). *Verh. Internat. Verein. Limnol.* 17: 1237–1239.

Burton HR (1981) Chemistry, physics and evolution of Antarctic saline lakes. In: Williams WD (ed.) *Salt Lakes*, pp. 339–362. London: Dr. W. Junk, Publ.

Carter JCH (1965) The ecology of the calanoid copepod *Pseudocalanus minutus* Kroger in Tessiarsuk, a coastal meromictic lake of northern Labrador. *Limnol. Oceanogr.* 10: 345–353.

Cooper S (Ecologist) Bryn Athyn, Pennsylvania (Personal Communication, 2007).

Dickman M (Limnologist) Wellland, Ontario, Canada (Personal Communication, 2007).

Dickman M, Krelina E, and Mott R (1975) An eleven thousand year history with indications of recent eutrophication in a meromictic lakes in Quebec, Canada. *Verh. Internat. Verein. Limnol.* 19: 2256–2259.

Duaime T (Geologist) Butte, Montana (Personal Communication, 2007).

Duthie HC and Carter JCH (1970) The meromixis of Sunfish Lake, southern Ontario. *J. Fish. Res. Bd. Canada* 27: 847–856.

Findenegg I (1935) Limnologische Untersuchungen in Karntner Seengebiete. Ein Beitrag zur Kenntnis des Stoffhaushaltes in Alpenseen. *Int. Rev. Hydrobiol.* 32: 369–423.

Findenegg I (1937) Holomiktische und meromiktische Seen. *Internationale Revue der gesamten Hydrobiologie und Hydrographie* 35: 586–610.

Findenege I (1968) Das Phytoplankton des Piburger See im Jahre 1966. *Ber. Nat.-med. Ver. Insbruck.* 56: 163–170.

Frey DG (1955) Langsee: a history of meromixis. *Mem. Ist. Ital. Idrobiol.* 8(Suppl.): 141–161.

Garcia-Gil LJ and Figueras JB (1993) Spatial heterogeneity of chlorophyll in Lake Vilar (Banyoles). *Verh. Internat. Verein. Limnol.* 25: 731–734.

Goehle KH and Storr JF (1978) Biological layering resulting from extreme meromictic stability, Devil's Hole, Abaco Island, Bahamas. *Verh. Internat. Verein. Limnol.* 20: 550–555.

Golebiowska D, Osuch M, Mielnik L, and Bejger R (2005) Optical characteristics of humic acids from bottom sediments of lakes with different mictic types. *Elec. J. Polish Agric. Univ.* 8(2): 1–9.

Gorham E (Ecologist) University of Minnesota, Minneapolis, Minnesota (Personal Communication, 2007).

Gorlenko VM and Chebotarev EN (1981) Microbiologic processes in meromictic Lake Sakov. *Mikrobiologiia* 50: 134–139.

Guerrero R, Abella C, and Miracle MR (1978) Spatial and temporal distribution of bacteria in a meromictic karstic lake basin: relationships with physicochemical parameters and zooplankton. *Verh. Internat. Verein. Limnol.* 20: 2264–2271.

Guilizzoni P, Marchetto A, Lami A, Brauer A, Vigliotti L, Musazzi S, Langone L, Manca M, Lucchini F, Calanchi N, Dinelli E, and Mordenti A (2006) Records of environmental and climatic changes during the late Holocene from Svalbard: palaeolimnology of Kongressvatnet. *J. Paleolimnology*.

Hakala A (2004) Meromixis as a part of lake evolution - observations and a revised classification of true meromictic lakes in Finland. *Boreal Environ. Res.* 9: 37–53.

Halbwachs M, Sabroux J-C, Grangeon F, Kayser G, Tochon-Danguy J-C, Felix A, Beard J-C, Villevielle A, Vitter G, Richon P, Wuest A, and Hell J (2004) Degassing the "Killer Lakes" Nyos and Monoun, Cameroon. EOS, Trans. *Am. Geophys. Union.* 85: 281–288.

Hammer UT (1994) Life and times of five Saskatchewan saline meromictic lakes. *Int. Revue ges. Hydrobiol.* 79: 235–248.

Hammer UT, Haynes RC, Lawrence JR, and Swift MC (1978) Meromixis in Waldsea Lake, Saskatchewan. *Verh. Internat. Verein. Limnol.* 20: 192–200.

Hand RM and Burton HR (1981) Microbiol ecology of an Antarctic saline meromictic lake. In: Williams WD (ed.) *Salt Lakes*, pp. 363–374. London: Dr. W. Junk, Publ.

Holler JD (2005) Lake management plan study for Island Lake, Lower Long Lake, Forest Lake, and Upper Long Lake in Bloomfield Hills, MI. 2005 Rept of Horne Engr. Services, Inc. for Property Owners Association, and Management Committee. 60 pp.

Hongve D (1974) Hydrographical features of Nordbytjernet, a manganese-rich meromictic lake in SE Norway. *Arch. Hydrobiol.* 74: 227–246.

Hutchinson GE (1957) *A Treatise on Limnology, vol. 1. Geography, Physics and Chemistry*, p. 1115. London: Wiley.

Janzen J (2004) *The diver's guide to Lake Wazee*. Inland Sea Corp, Andover, MN: 55304.

Kalff J (2002) Limnology, 529 pp. New Jersey: Prentice-Hall.

Krajick K (2003) Defusing Africa's killer lakes. *Smithsonian Magazine*. September.

Larson DW (1979) Turbidity-Induced Meromixis in an Oregon Reservoir: Hypothesis. *Wat. Resour. Res.* 15: 1560–1566.

Likens GE (1967) Some chemical characteristics of meromictic lakes in North America. Some Aspects of Meromixis, pp. 17–62. *Trans. Symp. Meromictic Lakes*, Fayettville, NY, Green Lakes, 23–24 April 1965 Dept. Civil Engr. Syracuse, NY. Dec 1967.

Li S and Li W (1999) Hydrochemistry in meromictic Lake Zige Tangco, Central Tibetan Plateau. *Asian J. Wat. Environ. Pollut.* 1(1–2): 1–4.

Li S, Wunnemann B, Yu S, Li W, Wu Y, and Xia W (2005) Holocene environmental and climatic change events derived from a meromictic lake sediment record on the Tibetan Plateau. *Poster Abstract: PAGES Second Open Science Meeting.* 10-12 Aug 2005. China: Beijing.

Löffler H (1975) The onset of meromictic conditions in Goggausee, Carinthia. *Verh. Internat. Verein. Limnol.* 19: 2384–2389.

Margaritora FG, Bazzanti M, Ferrara O, Mastrantuono L, Seminara M, and Vagaggini D (2003) Classification of the ecological status of volcanic lakes in central Italy. *J. Limnol.* 62 (Suppl.1): 49–59.

McNeely R (Ecologist) Geol. Survey of Canada, Ottawa, Ontario, Canada (Personal Communication, 2007).

Meybeck M, Martin JM, and Olive P (1975) Geochimie des eaux et des sediments de quelques lacs volcaniques du Massif Central francais. *Verh. Internat. Verein. Limnol.* 19: 1150–1164.

Miracle MR, Armengol-Diaz J, and Dasi MJ (1993) Extreme meromixis determines strong differential planktonic vertrical distribution. *Verh. Internat. Verein. Limnol.* 25: 705–710.

Newcombe CL and Slater JV (1950) Environmental factors of Sodon Lake - a dichothermic lake in southeastern Michigan. *Ecol. Monogr.* 20: 207–227.

Northcote TG and Halsey TG (1969) Seasonal changes in the limnology of some meromictic lakes in southern British Columbia. *J. Fish. Res. Bd. Canada* 26: 1763–1787.

Parkin TB, Winfrey MR, and Brock TD (1980) The physical and chemical limnology of a Wisconsin meromictic lake. *Trans. Wis. Acad. Sci. Arts Lett.* 68: 111–125.

Pendl MP and Stewart KM (1986) Variations in carbon fractions within a dimictic and a meromictic basin of the Junius Ponds, New York. *Freshwat. Biol.* 16: 539–555.

Psenner R (1984) Phosphorus release patterns from sediments of a meromictic mesotrophic lake (Piburger See, Austria). *Very Internat. Verein. Limnol.* 22: 219–228.

Psenner R (1988) Alkalinity generation in a soft-water lake: Watershed and in-lake processes. *Limnol. Oceanogr.* 33: 1463–1475.

Reviriego B, March J, Quetglan G, and Moya G (1993) Diel vertical activity of planktonic communities in the meromictic coastal lagoon of Cibollar (Mallorca, Spain). *Verh. Internat. Verein. Limnol.* 25: 1027–1030.

Richards FA, Cline JD, Broenkow WV, and Atkinson LP (1965) Some consquences of the decomposition of organic matter in Lake Nitinat, an anoxic fjord. *Limnol. Oceanogr.* 10: R185–R201.

Robarts RD and Allanson BR (1977) Meromixis in the lake-like upper reaches of a South African estuary. *Arh. Hydrobiol.* 80: 531–540.

Robdell D (Geologist) Union College, Schenectady, New York (Personal Communication, 2007).

Roberts N, Reed JM, Leng MJ, Kuzucuoglu C, Fontugne M, Bertaux J, Woldring H, Bottema S, Black S, Hunt E, and Karabiyikoglu M (2001) The tempo of Holcene climatic change in the eastern Mediterranean region: new high-resolution crater-lake data from central Turkey. *Holocene.* 11: 721–736.

Robertson D (Limnologist) US Geological Survery, Madison, Wisconsin. (Personal Communication, 2007).

Rodrigo MA, Vicente E, and Miracle MR (1993) Short-term calcite precipitation in the karstic meromictic Lake Cruz (Cuenca, Spain). *Verh. Internat. Verein. Limnol.* 25: 713–719.

Savvichev AS, Rusanov ll, Yu Rogozin D, Zakharova EE, Lunina ON, Bryantseva IA, Yusupov SK, Pimenov NP, Degermendzhi AG, and Ivanov MV (2005) Microbiological and isotopic-geochemical invesatigations of meromictic lakes in Khakasia in winter. *Microbiology.* 74: 477–485.

Sawayko P (Ecologist) Ann Arbor, Michigan (Personal Communication, 2007).

Schanz F and Friedl C (1993) Environmental effects on light utilization by phytoplankton in a meromictic alpine lake. *Verh. Internat. Verein. Limnol.* 25: 621–624.

Slusarczyk A (2003) Limnological study of a lake formed in limestone quarry (Krakow, Poland). I. Zooplankton Community. *Polish J. Environ. Studies.* 12: 489–493.

Stewart KM, Malueg KW, and Sager PE (1966) Comparative winter studies on dimictic and meromictic lakes. *Verh. Internat. Verein. Limnol.* 16: 47–57.

Stewart KM and Hollan E (1975) Meromixis in Ulmener Maar (Germany). *Verh. Internat. Verein. Limnol.* 19: 1211–1219.

Stewart KM and Platford RF (1986) Hypersaline gradients in two Canadian High Arctic lakes. *Can. J. Fish. Aquatic Sci.* 43: 1795–1803.

Strom K (1957) A lake with trapped sea water. *Nature* 180: 982–983.

Tims BV (1972) A meromictic lake in Australia. *Limnol. Oceanogr.* 17: 918–921.

Trolle A (1913) Hydrographical observations from the Danmark expedition. *Medd. Greenland* 41: 271–426.

Truper HG and Genovese S (1968) Characterization of photosynthetic sulfur bacteria causing red water in Lake Faro (Messina, Sicily). *Limnol. Oceanogr.* 13: 225–232.

Vicente E, Camacho A, and Rodrigo MA (1993) Morphometry and physico-chemistry of the crenogenic meromictic Lake EI Tobar (Spain). *Verh. Internat. Verein. Limnol.* 25: 698–704.

Walker KF (1974) The stability of meromictic lakes in central Washington. *Limnol. Oceanogr.* 19: 209–222.

Walker KF and Likens GE (1975) Meromixis and a reconsidered typology of lake circulation patterns. *Verhandlungen der Internationalen Vereinigung für Theoretische und Angewandte Limnologie* 19: 442–458.

Wetzel RG (2001) *Limnology. Lake and River Ecosystems*, 3rd Edition. New York, NY: Academic Press 1006 pp.

Yoshsimura S (1933) Chloride as indicator in dectecting the inflowing into an inland-water lake of underground water, possessing special physico-chemical properties. *Proc. Imp. Acad. Japan* 9: 156–158.

Ziegler W (Fisheries Biologist) Michigan DNR. Escanaba, Michigan (Personal communication, 2007).

# LAKES AND RESERVOIRS OF THE WORLD

Contents

## Origins of Types of Lake Basins

**D K Branstrator,** University of Minnesota Duluth, Duluth, MN, USA

### Introduction

Lake basins originate through a wide variety of natural and anthropogenic processes. Some of these processes are cataclysmic (volcanism), some are gradual and usually imperceptible (tectonic movement), while others (meteorite impact) are rare and extraordinary. To the human observer, most lakes are permanent features of the landscape. On a geologic time scale, by contrast, most lakes are fleeting and all are ultimately ephemeral. Although we have few opportunities to witness a lake's origin, the same set of lake-forming processes at work historically continue unimpeded today.

Three elements are common to the origin of every lake: (1) an environmental force, (2) a body of terrain reshaped by that force into a closed depression (basin), and (3) a water supply. These three elements have met on the landscape with immense frequency during the Earth's history and have given rise to an estimated 304 million natural lakes in existence today.

The first element in a lake's origin is an environmental force and is the facet of its natural history most commonly used by scientists to guide the general classification of lake basins. George Evelyn Hutchinson (1903–1991) provided one of the most extensive surveys available on the origins of lake basins in the first chapter of his four-volume series, *A Treatise on Limnology.* There he describes the formation of numerous distinct types of lake basins resulting from 11 principal environmental forces, including glacial, tectonic, volcanic, fluvial, organism behavior, chemical, wind, landslide, shoreline, meteorite, and organic accumulation.

The second element in a lake's origin concerns the specific process by which an environmental force reshapes terrain into a closed depression. This is the originating element conventionally used by scientists to name distinct types of lake basins. In the case of glacial force, for example, different names are given to basins scoured in bedrock versus those impounded by moraine. Scientists have also found it useful at times to organize lake basins quite broadly according to whether the terrain-shaping process works destructively, constructively, or obstructively. Destructive processes (glacial scouring) excavate depressions, constructive processes (moraine deposition) build rims that define depressions, while obstructive processes (landslide) build rims that barricade preexisting flows. An individual lake basin may be molded by more than one process. Because the Earth's terrain is so richly varied in its composition of soil, rock structure, and topographic relief, there are ceaseless opportunities for unique outcomes in the bathymetry and shoreline structure of lake basins.

A lake district refers to a set of lakes that share a principal originating force and that reside geographically in a defined setting on the landscape. Although individual basins in a lake district are generally of similar age and origin, they can differ exceptionally in size and shape. For example, the set of English Lake District basins shown in **Figure 1** all originated at about the same time by glacial scouring. Their differences in shape and bathymetry illustrate how the interplay between an environmental force and the terrain under influence can change rapidly across small geographic distances.

The third element in a lake's origin is its water supply. Unlike the first two elements, water itself is not always present at a lake basin's inception. Water derives variously from ice, rivers, precipitation, groundwater, wetlands, and preexisting lakes. Water sources to lakes may shift radically through time. For instance, most glacial lakes that were filled originally

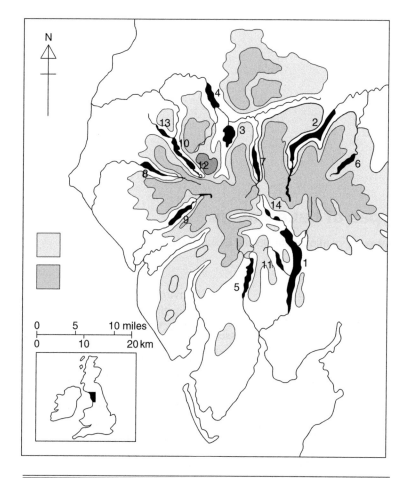

|  | Area (km²) | Length (km) | Max-depth (m) | Mean depth (m) | Volume (m³)(10⁶) | Area of drainage basin (km²) |
|---|---|---|---|---|---|---|
| 1. Windermere | 14.8 | 17 | 64 | 21.3 | 314.5 | 230.5 |
| 2. Ullswater | 8.9 | 11.8 | 62.5 | 25.3 | 223.0 | 145.5 |
| 3. Derwent Water | 5.3 | 4.6 | 22 | 5.5 | 29 | 82.7 |
| 4. Bassenthwaite Lake | 5.3 | 6.2 | 19 | 5.3 | 27.9 | 237.9 |
| 5. Coniston Water | 4.9 | 8.7 | 56 | 24.1 | 113.3 | 60.7 |
| 6. Haweswater | 3.9 | 6.9 | 57 | 23.4 | 76.6 | 26.6 |
| 7. Thirlmere | 3.3 | 6.0 | 46 | 16.1 | 52.5 | 29.3 |
| 8. Ennendale Water | 3.0 | 3.8 | 42 | 17.8 | 53.2 | 44.1 |
| 9. Wastwater | 2.9 | 4.8 | 76 | 39.7 | 115.6 | 48.5 |
| 10. Crummock Water | 2.5 | 4.0 | 44 | 26.7 | 66.4 | 43.6 |
| 11. Esthwaite Water | 1.0 | 2.5 | 15.5 | 6.4 | 6.4 | 17.1 |
| 12. Buttermere | 0.9 | 2 | 28.6 | 16.6 | 15.2 | 16.9 |
| 13. Loweswater | 0.6 | 1.8 | 16 | 8.4 | 5.4 | 8.9 |
| 14. Grasmere | 0.6 | 1.6 | 21.5 | 7.7 | 4.9 | 27.9 |

**Figure 1** Fourteen of the largest ice scour lakes in the English Lake District (shown blackened above) and their physical characteristics (associated table). Burgis MJ and Morris P (1987) *The Natural History of Lakes*. Cambridge: University of Cambridge. Reprinted by permission.

by melt water are now maintained by alternative supplies (precipitation, groundwater). Scientists find it useful to differentiate between lakes connected to rivers (drainage lakes) versus those whose hydrology is reliant exclusively on precipitation and groundwater (seepage lakes). On a regional scale, water supply can be unstable, resetting the hydrological origin of a lake long after its geological birth. One prominent example of this phenomenon is Lake Victoria (Africa), which originated several hundred thousand years before present but was entirely dry as recently as 12 400 years ago.

Certain types of lake-building processes can be expected to dominate over others at any given moment in the Earth's history. In today's inventory of larger lakes, those whose surface areas are greater than $0.01 \, km^2$ (1 ha), approximately 90% have basin origins that trace to glacial, tectonic, or fluvial forces (**Table 1**). Of these, glacial force far outweighs the importance of all others. This contemporary bias owes to recent and widespread glaciation during the Pleistocene when ice sheets covered nearly 25% of the Earth's continents.

## Lake Types

Scientists have long appreciated that a lake's physics, chemistry, and biological potential are predictable end products of its origin. In fact, trained scientists can infer much about a lake's current limnology by simply knowing its originating process. In the following paragraphs, 22 specific processes that originate distinct types of lake basins are enumerated and described. They are organized and presented by a principal environmental force as summarized in **Table 2**. Organic

**Table 1** Estimates of the global number and aerial expanse of lakes greater than $0.01 \, km^2$ in surface area organized by principal environmental force

| Principal force | Number of lakes | Percent of total lakes | Total lake area $(km^2)$ | Percent of total lake area |
|---|---|---|---|---|
| Glacial | 3 875 000 | 74 | 1 247 000 | 50 |
| Tectonic | 249 000 | 5 | 893 000 | 35 |
| Fluvial | 531 000 | 10 | 218 000 | 9 |
| Volcanic | 1000 | ≪1 | 3000 | ≪1 |
| Coastal | 41 000 | <1 | 60 000 | 2 |
| Miscellaneous | 567 000 | 10 | 88 000 | 4 |
| Total | 5 264 000 | ~100 | 2 509 000 | 100 |

A more recent study estimates that the Earth presently holds approximately 27 million natural lakes that are greater than $0.01 \, km^2$ in surface area, about five times more than the total number of lakes shown here. In addition, it is estimated that the Earth presently holds 277 million smaller natural lakes between 0.001 and $0.01 \, km^2$ in surface area, and 0.5 million reservoirs. Adapted from Kalff J (2002) *Limnology: Inland Water Ecosystems*. Upper Saddle River: Prentice Hall. Reprinted by permission of Pearson Education, Inc.

accumulation is the only principal environmental force discussed by G. E. Hutchinson in his *A Treatise on Limnology* that is not considered here.

## Glacial

Glaciers transform the Earth's surface through a variety of erosive and depositional processes resulting from their sheer physical constitution, their forward motion (advance), and their recession through melting (retreat). Glaciers form through the compaction and transformation of snow and other precipitation. During the Pleistocene, glaciers reached heights of 2 km above the Earth's surface, establishing enormous weight loads on the landscape. Under gravitational force imposed by their own mass, glaciers creep internally and slide along terrain, aided by, but not requiring, relief in the landscape. Rock debris is commonly incorporated into glacial ice through abrasion and quarrying (plucking) at the basal surface. Liquid water developing on or in glaciers is heavier than ice and tends to sink and layer along the glacial sole. This lubrication further aids their advance and erosive action.

**Ice scour** Ice scour lake basins are excavations in bedrock caused by the crushing and removal of loose debris. The depressions are generally carved during glacial advance, and deepened over cycles of retreat and readvance. In addition to the scouring effect of pure glacial ice, erosion is facilitated by protruding rock debris and melt water issuing through basal channels. Notable lakes with significant ice scouring in their origins include the Laurentian Great Lakes (Canada, USA), Great Bear Lake and Great Slave Lake (Canada), the fjord lakes in Norway, several lakes in the English Lake District (**Figure 1**), Lago Maggiore (Italy), Lac Léman (France, Switzerland), Lake Te Anau (New Zealand), and innumerable small lakes carved in the pre-Cambrian shield in Canada and Europe.

Glaciers tend to preferentially exploit weaknesses in rock structure and composition. As a result, the basin shorelines and subsurface contours of ice scour lakes often follow preexisting fracture and transitional zones in the bedrock. An acute example of bedrock control on scouring activity is provided by a set of lakes in Minnesota (USA) where glaciers excavated basins in soft slate layered between resistant columns of diabase sill (**Figure 2**). Quite impressive is that the long axes of these lake basins lie oblique to the southerly direction of glacial advance.

A stunning type of ice scour lake basin called a cirque (tarn) originates at the snow line in mountainous relief. Cirque basins derive from

**Table 2**    Common types of lake basins organized by principal environmental force

| Principal force | Lake type | Process of basin origination |
|---|---|---|
| Glacial | Ice scour | D – Glacier excavates a depression. |
| | Moraine dam | C – Glacier pushes or deposits terrain to make a rim. |
| | Kettle | D – Stagnant glacier block displaces soil to make a depression. |
| | Ice dam | O – Lobe or wall of glacier prevents drainage. |
| | Ice basin | D,O – Depression or cavity in glacier prevents drainage. |
| | Thaw | O,C – Permafrost prevents drainage and soils heave to make a rim. |
| Tectonic | Fault block | D,C – Fracture, faulting and warping define a depression and rim. |
| | Reverse drainage | O – Uplift and tilting redirect drainage. |
| | Newland | C – Uplift of ocean floor exposes a submarine depression. |
| Volcanic | Volcanic crater | D – Magma chamber empties to define a depression. |
| | Volcanic dam | O – Volcano or cooled lava barricades a flow. |
| Fluvial | Plunge pool | D – Waterfall excavates a depression. |
| | Oxbow | D,O – Coupled erosion and deposition close a river segment. |
| | Fluvial dam | O – Fluvial sediments barricade a flow. |
| Organism behavior | Beaver pond | O – Wood and mud barricade a flow. |
| | Reservoir | O – Human-constructed dam barricades a flow. |
| | Farm pond, Mine pit | D – Human excavates a depression. |
| Chemical | Solution | D – Bedrock dissolves to make a depression. |
| Wind | Deflation | D – Wind excavates a depression. |
| Landslide | Landslide dam | O – Gravity moves terrain which barricades a flow. |
| Shoreline | Coastal | C – Sediments deposited from longshore currents close a bay. |
| Meteorite | Meteorite crater | D – Meteorite impact excavates a depression. |

The terrain-shaping process at origination is coded as destructive (D), constructive (C), or obstructive (O).

relatively small glaciers and are characteristically bowl-shaped and bounded on the upslope shore by a steep headwall of rock. The erosive power of freezing and thawing on a seasonal basis is believed to enhance local corrosion of the basin floor and walls.

**Moraine dam**    Advancing glaciers push terrain at their leading edge whereas retreating glaciers deposit previously held debris as they melt. Both processes erect mounds of rock and soil on the landscape loosely referred to as moraines. Moraine dam lakes commonly reside in former river valleys with the moraine serving as a rim to complete the basin. Examples include Lake Mendota in Wisconsin and Mille Lacs Lake in Minnesota (USA), the latter lake being almost half bounded by moraine deposits (**Figure 3**). Ice scour and moraine building processes regularly labor in concert to forge glacial lake basins. As a result, many cirque lakes are impounded by a moraine at their downslope edge, and a vast number of lake basins categorized as ice scour depend to some degree on moraine rims to maintain their current depths. The Finger Lakes in New York (USA) were joint products of ice scour and moraine building processes.

**Kettle**    Advancing and retreating glaciers commonly fracture and strand ice blocks. Ice blocks that become partially or fully buried in soil or in the sediment of an outwash plain can originate kettle lake basins

(**Figure 4**). In this process, the dimensions and extent of inlay of the ice strongly dictate the lake basin's shape and bathymetry. Kettle lakes are characteristically deep relative to surface area and they can be multibasined where two or more blocks of ice strand adjacent to one another. Kettle lakes abound in North America, Europe, and Asia. Lakes in the prairie pothole region in Canada and the lakes surveyed by pioneering limnologists Edward A. Birge and Chancey Juday in Wisconsin (USA) are primarily kettles.

**Ice dam**    When glaciers themselves represent barricades that obstruct water flow they originate ice dam lakes. In regions of notable relief, an ice dam lake typically occurs where the lobe of a glacier extends down a main valley to barricade a tributary river entering from a lateral valley. The Märjelensee (Switzerland) is a well-known example. The basins of these lakes are highly transitory and can drain in a marked fashion if the dam hemorrhages. Lake Missoula (USA) was an ice dam lake of the Clark Fork River that once grew to a depth exceeding 600 m and covered an area the combined size of current day Lake Ontario and Lake Erie (Canada, USA). Scientists estimate that the lake drained within a week after the ice dam ruptured. Ice dam lakes also form on flatter terrain when the edge of a glacier prevents the drainage of its own melt water. Here the lake forms through ponding in front of the glacier

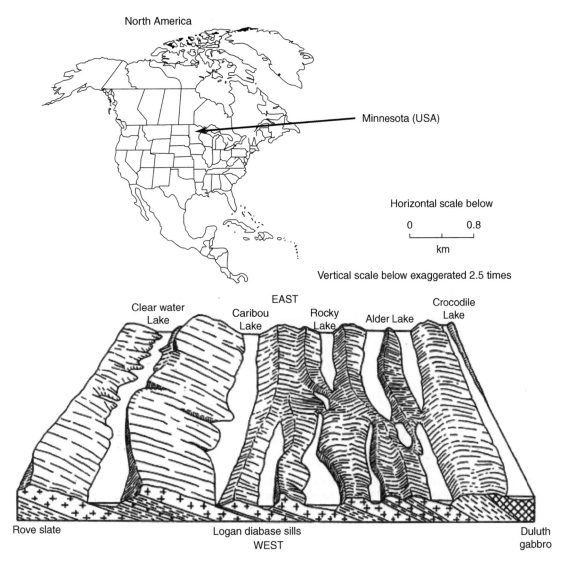

**Figure 2** Examples of some ice scour lakes in Minnesota (USA) that were carved in slate (metamorphic rock) resting between diabase sill (igneous rock). Adapted from Zumberge JH (1952) *The Lakes of Minnesota: Their Origin and Classification, University of Minnesota Geological Survey Bulletin 35*. Minneapolis: University of Minnesota. Continental map courtesy of Graphic Maps, Woolwine-Moen Group. Reprinted by permission.

and is more specifically termed a proglacial lake. Isostatic rebound of the recently uncovered terrain may tilt it toward the glacier and enhance the ponding effect. Lake Agassiz, the largest proglacial lake known, existed for some 4000 years and covered more than 350 000 km$^2$ during its life (**Figure 5**). The lake changed its configuration and reach many times as the glacier retreated northward. Vestiges of Lake Agassiz include Lake of the Woods (Canada, USA) and Lake Winnipeg (Canada).

**Ice basin** Ice basin lakes reside either on or in a glacier. They originate when melt water, obstructed from exiting a glacier, pools either in a surface depression or internally in a glacial cavity. The depressions and cavities form as a result of glacial movement,

fracture, fluvial erosion, and heat from the sun and the Earth. In the case of surface depressions, the entire lake is cupped in a basin of ice. In the case of internal cavities, the lake basin commonly resides at the floor of the glacier and is bounded by land underneath and by ice on the walls and ceiling. Lake Vostok (Antarctica) is an example of the latter, residing some 4000 m below the central Antarctic ice sheet see **Antarctica**. It is a huge lake with a liquid depth of at least 800 m and an area comparable to modern day Lake Ontario (Canada, USA).

**Thaw** Thaw (thermokarst, cryogenic) lakes have fascinated scientists for decades. These lakes cover vast coastal areas in the arctic regions of Eurasia and North America. They are best classified as

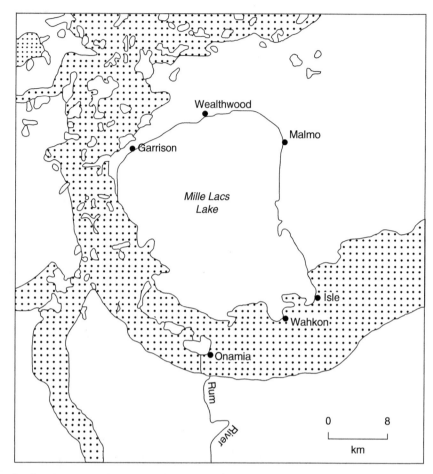

**Figure 3** An example of a moraine dam lake in Minnesota (USA). The stippled area shows moraine complex. Geographic reference as in **Figure 2**. Adapted from Zumberge JH (1952) *The Lakes of Minnesota: Their Origin and Classification, University of Minnesota Geological Survey Bulletin 35*. Minneapolis: University of Minnesota. Reprinted by permission.

periglacial because their origins depend on near-glacial conditions, but not glaciers themselves. A thaw lake originates when melt water in the surface layer of permafrost is prevented from draining downward by a deeper layer of frozen permafrost which serves as the basin floor. The surface area of a thaw lake may be quite small at first and polygon shaped. The soil rims that contain these lakes arise above fracture zones in the permafrost where annual freeze-thaw cycles lead to vertical expansion and soil upheaval. Contiguous thaw lakes will coalesce, resulting in large and small lakes in the same general area (**Figure 6**).

**Tectonic**

The Earth's exterior layer is comprised of a network of about a dozen relatively rigid, crustal plates that form a shell around the planet. The boundaries of these plates are zones of active slip, collision, and separation that generate what are called tectonic forces. Tectonic forces that translate upward to the

Earth's surface deform bedrock through fracture, rifting (separation), and warping (uplift and subsidence), resulting in the formation of mountains, ocean basins, and some of the world's largest, deepest, and oldest lake basins.

**Fault block** Fault block lake basins form where uplift and subsidence create vertical offset in adjacent blocks of fractured land. In cases where a single fault or fault complex is active, the process leads to a half-graben. This type of basin has characteristic steep-walled bathymetry on the fault side and an angled floor that slopes gradually to the opposite shore where vertical offset is minimal or nonexistent. Where multiple fault lines occur with wide parallel spacing, both sides of a land block can experience vertical offset and create a trough-shaped basin called a graben. Grabens generally contain precipitous bathymetric contours along both main shorelines (**Figure 7**). They are characteristically flanked by massive, steep escarpments that crest hundreds to

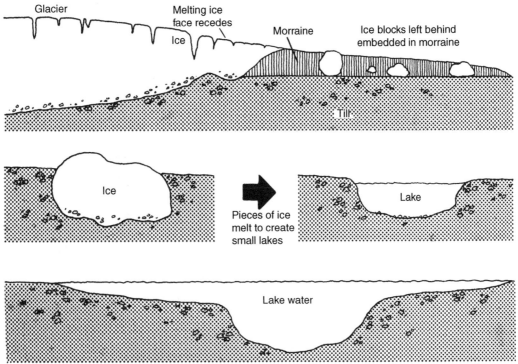

**Figure 4**   A schematic of how kettle lake basins originate. (a) Glacial retreat and ice block burial; (b) Post-melt; (c) General relationship between the size and shape of an ice block and a lake basin's bathymetry. Adapted from Hutchinson GE (1957) *A Treatise on Limnology. Volume 1: Geography, Physics, and Chemistry.* New York: Wiley. Exhaustive effort was made to secure permission.

thousands of meters above lake level (**Figure 8**). Well-known examples of fault block lakes include Lake Baikal (Russia), Lake Ohrid and Lake Prespa (Europe), Lake Issyk-Kul (Kyrgyzstan), Lake Tahoe (USA), and several lakes in the Central African Rift Valley District (**Figure 9**), including Lake Tanganyika, Lake Malawi, Lake Edward, Lake Albert, and Lake Rudolf. Some grabens that have been filled with water continuously for millions of years now house remarkably thick sediment layers such as those in Lake Baikal (~8 km thick) and Lake Tanganyika (~4.5 km thick). Such extraordinary sediment accumulations can only be explained if basin subsidence is ongoing. Evidence shows this to be the case which means that the longevity of these lakes depends in part on continual processes of origination.

**Reverse drainage**   Reverse drainage lakes result from uplift and tilting that redirect drainage. Two well-known examples include Lake Kyoga and Lake Victoria in Africa (**Figure 9**). These lakes were created when uplift around the plateau's western margin reversed flow in the Kafu, Katonga, and Kagera Rivers. All three major rivers historically flowed east to west across the plateau but now flow west to east

over much of their course, flooding what were once old river channels and riparian plains in the formation of these two lakes. The dendritic shoreline of Lake Kyoga and the angles of its bays with respect to the main arm of the lake remain vestiges to this day of an ancestral fluvial state and a history of drainage that once flowed east to west (**Figure 9**).

**Newland**   Newland lake basins originate when a submarine basin on the sea floor is uplifted and becomes exposed. The Caspian Sea, Aral Sea, and Lake Okeechobee (Florida, USA) are examples. Despite their marine ontogeny, the Caspian Sea and Aral Sea currently house salts derived overwhelmingly from terrestrial sources.

**Volcanic**

Volcanism is responsible for a variety of lake basin types that can be divided relatively naturally into two groups. One group includes those forming directly in the volcanic chamber where magma exited. The other group includes basins that result from obstruction imposed by the volcanic mountain itself or the expelled magma. Volcanic lakes are

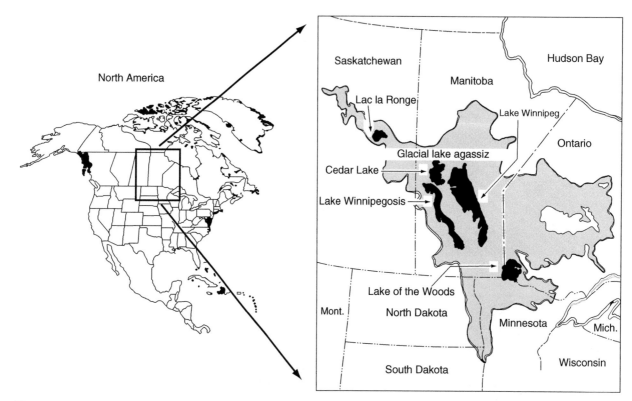

**Figure 5** The historical border of proglacial Lake Agassiz (stippled) and the current borders of five remnant extant basins (blackened) in North America. Adapted from Timms BV (1992) *Lake Geomorphology*. Adelaide: Gleneagles. Continental map courtesy of Graphic Maps, Woolwine-Moen Group. Reprinted by permission.

**Figure 6** Examples of thaw lakes in Alaska (USA). Joel Sartore / National Geographic Image Collection.

generally small but often deep and they comprise some of the world's most aesthetically pleasing and noteworthy ecosystems.

**Volcanic crater** Volcanic crater lake basins originate in the cavities from which magma was ejected. Small volcanic crater lakes (maars) and large ones (claderas) have representatives throughout the world, including many in the Eifel region (Germany), the Auvergne region (France), Indonesia, and central Africa. Because of their origin, these lakes generally have a small aspect ratio (maximum width:maximum depth), which can inhibit complete mixing (turnover) of the lake's water mass on an annual basis. Lake Nyos (Cameroon) is a maar with an aspect ratio of 9:1. In its recent history, Lake Nyos remained partially unmixed long enough to become supersaturated with carbon dioxide gas. A catastrophic episode of mass release of the gas in 1986 killed about 1700 humans and 3000 cattle. Crater Lake (USA) is a magnificent example of a caldera. At about 600 m depth, it is one of the top 10 deepest lakes in the world. Lake Tazawa and Lake Okama (Japan), and Lake Taupo (New Zealand) are other examples.

**Volcanic dam** Volcanic dam lakes originate as a result of drainage that is blocked by either a volcanic mountain or its expelled lava. One example is Lake Kivu, which lies on the western side of the Central African Rift Valley (**Figure 9**). In the origination of this lake, seven major volcanoes dammed a drainage pattern that historically flowed north into Lake Edward. Drainage in the watershed now accumulates in Lake Kivu, with excess water in the lake flowing

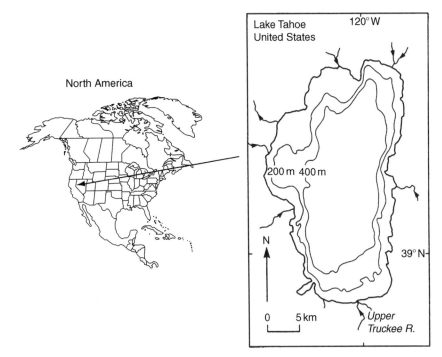

**Figure 7** The bathymetry of Lake Tahoe (USA), a graben. The lake's maximum depth is slightly over 500 m. Adapted from Hutchinson GE (1957) *A Treatise on Limnology. Volume 1: Geography, Physics, and Chemistry.* New York: Wiley. Exhaustive effort was made to secure permission. Continental map courtesy of Graphic Maps, Woolwine-Moen Group. Reprinted by permission.

**Figure 8** The Livingstone Mountains of Tanzania define the eastern fault in the north basin of Lake Malawi (Africa), a half-graben. By kind permission of Tom Johnson, University of Minnesota Duluth.

south to Lake Tanganyika. Lake Lanao (Philippines) is another example of a volcanic dam lake.

### Fluvial

Running water plays a profound role in sculpting the Earth's surface. Its dual ability to erode and construct, akin to glacial, tectonic, and volcanic forces, engender fluvial force with a wide range of originating processes.

**Plunge pool** As the name implies, a plunge pool lake basin originates at the base of a waterfall where the destructive energy of falling water excavates a hollow large enough to hold water long after the river has perished. A number of lakes in eastern Washington (USA) provide examples including Dry Falls Lake.

**Oxbow** An oxbow (billabong) lake basin originates through the coupled influence of erosion and deposition in what are often wide, gently sloping floodplains. The general process, illustrated in **Figure 10**, is one whereby a meandering loop in a river is eventually abandoned as the river cuts a newer, more direct path through the bank. The old river course is sealed at both ends with sediment deposits. Oxbows are commonly serpentine or crescent shaped, which reflects their position in the old river channel. Innumerable examples of oxbows exist worldwide that can best be appreciated from aerial views (**Figure 11**).

**Fluvial dam** Fluvial dam lakes originate when deposited silt creates a barrier that impounds drainage. Most common among this spectrum of lakes is a lateral lake that originates when a tributary is obstructed from entering a main river by a levee at the confluence. Lateral lakes are frequent on the Danube River (Europe) and the Yangtze River

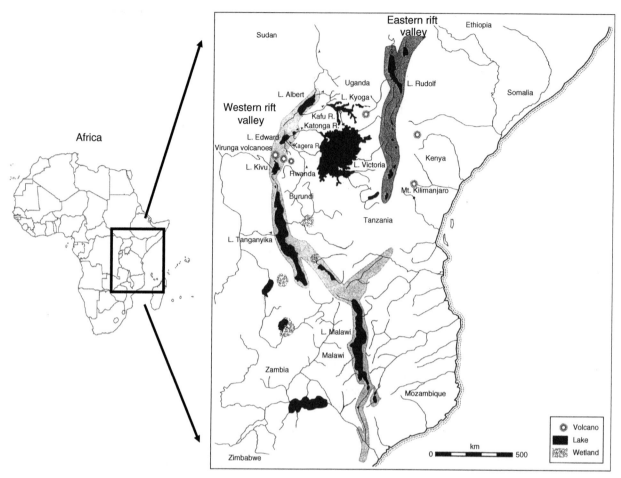

**Figure 9** Lakes in the Central African Rift Valley District (blackened). Adapted from Kalff J (2002) *Limnology: Inland Water Ecosystems*. Upper Saddle River: Prentice Hall. Reprinted by permission of Pearson Education, Inc. Continental map courtesy of Graphic Maps, Woolwine-Moen Group. Reprinted by permission.

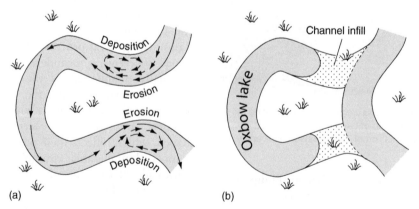

**Figure 10** Diagram of an oxbow lake during the (a) precursor phase and (b) late phase of origination. Kalff J (2002) *Limnology: Inland Water Ecosystems*. Upper Saddle River: Prentice Hall. Reprinted by permission of Pearson Education, Inc.

(China). A second type of fluvial dam lake is called a floodplain lake. This type originates when a levee develops along the edge of a main river and obstructs seasonal floodwater of the main river from reentering. The volume of a floodplain lake can shift by an order of magnitude on a seasonal basis in relation to rainfall. Floodplain lakes are common throughout low-latitude, riparian regions of South America.

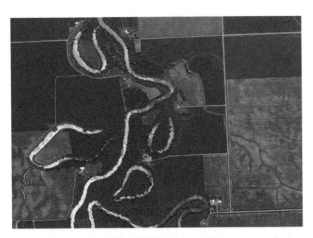

**Figure 11** Aerial photograph of four oxbow lakes flanking the Red River at the border of Minnesota and North Dakota (USA). Reprinted by permission of U.S. Department of Agriculture, Farm Service Agency, Aerial Photography Field Office.

### Organism Behavior

**Beaver pond**  Beavers are industrious ecosystem engineers that transform land surfaces from terrestrial to aquatic. Their dams are built of wood and mud for purposes of habitat expansion and predator protection. Beaver dams may reach 4 m in height and extend for up to 0.5 km in length, giving their aquatic impoundments the dimensions of small lakes. The European beaver (*Castor fiber*) was extirpated by trappers over most of its native range by 1900, and is now being reintroduced. A dire fate of similar proportions reduced the abundance of the American beaver (*Castor canadensis*) from an estimated 60–400 million individuals only two centuries ago to 6–12 million today. The remarkable numbers of beaver at one time suggest that their ponds may have once contributed significantly to lake numbers on a global scale.

**Reservoir**  Reservoirs are human-made impoundments that block the natural flow of rivers and submerge formerly terrestrial surfaces. They are generally built for purposes of flood control, water supply, power generation, navigation, fish production, or recreation. Modern, highly engineered reservoirs are capable of retaining enormous volumes of water and controlling its passage at the outlet with great precision. The numbers and sizes of reservoirs have been growing worldwide at rapid rates since World War II. One of the largest is Bratsk Reservoir (Russia) with a volume that exceeds Lake Tahoe (USA).

**Farm pond, Mine pit**  Humans have long been excavating basins to retain water for agricultural purposes. These basins, called farm ponds, are widespread globally and may be more important in their contribution to the total surface area of freshwater than once thought. Humans also excavate basins during mining operations for rocks, metals, and gems. These depressions are called mine pits and once abandoned they fill naturally with groundwater. Some mine pit lakes are remarkably deep. The Portsmouth Mine Pit Lake in Minnesota (USA) has a maximum depth of 150 m and a surface area of 0.5 km$^2$, making it the state's deepest inland lake.

### Chemical

**Solution**  Solution (karst, doline) lake basins form through a process of chemical dissolution of bedrock. The breakdown of limestone ($CaCO_3$) by natural levels of acidity in the groundwater is the most common chemical reaction involved. A solution lake basin generally originates as a subsurface cavern which progressively collapses under the strain of overlying soils. Solution basins have been known to appear suddenly, and disastrously, where large underground cavities collapse all at once. Subsurface outlets and cracks in solution basins may be sealed by residual rock, soil, or the hydrostatic pressure of the water table, including the ocean in coastal locations. They are common in the Balkan Peninsula, the European Alps, and Florida (USA).

### Wind

**Deflation**  Deflation (playa, pan) lake basins originate through the erosive force of wind that removes loose terrain. The process is facilitated by an arid climate and a lack of vegetative cover, and may be aided further in some instances by intermittent fluvial erosion and animal occupation (ungulates) which can help loosen sediment and reduce its grain diameter. Deflation lakes may dry up on a seasonal basis if precipitation and runoff are unable to maintain their evaporative losses. They are common throughout arid regions in Australia, Africa, and North America. One notable example is Lake Alablab (Kenya).

### Landslide

**Landslide dam**  A landslide is a gravitationally pushed, mass movement of debris. When one obstructs the passage of a river it originates a landslide dam lake. Because landslide debris is typically unconsolidated it erodes rapidly. These lakes are generally short-lived compared to other lake types. Two examples include Janet Lake in Glacier National Park (USA) and Lake Waikaremoana (New Zealand).

**Figure 12** Aerial photograph of Stoney Lake (USA), a coastal lake of Lake Michigan (background). Reprinted by permission of Aerial Graphics, L.L.C., Grand Rapids, Michigan (USA).

## Shoreline

**Coastal**    Coastal lakes originate when a bay or indentation in the shoreline of a lake or ocean becomes closed to the main body of water by a bar (spit) of sediment deposited through longshore currents. It is a process similar to that which creates fluvial dam lakes. Maritime examples of coastal lakes are common in France, Australia, and New Zealand. Lake Nabugabo (Uganda) is an example of a freshwater coastal lake which was cut off from Lake Victoria. The western edge of the state of Michigan (USA) is rich with coastal lake basins sealed off from Lake Michigan (**Figure 12**).

## Meteorite

**Meteorite crater**    The most bizarre of all originating events, and the rarest at this moment in the Earth's history, is that related to the impact of a meteorite. In this process, the catastrophic destruction and dispersal of terrain leaves a hollow called

a meteorite crater lake basin. Examples include Pingualuit Crater Lake in Québec (Canada) and Laguna Negra (Argentina).

## Conclusion

Most lake basins originate through nonliving processes but they occasionally involve the behavior of animals such as beaver, ungulates, and humans. Although a relatively limited number of recognized processes originate lake basins, the path followed by any particular lake will always be unique and inherently richer and more variable than a classification scheme can convey. Understanding the origin of a lake basin not only serves to document the natural history of its landscape but also provides valuable context for the investigation of its past and contemporary limnology.

*See also:* Abundance and Size Distribution of Lakes, Ponds and Impoundments; Africa: South of Sahara; Antarctica; Aquatic Ecosystem Services; Arctic; Geomorphology of Lake Basins; Mixing Dynamics in Lakes Across Climatic Zones; North America.

## Further Reading

Beadle LC (1974) *The Inland Waters of Tropical Africa*. London: Longman.

Hutchinson GE (1957) *A Treatise on Limnology. Volume 1: Geography, Physics, and Chemistry*. New York: Wiley.

Kalff J (2002) *Limnology: Inland Water Ecosystems*. Upper Saddle River: Prentice Hall.

Meybeck M (1995) Global distribution of lakes. In: Lerman A, Imboden D, and Gat J (eds.) *Physics and Chemistry of Lakes*, 2nd edn., pp. 1–35. Berlin: Springer-Verlag.

Sugden DE and John BS (1976) *Glaciers and Landscape: A Geomorphological Approach*. London: Edward Arnold.

Timms BV (1992) *Lake Geomorphology*. Adelaide: Gleneagles.

Wetzel RG (2001) *Limnology: Lake and River Ecosystems*, 3rd edn. San Diego: Academic.

Zumberge JH (1952) *The Lakes of Minnesota: Their Origin and Classification, University of Minnesota Geological Survey Bulletin 35*. Minneapolis: University of Minnesota.

# Geomorphology of Lake Basins

**B Timms,** University of Newcastle, Callaghan, NSW, Australia

## Introduction

Mapping is basic human activity, born in antiquity and developed today to embrace complex, multilayered geographic information systems. Maps of lakes are of more intrinsic value than terrestrial maps, because they show features below the water surface the eye cannot see. Moreover, they can be used to compute various parameters of value in understanding the structure and functioning of lakes. This chapter is devoted to explaining these parameters and to expand upon the role of lake geomorphology in limnology.

## Morphometric Parameters

In the past, mapping lakes was a laborious task. Lake shorelines had to be established by such methods as transverse, stadia or plane table surveys, all time consuming, and spot depths by lead lines at fixed positions. Today, outlines of shores are easily obtained from aerial or satellite images and depth profiling done by echosounders. Exceptions include shallow ephemeral lakes, like many in Australia, which are best mapped when dry using modern terrestrial surveying equipment. Once, the basic data had to be processed by hand to produce a map, but nowadays almost all stages can be computerized. In all cases scale is important, the finer the scale, the more accurate the map. The parameters derived from these maps can be divided into those taken directly or indirectly from the map and those that are derived by computation from the primary parameters.

(a) Primary Parameters

   (i) *Lake length.* This is the length of the line connecting the two most remote points on the shoreline. It should not cross land (so in an oxbow lake the line is curved), but it may cross islands. Such measurements are of intrinsic geomorphic value only. Of more use limnologically is the *maximum effective length*, which is defined by the longest straight line over water on which wind and waves can act. While the two values are similar in large deep lakes like Africa's Lake Victoria, in island-studded lakes like Sweden's L. Vänern, the maximum effective length is 69% of the maximum length. Knowledge of this parameter is important in geomorphic studies on shorelines and in studies on seiches and stratification in physical limnology.

   (ii) *Lake width* is defined by a straight line at right angles to the maximum length and connecting the two most remote shorelines. Again, of more value is the maximum effective width, which does not cross land. This, and the mean width are of use mainly in hydromechanical studies.

   (iii) *Lake depth* is the maximum known depth of a lake, and is the single most important geomorphic parameter of a lake. The world's deepest lake is Lake Baikal in Siberia (1620 m deep) and many large glacial, volcanic, and tectonic lakes exceed 200 m in depth. By contrast, most lakes on flood plains or formed by wind are shallow, rarely exceeding 5 m in depth. The limnological consequences of this are paramount and discussed later. *Mean depth* (lake volume/lake area) is also worth noting and has been used many times to explain varying productivities between lakes (e.g., the deeper the lake the less productive it is). Of the various other depth-related parameters, *relative depth*, which is the ratio of the maximum depth to the mean diameter of the lake, is useful in explaining stability of stratification in lakes. For large shallow lakes like the wind-stirred Lake Corangamite in Victoria, Australia, the value is 0.09%, while the nearby meromictic West Basin Lake in a volcanic crater, the relative depth is 3.0%.

   (iv) *Direction of major axis.* It is important to know this in geomorphic and hydrodynamic studies as lakes may be aligned to dominant wind direction and hence be more subject to wind than others. For instance, in the eastern inland Australia, only those lakes with an axis near N-S grow spits under the influence of winds from the NW that close off the southeast corner (see later).

   (v) *Shoreline length* is easy enough to measure by a map measurer, but is very much influenced by map scale. It is used to calculate shoreline development, a parameter used in littoral studies.

   (vi) *Lake area*, once determined by planimetry, but now easily done with a computer, and hardly affected by map scale, is another of the most basic lake parameters. The world's largest lakes are of tectonic and of glacial erosion origin. At the individual country scale, lakes are often listed by area with an explanation for any pattern based on geomorphic distinctiveness of the district/mode of origin. For instance in New Zealand, geology and climate explain the dominance of large piedmont glacial lakes on the South Island, and many somewhat smaller volcanic lakes on the North Island. If the area occupied by a lake fluctuates (because it is terminal or used for water supply) then it is useful to know the area at any depth and this is visualized in a *hypsographic curve* (**Figure 1(a, c)**).

(vii)  *Lake volume* is calculated from summing the volumes between each contour, though for increased accuracy different formulae are used according to lake form. This is another parameter often quoted for lakes, particularly if they are large. Most impressive in this instance is Lake Baikal's massive volume of $23\,000\,\text{km}^3$, representing one-fifth of the world's fresh water. Visualization of volume at any depth, particularly useful in lakes and reservoirs that fluctuate in depth, is achieved by a *volume/depth curve* (**Figure 1(b, d)**). In this respect, volume percentages derived from cumulative curves for reservoirs are widely quoted in the media in dry countries like Australia, where water reserves are precarious and precious.

(viii) *Insulosity* is the percentage of the lake area occupied by islands. Though some lakes have minor islands, like subsidiary cones in crater lakes, their insulosity values are of no consequence. It is mainly in glacial ice scour lakes and other lakes with highly irregular shorelines where there are many islands that this parameter exceeds ca. 10% and assumes importance. While its worth is intrinsically geomorphic, it is sometimes equated to the value of a lake for recreation, where humans

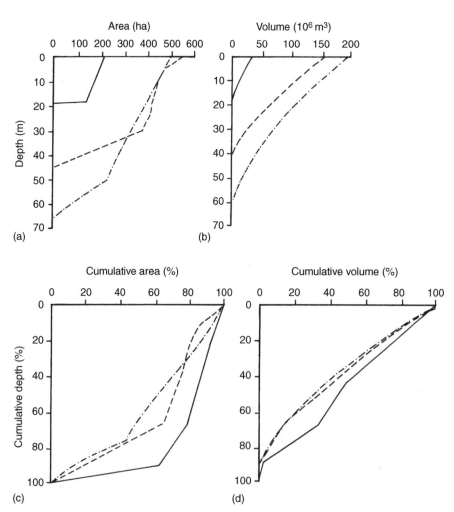

**Figure 1**  Absolute hypsographic (a) and volume curves (b), and relative hypsographic (c) and volume curves (d) for three maar lakes in Victoria, Australia.

require high shoreline–waterway interaction, as in Swedish lakes Vänern and Mälaren. On the other hand, sailors and waterskiers require open water for their recreation activities, so small, lakes without islands are favored. Thus, there is no relationship between insulosity and recreational use of a lake!

(ix) *Watershed to lake area/volume.* This is easy to calculate from maps and is used as an estimate of terrestrial inputs as in eutrophication studies and also as water renewal ratios for some in-lake processes.

(b) Common Derived Parameters

(i) *Shoreline development* ($D_s$) is a measure of the irregularity of the shoreline; its accuracy is dependent on map scale. Essentially, it is the ratio of the length of the shoreline to the length of the circumference of a circle of area equal to that of the lake. It is an index of the potential importance of littoral influences on a lake. Perfectly circular lakes have a $D_s$ of 1.0, average lakes have values between 1.5 and 2.5, while lakes and reservoirs with much indented shorelines have values exceeding 3 (**Figure 2**). Expressed in relation to lake types, volcanic vent lakes have minimal $D_s$s, and lakes in dammed valleys and in ice-scoured terrain have the highest values.

(ii) *Volume development.* This index ($D_v$) is used to express the form of a basin, and is defined as the ratio of the volume of a lake to that of a cone of basal area equal to the area of the lake and depth equal to the maximum depth of the lake. Lakes with a $D_v$ of 1.0 are perfectly cone-shaped and uncommon. Values <1.0 indicate a trumpeted-shaped lake and are rare

and values >ca. 2.5 indicate a beaker-shaped lake and are also uncommon (**Figure 2**). Most lakes have values somewhat greater than unity. The unusual values may be related to lake type, e.g., doline lakes usually have $D_v$s near 1 and claypans and maar lakes may have $D_v$s near 3, but sometimes $D_v$s may be the result of peat growth/marl deposition/shoreline slumping/erosion in the littoral, leading to values near 1. In all cases the index is a surrogate for the role of the sublittoral in a lake's limnological processes, the lower the value the greater the sublittoral influence.

(iii) *Slope* is the angle, usually expressed as a percentage of repose of the bottom sediments and can be determined directly from an echogram or from a contour map using appropriate measurements and formulae. Slopes of near 0% (profundal areas) to 20% (sublittoral slopes) are common, but occasionally values approach the maximum of 90% in a steep-shored crater, glacial erosion and lakes due to earth movements. This parameter is used mainly in sedimentological studies and to lesser extent in benthic studies in choosing suitable stations. *Mean slope* is an average for a whole lake and is derived from a contour map and application of formulae. Not surprisingly, mean values are much more subdued than slopes at designated contours and commonly range from <1% to 10%, but may exceed 25% in some lakes, e.g., Lake Barrine, a small maar in north Queensland, Australia, has a mean slope of 30%. Perhaps in interlake comparisons, visual inspections of comparative hypsographic curves (**Figure 1**) are just as instructive as are figures for mean slope, and far less troublesome to prepare.

(iv) Sometimes a derived parameter can be established for a special need. For example *Ratio of epilimnion sediment area to epilimnion volume* has been used more effectively than parameters based on whole lake volume in eutrophication studies because it more accurately accounts for nutrient recycling in a lake.

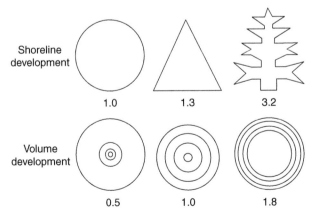

Shoreline development

1.0    1.3    3.2

Volume development

0.5    1.0    1.8

**Figure 2** Graphical representation of shoreline development for three lakes of different shape, and of volume development for another three lakes of different volume distribution. In the latter, concentric lines represent depth contours.

## Lake Shapes

Many lakes have characteristic shapes determined by their mode of origin that are only partly described by their parameters. The ratio of length over width can give information on whether a lake is rectangular, circular or ellipitical and high $D_s$s can indicate a dendritic lake, but a simple descriptive word is usually better. *Circular* lakes ($D_s$ near 1) are uncommon, but can be found in volcanic vents, in some deflation basins and in lakes due to meteoric impact. *Subcircular* lakes are more common and are associated with a variety of origins: volcanic craters, deflation basins, dolines, cirques, kettle-holes to name the more common ones. *Ellipitical* lakes are a special subgroup of these and are generally associated with wind deflation. *Subrectangular* elongate lakes generally are of structural origin or result for glacial erosion in valleys (piedmont lakes). Markedly *dendritic* lakes result from flooding of dissected valleys as in some coastal lagoons, some landside lakes and in many reservoirs. It is these in which $D_s$ is high (>3). *Triangular* lakes also usually arise from flooding, usually either on floodplains or along coasts. These may be confused with circular lakes on length/width ratios, but not by their descriptor. *Lunate* (or new-moon shaped) lakes result from river meandering to form oxbows, and sometimes from asymmetrical placed subcones in volcanic craters. Highly *irregular* lakes are possible from the fusion of lake basins and in glacially scoured areas. These also may have high $D_s$s.

## Changes with Time

Over geological time, lakes are temporary landscape features (but see later), filling with sediment and/or eroding at the outlet. Consequently lake morphometry changes with time, imperceptibly in well-watered areas, but at the other extreme, oscillating widely in terminal lakes in arid areas. The most obvious changes are lake level fluctuations and their concomittent shoreline changes. Sedimentation is less obvious, except for delta construction.

Lowered lake levels, of whatever cause, result in stranded beaches, spits and deltas. Worldwide, piedmont glacial lakes provide many examples, perhaps none better than New Zealand's Lake Wakatipu with its 10 strandlines cut into a major bench and deltaic fronts at 50 m above present lake level The terminal lakes of Tibet have numerous elevated shorelines attesting to a more fluvial past, as do most of the world's large lakes in endorheic regions. Interesting as these landforms may be, and useful in aging a lake, it is the changes with decreasing depth in physicochemical processes in the lake such as thermal and chemical stratification, that are important limnologically.

In some lakes, levels have been elevated since initial formation, drowning shoreline features and river channels. Coastal marine lagoons, tied to changing sea levels, provide many examples including Lake Macquarie near Sydney, Australia, where drowned river channels and spits are clearly discernable (**Figure 3**). As in lakes with lowered levels, raised levels may change physicochemical processes, and in addition, there may be hydrodynamic changes associated with changed inflows.

All lakes accumulate sediment, either mainly from stream inflows, or biological production in the lake itself, or occasionally from overland flows and rarely by showering ash from volcanic eruptions (as in lakes around Taupo and Rotorua in New Zealand). Rates

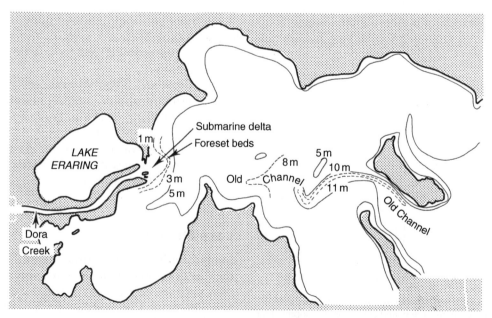

**Figure 3** The western bay of Lake Macquarie, near Sydney, Australia, a drowned coastal valley, showing the former river course and a lobate delta.

are highly variable and depend on geomorphic type, catchment size and erodability, and trophic status (and proximity to a volcano!). The most rapid accumulation is by delta building in lakes with large inflows in mountainous areas. Well-known examples include the deltaic plain at the head of Lake Geneva and the surficial sediments that divide Thuner See and Brienzer See at Interlaken, both in Switzerland. Deltaic form depends on the relative densities of lake and river water and lake currents. Arcuate Gilbert deltas due to homopycnal flows are perhaps the most common type in piedmont glacial lakes, lobate deltas associated with hypopycnal inflows in coastal marine lagoons (**Figure 3**), and alluvial fans in shallow desert lakes. No matter the deltaic form, lake surface area and volume are reduced, but probably with minimal influence on limnological processes in large lakes.

On the other hand, lake floor sedimentation due to submarine delta building by hyperpycnal inflows (as in Lake Pukaki in New Zealand) and by chemical or biological deposition decrease lake depth and hence may influence physicochemical processes, particularly if the sediments of productive lakes consume oxygen during stratification. Rearrangement of bottom sediments due to slumping or bioturbidation are of little geomorphic consequence, but may upset layering and hence interpretations of lake/catchment history recorded in sediment cores.

## Ancient Time plus Size

While differing geomorphology, as measured by the above parameters, influences lake ecology, there are many other factors involved, one of which is geological time. Most lakes exist for hundreds to thousands of years, perhaps even a few hundred thousand years, but some persist for many millions of years, and cognisance of this, together with their size (volume, perhaps area), allows greater understanding. Scale is indeed important in limnology.

By considering a new coefficient, Touchart divides lakes into four broad groups. He estimates the scale of a lake, by taking the logarithm of the product of the age (expressed in millennia) by its volume (in cubic kilometres). Lakes with values 4–8 are almost exclusively large old structural lakes such as Baikal, Caspian, Tanganyika (values of 8), Victoria (7), Issyl-Kul, Aral, Titicaca (6), Balkhash, Tahoe (5), and Biwa, Toba, Taupo (4), among others. Most of these are largely independent of morphoclimatic hazards and have high biological endemism. For instance in Lake Baikal there are 255 species of gammarid amphipods in 35 genera, 34 of which are endemic. New Zealand's Lake Taupo is an exception, going back perhaps only a million years, but subject to many catastrophic reincarnations as recent as 1800 years ago). Biodiversity is low as a consequence of this Recent age, and also it small size (623 km$^2$) and being on an isolated island.

The next group usually have values 1–4 and are medium sized, largely of morphoclimatic heritage, i.e., shaped largely by the Würm glaciation. Examples include, Lakes Superior, Great Bear (4), Ontario, Winnipeg (3), Constance, Geneva (2), Te Anau, Wakatipu (both NZ)(1) among a host of others. They are often large (>10 000 km$^2$) and deep (200–500 m), and formed by glacial erosion. Life expectancy is in the order of only thousands/tens of thousands of years. Biodiversity is lower and morphodynmaic processes (delta-building, sedimentation) in their basin are relative important in limnological processes. These are the lakes most common in the Temperate Zone of the Northern Hemisphere where limnology developed, so they are the 'standard' lakes of text books. To a large degree both of these classes of lakes are dominated by their shear size, so that morphometric parameters other than area and depth are generally unimportant in influencing limnological processes.

A third group of lakes are small (Touchart coefficients of 1 or less) and largely influenced by current morphodynamic processes or recently inherited ones. These include fluviatile lakes, karstic lakes, aeolian lakes, and landside lakes in the first subgroup and many small glacial and volcanic lakes in the second. All are subject to changing shape and dimensions and many have a particularly precarious existence (e.g., landslide lakes). It is in these lakes where key abiotic variables (e.g., Secchi depth) are related in varying degrees to lake morphology and catchment area.

Touchart's fourth group of lakes also have coefficients of 1 or less; they are the lakes (reservoirs) of human origin. They are of variable size (in area, some challenge lakes with coefficients of 4), but are recent in origin and have a short life span. Moreover, their morphometry is forever changing and their catchments dominate their limnological processes.

Touchart largely omits consideration of short-term changing morphology, partly seen in reservoirs, but absolutely characteristic of intermittent lakes. Such lakes are uncommon where most limnologists live in Europe and North America, but prevalent in the drier parts of Asia, Africa, South America and especially Australia. Some are of very ancient lineage, e.g., Lake Chad, Lake Eyre, but lack ancient/endemic biodiversity because intermittent lakes are useless evolutionary loci. Others (e.g., seasonal lakes due to flooding; shallow lakes due to wind deflation) are temporary landscape features and if they lie in a distinct basin, are best depicted with inverse contours (i.e., bottom at 0 m, highest level at x metres/centimetres to account for their fluctuations in water level (**Figure 5**). Most abound in geomorphic expressions of their shallowness and variability, so that a consideration of their geomorphic parameters together with their hydrologic variability can explain some ecological features.

## Parameters of the Geomorphic Lake Types

### Structural (Tectonic) Lakes

As shown above, many of these like Lakes Baikal and Tanganyika, are especially large, deep and old, so that they have special limnological features, that are expressed most meaningfully by their Touchart coefficients. Such lakes have $D_v$s a little above unity (1.2–1.5) and moderate $D_s$ (1.3–3.4). Many older tectonic lakes of moderate depth (e.g., Caspain Sea, Aral Sea) have $D_v$ s <1, while the shallow intermittent ones vary according to hydrological condition. For instance in Lake Eyre $D_v$ decreases from 1.68 when 'full'(=5.7 m deep) to 1.55 at 4 m deep to 1.43 at 1 m deep, while $D_s$ decreases from 4.7 when full (it spreads up entering creeks to give a highly indented shoreline) to much lower values when contained well within the

vast salt flats. The limnology of Lake Eyre is different when full to when near empty, largely because of differing water salinities but the consequences of its differing geomorphic parameters contribute too. Lake Chad also has differing geomorphologies at different water levels, and probably so do many other shallow arid zone lakes. In addition, tectonic lakes in terminal basins have many stranded beach features useful in deducting past histories, as in many Tibetan lakes to give one set of numerous possible examples.

### Meteoritic Impact Lakes

These are rare, but impressive because of their roundness ($D_s = 1$). Depths vary, but Lake Ungava (=Chubb) in Quebec Province, Canada, is 251 m deep and diameter about 3 km. Intermittent Lake Acraman in South Australia, although not quite round is notable because of its huge size (diameter 22 km).

### Volcanic Lakes

Like structural lakes, there is no one set of parameters which characterize all volcanic lakes, though some subtypes have characteristic features. One group are the maars and calderas; generally these have low $D_s$s (near 1), high $D_v$s (near 2), high littoral slopes and low profundal slopes. In an analysis of 15 Australian maars, $D_s$ averaged 1.14 (lowest 1.01) and $D_v$ 1.91 (highest 2.8) (**Figure 4**). Such lakes are

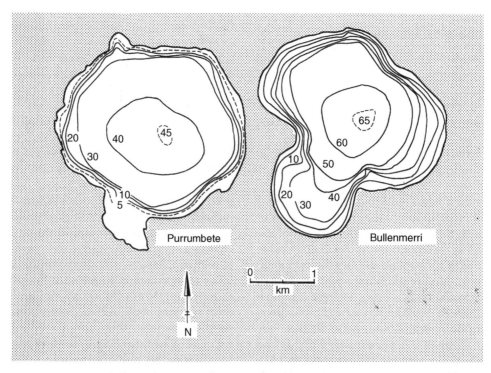

**Figure 4**    Bathymetric map of two maar lakes in western Victoria, Australia. Lake Purrumbete lies almost wholly within its crater, while Lake Bullenmerri lies within three adjoining craters. Both have a $D_s$ a little >1 and a $D_v$ near 2.

typically dominated by their limnetic zone and have physicochemical processes not too dissimilar to the classic large lakes of limnological literature. On the other hand shallower lakes on Victorian lava fields are not that different (except for their saline waters) to shallower lakes elsewhere. These are dominated by shallow water and littoral processes due to low maximum and mean depths, low littoral slopes and often high shoreline developments. Again, the basic geomorphic differences between the various types of lakes are expressed in depth and size, and perhaps there is scope in the Touchart system to differentiate lakes with coefficients <1 within his third type.

### Glacial Lakes

Again there is a wide variety of limnologies better explained by the size-age coefficient of Touchart (group 2, see above) than by geomorphic parameters alone. Nevertheless for the piedmont lakes due to glacial erosion, waters are deep, and $D_v$s are high. For 10 such large lakes on the South Island of New Zealand, 7 are deeper than 200 m and the average $D_v$ is 1.75. Shore and littoral features are unimportant compared with the contribution of large and deep limnetic region. Cirque lakes for another group, but their geomorphic parameters are hardly unique and influencial. The smaller kettles and the like are dominated by their small overall size and hence littoral processes.

### Fluviatile Lakes

Almost all of these are small, and even if >10 km$^2$ are still shallow and hence belong to Touchart's group 3 and so littoral dominated. Two common types have some characteristic parameters: most oxbows are lunate-shaped with a curving maximum length and moderate $D_s$s (<2.5), while most blocked valley lakes are dendritic with higher $D_s$s (2–4). Some, e.g., the bog lakes of the Waikato, New Zealand have shorelines much smoothed by growth of littoral vegetation so that $D_s$s are low, as are $D_v$s.

### Solution Lakes

As for fluviatile lakes most of these, especially dolines, are small and shallow, and though uvalas and poljes may be larger in area, they are still shallow. Perhaps the parameter of most interest is their $D_v$ – this is often 1 or less, indicating a cone or trumpet-shaped basin. Unlike the previous fluviatile lakes, but like kettles, they may have multiple deeper subbasins.

### Aeolian Lakes

Lakes formed, or largely moulded by wind action, are generally shallow (<5 m, often <1 m deep), relatively small (<100 km$^2$, often <1 km$^2$) and have sandy shores, so that shore processes are most important in their limnology. Hence spits, beaches, islands, and deltas feature in their evolution and their basin orientation is both a consequence of their development and/or a causal driver in their further evolution. Orientation to wind and effective maximum lengths for wave generation are basic controllers of their geomorphology. Many lakes in southwestern Western Australia and elsewhere are rounded or elliptical and difference in orientation and size between areas is related to different wind directions and/or rainfall distribution. On the other side of Australia, only those lakes orientated with their major axis N-S, develop spits in the southeastern corner in an attempt at lake compartmentalization (see later) (**Figure 5**). These spits increase $D_s$, which increases the habitat for shorebirds. Under unidirectional or bidirectional winds, shallow lakes may completely segment (i.e., form separate compartments), creating smaller lakes with more shorelines and increasing the important of littoral processes.

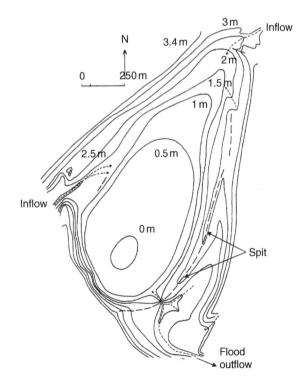

**Figure 5** Bathymetric map of Lake Yumberarra, Outback Queensland, Australia. Key: beach ridges long dashes; creek channels, short dashes. Depth contours inverted from normal pattern, as this is more convenient in shallow intermittent lakes with highly variable water levels.

### Coastal Lakes

Although most coastal lakes are due to sea-level rise, their geomorphology is variable due to different inherited features. A few are in fjords, so may be deep and steep-sided, while most are in drowned valleys, so dendritic or triangular, and those on low coasts may be elliptical. Australia's southeast coast has numerous coastal lakes of many subtypes, as does the Landes coast of France, the Cape Cod coast of North America, the Baltic coast to name a few others. Their active depositional environment encourages delta and spit formation, and lake compartmentilization in low coasts, so that shorelines are longer than those inherited. Given most lie in windy environs, basin orientation, and effective lengths are particularly important for those lake processes dependent on wave generation.

### Glossary

**Arcuate Gilbert delta** – A delta arched in plan and composed of a wide block of sediments up to the water surface.

**Homopycnal inflows** – Incoming water and lake water of same density, so thorough mixing of the two.

**Hyperpycnal inflows** – Incoming water more dense than lake water, so it flows along the bottom of the lake.

**Hypopycnal inflows** – Incoming water less dense than lake water, so it flows on the surface.

*See also:* Abundance and Size Distribution of Lakes, Ponds and Impoundments; Antarctica; Australia and New Zealand; Origins of Types of Lake Basins.

### Further Reading

Håkanson L (1981) *A Manual of Lake Morphometry.* Berlin: Springer.

Håkanson L (2005) The importance of lake morphometry and catchment characteristics in limnology – Ranking based on statistical analyses. *Hydrobiologia* 541: 117–137.

Hutchinson GE (1957) *A Treatise on Limnology* Vol. 1). New York: Wiley.

Kotwicki V (1986) The Floods of Lake Eyre. Adelaide: Engineering and Water Supply Department.

Lowe DJ and Green JD (1987) Origins and development of the lakes. In: Viner AB (ed.) *Inland Waters of New Zealand*, pp. 1–64. Wellington: DSIR Bulletin 241. DSIR Science Information Publishing Centre.

Timms BV (1992) Lake Geomorphology. Adelaide: Gleneagles Publishing.

Timms BV (2006) The geomorphology and hydrology of saline lakes of the middle Paroo, arid-zone Australia. *Proceedings of the Linnean Society of New South Wales* 127: 157–174.

Touchart L (2000) Les Lacs L'Harmattan. Paris.

# Reservoirs

**C Nilsson,** Umeå University, Umeå, Sweden

## Introduction

The Merriam-Webster Unabridged Dictionary defines the term reservoir as 'an artificial lake in which water is impounded for domestic and industrial use, irrigation, hydroelectric power, flood control, or other purposes.' The size and water content of a reservoir are controlled by a dam. The reservoir volume is usually defined by its live or dead storage capacity. Live storage capacity is the entire volume that can be withdrawn from the reservoir, whereas dead storage is the volume of water remaining in the reservoir when it is emptied to its (legislative) low-water level. Live storage capacity can be expressed as a degree of regulation, i.e., the proportion of a river's mean annual discharge that can be stored in a reservoir. Most reservoirs have a fairly low degree of regulation – usually far below 100% – but Lake Volta in the Volta River in Ghana has the world record of 428%, implying that more than 4 years of average discharge can be stored in the reservoir without releasing any water downstream of the dam. Apart from storing water, the function of a reservoir is to raise the level of the water to be diverted into a canal or pipe or to increase the hydraulic head. The head is an expression of water pressure that can be measured as the difference in height between the surface of a reservoir and the river downstream. Hydroelectric stations convert this pressure to electricity.

## Global Distribution of Reservoirs

There are nearly 50 000 dams in the world with heights above 15 m – defined as large dams – and an almost innumerable number of small dams built for farm ponds and other tiny impoundments. These dams can retain >6500 km$^3$ of water, which represents >15% of the annual global runoff (**Figure 1**). The area of former terrestrial habitat inundated by all large (>$10^8$ m$^3$) reservoirs in the world is comparable to the area of California or France. The environmental values that were lost as a result of this inundation are only known sporadically. It is not even known how many people were forced to move because of the reservoirs. The estimated number is 40–80 million people. Given that California has a population of approximately 37 million and the French population is 64 million, the magnitude seems accurate.

The project resulting in the highest number of forced resettlement – 1.3 million people – is the Three Gorges Dam on the Chang Jiang (Yangtze) in China. This estimate may increase because landslides in the margins of the filled reservoir threaten populated areas. The relative abundance of various sizes of lakes and reservoirs is discussed elsewhere.

Although dams were probably used much earlier, the first dams for which remains have been found were built about 3000 BC in modern Jordan. Prior to 1950 there were only about 5000 dams in the world, implying that many other dams were built after 1950 (**Figure 2**). China, which by far has the largest number of dams of the world's nations, shows an even later expansion in this respect. During the revolution in 1949 there were only eight large dams in China, but 50 years later the number had increased to around 22 000. The second most dam-rich country – the United States of America (USA) – lags far behind with only about 6600 dams, followed by India (about 4300 dams), Japan (about 2700 dams), and Spain (about 1200 dams).

The peak in number of dams and reservoirs created per year was reached in the 1970s. Although the rate of dam building has decreased, new dams are continuously being added. A few years ago, on average one new dam was completed every day after an average construction time of 4 years, implying that around 1500 dams were under construction. In 2004, large dams were planned or under construction on 46 of the world's largest rivers, with anywhere from 1 to 49 new dams per basin. Forty of these rivers are in nations not belonging to the Organization for Economic Cooperation and Development (OECD), indicating that future dam development does not depend on strong national economies. Almost half of the new dams are located on just four rivers, i.e., 49 on the Chang Jiang, 29 on the Rio de la Plata in South America, 26 on the Shatt Al Arab in the Middle East, and 25 on the Ganges–Brahmaputra in south Asia. New dams are also planned for several unaffected large river systems, including the Jequitinhonha in South America, and the Cá, Agusan, Rajang, and Salween in Asia. Many of these dams provide a serious threat to many species and habitats. For example, very large hydroelectric reservoirs in the tropics are especially likely to cause global species extinctions although such losses are rarely documented because scientific data are lacking.

**Figure 1**   There is yet no complete database of the world's water reservoirs. This map shows the location of about 1600 reservoirs from the *Global Lakes and Wetlands Database* [Lehner and Döll (2004) Development and validation of a global database of lakes, reservoirs and wetlands. *Journal of Hydrology* 296, 1–22]. The total storage capacity of these reservoirs amounts to approximately 5100 km³. This database will soon be superseded by *GRanD* (Global Reservoir and Dam Database) – a much more comprehensive database produced under the auspices of the Global Water System Project. Figure credit: Bernhard Lehner.

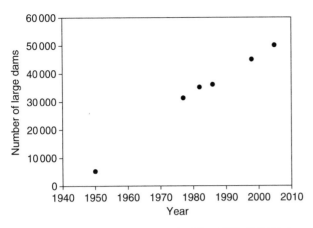

**Figure 2** Global rate of large-dam building 1950–2005. Data are taken from McCully P (2001) *Silenced Rivers: The Ecology and Politics of Large Dams.* London: Zed Books; WCD (2000) *Dams and Development: A New Framework for Decision-Making.* London: Earthscan; Scudder T (2005) *The Future of Large Dams: Dealing with Social, Environmental, Institutional and Political Costs.* London: Earthscan.

While new dams and reservoirs are added, some of the old ones are also taken out. In fact, removal of dams and reservoirs has become increasingly common, especially in western societies where old dams are widespread. The reasons for taking down a dam can vary, but one important cause is that a large dam is no longer safe. Another is that the dam is not needed; it might lack owners and be abandoned, or its owners cannot find funds to maintain it. Yet another is that the dam owners fail in the relicensing of the dam, in most cases because the dam has too large an environmental impact. For example, dams that interfere with migration of ecologically important fish stocks are key targets for removal.

## Functions of Reservoirs

Although all reservoirs are built to store water, their functions may differ. There are at least five major types of reservoirs but many are designed for multiple purposes. The management of multipurpose reservoirs builds upon compromises because it is nearly impossible to operate each function at its maximum level.

### Irrigation Reservoirs

This is the most common type of reservoir, built especially in dry parts of the world. Water in irrigation reservoirs is released into networks of canals mainly for use in farmlands. Water in irrigation reservoirs is generally not used for drinking. Thirty to forty percent of the 271 million ha of agricultural land irrigated worldwide rely on reservoirs. In other cases, water is extracted directly from rivers or from groundwater systems.

### Hydroelectric Reservoirs

A hydroelectric power station consists of turbines that rely on a gravity flow of water from the dam to turn a turbine to generate electricity. The water can be either released to the river downstream of the dam or pumped back into the reservoir and reused. Generally, hydroelectric dams are built specifically for electricity generation and are not used for drinking or irrigation water. Hydropower provides 19% of the world's total electricity supply, and is used in over 150 countries, with 24 of these countries depending on it for 90% of their supply. The countries producing most hydroelectric energy are Canada, USA, Russia, Brazil, and China.

### Standard Reservoirs

'Raw water' means surface water or groundwater, which, because of its bacteriological and chemical quality, turbidity, color, or mineral content, is unsatisfactory as a source for a community water system without treatment. Raw water reservoirs are used primarily for storing raw water to be treated and used in a community water system.

### Flood Control Reservoirs

Commonly known as 'attenuation' reservoirs, these are used to prevent flooding of lower lying lands. Flood control reservoirs collect water at times of unseasonally high rainfall, and then release it slowly over the course of the following weeks or months. To solve problems of hydrographic imbalance, transvasements, i.e., artificial crossings of water from one river basin to another, are used to decrease floods or to move water to lands with droughts. In countries with timber-floating, such reservoirs were used to even out the spring flood peak and make rivers floatable over a longer time.

### Recreational Reservoirs

Rarely, reservoirs are built solely for recreation. Most reservoirs are built for a civic purpose, but still allow fishing, boating, and other activities. At some reservoirs, special rules may apply for the safety of the public.

## Size of Reservoirs

Lake Volta in western Africa is known as the largest human-made lake in the world – by area as well as by volume. It covers 8502 km² or 3.6% of Ghana's area and has a maximum length of 520 km from its northernmost point at the town of Yapei downstream to the Akosombo dam. The volume of Lake Volta is 148 km³. When the reservoir was formed in 1965 about 78 000 people in 740 villages were relocated to new townships, along with 200 000 animals belonging to them. The reservoir is a major fishing area and provides irrigation for farmlands in the Accra plains. It is also important for transportation – generally by ferries and cargo boats. The Akosombo Dam controls the reservoir and is 660 m long at crest and 114 m high from crest to base. Its hydropower station produces electricity for much of the Ghana nation.

Although the Akosombo Dam is impounding the world's largest reservoir, it is not the largest dam in the world. The Three Gorges Dam – a dam which is 2309 m long and 185 m high but with a reservoir that is considerably smaller than Lake Volta – holds this position. Upon completion, the Three Gorges Dam will flood 632 km² of land to create a reservoir about 644 km long and 112 km wide and with a live capacity of 39 km³ of water. The dam will have the largest hydroelectric capacity of the world's dams, reaching 18 200 MW, thus surpassing the former record holder, the Itaipu dam on the Upper Parana River at the Brazil–Paraguay border, by more than 6000 MW. Although the reservoir surface area will reach only 1080 km², the entire Three Gorges Reservoir area covers 58 000 km², an area 16 710 km² larger than Switzerland. Consequently, several dozens to more than 100 mountain tops may become modern landbridge islands in this reservoir landscape.

The highest dam in the world, Nurek on the Vakhsh River in Tajikistan, is 300 m high. Its reservoir stores water for irrigation and hydropower generation. A yet higher dam, Rogun, on the same river has been under construction since 1976 and will be 335 m high upon completion. It is possible to get an idea of the size of these dams by comparing them to the Eiffel Tower in Paris, which has a height from the ground to the tip of the aerial of 324 m.

## Hydrology of Reservoirs

Among reservoirs, those built for generating hydroelectricity usually have the most pronounced fluctuations in water level (**Figure 3**). These fluctuations result from variations in the demand for electricity.

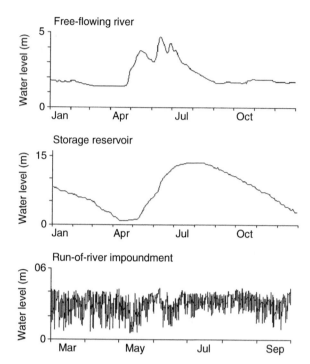

**Figure 3**  Typical annual water-level fluctuations in boreal free-flowing rivers and in the two major types of impounded waters. Note that the range of fluctuations differs between water bodies and that the storage reservoir has reversed hydrological conditions during summer, with early low-water and a late flood. From Jansson *et al.* (2000b) Fragmentation of riparian floras in rivers with multiple dams. *Ecology* 81: 899–903, with permission form Ecological Society of America.

Principally, there are two kinds of reservoir operation: storage reservoirs and run-of-river impoundments. Storage reservoirs are built primarily to sustain flow in the river downstream and level out ordinary fluctuations in discharge. In regions exhibiting seasonal climate variations and where rivers show large natural fluctuations in flow within or between years, storage reservoirs need to be large to fulfill this task. For example, in Norway there are two storage reservoirs with maximum legislated water-level fluctuations of 125 and 140 m, respectively.

Run-of-river impoundments are built primarily to balance daily variations in the demands for water for electricity production, and to provide hydraulic head. A river completely developed for hydropower production forms stairs of dams and impounded water surfaces without leaving any runs or rapids in between. Normally, power stations process more water during the day than during the night, implying that impounded water-levels are lowered during the day and raised during the night. This variation is commonly rather small to avoid loss of hydraulic head and thus reductions in power production.

Therefore, on a monthly or annual basis water levels could become more stable, but on a daily basis there could be frequent variation.

The numbers and locations of reservoirs vary between rivers. Many impounded rivers have big reservoirs in the upstream reaches, but long unimpounded reaches downstream. The flow hydrograph of such downstream reaches is affected by the reservoir. It has been estimated that, on average, about 5% of the water in a reservoir seeps into the ground and another 3.5% evaporates each year. The rest may be extracted or released to the river downstream. Usually, annual variations in flow are reduced so that floods and low-water periods become less dramatic. Additionally, the timing of the flood events that do occur is frequently changed. For example, in northern regions in which the spring floodwater is stored in reservoirs, floods may occur during summer or fall if reservoirs are filled in the spring and heavy rains continue through summer. In winter, storage reservoirs usually freeze over at a high water-level. Along with the emptying of the reservoir during winter, ice settles on the shorelines. Reservoir operation does not only change flow patterns, reservoirs also affect temperature, dissolved gases, and concentrations of waterborne material in the river downstream of the reservoir.

In recent years, an increasing number of ecologists advocate modified flow releases from such reservoirs, in order to attain more natural hydrologic conditions in the river downstream. This concept of dam reoperation attempting to find a compromise between human and environmental needs without sacrificing one or another is called environmental flows. It has been developed in water-poor areas such as South Africa and Australia where wise sharing of available water resources has become a chief issue, but the spirit of this concept is applicable to regulated rivers all over the world.

## Geology of Reservoirs

Humans have stored so much water in reservoirs that it has been suggested to have subtly altered the planet's rotation. Water impoundment is thought to have shortened the length of the day and shifted the Earth's axis by tiny amounts. No other human activity has been big enough to cause any appreciable alteration in these global phenomena. Large reservoirs can also cause earthquakes because of their addition of heavy mass. The largest reservoir-induced earthquake occurred in 1967 in India and had a magnitude of 6.3 or 6.8 on the Richter scale. Reservoirs are also recognized for their ability to collect sediments. All rivers erode their beds and carry sediments

downstream. Large reservoirs trap most of these sediments because currents slow down and drop their sediment loads when entering the reservoirs. Relatively sediments-free water is then released from the reservoir, in turn eroding new sediments from the channel below the dam and carrying them downstream. Because no sediments are deposited where this erosion is happening below the dam, the erosive capacity of the water may lower the riverbed by several meters until it is armored by stones and boulders and free from available fine-grade material. All reservoirs are, however, not effective sediment sinks. For example, in small, narrow reservoirs with low water residence time sediment deposition may be negligible.

The continuous addition of sediments to reservoirs successively reduces their storage capacity. Small reservoirs in sediment-rich regions can be completely filled within a few decades whereas large reservoirs in rivers running over coarse substrates have an expected life span of several centuries. It has been estimated that almost 30% or more than 100 billion metric tons of the global sediment load is trapped behind dams. This development is one of the factors eliciting demands for construction of new reservoirs. Nutrients, pesticides, and heavy metals that are transported with sediments are also trapped in the reservoirs. Such loadings make it more difficult to restore a river stretch should the dam and the reservoir be regarded as no longer necessary at some time in the future.

## Ecological Development of Reservoirs

Early environmental researchers identified three phases in the biological development of reservoirs: a breakdown of the original plant and animal communities and a recolonization by different species; a stage of increased primary and secondary production, mainly of plankton, and the establishment of temporary biotic communities; and, finally, a period of relative stabilization (**Figure 4**). This succession holds true for most reservoirs, but the levels of available nutrients at the more final stages of development can vary quite substantially. While some reservoirs stabilize at nutrient levels far below those of original water bodies in the reservoir area, others may experience increasing eutrophication because of extensive nutrient inputs.

The most apparent environmental alteration during the filling phase of a reservoir is that terrestrial habitats are becoming inundated and their vegetation destroyed. There are many examples of reservoir formation where forests and even villages were inundated and left to deteriorate. The decomposition of submerged vegetation and organic soils may

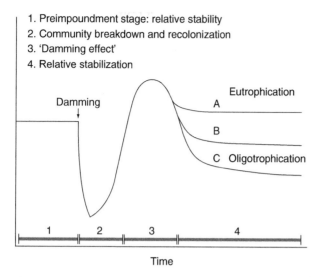

1. Preimpoundment stage: relative stability
2. Community breakdown and recolonization
3. 'Damming effect'
4. Relative stabilization

**Figure 4**   Hypothesized ecological successions in reservoirs. A–C denote alternative scenarios where A represents a eutrophication and B–C an oligotrophication of the reservoir compared with preimpoundment nutrient levels.

**Figure 5**   Extensive shoreline erosion in the Gardiken Reservoir as a result of impoundment of former terrestrial ground in the upper Ume River in northern Sweden. The disturbance of such areas makes establishment of plant and animal communities almost impossible. The woody debris is uprooted stumps from inundated forest land. Note that lower parts of the reservoir are ice-covered. Photo: Christer Nilsson.

produce huge amounts of greenhouse gases such as methane ($CH_4$) and carbon dioxide ($CO_2$) – especially when reservoirs are created in tropical rainforests and boreal peat lands. It is not known how many reservoirs are net C emitters, but some of the ∼1–3 billion metric tons of carbon that is sequestered in reservoirs is also converted into $CH_4$ and $CO_2$. It has been estimated that water reservoirs release 20% and 4%, respectively, of the world's annual anthropogenic emissions of $CH_4$ and $CO_2$. Other reservoir constructions are preceded by clearings of the areas to become inundated – to avoid timber and other vegetation floating around in the reservoir.

As a reservoir fills with water, the inundation and erosion of previously unflooded land and vegetation release nutrients and vegetative debris into the water. The increase in nutrient levels together with the increased light penetration in the reservoir cause the phytoplankton to multiply rapidly. As the phytoplankton respond to the new conditions, zooplankton and macroinvertebrates react in much the same way, resulting in an overall short-term increase in the productivity of the reservoir ecosystem. But once the nutrients from the flooded soils are depleted, the plankton population might either decrease or increase, depending on the inflow of nutrients to the reservoir.

Along with the erosion on new reservoir shorelines there is recolonization of plant species that are more adapted to the new conditions. Weed species exploit this situation in large reservoirs. The development of new plant communities very much depends on the stabilization of the substrate. At sites exposed

to wave and ice action, new shorelines can be completely deprived of their fine sediments, leaving barren ground (a 'bathtub ring') that does not offer suitable habitat for plants (**Figure 5**). Therefore, new reservoirs may exhibit a gradual increase in the cover of shoreline plants during the first years, succeeded by a gradual decrease as substrates become depleted.

In contrast, reservoirs in regions with warm climates and with small fluctuations in water level can develop luxuriant vegetation. For example, eutrophicated reservoirs in the tropics may be invaded by dense mats of floating aquatic plants, such as the water hyacinth (*Eichhornia crassipes*), giant Salvinia (*Salvinia molesta*) and Nile cabbage (*Pistia stratiotes*). These mats block light from reaching submerged vascular plants and phytoplankton, and often produce large quantities of organic detritus that can lead to anoxia and emission of gases, such as $CH_4$ and hydrogen sulfide ($H_2S$). The material derived from these plants is usually of low nutritional quality and is not typically an important component of the food for zooplankton or fish. Accumulations of aquatic macrophytes can restrict access for fishing or recreational uses of reservoirs and can block irrigation and navigation channels and intakes of hydroelectric power plants.

Most aquatic animals are confined to the surface layers of reservoirs because of the general decrease in oxygen with depth. The relative productivity of a reservoir is therefore proportional to its area rather than to its volume. As for plants, the abundance of animals depends on the range of water level fluctuations; reservoirs with large fluctuations having a much poorer fauna. For example, many benthic invertebrates have little motility and die as a result of rapid water-level drawdown. This has direct negative consequences for benthic-feeding fish. In reservoirs without large aquatic plants along the margins, the basic productivity will rely on plankton, and fish communities will be dominated by planktivorous species. The value of such reservoirs for waterfowl will be seriously curtailed because food resources will be poor and nesting places few.

Just as disturbance makes a landscape susceptible to invasion by alien plant species, the construction of water reservoirs around the globe is also considered to contribute to the accelerating spread of exotic aquatic species. One reason is that reservoir habitats are more homogeneous than those of streams, more disturbed and often more connected to other water bodies. Another is that reservoirs typically contain unstable, recently assembled communities of stocked fish. There are several reasons for introducing fish in reservoirs, such as utilizing ecological niches to which none of the existing species are adapted, increasing fishing success, providing more food fish and controlling aquatic weeds. Linked systems of reservoirs have faced increased invasion of exotic species such as the cladoceran *Daphnia lumholtzi* and the zebra mussel (*Dreissena polymorpha*). Reservoirs have also been linked to parasitic disease. For example, reservoirs have facilitated the spread of the *Schistosoma* parasite (including five species of flatworms) by greatly expanding its habitat, thus causing an increased incidence of deadly schistosomiasis or bilharzia throughout the tropics, where it affects about 200 million people.

When reservoirs serve as sources for the spread of exotic or generally invasive species into surrounding landscapes they are having landscape ecological effects, i.e., effects beyond the reservoirs themselves. Reservoirs also have other such effects. One example is their ecological fragmentation of rivers, e.g., by stopping the downstream drift of plant propagules and by providing obstacles to animal migration, thus fostering discontinuities in the riverine flora and fauna. For example, while free-flowing rivers in northern Sweden show continuous downstream changes in species composition of their riparian plant communities, chains of reservoirs and run-of-river impoundments demonstrate series of distinct assemblages with shifts from one to another at each dam (**Figure 6**). Otherwise, the most well-known ecological fragmentation effect dams and reservoirs have on rivers is the hindering of fish runs. Fish ladders, bypass channels, and detouring of fish by trucks are examples of measures to get the fish past the dam. If the fish are anadromous, it is required that there is spawning ground upstream of the dam and reservoir for the bypass to have an effect. Yet another landscape ecological impact of reservoirs is their retention of silicon that modifies the silicon:nitrogen:phosphorus ratio and causes dramatic shift in phytoplankton communities in the sea.

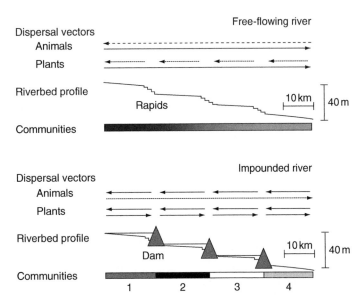

**Figure 6** Hypothesized relationships between vectors for animal and plant dispersal and migration, riverbed profile, and riverine communities in free-flowing vs. impounded rivers. The organisms in the free-flowing river are hypothesized to describe a gradual change downstream, whereas in the regulated river, each impoundment or reservoir is expected to develop individual organism communities (denoted 1–4). Note that animals in general have better capacities than plants do to move upstream rapids and downstream through dams. Scales are approximate. From Jansson *et al*. (2000a) Effects of river regulations on river-margin vegetation: a comparison of eight boreal rivers. *Ecological Applications* 10: 203–224, with permission from Ecological Society of America.

A reservoir does not only affect its surroundings but is also a product of its catchment. Practically all kinds of activities in the catchment can be transferred to the reservoir, in most cases through materials (such as sediments and various 'pollutants') carried by moving water or through a change in the hydrograph of the water draining the catchment. For example, a catchment that is cleared from forest or that has impervious surfaces or trenches shows more dramatic responses to rainstorms, the overflowing water is erosive and the catchment thus increases its footprint in the reservoir. For obvious reasons, there are cases when this collecting effect can be the primary motive for reservoir creation, and as mentioned above it is also a major obstacle when dams are planned for removal.

## Reservoirs and Global Warming

Reservoirs play several roles in the ongoing global warming. One suggested role is advancing the increase of carbon emissions to the atmosphere, caused by the generation of greenhouse gases resulting from decomposition of impounded organic matter. However, given that some reservoirs are net C emitters and others not, the jury is still out on the total figure. Another role is masking the ongoing rise of ocean water levels following melting of inland ice, groundwater extraction, and the thermal expansion of the oceans. This happens because of the constant addition of new reservoirs that increases the proportion of inland vs. ocean waters. In 1997, 13.6 mm of sea level equivalent was estimated to be impounded in reservoirs. Simultaneously, however, the trapping of sediments in reservoirs makes coastal deltas shrink, thus threatening some 300 million people who inhabit these deltas.

The accelerating global warming will also change global water cycling and water use by humans. By 2050, some large rivers are predicted to have doubled their discharge whereas others will face a halving of their flow. Impounded rivers are considered less able to handle these changes than the free-flowing ones because they have not been designed for discharges far beyond their present range of variability. For example, impounded rivers experiencing increased flow will run the risk of megafloods and dam failures. Our history includes many examples of spectacular dam-bursts, the worst probably being the catastrophe in the Henan province in China in 1975 when as many as 230 000 people may have died. In contrast, reservoirs facing increased temperatures and reduced discharge will lose even more water by evaporation and face worsened water stress. The ecological and societal costs of such changes may be momentous, calling for radical measures that restore the natural capacity of rivers to buffer climate-change impacts. This demand represents a major challenge for the scientific understanding of the factors involved in regulating the global water system and the ways in which humans are transforming it.

*See also:* Abundance and Size Distribution of Lakes, Ponds and Impoundments; Mixing Dynamics in Lakes Across Climatic Zones.

## Further Reading

Avakian AB and Iakovleva VB (1998) Status of global reservoirs: The position in the late twentieth century. *Lakes and Reservoirs: Research and Management* 3: 45–52.

Bednarek AT (2002) Undamming rivers: A review of the ecological impacts of dam removal. *Environmental Management* 27: 803–814.

Chao BF (1995) Anthropological impact on global geodynamics due to water impoundment in major reservoirs. *Geophysical Research Letters* 22: 3533–3536.

Dynesius M and Nilsson C (1994) Fragmentation and flow regulation of river systems in the northern third of the world. *Science* 266: 753–762.

Ericson JP, Vörösmarty CJ, Dingman SL, Ward LG, and Meybeck M (2006) Effective sea-level rise and deltas: Causes of change and human dimension implications. *Global and Planetary Change* 50: 63–82.

Fourniadis LG, Liu JG, and Mason PJ (2007) Landslide hazard assessment in the Three Gorges area, China, using ASTER imagery: Wushan–Badong. *Geomorphology* 84: 126–144.

Havel JE, Lee CE, and Vander Zanden MJ (2005) Do reservoirs facilitate invasions into landscapes? *BioScience* 55: 518–525.

Humborg C, Conley DJ, Rahm L, Wulff F, Cociasu A, and Ittekkot V (2000) Silicon retention in river basins: Far-reaching effects on biogeochemistry and aquatic food webs in coastal marine environments. *Ambio* 29: 45–50.

Jansson R, Nilsson C, Dynesius M and Andersson E (2000a) Effects of river regulation on river-margin vegetation: a comparison of eight boreal rivers. *Ecological Applications* 10: 203–224.

Jansson R, Nilsson C, and Renöfält B (2000b) Fragmentation of riparian floras in rivers with multiple dams. *Ecology* 81: 899–903.

Lehner B and Döll P (2004) Development and validation of a global database of lakes, reservoirs and wetlands. *Journal of Hydrology* 296: 1–22.

Matcharashvili T, Chelidze T, and Peinke J (2008) Increase of order in seismic processes around large reservoir induced by water level periodic variation. *Nonlinear Dynamics* 51: 399–407.

McCully P (2001) *Silenced Rivers: The Ecology and Politics of Large Dams.* London: Zed Books.

Nilsson C, Jansson R, and Zinko U (1997) Long-term responses of river-margin vegetation to water-level regulation. *Science* 276: 798–800.

Nilsson C, Reidy CA, Dynesius M, and Revenga C (2005) Fragmentation and flow regulation of the world's large river systems. *Science* 308: 405–408.

Palmer MA, Reidy Liermann CA, *et al.* (2008) Climate change and the world's river basins: Anticipating management options. *Frontiers in Ecology and the Environment* 6: 81–89.

Scudder T (2005) *The Future of Large Dams: Dealing with Social, Environmental, Institutional and Political Costs.* London: Earthscan.

St Louis V, Kelly CA, Duchemin É, Rudd JWM, and Rosenberg D (2001) Reservoir surfaces as sources of greenhouse gases to the atmosphere: A global estimate. *BioScience* 50: 766–775.

Syvitski JPM, Vörösmarty CJ, Kettner AJ, and Green P (2005) Impact of humans on the flux of terrestrial sediment to the global coastal ocean. *Science* 308: 376–380.

Tharme RE (2003) A global perspective on environmental flow assessment: Emerging trends in the development and application of environmental flow methodologies for rivers. *River Research and Applications* 19: 397–441.

World Commission on Dams (WCD) (2000) *Dams and Development: A New Framework for Decision-Making.* London: Earthscan.

## Relevant Websites

http://www.sandelman.ottawa.on.ca/dams/ – Dam-Reservoir Info and Impact Archive.

http://www.icold-cigb.net/ – International Commission on Large Dams.

http://www.irn.org/ – International Rivers Network.

http://npdp.stanford.edu/index.html – National Performance of Dams Program.

http://www.dams.org/ – The World Commission on Dams.

http://gcmd.nasa.gov/records/GCMD_UNH_WWRDII_DAMS.html – Dams, Lakes and Reservoirs Database for the World Water Development Report II.

http://www.gwsp.org/index.html.

# Abundance and Size Distribution of Lakes, Ponds and Impoundments

**J A Downing,** Iowa State University, Ames, IA, USA
**C M Duarte,** IMEDEA (CSIC-UIB), Esporles, Islas Baleares, Spain

## Abundance and Size of Lakes Dictated by Climate and Geology

A global understanding of the role of lakes and impoundments in the functioning of any region of the Earth requires quantification of their number and size distribution. Geology, tectonics, and climatic processing have made the Earth's surface an undulating and tilted surface with a hypsometry defined by the amount of relief and the irregularity of the underlying materials. If this undulating surface is cut through by a plane (tilted or not) that represents the groundwater and surface-water table, the water will outcrop in a manner that fills some large depressions and more small ones with water. Depending upon the net rate of replenishment of water by precipitation, the open surface waters of a region may cover much of the land surface or little of it, but the size and shape of depressions in the Earth will determine the regional morphometry or hypsometry. In this way, hypsometry and climate combine to determine the abundance and size distributions of natural lakes and depressions that can be made to hold water through the addition of dams.

## Historical Estimates of Limnicity

The 'limnicity' of land surfaces has been speculated upon since the early 1900s. In 1925, August Thienemann[1] concluded from map data summarized in 1914 by Wilhelm Halbfass, that "... around 2.5 million km$^2$, that is about 1.8% of dry lands, are covered with freshwaters; in Germany, the area of all lakes covers about 5200 km$^2$, or about 1% of the land surface. The largest lake on Earth is the Caspian Sea which has a surface area of 438 000 km$^2$." Several modern assessments of the global area covered by lakes and ponds have been made but were probably underestimates. These estimates were also made using similar map-based methods and have ranged from $2 \times 10^6$ to $2.8 \times 10^6$ km$^2$. Even 75 years after Thienemann's work, based on sometimes poor and incomplete maps, lakes and ponds were still thought to constitute only 1.3–1.8% of the Earth's non-oceanic area.

---

[1] Text located by Lars Tranvik. Translation supplied by Lars Tranvik and JAD.

The historical stability of estimates of limnicity of land surfaces is due to the prevailing assumption that large lakes make up the majority of Earth's lake area so that map inventories of the world's largest lakes would offer an estimate deficient in only a minor area represented by small water bodies. This idea was substantiated by analyses of the size distribution of lakes. One of the earliest of these was performed by R.D. Schuiling of the Netherlands in 1977. Schuiling plotted the number of lakes found in various size categories of world lakes and European lakes against the surface area of these lakes and found a consistent pattern. These data and others are shown expressed as measures of limnicity ($d_L$) of different size ranges of lakes in **Figure 1**. On the basis of such analyses, Schuiling and others concluded that lakes are numerically dominated by small water bodies but globally dominated in area by a small number of large lakes. The only dissenting voice in this dogma was Robert Wetzel who designed a graph of world lake abundance (reportedly, first on the back of a napkin) that showed a disproportionately great density of small lakes in the world, relative to large ones. Although meant to be a conceptual analysis rather than a quantitative one, Wetzel's perceptions concerning the relative abundance of different sizes of lakes in the world were remarkably accurate (**Figure 1**). He postulated that small lakes dominate the area of land surface covered by water.

The great stability from 1925 to 2002 in the estimate of the cumulative continental area occupied by lakes and other water bodies is due to agreement in the area of the world's largest lakes shown on maps. **Table 1** shows some characteristics of world lakes of greatest area, down to a size of 2000 km$^2$. The Caspian Sea, the lake with the largest area, alone makes up about 15% of the 2.5 million km$^2$ represented by the largest mapped lakes of the world. Lake Superior, the world's second largest lake, makes up 3.3% of the area of these lakes, while it takes the second through seventh largest lakes (Superior, Victoria, the Aral Sea, Lake Huron, Lake Michigan, and Lake Tanganyika) to have a cumulative area as large as the Caspian Sea. Lakes of progressively smaller sizes are represented by exponentially increasing numbers of lakes. The area distribution thus appears to approximate the Pareto distribution (explained later), which is precisely the type of distribution one would expect for objects created by random or directed processes in fields as diverse as linguistics and astrophysics. In fact, the word-frequency distribution in this Encyclopedia very likely follows a similar distribution.

The large area lakes of the world are interesting phenomena in themselves. They are found across a broad range of latitudes and elevations and represent lakes of divergent shape, receiving input from small and large watersheds. Of the world's lakes with greatest area (**Table 1**), the most northerly is Lake Taymyr at 74.5° N on the Taymyr Peninsula in Krasnoyarsk Krai, Russia. It is ice-covered much of the year, with a brief ice-free season from June to September. The most southerly of the large lakes is Lake Buenos Aires (also known as Lake General Carrera) at 46.5° S in Argentina and Chile, which is large enough to create its own climate in the Patagonia region. The large lake at the highest elevation is Lake Nam Co in Tibet, a holy lake large enough that it takes pilgrims 10 days to walk its circumference. Other well-known high-elevation lakes are Lake Titicaca on the border between Peru and Bolivia and Lake Tahoe in the USA. Some lakes are found below sea level, and many of these are very saline. The Caspian Sea is found at −28 m and several other somewhat smaller large lakes are also found below sea level (e.g., Lake Eyre in Australia, the Dead Sea in Israel and Jordan, the Salton Sea in the USA, and Lake Enriquillo in the Dominican Republic). Although many large lakes have somewhat rounded shapes (**Table 1**; shoreline development ratios <2), Lake Saimaa, Finland's largest lake, has shores that are extremely convoluted (development ratio >60). Some large lakes of the world have watersheds that are quite small relative to the lake area (i.e., 2–10; **Table 1**), whereas others such as Lake Chad and Lake Eyre drain areas >100-times their size.

Large lakes have a remarkable range of depths (**Table 2**), owing to regional hypsometry and the diverse geologic age and composition of their watersheds. The world's deepest lake is Lake Baikal in southern Siberia, which has a maximum depth of >1700 m. With Lake Tanganyika and the Caspian Sea, it is one of three lakes with a maximum depth >1000 m. In contrast, Lake Patos, a floodplain lake in Brazil the size of which is about 30% of the surface area of Lake Baikal, has a maximum depth of only about 5 m. Shallow lakes often are subject to extreme fluctuations in area owing to flooding or drying during periods of climatic variation. Despite the likelihood that the largest lakes might be expected to cover the greatest range of topography, there is only a weak general relationship between lake size and maximum depth in lakes larger than about 400 km$^2$ (**Figure 2**). For lakes >5000 km$^2$, however, there is a relationship between the minimum size of maximum depth observed in a size class of lakes and their areas. This relationship tells us that very large contiguous areas of the Earth's surface are unlikely to lack some level of significant relief.

The main problem with analyses of lake size distributions performed prior to the current decade is that small water bodies have been omitted from, or are poorly represented on, many maps. Therefore, the resolution of maps has dictated the perceived relative abundance of small lakes. This can be illustrated using modern satellite imagery by comparing the three panels of **Figure 3**, all with the same geographic center, but shown at three levels of spatial resolution. In the top panel, we perceive only the largest of the lakes (the North American Great Lakes and a few others). At intermediate resolution, perhaps similar to the resolution seen on maps in the early part of the twentieth century, we see large lakes and many of intermediate size. In the bottom panel, which still has inferior resolution to modern GIS coverages, a myriad of water bodies appear. These likely equal or exceed the area of larger lakes when summed over great land areas. In fact, if one assumes that some random process created pits and bumps in landscapes that were then filled with water, the aspect of lake regions and the size distribution of lakes is very similar to that observed in high-resolution satellite images

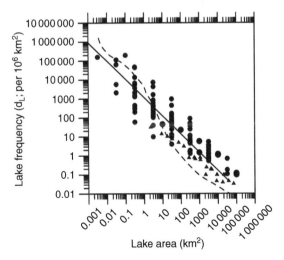

**Figure 1** Relationship between lake surface area and areal frequencies of different sized lakes measured as d$_L$ (number of lakes per $10^6$ km$^2$). The filled triangles indicate the frequencies of lake sizes digitized from Schuiling's work. The dashed line represents the hypothesis advanced by Wetzel, digitized from Figure 5 of his 1990 publication. All data represented by filled circles are from Meybeck's publications.

**Table 1**   The world's lakes >2000 km$^2$ in area, arranged in decreasing order of lake area

| Name | Latitude | Continent | Lake area (km$^2$) | WA:LA | Elev. (m) | $Z_{mean}$ (m) | Dev. ratio |
|---|---|---|---|---|---|---|---|
| Caspian | 42.0 | Asia | 374 000 | 10 | −28 | 209 | 2.8 |
| Superior | 47.6 | N. America | 82 100 | 2 | 183 | 149 | 4.7 |
| Victoria | −1.0 | Africa | 68 460 | 4 | 1 | 40 | 3.7 |
| Aral | 45.0 | Asia | 64 100$^a$ | 25 | 53 | 16 | 2.6 |
| Huron | 45.0 | N. America | 59 500 | 2 | 177 | 59 | 5.9 |
| Michigan | 44.0 | N. America | 57 750 | 2 | 177 | 85 | 3.1 |
| Tanganyika | −6.0 | Africa | 32 900 | 8 | 774 | 574 | 3.0 |
| Baikal | 54.0 | Asia | 31 500 | 21 | 456 | 730 | 3.5 |
| Great Bear | 66.0 | N. America | 31 326 | 5 | 156 | 76 | 4.3 |
| Great Slave | 61.8 | N. America | 28 568 | 34 | 156 | 73 | 3.7 |
| Erie | 42.2 | N. America | 25 657 | 2 | 171 | 19 | 2.4 |
| Winnipeg | 52.5 | N. America | 24 387 | 40 | 217 | 14 | 2.5 |
| Nyasa | −12.0 | Africa | 22 490 | 3 | 475 | 273 | 2.8 |
| Ontario | 43.7 | N. America | 19 000 | 4 | 75 | 86 | 2.4 |
| Balkhash | 46.0 | Asia | 18 200$^a$ | 10 | 343 | 6 | 5.0 |
| Ladoga | 61.0 | Europe | 17 700 | | 4 | 52 | 2.0 |
| Chad | 13.3 | Africa | 16 600$^a$ | 151 | 240 | 3 | 2.2 |
| Maracaibo | 9.7 | S. America | 13 010 | 7 | 0 | 22 | 1.5 |
| Patos | −31.1 | S. America | 10 140 | | 0 | 2 | 2.7 |
| Onega | 61.5 | Europe | 9700 | | 33 | 30 | 3.3 |
| Rudolf | 3.5 | Africa | 8660 | | 427 | 29 | 1.6 |
| Nicaragua | 11.5 | N. America | 8150 | | 32 | 13 | 2.4 |
| Titicaca | −15.8 | S. America | 8030 | 8 | 3809 | 103 | 3.5 |
| Athabasca | 59.2 | N. America | 7935 | 20 | 213 | 26 | 2.8 |
| Eyre | −28.5 | Oceania | 7690$^a$ | 146 | −12 | 3 | 4.5 |
| Reindeer | 57.3 | N. America | 6640 | 10 | 337 | 17 | 5.3 |
| Issykkul | 42.4 | Asia | 6240 | | 1608 | 277 | 2.7 |
| Tungting | 29.3 | Asia | 6000$^a$ | | 11 | 3 | 1.5 |
| Urmia | 37.7 | Asia | 5800$^a$ | 9 | 1275 | 8 | 1.8 |
| Torrens | −31.0 | Oceania | 5780$^a$ | 12 | 30 | <1 | 2.7 |
| Vanern | 58.9 | Europe | 5648 | 8 | 44 | 27 | 7.3 |
| Albert | 1.7 | Africa | 5590 | | 617 | 27 | 1.8 |
| Netilling | 66.5 | N. America | 5530 | | 30 | | 3.8 |
| Winnipegosis | 52.6 | N. America | 5375 | | 3 | 3 | 3.7 |
| Bangweulu | −11.1 | Africa | 4920$^a$ | 20 | 1140 | 1 | 2.0 |
| Nipigon | 49.8 | N. America | 4848 | | 320 | 63 | 2.9 |
| Gairdner | −31.6 | Oceania | 4770$^a$ | 2 | 34 | <1 | 2.3 |
| Manitoba | 50.9 | N. America | 4625 | | 248 | 3 | 3.4 |
| Taymyr | 74.5 | Asia | 4560$^a$ | | 3 | 3 | 3.7 |
| Koko | 37.0 | Asia | 4460 | | 3197 | 14 | 1.5 |
| Kyoga | 1.5 | Africa | 4430 | | 1036 | 6 | 7.5 |
| Saimaa | 61.3 | Europe | 4380 | 14 | 76 | 14 | 63.3 |
| Great Salt | 41.2 | N. America | 4360 | 12 | 1280 | 4 | 2.1 |
| Mweru | −9.0 | Africa | 4350 | | 922 | 7 | 1.5 |
| Woods | 49.3 | N. America | 4350 | | 323 | 8 | 4.9 |
| Peipus | 57.3 | Europe | 4300 | 11 | 30 | 6 | 2.0 |
| Khanka | 45.0 | Asia | 4190$^a$ | | 69 | 5 | 1.5 |
| Dubawnt | 63.1 | N. America | 3833 | | 236 | | 3.5 |
| Mirim | −32.8 | S. America | 3750 | 17 | 1 | 5 | 2.7 |
| Van | 38.6 | Asia | 3740 | 4 | 1646 | 55 | 2.3 |
| Tana | 12.2 | Africa | 3600 | | 1811 | | 1.5 |
| Poyang | 29.0 | Asia | 3350 | | 10 | 8 | 6.4 |
| Uvs | 50.3 | Asia | 3350 | | 759 | 1 | 1.6 |
| Amadjuak | 64.9 | N. America | 3115 | | 113 | | 3.5 |
| Lop | 40.5 | Asia | 3100 | | 768 | 2 | 5.4 |
| Melville | 53.8 | N. America | 3069 | | 0 | 97 | 2.7 |
| Rukwa | −8.0 | Africa | 2716$^a$ | 28 | 793 | <1 | 1.9 |
| Hungtze | 33.3 | Asia | 2700 | | 15 | | 1.9 |
| Wollaston | 58.2 | N. America | 2690 | 9 | 398 | 17 | 5.6 |
| Alakol | 46.2 | Asia | 2650 | | 347 | 22 | 1.8 |

Continued

**Table 1** Continued

| Name | Latitude | Continent | Lake area (km²) | WA:LA | Elev. (m) | $Z_{mean}$ (m) | Dev. ratio |
|------|----------|-----------|-----------------|-------|-----------|----------------|------------|
| Hovsgol | 51.0 | Asia | 2620 | | 1624 | 183 | 2.1 |
| Iliamna | 59.5 | N. America | 2590 | | 15 | 123 | 2.2 |
| Chany | 54.8 | Asia | 2500 | | 105 | 2 | 4.1 |
| Nam | 30.8 | Asia | 2500 | | 4627 | | 1.6 |
| Sap | 13.0 | Asia | 2450[a] | 33 | 1 | 4 | 2.2 |
| From | −30.7 | Oceania | 2410 | 35 | 49 | <1 | 1.5 |
| Kivu | −2.0 | Africa | 2370 | | 1460 | 240 | 3.3 |
| Mistassini | 50.9 | N. America | 2335 | 8 | 372 | 75 | 4.5 |
| Mai-Ndombe | −2.0 | Africa | 2325[a] | | 340 | 5 | 2.6 |
| Nueltin | 60.2 | N. America | 2279 | | 278 | | 2.5 |
| South Indian | 57.1 | N. America | 2247 | | 254 | 7 | 5.7 |
| Buenos Aires | −46.5 | S. America | 2240 | | 217 | | 1.6 |
| Tai | 31.3 | Asia | 2210 | | 3 | 2 | 2.2 |
| Edward | −0.4 | Africa | 2150 | | 912 | 35 | 1.7 |
| Ilmen | 58.3 | Europe | 2100[a] | 28 | 18 | 6 | 1.5 |
| Helmand | 31.0 | Asia | 2080[a] | 168 | 510 | 4 | 2.9 |
| Michikamu | 54.1 | N. America | 2030 | | 460 | 33 | 3.9 |

A superscripted 'a' indicates that a lake's area is variable in time and that the area may be nominal. WA:LA is the ratio of watershed to lake area, 'Elev.' Is the elevation of the lake above mean sea level, $Z_{mean}$ is the average depth, and 'Dev. ratio' is the shoreline development ratio (high numbers mean less circular). Latitudes are expressed in decimal degrees with latitudes in the southern hemisphere expressed as negative numbers. Data are after Herdendorf's work.

(**Figure 4**). Therefore, analyses of limnicity based on maps drew faulty conclusions about the amount of land surface covered by lakes and ponds, as well as the relative areal importance of small and large lakes.

## Modern Analyses of Limnicity and Lake Size Distribution

Nearly concurrently, Lehner and Döll and an international team of scientists working with Downing at the US National Center for Ecological Analysis and Synthesis applied modern GIS methods and updated geographic imagery to updating inventories of world lakes. These two efforts used divergent approaches but both had the objective of using new technologies to provide a more accurate estimate of the global extent and distribution of lakes and other water bodies.

The approach used by Lehner and Döll was to combine many data sources to create a global database of lakes and wetlands. This database was created by combining data sources, registers, and inventories focusing on descriptive attributes with analog or digital maps that show the spatial extent and locations of lakes and impoundments. This important step replaced the >13 published, list-based attribute tables with a GIS approach, which allows evaluation of area, shape, and location of lakes and impoundments. These two data sources allow some historical perspective, as well, since modern sources such as satellite images may not include long-term variations that can only be derived using local information. For

example, the Aral Sea has decreased in area to less than 30% of its former area. Although the database was compiled to represent lakes ≥0.1 km² in area, Lehner and Döll judged that even high-resolution satellite imagery under-samples lakes <1 km². Therefore, this approach did not effectively inventory lakes smaller than about 100 ha in surface area but offered the most substantial advance in the inventory of world lakes since Halbfass's compilation of 1914.

Lehner and Döll confirmed previous estimates of both the size distribution (as area) of moderate-to-large lakes as well as the surface area of the Earth they occupy. They showed quantitatively that the number of lakes in a size category increases as a power function of decreasing area. They showed, for example, that although there are $10^5$ lakes >1 km², there are only about 100 lakes >1000 km² in area. They also confirmed the previous estimates of the global area covered by moderate-to-large lakes as being near to 2.5 million km². More importantly, however, their use of GIS and satellite imagery allows a greatly improved understanding of how global lake area is distributed geographically (**Figure 5**) and how lake area is distributed compared with that occupied by rivers and impoundments. In this analysis, rivers were distinguished from lakes and impoundments based on their development ratio so that very long and narrow water bodies were called 'rivers'. Further, this analysis includes only medium-to-large lakes and likely includes only the largest of rivers (i.e., with a breadth of 100 m or more; Strahler order >5). The distribution of lake area is strongly skewed toward the north temperate zone (35–70° N) with a small increase also

**Table 2**  Some of the world's large lakes that are of great depth, arranged in decreasing order of maximum depth

| Name | Latitude | Continent | $Z_{max}$ (m) | $Z_{mean}$ (m) | Lake area (km$^2$) |
|---|---|---|---|---|---|
| Baikal | 54.0 | Asia | 1741 | 730 | 31 500 |
| Tanganyika | −6.0 | Africa | 1471 | 574 | 32 900 |
| Caspian | 42.0 | Asia | 1025 | 209 | 374 000 |
| Nyasa | −12.0 | Africa | 706 | 273 | 22 490 |
| Issykkul | 42.4 | Asia | 702 | 277 | 6240 |
| Great Slave | 61.8 | N. America | 625 | 73 | 28 568 |
| Toba | 2.6 | Asia | 529 | 249 | 1150 |
| Tahoe | 39.1 | N. America | 501 | 249 | 500 |
| Kivu | −2.0 | Africa | 480 | 240 | 2370 |
| Great Bear | 66.0 | N. America | 452 | 76 | 31 326 |
| Fagnano | −54.6 | S. America | 449 | 211 | 590 |
| Nahuel Huapi | −41.0 | S. America | 438 | 206 | 550 |
| Dead | 31.5 | Asia | 433 | 184 | 1020 |
| Superior | 47.6 | N. America | 407 | 149 | 82 100 |
| Llanquihue | −41.1 | S. America | 350 | 133 | 800 |
| Geneva | 46.4 | Europe | 310 | 153 | 580 |
| Titicaca | −15.8 | S. America | 304 | 103 | 8030 |
| Aregentino | −50.2 | S. America | 300 | 120 | 1410 |
| Iliamna | 59.5 | N. America | 299 | 123 | 2590 |
| Atlin | 59.5 | N. America | 283 | 86 | 774 |
| Michigan | 44.0 | N. America | 282 | 85 | 57 750 |
| Hovsgol | 51.0 | Asia | 270 | 183 | 2620 |
| Melville | 53.8 | N. America | 256 | 97 | 3069 |
| Constance | 47.6 | Europe | 252 | 90 | 540 |
| Ontario | 43.7 | N. America | 245 | 86 | 19 000 |
| Ladoga | 61.0 | Europe | 230 | 52 | 17 700 |
| Baker | 64.2 | N. America | 230 | 93 | 1887 |
| Huron | 45.0 | N. America | 229 | 59 | 59 500 |
| Reindeer | 57.3 | N. America | 219 | 17 | 6640 |
| Mistassini | 50.9 | N. America | 183 | 75 | 2335 |
| San Martin | −48.9 | S. America | 170 | 68 | 1010 |
| Nipigon | 49.8 | N. America | 165 | 63 | 4848 |
| Taupo | −38.8 | Oceania | 165 | 97 | 616 |
| Van | 38.6 | Asia | 145 | 55 | 3740 |
| Vattern | 58.4 | Europe | 128 | 40 | 1856 |
| Onega | 61.5 | Europe | 127 | 30 | 9700 |
| Champlain | 44.6 | N. America | 122 | 49 | 1100 |
| Athabasca | 59.2 | N. America | 120 | 26 | 7935 |
| Edward | −0.4 | Africa | 117 | 35 | 2150 |
| Flathead | 47.9 | N. America | 113 | 50 | 500 |
| Sakami | 53.3 | N. America | 110 | 52 | 592 |
| Grand | 48.9 | N. America | 110 | 52 | 537 |
| Vanern | 58.9 | Europe | 106 | 27 | 5648 |
| Biwa | 35.3 | Asia | 103 | 41 | 688 |
| Pyramid | 40.0 | N. America | 101 | 54 | 510[a] |

A superscripted 'a' indicates that a lake's area is variable in time and that the area may be nominal. $Z_{max}$ is the maximum depth and $Z_{mean}$ is the average depth. Latitudes are expressed in decimal degrees with latitudes in the southern hemisphere expressed as negative numbers. Data are after Herdendorf's publications.

near the equator (**Figure 5**). This is consistent with what is known about the distribution of large lakes (North American Great Lakes and the Caspian Sea), the balance of precipitation and evaporation, and agrees well with satellite scans of open water (except where large numbers of small lakes could not be inventoried). This analysis suggested that lakes make up about 1.8% of the land area (2 428 000 km$^2$), impoundments about 0.2% (251 000 km$^2$), and rivers about 0.3% (360 000 km$^2$).

The perennial problem in lake inventories has been that small water bodies have gone un-inventoried. This has been assumed, somewhat tautologically, to be of little consequence because small water bodies have been found to be numerically dominant but inconsequential in terms of the area of land surface they cover. On the one hand, as shown by Lehner and Döll, no worldwide GIS coverage exists that has high enough spatial resolution to resolve all of the world's small lakes. On the other hand, we have

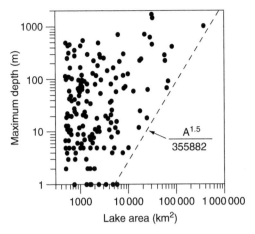

**Figure 2**   Relationship between lake area and maximum depth taken from data compiled and published by Herdendorf. The dashed line is an empirically derived lower limit to maximum depth (m) for lakes larger than 5000 km², where A is the lake area in km².

long-standing information on the size distribution of the world's lakes down to a size of 1–10 km² as well as high-resolution GIS coverage on many regions that can be used to examine and characterize the size distribution of lakes down to the smallest sizes.

Downing and coworkers improved estimates of lake abundance and area by working on the small-lake under-sampling problem. They estimated the world abundance of natural lakes by characterizing the size distribution using a nearly universally applicable distribution function, testing its fit to data down to the smallest sizes of lakes, anchoring it in empirical data for large lakes, and solving the distribution function to calculate the world abundance of large and small lakes.

Because the relationship between $d_L$ and lake area (A) in **Figure 1** appears to fit a power function, Downing and coworkers and Lehner and Döll suggested that lake-size distributions fit a size–frequency function of the form:

$$N_{a \geq A} = \alpha A^{\beta} \qquad [1]$$

where $N_{a \geq A}$ is the number of lakes of great or equal area (a) than a threshold area (A), and $\alpha$ and $\beta$ are fitted parameters describing the total number of lakes in the dataset that would be of one unit area in size and the logarithmic rate of decline in number of lakes with lake area, respectively. This model corresponds to a Pareto distribution that is particularly versatile and used in fields from linguistics to engineering. It has also been found useful in describing lakes' size–frequency distributions, as long as the data are not truncated or censored.

**Figure 3**   Illustration of the influence of scale of observation on the perceived size distribution of lakes. Source: earth.google. com. The images are centered on the same point but have improved spatial resolution from top to bottom.

To test lake size distributions for general fit to the Pareto distribution, Downing and coworkers collected exhaustive inventories of all lakes within a variety of geographical settings representing divergent topography and geology. **Figure 6** shows that there are interregional similarities among slopes of

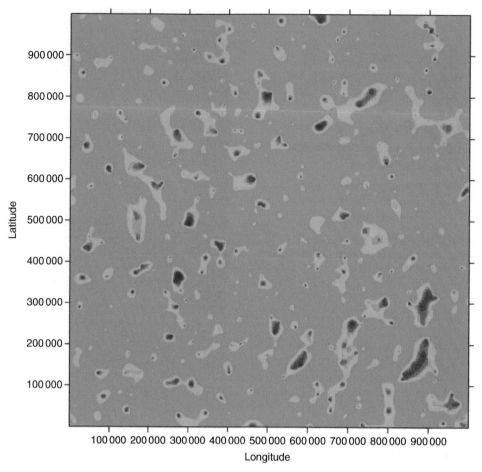

**Figure 4** A simulated lake region generated using a random number generator in Microsoft Excel™ to determine 2000 positions of pits and bumps in a $10^6 \times 10^6$ land-unit area. Pit or bump height was determined using the same random number generator to create relief features with an elevation of 0–100 units. The topography of the land surface was created using kriging and the lake shores were arbitrarily set at 27 units of elevation.

Pareto curves. These curves show similar rates of decline in abundance with lake size among many regions of the Earth.

Because the size–frequency distributions of lakes follow a Pareto distribution in many regions down to very small lake sizes (**Figure 6**), exhaustively censused (canonical) data on the abundance of the world's largest lakes allows eqn. [1] to be anchored at the upper end to permit estimation of the world-wide abundance of lakes across the full range of lake sizes. Two collections of exhaustively censused lakes are shown plotted on **Figure 6**. Considering only the 17 357 natural lakes $>10\,km^2$ in area, eqn. [1] can be fitted by least squares regression as

$$N_{a \geq A} = 195560 A^{-1.06079}, r^2 = 0.998;$$
$$n = 17357;\ SE_\beta = 0.0003 \qquad [2]$$

Since the shape of Pareto distributions is similar among diverse regions of the Earth (**Figure 6**) and

the parameters of this distribution are estimatable from the canonical data sets, Downing and coworkers provided a means of estimating the global extent of ponds and lakes down to very small sizes.

Analyses of lake-size distributions based on the Pareto distribution (**Table 3**) reveal that, contrary to others' predictions, small lakes represent a greater lacustrine area than do large ones. The large lakes $\geq 10\,000\,km^2$ in individual lake area make up only about 25% of the world's lake area. Together, the two smallest size categories of lakes in **Table 3** make up more area than the three top size categories. Thus, undercounting small lakes has resulted in the significant underestimation of the world's lake and pond area over the last century. World lakes and ponds account for roughly $4.2 \times 10^6\,km^2$ of the land area of the Earth. This is more than double the historical estimates. Natural lakes and ponds $\geq 0.001\,km^2$ make up roughly 2.8% of the non-oceanic land area; not 1.3–1.8% as assumed since the early 1900s.

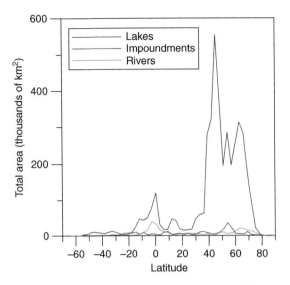

**Figure 5** Measured latitudinal distribution of area of lakes, impoundments, and streams redrawn from Lehner and Döll's Figure 3. Areas of water bodies are summed over 3° increments of latitude. Lake areas are corrected following the work of Downing and coworkers by assuming that previous underestimates of lake areas have been distributed evenly across all latitudes. Areas of impoundments are likely to be correct but river areas are probably underestimated.

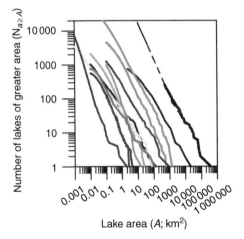

**Figure 6** Plots of data on the axes implied by eqn. [1]. Colored lines are high resolution GIS analyses of geographically dissimilar regions. The black lines represent canonical (complete) censuses of world lakes taken from publications by Herdendorf and by Lehner and Döll. This figure is redrawn from a publication by Downing and coworkers.

## Impoundments

Parallel with the analysis of natural lakes above, one can fit the Pareto distribution to the sizes of large impoundments. Downing and coworkers did this using data from the International Commission on Large Dams who publish data on dams around the

world that are of safety, engineering or resource concern (e.g., **Figure 7**). These data are purposefully biased toward large dams so the data provide the most accurate estimate of impoundments with the largest impounded areas, and progressively less complete coverage of small ones. Fitting eqn. [1] to data on the 41 largest impoundments from the largest impoundment (13 500 km²) down to 1000 km² yields:

$$N_{a \geq A} = 2922123 A^{-1.4919}, r^2 = 0.97, n = 41 \qquad [3]$$

This equation implies that the smallest of the large impoundments make up more surface area than the largest of them because the exponent is strongly negative. This dataset is intentionally biased to large dams so the exponent increases if small impoundments are included. If one considers all of the ICOLD impoundments down to 1 km², the result will be biased toward underestimation of impoundment area, but is

$$N_{a \geq A} = 20107 A^{-0.8647}, r^2 = 0.97, n = 9604 \qquad [4]$$

This equation results in an underestimate of the area covered by impoundments because it ignores many impoundments formed by small dams. Calculations using this function and the Pareto distribution, however, show that there are at least 0.5 million impoundments $\geq 0.01$ km² in the world and they cover >0.25 million km² of the Earth's land surface (**Table 3**). This calculation using the Pareto distribution yields a smaller area than estimates based on extrapolation but nearly identical to GIS-based estimates. Large impoundment data suggest that small impoundments cover less area than large ones (**Table 3**).

Most analyses of impoundment size distributions have ignored small, low-tech impoundments created using small-scale technologies. Farm and agricultural ponds are growing in abundance, world-wide, and are built as sources of water for livestock, irrigation, fish culture, recreation, sedimentation, and water quality control.

**Figure 8** shows that the area of water impounded by agricultural ponds in several political units expressed as a fraction of farm land varies systematically with climate. In dry regions, farm ponds are rare, but up to about 1600 mm of annual precipitation, farm ponds are a rapidly increasing fraction of the agricultural landscape. In moist climates, farm ponds make up 3–4% of agricultural land. Downing and coworkers used this relationship (**Figure 8**) with data on area under farming practice, pond size, and estimates of annual average precipitation to estimate the global area covered by farm pond impoundments. They found that 76 830 km², worldwide, is covered with farm ponds. These small impoundments are growing in importance at annual rates of from 0.7%

**Table 3**   The frequency of lakes and impoundments of different sizes worldwide, estimated following Downing and coworkers' method employing the Pareto distribution

| Area range (km²) | Number of lakes | Total area of lakes (km²) | Number of impoundments | Total area of impoundments (km²) |
|---|---|---|---|---|
| 0.001–0.01 | 277 400 000 | 692 600 | 76 830 000[a] | 76 830[b] |
| 0.01–0.1 | 24 120 000 | 602 100 | 444 800 | 12 040 |
| 0.1–1 | 2 097 000 | 523 400 | 60 740 | 16 430 |
| 1–10 | 182 300 | 455 100 | 8 295 | 22 440 |
| 10–100 | 15 905 | 392 362 | 1 133 | 30 640 |
| 100–1000 | 1 330 | 329 816 | 157 | 41 850 |
| 1000–10000 | 105 | 257 856 | 21 | 57 140 |
| 10000–100000 | 16 | 607 650 | 3 | 78 030 |
| >100000 | 1 | 378 119 | | |
| All water bodies | 304 000 000 | 4 200 000 | 77 345 000 | 335 400 |
| Percent of land area | 2.80% | | 0.22% | |

Superscript 'a' indicates that the calculation was based on an average farm pond size of 0.001 km². 'b' indicates an estimate of the global area of farm ponds.

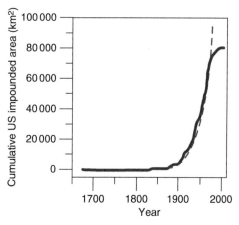

**Figure 7**   Rate of change in impounded area in the United States for all impoundments listed by the United States Army Corps of Engineers with dams listed as potential hazards, or low hazard dams that are either taller than 8 m, impounding at least 18 500 m³, or taller than 2 m, impounding at least 61 675 m³ of water (after Downing and coworkers). All data were ignored where the date of dam construction was unknown (ca. 12% of impounded area) or natural lakes (e.g., Lake Superior) were listed as impoundments. The dashed line shows a semi-log regression ($r^2 = 0.95$) based on the exponential phase of impoundment construction.

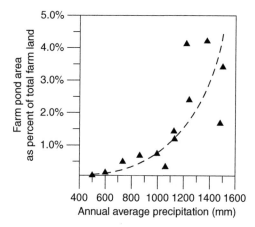

**Figure 8**   Relationship between the surface area of farm ponds and the annual average precipitation in several political units (after Downing and coworkers). The line is a least squares regression ($r^2 = 0.80$, $n = 13$) where the area of farm ponds expressed as a percentage of the area of farm land (FP) rises with annual average precipitation (P; mm) as $FP = 0.019\,e^{0.0036P}$.

per year in Great Britain, to 1–2% per year in the agricultural parts of the United States, to >60% per year in dry agricultural regions of India.

## Conclusions

Although data have been available on the world's largest and most spectacular lakes since the early 1900s, global estimates of the abundance and size distribution have underestimated the abundance and area of small natural lakes. When appropriate GIS

methods and mathematical models are used, the area of land surface covered by lakes has been found to be 2.8% of the terrestrial surface area of the Earth, nearly double some historical estimates. There are 304 million natural lakes in the world that cover 4.2 million km². More than 207 million lakes are <0.01 km² in individual area and the area covered by lakes of different sizes is greater for small lake size-classes than for large ones. Inventories of natural lakes consider only lakes with permanent open water, including lakes, ponds, and bogs, but other systems, including flood lakes, temporary water bodies, and wetlands, would add substantially to the area occupied by continental aquatic systems. The land area covered by constructed lakes and impoundments is smaller than that of natural lakes but 77 million world impoundments cover 335 000 km² or about

0.22% of the land surface. The area covered by small, low-tech farm ponds is about the same as that covered by the world's three largest impoundments. Impoundment area is generally underestimated, however, by poor availability of data on small and moderately sized basins and the number of such low-tech structures is increasing rapidly where agricultural needs are great.

*See also:* Aquatic Ecosystem Services; Geomorphology of Lake Basins; Mixing Dynamics in Lakes Across Climatic Zones; Origins of Types of Lake Basins.

## Further Reading

Dean WE and Gorham E (1998) Magnitude and significance of carbon burial in lakes, reservoirs, and peatlands. *Geology* 26: 535–538.

Downing JA, *et al.* (2006) The global abundance and size distribution of lakes, ponds, and impoundments. *Limnology and Oceanography* 51: 2388–2397.

Hamilton SK, Melack JM, Goodchild MF, and Lewis WM Jr. (1992) Estimation of the fractal dimension of terrain from lake size distributions. In: Petts GE and Carling PA (eds.) *Lowland Floodplain Rivers: A Geomorphological Perspective*, pp. 145–164. John Wiley.

Herdendorf CE (1984) *Inventory of the Morphopmetric and Limnologic Characteristics of the Large Lakes of the World.*

*Technical Bulletin*, pp. 1–54. The Ohio State University Sea Grant Program.

Kalff J (2001) *Limnology: Inland Water Ecosystems.* Prentice Hall.

Lehner B and Döll P (2004) Development and validation of a global database of lakes, reservoirs and wetlands. *Journal of Hydrology* 296: 1–22.

Meybeck M (1995) Global distribution of lakes. In: Lerman A, Imboden DM, and Gat JR (eds.) *Physics and Chemistry of Lakes*, pp. 1–35. Springer-Verlag.

Schuiling RD (1977) Source and composition of lake sediments. In: Golterman HL (ed.) *Interaction Between Sediments and Fresh Water*, pp. 12–18. Dr. W. Junk B.V. Proceedings of an international symposium held at Amsterdam, the Netherlands, September 6–10, 1976.

Shiklomanov IA and Rodda JC (2003) *World Water Resources at the Beginning of the Twenty-First Century.* Cambridge University Press.

Smith SV, Renwick WH, Bartley JD, and Buddemeier RW (2002) Distribution and significance of small, artificial water bodies across the United States landscape. *Science of the Total Environment* 299: 21–36.

St. Louis VL, Kelly CA, Duchemin E, Rudd JW, and Rosenberg DM (2000) Reservoir surfaces as sources of greenhouse gases to the atmosphere: A global estimate. *BioScience* 50: 766–775.

Tilzer MM and Serruya C (eds.) (1990) *Large Lakes: Ecological Structure and Function.* Springer-Verlag.

Wetzel RW (1990) Land-water interfaces: Metabolic and limnological regulators. *Verhandlungen der Internationale Vereinigung der Limnologie* 24: 6–24.

Wetzel RW (2001) *Limnology: Lake and River Ecosystems.* Academic Press.

# Saline Inland Waters

**M J Waiser,** Environment Canada, Saskatoon, SK, Canada
**R D Robarts,** UNEP GEMS/Water Program, Saskatoon, SK, Canada

## Introduction

Lakes, unlike rivers, are mainly storage bodies. They are dynamic ecosystems and in addition to their storage function, are the source of food and recreation for humans, support a large range of biodiversity and may provide the foundation for people's livelihoods. Unfortunately, lakes are also among the most vulnerable and fragile aquatic ecosystems as they are a sink for a wide range of dissolved and particulate substances. Lakes, therefore, serve not only as sensitive indicator systems but recorders of effects of human and natural disturbances inside, and sometimes outside, their drainage basins. In addition, effects of climate change can interact with other direct human (e.g., pollution, water diversion) impacts with both positive and negative consequences for specific lake systems.

Historically, saline lakes have unfortunately been perceived as being unimportant, of less utility, and less abundant than fresh waters. They are, however, distributed worldwide in semi-arid or arid climatic zones and are often close to large population centers. On a worldwide basis they have accounted for only slightly less water volume (0.008%) than their freshwater counterparts (0.009%) (**Figure 1**). Indeed, in spite of their salinity, most of these lakes have important values and uses (except as drinking water) for both society and the environment in semi-arid and arid regions of the world. As with freshwater lakes, more attention needs to be directed to the protection, management, and conservation of the world's saline lakes.

## Where Do Saline Lakes Occur?

Most saline lake basins are found between about 400 and 1500 m above sea level (masl) and were formed by tectonic action. Some were formed during volcanic activity (maars and craters) while others were created during recent glacial activity.

True saline lakes (athalassic) occur mainly in arid (precipitation 25–200 mm) and semi-arid (precipitation 200–500 mm) climatic zones (between 53° N and 30° N in the Northern Hemisphere and 3° N and 42° S in the Southern Hemisphere; and south of 77° S in Antarctica) where net evaporation exceeds precipitation. They are usually located in hydrologically closed (endorheic) basins and consequently are the termini of inland drainage basins (inflows but no outflows).

Some saline lakes (thalassic), however, due to their proximity to marine environments, exhibit features (ionic composition and biology) closer to marine environments than inland saline lakes.

Sources of incoming water include precipitation, groundwater seepage, underwater springs (mostly in karst environments), creeks, and rivers. Approximately 1/10 of the Earth's surface area is made up of such closed or endorheic drainage basins (**Figure 2**). The location of saline lakes within semiarid and arid climatic zones and endorheic drainage basins means that saline lakes are extremely responsive to climate, both within and between years. During times of decreased precipitation and/or drought, or changes in land use patterns within the basin, water levels and volumes can decrease markedly with concomitant increases in salinity. For example, from 1904 to the 1990s, water levels in Redberry Lake, a $MgSO_4$ dominated lake in south central Saskatchewan, dropped from 737 to 728 masl. At the same time total dissolved solids (TDS) rose from 12 000 to 19 000 mg $l^{-1}$ (12–19‰) (**Figure 3**). These changes are thought to be a direct result of land use changes in the basin as well as periods of drought. Small shallow North American prairie wetlands can show this kind of decrease in water level and increase in salinity within one season (**Figure 4**). Because small climatic changes can be amplified in saline lakes and wetlands, they are very sensitive to climate change.

## Salinity

Salinity is usually defined as the sum of ionic compounds dissolved in water and can be measured in several ways. Specific conductivity quantifies the relative ease with which an electrical current passes through a water sample and is usually expressed as milli or micro Siemens per centimeter (mS or µS $cm^{-1}$). Because the conductivity of NaCl increases 2% for every degree increase in temperature, specific conductivity is always measured at 25 °C.

Salinity can also be expressed as milligrams per liter of total dissolved solids (mg $l^{-1}$ TDS). For TDS, water is filtered through a 2 µm pore-size filter, evaporated to dryness at <100 °C, and then weighed. Lakes with TDS $\geq$3000 mg $l^{-1}$ (3‰ or 5500 µS $cm^{-1}$) are commonly referred to as saline and this is the salinity at which most people start to taste salt.

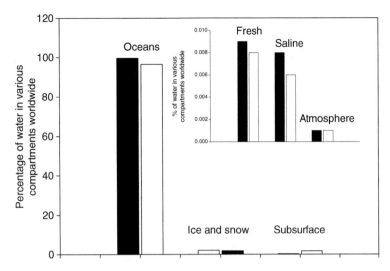

**Figure 1** Percentage of water in various compartments worldwide. Data from Wetzel RG (1983) Limnology 2nd Edn. Philadelphia: Saunders (black bars) and from Shiklomanov IA (1990) Global water resources. *Natural Resources* 26:34–43 (white bars) as cited in Williams WD (1996) The largest, highest and lowest lakes of the world: saline lakes. *Verhandlungen Internationale Vereinigung Limnologie* 26: 61–79.

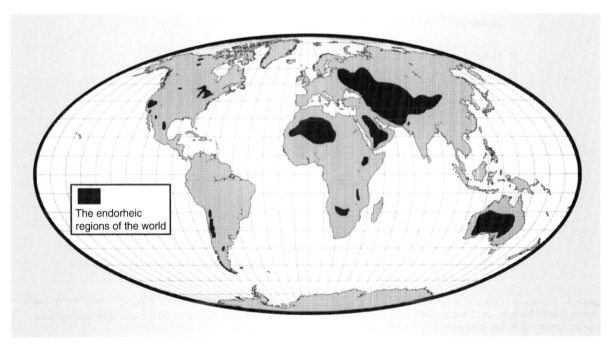

**Figure 2** Endorheic drainage basins of the world. With permission from Shiklomanov IA (1998) *World water resources. A new appraisal and assessment for the 21st century.* Paris: UNESCO.

Freshwater species start to disappear at salinity above 3000 mg l$^{-1}$ and below this higher salinity biota are not found. Saline lakes can be further differentiated into hyposaline, mesohaline, and hypersaline (**Table 1**). Freshwater lakes have TDS of 500 mg l$^{-1}$ while subsaline lakes are those whose TDS ranges from 500 to 3000 mg l$^{-1}$ (**Table 1**). In Africa and Australia, lakes lie at the extreme end of the salinity gradient with salinities up to 270 000 μS cm$^{-1}$. Some Antarctic saline lakes have recorded salinities as high as 790 000 μS cm$^{-1}$. Such high salinities result not only from evaporation but the fact that salts are frozen out of the ice cover. In the North American north Temperate Zone, however, specific conductivities are more typically in the range of 20–600 μS cm$^{-1}$. Across the semiarid prairie pothole region of Saskatchewan,

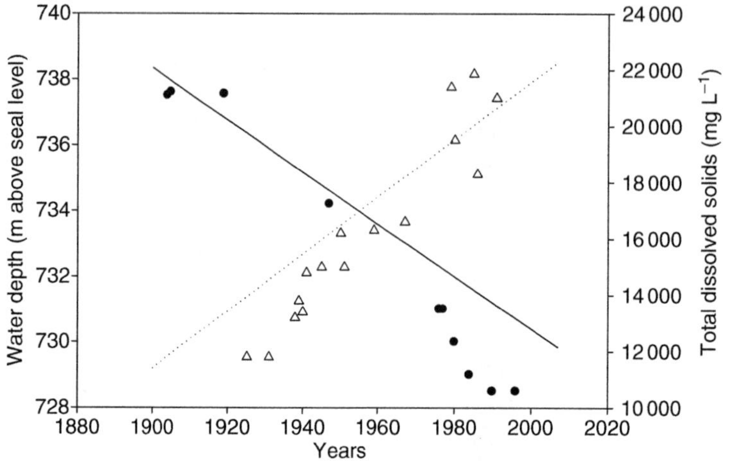


**Figure 3** Declining water levels (predicted: solid line; actual: solid circles); and increasing salinity (predicted: broken line and actual: triangles) in Redberry Lake, SK, Canada, as a result of drought and land use changes in the basin during the period 1904–2003. Data from G. Van der Kamp and M.J. Waiser (unpublished data).

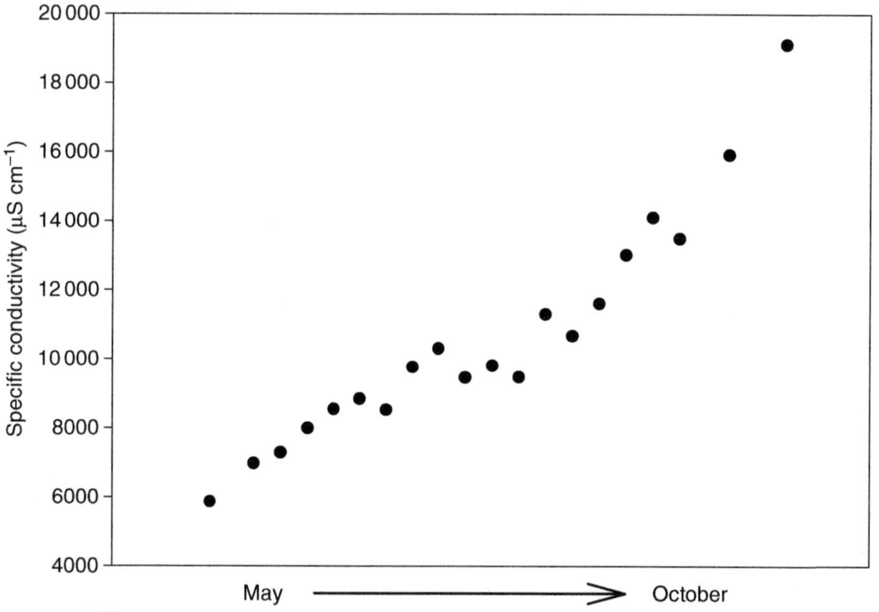


**Figure 4** Increases in specific conductivity from May to October, 2000, in a shallow saline prairie wetland – Pond 50, St. Denis National Wildlife Refuge, SK, Canada. Data from Waiser MJ (2001) *The Effect of Solar Radiation on the Microbial Ecology and Biogeochemistry of Prairie Wetlands*. PhD thesis, Edinburgh: Napier University.

Canada specific conductivities as high as 72 000 $\mu$S cm$^{-1}$ have been recorded in some saline wetlands.

Although ions that comprise salinity can originate from marine aerosols deposited on lake surfaces (those lakes found near to marine environments and closely related hydrologically and biologically to the sea are known as thalassic lakes), they usually originate in soils or rocks within a lake catchment (such lakes are known as athalassic or true inland saline lakes). Surface catchment geology, therefore, is one factor that will determine whether receiving water bodies are saline or not. For example, igneous rocks that characterize the Precambrian Shield of North America and areas of Scandinavia, Russia as well as the Amazon and Congo River Basins, are of low solubility. Their receiving waters will consequently be low in salinity and precipitation will determine lake salinity. On the other hand, soils formed after retreat of the last glacier (i.e., Pleistocene in the Prairie Pothole region) tend to have higher solubility and therefore receiving waters will be higher in salinity. Aside from air borne salts, and weathering of rocks, the other major source of salts is from springs rich in minerals leached from underground rocks and sediments.

Within saline lakes, however, ionic composition tends to differ greatly from incoming waters and geochemistry of catchment soils. This is due to the fact that salinity arises from differential ion precipitation as water evaporates from the lake. In general terms, $CaCO_3$ is precipitated first during evaporation and results in relative enrichment of Na, Mg, Cl, and $SO_4$ in remaining water. Next, $MgCO_3$ precipitates out as dolomite $[CaMg(CO_3)_2]$ in waters where concentration of Ca and Mg is high. If concentration of Ca is high enough then gypsum will precipitate $(CaSO_4·2H_2O)$. After all of this, precipitation has occurred, the remaining water tends to be enriched in Cl relative to other anions, and Na relative to other cations. For this reason there are more saline lakes with Na and Cl as their major ions owing to the high solubility of Na and Cl.

Saline lakes exhibit a diverse ionic composition (Table 2). Globally, NaCl lakes are particularly common in Australia, South America, and Antarctica. In East Africa, however, waters are enriched in $HCO_3$, but poor in Ca and Mg and these are the so-called soda (sodium carbonate) lakes of the African Great Rift Valley (Lakes Nakuru and Magadi). Across the prairie pothole region of North America, many of the lakes are dominated by Mg and $SO_4$ (e.g., Redberry Lake).

Major determinants of salinity, therefore, are local geology, climate, soil age, and distance from the sea (ion deposition).

## Global Distribution

Saline lakes are found on all continents and are often close to large population centers. Some of the largest lakes in the world are saline, such as the Caspian Sea. (Table 3). Recent major declines in the volume of saline lakes such as the Aral Sea (75%) may have greatly altered this ratio but this has not yet been estimated.

### South America

In South America there are two major saline lake areas. The first is the Bolivian Altiplano region and its northern extension into Peru, which contains many shallow permanent saline lakes as well as ephemeral salt pans. Some of the highest saline lakes in the world are located here (4000–4500 masl). The second region is found in the Pampas of central Argentina and stretches into northern Patagonia. Its

**Table 1** Classification of saline lakes according to total dissolved solids (TDS), specific conductivity and salinity

| Lake type | TDS (mg l$^{-1}$) | Specific conductivity ($\mu S\ cm^{-1}$) | Salinity ‰ (ppt) |
|---|---|---|---|
| Fresh | <500 | <850 | 0.5 |
| Sub-saline | 500–3000 | 850–5500 | 0.5–3 |
| Hyposaline | 3000–20 000 | 5500–30 000 | 3–20 |
| Mesohaline | 20 000–50 000 | 30 000–70 000 | 20–50 |
| Hypersaline | >50 000 | >70 000 | >50 |
| Seawater | 35 000 | 53 000 | 35 |

Freshwater and seawater values are included for comparison.

**Table 2** Divergent chemical composition of selected Saskatchewan and African saline lakes

| Lake | Salinity | Ca | Mg | Na | K | HCO$_3$ | CO$_3$ | Cl | SO$_4$ |
|---|---|---|---|---|---|---|---|---|---|
| Muskiki[a] | 370.2 | 145 | 29 000 | 62 200 | 1600 | 1469 | 0 | 11 400 | 237 000 |
| Patience[a] | 219.3 | 1880 | 2625 | 55 200 | 25 200 | 102 | 24 | 105 000 | 17 200 |
| Little Manitou[a] | 95.9 | 497 | 9518 | 12 300 | 890 | 776 | 209 | 18 000 | 39 600 |
| Deadmoose[a] | 27.1 | 268 | 1592 | 5390 | 324 | 429 | 122 | 6000 | 11 000 |
| Redberry[a] | 19.4 | 99 | 2271 | 1860 | 178 | 551 | 125 | 220 | 12 500 |
| Reflex[a] | 8.1 | 10 | 65 | 2900 | 67 | 1220 | 396 | 3500 | 480 |
| Victoria[b] (Africa) | 118 | 5.6 | 2.6 | 10.4 | 3.8 | 56.1 | | 3.9 | 2.3 |
| Tanganyika[b] (Africa) | 460 | 9.2 | 1.6 | 57 | 18 | 6.0 | | 20.9 | 7.2 |
| Malawi[b] (Africa) | 200 | 16.4 | 4.7 | 21 | 6.4 | 2.4 | | 3.6 | 5.5 |
| Eyre[c] (Australia) | 115 940 | 910 | 300 | 43 780 | 10 | 40 | | 67 960 | 2940 |
| Magadi[d] (Africa) | n/a | n/a | n/a | 162 400 | 2340 | 0 | 121 800 | 74 100 | 2480 |

Salinity in ppt (‰) and ions in mg l$^{-1}$ (ppm). n/a – not available.

[a]From Hammer UT (1978) The saline lakes of Saskatchewan III: Chemical Composition. *Int. Revue Ges. Hydrobiol.* 63: 311–335.

[b]From JF and Talling IB (1965) The chemical composition of African lake waters. *Int. Revue ges. Hydrobiol.* 50: 421–463.

[c]From Williams WD (1984) Chapter 20: Australian Saline Lakes. pp. 499–519. In Taub F (ed.) Ecosystems of the World 23: Lakes and Reservoirs. Elsevier, New York.

[d]From Livingstone DA and Melack JM (1984) Chapter 19: Lakes of Sub-Saharan Africa pp. 467–497. In Taub F (ed.) Ecosystems of the World 23: Lakes and Reservoirs. Elsevier, New York.

**Table 3** Area, volume, mean, and maximum depth of the 10 largest saline lakes in the world

| Lake | Area (km²) | Volume (km³) | Mean depth (m) | Maximum depth (m) |
|---|---|---|---|---|
| Caspian | 422 000 | 79 000 | 187 | 1072 |
| Aral[1] | 66 000* | 1064* | 16 | 69* |
|  |  | 304[†] |  | 37.8[†] |
|  | 28 687** |  |  |  |
|  | 17 160*** |  |  |  |
| Balkash | 22 000 | 122 | 6 | 27 |
| Eyre, North[2] | 7000 | 23 | 3 | 6 |
| Issyk-kul | 6300 | 1730 | 275 | 702 |
| Urmia | ~5000 | 25 | 5 | 16 |
| Qinghai Hu | 4600 | 85 | 17.5 | 27 |
| Great Salt Lake | 4400 | 19 | 4 | 10 |
| Van | 3600 | 191 | 53 | 550 |
| Dead Sea | 940 | 136 | 145 | 330 |

[1] *1969; **1998; ***2004; [†]1990.
[2]When full. Sources: Williams WD (1996) The largest, highest and lowest lakes of the world: Saline lakes. *Verhandlungen Internationale Vereinigung Limnologie* 26: 61–79. Glazovsky NF (1995) The Aral Sea Basin. In: Kasperson J, Kasperson R, and Turner BL (eds.) *Regions at Risk: Comparisons of and Threatened Environments*. New York: United Nations University Press. Nikolyayeva RV (1969) *The main morphological characteristics of the Aral Sea. The problems of the Aral Sea.* pp 25–38. Moscow: Nauka. Jellison R, Zadareev YS, DasSarma PA, *et al.* (2004) *Conservation and Management of Saline Lakes: A Review of Five Experience Briefs.* http://www.worldlakes.org/uploads/S... Kes%20thematic%20Paper%2022Jun04.pdf.

largest and most permanent saline lake is Mar Chiquita while one of the world's largest complexes of saline lakes, Salinas Grandes, is also located here. Other areas containing saline lakes include the Pantanal (southwestern Brazil), where round saline ox bow lakes are known locally as salinas, as well as along the eastern and southern coasts of Brazil and Chile, respectively.

### North America

Saline lakes are widely distributed throughout the western half of the North American continent where six regions can be distinguished including the Great Plains, the North West, the Great Basin, the Mid Continental Region, the Southwest Region, and the Chihuahuan region of North Mexico. One of the best studied is the Great Plains or Prairie Pothole Region. This area extends from the Dakotas and Western Minnesota into Southern Manitoba, Saskatchewan, and Alberta. No other area in the world has the concentration or diversity of saline lake environments seen here. Estimates of saline lake numbers range from 1 to greater than 10 million and density can be as high as 120 lakes per km². Every water chemistry is represented in this region. In fact, because of this, over 40 species of endogenic precipitates have been identified in lake sediments. Types of lakes range from large, ephemeral salt pans (e.g., Muskiki Lake, Saskatchewan) to deep permanent lakes (e.g., Deadmoose Lake Saskatchewan where some 'holes' are >60 meters deep).

The Great Basin region, which includes all of Nevada, the western half of Utah, southeastern Oregon and a large area of southeastern California, contains some of the most well known lakes in North America. These include Mono, Pyramid, Walker, and Great Salt Lakes as well as the Salton Sea (**Figure 5**).

### Europe (Excluding Russia)

Although there are many athalassic lakes in Europe, true athalassic lakes are found only in hydrologically closed basins which typify Spain and southeastern Europe. In Spain, saline lakes are quite common, with the largest and most permanent being Lake Gallocanto. Some saline lakes are also found in Austria, Hungary, eastern Romania, Greece, Sardinia, Sicily, and Cyprus.

### Africa

Saline lakes on the African continent are located in four major areas which include the northern half of the continent (Saharan and semiarid region bordering the Mediterranean), East Africa, the Western Rift Valley and South Africa. Perhaps the best known African lakes are the so-called soda lakes of East Africa. These sodium bicarbonate dominated lakes occur in a narrow strip extending from the Red Sea south through Ethiopia and Kenya into Central Tanzania and include small deep permanent lakes as well as shallow salt pans. Salinities here range from 3000 to 300 000 mg l$^{-1}$. A number of lakes (e.g., Lakes Nakuru and Magadi) are major staging and feeding areas for pink flamingoes.

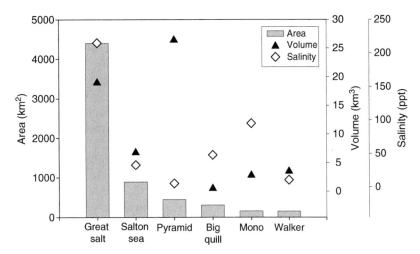

**Figure 5**   Salinity, volume and area of some large North American saline lakes. Data from: Williams, WD (1996) The largest, highest and lowest lakes of the world: Saline lakes. *Verhandlungen Internationale Vereinigung Limnologie* 26: 61–79.

Much less well known are coastal saline lakes along the coast of South Africa, e.g., Swartvlei in Cape Province. These lakes are formed when the action of long shore currents creates sand bars preventing drainage to the sea. They are re-opened when the water level in a lake flows over the bar and erodes it. However, in some cases these are artificially destroyed when levels get too high and start to flood houses or businesses built close to the lake shore. For part of a year the upper water column may become fresh water as riverine inflows spread over the denser saline waters until the breakdown of the sand bar allows the water to flow into the ocean. The bottom part of the water column remains saline throughout the year except under exceptional hydrological and climatic circumstances. Similar coastal saline lakes also occur on the coasts of Japan, Australia, and Korea.

### Russia, China, Iranian Plateau, and the Middle East

Large areas of southern Russia and Central Asia are endorheically drained. Saline lake regions here include one associated with the Black Sea and another within the Aral-Caspian depression. The Aral-Caspian depression contains two of the largest saline lakes in the world, the Caspian Sea and the now severely depleted Aral Sea. Northeast is Lake Tengiz, a major habitat for the greater flamingo and located in the Baraba-Kulunda steppe region. Southeast of Lake Tengiz is the Balkhash region whose largest saline lake is Lake Balkhash. Further south in the Tien Shan Mountain area is another large saline lake, Issyk-kul, the 11th largest lake in the world (by volume) and amongst the most ancient (~25 million years).

In China, saline lakes account for 55% of the area of all lakes and include the largest Chinese lake,

Qinghai Hu, which is also the seventh largest saline lake worldwide. The five major saline lake regions in China include Inner Mongolia, the Qaidam basin, Quinghai Hu, the Tibetan plateau (which includes the highest salt lake known – Nan Tso located at 4718 masl) and the Tarim Basin. Lake Urmia, the sixth largest saline lake in the world, is located on the Iranian plateau and is an important flamingo habitat. In the Middle East, the largest and best known saline lake is the Dead Sea. Located at 400 m below sea level it is not only the lowest lake worldwide, but the lowest place on earth. Unfortunately, it is under threat due to reduced water inflow from the Jordan River as a result of diversion and reservoir construction. The surface area has been reduced by more than one third resulting in a drop of over 20 m since development of the region started early in the 20th century.

### Australia

There are three major salt lake regions in Australia: a large central one, as well as ones in the southeast and southwest. Australian saline lakes are typically dominated by NaCl and exhibit a high degree of biological diversity. The largest lakes include Lake Eyre (Central region), and Lakes Corangamite and Bullen Merri in the southeast.

### Antarctica

Saline lakes characterize a number of regions in Antarctica including the Vestfold, Bunger and Larsemann Hills as well as the McMurdo Dry Valley and in the west on the Antarctic Peninsula.

## Flora and Fauna

Generally speaking, freshwater species are routinely found in waters with <1000 mg l$^{-1}$ TDS (1‰; 1600 µS cm$^{-1}$), while few are found at concentrations greater than this. At low salt concentrations, organisms tend to show flexibility in their adaptation to various ion combinations. In saline lakes across Western Canada, for example, the rotifer *Branchionus plicatilis* and the harpactacoid copepod *Cletocamptus* spp., tend to prevail in lakes dominated by Cl while *Leptodiaptomus sicilis* and *Diaptomus nevadensis* dominate in SO$_4$ and CO$_3$ waters. Such flexibility, however, disappears as salinity increases. This is due to the fact that high salinity exerts an osmoregulatory stress on endemic organisms. The term osmoregulation, or osmotic regulation, refers to the regulation of the internal cell environment relative to the concentration in the external environment, across a semi-permeable membrane (cell membrane). Organisms adapted to life in salty environments (halophiles) do this by accumulating organic solutes like glycerol, sucrose and glycine within their cells. A great deal of energy is consequently expended to maintain properly functioning tissues. If cells did not do this, water would move across the cell membrane to the surrounding saline water (osmosis) in an attempt to equalize the salt concentration and cells would ultimately die. Because salt tolerant organisms expend a great deal of energy in osmoregulation, those halophiles that can produce energy in other ways (e.g., through photosynthesis) will tend to be favored in saline lakes. For example, the green alga *Dunaliella* spp. is present world wide as the main or sole primary producer in hypersaline environments.

### Plankton

**Microbial Community** Aquatic bacteria, algae, fungi, and protozoa have demonstrated high evolutionary adaptability to salinity changes and as a result, halophiles are widespread in all three domains – Archeae, Eubacteria, and Eukarya. In fact, some bacteria can exist up to NaCl saturation. A common benthic feature of many saline lakes is microbial mats formed by an association between cyanobacteria, bacteria, and algae. Cyanobacterial mats in the Salton Sea, California, contain several filamentous cyanobacteria, including *Oscillatoria* spp. and *Geitlerinema* spp. as well as unicellular cyanobacteria, small diatoms, and *Beggiatoa*, a sulfur oxidizing bacteria. It is thought that cyanobacteria, particularly the mat forms associated with *Beggiatoa*, may play an important role in carbon fixation and other biogeochemical processes in the Salton Sea.

Numerous algal species are highly adapted to saline conditions as well. In a study of 41 saline lakes across southern Saskatchewan, Canada, ranging in salinity from 3.2 to 428 g l$^{-1}$, algae in 7 phyla, 8 classes, 42 families, 91 genera and 212 species were identified. Fourteen species were restricted to hypersaline (50 g l$^{-1}$; 50‰) waters and eleven of these were diatoms. Generally, species diversity was inversely related to lake salinity.

Important community constituents over a wide salinity range were green algae, cyanobacteria, and diatoms. Commonly, green algae dominated when lake salinity exceeded 10.0 g l$^{-1}$ (10‰). In some saline lakes, models developed for temperate freshwater lakes that predict algal biomass from either ambient total phosphorus or nitrogen concentration routinely overestimate chlorophyll *a* and primary production. As well, bacteria in some saline lakes are phosphorus limited despite high measured concentrations of biologically available phosphorus. Some aspect of salinity and high levels of dissolved organic carbon, which may inhibit algae and/or bind phosphorus, may explain these anomalies.

### Zooplankton (Invertebrate)

Countless zooplankton species are also salt tolerant. For example, approximately 31 species are tolerant of hypersaline conditions and these include 20 copepods, 4 rotifers, 2 anostracans, and 5 cladocerans. One of the most ubiquitous of these, the brine shrimp *Artemia*, is found worldwide in fishless saline lakes and tolerates salinities >30 000 mg l$^{-1}$ (30‰). Many more zooplankton species are tolerant of mesosaline and low salinity waters compared to those which are hypersaline. In East African saline lakes, the number of rotifer, cladoceran and copepod species starts to decline at conductivity >1000 µS cm$^{-1}$ (~700 mg l$^{-1}$) falling further to only a few species above 3000 µS cm$^{-1}$ (~1800 mg l$^{-1}$).

### Other Invertebrates

Many other invertebrates characterize saline lakes and include annelids (worms), nematodes (roundworms), gastropods (snails), amphipods (side swimmers), corixids (water boatmen) and larval stages of chironomids (midges), anopheles (mosquitoes), and ephydra (brine fly).

### Fish

Saline lakes, especially those with lower salt concentrations and ion species dominated by Na and Cl, can have complex food webs and be extremely productive. In fact some saline lakes support robust fisheries (e.g., Lake Issyk-kul, Caspian Sea, Lake Balkhash, Lake Chany complex, Siberia). Lake Issyk-kul, a large mesohaline $(6000\,mg\,l^{-1})$ sodium-sulphate-chloride lake in northeastern Kirghizia has 299 species of phytoplankton, 154 zooplankton species, 176 species of benthic invertebrates and 27 fish taxa in five families. Although the Caspian Sea once supported 82% of the world's sturgeon population, recent water regulation and falling water levels have caused many spawning grounds to disappear and sturgeon catches to decline. Fish survival in saline systems is limited not only by salt concentration but also by specific types of ions present.

### Macrophytes

Although saline waters support numerous macrophytes, their major stress in saline waters is water level fluctuation. Differing macrophyte associations characterize waters of differing salinity. Angiosperms commonly found in saline lakes worldwide include various species of reeds (*Phragmites*), grasses/sedges (*Carex, Ruppia, Cyperus, Puccinellia, Triglochin*), cattails (*Typha*), rushes (*Juncus*), bullrushes (*Scirpus*), pondweed (*Potamogeton*), saltwort (*Salicornia*), and saltbush (*Atriplex*).

### Other

Worthy of mention as well is *Phoca caspica*, a seal endemic to the Caspian Sea, and one of only a very few worldwide, which make their home in inland waters.

## Effect of Salinity on Biodiversity

As salt concentration increases, however, biodiversity and species richness tends to decrease. At highest salinities, therefore, food webs become simplified. In highly saline Lake Nakuru, Kenya (TDS 10.0–120.0 g $l^{-1}$), for example, the food web consists of a top predator (the lesser flamingo – *Phoeniconaias minor*), one cyanobacterial species (*Oscillatoria* spp.), two zooplankton species (one copepod and one rotifer) and one introduced fish species. Salinity in the Chany Lake complex of Siberia ranges from 0.8 g $l^{-1}$, where two rivers enter the lake, to 6.5 g $l^{-1}$ at the point furthest from the river mouths. Along the same salinity gradient the number of aquatic vascular plant species declined from 16 to 12, phytoplankton species from 98 to 52 and zooplankton from 61 to 16.

## Bird Habitat

Saline lakes around the world are important nesting, feeding and staging areas for all types of waterbirds. The saline soda lakes of east Africa are renowned habitats for pink flamingoes. Lakes Natron and Bahi, for example, have distinctively large populations of flamingos. According to a recent aerial survey, Lake Natron has the highest concentration of flamingos in East Africa. Both the greater and lesser flamingo (*Phoenicopterus ruber* and *Phoeniconaias minor*) are found at these lakes, with the lesser flamingo outnumbering the greater by 100 to one.

Mono Lake, California is designated an International Reserve in the Western Hemisphere Reserve Network with nearly 2 million waterbirds (including 35 species of shorebirds, e.g., Western phalaropes, American avocets, Western and Least sandpipers, Snowy plovers, White-faced ibises, Dowitchers) using it as a food and rest stop for at least part of the year. Mono Lake is also home to the second largest California Gull rookery in North America (Great Salt Lake is the largest). Another major staging and feeding area for migratory and shore birds is the Salton Sea which supports the greatest diversity of birds of any U.S. National Wildlife refuge. Redberry Lake (Saskatchewan) has been designated a World Biosphere Reserve by the United Nations Educational, Scientific and Cultural Organization (UNESCO). This lake serves as a significant staging, nesting and feeding area for Canada Geese (*Branta canadensis*), Tundra Swans (*Cygna columbianus*), and about 30 000 ducks. Over 188 species of birds 'hang out' and nest here including the Piping Plover (*Charadrius melodus*), a small endangered shorebird. American White Pelicans (*Pelecanus erthrorhynchos*) are also summer inhabitants.

## Economic Uses

Saline lakes are a source of many evaporitic minerals which have a wide range of uses in manufacturing, construction, agriculture, medicine as well as chemical industries. Such evaporites include halite (NaCl), uranium, zeolites (hydrated alumino-silicate minerals with an 'open' structure that can accommodate a wide variety of cations, such as $Na^+$, $K^+$, $Ca^{2+}$, $Mg^{2+}$ and others, i.e., $Na_2Al_2Si_3O_{10}$–$2H_2O$ (natrolite) used as molecular filters and ion exchange agents), lithium (used in heat transfer applications and salts used in pharmacology as mood stabilizers), potash ($K_2O$ – fertilizer, glass, soap) and borax ($Na_2B_4O_7 \cdot 10H_2O$ – detergents, cosmetics, glass, pottery). Saline lakes are also a source of Glauber's salts ($Na_2SO_4$ – used in manufacturing of detergents, carpet fresheners, glass, paper and textiles) as well as Epsom salts ($MgSO_4 \cdot 7H_2O$ – agricultural and medicinal uses). Glycerol (used in dynamite and light industry) and β-carotene (used in food and medical industries) are both derived from *Dunaliella*, a green alga widely present in saline lakes. β-carotene is also derived from the salt tolerant cyanobacteria *Spirulina*. Finally, *Artemia* cysts, used as food materials for aquaculture (shrimp, fish, and crabs), are harvested commercially from saline lakes.

In some countries, thriving health spas are located adjacent to saline lakes; for example, the Dead Sea (Israel/Jordan) and Manitou Lake (Saskatchewan, Canada). The salty water appears useful in treating certain skin diseases (i.e., psoriasis) and easing pain associated with arthritis and other joint afflictions. Flourishing cosmetic industries, which incorporate salts from saline lakes, have also been established.

A number of lakes support important fisheries. In the Caspian Sea, the Beluga sturgeon (*Huso huso*) supplies 90% of the world's caviar. Fisheries in saline lakes, however, have been threatened by introduction of exotic fish species. In Lake Issyk-kul, for example, introduction of the Sevan trout led to a drastic decline in several endemic species and subsequently of the trout itself.

## Paleolimnology

Due to their hydrologically closed nature, saline lakes are very responsive to past and present climate change. Paleolimnologists can track these climatic changes by examining cores from lake sediments. Preserved in the core are biological, geological and chemical signals – essentially clues which can reveal not only the ecological history of the lake but the surrounding landscape as well. Paleolimnological examination of cores from saline lakes, for example, has been used to predict the duration and frequency of droughts on the Canadian prairies. During droughts, evaporation rates increase and in saline lakes, as water is lost, salinity increases. As salinity increases, the numbers of algal species able to tolerate this increase decline and species change. Scientists can track these changes by first cutting lake sediment cores into sections and dating them. Dating techniques for these core sections include use of the following: radioactive carbon-14 ($^{14}C$) and lead-210 ($^{210}Pb$) isotopes, pigment remnant analysis, plant pollen and spore analysis, algal microfossil analysis, plant macrofossil analysis and fossilized remains of Cladocera, ostracods, and midges. Because the outer shells (silica valves) of diatoms are so well preserved in sediments, past changes in their communities can be related to historical changes in nutrient chemistry and salinity due to variations in climate and lake

levels. Also, by-products of plant pigments, are often well preserved in sediments and so the stratigraphy of chlorophyll and carotenoid (plant pigments) degradation products can also be used to determine historical lake productivity.

## Threats to Saline Lakes

Perhaps the greatest threat to saline lakes is anthropogenic or secondary salinization – in other words increased lake salinity as a result of human activity (industry, agriculture, construction) in the lake basin (**Figure 6**). In fact, W.D. Williams has stated that 'in some countries, anthropogenic salinization represents the most important threat to water resources.' For example, the disruption of the hydrological cycle by agriculture or diversion or damming of lake inflows can lead to freshwaters becoming saline and saline waters becoming more saline. On the Murray River flood plain in Australia, for example, agricultural clearing and land irrigation have caused the saline aquifer to rise. Concomitant salinization of floodplain wetlands has led to the disappearance of wetland macrophytes and riparian trees. Their disappearance is testament to the fact that small increases in salinity can have a large effect on biota due to the narrow salinity tolerance range for freshwater organisms. In Australia, it is estimated that anthropogenic salinization costs in excess of $50 million US annually.

### The Aral Sea

The best known case of anthropogenic salinization is that of the Aral Sea. For many years, lake levels, areas, volumes and salinity of this lake were stable. Lake area was ~68 000 $km^2$, the maximum depth was 69 m, volume 1000 $km^3$ and salinity 10‰. During the 1950s, however, planners decided to intensify cotton production in the lake basin. As a result, the irrigation area surrounding the lake increased from 5 to 7.4 million hectares and most of the water flowing into the Aral Sea (from the Syr Darya and the Amu Darya Rivers) was diverted for this purpose. So much water was diverted in fact that in some years (e.g., 1980s) there was no inflow to the lake. As a result, the lake is only 50% of its former area and 25% of its former volume. Salinity tripled from 10 000 mg $l^{-1}$ (10‰) in the 1960s to over 30 000 mg $l^{-1}$ (30‰) and currently exceeds that of the ocean. Salt deflation (removal of salts by wind action) from the exposed basin and subsequent deposition in surrounding agricultural land caused a marked reduction in agricultural activities. Subsequent land loss has topped 10 000 $km^2$. Pesticides used in agricultural activities in the basin have wiped out microbial and

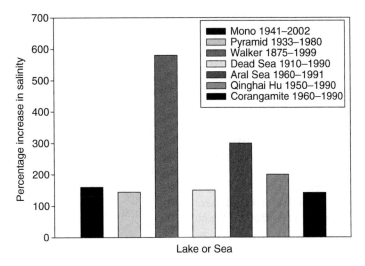

**Figure 6** Percent increase in salinity due to secondary salinization in some of the largest lakes in the world. Data from: Williams, WD (1996) The largest, highest and lowest lakes of the world: saline lakes. *Verhandlungen Internationale Vereinigung Limnologie* 26: 61–79 and from Williams WD (2002) Environmental threats to salt lakes and the likely status of inland saline ecosystems in 2025. *Environmental Conservation* 29: 154–167.

zooplankton communities in the lake and fish stocks have consequently declined markedly. Originally 20 fish species were present in the Aral Sea and of these, 12 were economically important. Destruction of the fishing industry due to increased salinization occurred during the 1980s and by 1999, only five species remained. In the near future, the only animal able to inhabit the Aral Sea may be the halotolerant brine shrimp, *Artemia*. Deflation of salts from the basin has also led to increased human health problems. Salinization of the Aral Sea has been an ecological disaster of gargantuan proportions. In terms of remediation, although efforts have been made, the lake and its surrounding basin have not yet recovered.

### Mono Lake

In 1941, the city of Los Angeles diverted water from the Owens River and three of four tributaries that fed Mono Lake in an effort to meet its increasing water demands. At the time Mono Lake was 2105 masl and salinity was $50\,000\,\mathrm{mg}\,\mathrm{l}^{-1}$ (50‰). By 1982, however, lake levels had fallen to 2091 masl and salinity had almost doubled to $99\,000\,\mathrm{mg}\,\mathrm{l}^{-1}$ (99‰) due to the water diversion. Shrinking water levels caused landbridges to form between the mainland and islands, allowing coyotes and other predators' access to California Gull nesting sites. As well, due to increasing salinity, Mono Lake's brine shrimp (*Artemia monica*) was being considered for listing as a federally threatened species by the U.S. Fish and Wildlife Service.

Lowered water levels also exposed tufas, irregular shaped thick towers of calcium carbonate, endemic to this lake. Tufa rock is the primary substrate on which the aquatic larvae and pupae of *Ephydra hians* (alkali fly) are found. These flies are a dietary staple for many migratory birds that spend time at the lake. Falling lake levels also created a smaller shallow area around the lake (littoral zone). This meant less suitable habitat for fly larvae and pupae, declining fly numbers and therefore less food for migratory birds. Alkali flies, like brine shrimp, were threatened by increasing salinity as well. Concerted conservation efforts by the Mono Lake Committee, however, paid off. In 1983, the California Supreme Court ruled that under the public trust doctrine of the State's constitution, the State of California must protect natural resources like Mono Lake. As a result, Los Angeles currently takes only 16% of the original water diversion. Today, the lake stands at 2095 masl and in another eight years the water level is expected to rise another two meters. Rising water levels have covered the old land bridges preventing coyote access to gull nesting sites and decreasing salinity means the lake is once again teeming with brine shrimp and alkali flies which feed migratory birds. Mono Lake is an ecological success story that hopefully will be repeated worldwide.

### Other Saline Lakes

There are many other saline lakes similarly affected by water diversions and threatened by secondary salinization. These include the Dead Sea, Lake Corangamite, Lake Balkhash, Pyramid and Walker Lakes, Qinghai Hu, and the Caspian Sea (**Figure 6**) Elevation of Lake Walker, for example, dropped 40 m from 1882 to 1996 and salinity increased from 2.6 to $12.4\,\mathrm{g}\,\mathrm{l}^{-1}$ as a result of water diversion in the basin.

This secondary salinization led to a decrease in tui chub minnow (*Gila bicolor*) numbers, which then threatened survival of its main predator, the Lahontan cutthroat trout (*Oncorhyncus clarki henshawi*).

Other threats to saline lakes include introduction of exotic fish species leading to a decline in endemic species (Lake Issyk-kul, Lake Balkhash); heavy metal contamination (Lake Poopó – Bolivia); toxic and radioactive waste contamination (Lake Issyk-kul); and eutrophication (Lake Poopó and the Caspian Sea).

*See also:* Africa: South of Sahara; Antarctica; Asia; Australia and New Zealand; Europe; Meromictic Lakes; North America; Origins of Types of Lake Basins; South America.

## Further Reading

Ellis WS (1990) A Soviet sea lies dying. *National Geographic* 177: 73–93.

Ferguson R (2003) *The Devil and the Disappearing Sea.* Vancouver: Raincoast Books.

Hammer UT (1986) *Saline Lake Ecosystems of the World.* Dordrecht: Dr. W. Junk Publishers.

Hammer UT (1990) The effects of climate change on the salinity, water levels, and biota of Canadian prairie saline lakes. *Verhandlungen Internationale Vereinigung Limnologie* 24: 321–326.

ILEC (2003) *World Lake Vision: A Call to Action.* Kusatsu, Japan: International Lake Environment Committee and UNEP-IETC.

Jellison R, Zadereev YS, DasSarma PA, *et al.* (2004) *Conservation and Management of Saline Lakes: A Review of Five Experience Briefs.* http://www.worldlakes.org/uploads/S... kes%20Thematic%20Paper%2022Jun04.pdf.

Kalff J (2003) *Limnology.* New Jersey: Prentice-Hall.

Micklin PP (1988) Desiccation of the Aral Sea: A water management disaster in the Soviet Union. *Science* 241: 1170–1176.

Patten DT, *et al.* (1987) *The Mono Basin Ecosystem. Effects of Changing Lake Level.* Washington, D.C: National Academy Press.

Williams WD (1993) The conservation of salt lakes. *Hydrobiologia* 267: 291–306.

Williams WD (2001) Anthropogenic salinization of inland waters. *Hydrobiologia* 466: 329–337.

Williams WD (2002) Environmental threats to salt lakes and the likely status of inland saline ecosystems in 2025. *Environmental Conservation* 29: 154–167.

Williams WD and Aladin NV (1991) The Aral Sea: Recent limnological changes and their conservation significance. *Aquatic Conservation* 1: 324.

# Antarctica

**J C Priscu and C M Foreman,** Montana State University, Bozeman, MT, USA

## Introduction

The evolutionary history of Antarctic lakes reflects the history of the continent itself. More than 170 Mya, Antarctica was part of the supercontinent Gondwana. Over time Gondwana broke apart and Antarctica, as we know it today, was formed around 25 Mya. During its evolution, the continent underwent numerous climate shifts. Around 65 Mya, Antarctica still had a tropical to subtropical climate, complete with an Australasian flora and fauna. Ice first began to appear around 40 Mya. The opening of the Drake Passage between Antarctica and South America around 23 Mya resulted in the Antarctic Circumpolar Current, which effectively isolated the advection of lower latitude warm water to the region, leading to continent-scale glaciations that now typify Antarctica. The period between 14.8 and 13.6 Mya (mid Miocene) saw an important change in the landscape evolution. During this time, the linked climate-and-glacial system changed from one dominated by intermittent fluvial erosion and wet-based glaciation, to one featuring a largely cold-based ice sheet, with cold-based alpine glaciers in the hyperarid, cold-desert conditions of the Transantarctic Mountains. The last Antarctic glaciation reached a maximum around 18 000 years ago, a period when the present ice sheet was much thicker and extended out to the edge of the continental shelf. The icecaps of offshore islands were similarly more extensive. These extensive ice sheets retreated during the late Pleistocene and have remained relatively stable during the current Holocene epoch. As a result of this temporal evolution, we now see lakes distributed on maritime islands, along the margins of the continent in ablation regions, and subglacially, beneath the thick ice sheet. All these lakes reflect, to varying degrees, the legacy left by past geological and climatological conditions.

This article describes the formation, distribution, and diversity of lakes in selected regions in Antarctica where focused research efforts have occurred. Although no subglacial lakes have been sampled directly, we present an overview of what is known about them, with a focus on Lake Vostok, the largest of these lakes.

## The Antarctic Continent: An Overview of Lake Regions

Antarctica comprises more than $14 \times 10^6 \, km^2$, making it the fifth largest continent. Physically, it is divided into West Antarctica and East Antarctica by the Transantarctic Mountains. Antarctica is the coldest place on Earth with about 98% of the continent covered by permanent ice, which averages 2.5 km in thickness. The continent holds 70% of all the fresh water on Earth, in the form of ice. Average winter temperatures approach $-75\,°C$ in the continental interior and $-25\,°C$ along the margins. The average temperatures during summer are considerably warmer, reaching $-35$ and $-10\,°C$ for the same regions. Despite the subzero temperatures and thick ice sheets, numerous lakes exist on the Antarctic continent (**Figure 1**). Except for the subglacial lakes, Antarctic lakes are confined to the ice free coastal regions. The surface lakes range in latitude from $60°43'$ S (Signy Island) to $77°30'$ S (McMurdo Dry Valleys) (**Table 1**). The air temperatures and ice cover characteristics of the lakes reflect this latitudinal location with lakes at lower latitudes having thinner and shorter seasonal ice covers and higher productivity.

### Subglacial Lakes

**Location** The earliest evidence of subglacial lakes was from Russian aircraft pilots flying missions over the Antarctic continent, claims subsequently verified by airborne radio–echo sounding during the 1960s and 1970s. We now know that more than 150 lakes exist beneath the Antarctic ice sheet (**Figure 1**), many of which may be connected by large subglacial rivers. Approximately 81% of the detected lakes lie at elevations less than $\sim$200 m above mean sea level, while the majority of the remaining lakes are 'perched' at higher elevations. Sixty-six percent of the lakes lie within 50 km of a local ice divide and 88% lie within 100 km of a local divide. The high density of lakes in the Dome-C region implies that they may be hydrologically connected within the same watershed and would be an important system to study from the standpoint of subglacial hydrology and biological and geochemical diversity.

**Formation and diversity** The association of subglacial lakes with local ice divides leads to a fundamental question concerning the evolution of subglacial lake environments:

Does the evolving ice sheet control the location of subglacial lakes or does the fixed lithospheric character necessary for lake formation (e.g., basal morphology,

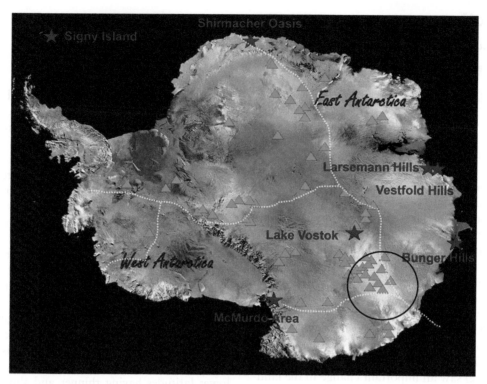

**Figure 1** Locations of the lakes discussed in the text. Red stars show the locations of specific lakes or regions; triangles denote the location of known subglacial lakes; yellow dashed lines represent the approximate location of several of the major ice divides on the continent; blue circle denotes the Dome-C subglacial lake cluster.

**Table 1** Location, air temperature, and ice cover characteristics of selected lake regions discussed in this article

| District | Latitude | Mean temperature | | | Ice cover | Notes |
|---|---|---|---|---|---|---|
| | | Annual | Summer | Winter | | |
| *Subantarctic Islands* | | | | | | |
| Signy Island | 60°43′ S | −4.2 | 1.4 | −9.6 | Variable (8–12 mo, 1–2 m) | Maritime climate zone |
| *Continental ice-free zones* | | | | | | |
| Shirmacher Hills | 70°45′ S | −10.8 | −2.0 | −16.1 | Permanent (~3 m) | Freshwater and epishelf lakes |
| Larsemann Hills | 69°03′ S | −10.5 | 4.0 | −15.0 | Variable (8–10 mo), ~2 m | Freshwater lakes |
| Vestfold Hills | 68°33′ S | −10.0 | −0.9 | −16.9 | Variable (8–10 mo), 0.5–2 m | Saline lakes formed by isostatic rebound |
| Bunger Hills | 66°17′ S | −9.0 | 0.4 | −16.3 | Variable (8–12 mo, 2–4 m) | Many tidally influenced epishelf lakes |
| McMurdo Dry Valleys | 77°30′ S | −17.7 | −3.1 | −25 | Permanent (4–19 m) | Many chemically stratified with ancient bottom water brines |

Sources

Jacka TH, Budd WF, and Holder A (2004) A further assessment of surface temperature changes at stations in the Antarctic and Southern Ocean, 1949–2002. *Annals of Glaciology* 39: 331–338.

Simmons GM, Vestal JR, and Wharton RA (1993) Environmental regulators of microbial activity in continental Antarctic lakes. In: Friedmann I (ed.) *Antarctic Microbiology*, pp. 491–451. New York: Wiley-Liss.

Heywood RB (1984) Inland waters. In: Laws RM (ed.) *Antarctic Ecology*, vol. 1, pp. 279–334.

Gibson JAE and Anderson DT (2002) Physical structure of epishelf lakes of the southern Bunger Hills, East Antarctica. *Antarctic Science* 14(3): 253–261.

geothermal flux or the nature of sub-ice aquifers) constrain the evolution of ice sheet catchments? With the exception of central West Antarctica (where lakes are few), we know little about either the lithospheric character along these catchment boundaries or the history of their migration, given by layering within the ice sheet. Subglacial lake environments rest at the intersection of continental ice sheets and the underlying lithosphere. This unique location sets the stage for generating a spectrum of subglacial environments reflective of the complex interplay of ice sheets and the lithosphere.

Antarctic subglacial lakes have been categorized into three main types: (1) lakes in subglacial basins in the ice-sheet interior; (2) lakes perched on the flanks of subglacial mountains; and (3) lakes close to the onset of enhanced ice flow. The bedrock topography of the ice-sheet interior involves large subglacial basins separated by mountain ranges. The lakes in the first category are found mostly in and on the margins of subglacial basins. These lakes can be divided into two subgroups. The first subgroup is located where subglacial topography is relatively subdued, often toward the center of subglacial basins; the second subgroup of lakes occurs in significant topographic depressions, often closer to subglacial basin margins, but still near the slow-flowing center of the Antarctic Ice Sheet. Where bed topography is very subdued, deep subglacial lakes are unlikely to develop. Lake Vostok (surface area of $\sim$14 000 km$^2$, maximum depth $\sim$800 m; volume $\sim$5400 km$^3$) is the largest known subglacial lake and the only one that occupies an entire section of a large subglacial trough. Theoretical models reveal that the subglacial environment may hold $\sim$10% of all surface lake water on Earth, enough to cover the whole continent with a uniform water layer with a thickness of $\sim$1 m. These models further reveal that the average water residence time in the subglacial zone is $\sim$1000 years.

Much attention is currently focused on the exciting possibility that the subglacial environments of Antarctica may harbor microbial ecosystems isolated from the atmosphere for as long as the continent has been glaciated (20–25 My). The recent study of ice cores comprised of water from Lake Vostok frozen to the overlying ice sheet has shown the presence, diversity, and metabolic potential of bacteria within the accreted ice overlying the lake water. Estimates of bacterial abundance in the surface waters range from 150 to 460 cells ml$^{-1}$ and small subunit rDNA gene sequences show low diversity. The sequence data indicate that bacteria in the surface waters of Lake Vostok are similar to present day organisms. This similarity implies that the seed populations for the lake were incorporated into the glacial ice from the atmosphere and were released into the lake water following the downward transport and subsequent melting from the bottom of the ice sheet. Subglacial lakes present a new paradigm for limnology, and once sampled, will produce exciting information on lakes that have been isolated from the atmosphere for more than 10 My.

## Signy Island, South Orkney Islands

**Location** Signy Island (60°43′ S, 45°36′ W) is a 20 km$^2$ island in the South Orkney archipelago. It lies at the confluence of the ice-bound Weddell Sea and the warmer Scotia Sea, and its climate is influenced by the cold and warm air masses from these two areas. The lakes on Signy Island share characteristics of Antarctic and subAntarctic environments, a fact reflected in the diverse flora and fauna. This region of the continent has been ice free for the past 6000 years and is referred to as the maritime Antarctic zone, with an annual mean air temperature near −4 °C. The island is small ($7 \times 5$ km$^2$; surface area 19.9 km$^2$) and has relatively little relief (maximum elevation 279 m). Most of the 17 lakes on Signy Island lie in the valleys and plains of the narrow, coastal lowland, which is usually snow free during the summer. The lakes on Signy Island that have received extensive study include Heywood, Sombre, Amos, and Moss. Radiocarbon dates on basal sediment from the lakes on Signy Island show that these lakes did not exist more than $\sim$12 000 years ago, making them similar in age to many of the continental lakes.

**Formation and diversity** These lakes share a common geology but cover a wide range of physical, chemical, and biological properties. Sombre lake is an ultraoligotrophic system receiving no significant nutrient input. The inflows consist primarily of snow melt streams which are frozen for 8–9 months in a year. As a result, the water column is relatively clear and most of the primary production comes from benthic cyanobacterial mats. Many of the lakes are suffering animal-induced (fur seal) eutrophication with Amos and Heywood lakes being the most severely affected. Unlike Sombre lake, the water column of Amos lake develops a dense phytoplankton bloom during spring and summer in response to elevated nutrient enrichment.

A summary of data averaged over a 6-year period (**Table 2**) reveals that the pH in these lakes is circumneutral, whereas the conductivity varies from 40 to 233 μS cm$^{-1}$, reflecting waters of extremely low to moderate ionic strength. The chlorophyll $a$ levels generally increase with conductivity and represent waters ranging from mesooligotrophic to mesoeutrophic. Levels exceeding 10 μg l$^{-1}$ chlorophyll $a$ in Heywood lake result from wildlife-induced nutrient loading. Dissolved inorganic nitrogen (DIN; $NO_3^- + NH_4^+$) and soluble reactive phosphorus (SRP) concentrations also vary considerably across the lakes on Signy Island, reflecting various degrees of nutrient loading and biological consumption. The ratio of DIN:SRP ranges from 3.6 to 445.3 revealing a wide range of potential

**Table 2** Selected chemical and biological properties from 14 lakes on Signy Island

| Lake | Depth (m) | pH | Conductivity ($\mu S\,cm^{-1}$) | Chl a ($\mu g\,l^{-1}$) | $Cl^-$ ($mg\,l^{-1}$) | $NO_3^-$–N ($\mu g\,l^{-1}$) | $NH_4^+$–N ($\mu g\,l^{-1}$) | SRP ($\mu g\,l^{-1}$) | DIN:SRP (g:g) |
|------|-----------|-----|------------|-------|------|--------|--------|-----|---------|
| Amos | 4.3 | 8.12 | 120 | 4.11 | 33.9 | 520.6 | 214.4 | 20.3 | 3.6 |
| Bothy | 2 | 6.82 | 233 | 1.24 | 44.2 | 570.7 | 94.8 | 4.1 | 12.2 |
| Changing | 5.4 | 6.82 | 94 | 2.43 | 25.3 | 146.4 | 8.5 | 4.2 | 44.3 |
| Emerald | 15 | 6.62 | 67 | 1.33 | 23.4 | 97.3 | 10.3 | 2.1 | 33.6 |
| Heywood | 6.4 | 6.92 | 134 | 10.06 | 42.3 | 327 | 56.1 | 5.9 | 11.9 |
| Knob | 3.5 | 7.32 | 62 | 8.70 | 18.7 | 123.3 | 16.3 | 3.1 | 155.1 |
| Light | 4.4 | 6.82 | 121 | 9.21 | 44.1 | 33.4 | 11.2 | 4.7 | 31.9 |
| Moss | 10.4 | 6.82 | 40 | 1.85 | 23.2 | 111 | 5.6 | 1.4 | 50.7 |
| Pumphouse | 4 | 6.92 | 86 | 3.02 | 18.4 | 63.6 | 7.1 | 2.5 | 20.2 |
| Sombre | 11.2 | 6.82 | 78 | 3.99 | 25.8 | 181.9 | 12.9 | 4.6 | 74.9 |
| Spirogyra | 1.5 | 7.42 | 60 | 1.51 | 19.4 | 116.2 | 17.4 | 4.7 | 445.3 |
| Tioga | 4 | 7.42 | 134 | 4.29 | 20.9 | 105.7 | 53.3 | 6.1 | 5.7 |
| Tranquil | 8 | 6.92 | 52 | 1.53 | 16.7 | 80.7 | 1.8 | 2.7 | 275.0 |
| Twisted | 4 | 6.82 | 92 | 2.43 | 28.7 | 65.6 | 2.5 | 3.4 | 26.2 |

Adapted from Jones VJ, Juggins S, and Ellis-Evans C (1993) The relationship between water chemistry and surface sediment diatom assemblages in maritime Antarctic lakes. *Antarctic Science* 5(4): 339–348. Data averaged from collections during early winter, spring, summer open water between 1985 and 1991.

nutrient limitation (a ratio of $\sim$7 represents balanced growth of phytoplankton). Marked seasonal variations in the abundance, composition, and productivity of phytoplankton, bacterioplankton and protozooplankton within Heywood lake have been shown to be correlated with the seasonality of both physical factors and nutrient levels. The biota within the lakes on Signy Island can be expected to continue changing in response to animal-induced nutrient loading.

In addition to the well documented increases in nutrient loading in many of the lakes on Signy Island, the air temperatures in this region have been rising by $\sim$0.25 °C per decade, which has led to increasing lake water temperatures (0.6 °C per decade) and more ice-free days each year. The higher air temperatures have caused elevated ice melt in the lake catchments, leading to elevated phosphorus loading from soil leaching. Measurements of dissolved phosphorus and chlorophyll *a* increased fourfold between 1980 and 1995 in response to climate warming. These data clearly point to the sensitivity of polar lakes to climate change, where small changes in air temperature result in extreme ecological change.

### Shirmacher Hills, Queen Maud Land, East Antarctica

**Location** Schirmacher Hills, located in the Central Queen Maud Land of East Antarctica, is an ice free area, bounded by a continental ice sheet on the south and the Fimbul ice-shelf on the north. The region is 17 km in length, 2–3 km wide and lies 100 km from the East Antarctic ice-sheet. Constituted by low lying hills that run East–West, the area varies widely in terms of elevation above sea level. The southern portion of the area is overlain by continental ice-sheet with recessional moraines and ice sheets descending onto bare rock. In contrast, the northern area exhibits steep escarpments all along its length in contact with the ice-shelf, an area often containing epishelf or tidal freshwater lakes similar to those that occur in the Bunger Hills. There are over 100 lakes and ponds of varying depths and melt-water streams in the area.

**Formation and diversity** The Schirmacher Hills contains several freshwater lakes formed by ice erosion and relict saline lakes formed as a result of the isolation of sea inlets and lagoons. The lakes are fed by melt water from snow beds and ice slopes during the summer.

Lakes Verkhneye, Pomornik, Glubokoye, and Stancionnoye lie in the eastern portion of the Shirmacher Hills and originate from glacial ice-scour. These lakes range in depth from 3 to 35 m. Lakes Untersee and Obersee are the two most studied lakes in the region and lie at elevations between 650 and 800 m. Lake Untersee (11.4 km$^2$) is a perennially ice-covered ($\sim$3 m ice cover), ultra-oligotrophic lake with a maximum depth of 169 m. About 2% of the incident sunlight reaches 145 m, making Lake Untersee one of the most transparent water bodies in the world. The water column throughout the majority of the lake is well mixed; however, there is a trough in the south-eastern portion of the lake that is physically and chemically stratified. The surface waters are supersaturated with oxygen, while water below 80 m is anoxic. High pH values (10–11) occur above 75 m, but decrease to $\sim$7 at 100 m. Methanogenesis, methane oxidation, and sulphate reduction have all recently been measured in the deeper waters of this lake. Although smaller, Lake Obersee (3.4 km$^2$) shares many of the same physiochemical features (i.e., high pH, oxygen supersaturation, and high transparency of ice) with Lake Untersee.

## Larsemann Hills

**Location** The Larsemann Hills (69°24′ S, 76°20′ E) are a series of rocky peninsulas and islands, located midway between the eastern extent of the Amery ice shelf and the southern boundary of the Vestfold Hills, in Prydz Bay, East Antarctica. Occupying an area of 50 km$^2$, the Larsemann Hills consist of two main peninsulas, Broknes to the east and Stornes to the west, as well as several off-shore islands. More than 150 freshwater lakes exist in the area, ranging from small ephemeral ponds to lakes exceeding 0.13 km$^2$ (Lake Nella) and 38 m deep (Progress Lake).

**Formation and diversity** The lakes in Larsemann Hills are thought to have formed from the exposure of basins after the retreat of the continental ice cap or after isolation due to the isostatic uplift following deglaciation. The lakes are connected to the coast by steep-sided valleys that dissect the area. During the summer months these lakes are partially to fully ice free and are well mixed by the strong easterly katabatic winds, while during the winter they lie under ice covers ~2 m thick. In general, the lakes on Stornes Peninsula have lower conductivities than those on Broknes Peninsula; these differences are believed to reflect the amount of freshwater input or the time elapsed since deglaciation. There is more snow cover on Stornes than Broknes Peninsula, which results in a cooler microclimate. Lake Reid (69°22′ S, 76°23′ E) on Broknes Peninsula is a relatively small (0.2 km$^2$), shallow lake (4 m deep), which is seasonally stratified and brackish. During winter the lake becomes anoxic, acquiring oxygen after ice melt and mixing in summer.

## Vestfold Hills

**Location** The Vestfold Hills lie along the coast of Princess Elizabeth Land (68°25′–68°40′ S, 77°50′–77°35′ E) and occupy an area of ~410 km$^2$. The region and several outlying islands are typically snow free and contain ~150 freshwater and saline lakes, which account for 8% and 2% of the total area, respectively.

**Formation and diversity** Following the retreat of the continental ice sheet after the last glacial maximum ~12 000 years ago, isostatic rebound occurred at a faster rate than sea level rise. As the land rose, it cut off fjords and trapped pockets of seawater, creating lakes. The freshwater lakes include supraglacial and proglacial lakes, while the saline lakes range from brackish to hypersaline (6× seawater) and include both permanently stratified and seasonally mixed lakes. The lakes closest to the ice sheet are typically fresh, while those closer to the coast tend to be saline or hypersaline. The stratified lakes can be divided into different geographical regions: Long Peninsula, Broad Peninsula, Ellis Fjord, and Mule Peninsula.

Ace Lake, lying in the northern portion of the Vestfold Hills on Long Peninsula, 10 km from the ice sheet and 150 m from the sea, went through a complicated evolution, beginning as freshwater, which was then inundated by seawater, dried down and refilled by glacial melt water. The bottom saline waters and the upper fresh waters do not mix and the lake is permanently stratified. The lake is oligotrophic with a depth of 25 m, an area of 0.16 km$^2$ and a 1.5–2 m thick ice cover. The water below 12 m is permanently anoxic and contains sulfate reducing and methanogenic bacteria. Extensive studies have been carried out on Ace Lake and have shown that there are significant interannual variations in biological activity related to the ice cover and local meteorology.

## Bunger Hills

**Location** The Bunger Hills is a rocky, ice-free area located in eastern Antarctica (66°17′ S) and is surrounded on all sides by glacial ice. This region is thus different from Vestfold Hills, which has the open ocean as one of its boundaries, but similar to Schirmacher Hills, which is also surrounded by glaciers. The total area is about 950 km$^2$, of which 420 km$^2$ is exposed rock. Most lake studies have occurred in the southern Bunger Hills, which has a maximum relief of about 160 m and is dissected by many steep valleys filled by lakes. Lakes also occur in the till-covered lowlands, but are generally smaller and more circular in shape. The climate of the Bunger Hills is similar to that of other rocky coastal areas of the East Antarctic coastline and most closely related to that of the Vestfold Hills.

**Formation and diversity** Over 200 water bodies occur in the Bunger Hills, ranging in size from small, shallow ponds that freeze to the bottom during winter, to Algae Lake, one of the largest and deepest surface freshwater lakes in Antarctica (14.3 km$^2$, 143 m deep). The conductivity of the lakes ranges from ultrafresh (<20 mg l$^{-1}$ total dissolved solids (TDS)) for those which receive melt water from the Antarctic plateau, to highly mineralized (>80 g l$^{-1}$ TDS).

Russian scientists working on these lakes characterized four types depending upon their hydrological and chemical characteristics (**Table 3**).

The lakes of the Bunger Hills differ from other coastal Antarctic lakes in the presence of a relatively high number of epishelf lakes. Owing to their position between land and a floating ice shelf or glacier, these lakes are tidal, containing a layer of freshwater overlying the water derived from the adjacent marine environment (**Figure 2**). The degree to which marine incursion occurs depends upon the actual hydraulic potential of the lake. Higher inflows produce hydrostatic pressure that minimizes the movement of marine water into the lake, whereas lower inflows allow marine water to enter the lake basin. Only about 10 known examples of this type of lake exist and are believed to have provided important refuges for aquatic organisms during glacial periods.

The biology of lakes in the Bunger Hills has received relatively little study to date. Research by Russian and Australian scientists revealed that the larger lakes have low water column primary production (0.08–326 mg carbon per cubic meter per day) and phytoplankton biomass (chlorophyll $a$ is typically $<2 \mu g l^{-1}$). The highest production occurs in lakes with the highest salinities. Despite low water column primary productivity, microbial mats exist in nearly all lakes of the Bunger Hills and consist of cyanobacteria, chlorophytes, diatoms, mosses, and heterotrophic grazers.

## The McMurdo Region

### Inland Lakes

**Location** The McMurdo Dry Valleys (MCM) (77°30′ S) of southern Victoria Land have a surface area of ~4000 km², representing the largest and most southerly ice-free area on the Antarctic continent (**Figure 3**). The MCM comprise three large valleys (Victoria, Wright, Taylor) along with many adjoining areas and consist of a mosaic of landscape features including glaciers, ephemeral streams, perennially ice-covered lakes, and exposed bedrock and soils. The region is ice-free because the Transantarctic Mountains block the flow of ice from the Polar Plateau and the warm katabatic winds flowing from the Polar Plateau to the sea through the east–west trending valleys lead to relatively high rates of ablation and associated ice loss. The largest lakes in this

**Table 3** Characteristics of the major lakes within the Bunger Hills based on watershed type, and the geochemistry and physical properties within the lakes

| Lake type | Lake character | Examples |
|---|---|---|
| Low water retention resulting from input from glacial melt | Low conductivity, typically isothermal | Algae Lake, Lake Dalekoje |
| Lakes of glacial origin with some through flow | Low to moderate conductivity, melt water dominated by land sources | Lake Dolgoe, Lake Dolinnoje |
| Lakes with a marine origin, often isolated from other hydraulic input | High conductivity, generally closed | Lake Polest, Lake Vostochnoye |
| Lakes with marine incursions (epishelf lakes) | Low salinity, tidal | Transkriptsii Gulf, Lake Pol'anskogo |

Modified from Klokov V, Kaup E, Zierath R, and Haendel D (1990) Lakes of the Bunger Hills (East Antarctica): Chemical and ecological properties. *Polish Polar Research* 11: 147–159.

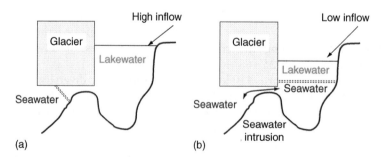

**Figure 2** Conceptual diagram of the mixing dynamics in an epishelf lake. (a) High inflow rates keep seawater from entering the lake basin; (b) Low inflow allows seawater to enter the lake basin producing a two layer system with freshwater overlying a marine layer. The double dashed line depicts the interface between seawater and freshwater. Modified from Gibson JAE and Anderson DT (2002) Physical structure of epishelf lakes of the southern Bunger Hills, East Antarctica. *Antarctic Science* 14(3): 253–261.

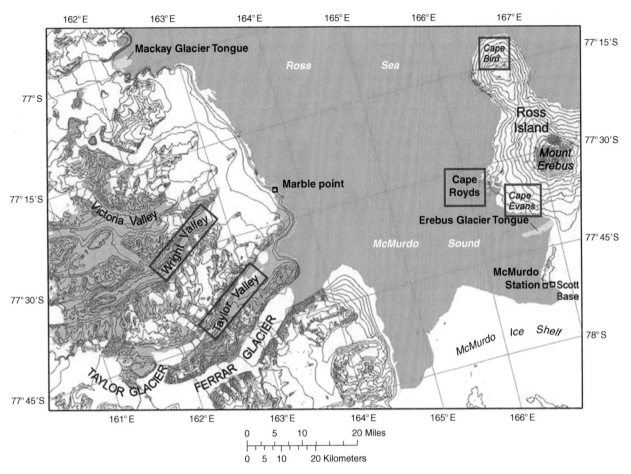

**Figure 3**  Locator map of the McMurdo Sound area showing the McMurdo Dry Valleys to the east (brown areas) along with the locations of the Wright and Taylor Valleys. The coastal lakes areas on Ross Island (Capes Byrd, Evans and Royds) are shown to the west. The red boxes denote regions containing lakes discussed in the manuscript.

area lie at the bottom of these valleys. The average annual temperature ($\sim-18\,^{\circ}$C) and precipitation (<5 cm water equivalents per year) make this ecosystem the coldest and driest of all lake regions in Antarctica. As a result, all lakes in the area are permanently ice-covered, except for a single pond (Don Juan Pond) where salinity is more than 18 times that of seawater.

**Formation and diversity**  The McMurdo Dry Valleys contain eight relatively large lakes (Vida, Vanda, Fryxell, Hoare, Bonney, Joyce, Miers, and Trough) all of which have permanent ice covers ranging in thickness from $\sim$4 to 7 m (Vida is an exception with a 18 m thick ice cover overlying saline – 5× seawater – water) and possess unique physical and geochemical attributes. Except for Lake Miers, all these lakes have closed basins with no surface outflow. Studies on these lakes began in 1957 as part of the IGY, and today, lakes in the Taylor Valley (78° S) form the centerpiece for the US National Science

Foundation's MCM Long-Term Ecological Research (LTER), which has collected an extensive set of data on three of the lakes (Fryxell, Hoare, and Bonney) and the surrounding ecosystem, since 1993.

The lakes in the present day MCM evolved as the result of changing climate conditions since the last glacial maximum. Data from the Taylor Valley have shown that over the past $\sim$20 000 years, lake levels within the MCM have varied considerably in relation to climate. Glacial Lake Washburn filled the entire Taylor Valley from the last glacial maximum to the early Holocene as a result of an ice dam formed at the base of the valley by the advancing West Antarctic Ice Sheet. As the climate became warm, the West Antarctic Ice Sheet retreated and Lake Washburn drained to McMurdo Sound, leaving behind smaller lakes in the lowest portions of the valley. Little is known about the limnology of these lakes but recent isotopic measurements show that many of the lakes in the MCM lost their ice covers and evaporated to small brine ponds or disappeared completely

~1200 years ago. A warmer climate since this period produced a flush of glacier melt that overflowed the brine ponds, producing the chemically stratified lakes we see today. The legacy of the ancient lake stands is now evident and relict resources left behind by these systems drive many of the biological processes within the MCM. For example, as the lakes rose, the soils inundated and became organic rich lake sediments, which became part of the terrestrial landscape as the lakes fell. The organic matter deposited during the period of inundation fuels much of the present day heterotrophic activity in the dry valley soils. These ancient, climate-driven lake-level changes also led to concentrated brine pools containing high levels of dissolved organic carbon, inorganic nitrogen, and inorganic phosphorus, which now form in the deep waters of many of the present lakes. The upward diffusion of these ancient nutrients has been shown to drive contemporary phytoplankton and bacterio-plankton productivity. Owing to low rates of annual primary production resulting from the long polar night and low light penetration through the thick permanent ice covers during the austral summer, annual primary production to respiration ratios in Lake Bonney (and presumably other lakes in the area) are less than unity. Hence, these ancient nutrient pools are essential to contemporary life in the lakes – without them, biology would cease.

The deep-water salts in the lakes and ponds of Wright Valley (e.g., Lake Vanda, Don Juan Pond) are comprised of $CaCl_2$ whereas NaCl dominates the brines of lakes in the other valleys. The large differences in salinity and ionic composition of the lakes (**Figure 4**) are related, in part, to how the lakes have responded to temperature changes through the Holocene. Specifically, the difference in brine composition among lakes is related to the eutectic properties of NaCl and $CaCl_2$. The permanent ice-covers, low advective stream inflow (stream flow is low and exists for 4–6 weeks each year), and strong vertical chemical gradients that result from relatively young freshwater overlying ancient brines suppress vertical mixing in these lakes to the level of molecular diffusion. As a consequence, they have not mixed completely for thousands of years. The deep saline waters also trap and store solar energy in the chemically stratified lakes, producing deep warm waters that exceed 20 °C in Lake Vanda.

Biological measurements on these lakes reveal a truncated food web with relatively few metazoans

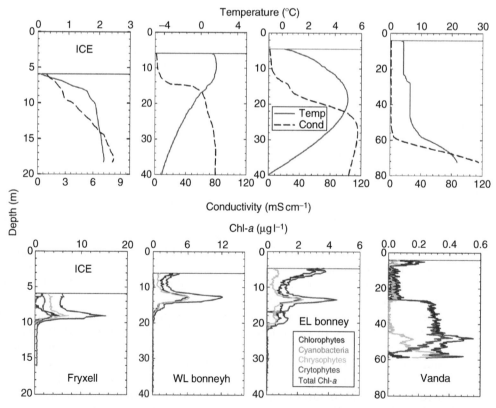

**Figure 4** Temperature, conductivity, and chlorophyll *a* profiles for Lakes Fryxell, Bonney (west and east lobes), and Vanda in the McMurdo Dry Valleys. The chlorophyll *a* profiles, obtained with a spectral fluorometer, depict values for four algal groups as well as total chlorophyll *a*.

(primarily rotifers); the lakes completely lack crustaceous zooplankton and fish. The vertical zonation of phytoplankton reflects the lack of vertical mixing and the presence of strong chemical gradients. **Figure 4** shows the strong vertical stratification of biomass and species composition through the water column, and the relationship between the deep-chlorophyll maxima and upward diffusion of ancient nutrient pools. A statistical comparison of the phytoplankton diversity among the lakes reveals that surface groups differ from the deep-living populations, and that the phytoplanktons in the lakes of the Taylor Valley differ from those in Lake Vanda in the Wright Valley. This pattern reflects the chemical evolution of these lake ecosystems.

Unlike the temperature regime on Signy Island, air temperatures in the MCM have come down at an average rate of 0.8 °C per decade over the past two decades. This cooling trend has led to the formation of thicker ice covers on the MCM lakes and decreased light penetration to the water column (0.055 mol photons per square meter per day annually in west lobe Lake Bonney). Because phytoplankton photosynthesis in the lakes is light limited, phototrophic primary production has decreased by 50% over the past 10 years in response to higher light attenuation by the thicker ice covers (**Figure 5**). The increasing trend in chlorophyll *a* after 2001 is the result of nutrient enrichment following an unusually warm year. Continued cooling in this region will clearly

produce a cascade of ecological changes within the MCM lake ecosystems as phototrophic primary production decreases even further.

### Coastal Ponds

**Location**  Coastal lakes and ponds are distributed around the margins of the Antarctic continent, and are particularly abundant around the ice free areas of McMurdo Sound. These shallow coastal aquatic systems typically freeze solid over the winter months and cryoconcentrate the organisms in the ice, forcing the majority of the organisms, gases, and dissolved organic matter to the bottom (**Figure 6**). Most of the lakes studied are on Ross Island, east of McMurdo Sound near capes Evans, Royds and Bird.

**Formation and diversity**  The coastal lakes and ponds occupy ice gouged areas in close proximity to McMurdo Sound (**Figure 3**). The main glacial groove in which the majority of the larger lakes at Cape Royds lie runs from Backdoor Bay along Blue Lake and branches off towards the coast with Clear, Coast, and Pony lakes. Two lakes at Cape Barnes, Sunk Lake, and Deep Lake, are found in a continuation of this same groove. These lakes range from freshwater to brackish with the ionic content enriched by precipitation, wind blown salt spray from the nearby sea, leaching from the volcanic rocks, and biologic inputs from penguins and sea birds. During the summer months, these ponds undergo varying degrees of melt depending upon snowpack and climatic conditions. The organisms that live in these coastal ponds experience extreme and often abrupt changes in light, temperature, water availability, salinity, and nutrients in contrast to those that live in the more stable inland meromictic lakes (**Table 4**).

Pony Lake is a shallow, eutrophic coastal pond located on Cape Royds. This lake is ice-covered except in mid-summer, when strong winds cause complete mixing of the water column. The lake is ~120 m long and 70 m wide, and 1–2 m deep. The source of water is snow, and water is lost by both ablation (from the snow and ice cover) and evaporation in mid-summer. Ice cover typically persists until late December. The pond is saline (~0.21 ppt), as a result of proximity to the sea and accumulation of salts by ablation. Phytoplankton are abundant (Chl *a* 28–140 µg l$^{-1}$) with the dominant alga being the chlorophyte *Chlamydomonas intermedia*. There is a penguin rookery along the eastern shore and the lake has high nutrient concentrations. DOC concentrations range from 10 mg carbon per liter during the early season to as high as 110 mg carbon per liter during the height of the algal blooms.

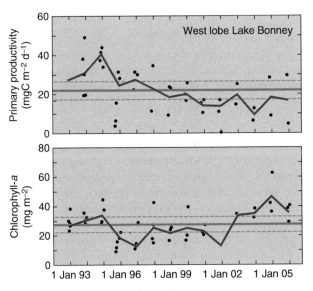

**Figure 5**  Long-term trends in depth integrated primary productivity and chlorophyll-*a* in the west lobe of Lake Bonney. Black circles represent data from all dates where measurements were made; the solid red line shows the long-term trend in average values for November and December. The solid and dashed green lines denote the mean and 95% confidence intervals around the November and December trend.

**Figure 6** Ice core being collected from Pony Lake, with inset on left showing the bottom portion of the core where organisms, gases and organic matter cryo-concentrate.

**Table 4** Comparison of ecological parameters in phytoplankton communities of Antarctic coastal ponds (i.e., Pony Lake, Blue Lake) and inland meromictic lakes (i.e., Lake Bonney, Lake Vanda)

| Parameter | Coastal lakes and ponds | Meromictic lakes |
| --- | --- | --- |
| Habitat stability | Low, experiences catastrophic changes | High, predictable |
| Growth season | Weeks | Months |
| Nutrient supply | High, C, N, P in excess | Low, growth limiting |
| Strategies | $r$-Selection (opportunistic, fast growing, rapid population shifts) | $K$-selection (in equilibrium highly efficient, slow population shifts) |
| Niche breadth | Wide, broad tolerances | Narrow, specialists |
| Species diversity | Low | High |
| Productivity | High | Low |

Adapted from Vincent WF and Vincent CL (1982) Response to nutrient enrichment by the plankton of Antarctic coastal lakes and the inshore Ross Sea. *Polar Biology* 1: 159–165.

## Conclusions

Despite temperatures well below the freezing point, the Antarctic continent possesses a wide variety of lakes (**Figure 7**) that contain liquid water throughout the year, making them an oasis for life in what would otherwise appear to be an uninhabitable environment. Interestingly, owing to their sensitivity to climate changes (small climate changes produce a magnified cascade of physical, chemical and biological changes), these lakes receive relatively little discussion in limnology textbooks. Antarctic lakes contain reservoirs of the evolutionary history of the continent and the contemporary activity we measure today often reflects the legacy of resources deposited in the environment during the evolution of the lake basins. Subglacial lakes, which have been isolated beneath the Antarctic ice sheet for >10 My, present an exciting new paradigm for limnologists. Recent calculations show that the subglacial lakes and rivers beneath the Antarctic ice sheet contain ~10 000 km³ of liquid water. This volume is ~10% of all lake water on Earth, enough to cover the whole continent with a uniform water layer with a thickness of ~1 m. Once sampled, information from these systems will change the way we view global water and biological reservoirs on our planet. Importantly, lakes in Antarctica provide clues on ecosystem properties and metabolic lifestyles that may exist on other frozen worlds such as Mars and Europa. As such, they provide us with a logical step in our search for extraterrestrial life.

**Figure 7**   Examples of different Antarctic lake types: (a) Lake Bonney an epiglacial lake, located in the McMurdo Dry Valleys is bound by both rock and glacier; the Taylor Glacier forms the lower boundary of the lake in the photo; (b) Radarsat image showing the flat area on the surface of the East Antarctic Ice sheet where the ice sheet 'floats' as it passes over the waters of Subglacial Lake Vostok; (c) Pony Lake is a shallow coastal lake at Cape Royds that freezes to the bottom in winter. Radarsat image is courtesy of NASA-GSFC. Shadows on the mountain side appear as dark blue areas in panel (a) and should not be confused with liquid water or dark mineral deposits.

*See also:* Effects of Climate Change on Lakes; Meromictic Lakes; Origins of Types of Lake Basins; Saline Inland Waters.

## Further Reading

Burgess JS, Spate AP, and Shevlin J (1994) The onset of deglaciation in the Larsemann Hills, Eastern Antarctica. *Antarctic Science* 6(4): 491–498.

Christner BC, Royston-Bishop G, Foreman CM, *et al.* (2006) Limnological Conditions in Subglacial Lake Vostok, Antarctica. *Limnology and Oceanography* 51: 2485–2501.

Doran PT, McKay CP, Clow GD, *et al.* (2002) Valley floor climate observations from the McMurdo dry valleys, Antarctica, 1986–2000. *Journal of Geophysical Research* 107(D24): 4772–4784.

Foreman CM, Wolf CF, and Priscu JC (2004) Impact of episodic warming events on the physical, chemical and biological relationships of lakes in the McMurdo Dry Valleys, Antarctica. *Aquatic Geochemistry* 10: 239–268.

Gibson JAE (1999) The meromictic lakes and stratified marine basins of the Vestfold Hills, East Antarctica. *Antarctic Science* 11(2): 175–192.

Gibson JAE, Wilmotte A, Taton A, Van de Vijver B, Beyens L, and Dartnall HJG (2006) Biogeographic trends in Antarctic lake communities. In: Bergstrom DM, Convey P, and Husikes AHL (eds.) *Trends in Antarctic Terrestrial and Limnetic Ecosystesms: Antarctica as a Global Indicator*, pp. 71–98. The Netherlands: Kluwer Academic Publishers.

Green WJ and Friedmann I (eds.) (1993) *Physical and Biogeochemical Processes in Antarctic Lakes*. Washington, DC: American Geophysical Union.

Heywood RB (1984) Inland waters. In: Laws RM (ed.) *Antarctic Ecology*, vol. 1, pp. 279–334. Academic Press.

Lyons WB, Laybourn-Parry J, Welch KA, and Priscu JC (2006) Antarctic lake systems and climate change. In: Bergstrom DM, Convey P, and Husikes AHL (eds.) *Trends in Antarctic Terrestrial and Limnetic Ecosystems: Antarctica as a Global Indicator.* The Netherlands: Kluwer Academic Publishers.

Lyons WB, Howard-Williams C, and Hawes I (eds.) (1997) *Ecosystem Processes in Antarctic Ice-Free Landscapes.* Rotterdam, Netherlands: A.A. Balkema Press.

McKnight DM, Andrews ED, Spaulding SA, and Aiken GR (1994) Aquatic fulvic acids in algal-rich Antarctic ponds. *Limnology and Oceanography* 39(8): 1972–1979.

Priscu JC (ed.) (1998) *Ecosystem Dynamics in a Polar Desert: The McMurdo Dry Valleys, Antarctica.* Antarctic Research Series. vol 72. Washington, DC: American Geophysical Union.

Priscu JC, Wolf CF, Takacs CD, *et al.* (1999) Carbon transformations in the water column of a perennially ice-covered Antarctic Lake. *Bioscience* 49: 997–1008.

Priscu JC, *et al.* (2008) Antarctic subglacial water: Origin, evolution and ecology Polar Lakes and Rivers – *Limnology of Arctic and Antarctic Aquatic Ecosystems.* Oxford University Press, pp. 119–136.

Vincent WF and Laybourn-Parry J (2008) *Polar Lakes and Rivers – Limnology of Arctic and Antarctic Aquatic Ecosystems.* Oxford University Press, pp. 119–136.

Wand U, Samarkin VA, Nitzsche H-M, and Hubberten H-W (2006) Biogeochemistry of methane in the permanently ice-covered Lake Untersee, Central Dronning Maud Land, East Antarctica. *Limnology and Oceanography* 51(2): 1180–1194.

## Relevant Websites

http://www.aad.gov.au – Australian Antarctic Division.
http://www.scar.org – Scientific Committee on Antarctic Research.
http://www.mcmlter.org – McMurdo Dry Valley LTER.
http://scarsale.tamu.edu – Subglacial Antarctic Lake Environments.
http://www.homepage.montana.edu/~lkbonney/ – Priscu Research Group.
http://www.Ideo.columbia.edu/~mstudinger/Vostok.html – M. Studinger's Vostok Webpage.

# Arctic

**G W Kling,** University of Michigan, Ann Arbor, MI, USA

## Introduction

Arctic lakes contain key resources for arctic communities in terms of food, water, transport, and industrial water for hydropower or mining. The variety of lakes in the Arctic is surprisingly large, and the ranges in lake type, water chemistry, and physical mixing characteristics rival those in other regions of the world. But one universal and distinguishing feature is the long period of ice cover in winter months. Another distinguishing feature is a long period of complete darkness in the winter, coupled with several months of 24-h light during the summer. A final distinction of arctic lakes that has gained importance recently is their responsiveness to climate change. These lakes of the far north will be our beacon signals of the changes expected in potentially all aquatic ecosystems in this century.

## The Arctic

The arctic region is defined geographically as lying above the Arctic Circle at 66° 33′N latitude. However, similar to the distinction between montane (trees) and alpine (treeless) zones, the more typical definition of the arctic region is that it is treeless, which roughly corresponds to the 10 °C July isotherm (**Figure 1**). A similar definition is that the arctic region is bounded by the southern limit of discontinuous permafrost (ground frozen for more than 2 years running). Whether the boundary is a temperature limit, or frozen ground, or the presence of trees, the broad ecotone between the Arctic and Boreal is called the subarctic region. Permafrost can extend to great depth (hundreds of meters), and underlies a surface active layer that thaws shallowly in summer to depths of only 30–100 cm. Because the permafrost acts as a seal to drainage and deep groundwater flow, this shallow active layer is often waterlogged when thawed, resulting in the Arctic being the world's largest wetland.

## Climate

The arctic climate has high spatial variability, extreme conditions of solar radiation, and is characterized by both polar maritime and continental climates. Maritime conditions prevail over the Arctic Ocean, coastal Alaska, Iceland, northern Norway and adjoining parts of Russia, and result in cold winters and summer average temperatures of ~10 °C. Continental climates of the interior have more severe winters with lower precipitation, and while frost may occur year-round the average summer temperatures are >10 °C. Winter weather is dominated by the semipermanent Icelandic and Aleutian Low and the Siberian High pressure systems. These weaken in summer and weather patterns are governed by the movement of cyclones (low pressure) across Siberia and into the Arctic Basin, or cyclones generated locally along the arctic or polar fronts. The importance of these patterns for lakes is that first the location or origin of the air masses affects their moisture content, and second the passage of fronts strongly impacts mixing dynamics in lakes and runoff into lakes from storm events in summer.

## History

As with human exploration of the Earth's poles in general, scientific expeditions to the Arctic have been relatively recent. In the 1800s there were casual observations of freezing and thawing limited in extent to lakes found along seafaring routes, such as the discovery of and records made for Lake Hazen by the Greely expedition to Ellesmere Island in 1882. Ancillary information on lakes from such expeditions continued through the early 1900s, but following World War II several permanent bases were established in the Arctic that increased the number of scientific observations on lakes. There was a big push to understand arctic ecosystems from the international community during the 1960s and especially in the 1970s with the establishment of the International Biological Programme (IBP). The IBP had two aquatic sites in the Arctic (Point Barrow, Alaska and Char Lake, Canada) and paved the way for several intensive research programs, notably the Long-Term Ecological Research program (LTER) based initially in the U.S. and now extending internationally. With the ramp-up of interest in climate change in the 1990s, and the early indicators of its potential global impact, the lack of earlier investigations of arctic lakes has been compensated by a flurry of recent studies and reports, including the joint international project of Arctic Climate Impact Assessment. This interest continues today, and for those interested there are several more in-depth reviews of arctic limnology by John Hobbie and Warwick Vincent, and of the past history of arctic lakes by John Smol and colleagues that provide an entry point to current topics and the primary literature.

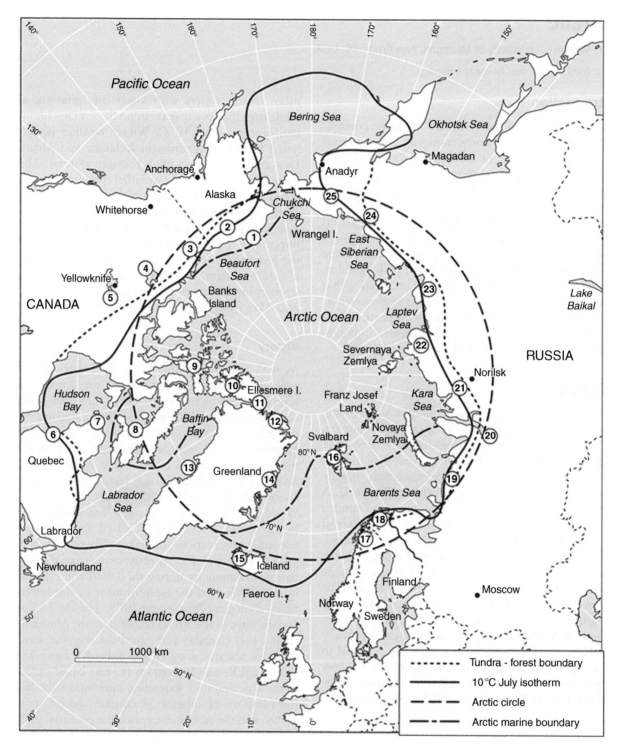

**Figure 1**    Map of the Arctic showing the Arctic Circle (long-dashed line), treeline (short-dashed line), and July 10 °C isotherm (solid line). Adapted from Vincent *et al.* (2008).

## Arctic Lake Ecosystems

Limnology is classically concerned with lake formation and geomorphology, and with lake physics, chemistry, and biology. Of the over 75 documented ways there are to form a lake, the Arctic contains examples of nearly all of those processes and certainly of lakes formed by the major categories of volcanic, glacial, and tectonic forces. But one unique aspect of arctic lakes is the role of thermokarst processes, which contribute to the most numerous lake forms on the planet. These shallow

**Figure 2**   The marked surface coverage of lakes shown on the coastal plain of Alaska near Barrow (a – top left, 71.3° N), in western Siberia near the Pur River (b – top right, 67° N), the Mackenzie River Delta (c – bottom left, 69°N), and ice-wedge polygons on Bylot Island (d – bottom right, 72.8°N). Photos a and b by G Kling; c by L Lesack; d by W Vincent.

ponds are formed inside ice-wedge polygons or from the melting of shallow ground ice to form thermokarst lakes. The numbers of these thaw lakes and ponds are staggering; ~45 000 have been mapped on the Mackenzie River delta and floodplain, and ~200 000 are found on the Yukon River delta alone. The Arctic also stands out as typically having the greatest water coverage of any terrestrial surface (**Figure 2**), which has implications for climate feedbacks and for the role of lakes in regional carbon budgets. An interesting feature of the thermokarst lakes is that they are typically oblong and oriented with their main axis perpendicular to the prevailing wind, apparently caused by surface currents that are strongest at the leeward corners of the lake.

## Lake Physics

The physics of arctic lakes are controlled by the prevalence of ice and the extremes of solar radiation. Some high-latitude arctic lakes are permanently covered with ice up to 4 m thick, and nearly all are ice-covered for at least 8–11 months of the year. Even during the ice-free summer months average temperatures in surface waters rarely exceed 18 °C and are more commonly less than 10 °C. The bottom waters in some deep lakes stay cold year-round and may never reach 4 °C, yet there exist lakes whose bottom waters are saline and whose temperatures are abnormally high owing to the trapping of solar heat penetrating through permanent ice covers.

### Solar Radiation

The Arctic also enjoys 24-h daylight through the summer months, and is blanketed by 24-h darkness for several winter months depending on latitude. Halfway between the Arctic Circle and the North Pole (78° N), the sun sets on October 27th and rises again on February 15th, resulting in quite low annual inputs of solar radiation. However, starting in mid-April the sun never sets for the next 4 months, and thus summer daily values of insolation received by lakes can be ~800 W m$^{-2}$ (~40 MJ day$^{-1}$), which is equivalent to what lakes receive in the temperate zone. The extinction of light by dissolved and particulate material in arctic lakes is similar to that found in temperate or tropical regions; extinction coefficients can range from quite low ($\sim 0.1\,\mathrm{m^{-1}}$) in some ice-covered lakes up to $2.5\,\mathrm{m^{-1}}$ in lakes clouded by glacial flour. Although recent increases in UV radiation are small in the Arctic compared with the large increases in Antarctica driven by the ozone hole, shallow arctic lakes may receive more UV in the future, especially if there are reductions in winter snow cover, which blocks UV much more effectively than does ice.

### Patterns of Lake Mixing

Energy inputs and the cover of ice control the stratification and mixing patterns in arctic lakes. Ice cover isolates the lake from the mixing energy of the wind, which in lakes that lack an ice-free period tends to result in permanent inverse stratification. Because freshwater is most dense at $\sim 4\,^{\circ}\mathrm{C}$ the temperature profile in these lakes is inverted, with the coldest and least dense water near $0\,^{\circ}\mathrm{C}$ at the surface just under the ice; temperature and density increase toward the bottom. Although inputs of solar radiation through the ice may warm surface water and create convection currents and mixing, these energy inputs tend to be weak and in general the vertical mixing rates of solutes under ice are very slow, often on the order of molecular diffusion. In winter the sediments of arctic lakes may release heat accumulated during the ice-free summer, warming the adjacent water toward $4\,^{\circ}\mathrm{C}$ and creating density currents that slowly move oxygen-reduced water toward the deep basins. A second category of lakes that are permanently stratified (meromictic – never fully mix top to bottom) also is represented in the Arctic, although as in other parts of the world they are rare. These lakes may or may not have permanent ice cover, but have salty, high-density water in the bottom layers, which requires too much energy to be lifted and mixed with surface waters. The result is a continuous separation and varied trajectories of evolution of water layers and their chemical and biological contents.

In contrast to these poorly mixed lakes, ice-free shallow lakes that mix continuously throughout (polymictic) are most common in the Arctic. In slightly deeper lakes ($>4$–$5\,\mathrm{m}$ maximum depth) a typical stratification pattern occurs where during the ice-free summer period a mixing and warmer upper layer (epilimnion) overlies a middle transition zone where temperature drops rapidly with depth (metalimnion), which is in turn underlain by a poorly mixed bottom layer (hypolimnion), where salts or particulate materials accumulate. When ice-covered, these lakes exhibit the inverse stratification described earlier. In the cold temperatures of the Arctic, even when lakes are ice-free the balance between stratification and mixing is adjusted by energy inputs.

In the coldest lakes during the ice-free period, if solar heating cannot raise temperatures to the density maximum of $4.0\,^{\circ}\mathrm{C}$, then the lakes will have no summer stratification.

Once lakes stratify, the isolation of water masses strongly impacts chemical and biological processes. Various mechanisms of mixing such as convection, wind stirring, density currents, or breaking of internal waves may increase the rate and extent of vertical mixing of oxygen, nutrients, and organisms. In addition, stream water inputs during storms, especially to smaller lakes, can strongly modify water column structure and mixing rates, again impacting chemistry and biology. Determining the relative importance of these various mixing processes and their impacts on lake ecosystem function is a current puzzle and theme of research in the physical limnology of arctic lakes.

## Lake Chemistry

Arctic lakes exhibit a full range of chemistries, from very dilute waters with electrical conductivities approaching rainwater to waters concentrated by evaporation to beyond the salinity of seawater. Within this range, however, the majority of lakes are relatively dilute (conductivity $<300\,\mu\mathrm{S\,cm^{-1}}$); this is due in large part to the underlying permafrost, which isolates surface waters and soils from weathering interactions with deeper mineral soils and rocks. The unfrozen zone beneath lakes (talik) may extend for many meters, but the impact of weathering in this zone on lake chemistry is almost entirely unknown. In the rest of the catchment, weathering reactions are confined to the very shallow unfrozen layers at the surface. In addition, weathering rates are slowed by cold temperatures and essentially stopped by lack of liquid water, and thus rivers that drain arctic tundra and feed lakes tend to be very dilute. Finally, the particular chemical compositions of lakes are dependent on several factors, including proximity to the ocean and the land surface age and geology for inorganic materials, and for organic materials compositions depend on the extent and type of terrestrial vegetation in the surrounding catchment.

### Major Ions

Tremendous numbers of coastal lakes and ponds in the Arctic are influenced by sea spray, particularly within $\sim 20$–$30\,\mathrm{km}$ of the coast. The ratios of major ions such as $Na^+$, $Cl^-$, and $K^+$ in these lakes are similar to ratios found in seawater, with the exception that $Ca^{2+}$ and $HCO_3^-$ tend to be enriched over seawater owing to inputs from weathering or most often from blowing soil (loess). Atmospheric inputs have little influence on the more inland lakes, however, which in all landscapes tend to be enriched in bicarbonate and cations (relative to $Cl^-$) compared with ratios in rain.

The second major factor influencing water chemistry is the time since last glaciation of the landscape, broadly characterized by younger soils and more recently exposed rocks with higher carbonate mineral content and pH, versus older, more weathered rocks and acidic soils. For example, Ca:Cl and Mg:Cl ratios are typically 3–5 times higher in lakes on younger than on older glacial surfaces. This influence is found circumpolar, and reflects the geographic asymmetry in Pleistocene glaciations, where much of the North American and European arctic had young land surfaces exposed by recession of the last continental or mountain-glacier ice sheets (<15 000 a BP), whereas much of Siberia to far northwestern Canada had older surfaces (>100 000 a BP) unglaciated during the late-Quaternary.

### Nutrients

The total amounts of major nutrients (N and P) in lakes are usually set by the surrounding geology, rates of weathering, and geochemical reactions, but the observed concentrations of inorganic nutrients dissolved in lakes tend to be strongly controlled by biological processes. Arctic systems follow this generality as well, but what is special in these lakes is that the supply rates, geochemistry and biological controls are often at the extremes of typical behavior. Because of the isolation of mineral soils by permafrost there are low amounts of P weathered from rocks, and because of low temperatures and shallow soil depths the generation of N and P through decomposition of organic matter on land is also low. Add to this the fact that tundra plants are critically limited by both N and P, and the result is a low release rate of nutrients from land to surface waters. Consider that an average boreal forest may retain and recycle 60–70% of the N needed to support annual primary production, while in low-arctic tundra this retention is likely much greater than 90%. Within at least some arctic lakes, a further governor is the geochemistry of sediments exceptionally rich in iron and manganese, which operates to bind P and both scavenge it from the overlying water and prevent its return. The net consequences of these extremes are that inorganic nutrient concentrations are extremely low in arctic (and alpine tundra) lakes, and one implication is that arctic lake primary production is more often co-limited by both N and P than is found in other regions.

### Land–Water Interactions

Lakes everywhere are intimately tied to inputs of materials and nutrients from land, but this terrestrial connection is particularly strong in the Arctic for organic matter in dissolved and particulate forms. In terms of C (carbon), N, and P, the dissolved organic matter (DOM) concentrations typically represent 70–90% of the total C, N, and P in lake water or entering the lakes. This terrestrial connection is so strong mainly because the primary production of terrestrial environments is much greater than that found in aquatic habitats; as an example, Toolik Lake in northern Alaska is a typical deep-water lake (68.38° N, 1.5 km$^2$ surface area, 25 m maximum depth) and has an annual primary productivity of ~10 g C m$^{-2}$, whereas in the surrounding tundra the productivity is not only 10 times that amount but the storage of organic C in soils is 1000 times that amount (~10 kg C m$^{-2}$ in the upper thawed layer alone). There is, however, a distinct gradient of terrestrial influence on lakes in the Arctic, because as one travels further north or to higher elevation the climate is colder, the terrestrial vegetation becomes scarce, the organic matter in soils is much reduced, and there is less precipitation in the form of rain to flush DOM from soils into surface waters (**Figure 3**). The endpoint of this gradient is found in antarctic lakes, and consists of a full reversal of the pattern – there, in some areas the organic matter produced in lakes is blown onto land and may constitute the dominant C input to soils.

The strong terrestrial–aquatic linkage affecting arctic lakes is also illustrated by the case of landscape-level carbon budgets. Lakes with ice-free periods in the Arctic, as in most other areas of the world, tend to release more $CO_2$ to the atmosphere than they take up in photosynthesis on an annual basis (net C sources). This supersaturation with respect to atmospheric $CO_2$ is driven by the export from land of $CO_2$ and of organic carbon that is respired to $CO_2$ and methane in the lakes. What is different about this landscape integration of terrestrial and aquatic ecosystems in the far north is that the surface area of water is so great that the loss of C back to the atmosphere is a significant term in the carbon balance of the entire Arctic (~20% of the net C exchange with the atmosphere; **Figure 4**).

### Seasonal Variability

The forcing functions of seasonal variations in lake chemistry fall into two categories. First is the concentration of salts, gases, and organic matter by evaporation in the summer or ice formation in the

**Figure 3** High-arctic Lake Hazen (*left*, 81.8° N) and low-arctic Toolik Lake (*right*, 68.38° N) showing the comparison of terrestrial vegetation in the catchment, leading to reduced inputs of organic matter from land into the high-arctic versus the low-arctic lake. (Photos: left – W Vincent; right – G Kling).

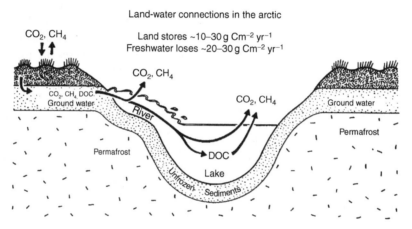

**Figure 4** Regional and circumpolar carbon budgets are influenced strongly by lakes acting as conduits of $CO_2$ and $CH_4$ to the atmosphere. On a per area basis, the annual evasion of gases to the atmosphere roughly balances the net long-term C storage on land. Shallow, organic rich soils lying above permafrost allow the products of plant and soil respiration to accumulate, which are then moved into streams and lakes and returned to the atmosphere or exported to the sea.

winter; although these processes occur worldwide where lake ice forms, the generally dry conditions and development of thick ice in the Arctic accentuate the magnitude of effects. As water freezes the crystals exclude salts and dissolved organic materials, and the chemical concentrations increase in the unfrozen water just below the ice surface. This cryo-concentration can affect large volumes of surface water even in deep lakes, but the most dramatic impacts are in the many shallow lakes that freeze completely or nearly to the bottom. Once the ice melts, evaporative concentration in these shallow ponds may also alter water chemistry progressively through the ice-free season. In summer, flat coastal plain areas can have high evaporation and especially transpiration of water in wetlands surrounding thermokarst ponds, which can actually reverse the hydrological gradient

so that water flows out of the ponds and onto land. Although this transfer does not change water chemistry directly, it can impact the chemical loading and budgets of lakes.

The second category of processes that change lake chemistry seasonally is the external input from snowmelt and summer storms. Little happens chemically during winter once the maximum ice thickness is established, but warming temperatures in spring cause snowmelt and result in the largest, periodic mass flow in the hydrological cycle. This spring snowmelt and associated runoff can completely flush smaller lakes with relatively dilute but often nutrient-rich water. Because soils are still mostly frozen at the time of snowmelt, runoff water is short-circuited from the normal terrestrial flow path where it is exposed to soil microbes and plant roots; thus it can be rich in nutrients or labile organics leached from the surface mat of vegetation. Therefore even in larger-volume lakes the runoff water can greatly stimulate bacterial activity, and can provide the majority of inorganic nutrients used by algae in the lake throughout the summer growing season. Snowmelt typically occurs when the deeper lakes are still ice-covered, and because the inflow water is close to ~1 °C the runoff enters the lake and floats just below the lower ice surface in close proximity to surface algae and bacteria, increasing the impact of this seasonal change in water chemistry on other aspects of ecosystem function.

Large hydrological excursions during summer storm events can similarly alter the chemistry of lakes, especially through inputs of N, P, and DOM. These alterations can persist for long periods of time (many weeks) depending on the volume and character of the inflow. One important distinction from snowmelt is that summer stormwaters have variable temperature and density and can produce overflows (as occur under ice), but more often this water forms an underflow and intrudes at depth in the lake. These intrusions may remain in thin layers but can also thicken and mix nutrients into both underlying and overlying waters. The specific placement of the intrusions relative to water column structure, organism distribution, and the availability of light for photosynthesis or photochemical reactions can determine the ultimate impact of the inflow on lake chemistry and biology. In fact, determining the dynamics of these events and their impacts on arctic lakes is a critical area of current and future research.

## Lake Biology

### Species Diversity

For the most part lakes in the Arctic (1) are relatively depauperate with respect to biodiversity compared to temperate and tropical zones, (2) are populated mainly by cold adapted or cold tolerant species, and (3) have low productivity and simplified food webs. This relative impoverishment increases moving further north or to higher elevation, and it increases moving from lower to higher trophic levels in the food web (**Table 1**). That is, the ratio of arctic diversity to temperate or tropical diversity is much lower for fish than for algae (early indications suggest that this pattern holds for bacterial diversity as well). Additionally, the species-area curves in arctic regions reach an asymptote very quickly compared to regions further south. To illustrate these points consider the well-studied lakes near the NSF-LTER site of Toolik Lake on the North Slope of Alaska (low arctic, 68.38° N). There are ~150 species of phytoplankton in Toolik and surrounding lakes, and perhaps 50–100% more could be found in a similar number of lakes in one area of the temperate zone. Strikingly, most of the phytoplankton species are extremely small (nanoplankton <20 μm in size), with very few larger species that dominate diversity in lakes of regions further south. For the zooplankton, surveys of 100s of lakes near Toolik and extending northward to the Arctic Ocean (70.3° N) found a total of 13 macrozooplankton and ~15 rotifer species, whereas 40 and 60 species, respectively, were found in just 39 similar lakes in tropical Cameroon. The greatest contrast occurs at the top of the food chain where only 6 fish species exist in and around Toolik (and perhaps only 10 genera and 15 freshwater species in lakes throughout the Arctic), and yet the U.S. state of Michigan alone has 157 species of fish. Of course there is some variability in these patterns (e.g., 33 rotifer species occur in only 5 small ponds on Ellesmere Island, 78.37° N), and few detailed taxonomic surveys of arctic lakes exist, especially for phytoplankton. Overall the generalities of patterns in diversity are quite strong, and are consistent with the explanation that while autotrophs or decomposers may generate sufficient energy to fill many available niches in lakes, the minimal energy available to subsequent trophic levels severely truncates the species richness.

### Productivity

In line with the low species diversity, the productivity of arctic lakes is generally classed as oligotrophic (unproductive) to ultra-oligotrophic. In unpolluted lakes the biomass of algae measured as chlorophyll $a$ is often much less than $1\,mg\,m^{-3}$ in the high arctic, and may increase to $3$–$5\,mg\,m^{-3}$ in low arctic lakes just before or after ice-out. The rates of primary production in the water column can be variable but are equally low, typically from $1$–$200\,mg\;C\,m^{-3}$ $day^{-1}$, and integrated annual production may be $\sim10\,g\,C\,m^{-2}$ in the productive lakes but is usually much less. On an annual basis this low productivity is a result of insufficient light and a short growing season. When sunlight is ample during summer, the controls on primary productivity shift to the scarcity of nutrients and the rapidness of light extinction with depth in the lake. Because of the large terrestrial contributions of DOM to arctic lakes, bacterial production at least in low-arctic lakes is roughly comparable to other lakes worldwide.

In shallow arctic lakes, algae and cyanobacteria exist in these barren conditions by forming benthic mats that offer both stability from frequent mixing and provide proximity to nutrient sources in the sediments. The resulting rates of productivity can be

**Table 1** Selected studies reporting the diversity of phytoplankton, benthic algae, zooplankton, benthos, and aquatic macrophytes in arctic lakes and ponds

| Source | Site | Species, Genera (N) | Lakes (N) | Notes |
|---|---|---|---|---|
| | **Phytoplankton; Algae** | | | |
| 9 | Linnevatnet, Spitsbergen | 150 | – | E Willen (unpublished), cited in 1 |
| 9 | Peters Lake area, Alaska | 221 | – | Kinney et al. (1972), cited in 1 |
| 9 | Colville River drainage | 172 | 10 | Kinney et al. (1972), cited in 1 |
| 13 | Toolik Lake, Alaska | 142 | 1 | H Kling, cited in 2 |
| 1 | Barrow, Alaska | 105 | 5+ | Tundra ponds |
| 1 | Barrow, Alaska | 45+ | 1 | Epipelic algae, tundra pond |
| 17 | Char Lake, Canada | 155 | 1 | H Kling, cited in 16 |
| 3 | Southwest Greenland | 13 | 2 | Genera |
| 5 | Lake El'gygytgyn, Siberia | 2 | 1 | Planktonic diatoms only |
| 5 | Lake El'gygytgyn, Siberia | 113 | 1 | Sediments, benthic diatoms |
| 20 | Bylot Island, NWT, Canada | ~37 | 9 | Mixture of planktonic and benthic cyanobacteria and diatoms |
| 7 | Lac du Cratere, Canada | 101 | 1 | Sediments, benthic diatoms |
| 11 | Banks Island, Canada | 109 | 36 | Sediments, benthic diatoms |
| 6 | Ellesmere Island | 13 | – | Cyanobacteria |
| 2 | Ward Lake, Canada | 11+ | 1 | Benthic mat cyanobacteria, diatoms, etc. |
| | **Zooplankton** | | | |
| 10 | North Slope of Alaska | 12 | 45 | Macrozooplankton |
| 12 | Ellesmere Island, NWT | 33 | 1 | Rotifers only |
| 14 | North Slope of Alaska | 8 | 105 | Macrozooplankton |
| 16 | Cornwallis Island, NWT | ~7 | 5 | Macrozooplankton |
| 4 | Circum-arctic | 165 | 212 | Rotifers; arctic and subarctic |
| 4 | Bathurst Island, NWT | 21 | 38 | Rotifers only |
| 4 | North Slope, Alaska | 87 | 38 | Rotifers only |
| 4 | N. Yukon and NWT, Canada | 81 | 22 | Rotifers only |
| 18 | Toolik Lake, Alaska | 9 | 1 | Rotifers only |
| | **Benthos; Macrophytes** | | | |
| 10 | North Slope of Alaska | 32 | 45 | Chironomids |
| 15 | Lake 18, NWT, Canada | 7+ | 1 | Macrophytes and mosses |
| 8 | Barrow Ponds, Alaska | 2 | 5+ | Macrophytes |
| 19 | Faroe Islands | 24 | 6 | Aquatic macrophytes |
| 1 | Barrow, Alaska | 3 | 5 | Rooted aquatic macrophytes |

All diversity numbers are for species unless noted. A '+' indicates the number (of species or lakes sampled) is a minimum estimate.
Sources

1. Alexander V, et al. (1980) Primary producers. In: Hobbie JE (ed.) *Limnology of Tundra Ponds: Barrow, Alaska*, pp. 179–250, 514 pp. Stroudsberg PA: Dowden, Hutchinson, & Ross.
2. Bonilla S, et al. (2005) Benthic and planktonic algal communities in a high arctic lake: Pigment structure and contrasting responses to nutrient enrichment. *Journal of Phycology* 41: 1120–1130.
3. Brutemark A, et al. (2005) An experimental investigation of phytoplankton nutrient limitation in two contrasting low arctic lakes. *Polar Biology* 29: 487–494.
4. Chengalath R and Koste W (1989) Composition and distributional patterns in arctic rotifers. *Hydrobiologia* 186/187: 191–200.
5. Cremer H and Wagner B (2003) The diatom flora in the ultra-oligotrophic Lake El'gygytgyn, Chukotka. *Polar Biology* 26: 105–114.
6. Croasdale H (1973) Freshwater algae of Ellesmere Island, N.W.T. *National Museums of Canada Publications in Botany* 3: 1–131.
7. Grönlund T, et al. (1990) Diatoms and arcellaceans from Lac du Cratère du Nouveau-Québec, Ungava, Québec, Canada. *Canadian Journal of Botany* 68: 1187–1200.
8. Hobbie JE (1980) *Limnology of Tundra Ponds: Barrow, Alaska*, 514 pp. Stroudsburg, PA: Dowden, Hutchinson and Ross.
9. Hobbie JE (1984) Polar limnology. In: Taub F (ed.) *Lakes and Reservoirs*, pp. 63–105. Amsterdam: Elsevier.
10. Kling GW et al. (1992) The biogeochemistry and zoogeography of lakes and rivers in arctic Alaska. *Hydrobiologia* 240: 1–14.
11. Lim DSS, Smol JP, and Douglas MSV (2007) Diatom assemblages and their relationships to lakewater nitrogen levels and other limnological variables from 36 lakes and ponds on Banks Island, N.W.T., Canadian Arctic. *Hydrobiologia* 586: 191–211.
12. Nogrady T and Smol JP (1989) Rotifers from five arctic ponds (Cape Herschel, Ellesmere Island, N.W.T.). *Hydrobiologia* 173: 231–242.
13. O'Brien WJ, et al. (1997) The limnology of Toolik Lake. In: Milner AM and Oswood MW (eds.) *Freshwaters of Alaska: Ecological Syntheses*, pp. 61–106. New York: Springer.
14. O'Brien WJ, et al. (2004) Physical, chemical, and biotic effects on arctic zooplankton communities and diversity. *Limnology and Oceanography* 49 (4, part 2): 1250–1261.
15. Ramlal PS, et al. (1994) The organic carbon budget of a shallow Arctic tundra lake on the Tuktoyaktuk Peninsula, N.W.T., Canada. *Biogeochemistry* 24: 145–172.
16. Rautio M, Vincent WF (2006) Benthic and pelagic food resources for zooplankton in shallow high-latitude lakes and ponds. *Freshwater Biology* 51: 1038–1052.
17. Rigler FH (1974) *Char Lake Project PF-2. Final Rep. 1974*, 96 pp. Toronto: Canadian Commitee for International Biological Programme.
18. Rublee PA (1992) Community structure and bottom-up regulation of heterotrophic microplankton in arctic LTER lakes. *Hydrobiologia* 240: 133–141.
19. Schierup H-H, et al. (2002) Aquatic Macrophytes in Six Faroese Lakes. *Annales Societas Færoensis Supplement* 36: 47–58.
20. Vezina and Vincent W (1997) Arctic cyanobacteria and limnological properties of their environment: Bylot Island, Northwest Territories, Canada (73° N, 80° W). *Polar Biology* 17: 523–534.

10–15 times higher per area than in the shallow water column, and thus account for the majority of newly produced organic matter. Mosses and emergent macrophytes such as *Arctophila* and *Hippurus* can also dominate the shallows ($\sim$<1.5 m depending on light extinction) of any lake.

Arctic phytoplankton species are often motile (cryptophytes, dinoflagellates, crysophytes) and in the deeper parts of lakes they can separate into distinct layers. Similar thin layers of bacteria have been studied in permanently or strongly stratified lakes, but there is little or no research on the nature and productivity of these thin layers of phytoplankton in lakes with weak or variable stratification. How these thin layers form, rely on, and interact with the physics and chemistry of horizontal intrusions of nutrient-rich water from stream inputs, or of horizontal jets formed by breaking internal waves, is a fascinating question and an area of research relevant not only to the Arctic but to limnology and oceanography broadly.

### Food Webs

The food web structure of arctic lakes is simplified and contains a limited number of species overall and often a truncated number of trophic levels. For example, in many lakes the top predators are zooplankton or even insects, such as the larvae of dytiscid beetles. Even when fish are present it is rare to have the classic (yet simply) structure of piscivorous top predators feeding only on forage fish which feed only on zooplankton. Typically fish feed on a variety of prey, and their presence can strongly impact the structure of the food web. If a lake is shallow enough to freeze to the bottom, or if the outlet of the lake is too steep to be negotiated by fish, the fish-less lake food web will be distinguished by very abundant and large-bodied zooplankton. Lakes with fish tend to have a small-bodied zooplankton community or sometimes a mixture of the two communities in larger lakes containing greater habitat diversity.

The energy flow in arctic lake food webs is most often driven by benthic production. Starting as early as the 1950s it was noticed that zooplankton existed in unusual abundances where the pelagic (open water) primary production was insufficient to support such high biomass. In turn, the level of fish biomass in arctic lakes requires energy subsidies from benthic organisms such as mollusks and chironomids in addition to open-water zooplankton food. Despite the well-known importance of benthic organisms in arctic lakes, only $\sim$2–3 complete energy budgets of arctic lakes exist, meaning that to date we know very little about the energy flows and ecosystem dynamics in lakes of the Arctic.

## The Future of Arctic Lakes

It is now commonly understood from paleo records that much of the Arctic began to warm in the mid-19th century, that the warming will likely intensify in the future, and that arctic ecosystems are extremely sensitive to such warming and its associated ecological effects. This sensitivity is due mainly to the fact that the Arctic lies at the cusp of change from solid to liquid water. The indirect, physical consequences of even slight increases in temperature that will melt lake ice, thaw permafrost, and increase evaporation are certainly grander in their effects on organisms and system function than is the temperature change alone.

The first threshold we are crossing is the melting of ice cover, which immediately alters the mixing patterns

**Figure 5** Recently dried lakes on the North Slope of Alaska (*left*, shallow thermokarst pond $\sim$69° N; *right*, small lake in glacial till near Toolik Lake, 68.4° N). Photos by G Kling.

of lakes, increases the growing season, and greatly enhances the UV exposure of aquatic organisms; these changes have led to increased productivity and shifts in the community structure of many lakes, especially in the high arctic. If water temperatures warm above 4 °C in summer, lakes that now mix continuously in the ice-free season may start to stratify. One biological consequence is that under warmer temperatures lakes that now stratify may have increased oxygen depletion in bottom waters, and if coupled with warmer surface waters this could physiologically stress fish and shrink their effective habitat.

A second threshold includes the thawing of permafrost and the net balance of precipitation and evaporation that allows lakes and ponds to persist. There is evidence of both pond desiccation due to increasing temperatures and evaporation, and of the drainage of lakes assisted by melting permafrost and altered hydrological patterns (**Figure 5**). Thermokarst ponds have a relatively short life span, and are constantly draining and reforming on time scales of hundreds of years mainly as a result of altered drainage or river capture. Permafrost thawing may initially increase thermokarst activity and pond formation, but in the long term there appears to be a distinct shift in the balance of these processes that has allowed thousands of lakes to shrink or disappear completely just in the last 30–40 years.

The melting of massive ground ice will have other potentially marked impacts on aquatic ecosystems,

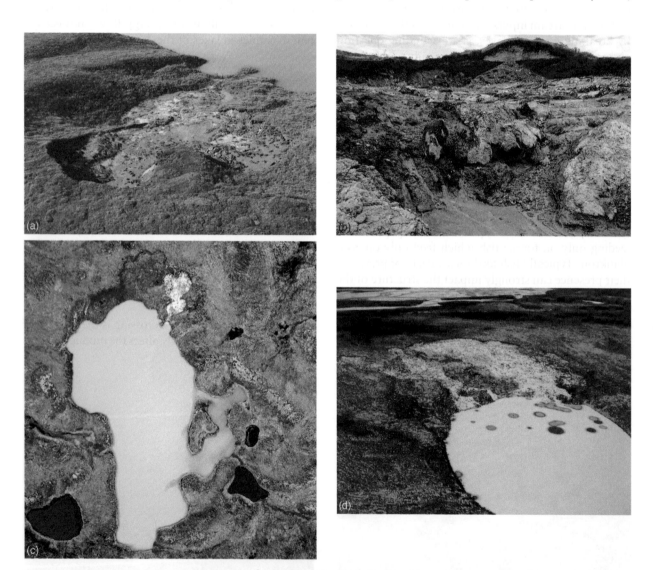

**Figure 6**   *Left:* Aerial views of Lake NE-14 (0.24 km² area, 68.41° N) near Toolik Lake, Alaska with a glacial thaw slump on shore adding sediment to the lake; prior to the slump in 2006 the lake color was the same as the lake in the lower left of the photograph. Note the visual evidence for previous slumping scars to the left of the current scar (lower left picture). *Top right:* Ground view of massive land failure associated with the slump. *Bottom right:* Aerial view of a thermokarst slump on the shore of a lake near the Sagavanirktok River, Alaska (~69° N). Photos by B Bowden (a – top left and d – lower right), A Balser (c – bottom left), G Kling (b – top right).

such as land failure and catastrophic slumping of soils into lakes (**Figure 6**). Even without such slumps, the implications of permafrost thawing and deepening of the surface active layer include: (1) chemical changes such as increased carbonate minerals and P released from weathering of previously frozen mineral soils; (2) biological changes in vegetation, demonstrated by increased shrubs in low-arctic tundra, which can alter the amount and character of DOM exported from land to lakes; and (3) warming of lake sediments and soils, some of which are extremely C rich (especially the Yedoma soils in Siberia) and could substantially increase the source of both $CO_2$ and methane to the global atmosphere. This C has been locked up in permafrost and not participating in the global C cycle, but the tremendous stores of C in wetlands and sediments of the Arctic (estimated as up to one-third the mass of C in the entire atmosphere) may provide strong positive feedbacks to climate warming in the future. Finally, there are indications that fire frequency and magnitude on the arctic tundra has recently increased (**Figure 7**), and although the implications for both tundra C budgets and lake ecosystems could be considerable, they are currently unknown.

While the island continent of Antarctic suffers from extreme biogeographical isolation, the Arctic faces more serious threats from both species invasions from the south and a truncated northern border of ocean. Unlike current boreal species with the opportunity to survive climate warming by shifting distribution patterns northward, the arctic species have nowhere to go. This presents an interesting aspect of research, especially for the future and in terms of arctic biodiversity and its controls, and determining which species have enough genetic flexibility to adapt to the rapidly changing (with respect to evolutionary time scales) environment.

## Nomenclature

| | |
|---|---|
| g | gram |
| mg | milligram |
| m | meter |
| km | kilometer |
| $km^2$ | square kilometer |
| μm | micrometer |
| day | day |
| a | annum (year) |
| BP | before present |
| a BP | years before present |
| °C | degrees Celsius |
| °N | degrees North latitude |
| J | Joules |
| MJ | megajoules |
| Ca | calcium |
| Mg | magnesium |
| Na | sodium |
| Cl | chloride |
| $HCO_3^-$ | bicarbonate |
| $CO_2$ | carbon dioxide |
| C | carbon |
| N | nitrogen |
| P | phosphorus |
| DOM | dissolved organic matter |

**Figure 7** Aerial views of two tundra fires on the North Slope of Alaska. Top (a) shows a small fire in 2004 near the Sagavanirktok River (photo by R Flanders), while bottom (b) shows a gigantic fire (>1000 $km^2$) near the Anaktuvuk River that burned in August and September 2007 (photo R Jandt, BLM).

*See also:* Abundance and Size Distribution of Lakes, Ponds and Impoundments; Antarctica; Effects of Climate Change on Lakes; Geomorphology of Lake Basins; Meromictic Lakes; Origins of Types of Lake Basins; Saline Inland Waters.

## Further Reading

Arctic Climate Impact Assessment (2005) *ACIA Scientific Report*, 140 pp. Cambridge, UK: Cambridge University Press. ISBN-13: 9780521617789/ISBN-10: 0521617782.

Crump BC, Adams HE, Hobbie JE, and Kling GW (2007) Biogeography of bacterioplankton in lakes and streams of an Arctic tundra catchment. *Ecology* 88: 1365–1378.

Hobbie JE (ed.) (1980) *Limnology of Tundra Ponds, Barrow, Alaska*, US/IBP Synthesis Series, vol. 13, 514 pp. Stroudsburg, USA: Dowden, Hutchinson and Ross. Available at. http://www.biodiversitylibrary.org/ia/arcticecosystemc00brow.

Kling GW, Kipphut GW, and Miller MC (1991) Arctic lakes and rivers as gas conduits to the atmosphere: Implications for tundra carbon budgets. *Science* 251: 298–301.

MacIntyre SM, Sickman JO, Goldthwait SA, and Kling GW (2006) Physical pathways of nutrient supply in a small, ultra-oligotrophic arctic lake during summer stratification. *Limnology and Oceanography* 51: 1107–1124.

O'Brien WJ (ed.) (1992) Toolik Lake: Ecology of an aquatic ecosystem in arctic Alaska, *Hydrobiologia* 1–269.

Pienitz R, Douglas MSV, and Smol JP (2004) *Long-Term Environmental Change in Arctic and Antarctic Lakes*, 562 pp. Dordrecht: Springer.

Schindler DW and Smol JP (2006) Cumulative effects of climate warming and other human activities on freshwaters of arctic and subarctic North America. *Ambio* 35: 160–168.

Smith LC, Sheng Y, MacDonald GM, and Hinzman LD (2005) Disappearing arctic lakes. *Science* 308: 1429.

Smol JP, Wolfe AP, Birks HJB, Douglas MSV, *et al.* (2005) Climate-driven regime shifts in the biological communities of Arctic lakes.

*Proceedings of the National Academy of Sciences USA* 102: 4397–4402.

Vincent WF and Hobbie JE (2000) Ecology of Arctic lakes and rivers. In: Nuttall M and Callaghan TV (eds.) *The Arctic: Environment, People, Policies*, pp. 197–231. UK: Harwood Academic Publishers.

Vincent WF and Laybourn-Parry J (eds.) (2008) *Polar Lakes and Rivers – Limnology of Arctic and Antarctic Aquatic Ecosystems.* Oxford, UK: Oxford University Press.

## Relevant Websites

http://www.arctic-council.org/ – The international Arctic Council.

http://www.amap.no/ – The international Arctic Monitoring and Assessment Program.

http://www.ipy.org/ – Website for the International Polar Year (2007–2008).

http://www.arcus.org/ARCUS/links/index.html – Very complete compendium of links to Arctic Research and Information.

http://www.spri.cam.ac.uk/resources/organisations/keyindex.html – Directory of Polar and Cold Regions Organizations (Scott Polar Research Institute).

http://www.arcus.org/ – The Arctic Research Consortium of the U.S.

http://www.arctic.gov/ – The U.S. Arctic Research Commission.

http://www.arctic.noaa.gov/ – Arctic news of the National Oceanic and Atmospheric Administration.

http://www.nsf.gov/div/index.jsp?div=ARC – Arctic science at the U.S. National Science Foundation.

# Africa: North of Sahara

**M Ramdani,** Mohammed V University, Rabat, Morocco
**N Elkhiati,** Hassan II University, Casablanca, Morocco
**R J Flower,** University College, London, UK

## Introduction

The Mediterranean climate of North Africa is characterized by hot dry summers and seasonally restricted rainfall. The western region experiences a subhumid Mediterranean climate with mild, moist winters from October through March/April, and hot, dry summers from May through September. Nevertheless, the climate is quite variable from year to year, and the North Atlantic oscillation has a large interannual influence on winter rainfall patterns, particularly in Morocco. The Saharan mountain chains running northeast also have a major influence on rainfall, and in Morocco the Atlas Mountains can receive an annual rain fall of over 200 cm. The Atlantic influence on rainfall diminishes eastwards, although mountains in northwest Tunisia can receive over 150 cm of the annual rainfall. Further to the east, across the northern parts of Libya and Egypt, rainfall <10 cm is received annually. The temporal and spatial distribution of rainfall in North Africa has important consequences for inland waters. Irregular rainfall from year to year and high evapotranspiration rates combined with an essentially unglaciated geology favor a diversity of standing water types, with a bias toward shallow ephemeral water bodies in lowland regions. Run-off is collected by a variety of depressions that range from small transitory ponds (often salinized in summer months) to permanent natural and predominantly freshwater lakes, which are sometimes of large dimension. Of the larger natural freshwater lakes, exceeding 1 km$^2$, most are restricted to Morocco and Algeria and the Nile delta region of Egypt, although substantial coastal lagoons are common in all the countries bordering the southern Mediterranean. There are several common local terms used to categorize North African lakes (including daya and chott) and use of these has become current practice by limnologists and hydrologists, beyond the southern Mediterranean region, especially in Mediterranean Europe.

## Formation, Diversity, and Distribution of Inland Waters North of the Sahara

North of the Sahara is defined as the southern Mediterranean region (SMR), an area surrounded by Atlantic Ocean in the west and Mediterranean Sea to the north. The SMR is located at the crossroads of three continents – Africa, Asia, and Europe – and displays much diversity in terrain, climate, and biodiversity. Both North Africa and the Middle East have considerable inland water resources with specialized aquatic flora and fauna. Internationally, many inland and costal lakes have particular value for bird life. Following the typical hot dry summer period, winter rains are an important feature of the climate in the Maghrebian countries (Morocco, Algeria, and Tunisia). In early spring, these rains bring about a burst of exuberant growth in and around standing water bodies. Wildflowers and wetland vegetation generally grow prolifically in and around most ponds and lakes usually from late February each year. Nevertheless, climate varies strongly in the SMR according to location and to terrain, hence rainfall decreases strongly in both southern and easterly directions. Lakes can be found across the climate spectrum ranging from subhumid to arid conditions and so the timing and seasonal diversity of aquatic life shows much variation. In the eastern SMR, the Nile delta lakes experience relatively small seasonal changes and much of the aquatic vegetation grows throughout the year. On the other hand, water bodies in mountainous areas have great temperature variations, with freezing winter and even snow in the Atlas Mountains. The Mediterranean coastal areas are more humid and mild than most inland areas, with temperatures ranging from 30 °C in August to 3 °C in December. The subdesert areas in the south are hot and dry all year round and lakes tend to be groundwater-fed systems. Average summer temperatures in these regions are high at around 42 °C in August and sometimes reaching 46 °C. Many parts of the southern areas receive only a few centimeters of rain annually, although severe rain storms do occur at irregular inter-annual intervals. During the three last decades, the SMR as a whole has received less rainfall than usual, and some regions face water shortages that are exacerbated by increasing water demand.

Natural standing water resources – ponds, oasis, and lakes – also have cultural significance in many areas. They are very important for sustaining the rural populations, especially in the north Sahara region where scarce water resources are exploited for crop irrigation, livestock, and for domestic uses. Many of these waters are highly developed but remain important for aquatic ecology, and, where salinity is low, many are key habitats for amphibians. Provision of water storage impoundments and sustainable use of natural freshwaters have been integrated into cultural practices over thousands of years (see the section on 'Oases' below). Although many of these water resources have been strongly developed and exploited since at least the Roman period, it seems that only within the last 100 years or so has the water usage expanded unsustainably.

A general map showing the distribution of major standing water bodies in northwestern part of Africa is given in **Figure 1**. In the west, water availability for lakes is usually from drainage of upland areas but in Egypt they exist by virtue of the Nile. A variety of local terms have been developed in North Africa to distinguish water bodies and most of them are used by water managers and researchers as well as by local people. They are summarized in the upcoming sections.

### Daya

The Arabic term 'Daya' indicates a shallow natural or sometimes artificial depression (usually <1.5 m deep) that contains water for at least some period of the year. These ponds or small lakes occur mainly in the lowland plains and are characterized by a certain morphological homogeneity and are very common in subhumid and humid areas of North Africa, especially where annual rainfall exceeds 20 cm. Water

levels vary seasonally and inter-annually and are usually recharged during October–November each year. The occurrence of these water bodies, in the least dry areas (northwestern parts of the Maghrebian countries), is usually semipermanent (type 3), whereas in the arid areas their presence is less irregular and some dayas can remain dry for years, depending on rainfall (types 1 and 2). A typical daya completely loses its water by evaporation and drainage during the summer period, and the site can be distinguished in the dry landscape by an area of green vegetation (types 3 and 4). In certain formations in the less dry areas, the daya can persist the summer as a saline pool of smaller dimensions. They are frequently encountered in the northern parts of Morocco, Algeria, and Tunisia and in the south of Spain and Italy, but are infrequent in the east of North Africa (see **Figure 1**). The duration of inundation can vary from 9 to 3 months in a year according to rainfall and local climate. Dayas usually have silty clay bottom sediments derived from local soil erosion. Vegetation in and around a daya can vary from one formation to another not only according to water availability but also to disturbance. *Isoetes velata*, *Ranuculus aquatilis*, and *Marsilea quadrifolia* can be common in dayas type 3 and type 4. The Notostraca *Triops concriformis mauritanicus* and *Lepidus apus* can be characteristics of dayas in North Africa. Livestock can exert a very strong pressure on soils and vegetation around dayas leading to deforestation and siltation. Those dayas are prone to large water level

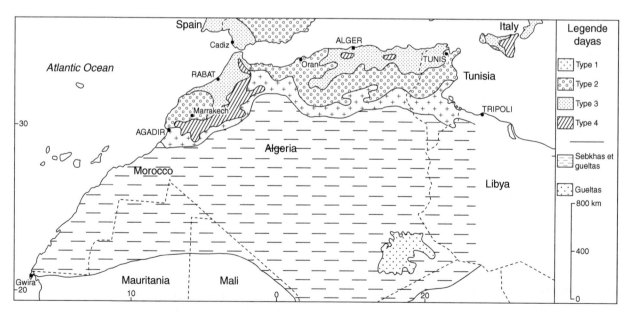

**Figure 1**  Distribution of ponds and lakes areas in North African Sahara and south of Spain regarding annual rainfall average. Dayas type 1 are ephemeral water body limited between Sahara and Atlas mountains. Dayas type 2 in the arid areas; dayas type 3 in subhumid and humid areas, and dayas type 4 in mountain areas.

changes that they support only a few marginal plants. Generally, in the wet period, water within a daya is constantly turbid as a result of suspended sediments arising from silt inwash and wave disturbance. Nevertheless, they can be valuable sites for aquatic diversity and water birds. In the areas where the annual average of rains do not exceed 30 cm and where the period of dryness reaches 9–10 months, water can be stored as groundwater and the daya quickly becomes viable again even after a small rainfall (**Figure 2**).

## Merja

The term *Merja* indicates a shallow surface sheet of water with extensive mud flats. Deep mud is typical (1–2 m) and immersion can vary from daily, where there is tidal influence, to seasonal to permanent water. They are often brackish, high in suspended matter, but can be productive systems. The water level can exceed 4 m depth usually in March. The receding water level during summer leaves a large expanse of usually bare mud and salt deposits around the merja. The inundation limits vary not only according to the season but also between years. In summer, desiccation of exposed mud occurs and colonization by ruderal (weed-like) plants, depending on surface salinity, often proceeds. Merjas are few in Morocco, and they are generally found in the coastal zone occupying depressions on the landward side of coastal dune systems; they are usually above sea level but exceptionally they do communicate directly with the sea (e.g., Merja Zerga). They are often characterized by extensive stands of emergent plants, including *Phragmites communis*, *Typha angustifolia*, and *Scirpus* with a broad marginal belt of *Juncus maritimus* and *Juncus acutus* (central parts being deprived of vegetation). Hence, the merjas are often associated with significant wetlands that may be extensive in the wet season. A typical merja receives stream runoff from the catchment but locally, springs often mark

the emergence of artesian water. Generally, groundwater in the region of a merja is very near to the surface and also contributes to the maintenance of water level. These sites are however strongly threatened by the expansion of human activities, especially agriculture (where pumping and draining of water for crop cultivation lowers the water table, causing the merja and any associated wetland to shrink). One merja, Merja Bokka (northern Morocco), was desiccated completely by these processes in 1998 during environmental monitoring. Merja Bokka was formerly a substantial lake that in the early 1990s was still ~250 m in diameter but very shallow. Before the 1980s, this permanent water body supported extensive emergent macrophyte communities. Aerial photographs showed several small low islands, which by the mid-1990s were covered with *Panicum repens* and *Typha angustifolia*. Several former higher lake-level shorelines occurred around the lake, indicating that past water depths were over 2 m. Since the 1980s, the inflow stream became redundant as water was taken for agriculture and the formerly common *Phragmites* was reduced by burning and drainage. By 1997, the water level was a few centimeters and the lake possessed only a few introduced small carps but many amphibians, especially the newt *Pleurodeles waltli* and the Chelonia *Mauremys leprosa*. In 1998, the entire lake basin was dry and was prepared for agriculture. Since the 1960s, many shallow merjas and other lowland water bodies in agricultural regions of all the Maghrebian countries have followed a fate similar to that of Merja Bokka.

## Garaet

In Tunisia, Garaet is a typical term for a large permanent body of water. An example for this is Garaet El Ichkeul, a large shallow brackish lake formerly famous for its water birds. It is surrounded by productive agriculture and major disturbances of its hydrological system occurred during the twentieth century. Following canalization, hydrological modifications were made to the five main inflowing rivers and all but one of these inflows has been dammed since 1984. Consequently, freshwater drainage inputs are now much reduced and the once extensive marginal *Phragmites* beds have greatly diminished. The lake bed is 1.5 m below sea level, which is the lake depth in summer. In winter, lake level increases are now attenuated and the former marsh zone is only partly inundated. In the past, water usually flowed out from the lake to the sea, but now seawater tends to inflow during dry periods. Since 1986, sluice gates on the Tinja River outflow have control over seawater inflow, but there are conflicts between fisheries

**Figure 2** Daya type 4 (Dayet Ifrah) in the Middle Atlas (Morocco). Note the typical circular shape and deforested landscape.

and biodiversity interests. In recent years, the salinity of Ichkeul has increased so that hypersalinity persisted in the mid-1990s. The Ichkeul lake was listed as a Ramsar site in 1971, designated a Biosphere Reserve in 1977, and an International Park in 1980. (The Convention on Wetlands, signed in Ramsar, Iran, in 1971, is an intergovernmental treaty, which provides the framework for national action and international cooperation for the conservation and wise use of wetlands and their resources.) The marshland and the lake used to receive some 200 000 overwintering and migratory birds annually, mainly ducks, coots, grebes cormorants, spoonbills, storks, herons, and waders; especially, the geese which declined in the 1990s, mainly because of the loss of marshland.

**Figure 3** Megene Chitane in Northern Tunisia, a shallow and relatively undisturbed upland lake showing a late spring abundance of aquatic plants. The lake is unusual because the water is slightly acidic (photo April 1997).

### Megene

This term is used in the Maghreb countries for describing a small lake in mountain areas. A typical example is Megene Chitane in North Tunisia, a small, usually clear, soft water lake located on the north slope of Chitane Mountain at 150 m altitude, overlooking the Mediterranean Sea to the north. The site is protected and was declared a nature reserve in 1993. The outflow is ephemeral and maximum water depth is normally ~1.2 m. The catchment includes cork oak with fairly undisturbed scrub understorey. A perimeter fence some 50–100 m from the lake shores protects the catchment within the vicinity of the lake. Upslope from the lake, subsistence farming is practised on a small area of valley mire. Water flowing through the mire is currently being exploited for crop irrigation. A small aquifer supplies two small freshwater springs that feeds the upper mire bog and eventually the lake. The sandstone aquifer and the valley mire confers acidity to the lake as they are both acidic. The lake water is only slightly acidic, which is most unusual for North African lakes, because nearly all of them nearly possess alkaline water (**Figure 3**). In 1998, the underwater lake sediment was observed to be covered everywhere with desiccation fissures marking the earlier loss of surface water (Elkhiati *et al.*, 2002). Loss of standing water probably resulted from overexploitation of the lake's natural acid spring water supply. Nevertheless, the lake still supports many aquatic plants, including *Nymphaea alba*, *Juncus heterophyllus*, and the rare *Isoetes velata*. The littoral zone is covered with terrestrial vegetation (*Cotula coronopifolia* and some grasses). Fish were introduced by Forestry authorities but these seem to have been lost following the mid-1990s desiccation period.

### Sebkha, Oglet, and Chott

These types of water bodies are typically ephemeral and usually possess brackish water. They occur in steppe or substep zones with arid climate and are typical in dryer parts of North Africa that receive seasonal runoff from higher ground. Their salinity is derived from soil-derived salts and evaporation; sodium and calcium with chloride and bicarbonate are usually the dominating ions controlling salinity, but sulfate can be locally important depending on hydrogeology. Both sebkhas and oglets exist as shallow basins often of large dimensions (10–30 km); the water is always shallow and is susceptible to wind effects, especially desiccation. They can be dry or very shallow in summer, but can offer useful habitats for birds, e.g., Sebkha Kelbia in Tunisia. The term sebkha is used in the South the term oglet in the East, for identical sites in the SMR.

Chotts are immense depressions (30–100 km or more) without defined outflow channel; some years they may remain dry, but when flooded following a wet period, they expand to a large expanse of silty water. By the late summer, open water will be extremely reduced or nonexistent and in some cases standing water exists for only a few days before it is removed by infiltration and evaporation. These formations when exposed are often largely covered by a steppe terrestrial vegetation (e.g., Salsolacées) on suitable ground. Elsewhere, the chott landscape is dominated by exposed alluvium (often sands and coarse gravels) with occasional local depressions with evaporites. In these areas a surface crust of gypsohalite-carbonates often rest on sands and silts. These hollows may be reflooded several times in a year depending on rainfall in the catchment area. Duration of immersion of more than 5 months would be

**Figure 4** Chott Iriqi (southeast Morocco) – a site that is dry in summer (left) but is flooded in the winter period (right) where it supports abundant microinvertebrates and attracts flocks of flamingos.

exceptional. A well-known example of a large chott is the Chott El Jerid in Tunisia, an interesting site that possesses occasional accumulations of marine molluscs (*Cerastoderma*). In southern Morocco, the Sebkha Zima (Chemaia) presents characteristics between a merja and a sebkha because of the absence of vegetation, but having an effluent drainage system in wet years. Located in south-east Morocco, Chott Iriqi is another interesting site for birds (**Figure 4**).

### Guelta (or Galta or Gueltat)

This term is used to indicate pockets of water left in drainage channels or wadis during the dry season; the dimension and the duration of immersion vary from one guelta to another and from one region to another. They are temporary but can persist through very dry winters depending on runoff, water flow, and/or groundwater. They can occupy meanders left by permanent rivers (cf. ox-bow lakes of north Europe) as well as in temporary wadis. In the arid and steppe areas, the gueltas occur in lowland rivers and their deltas where flow becomes greatly reduced such as by channel blockage by sand dunes. In the Moroccan Sahara, these formations often correspond to oases; Gueltat Zemmour in Moroccan Sahara near Smara is a typical example.

### Aguelmane

The Berber term 'Aguelmane' or 'Aguelmam' is used only within the Middle Atlas of Morocco. It means a large, permanent, major natural body of water (often 6–40 m deep) and they are often surrounded by upland forest. Of all the North African water bodies, they are the nearest equivalent of a European upland lake. However, sometimes the term daya is also used to indicate this type of lake formation. Lakes are rare in the High Atlas Mountains but in the Middle Atlas

**Figure 5** Aguelmane Azigza (Middle Atlas, Morocco) showing cedar forest at the east end of the lake and a large exposed littoral area resulting from several years of drought (photo June 1984).

they are common. They occupy limestone (Lias) basins and persist because of an impermeable clay bottom; Aguelmane Azigza (Green Lake in Berber translation) is a typical example (**Figure 5**, others are shown in **Figure 6**). It is fed by precipitation and is subject to large water level changes depending on rainfall quantity and drainage; being located in a karstic landscape, subterranean drainage is common. Some aguelmanes are deprived of vegetation while others support a proliferation of aquatic plants, largely controlled by siltation effects, and/or water levels changes. Where a surface outflow occurs, the water flow is commonly modified by construction of a small barrage to increase the depth and extent of the lakes. Hence, some sites are transformed into 'semiaguelmanes' or artificial lakes, for example, Dayet Aoua in the Middle Atlas of Morocco.

### Birket (or Birkat)

This a term used in the eastern part of North Africa to describe a pond or a lake. They can be either fresh or

**Figure 6**    Isli Lake (left) and Tislit Lake (right) – two substantial hard water lakes in the High Atlas Mountains (Morocco).

**Figure 7**    North shore of Lake (Birket) Qarun (April 2003) – a salty lake at the margin of the hyper-arid Egyptian western desert. Note the absence of both vegetation and run-off drainage features on this shore.

saline, and rainfall is not the main source of water. They can be independent water bodies without inflows or outflow channels (or both) and examples are the group of salty lakes at the Siwa Oasis in the western desert of Egypt, where they are sustained by artesian water supply. Other birkets are dependant on River Nile, and Birket Qarun, for example, is a substantial 40-km long lake on the margin of the Great Western Desert (**Figure 7**). The lake is fed by Nile water through the Bahr Yusef canal, and although it was formerly a larger and essentially a freshwater lake, hydrological and other changes in the last 100 years or so have induced strong salinization. The salinity of the lake water is around that of seawater (depending on season and location), and the site has developed an interesting part marine fauna, including molluscs (*Cerastoderma*), crustaceans, several fish species (*Solea*), and even a few sea anemones.

**Reservoirs**

Many of these artificial lakes were created in the SMR during the mid to latter part of the twentieth century to store water for a variety of purposes. Perhaps the most famous is Lake Nasser and the Aswan High Dam constructed in the 1960s. This large dam not only regulates the seasonal Nile floods but also enables hydroelectric power generation. Also, the function of most North African dams and reservoirs is to provide water for irrigation and for drinking water supply. They extend over large areas and exceed the size of natural lakes in both the Maghreb region and in Egypt. Being dependant on rainfall far away in Ethiopia and Central Africa, Lake Nasser is replenished annually by high Nile flow in late summer. In the western SMR, however, reservoir water levels are strongly controlled by local precipitation and water demand, but they typically rise in winter as a result of silt laden water inflows and fall strongly in summer. Depressions around their margins can, however, preserve water all the year and constitute permanent marshy zones. Hence, while the littoral shore of these reservoirs are largely devoid of life, especially aquatic plants, these wet marginal depressions can support considerable diversity and often with abundant hydrophyte communities. All these artificial water bodies face problems of high siltation through erosion of newly established shorelines and delivery of silt inflowing water.

**Oases**

Oases are characteristic of the hyperarid areas in or near desert areas. They are typically supplied by ground water or springs, and temporary rivers can also sustain some oases by conducting rainwater (above or below ground) from far away (e.g., the Nile and Draa rivers) to desert depressions. Where oases are fresh, aquatic vegetation and animal life can flourish. They are a foci for people, and human intervention for water use often disturbs standing water habitats in oases so that eutrophication and water withdrawal can cause sustainability problems. The more isolated oases are subject to a hostile climatic ambiance and are usually fairly species poor but are not without interest as specialized insects and amphibians are often abundant. However, where salinity is high, for example in the hypersaline lakes of the Siwa

(Egypt), aquatic life is extremely reduced. Natural oases exist in depressions in the surface of the deserts and can move according to sand dune migration. Some are artificial, and desert people have shown the ingenuity to use water sustainably in dry areas and many oases can be regarded as the product of human effort and will disappear without regular management. Today, traditional water management techniques are declining around oases as technological innovations and finance make it possible to artificially extend water supply to thousands of hectares and 'make the desert flower.' Nevertheless, in many regions overuse of this precious freshwater resource has lowered the groundwater table and exploitation of deep water aquifer resources is now being undertaken.

## The Biota of North African Lakes

Water quality and its availability as well as geography play important roles in influencing the occurrences and abundances of aquatic animals and plants in North African lakes. Superimposed on these factors are an array of disturbance activities caused by people, and so many lakes today are modified systems where typically nutrient enrichment and salinity changes have occurred as a result of agriculture and development generally. Nevertheless, many lakes retain significant biological interest and some support communities recognized for their high conservation value. Often, however, the focus for conservation is on the surrounding vegetation or on extent of water of value for birds; in fact it is the quality of the aquatic environment that underpins the conservational value. A brief review of some of the aquatic groups of interest in SMR lakes follows, and generally species are all characteristic of either hard or brackish water systems, and in a wide survey of SMR lakes between 1996 and 2004, only one acid water lake was found: Megene Chitane in Tunisia (water quality in selected North African lakes is indicated in **Table 1**).

Climate and water quality together influence the occurrence and abundance of aquatic organisms in the shallow SMR lake systems. Many organisms typically show adaptations to strong seasonal changes and fluctuating conditions, including receding water levels and water quality variations similar to those described elsewhere for desert river systems and temporary waters. Aspects of hydrobiology are considered as follow.

### Plankton

Primary production in open waters is driven by phytoplankton and many species have been described in the SMR since systematic research began earlier in the twentieth century (e.g., Gayral, 1954). The trend in availability of phytoplanktons related to the seasonal changes is comparable to that in European lakes with early and late summer peaks. However, high turbidity of some lakes can strongly depress phytoplankton growth despite an excess of nutrients. Green algae tend to be the most diverse group in all lakes except enriched lakes, where blue-greens (cyanophytes) can dominate, making the open water highly turbid and green. For diatoms, *Cyclotella* species such as *C. ocellata* are common in the hard water upland lakes but in the lowland sites, more salt tolerant taxa are found, including *Thalassiosira weissfloggi* and *Cyclotella meneghiniana*. In the Nile system, *Aulacoseira granulata* is usually the common plantkonic diatom. Phytoplankton was scarce in acidic Megene Chitane but some *Cyclotella pseudostelligera* and several acidophilous diatom species were recorded.

Being at the base of the food chain, the phytoplankton sustains the zooplankton as well as fisheries. Rotifers seem to dominate where organic pollution is high, such as in Lake Manzala (Egypt), but elsewhere crustaceans such as *Daphnia magna*, *Ceriodaphnia dubia*, *Moina micrura*, *Dunhevidea crassa*, *Chydorus sphaericus*, *Alona esteparica*, *Alona quadrangula*, *Leidigia ciliata*, *Dipatomus cyaneus admotus*, *Metadiaptomus chevreuxi*, *Hemidiaptomus maroccanus*, *Mixodiaptomus laciniatus atlantis*, and *Diacyclops bicuspidatus* are common. One characteristic of some of the SMR zooplankton species is their ability to withstand high salt and desiccation, and estivation

**Table 1**  Summary examples of mean values for some water chemistry attributes in selected North African lakes

|  | Sidi Bou Ghaba | Azizza | Zerga | Chitane | Korba | Manzala | Qarun |
|---|---|---|---|---|---|---|---|
| pH | 8.3 | 8.1 | 8.1 | 5.5 | 8.1 | 8.2 | 8.7 |
| TDS (g l$^{-1}$) | 5.3 | 0.7 | 15.8 | 0.5 | 41.7 | 0.9 | 34 |
| Ca (mg 1$^{-1}$) | 64 | 48 | 442 | 47 | 465 | 48 | 374 |
| Na (g 1$^{-1}$) | 2.9 | 0.01 | 2.7 | 0.9 | 12.9 | 0.2 | 7.7 |

Merjas Sidi bou Ghaba and Zerga are on the western coast of Morocco, Aguelmane Azizga in the Middle Atlas Mountains, Megene Chitane and Lac de Korba are on the north and east coast of Tunisia, respectively, and Lake Manzala and Birket Qarun are in Egypt. Note that the upland lake M. Chitane alone has acidic water and that both Zerga and Korba are linked to the sea by channels. Ca – calcium, Na – sodium, TDS – total dissolved salts.

in pond habitats (dayas types 1–4) is usual. Crustaceans showing summer dormancy are the Notostracha (*Triops concriformis mauritanicas*, *Lepidurus apus*), Anostraca (*Chirocephalus diaphanous*), Conchostracha (*Cyzicus* sp.), Cladocera (*Daphnia*, *Alona*, *Chydorus*, and *Leydigia*), and copepods (Diaptomidae *Neolovenula alluaudi* and *Eudiaptomus cyaneus*, Cyclopidae *Metacyclops minutus*, and Harpacticoida).

### Aquatic Plants

Emergent macrophytes are frequently much reduced in the Maghrebian region by overexploitation of water resources but many are still abundant in the Egyptian Delta lakes and in Moroccan mountain lakes. Emergent plants include *Juncus acutus*, *J. maritimus*, *J. rigidus*, *Scirpus* spp., *Typha angustifolia*, and *Phragmites australis*, and are so abundant in the Nile delta lakes that they not only provide sheltered habitats for many reed nest birds but also permit harvesting for a variety of domestic uses. In the predominantly fresh water lakes, submerged vegetation in many SMR lakes is commonly composed of abundant charophytes, *Tolypella glomerata*, *Nitella opaca*, *N. gracilis*, *N. hyalina*, *N. translucens*, *Lamprothamnium papulosum*, *L. succinctum*, *Chara hispida*, *C. vulgaris*, *C. aspera*, and *C. canescens*. Other occasionally common aquatic macrophytes, especially in the more brackish sites, are *Naias minor*, *N. major*, *Ruppia cirrhosa*, *R. maritima*, *Potamogeton*, *Myriophyllum*, and *Ceratophyllum*. *Isoetes velata* and *I. lacustris* are mainly common in freshwater (dayas types 3 and 4).

Floating aquatic plants such as the water ferns (*Lemna gibba*, *L. minor*, and *Azolla filiculoïdes*), water lettuce (*Pistia stratoites*), and the water hyacinth (*Eichhornia crassipes*) can form nuisance growths on surface waters and the latter two species are restricted to Egypt.

### Amphibians and Reptiles

The turtle *Mauremys leprosa is* very common in different water bodies in North Africa but *Emys orbicularis* is very rare and limited to the Rif and the Atlas mountain areas. Dayas, Merja, and Aguelmane are important habitats for *Bufo bufo spinosus*, *B. mauritanicus*, and *B. viridis*. *Bufo bongersmai* is more frequent in the desertic and arid areas (South west of Morocco and south of Algeria). *Hyla meridionalis*, *Rana saharica*, *Discoglossus pictus*, *Alytes obstetricans*, *Pelobates varaldii* are common in permanent water bodies. *Salamandra algira* is very rare in mountain water bodies (Rif and Atlas mountains). *Pleurodeles waltli* is common in dayas types 3 and 4.

### Fisheries

Barbel (*Barbus callensis*, *B. nasus*, *B. pallaryi*, and *Varicorhynus marocannus*), *Cobitis maroccana* and, in the Nile fed lakes, an array of *Tilapia* species (*T. nilotica* and *T. zillii*) represent the main resident species in inland waters of North of the Sahara. These fish are found in a variety of habitats, including oasis and delta lakes. Fish aquaculture is frequently developed for mullet and for *Tilapia* in the Nile delta lakes. A few species are endemic to the region in the Nile system and include such species as the Nile perch (*Lates niloctica*). Also in Morocco, there is a local population variety of brown trout (*Salmo trutta* f. *fario* and *S. trutta macrostigma* endemic in Atlas Mountains) in Ifni, Isli, and Tislit lakes (High Atlas, Morocco). Other fish species have been introduced recently to North African freshwaters and these include the common carp (*Cyprinus carpio*), pike (*Esox esox*), black bass (*Sander lucioperca*), rainbow trout (*Onchorynchus mykiss*), and tench (*Tinca tinca*). *Lepomis gibbosus*, *L. macrochirus*, *L. microchirus*, *Perca fluviatilis*, *Micropterus salmoide*, *Aristichthys nobilis*, *Cteropharangodon idellus*, *Hypophthalmichthys molitrix*, and *Scardinus erythrophthalmus* were introduced in Mountain lakes, reservoirs, and rivers for sportive fishery. *Gambusia affinis* is very common in different water bodies and was introduced as a control for mosquito larvae.

### Birds

Main migration routes are through Morocco, Tunisia, and eastern Egypt, and lakes on these routes are important habitats for these birds. Sites such as Merja Zerga, Ichkeul, Burrulus, Manzala are important resting areas for migrating birds. These wetland areas are inundated with migrant waders and waterfowl during each migration season and are supplemented by notable resident species such as the Coot (*Fulica atra*), Crested Coot (*Fulica cristata*), African Marsh Owl (*Asio capensis*), and Double-spurred Francolin (*Francolinus bicalcaratus*).

## Environmental Change in North African Lakes

Exploitation of water resources and intensity of land usage have increased rapidly throughout the SMR during the twentieth century. This has occurred especially in Morocco, Tunisia, Egypt, Lebanon, and Israel, and on the Mediterranean islands and has paralleled regional population growth. Lakes are especially vulnerable habitats and many of North Africa's natural standing waters are now degraded by human activities. Although past human impacts on North African landscapes began in a major way during the Roman period, water has been managed in Egypt for over 5000 years. Nevertheless, shallow lake SMR ecosystems may have been the most intensively disturbed during the last 100 years following the introduction of mechanization of land management. Large-scale hydraulic modifications of most North African river catchments have been carried out since the 1950s (e.g., Sebou River in Morocco, the Mejerda in Tunisia, the Nile by the Aswan High Dam in Egypt). In all cases, these modifications were carried out to control annual floods and/or to improve water availability for people and for agriculture. Although people have benefited, many small lowland ponds and wetlands, especially those in the Maghrebian countries, have been a major casualty of this program as natural fresh water availability has generally declined. An exception is the Nile delta lakes, where improved water availability, especially following completion of the Aswan High Dam, in the Delta has lead to increased agricultural returns entering the lakes, causing a freshening of receiving standing waters. Elsewhere, there is evidence that, at least initially, improvements in water supply for agriculture led to local freshening of lakes and wetlands but since the 1980s many sites are showing a salinization trend. Furthermore, local eutrophication and pollution are also impacting many inland waters of the SMR, and, despite some improvements in environmental management in the latter part of the twentieth century, water resources and their aquatic communities have been generally diminished. Land drainage, afforestation, and intensified crop production arising from development programs as well as from effluents and water abstraction continue to threaten lake water quality in most North African regions. Unfortunately, without active management, all the current problems that affect water quality and availability now undoubtedly will be exacerbated by global warming.

The upland regions, where there are no major groundwater reserves to exploit (such as in Libya and Mauritania), are particularly prone to water shortages brought about by climate change. In the Middle Atlas Mountains of Morocco, a karstic landscape makes most natural lakes very prone to major water level fluctuations as a result of the inter-annual periods of low rainfall. The level of one such lake (Aguelmane Azigza) declined by over 11 m between 1979 and 1984 following a period of low rainfall. Some recharge has occurred subsequently but this lake and others like it remain vulnerable to persistent climate change. At the other end of the spectrum of SMR lakes, many oases are declining because of a variety of reasons, including rural exodus and

reduced or overmanagement of the semiartificial sites, increased water deficit, and siltation by the progression of the dunes. In the lowland plains, shallow lakes (dayas and merjas) occupy natural closed depression and all are threatened by climate, by encroaching agriculture as well as by sea level rise. Locally, afforestation offers a special threat to dayas in the Maghreb. In Morocco, ancient forests of cork oak were very extensive in the Gharb region, where dayas were abundant. Several recent processes have lead to the loss of most of these dayas. During the periods 1980–1981 and 1992–1996, rainfalls were only two-thirds compared with the normal, and the period of dryness often extended over 9 months. These also contributed to the on-going decline of cork oak in the Mamora and Bin Abid forests. Forest openings expanded, permitting access by livestock and facilitating plantation of the *Eucalyptus* forest which began after the 1900s. This intense pressure has lead to the widespread disappearance of dayas from these Moroccan forest areas by the twenty-first century. Apparently both human- and climate-induced changes have strongly impacted North African lake systems, but it is in the lowland regions where landscape modifications have been most important in diminishing sites during the recent past. Depending on the type and degree of water abstraction and the relationship of a lake basin with the sea, hydrological modification can either bring about freshening (cf. Nile Delta lakes) or salinization (cf. Garaet Ichkeul, Tunisia), or complete desiccation (Merja Bokka). Even where there has been a freshening trend (Nile Delta lakes), it is unlikely that this will persist into the future since the Mediterranean Sea level is rising and the Delta is subsiding.

Many of the North African natural lakes are conservationally important habitats and some have been designated RAMSAR sites – internationally recognized bird reserves – and several support significant fisheries (such sites are Lake Burullus in Egypt, Merja Sidi Bou Ghaba (**Figure 8**), Aguelmane Afenourir, and Merja Zerga in Morocco, lakes Oubeira and Tonga in Algeria, and Ichkeul in Tunisia). Despite achieving this status many of these sites are already impacted by anthropogenic activities, but the extent of environmental stress is largely unknown except that habitat quality is generally declining and migrant bird populations are threatened. One way to track the development of a lake through time is by using palaeolimnology. Analysis of sediment cores for fossils and geochemistry is commonly used to answer questions about the rate and extent of changes in aquatic biota in wetlands and lake ecosystems on a variety of time scales. Sedimentary records can also help identify trends in pollution impacts and determine the rate of siltation. In the SMR, lake sediment

**Figure 8** Upper: The nature reserve and Ramsar site Merja Sidi Bou Ghaba, a relatively undisturbed coastal lake in north-west Morocco in spring. Lower: Sidi bou Ghaba in late summer showing desiccation of the south part of the lake (photo 2002).

cores have been used to infer historical changes in several lakes in Morocco, Tunisia, and Egypt. Results generally show that nutrient enrichment, salinization, metal and pesticide contamination, and soil erosion have all impacted the investigated lakes during the twentieth century. Further, these sediment archives demonstrate major changes that have occurred in biodiversity with species of some aquatic organisms, especially plants and invertebrates, being lost as salinity and water availability have changed. However, compared with boreal zone lake sediments, sedimentary preservation of some organism groups such as diatoms is often poor so limiting the potential of some palaeolimnological techniques for reconstructing past environmental changes. Probably, the frequently warm and strongly alkaline conditions in SMR lakes do not favor preservation. Hence, little is known about the resilience of some aquatic communities to recent environmental changes, but where lake monitoring has not been undertaken or only sporadically, palaeoecology and lake sediment core studies do offer the only way of tracking lake changes through time.

It is anticipated that detrimental effects of environmental changes on SMR lake ecosystems will increase during the twenty-first century as a result of the demands of local human populations and of accelerating climate change. To track how lake ecosystems are responding to these changes, a program of well-coordinated environmental monitoring is needed, which involves remote sensing and hydrological

modeling combined with on-site water and sediment quality assessments. Nevertheless, it is already clear that diminishing freshwater availability is the most widespread threat to the persistence of many lakes in the SMR. So while monitoring ecosystem health is essential, appropriate restoration and conservation measures need to be introduced. Not least is improved water management that takes advantage of both traditional methods as well as new innovations so that natural water resources can be used more efficiently. Such actions taken together will help inform and set policy for future management of freshwaters in North Africa. These actions must also take account of aquatic biodiversity as well as the water demand.

*See also:* Lakes as Ecosystems; Mixing Dynamics in Lakes Across Climatic Zones; Saline Inland Waters.

## Further Reading

Birks HH, Peglar SM, Boomer I, Flower RJ, Ramdani M, with contributions from Appleby PG, Bjune AE, Patrick ST, Kraiem M, Fathi AA, and Abdelzaher HMA (2001) Palaeolimnological responses of nine North African Lakes to recent environmental changes and human impacts detected by macrofossil and pollen analyses. *Aquatic Ecology* 35(3–4): 405–430.

Elkhiati N, Soulie-Marshe I, Ramdani M, and Flower RJ (2002) A study of the subfossil oospores of *Nitella opaca* (Charophyceae) from Megene Chitane (Tunisia). *Cryptogamie Algologie* 23(1): 65–73.

Dudley Williams D (2006) *The Biology of Temporary Waters.* Oxford: OUP.

Flower RJ and Foster IDL (1992) Climatic implications of recent changes in lake level at Lac Azigza (Morocco). *Bulletin of the Society of Geology France* 163: 91–96.

Flower RJ, Dobinson S, Ramdani M, *et al.* (2001) Recent environmental change in North African wetland lakes: diatom and other stratigraphic evidence from nine sites in the CASSARINA Project. *Aquatic Ecology* 35(3–4): 369–388.

Gayral P (1954) Recherches phytolimnologiques au Maroc. *Travaux Institut Scientifique Chérifien série botanique* 4: 1–308.

Hughes IMR, Ayache F, Hollis GE, Mamouri F, Avis CH, Giansante C, and Thompson I (1996) Inventaire préliminaire des zones humides tunisiennes. Doc. CEE (DGXII), Report by the Wetlands Research Unit, University College London, 473 pp.

Kingford R (ed.) (2006) *Ecology of Desert Rivers.* Cambridge: CUP.

Lamb PJ and Peppler RA (1987) North Atlantic Oscillation: concept and application. *American Meteorological Society* 68: 1218–1225.

Ramdani M (1988) Hydrobiology of sebkhas and gueltas in the Khnifiss-La'youne region. In: The Khnifiss Lagoon and its Surrounding Environment. *Trav. Inst. Sci. Rabat, Mém. hors série,* pp. 87–94.

Ramdani M (1988) Les eaux stagnantes au Maroc: études biotypologique et biogéographique du zooplancton. *Trav. Inst. Sci. Rabat, Zool* 43: 40.

Ramdani M, Elkhiati N, Flower RJ, Kraiem MM, Fathi AA, Birks HH, and Patrick ST (2001) Open water zooplankton communities in North African wetland lakes: the CASSARINA Project. *Aquatic Ecology* 35(3–4): 319–333.

Ramdani M, Flower RJ, Elkhiati N, Kraiem MM, Fathi AA, Birks HH, and Patrick ST (2001) North African wetland lakes: Characterization of nine sites included in the CASSARINA Project. *Aquatic Ecology* 35(3–4): 281–302.

Stanley DJ and Warne AG (1993) Nile Delta: Recent geological evolution and human impact. *Science* 260: 628–634.

Vita-Finzi C (1969) *The Mediterrean Valleys*. Cambridge: CUP.

Zahran MA and Willis AJ (2003) *Plant Life in the River Nile in Egypt*. Riyad: Mars.

# Africa: South of Sahara

**O V Msiska,** Mzuzu University, Luwinga, Mzuzu, Malawi

## Lakes

Lakes refer to large bodies of water that are enclosed by land, either partially or without drainage outlet as in endorheic ecosystems. Sub-Saharan Africa is endowed with numerous lakes some of which tend to attract interests of a wider scientific community owing to their ecological significance. Many communities depend on lakes for potable water, power generation, transport, fishing, and recreation activities.

## Formation

With the exception of glacial lakes, all other types of lake formation systems are represented in sub-Saharan Africa ranging from floodplain, depression and volcanic to rift valley. Among them, the most notable ones are the African Great Lakes found in the Eastern and Western Rift Valley regions of Africa (**Figure 1**). In contrast to the Laurentian Great Lakes, African Great Lakes are known to be very old. Using sediment-aging techniques, Lakes Tanganyika and Malawi/Nyasa/Niassa date their origin to the Miocene period 10–20 million years ago. Underlying the bed of these lakes is an accumulation of sediments covering depth of more than 4 km and enabling the reconstruction of paleolimnological and paleoclimatic events that provide the most compelling evidence about their genesis.

Owing to tectonic forces, these lakes were reformed at the end of the Pleistocene period when land transformations caused uplifting of portions of the western region to form Lake Victoria, now the second largest freshwater lake on earth. The recent volcanic eruptions recorded on 17 January 2002 in Lake Kivu provide evidence that volcanic activities are still active in the area.

Many other lakes are associated with a network of flood plains in the region, ranging from temporal pools to permanent wetlands that experience different flood regimes according to their location with respect to the equator (**Figure 2**). The majority of these lakes are found in Niger delta, Lake Chad, southern Sudan (Sudd), Zaire basin (Lake Bangweulu), and Cubango-Kalahari (Lake Ngami). The hydrological regime of Lake Chad and the associated river system suggests that this might be a remnant of once the greatest lake in the world, which has since undergone contraction owing to drought in the catchment's area as shown in Figure 3, the so-called Mega Chad estimated to have covered 300 000–400 000 km².

Geological records of ichthyofauna support the hypothesis of continental connections indicating that differentiation of some aquatic forms was already complete by the Mid-Jurassic period at a time when Gondwanaland was breaking up. The most ancient group of living bony fishes, Dipnoi, is represented by four freshwater species of *Protopterus* found in Africa that are believed to have originated during the Lower Devonian period. Similarly, the worldwide distribution of a group of Osteoglossidae fish, represented by *Heterotis* in Africa, *Arapaima* and *Osteoglossum* in South America, *Sclerophages* in Australia and fossil records found in Asia, Australia, and North America lends credence to connectivity of the continents, and the African origin of fish species.

## Diversity

**Physical and chemical factors** The lakes of sub-Saharan Africa can be divided into two broad categories of low and high relief conditions. The low relief areas occur in sedimentary basins and upland plains that are below 600 m above sea level and are represented by catchments of Nile, Chari, Niger, Senegal, Volta, and Zaire. Few mountainous regions such as the Guinean highlands occur here. The south and east are classified as high with relief of above 1000 m, and two major faultlines running NE/SW from Ethiopia to Zimbabwe are a major feature. The Western and Eastern Rift Valleys form trenches of about 1000 m below the top of escarpments bearing remains of intense volcanic activity. Such trenches are ideal traps of freshwater that harbor most of the lakes.

Owing to chemical constituents, Lake Tanganyika is classified as a soda lake, and is sometimes more closely associated with the Indian Ocean when compared with other adjoining lakes. The shallowness of Lake Victoria makes it one of the most productive as it undergoes complete mixing owing to current movement initiated by wind and internal seiches. Horizontal currents are influenced by wind speed and density-based buoyancy, resulting in upwelling and recycling of nutrients to sustain the food web. In deeper lakes, there is almost permanent stratification except for localized upwelling that sometimes generates catastrophic consequences as witnessed in August–September 1999 in central Lake Malawi,

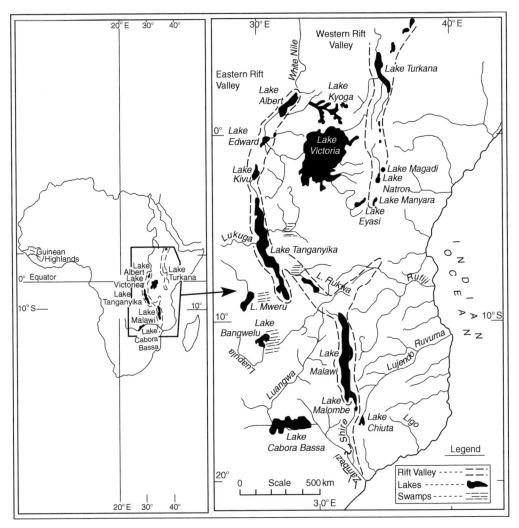

**Figure 1** The two Rift Valleys of eastern Africa and their associated lakes.

where thousands of fish died from chemically reduced anoxic gases such as $CO_2$, $CH_4$, and $H_2S$. A similar episode was reported from around Nkhata-Bay in northern Malawi in 1937.

## Biological Diversity

A combination of water level fluctuations, spatial heterogeneity, and isolation of lakes are principally known to have profound influence on aquatic fauna and flora, notably fish speciation in the aquatic systems of sub-Saharan African lakes.

## Algae

Algae types and primary production in natural lakes are dependent on several factors; the major ones are water levels, nutrient recruitment, subterranean rocks, and temperature. Primary production is relatively low in lakes Malawi and Tanganyika with

records of $0.73$ g.C.m$^2$ and $1.4$ g. C m$^2$ per day, respectively, recorded in surface waters only. Of the Great Lakes, Lake Victoria is more productive with chlorophyll-$a$ values of $8.8–51.4$ mg per m$^3$. Many species of diatoms (*Nitzschia, Stephanodiscus, Navicula, Rhopalodia, Cymbella + Encyonema, Ggyrosigma*), green algae (*Chlorella + Chlorocystis, Coelostrum, Chroococcus, Mougeotia, Euastrum*) and blue green algae (*Anabaena, Calothrix, Lyngbya, Tychonema, Schzothrix, Planktolyngbya, Chroococciods*) commonly occur in lakes Tanganyika and Malawi. *Melosira* is prevalent in Lake Malawi, where it is a keystone species of the pelagic zone. Apart from forming the base of the food chain, deposited lake-bed forms of dead algae provide reliable material for seismic and carbon dating studies.

Epilithic periphyton is significant to Lakes Tanganyika and Malawi, where it is grazed upon by the fast evolving rock-dwelling haplochromis species. In Lake Malawi, heterocystous Cyanobacteria and diatoms

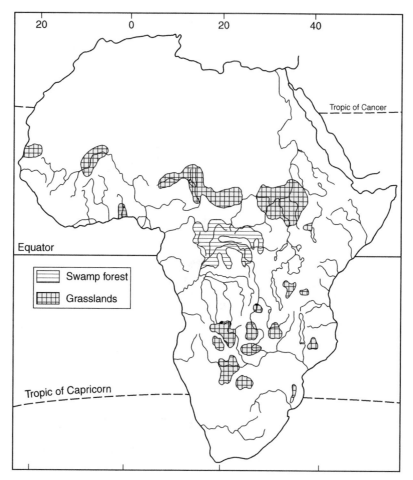

**Figure 2**  Situation of the major African wetlands.

are the dominant forms. They are endosymbionts that undertake nitrogen fixation. In some locations, there is a shift towards diatoms and chlorophytes owing to sediments and nutrient loading. *Calothrix* species dominate the littoral periphyton zone. It is stated that heterocytes decrease with depth in Cyanobacteria. The dominance of nitrogen fixing algae is indicative of N-limitation in rocky littoral zones. The mixture of algal species found in sediment is grouped under the common name 'Awfuchs' and are rather poorly taxonomically defined. Eutrophication has resulted in algal shifts dominated by *Anabaena* and *Microcystis* in most lakes.

### Zooplankton

The crustacean zooplankton, protozoa, or insect larvae are dominant in the secondary food chain of many African Lakes. While studies on their identification continue to provide insights into ecosystem structure and species succession, these are also important to an understanding of ecosystem-wide trophic functioning. This is especially true where a whole fish species flock may have undergone a decline due to (i) overexploitation and where (ii) foreign introductions into natural complex ecosystems have caused adverse effects as in Lake Victoria, or through (iii) natural succession.

Predatory fish and other organisms cause changes to zooplankton diversity as evident from studies conducted in Lake Chad: here a microzooplankton of 80 μm is selectively consumed by *Brachysynodontis batensoda* although nauplii and rotifers are also variably preyed upon. Some zooplanktons like cladoceran *Moina micrura* are incapable of avoiding predation by fish owing to low mobility in contrast to diaptomids.

Where predation is through passive selection, mean prey size and species composition of zooplankton are altered. In Lake Naivasha, zooplankton *Daphnia laevis* grows to 2.0–2.5 mm, *Diaphanosoma excism* attains 1.0–1.6 mm while *Tropodiaptomus neumanni* reaches 2.6 mm; this is attributed to low predation pressure. By contrast, in Lake Tanganyika intense

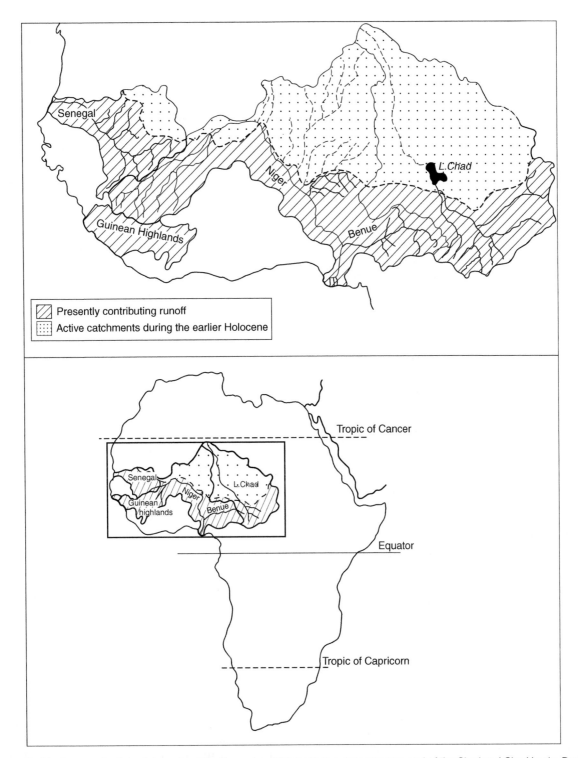

**Figure 3** Maximum potential catchments of the Senegal and Niger–Benue river systems, and of the Chari and Chad basin. During periods of maximum humidity, such as the early Holocene, the whole of the catchments were probably active. Today, only part of the catchment (the shaded areas) contributes to the runoff.

predation by endemic clupeids *Stolothrissa tanganyicae* and *Limnothrissa miodon* is thought to have led to the absence of cladocerans. It is thought that inefficiency of a planktivore fish of the pelagic zone of

Lake Malawi is responsible for the occurrence of *Bosmina longirostis, Diaphonosma excisum*, and *Daphnia lumhotzi*. On the other hand, passive feeding behavior exhibited by the pelagic *Engraulicypris sardella* on *Chaoborus* during daytime offers little pressure on this zooplankton. Thus *Chaoborus edulis* larvae are known to dominate zooplankton of Lake Malawi, and are also found in other rift valley lakes of similar trophic structure. However, *Chaoborus* are absent from Lake Tanganyika, where instead protozoa like *Strombidium cf. viride* and their symbionts are prevalent. In Lake Victoria, swarms of emergent insects have in recent years increased owing to changes in niche structure as a result of a population crash experienced with native cichlids. In turn, this has resulted in the population explosion of the sand martin, *Riparia riparia*.

Further experimental evidence of predation pressure was obtained in Lake Kariba, where the introduction of *Limonothrissa miodon* in 1967–68 led to reduction of *Ceriodaphnia*, *Diaphanosoma*, and *Diaptomus*.

### Benthos Communities

The basic composition of the epilithic fauna is dominated by ostracods, water mites, chironomids, and copepods, forming two-thirds of the total number of animals: this is a common feature of all large African lakes, including Turkana and Tanganyika. The effect of predation by fish on benthic organisms is evident from relatively low numbers of ostracods, hydracarina, chironomids, copepods, and Corixidae recorded in Lake Malawi.

### Aquatic Macrophytes

African Great lakes are fringed by sandy shorelines, only occasionally interspaced by macrophytes like reeds (emergent *Phragmites*), *Vossia cuspidata*, and sedges (*Typha* species). Few submergents like *Salvinia* and broad-leaved floating plants occur. In recent times, the water hyacinth *Eichhornia crassipes* has become a menace to many of the lakes. The biomass of Hippo grass *Vossia* species and papyrus *Cyperus papyrus* have recently increased in Lake Victoria. Papyrus is characteristic of swampy zones around lakes of central and eastern tropical Africa, suggesting a non-sustainable status. Even without alien introductions, new studies indicate trophic shifts in Lakes Malawi/Nyasa and Malombe attributed to overfishing and possibly also to climate change.

### Fish Fauna

**Species diversity**   By far, the most studied organisms in African lakes are Cichlid fish species. Because many are of recent origin, morphometric differences might be very subtle such that cladistic techniques are unable to separate them; hence advanced molecular genetic methods are increasingly being used. Studies of chromosome numbers and morphology have also been advanced. Chromosome number and/or chromosome arm numbers show great variation indicating that the karyotype is also species-specific. A range of chromosome numbers has been recorded ($n = 15$, 13, 11, 10, 9 in *Aphyosemion*, ($n = 17$, 18 in *Epiplatys*) and polyploidy in *Barbus* species ($n = 148$–150), indicating the great genetic variation within genera. Generally, a high chromosome number is associated with larger size, longer life, faster growth, better ecological adaptation and robustness.

Although there is still some confusion about ways of calculating Fundamental Number, which refers to long chromosome arms, comparative analysis of banding patterns for each chromosome has also been demonstrated to be species-specific. Among tilapias, whose origin is Africa, the C-heterochromatin distribution differs according to species. The diploid value in the Tilapiini tribe is relatively conservative with chromosome number of around 44; the most conservative being *Oreochromis alcalicus* ($2n = 48$) while *Oreochromis karongae* ($2n = 38$) in Lake Malawi is regarded to be of more recent origin. These results suggest a need for more research to refine characterization of stock for better formulation of conservation. Similarly, determination of variation is crucial where there is need to enrich farmed stocks as successfully demonstrated in *Oreochromis niloticus*.

The Siluriformes have relatively high diversity with respect to number and shape of chromosomes, exemplified by *Aeuchenoglanis* and *Bagrus* ($2n = 54$ or 56), *Chrysichthys* and *Clarotes* ($2n = 70$ or 72). At the population level, heterozygosity ($H = \sum h/r$) of *Chrysichthys maurus*, *Chrysichthys auratus*, and *Clarotes laticeps is* high. The chromosomes number is higher in these catfishes than in *Clarias gariepinus* from Asia and Africa that exhibits stable karyotype. Within the major groups of cichlids, there appears to be no relationship between karyotypic diversity and phylogeny.

The greatest change of fish species diversity occurred in Lake Victoria, where more than 13 native detrivorous/phytoplanktivorous haplochromids became extinct and were naturally replaced by the atyid prawn, *Cardina nilotica*. Similarly, more than 20 zooplanktivorous haplochromis were replaced by the cyprinid *Rastrineobola argentea*. The Nile Perch *Lates niloticus* is reported to have predated on more than 109 species of the original stock of haplochromids. The resulting cascading trophic interactions resulted in a bizarre phenomenon where *Lates niloticus* resorted to eating its own young.

## Genetic Diversity

By using molecular markers, genetic variation at population level can be discerned through a range of alleles representing genes. Mean genetic variability or heterozygosity among the populations has been measured using electrophoresis and DNA fingerprinting. Mostly this has been utilized for constructing evolutionary relationships among fishes. The taxonomy and management of cichlid species such as *Haplochromis* and tilapias have benefited from these techniques.

## Biodiversity and Conservation

Inland water bodies are generally patchy with respect to habitats lying within each catchments area. Recent mitochondrial studies have shown that each habitat, even within a range of 700 m, may be an island with endemic species arising through allopatric and sympatric speciation. The characterization of species and species complexes in African lakes is a challenge but also provide a living laboratory for scientists. Thus conservation of species flocks to uphold variation is tied to maintaining ecosystem integrity. In turn, this important for sustenance of rural livelihoods that depend on fishing. Some of the unique ecosystems have been designated heritage sites led, among freshwaters, by the establishment of the Lake Malawi National Park in 1982 and its subsequent attainment of status as a World Heritage Site. These sites have particular value to science and will serve future generations.

The inability of cichlid to cross valleys is exemplified by the proliferation of *Pseudotropheus* species and/or subspecies in Lakes Tanganyika and Malawi. It is contended that this might have contributed to species radiation, coupled with lake level changes.

**Table 1** Diversity of physical and chemical characteristics of selected sub-Saharan African Lakes

*African Great Lakes and other lakes*

| Parameter | Lake Chad | Lake Chilwa | Lake Kivu | Lake Albert | Lake Turkana | Lake Edward | Lake Victoria | Lake Tanganyika | Lake Nyassa/Malawi |
|---|---|---|---|---|---|---|---|---|---|
| Surface area (km²) | 2000–22 000 | 1858 | 2700 | 5300 | 6750 | 2325 | 68 800 | 32 600 | 29 500 |
| Max. Depth (m) | 9.5 | 2.1 | 240 | 58 | 109 | 112 | 79 | 1470 | 700 |
| Mean depth (m) | 3.9 | | 220 | 25 | 30.2 | 17 | 40 | 580 | 264 |
| Volume (km³) | 75 | – | 500 | 280 | 203.6 | 912 | 2760 | 18 900 | 7775 |
| Drainage area (km²) | $2.5 \times 10^6$ | 1160 | 7000 | – | 130 860 | 12 096 | 195 000 | 220 000 | 100 500 |
| Altitude (m.asl) | 1463 | 622 | 1500 | 615 | 360 | – | 1134 | 774 | 474 |
| River inflow (km³ year⁻¹) | | | | – | – | – | 20 | 14 | 29 |
| River outflow (km³ year⁻¹) | | | | – | – | – | 20 | 2.7 | 12 |
| Rainfall (mm year⁻¹) | | | | – | <250 | – | 100 | 29 | 39 |
| Evaporation (mm year⁻¹) | | 1757 | | – | 2335 | – | 100 | 50 | 57 |
| Residence time (years) | | – | | – | 12.5 | – | 23 | 440 | 114 |
| Na⁺ (μmol l⁻¹) | 0.5 | 189–780 | 5.70 | 3.96 | | 4.78 | 450 | 2700 | 840 |
| K⁺ (μmol l⁻¹) | 0.2 | 3.7–23.8 | 2.17 | 1.67 | | 2.32 | 97 | 820 | 150 |
| Ca²⁺ (μmol l⁻¹) | 0.8 | 7.0–18 | 1.06 | 0.49 | | 0.57 | 140 | 270 | 450 |
| Mg²⁺ (mg l⁻¹) | 0.3 | 5.2–8.6 | 7.00 | 2.69 | | 3.98 | 110 | 1650 | 300 |
| Cl⁻¹ (mg l⁻¹) | 0.0 | 182–515 | 0.89 | 0.94 | | 1.03 | 110 | 750 | 100 |
| SO₄²⁻ (mg l⁻¹) | 0.005 | – | 0.33 | 0.76 | | 0.89 | ≤24 | 37 | 30 |
| Alkalinity (mg l⁻¹) | 0.11 | 6.7–19 | 16.4 | 7.33 | | 9.85 | 0.92 | 6.52 | 2.3 |
| Conductivity (μS cm⁻¹) | 180 | 800–2500 | 1240 | 735 | 3300 | 925 | 97 | 610 | 230 |
| pH | 8.0–8.5 | 7.6–9.5 | 9.1–9.5 | 8.9–9.5 | 9.5–10.5 | 8.8–9.1 | 7.1–8.5 | 8.0–9.0 | |
| Cations | 1.8 | | 15.93 | 8.81 | | 11.65 | 1.02 | 7.46 | 2.46 |
| Anions | 1.9 | | 17.62 | 9.03 | | 11.77 | 1.05 | 7.62 | 2.59 |

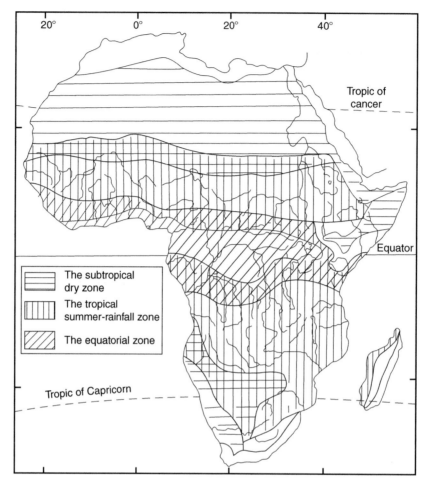

**Figure 4**   The major climatic zones of tropical Africa (modified from Walter *et al.,* 1975).

Existing as either relics of a large body of water contracted by siltation and other changes in the earth crust, as is the case with Lake Chad or by uplifting of the Earth's crust as in Lake Victoria, there is a large diversity of lakes. Some of the lakes have become internal drainage systems that undergo fluctuating salt changes during periods of drying and refilling, as is the case with Lakes Chala, Chilwa, Nakuru, Magadi or Turkana among the larger forms and Mombolo, Rombou or Latir of the smaller lakes.

Spatial and temporal structural heterogeneity of lake habitats is a major ecological factor influencing the existence and sustainability of aquatic ecosystems in Africa. As a result, there is a relatively high proportion of endemism among fish species, which have evolved independently in the different freshwater systems in Africa.

While the African Great Lakes harbor diverse and unique flora and fauna, it is the high endemism in fish species that attracted most interest. Among the Great Lakes, Lake Malawi has the most species exceeding 850, nearly all of them being endemic, despite the fact that it is of more recent origin than Lake Tanganyika.

It is generally accepted that reported numbers are an underestimate and 11 families have been identified with Cichlidae being the most speciose. The Lake Tanganyika fauna is less speciose, with a total of 16 families and more than 200 cichlid species. Lake Victoria is estimated to have been inhabited by more than 500 species, which have since declined, some becoming completely extinct through predation and competition from introduced Nile perch (*Lates niloticus*) and *Oreochromis niloticus*, respectively. The introduction of *Lates niloticus* and *Oreochromis niloticus* into Lake Victoria has led to the disappearance of more than 200 endemic species within a relatively short space of time of ten years. Arguably, this represents the greatest demise of the vertebrate group recorded in modern times.

### Distribution

The highest concentration of large lakes is found in East Africa. In addition to the African Great Lakes, there are smaller ones such as Lakes Turkana, Baringo, Nakuru, Naivasha, Magadi, Natron, and

Manyara on the Western Rift Valley and Albert, George, Edward, Kivu, and Rukwa on Eastern Rift Valley (**Figure 1**). Occupying a surface area of $69\,000\,km^2$, Lake Victoria is the largest freshwater lake in Africa and the second on earth. Depth-wise, Lake Tanganyika is second only to Lake Baikal, followed by Lake Malawi (**Table 1**). Lake Victoria is shallower and occurs between the Two Rift Valleys lakes of Kyoga and Eyasi. The shallower Lake Victoria occupies an uplifted region between the western and eastern arms of the Rift Valley. Other notable lakes of eastern Africa that lie close to the Rift Valleys are Mweru, Bangweulu, Chilwa, and Chiuta.

Climatic factors influence the location of aquatic systems and associated lakes in the Africa region. Their occurrence stretch across approximately $35°$ of latitude on either side of the equator. Temperature, rainfall, and wind have strong bearing on the biodiversity of resident aquatic ecosystems as shown in **Figures 4, 5** and **6**. Since the climatic pattern does not exactly follow latitude, the northern and southern dry desert belts interrupt the limits of intertropical zone. In the tropical zone, there are (i) hot, humid areas with two rainy seasons, (ii) hot belt with summer rain, and (iii) hot and arid areas. The highest rainfall is found in equatorial Zaire and in the littoral zone of Gulf of Guinea. On the other hand, desert areas to the north and south of the equatorial belt experience little and unpredictable rains, thus fewer lakes exist there. Generally, annual and within-year rainfall variations are reflected in river flows and lake levels, which influence occurrence of aquatic systems within the respective environs. A summation of the physico-chemical characteristics of the lakes is given in **Table 1**.

## Reservoirs

In respect of the diverse water bodies, reservoirs refer to areas where water is stored for purposes of drinking, irrigated agriculture, hydropower generation, recreation or fisheries, some of which are human-made (**Table 2**). These water bodies vary greatly in size ranging from stock dams for livestock drinking to lakes that cater for the demands of urbanization and industrialization.

### Formation

The rising demand for electric power, potable water, irrigation, fisheries products and ecotourism are

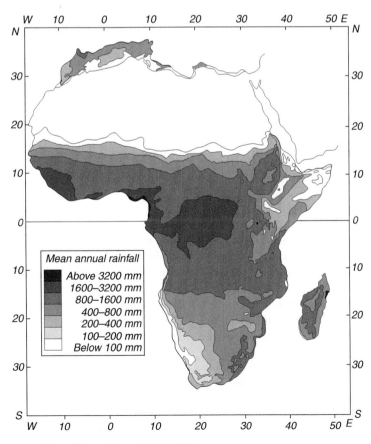

**Figure 5** Mean annual rainfall in Africa (Adapted from Balek, 1977).

**Figure 6**   The distribution of river systems in Africa.

linked to urbanization and industrialization, which has led to the construction of dams and storage reservoirs in sub-Saharan Africa. Thus, there is a proliferation of small dams, which have essentially altered the ecology and functional webs of rivers and lakes, creating lotic and lentic environments. The changed flood regime has elicited faunal and flora dynamics that exert influence over a broader expanse of the drainage basins.

Impoundment of streams and rivers has led to formation of many artificial water bodies in sub-Saharan Africa, ranging from small dams used for irrigated agriculture, flood control, small scale power generation, drinking water supply to large reservoirs which ecologically integrate functions of rivers and lakes.

Many combine characteristics of lotic and lentic environments, thus diverse biotopes are intertwined and vary seasonally with respect to flood dynamics. For example, Lake Volta was initially characterized by depletion of oxygen. Despite the oligotrophic nature, plankton populations developed well while algal blooms were frequent and aquatic macrophytes were persistent. Likewise aquatic organisms are adapted with respect to ecological functional heterogeneity of habitats, both spatially and temporally.

### Diversity

The proliferation and variation in the reservoir size, purpose for their construction and the hydrological

**Table 2** Physical and chemical characteristics of selected reservoirs of sub-Saharan Africa

| Parameter | Volta (Ghana) | Kossou (Cote d'Ivoire) | Lagdo (Cameroon) | Jebel Aulia (Sudan) | Mwandingusha (Zaire) | Kainji (Nigeria) | Kariba (Zambia/ Zimbabwe) | Cabora Bassa (Mozambique) |
|---|---|---|---|---|---|---|---|---|
| Surface area ($km^2$) | 8502 | 1600 | 700 | 600 | 446 | 1270 | 6000 $km^2$ | 2739 $km^2$ |
| Max. depth (m) | 75 m | 60 | 29 | – | 14 | 60 | – | 157 |
| Mean depth (m) | 18.8 m | 14.3 | 11 | – | – | 11 | – | 20.9 |
| Volume ($km^3$) | 148 | 20.5 | 7.7 | – | 1063 | 13.97 | – | 55.8 $km^2$ |
| Drainage area ($km^2$) | 385 185 | 33 000 | 216 | – | – | $1.6 \times 10^6$ | – | 56 925 $km^2$ |
| Altitude (m, asl) | 85 m | 204 | 216 | 378 | 1105.75 | 142 | 485 m | 314 |
| Usage | Hydropower, transport, potable water, fisheries | Fisheries and aquaculture, hydropower | Hydropower | Regulation of Nile irrigation | Hydropower, fisheries | Hydropower, fisheries | Power generation, irrigation, potable water, recreation, transportation, fisheries | Power generation, irrigation, potable water, recreation, transportation, fisheries |
| Year of closure | 1964 | 1971 | 1982 | 1937 | 1947 | 1968 | 1958 | – |
| Rainfall (mm $year^{-1}$) | – | 1150–1400 | – | – | – | – | | |
| Evaporation ($km^3 . year^{-1}$) | – | 2.4, 0.73 | – | – | – | – | | |
| Displacement time (years) | 4.3 | 5 | – | – | – | – | | 0.54 |
| $Na^+$ (mg $l^{-1}$) | | 5.8–6.4 | 3.9 | – | 3.07–6.0 | 1.8–5.2 | | |
| $K^+$ (mg $l^{-1}$) | | 1.6–2.5 | 3.0 | – | 0.78–1.56 | 1.4–3.6 | | |
| $Ca^{2+}$ (mg $l^{-1}$) | | 4.4–5.0 | 9.0 | 8–11 | 11.04–49.07 | 3.0–11.2 | | |
| $Mg^{2+}$ (mg $l^{-1}$) | | 2.0–2.2 | 2.0 | – | | 2.6–3.3 | | |
| $Cl^-$ (mg $l^{-1}$) | 3–14 | 1.0–35.0 | 3.0 | 3–7 | 0.71–2.49 | – | 2.2 | |
| $SO_4^{2-}$ (mg $l^{-1}$) | | 0.2–1.0 | – | 1 | 0.19–5.00 | – | | |
| Alkalinity (mg $l^{-1}$) | | 24–36 | 55 | 65–70 | – | 0.45–0.69 | | |
| $PO_4^-$ (mg $l^{-1}$) | 0.08–0.68 | 0.2–0.8 | 0 | 0.01–25 | 9.0–18.0 | 0.02–10.6 | | |
| Conductivity (μS $cm^{-1}$) | | 70–180 | 80 | 160–330 | 128–309 | 46.6–99.6 | | |
| pH | 6.9–8.1 | 7.7–9.0 | 7.2 | 8.0–9.8 | 7.1–8.4 | 6–7.6 | 7.5–8.5 | 6.9–8.7 |

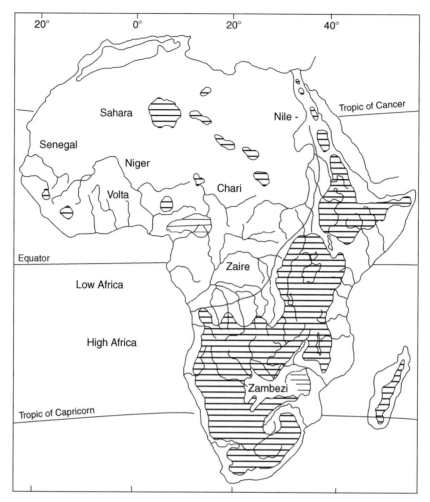

**Figure 7**  The general topography of Africa showing the main watersheds, the areas over 1000 m, and the approximate division between High and Low Africa (Adapted from Beadle, 1981).

regime are so high that it is not possible to discuss each one of them separately. Thirteen reservoirs of over 300 km² have been reviewed with respect to fisheries potential and the physico-chemical characteristics of some of them are given in **Table 2**.

Aquatic organisms have likewise adapted to spatial and temporal changes in the new habitats. The discontinuities between a riverine regime and lacustrine conditions are best illustrated in Jewel Aulia.

### Algae

Few reservoirs have received thorough investigations for algae types, but good accounts are available from studies of Lakes Volta and Kariba. Following filling up of the dams, colonization by algae was dominated by chlorophytes later succeeded by diatoms, desmids, cyanophytes and euglenophytes. The density rises after closure of the dam and fluctuates around

a peak. The growth of phytoplankton is reflected by a rise in pH, a decline in phosphate during initial stages and subsequent rise in silicate concentration. Seasonal periodicity, primary production, and community structure are influenced by depth, latitude, nutrient inflow, temperature and sediment type. Eutrophication is common in small dams, which are close to sewerage systems or big towns and are usually dominated by *Chlorocystis* and cyanophytes.

Aquatic macrophyte communities range from small plants like *Azolla, Lemna, Spirodela* and *Salvinia*, to large floating *Pistia, Elodea* and the nuisance unintentionally introduced water hyacinth *Echhornia crassipes. Ceratophyllum* and *Utricularia* represent submerged aquatic forms. Emergent plants occur on the fringes of the reservoirs and are represented by *Jussiaea, Echinochloa,* and *Alternanthera.* The marginal elephant grass *Vossia cuspidata* colonizes dams within one year of filling while semi-aquatic

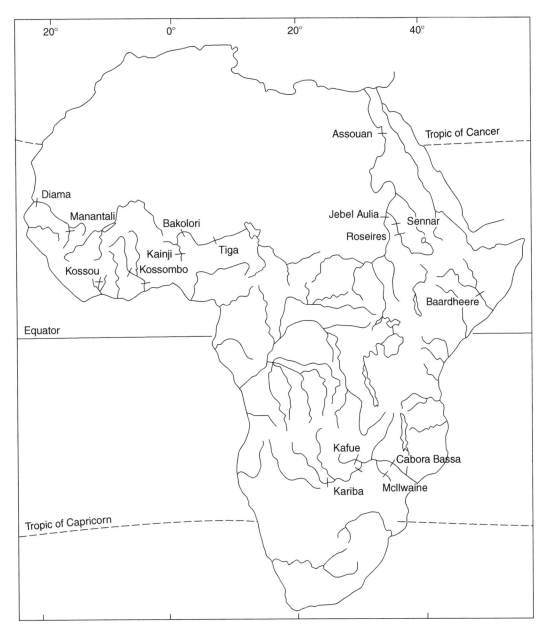

**Figure 8** Location of major human-made lakes in Africa.

vegetation includes *Typha, Phragmites, Echinochloa, Ficus, Scirpus* and *Leersia*. Generally, weed infestation is related to the annual water level changes and influx of nutrients, and has created problems for transport, fishing and water supply.

## Zooplankton

Expansion of the water column provides changes in chemical characteristics that favor establishment of invertebrates. As reservoirs mature, there is generally an increase in the biomass of larval forms of terrestrial insects like *Mansonia*, mayfly nymphs,

dragonfly, caddisfly, and truly aquatic insects like hemiptera and colleoptera, and mollusks. Crustacean zooplanktons colonize reservoirs according to feeding behavior; filter feeders being first and predators are last. There are differences between upper and lower sections of the reservoir with copepods being dominant in the former. As the dam fills, there is a succession of crustacean zooplankton from rotifers (*Brachionus, Keratella, Filinia, Polyartha*) followed by daphnids (*Bosmina, Diaphanosoma, Ceriodaphnia, Daphnia*) and lastly copepods (*Mesocyclops leurkarti*). Due to predation pressure, a shift might occur in community structure affecting diversity.

For example, the introduction of *Limnothrissa miodon* into Lake Kariba in 1967–1968 led to the decline of large zooplankton by 1968–1974, notably *Ceriodaphnia, Diaphanosoma* and *Diaptomus*. Fish tend to select *Cladocera,* so copepods are more cosmopolitan.

In Lake Kariba, the benthos fauna was studied during a complete flood regime. Recruitment of organic and allochthonous matter from the terrestrial zone during high floods enhanced production of invertebrate biomass. The biomass of fish follows a prey–predatory relationship.

### Fish Fauna

The creation of lacustrine systems in artificial reservoirs provides a survival challenge to riverine species some of which fail to adapt, being unable to utilize organisms of the pelagic zone and to reproductive in new habitats. Clupeids have successfully occupied the pelagic zone and become the dominant group, whether through natural colonization as in Lake Volta or as introductions as in Lake Kariba and Cabora Bassa. Cichlidae have also been successful and so have Cyprinidae, Bagridae, Claridae, Citharinidae, Centropomidae, Mochokidae, Mormyridae, Osteoglossidae, Polyteridae, Schilbeidae, and Characidae, depending on their occurrence in the catchment area. The introduction of *Limnothrissa miodon* and *Stolothrissa tanganyikae* into Lakes Kariba and Cabora Bassa is generally hailed as a success story. On the contrary, the introduction of *Lates niloticus* and *Oreochromis niloticus* into Lake Victoria has had adverse impacts on native fish species and has received bad publicity.

### Distribution

Just like natural lakes, big reservoirs occur close to main rivers. **Figures 7** and **8** show the association between major reservoirs and key wetlands of sub-Saharan Africa.

The biogeography of sub-Saharan fauna and flora shows patterns that can be reconstructed from the sequence of events which commenced with drifting apart of the continents. Within the sub-Saharan region, there are some keystone species that evidently connect water catchments. From the Nile to West Africa and southward to Zaire, *Sarotherodon galilelaeus, Lates niloticus, Hydrocynus forskalii,* and *Tilapia zillii* are represented. These species occur in reservoirs as far apart as Kossou in Cote d'Ivoire, Ladgo in Cameroun, and Jebel Aulia in Sudan. The connection between Nile and the western reservoirs that excludes the south is suggested by occurrence of

*Alestes dentex, Citharinus citharus, Oreochromis aureus* and *Oreochromi niloticus*. The Chad basin's connection to West Africa and Zaire is indicated by presence of *Hepseus odoe, Labeo parvus* and *Gnathonemus petersii*.

Fish are also good indicators of environmental conditions because they cannot move overland from one aquatic system to another, hence they have to adapt to changes within a given system or face extinction. In addition, they impact on the distribution and abundance of other aquatic organisms through the food chain or competition.

## Further Reading

Bootsma HA and Hecky RE (2003) A comparative introduction to the biology and limnology of the Great Lakes. *Journal of Great Lakes Research* 29(Suppl. 2): 3–18.

Cohen AS, Soreghan MJ, and Scholz CA (1993) Estimating the age of ancient lakes: An example from Lake Tanganyika, East African Rift system. *Geology* 21: 511–514.

FAO (1990) *Source Book for Inland Fishery Resources of Africa (3 Vols.)* Committee for Inland Fisheries of Africa Technical Papers.

FAO (1994) *Status of Fish Stocks and Fisheries of Thirteen Medium sized African Reservoirs* by M. Van der Knaap. Committee for Inland Fisheries of Africa Technical Paper 26. Rome: FAO.

Habernyan KA and Hecky RE (1987) The late Pleistocene Holocene stratigraphy and paleolimonology of Lakes Kivu and Tanganyika. *Palaeogeography, Palaeoclimatology and Palaeoecology* 61: 169–197.

Johnson TC (1996) Sedimentary processes and signals of past climatic change in the large lakes of the East African Rift Valley. In: Johnson TC and Odada EO (eds.) *Limnology, Climatology and Paleoclimatology of the East African Lakes*, pp. 367–412. Amsterdam: Gordon and Breach.

Karenge LP and Kolding J (1994) On the relationship between hydrology and fisheries in the Lake Kariba, Central Africa. *Fisheries Research* 22: 205–226.

Kodings Ad (1998) *Tanganyika Cichlids in their Natural Habitat.* USA: Cichlid Press.

Leveque C (1997) *Biodiversity Dynamics and Conservation: The Freshwater Fish of Tropical Africa.* Cambridge, UK: Cambridge University Press.

Livingstone DA and Melack JM (1984) Some lakes of sub-Saharan Africa. In: Taub FB (ed.) *Ecosystems of the World 23: Lakes and Reservoirs.* Oxford, UK: Elsevier Science.

Ogutu-Ohwayo R (1990a) The decline of the native fishes of lakes Victoria and Kyoga (East Africa) and the impact of introduced species, especially the Nile perch, *Lates niloticus,* and Nile tilapia, *Oreochromis niloticus. Environmental Fish Biology of Fishes* 27: 81–96.

Ogutu-Owhayo R (1990b) The reduction in fish species diversity in lakes Victoria and Kyoga (East Africa) following human exploitation and introduction of non-native fishes. *Journal of Fish Biology* 37: 207–208.

Owen RB, Crossley R, Johnson TC, *et al.* (1990) Major low levels of Lake Malawi and their implications for speciation rates in cichlid fishes. *Proceedings of the Royal Society of London* B 240: 519–553.

Rosendahl BR (1987) Architecture of continental rifts with special reference to East Africa. *Annual Review of Earth Planet Science* 15: 445–504.

Tiercelin J-J and Mondeguer A (1991) The geology of the Tanganyika trough. In: Coulter GW (ed.) *Lake Tanganyika and its Life*, pp. 7–48. Oxford, UK: Oxford University Press.

Welcomme RL (2001) *Inland Fisheries: Ecology and Management.* Oxford, UK: Fishing News Books Limited.

## Relevant Websites

http://www.springerlin.com.
http://na.unep.net/AfricanLakes/.
http://en.wikiprdia.org/wiki/African_Great_Lakes.
http://www.ilec.or.jp/database/map/regional/africa.html.

# Asia

**B Gopal and D Ghosh,** Jawaharlal Nehru University, New Delhi, India

## Lakes and Reservoirs of Asia (Formation, Diversity, Distribution)

Asia, the world's largest continent (44.39 million $km^2$), lies entirely in the Northern Hemisphere extending from the Equator to the Arctic Circle. The Ural and Caucasus mountains, and the Caspian Sea are accepted by some as the border with Europe, others extending the border to include the Black Sea, the Aegean Sea, and the Bosphorus and Dardanelles straits. The continent is separated from Africa by the Suez Canal and the Red Sea, and from North America by the Bering Strait.

Geologically, Asia consists of Precambrian landmasses, the Arabian and Indian peninsulas in the south and the central Siberian plateau in the north, as well as areas of very recent origin. The Asian continent however owes its present geomorphological features to the geological events of the past 50 million years. After the breaking apart of the Gondwanaland, the Arabian and Indian plates drifted north and northeastwards. The thrust of the Indian plate under the Asian plate obliterated the Tethys Sea and resulted in the rise of Himalayan ranges and the Tibetan Plateau. The collision of the Arabian plate with west Asia pushed up the Caucasus mountains and reduced the Tethys sea to the brackish Sarmantian Lake. The subsequent epeirogenic processes (large scale upliftment and subsidence of land) and volcanic and glacial activities have resulted in a highly complex pattern of landforms.

Physiographically, Asia comprises of (a) the vast central zone of high plateaus, rising to more than 4500 m and the surrounding high mountain ranges, (b) the northern lowlands of Central Asia and Siberia, (c) the eastern lowlands of eastern China and southeast Asia drained by mighty rivers such as Yangtse, Hwangho, and Mekong, (d) the southern peninsular plateaus of India and Arabia, and (e) vast plains of major rivers (Tigris-Euphrates, Indus, Ganga, and Brahmaputra) that are separated by various spurs of mountain ranges (**Figure 1**). Most important of the highland plateaus is the Pamir Plateau (often referred to as the Pamir Knot) from which high mountain ranges radiate both eastwards and westwards. The Hindukush, Koh-i-Baba, and Suleiman mountains radiate westwards, enclose the high plateaus of Afghanistan, Baluchistan, and Iran, and later coalesce in the Armenian Knot. Eastwards, Kunlun, Karakorum, and the Himalaya, which are the youngest (Tertiary period), enclose between them the large

Tibetan Plateau. The most northerly range, Tien Shan is the oldest and arches over a depression, the Tarim basin, which is bordered on the south by the Kunlun range.

The Himalaya, after a sharp southward turn, extends eastwards in a large arc across north India, and includes the world's highest peaks. The Himalayan range again turns southwards to continue as Arakan Yoma through Myanmar, and can be traced further south in Andaman-Nicobar islands and into Sumatra and Java Islands.

The eastern lowlands of China and southeast Asia are intersected by a series of relatively low ranges running in the north-south direction. The Deccan Plateau of the Indian subcontinent is separated from the Ganga–Brahmaputra plain by the Vindhyan and Satpura ranges and is also rimmed on the west and the east by a series of low mountain ranges (the Western Ghats and Eastern Ghats) that coalesce in the south.

The climate in Asia varies between extremes of both temperature and precipitation. It is governed by the monsoon winds and influenced by the high mountain ranges. After the temperatures start soaring during the summer (April onwards), the southwest monsoon brings very large amounts of precipitation over the south and southeast Asia. The large mountain ranges influence the spatial distribution of rainfall, resulting in sharp gradients. During winter, monsoons from the north move into East Asia and cause cold dry weather. Bitter cold polar winds keep part of northern Siberia frozen throughout the year. More than one-third of the Asian continent is therefore arid (annual precipitation <200 mm), semi-arid (200–500 mm), or even hyper-arid (<25 mm), and is either too hot or too cold. This includes most of the Middle East, Central Asia, Mongolia, western China (much of Qinghai, Xinjiang, and Xizang (=Tibet) provinces), and parts of the Indian subcontinent (most of Afghanistan and Pakistan, western India and parts of peninsular India, and even Sri Lanka). Interestingly, these dry regions have the largest concentration of natural, though small, lakes.

Thus, the continent is remarkably unique for its great contrasts and extremes because it includes the highest mountain peak (Mt Everest, 8850 m) and the lowest place on land (Dead Sea, 394 m below sea level), most rainy and most arid areas, as well as the hottest and coldest places on the Earth. Not only did innumerable plants and animals originate in Asia, but

**Figure 1** Map of Asia (from http://upload.wikimedia.org/wikipedia/commons/a/a8/Asia-map.png).

the human species and human civilization also evolved here. Today, Asia is the home to three-fifths of the world's human population, which exhibits as high diversity of human cultures as the continent's biological diversity.

## Natural Lakes

Commensurate with its long eventful geological history and complex geomorphology, Asia has a large diversity of natural lakes that vary in their age, area, depth, hydrology, water quality, and biodiversity. Asia has the world's largest (Caspian Sea), oldest and deepest (Lake Baikal) lakes as well as those lying at the highest (~6000 m) and lowest elevation (Dead Sea, >400 m below sea level), and with highest salinity (>400‰). Lakes larger than 1000 km$^2$ and those more than 100 m deep are listed in **Tables 1** and **2**. Some regions such as the central Asian highland and Indonesia have a very high concentration of lakes: Mongolia has >3500 lakes (>10 ha), most of which are brackish or saline and small. Only 12 lakes have an area of more than 100 km$^2$. There are 2800 lakes (>100 ha area) in China and of these more than half are in Tibet. In central Asia, Kyrgyzstan has 1923 lakes covering about 684 km$^2$. Distributed over all major islands of Indonesia, there are about 520 lakes of which three are among the world's 20 deepest lakes. Within Himalaya, some regions have a very high concentration of lakes: there are 227 lakes in Sikkim (about 7000 km$^2$ area) at altitudes rising up to 6000 m; and 90 lakes lie among the glaciers at 4460 m to 5645 m altitude within 650 km$^2$ of Khumbu Himal (East Nepal) region. The Indian subcontinent (excluding Himalaya) has very few natural lakes of which most are of fluvial origin (floodplain lakes). There is no natural lake in the Arabian Peninsula and in Sri Lanka.

Most of the natural lakes originated variously from tectonic activities over the past several million years. Lake Baikal (Russia) originated in a rift valley about 25 million years ago. The rift valley is still seismically active, and reported to have become deeper by 9 m after the 1959 earthquate, which had its epicenter in the middle of the lake. The 636 km long and 80 km wide lake with a surface area of 31 494 km$^2$ and a maximum depth of 1671 m, has its floor 1285 m below sea level. It is the largest freshwater lake in Asia that holds 23 600 km$^3$, or about 20% of the world's total freshwater resources. It is fed by 300 streams but the only outflow is the River Angara. The lake is also unique in its water quality (very low ion concentration, well mixed water column despite its depth) and high

**Table 1**  Natural lakes in Asia that are more than 100 m deep

| Lake | Characteristic | Depth, m |
|---|---|---|
| Lake Baikal, Russia | Tectonic | 1637 |
| Caspian Sea, Russia | Endorheic saline | 1025 |
| Lake Dieng Plateau, Indonesia | Crater | 884 |
| Lake Issyk-Kul, Kazakhstan | Endoheric/ monomitic | 668 |
| Lake Matano, Indonesia | Tectonic | 590 |
| Lake Toba, Indonesia | Volcanic/tectonic | 529 |
| Sarez Lake, Tajikistan | Tectonic | 505 |
| Lake Van, Turkey | Saline | 451 |
| Lake Poso, Indonesia | Tectonic | 450 |
| Lake Tazawa, Japan | Volcanic, caldera | 423.4 |
| Heaven lake, China | Crater lake | 384 |
| Lake Chonji/Tianchi, China | Crater | 384 |
| Lake Shikotsu, Japan | Crater | 360.1 |
| Lake Wisdom, New Guinea | Volcanic, caldera | 360 |
| Dead Sea, Israel | Endorheic/ hypersaline | 330 |
| Lake Towada, Japan | Volcanic, caldera | 326.8 |
| Lake Teletskoye, Russia | Tectonic | 325 |
| Lake Dibawah, Indonesia | Tectonic | 309 |
| Lake Singkarak, Indonesia | Tectonic | 268 |
| Lake Khuvsgul, Mongolia | Freshwater | 262 |
| Lake Kara Kul, Tajikistan | Glacial/tectonic lake | 240 |
| Lake Segara Anak, Indonesia | Crater | 230 |
| Lake Ranau, Indonesia | Tectonic/volcanic | 229 |
| Lake Mainit, Philippines | Freshwater | 223 |
| Lake Karakul, Tajikistan | Meteoritic impact | 220 |
| Lake Mashu, Japan | Endoheric | 211.4 |
| Lake Towuti, Indonesia | Tectonic | 203 |
| Lake Toya, Japan | Volcanic caldera | 179.7 |
| Lake El'gygytgyn, Russia | Meteoritic impact | 175 |
| Lake Maninjau, Indonesia | Caldera | 169 |
| Lake Chuzenji, Japan | Volcanic | 163 |
| Lake Pakis, Indonesia | Crater | 158 |
| Lake Okutama, Japan | Reservoir | 142 |
| Lake Klindungan, Indonesia | Crater | 134 |
| Lake Basum Tso, China | Tectonic | 120 |
| Lake Kussharo, Japan | Volcanic, caldera | 117.5 |
| Lake Ritsa, Georgia | Tectonic/freshwater | 116 |
| Lake Umbozero, Russia | Tectonic/freshwater | 115 |
| Lake Buhi, Philippines | Rift lake | 112 |
| Lake Lanao, Philippines | Rift lake | 112 |
| Lake Burdur, Turkey | Salt | 110 |
| Lake Biwa, Japan | Freshwater | 103.8 |
| Lake Ashi, Japan | Caldera | 103.6 |
| Lake Lindu, Indonesia | Tectonic | 100 |

biodiversity (1085 plants and 1550 animal species) of which 60% is endemic. An interesting feature of Lake Baikal is the endemic fauna adapted to deep water (e.g., the long-finned, translucent, Baikal oil fish (*Comephorus baicalensis* and *C. dybowskii*) living at a depth of 700–1600 m) and the large diversity of ostracods and flatworms. Lake Baikal is also known for the only freshwater seal, the Baikal Seal (*Phoca sibirica*).

**Table 2**   Natural lakes in Asia that are larger than 1000 km²

| Lake | Characteristic | Area, km² |
|---|---|---|
| Lake Songkhla, Thailand | Natural/brackish | 1040 |
| Lake Toba, Indonesia | Volcanic/tectonic | 1130 |
| Sevan, Armenia | Tectonic | 1236 |
| Lake Vygozero, Russia | Freshwater lake | 1250 |
| Kirov Bays, Azerbaijan | Coastal, brackish to saline | 1325 |
| Lake Tengiz, Kazakhstan | Salt | 1382 |
| Lake Hyargas, Mongolia | Saline, deep | 1407 |
| Lake Tuz, Turkey | Salt/shallow | 1500 |
| Lake Hamun-i-Helmand, Afghanistan | Marshy salt lake | 1600 |
| Lake Zaysan, Kazakhstan | Freshwater | 1810 |
| Lake Ogan-Komering, Indonesia | Floodplain | 2000 |
| Lake Taihu, China | Riverine, floodplain | 2428 |
| Lake Hulun, China | Freshwater | 2339 |
| Lake Namtso, China | Salt | 2470 |
| Lake Alakol, Kazakhstan | Salt | 2650 |
| Lake Dongting, China | Shallow, riverine | 2820–20 000 |
| Lake Aydar, Uzbekistan | Brackish/salt | 3000 |
| Lake Uvs Nuur, Mongolia | Salt | 3350 |
| Lake Peipus, Estonia | Freshwater lake | 3555 |
| Lake Van, Turkey | Saline | 3755 |
| Lake Xingkai Hu (=Lake Khanka) China/Russia | Riverine | 4380 |
| Lake Taymyr, Russia | Freshwater | 4560 |
| Lake Urmia (Iranian Azerbaijan) | Salt | 5200 |
| Lake Qinghai, China | Drainless lake/ saline | 5694 |
| Lake Issyk-Kul, Kazakhstan | Endoheric/ monomitic | 6236 |
| Lake Balkhash, Kazakhstan; China | Endorheic/saline | 16996 |
| Aral Sea, Kazakhstan/ Uzbekistan | Glacial/saline | 17158 |
| Lake Baikal, Russia | Tectonic | 31500 |
| Dead Sea, Israel | Endorheic/ hypersaline | 40650 |
| Naujan lake, Philippines | Volcanic | 81250 |
| Caspian Sea, Russia | Endorheic saline | 371 000 |
| Tonle Sap, Cambodia | Riverine | 2500–10 000 |
| Lake Poyang, China | Riverine, floodplain | 3585 |
| Lake Hawr al Hammar, Iraq | Seasonal flooding | 2500 |
| Lake Tharthar, Iraq | Seasonal flooding | 2700 |
| Lake Dibbis, Iraq | Seasonal flooding | 1985 |

Lake Issyk-kul (=Ysyk-Köl) in Kyrgyzstan is also ca. 25 Ma old lake, located at 1606 m elevation in an intermontane valley in the north of Tien Shan mountains. The 668 m deep lake is 182 km long and up to 60 km wide, and covers an area of 6236 km². The endorheic lake basin is surrounded by high glacier-capped mountains and receives more than 100 rivers

of which Djyrgalan and Tyup are the largest. The water is slightly saline and never freezes. The lake supports relatively few plant and animal species many of which are endemic. The native fisheries have declined in recent years after the introduction of the Sevan trout from Armenia. The freshwater Lake Biwa (Honshu, Japan) is another very old (2 Ma) lake that occupies an oblong tectonic rift basin (104 m deep).

The origin of some old lakes is traced back to the geological events that gave birth to the Asian continent. The Caspian Sea, the world's largest lake (surface area about 394 000 km² and total volume 78 000 km³), which lies east of the Caucasus and north of the Elburz Mountains, and 28 m below the sea level, evolved over the past >7 M years, passing through phases of constriction and expansion and large water level changes. The collision of the Arabian Peninsula with west Asia pushed up the Elburz-Kopet Dag and Caucasus Mountains and formed the Sarmatian Lake, composed of the present Black Sea and south Caspian. Further orogenic events separated the Black Sea from South Caspian which started sagging. A mountain arch rose across the south basin dividing it into the Khachmas and Lenkoran Lakes. Subsidence of this land bridge due to continued sagging reunited the two basins (Balakhan Lake). About 3–2 million years BP the lake expanded to more than three times its present area – reconnecting with Black Sea and Aral Sea. The lake surface was at +50 m. During the Pleistocene (2–0.7 Ma BP) the lake level rose and fell in a series of events but the lake level sank deeper each time, reaching its lowest level of −120 m from sea level. These changes were evidently linked to glaciation cycles, and the retreat of glaciers in Russia. In recent years, the water level has fluctuated by a few meters in response to the periodic changes in the course of the Amu-Darya River. The lake level dropped by 3 m between 1930 and 1977 but rose rapidly after 1978 to −26 m mark.

The lake comprises of three basins: a large shallow northern basin (max. 20 m deep), a small middle basin (max. 788 m deep) and a large deep (maximum 1006 m) south basin. The salinity varies from 0.1% to 1.2% along a north-south gradient. The lake is fed by River Volga, which forms a delta on the lake's northern shore, discharging annually an average 237 km³ fresh water (range: 200–450 km³) into the lake. River Kura (16.8 km³) and the Ural River (8.1 km³) provide other major inflows. During the past few years, water level has again started declining because of the exploitation of R. Volga waters.

Lake Baskunchak, north of the Caspian Sea, is another saline (>30% salinity) lake that lies 21 m below sea level. It also originated by getting isolated from the sea during the Pleistocene.

Two interesting lakes in Israel: Lake Kinneret in the north and Dead Sea in the south, lie in the Jordan rift valley which is a part of the Great Rift Valley system extending from Turkey to Zambezi in Africa. About 3 Ma BP, the valley was flooded from the Meditrerranean Sea that deposited thick layers of salts. The orogenic processes raised the land between the rift and the Mediterranean, separating the lakes from the Sea. Further tectonic processes, linked with the northward shift of the Arabian peninsula, resulted in sinking of the southern part deeper than in the north, and the deposition of sediments pushed up the salts as the Mt Sedom on the southwest side. Today, the Dead Sea has its surface 418 m below sea level whereas Lake Kinneret (also known as L. Tiberias or the Sea of Galilee) is 211 m below sea level. Dead Sea is the deepest (max. depth 330 m) hypersaline lake with average 310% salinity. Interestingly, the salts are mostly magnesium and potassium chloride (52% and 37%) and sodium concentration is low. During the past 70 k years, water level of the Dead Sea has fluctuated greatly by several hundred meters, and started dropping in the past few thousand years. High ariditry (<25 mm rain) is also responsible because the inflow from L. Kinneret has declined. L. Kinneret is nearly freshwater though the ionic composition varies at different depths. It is fed by R. Jordan which used to flow through Hula marshes before entering the lake but the river has now been regulated and a part of its flow used for irrigation.

Another large lake, Aral Sea, lies between Kazakhstan and Uzbekistan, in the Aral-Sarykamysh Depression which was formed by tectonic processes and wind erosion, ca. 3 Ma BP. The saline lake is fed by Amu Darya and Syr Darya rivers, which originate in the Pamir ranges and discharge together about 110 km$^3$ of water. However, the diversion of river flow for irrigation has resulted in rapid shrinking of the lake from 68 000 km$^2$ in 1960 to 17 160 km$^2$ in 2004 (**Figure 2**), and consequently a sharp increase in salinity.

Lake Balkhash in eastern Kazakhstan, lies about 1600 km east of Aral Sea, in another tectonic depression surrounded by mountain ranges. The 600-km long and 5–70-km wide lake covers an area of about 16 000 km$^2$. It is an endorheic lake, with its western half being shallow (6 m deep) fresh to brackish water and eastern half being deep (26 m) and saline (up to 7%). The lake receives about 80% of its water from R. Ili flowing from Tien Shan mountain ranges in China. Excessive water diversion for cotton cultivation has lowered the lake area and increased its salinity.

Several other tectonic lakes in central Asia are large and deep; e.g., Lake Hövsgöl (Khuvsgul) in Mongolia is a 262 m deep, 2 Ma old lake with freshwater of pristine quality whereas Lake Alakol at 347 m altitude in Kazakhstan is 54 m deep saline lake with an area of 2 650 km$^2$. Lake Zaysan in eastern Kazakhstan, however, is a large 106-km long (1,810 km$^2$) freshwater lake at 420 m elevation, and is the source of Irtysh River. Khar-us lake (1157 m altitude) is another large (1 852 km$^2$) but shallow (only 4 m) freshwater lake. Lake Teletskoye, the largest lake (233 km$^2$) at 434 m altitude in the Altai Mountains is 325 m deep tectonic lake.

In Qinghai-Xizang plateau, lakes have developed under the influence of the vertical, transversal, and oblique tectonic movements resulting in a lattice-like network of lakes. Most lakes are therefore long and narrow. Many fault valleys were later dammed by material deposited by landslide or moraine carried by retreating glaciers or by floods. Other lakes were created by glacial activity. Such lakes are distributed mainly at the periphery of the plateau; they are numerous but small. Most of these lakes are saline and lie at elevations of more than 3000 m. In most cases, the depth has not been determined yet. L. Qinghai (5 694 km$^2$) lying at 3205 m between Hainan and Haibei provinces, is the largest of them. During the past few decades, following the drying up of most of the tributary streams, the water level has gone down and several smaller lakes have separated from it near its periphery. The second largest lake in the region is Nam Co (2 470 km$^2$) lying close to Lhasa, at 4718 m altitude. Lake Pangong Tso (4 250 m altitude), a transboundary lake with about two-thirds of its 134 km length in China and the rest in India, also lies in a dammed fault valley. Lake Mansarovar (=Mapham Yutso) is a freshwater lake (320 km$^2$) at 4556 m altitude near Mt Kailash. Together with the adjacent Rakshastal lake (=La'nga Co), it is not only of great religious importance but four major rivers Sutlej, Brahmaputra (=Tsangpo), Indus and Karnali, have their source in its vicinity. Other important lakes in Tibet region are: Puma Yumco (5030 m elvation), Yamdrok (620 km$^2$) and Dagze Co (260 km$^2$).

In Yunnan province (China) there are numerous rift-valley lakes that are deep and developed during the late Tertiary or early Quaternary, along tectonic fracture belts. For example, Lakes Chehu, Yang Zong Hai, Fu Xian Hu (155 m deep), Xing Yun Hu, and Tong Hai lie along the Xiao Jiang fracture zone and Lakes Lugu Hu (93 m deep), Jian Chuan Hu, Er Hai, etc lie along the Hong He (Red River) fracture zone. Most of the important lakes in Inner Mongolia (Lake Hu Lun Hu and Lake Dai Hai), Xinjiang (Lakes Bosi

**Figure 2** (Continued)

**Figure 2** Changes in the area of Aral Sea between 1973 (a) and 2004 (b) following the diversion of river flows for cotton cultivation in arid Central Asia. (From URL: na.unep.net/digital_atlas2/webatlas.php?id=11.)

Teng Hu, Saili Mu Hu, Bulun Tuo Hu, Aibi Hu) and northern Hebei province are also relatively large and of tectonic origin. Tectonic lakes in Indonesia include Lakes Diatas, Dibawah, Lindu, Mahalona, Poso, Ranau Rawa Danau, Singkarak, Tawar Laut, Towuti and Yamur, and in Japan the better known are L. Akoi, L. Kizaki, and L. Nakatsuna. Major lakes in Nepal lie in deep gorges (e.g., Lake Rara; 167 m deep) or in blocked valleys (e.g., Lake Phewa).

Numerous lakes developed under the influence of glaciers either solely or in combination with the tectonic processes, throughout the Himalayan ranges from Kashmir in the west to Arunachal Pradesh in the east, particularly at elevations beyond approx. 2000 m. These lakes are generally small and relatively shallow. Many of them occupy fault depressions with blockages created by moraine or landslides. The outflows from most of the high altitude glacial lakes gradually turn into major rivers. Noteworthy among these lakes are Gangabal, Nilnag, Kounsarnag, Kishensar and Alipather in Kashmir, Tso Moriri (a large, 75 m deep brackishwater lake), Tso Kar and Statsapuk Tso, Yaye Tso, Khagyar Tso, Pangur Tso in Ladakh, Chandratal, Nainital and Bhimtal in Himachal and Uttarakhand, and Lake Tsomgo in Sikkim. In Nepal and Sikkim Himalaya, the lakes occur also at elevations beyond 5000 m; for example, Lake Sonlaphyo Tso at 6230 m in Sikkim.

Both caldera and crater lakes are common in southeast Asia, particularly Indonesia. The caldera lakes are formed by volcanic eruptions followed by the collapse of the dome causing large and deep depressions, the caldera. Crater lakes develop by filling in of the extinct craters. Lake Toba, in the middle of north Sumatra (Indonesia) is the world's largest caldera lake (100 km long and 30 km wide) at an elevation of 900 m. It lies near a fault line running along the Sumatra Fracture Zone, and was formed about $71\,500 \pm 4000$ years ago, by a huge eruption, the largest within the last two million years. The volcano spewed out about 2000 km$^3$ of lava flow and 800 km$^3$ of ash which was carried by the winds westwards up to the Indian subcontinent and deposited there as a 15-cm thick layer (up to 6 m thick at one site in central India). The volcanic dome in the middle of the lake formed a large island in the lake. The area is seismically active with frequent earthquakes and smaller eruptions near the margin of the caldera and lava domes. Some parts of the caldera have been uplifted (for example, Samosir Island by 450 m). Within Sumatra, Lake Maninjau is another caldera lake formed by a volcanic eruption ca. 52 000 years BP. The caldera measuring $20 \times 8$ km is occupied by the 99.5 km$^2$ lake with a maximum depth of 165 m. The lake outflows on its west into the Antokan river

which has been used for hydropower generation. Lake Wisdom (360 m deep) and Lake Dakataua (120 m deep) are two major caldera lakes in Papua-New Guinea. Lake Yak Loum in Ratanakiri province of Vietnam is another large caldera (about 75 km in diameter and 48 m deep). The deepest lake (373 m) in China, Lake Tian Chi is also a caldera lake, in the Bai Tou Shan (=Changhai) mountain at 2194 m altitude (area 10 km$^2$). Volcanic eruptions have recurred here quite frequently (in 1597, 1668, and 1702). Lake Kurile is a large caldera (formed 6440 BC) in the southern Kamchatka Peninsula. It has an area of 77 km$^2$ and an average depth of 176 m (maximum 306 m). Lake Taal of Philippines (234 km$^2$, 172 m deep) is a large caldera lake in a volcanically active region and its basin has been modified several times. An eruption in 1754 created an outflow (R. Pansipit) on its southwest corner and another eruption in 1911 in the middle of the lake created a 45 km$^2$ island with a small crater lake on it. In Japan, the 327-m deep Lake Towada, 233-m deep Lake Ikeda and the 179-m deep Lake Toya are major caldera lakes.

The largest freshwater lake in the Philippines, Laguna de Bay (949 km$^2$) was formed by a combination of volcanic and tectonic activities which separated it from the Manila Bay. Lake Lanao (340 km$^2$, 112 m deep) in Philippines was formed by the collapse of a volcano and tectonic-volcanic damming of the basin between two mountain ranges and similarly Lake Buhi (18 km$^2$) was formed following the collapse of Mt Iriga caused by an earthquake in 1641 AD. Lava flows often create large dams blocking the tectonic valleys or even rivers. Lake Sevan (Armenia) located at 1950 m elevation is one such large lake. The lake is nearly triangular and divided into two basins, a small, 95 m deep northwestern part and a large shallower southeastern part. The freshwater lake is nearly endorheic because only <10% of total inflow is released through the only stream Razdan. In an effort to reduce evaporation losses and increase the flow in the river for irrigation, the lake area has been reduced from 1416 km$^2$ to 1241 km$^2$ and water level lowered by about 20 m. Lake Van (=Van Gölü) in Turkey near its border with Iran is another large endorheic brackishwater lake that was formed by lava flow from the Nemrut volcano blocking the natural drainage to the Murat River. The basin of Lake Urmia, another large shallow (maximum 16 m) saline lake in Iran is also encircled by lava. Lake Jing Po Hu (area 95 km$^2$, maximum depth 62 m) of Heilongjiang province in China was formed by Quaternary basalt lava dam blocking the outlet of River Mu Dan Jiang. Similarly, Lake Wu Da Lian Chi (18.47 km$^2$) of Dedu county (China) was formed when River Bai

He was blocked by lava from a volcanic eruption in 1719–1721.

Indonesia has many lakes formed in the craters of extinct volcanoes. For example, Klindungan (134 m), Pakis (158 m), Segara Anak (230 m), Telaga Ngebel (46 m) and Telaga Pasir (35 m), Tigawarna (60 m) and Tondano (20 m). Several lakes in Japan are deep crater lakes; e.g., Shikotsu (360 m), Mashu (211.5 m).

Maar lakes are extensively distributed in volcanic areas of North-East China, South-East China and especially in Leizhou Peninsula of Guangdong and in Huinan district of Jilin, and in Indonesia.

There are several interesting lakes formed by meteoritic impacts. Lake Kara-Kul (25-km diameter, >220 m deep) in Tajikistan at 3,900 m elevation in the Pamir ranges was formed by a meteorite impact about 5 Ma BP. Lake El'gygytgyn (about 15 km diameter and 175 m deep) in northeast Siberia sits in a meteoritic impact crater created 3.6 million years ago. Lake Lonar in western India is a rather small crater lake (1.5 km diameter) whereas Lake Khajjiar (about 3 ha) in Himachal Pradesh (India) is also believed to have been formed by a meteoritic impact.

The next most common lakes are riverine lakes or of fluvial origin. The Asian continent has some of the world's major river systems draining its landscape. Most of these originate from the central highland and meander through the vast floodplains before discharging into the oceans. There are thousands of fluvial lakes in the floodplains of Indus, Ganga and Brahmaputra in south Asia, Yangtse and Huanghe in eastern, northwestern and Jianghan plains in China, Tigris and Euphrates in Iraq, Irrawady in Myanmar, and the Mekong in southeast Asia. The most notable example is the Tonle Sap in Cambodia which shrinks and expands dramatically from 2500 km$^2$ to 10 000 km$^2$ as it drains into the Mekong and gets flooded from the latter during the period of high precipitation (**Figure 3**). Lake Loktak in Manipur (NE India) was a similar smaller lake that alternately got flooded from or drained into the Imphal river until its water level was regulated by a barrage. In the Yangtse river basin, tectonic activities and fluvial processes have interacted to create a large number of lakes. Lake Dongting spreads from 2820 to 20 000 km$^2$ during annual floods from Yangtse and four other rivers. It has been shrinking rapidly because of high sediment load received from the R. Yangtse. Lake Poyang connects with R. Yangtse through Han and Xiu rivers. Another large shallow lake Taihu (max. depth 2.6 m) lies in the delta of Yangtse with which it is connected by many streams but the water flux is heavily regulated. The Amur (Heilongjiang), Sungari and Ussuri rivers, poor drainage has resulted in many large and shallow lakes such as the Lake Xing Kai Hu (Lake Khanka) on the

borders of China and Russia (maximum depth 10.0 m). The only natural lakes in Vietnam, Be Be (4.5 km$^2$) and Ho-Tay (or West Lake, 4.13 km$^2$) have a maximum depth of 3.5 m.

The floodplain lakes (=ox bow lakes) which have either become isolated from the rivers or are periodically connected with them during flood period are known variously in different countries. These are Tals in northern India, beels in eastern India: for example, Kabar Tal, Suraha Tal, Deepor beel and Sareshwar beel. In Bangladesh, these are haors (generally formed in tectonic depressions) or baors (cut off meander loops) such as Hakaluki haor (344.4 km$^2$), Tangua haor (255 km$^2$), Baluhar baor (2.8 km$^2$), Marjat baor (2.5 km$^2$) and Gobindapur baor (2.2 km$^2$). In Iraq, these are called Hawr or Hor. Tasek Bera, Tasek Cini and Ulu Lepar in Pahang (Malaysia) are also shallow foodplain lakes.

In Iraq, the seasonally flooded shallow lakes such as Tharthar, Habbaniya, Hawr al Hammar, Dibbis, Hawr Dalmaj, Hawr Ibn Najim, Hawr al Suwayqiyah, Hawr as Saniyah, and Hawr Hawizeh are spread over hundreds of square km areas.

The majority of the lakes in the arid regions are saline to brackish. In the cold-arid regions of central Asia, they are usually endorheic and depend largely upon snowmelt from surrounding glacier-covered mountain peaks. High rates of evaporation result in increasing salinity. Innumerable lakes in the Qinghai Tibet Plateau region were formed by the epeirogenic processes followed by glacial activity. In Mongolia, some of the very large, though shallow, saline lakes include Uvs Nuur (3350 km$^2$), Urureg (238 km$^2$), Achit (290 km$^2$), Khar Nuur (575 km$^2$), Durgun (305 km$^2$), Hyargas (1407 km$^2$), Telmen (194 km$^2$), Sangiyn Dalay (165 km$^2$), Oygon (61.3 km$^2$), Khyargas Lake (140 km$^2$), Boon-tsagaan (252 km$^2$) and Orog (140 km$^2$). In Kazakhstan, Lake Tengiz is a large shallow (max. 6.5 m) saline lake. Kazakhstan shares with Uzbekistan a group of shallow brackish water lakes (Aydar Kul, Arnasay and Tuzkan), which cover together about 4000 km$^2$.

In hot deserts, lakes have developed due to poor drainage that is caused by impervious subsurface layers formed by the precipitation of calcium salts. In Syria, the only shallow salt lake is Sabkat al Jabboul. Azraq oasis in eastern Jordan is a large (about 100 km$^2$) complex of marshes fed largely by springs. Iran has a large network of lakes and lagoons south of the Caspian Sea. Anjali wetlands (Anjali Murdab) are among the most significant. In northern Azarbayjan province lies a large (4830 km$^2$) hypersaline lake Orumiyeh (=Urmia). Numerous saline, brackish (Perishan, Maharloo, Bakhtegan, Tashk) and freshwater lakes (Dasht-e-Arjan and Haft Barm lakes) occur in the central Fars province.

The Neiriz (=Niriz or Neyriz) lake alone, when flooded during periods of heavy rainfall, covers some 1800 km² but gets fragmented into many smaller shallow water bodies/marshes during the dry period. Another complex of freshwater lakes (Hamoun-i-Puzak, Hamoun-i-Sabari, and Hamoun-i-Helmand) lies in the Seistan basin on the east. Important lakes in Afghanistan include Jamoun-i-puzak, Lake Hashmat (near Kabul), Dasht-e-Nawar, and Abi-Estada.

Eolian lakes are also formed in deserts where huge volumes of sand transported by wind are deposited to block the flow of natural channels. These generally small, shallow, and saline lakes are common in desert areas of India (e.g., Sambhar, Deedwana, and Pachpadra lakes in Rajasthan), Afghanistan (e.g., Hamun-e-Helmand), Iran, Middle East and China. Several lakes in region east to River Heihe for example, Lake Yihezagede Haizi of Inner Mongolia is between the sand dunes.

Karst lakes, which are formed by the dissolution of carbonate rocks or salt deposits, are relatively few. Major karst lakes of importance occur in southwest China. The Phong Nha–Ke Bang National Park (Vietnam) is the oldest major karst area in Asia, and has been subjected to massive tectonic changes. It includes

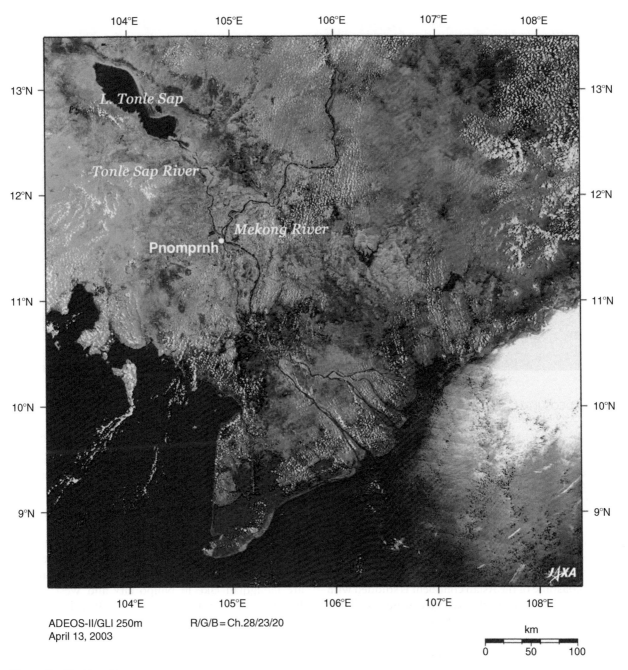

ADEOS-II/GLI 250m
April 13, 2003

R/G/B = Ch.28/23/20

km

0   50   100

**Figure 3** (Continued)

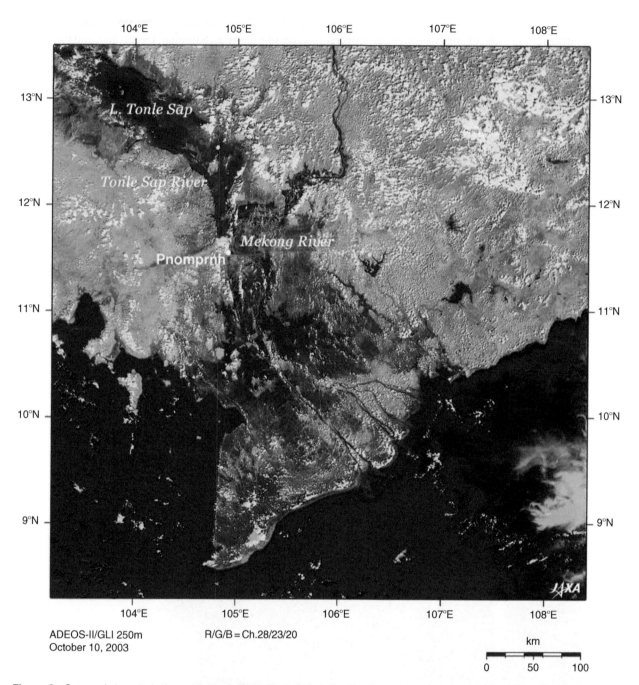

**Figure 3** Seasonal changes in the areal extent of lake Tonle Sap during the dry and rainy season. Portions shown in green represent vegetation areas, portions shown in red or brown, nonvegetation areas; and portions shown in black or blue are water areas. April 2003: Fig. 3b and Oct. 2003: Fig. 3b. (from http://www.eorc.jaxa.jp/en/imgdata/topics/2004/img/tp040419_03.jpg and http://www.eorc.jaxa.jp/en/imgdata/topics/2004/img/tp040419_02.jpg).

65 km of caves and underground rivers. In Papua-New Guinea, Wongabi and Louise are karst lakes. Also in Turkey, there are several karst lakes in western Taurus.

The coastline of the Asian continent is studded with numerous lagoons which exchange water with the sea through a narrow mouth. Some of these lagoons have in recent years turned into completely closed freshwater bodies. Depending upon their size and orientation, most of the lakes exhibit zonation along a salinity gradient, a part of the lake distant from the sea remains freshwater. Important examples are Dongqian lake in Ninpo city and West lake in Hangzhou city in China, Lakes Chilika, Pulicat and Vembnad in India, and Lake Songhkla in Thailand. Lake Kolleru on the east coast of India was a lagoon that gradually turned into a freshwater lake though it maintained a sea connection until the last century. There are numerous small lagoons along the coastline of Sri Lanka and along the Mediterranean coast in Turkey. Interestingly, most of these lagoons have a very high biodiversity and secondary productivity.

## Human-Made Lakes (Reservoirs)

Human-made lakes are an important and prominent feature of the Asian landscape. Their construction dates back to antiquity in response to the vagaries of the monsoon, which causes frequent floods and droughts often in the same area, and the need of water for domestic supplies as well as agriculture. In recent times, huge multipurpose reservoirs have been constructed for irrigation and hydropower generation as well as flood control. Some of the world's largest and highest dams have been built in Russia (Kuybyshev, Bratsk, Krasnoyarsk, Rybinsk Reservoirs), Kazakhstan (Bukhtarma Reservoir), China (Three Gorges), and India (Tehri, Sardar Sarovar and Hirakud). Nurek (Norak) Dam in Tajikistan is only 95 km$^2$ in area but is the highest (300 m) in Asia.

There are 86 881 reservoirs in the People's Republic of China. Of these, 328 are large (>10 million m$^3$); 2333 median (1–10 million m$^3$); 14 232 (0.1–1 million m$^3$); and the rest are very small. India has more than 100 000 human-made water bodies, most of which are concentrated in the southern and western semi-arid regions. There are about 20 000 reservoirs with an area of >10 km$^2$ (**Table 3**). Sri Lanka has a

very high density of human-made water bodies but only a few of them are perennial. The large reservoirs include Udawalwe Wewa (32.8 km$^2$), Parakrama Samudra (22.6 km$^2$), Hurulu Wewa (21.95 km$^2$), and Kala Wewa (26 km$^2$). Bangladesh has only two reservoirs, Kaptai and Feni, constructed in recent years for hydropower and flood control, respectively.

In southeast Asia there are also numerous reservoirs. Malaysia has over 50 reservoirs though most of them are small. The larger ones are Kenyir dam (370 km$^2$), Temenggor dam (152.1 km$^2$), and Subang dam (104.5 km$^2$), and Pedu dam (64 km$^2$). Laos has huge artificial lakes such as the Nam Ngum dam near Vientiane. Larger reservoirs in Indonesia constructed during the past 40 years include Jatiluhur (83 km$^2$), Saguling (56 km$^2$), Cirata (62 km$^2$), and Riam Kanan (92 km$^2$).

Major reservoirs in Iraq include Samarra, Kut, Ramadi, Hindiya, Yao, Meshkab, Hafar, and Akaika. Lebanon also has two important human-made lakes – Tanayel and Qaroun (the largest in the country, constructed on R. Litani in 1962).

*See also:* Abundance and Size Distribution of Lakes, Ponds and Impoundments; Europe; Geomorphology of Lake Basins; Origins of Types of Lake Basins; Reservoirs; Saline Inland Waters.

## Further Reading

Duker L and Borre L (2001) Biodiversity Conservation of the World's Lakes: A Preliminary Framework for Identifying Priorities. *LakeNet Report Series, No. 2.* Annapolis, MD: Monitor International.

Dulmaa A (1979) Hydrobiological outline of the Mongolian lakes. *Internationale Revue der gesamten Hydrobiologie* 64: 709–736.

Gopal B and Wetzel RG (eds.) (1995–2004) *Limnology in Developing Countries,* Vols. 1–4. USA: International Association of Theoretical and Applied Limnology.

International Lake Environment Committee (1993) *Data Book of World Lake Environments: A Survey of the State of World Lakes. 1. Asia and Oceania.* Japan: ILEC, Kusatsu.

Löffler H (1956) Limnologische Untersuchungen an Iranischen Binnengewässern. *Hydrobiologia* 8: 201–278.

Mirabzadeh and Parastu A (1999) *Wetlands in Western Asia. Background Paper for 'Conservation and Wise Use of Wetlands in Western Asia'.* Ramsar Bureau, Gland, Switzerland. 18 pp. http://www.ramsar.org/about_western_asia-bkgd.htm.

Petr T (ed.) (1999a) *Fish and Fisheries at Higher Altitudes: Asia,* FAO Fisheries Technical Paper 385. Rome: Food and Agricultural Organization of the United Nations.

Sugunan VV (1998) *Reservoir Fisheries of India.* FAO Fisheries Technical Paper 345. Rome: Food and Agricultural Organization of the United Nations.

Wang H (1987) The water resources of lakes in China. *Chinese Journal of Oceanography and Limnology* 5: 263–280.

Williams WD (1991) Chinese and Mongolian saline lakes: A limnological overview. *Hydrobiologia* 210: 33–66.

**Table 3** Major reservoirs in India

| Reservoir | Area, km$^2$ |
| --- | --- |
| Hirakud | 719.6 |
| Nagarjunasagar | 284.7 |
| Srisailam | 512 |
| Rana Pratap Sagar | 150 |
| Stanley | 153.5 |
| Krishnarajasagar | 132 |
| Govindsagar (=Bhakra) | 168.7 |
| Pochampalli | 448 |
| Pong | 246 |
| Sriramsagar | 378.3 |
| Nizamsagar | 128 |
| Sardar Sarovar | 370 |
| Bhavanisagar | 78.7 |
| Idukki | 61.6 |
| Tungbhadra | 378.1 |
| Sharavathi | 405 |
| Rihand (=GBPant Sagar) | 465.4 |
| Gandhisagar | 660 |
| Bergi | 273 |
| Tawa | 200.5 |
| Vallabhsagar (=Ukai) | 520 |
| Kakarapur | 442 |

# Australia and New Zealand

**J D Brookes,** The University of Adelaide, SA, Australia
**D P Hamilton,** University of Waikato, Hamilton, New Zealand

## Introduction

Australia is the driest continent on Earth, other than Antarctica, but has mean annual rainfall ranging from 3200–4000 mm in the tropical north-east and south-west Tasmania to <200 mm in the dry interior (**Figure 1**).

Rainfall in most areas of New Zealand (NZ) varies from 600 mm to 1600 mm, but extremes of >6000 mm are noted on the mountainous west coast of South Island (**Figure 2**).

Although these Southern Hemisphere neighbors have different water resource issues many of their lakes and reservoirs have undergone similar changes since European colonization, including modified hydrology and catchment land use, and the ensuing changes in timing and magnitude of water delivery, as well as deterioration of water quality.

The Australian continent is characterized by enormous climatic diversity and is also subject to highly variable seasonal and interannual climatic conditions. Rainfall and streamflow variability in Australia are greater than elsewhere in the world, except for South Africa, which shares similar hydrology. By contrast, New Zealand has a maritime climate characterized by small seasonal and interannual variations in rainfall, and lower amplitudes of variation in temperature and humidity. The maritime climate produces deeper surface mixing in lakes in NZ than those of comparable latitudes in the Northern Hemisphere.

A major influence on climate in the southern part of Australia is El Niño. The El Niño/Southern Oscillation (ENSO) phenomenon results from interactions between large scale ocean and atmospheric circulation processes in the equatorial Pacific Ocean. El Niño is used to describe the appearance of warm water along the coast of Peru and Ecuador, and the Southern Oscillation is the alternation of atmospheric pressure differences between the Australia–Indonesia region and the eastern tropical Pacific Ocean. El Niño conditions typically result in low rainfall throughout southern Australia and tend to be associated with Australia's worst droughts. The opposing condition is La Niña, in which above average rainfall would be expected. The ENSO effect accounts for only a small fraction of annual variations in climate in NZ, with El Niño tending to increase the intensity of the predominant westerly winds, which reduces rainfall in the more arid eastern regions of the country.

## Features and Origins of Natural Lakes

Australia is an ancient, weathered continent and the landscape has undergone considerable environmental changes since stretching and rifting from the supercontinent Gondwana in the Jurassic (150 m.y.a.) and traveling northward with separation off of NZ 80–60 m.y.a, and finally separation of Antarctica 45 m.y.a. Large parts of the north-east and south-west of NZ are remnants of the former eastern margin of Gondwanaland, but the NZ landscape has also been strongly influenced by recent tectonic and volcanic activity, as well as glaciations during the Pleistocene.

The Australian climate has generally been warm and moist, even when joined to Antarctica, and the drying of the continent probably began about 35–24 m.y.a. Drying continued as Australia drifted northward and temperature gradients between the equator and the pole increased until about 2.5 m.y.a. when continent-wide climatic conditions became similar to the present day.

Lakes in both countries have varied form and origin, reflecting the predominant geological processes, hydrology, and climate. Australia has an arid interior with eco-regions spanning arid, temperate, alpine, subtropical, and tropical. Standing waters include freshwater lakes, saline lakes, coastal lagoons, and drinking supply reservoirs in dammed river basins. The New Zealand landscape is young and mountainous, leading to tremendous diversity of lake types with at least ten different possible mechanisms of formation. Most of the lakes have been formed comparatively recently, within the past 20 000 years.

Accompanying the aridification of Australia has been the formation of saline lakes. The largest lake in Australia is Lake Eyre, which fills intermittently following high rainfall in the humid subtropical catchments of the tributary rivers, including the Diamantina-Warburton Rivers and Cooper Creek. The Lake Eyre basin, covering 1.3 million km$^2$, was formed via upwarping of the Central Lowlands near the South Australian Gulfs. Instead of draining to the sea, water was directed inland to the lake which has a current elevation of 15 m below sea level. The climate in the Basin is hot and dry, and evaporation exceeds rainfall in all months over the entire Basin.

Saline lakes are a common feature of the arid centre and southern coastal regions of Australia. The mechanisms by which salt entered the lakes

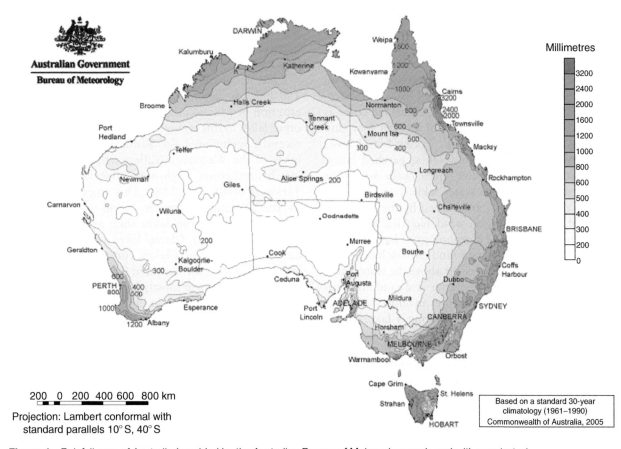

**Figure 1** Rainfall map of Australia (provided by the Australian Bureau of Meteorology and used with permission).

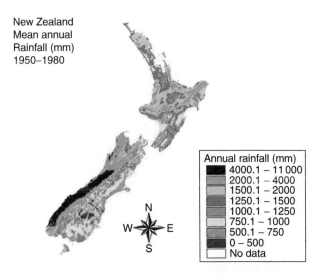

**Figure 2** Rainfall map of New Zealand (courtesy of Wei Ye, International Global Change Institute, University of Waikato).

include (1) evaporation of relict sea water from incursions of the ocean during the Tertiary and Quaternary and subsequent precipitation of dissolved solutes; (2) dissolution from rocks of marine origin; (3) additions of ocean sprays, continental dusts, and atmospheric

deposition; and (4) weathering of terrestrial rocks and minerals. In contrast to Australia, NZ has only one natural saline lake – Sutton Lake in arid Central Otago. The lake's elevated salinity is sustained by the windy maritime climate in which the rate of evaporation (700 mm per annum) exceeds rainfall (500 mm per annum).

Lakes of volcanic origin occur in the Atherton tablelands in north Queensland and in the south-east of Australia. The Southern Volcanic Hills in the southeast of South Australia and western Victoria are generally considered to be less than 10 000 years old. There is an older group of volcanoes to the north-west of Blue Lake that may have erupted between 1 000 000 to 20 000 years ago. The most famous of the south-east crater lakes is Blue Lake in Mount Gambier, so named because it is brilliant blue year-round. Blue Lake is fed by two major aquifers, has a mean depth of 70 m, and recent dating of sediments suggests that it was formed more than 28 000 years ago.

The Atherton Tablelands has several maar lakes, including Lake Barrine, Lake Eacham, and Lake Tinaroo, which are now included within the Wet Tropics World Heritage Area. Volcanic activity in the Atherton area occurred between 10 000 (Pleistocene) and 2 million (Pliocene) years ago. During the explosion that formed Lake Barrine, the aboriginal stories describe the landscape as open scrub. The lake is now surrounded by tropical rainforest, and sediment pollen records confirm that the rainforest was formed on the Atherton Tableland only around 7600 years ago.

Fraser Island, an island off the Queensland coast, is the world's largest sand island of 124 km long and covering an area of 1630 km². Fraser Island hosts three types of lakes: window lakes, perched lakes, and barrage lakes. Window lakes occur when the ground drops below the water table. Barrage lakes, as their name suggests, are formed by the damming action of sand blown by the wind, blocking the waters of a natural spring. The morphology of barrage lakes is dynamic, depending upon sand encroachment into these lakes. Perched lakes occur above the water table and are formed as organic matter and sand accumulate in shallow depressions to form an impervious base and eventually capture rain, creating a lake. Fraser Island has half of the world's perched dune lakes, including the world's largest perched dune lake – Lake Boomanjin, covering an area of 190 km².

Volcanic lakes in NZ occur in three main areas of the North Island: the Taupo Volcanic Zone, the Auckland Volcanic District, and the Bay of Islands-Kaikohe District. Lake Taupo (with an area of 616 km²), the largest lake in New Zealand, was formed from a series of explosive volcanic eruptions, with the last Taupo Pumice eruption considered to have occurred around 186 AD Many lakes of the Rotorua region (**Figure 3**) were formed through collapsed calderas following massive eruptions, often acting in combination with lava flows blocking outflows.

Lake Rotorua, for example, was formed around 140 000 years ago from a collapsed caldera following eruption of Mamaku Ignimbrite. Nevertheless many other lakes in this region have been formed relatively recently, including Lake Rotomahana and smaller lakes of the Waimangu Valley, which filled following hydrothermal explosions associated with the Tarawera eruption of 1886. This eruption is evident as the volcanic tephra is found in the bottom sediments of Rotorua lakes such as Tarawera and Okareka (**Figure 4**). Glacial lakes are the most common of New Zealand's lakes, comprising the deepest lakes due to their formation in former glacial valleys, with outlets dammed mostly by moraine and alluvial outwash. Lake Hauroko in the south of South Island is the deepest New Zealand lake (462 m) but Lakes Manapouri and Te Anau also exceed 400 m depth.

## Lake Facts in Brief

- New Zealand has 776 lakes having a length greater than 0.5 km.
- The largest lakes in Australia include Lake Eyre (9500 km²), Lake Torrens (5900 km²), and Lake Gairdner (4300 km²), which are all in South Australia. The largest lake in North Island of

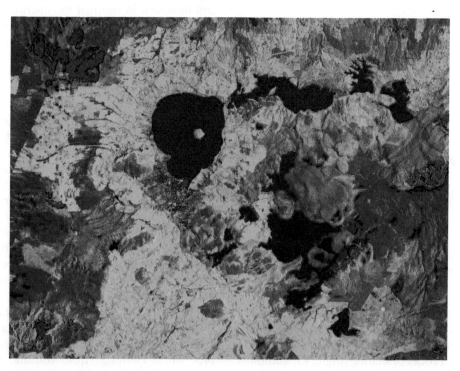

**Figure 3**   False color satellite image showing the Rotorua lakes' region (source: Google Earth).

**Figure 4** Mt. Tarawera (background) rising above Lake Tarawera and Lake Okareka (foreground). (Source: Tourism Rotorua.)

New Zealand is Lake Taupo (616 km²) and in South Island, Lake Te Anau (348 km²).
- The largest artificial lake in Australia is Lake Argyle (700 km²) in Western Australia. New Zealand's largest artificial lake is Lake Benmore (74 km²), retained by a large earth dam on the Waitaki River system in South Island.
- The deepest lake in Australia is Lake St Clair in Tasmania, at 174 m. The deepest lake in New Zealand is Lake Hauroko at 462 m.

## Largest Australian and New Zealand Lakes by State/Territory/Region

The largest waterbodies, whether natural or artificial, on the Australian continent and in New Zealand are given in **Table 1**.

## Artificial Lakes

### Drinking Supply Reservoirs

The population of Australia is highly concentrated in large coastal cities. Most Australian cities depend predominantly on drinking water reservoirs that capture and store surface water, although Perth in Western Australia uses about 40% groundwater for potable supply. Similarly, some larger cities in New Zealand

are reliant upon small water supply reservoirs, despite large river systems, such as the Waikato River, being important potable water sources for major cities in New Zealand.

### Reservoirs for Hydro Power and Irrigation

Most artificial lakes used for generation of hydroelectric power or securing water for irrigation in Australia and NZ have been developed since the 1940s. Schemes of note in Australia include the Snowy River Scheme for hydroelectric power in the high country New South Wales, the hydroelectric network in Tasmania, the Hume and Dartmouth Dams in the headwaters of the Murray River, and the Ord River Scheme for controlling water release for irrigation requirements in Western Australia. Hydro-electric power constitutes about two-thirds of New Zealand's electrical energy supply, mostly from a series of dams along major river and lake systems (e.g., Lake Ruataniwha, **Figure 5**).

### The Snowy River Scheme, Australia

The Snowy Mountains scheme is the largest engineering project undertaken in Australia, beginning in 1949 and reaching completion in 1974. The scheme, in the High Country of Southern New South Wales, involved redirecting water from the east-flowing Snowy River through a series of tunnels and dams across the range to produce ten percent of New South Wales' power needs and deliver the water to the Murray and Murrumbidgee Rivers. The scheme has sixteen major dams, seven power stations and 225 kilometers of tunnels, pipelines and aqueducts. The scheme has left the Snowy River with extremely low flows (approximately one percent of original) and plans are now in place to reinstate 21% of original flow to the river to improve its health.

### Hydro-Electricity in Tasmania, Australia

The entire electricity needs of Tasmania are met by hydroelectric power from 27 power stations fed by 52 storages. The landscape changes to meet this need are great and are summarized below:

- More than 1100 km² of river valleys, wetlands and pre-existing lakes have been inundated for Hydro development.
- The catchment area feeding into Hydro storages takes up 21 500 square kilometres, which equates to 36% of the area of Tasmania.
- 107 dams and weirs have been constructed, affecting 21 creeks, 25 rivers, and 7 estuaries.

**Table 1**  Inventory of the largest lakes in each state or territory of Australia and New Zealand

| State/territory | Lake | Area (km²) | Maximum depth (m) | Origin |
|---|---|---|---|---|
| South Australia | Lake Eyre (saline) | 9500 | 3–6[a] | Upwarping of adjoining lowlands |
| Western Australia | Lake Mackay (saline) | 3494 | Shallow, frequently dry | Deflated lake basin |
| Northern Territory | Lake Amadeus (saline) | 1032 | Shallow, frequently dry | Deflated lake basin (>5 M years old) |
| New South Wales | Lake Garnpung | 542 | Typically dry | Deflated lake basin |
| Tasmania | Lake Gordon | 270 | 140 | Dammed hydroelectric reservoir |
| Queensland | Lake Dalrymple | 220 | 30 | Dammed river |
| Victoria | Lake Corangamite (saline) | 209 | 6 | Depression in volcanic terrain |
| Australian Capital Territory | Lake Burley Griffin | 7.2 | 18 | Dammed river |
| North Island (NZ) | Lake Taupo | 616 | 163 | Volcanic |
| South Island (NZ) | Te Anau | 348 | 417 | Glacial |

[a]Depth varies from completely dry to maximum during extreme floods that periodically fill the lake.

**Figure 5**  Lake Ruataniwha in MacKenzie Country, central South Island, a major rowing venue in NZ and a controlled lake (for hydroelectric storage). The blue color is enhanced by light-scattering glacial flour, which decreases the optical depth. (Photo: J. Brookes.)

- At least 1200 kilometres of natural creeks and rivers are affected by diversion or addition of water, and modification of the natural flow regime.

### Lake Argyle, Australia

Lake Argyle in the North West of Australia was created in 1972 by damming the Ord River. The lake has a capacity of $10.76 \times 10^9 \, m^3$ and provides water for the Ord River Irrigation scheme.

### Lake Manapouri, New Zealand

Hydroelectric power was first generated from an underground station in Lake Manapouri in the Fiordland National Park of South Island, NZ, in 1969. Water discharged from the power station travels via two 10 km tunnels into a fiord on the Fiordland west coast. This station generates around 5400 GWh (billion Wh) of electricity annually, which is used almost exclusively for smelting aluminum at Tiwai Point on the southern coast of South Island. Original plans to raise the lake level by up to 30 m to increase power generation prompted a public campaign 'Save Manapouri' which highlighted the loss of pristine native forest, rare flora and beach subsidence, as well as identifying water hazards associated with the projected flooded forest. An initial controlled lake level 8.4 m above the natural mean level has subsequently been altered to maintain lake level within the natural range.

## Hydro-Electric Developments on Large Rivers in New Zealand

The Waikato River (North Island), Waitaki and Clutha Rivers (South Island) have all been extensively dammed. The Waikato River begins at the outlet of Lake Taupo, the Waitaki River receives water from three main glacial-formed lakes where water levels are controlled, and the main inputs to the Clutha River also arise via glacial-formed lakes; Wanaka, Wakitipu, and Hawea. Some statistics relating to these three hydroelectric dams are given below.

- Eight dams in series on the Waikato River, New Zealand's longest river system.
- Eight power stations on the Waitaki River including three artificial lakes, three natural lakes where water levels are regulated artificially for power generation, and a series of artificial canals.
- Two dams on the Clutha River.

## Largest Constructed Reservoirs by State/Territory/Region

The largest waterbodies serving water and power supply needs in Australia and New Zealand are shown in **Table 2**.

# Culturally Significant Lakes and Their History

Lakes and reservoirs in Australia and New Zealand have played important roles in ancient and recent cultural history. Aboriginal and Māori settlements have focused around lakes and rivers. The arid interior of Australia also hosted large populations of aboriginals who relied on rock holes and native wells for water. These formations were typically fed by groundwater, and many rock holes could best be described as windows to the groundwater via limestone tunnels. The culture of Māori is inextricably linked with 'wai' (water) to such an extent that in addressing someone in Māori with 'ko wai to ingoa' (what is your name) the reference to wai equates to the essence of their existence. According to Māori, lakes and rivers contain a taniwha (serpent) that acts as a guardian for the mauri (spiritual life force) of the waterbody. If the taniwha departs the waterbody as a result of contamination then the mauri may be stolen.

## Willandra Lake and Aboriginal History

The Willandra Lakes is a World Heritage region and covers $2400 \, km^2$ of semi-arid landscape in far South-Western New South Wales. The region contains a system of Pleistocene lakes formed over the last two million years. The original source for the lakes was a creek that flowed from the Eastern Highlands to the Murray River but this has dried over the past several thousand years and become progressively more saline. Aborigines have lived on the shores of the Willandra Lakes for more than 60 000 years. Lacustrine freshwater shells, burnt animal bones and stone artefacts indicate the presence of humans around 32 750 years ago and there is abundant evidence in lake sediments of humans living in the area over the last 10 000 years. In 1968, excavation uncovered a cremated woman in the dunes of Lake Mungo, making this 60 000-year-old cremation site the oldest in the world. In 1974, the ochred burial site of a man was found nearby. Radiocarbon dating has established that these are some of the earliest evidence of modern humans in the world and are least 60 000 years old.

## Lake Pedder and the Fight for Conservation

The economic development of Australia has been associated with environmental impacts and in some cases has generated considerable public ire. The 'green movement' was initially strongest in Tasmania with

**Table 2** Largest constructed reservoirs retained by dams by state, territory, or region

| State/Territory | Name | Reservoir | Capacity ($\times 10^9$) ($m^3$) |
|---|---|---|---|
| Tasmania | Gordon | Lake Gordon | 12.450 |
| Western Australia | Ord River | Lake Argyle | 10.760 |
| New South Wales | Eucumbene | Lake Eucumbene | 4.798 |
| Victoria | Dartmouth | – | 4.000 |
| Queensland | Burdekin Falls | Lake Dalrymple | 1.860 |
| Northern Territory | Darwin River | – | 0.259 |
| Australian Capital Territory | Corin | – | 0.0755 |
| South Australia | Mount Bold | Mount Bold | 0.0459 |
| North Island (NZ) | Arapuni | | 0.144 |
| South Island (NZ) | Benmore | | 1.680 |

campaigns to preserve wilderness areas, prevent logging of natural forests and arrest the building of dams for the generation of hydroelectricity. In the 1970s, the Gordon River was dammed, flooding Lake Pedder to unnaturally high levels. Although the battle to preserve Lake Pedder and its distinctive surrounding landscape failed, it stimulated the public into protesting against environmental destruction, and this fight stimulated the birth of the modern conservation movement in Australia. A decade later a proposal to dam the Franklin River was quashed by the Environmental movement. A key figure in this campaign was Bob Brown who founded the Green Party in Tasmania. He was jailed for three weeks for his protest but was enlisted as a parliamentarian less than 24 h after his release.

### The Rotorua Lakes

The shores of Lake Rotorua were a prominent site of battles between the Māori tribe Te Arawa and early European settlers. The European immigrants sought land from around the Rotorua lakes that was highly prized by Māori as a source of food and where many Māori settlements had developed. In 1922, the Crown advised Te Arawa Māori that because of the recession it was unable to afford a proposed settlement on sharing the lakes as a resource and instead offered to negotiate an annuity for 14 of the lakes, with just one of the lakes remaining vested with its Māori guardians. This annuity remained largely unaltered until 2004 when the Crown offered Te Arawa Māori a settlement that included title of the lake beds, a strategic management role in the lakes, financial and cultural redress, and a formal apology. The period prior to return of the lake beds to Te Arawa Māori has been characterized by environmental degradation of many of the lakes, including introductions of invasive weeds, decimation of native fish by introduced salmonids, and eutrophication.

### Lake Ngaroto

The Battle of Hingakaka took place around 1803 on the shores of Lake Ngaroto, a riverine lake in central North Island. An army of around 10 000 Māori warriors of several tribes attempted to settle a long-running grievance by destroying a settlement of 3000 men, women and children, mostly from the Ngati Maniapoto tribe, who lived on the shores and on an island of Lake Ngaroto. The shoreline dwellers used great stealth as well as familiarity with their local wetland and lake environment to kill many of the invaders, some of whom were forced to swim for safety in the lake, only to drown or be dispatched of as they attempted to return to shore. Around one century later, as the wetlands around the lake were drained and the lake water level decreased to support agricultural development, not only were the bones of many who drowned in the lake uncovered but also a carving was discovered in which a stone was embedded. This stone is reputed to hold one of the traditional gods of the Māori people, Uenuku, and was brought across the Pacific Ocean to New Zealand in a canoe by some of the first Māori settlers of the Tainui tribe. Uenuku had apparently been placed in Lake Ngaroto for safe-keeping during the battle of Hingakaka and is today housed in a local museum.

## Lake Ecology

Lake biological communities in New Zealand and Australia appear to be structured quite differently to those of Northern Hemisphere lakes. At the base of the aquatic food chain the paradigm of nutrient limitation by phosphorus cannot be generalized to lakes in the Southern Hemisphere. Nutrient limitation by nitrogen has often been observed in open waters of many New Zealand lakes, particularly those of the Central Volcanic Plateau region of the North Island. Relatively recent land development and intensification of farming activities present a significant challenge for managing lakes in this region due to increased nitrogen leaching to the groundwater aquifers that contribute most of the lakes' nutrient load.

Phytoplankton succession in the temperate lakes follows the typical succession from diatoms and green algae to cyanobacteria and motile algae as turbulent mixing decreases. In Northern Australia the cyanobacteria can dominate lake ecosystems for much of the year. *Cylindrospermopsis* is notable as a nitrogen-fixing cyanobacterium that has relatively recently become widespread in both Australia and New Zealand, having mainly dominated in (sub) tropical systems in the past.

In the littoral region native submerged macrophyte communities tend to be low-growing turf communities that are particularly susceptible to being outcompeted by some of the taller canopy-forming exotic species such as *Lagarosiphon major* and *Egeria densa*, with the latter species able to still be highly productive under nutrient-enriched conditions. Only a very limited number of macroinvertebrate species in New Zealand can actually consume macrophytes directly, so again there is high susceptibility to invasions by the tall, canopy-forming species which are grazed minimally and contribute little to carbon cycling through the aquatic food chain.

Susceptibility to eutrophication may also be significantly enhanced in New Zealand and Australia, compared with the Northern Hemisphere, as the zooplankton communities are considered to be depauperate, without many of the large planktivorous species that are characteristic of North Hemisphere lakes. Furthermore there are few if any obligate planktivorous fish species that may exert direct control by grazing on phytoplankton populations, though some of the smaller native fish species (e.g., bullies and kaoro) have planktivorous larval stages. Iconic salmonids (e.g., brown trout and rainbow trout in particular) and perch have been extensively introduced into New Zealand and the southern temperate region of Australia and likely play a major role in top-down control of zooplankton populations, thereby influencing trophic structure and once again weakening the capacity for zooplankton to exert top-down control on phytoplankton populations. Introductions of coarse fish (tench, rudd, catfish, and koi carp), mostly liberated and spread illegally, have also been linked to declining water quality in lakes in the Auckland and Waikato regions of New Zealand.

For shallow lakes there is, however, a strong similarity between Australasia and the Northern Hemisphere related to the existence of alternative stable states; a clear-water, macrophyte dominated phase and a turbid, phytoplankton dominated phase. In both cases nutrient supply may play an important role in effecting switches from one state to another, but bio-manipulations involving fish removal in some Northern Hemisphere lakes have also been highly effective in restoring a clear-water phase. It is less likely that biomanipulation would meet with the same degree of success in counterpart Australasian lakes under the reduced top-down structuring, and physical factors such as strong winds in the prevailing maritime climate of New Zealand and coastal Australia and anthropogenic salinization in Australia, as well as the ephemeral nature of many Australian inland waters, greatly complicate application of alternative stable state theory to shallow lakes of Australia and New Zealand. Examples of lakes that have shown alternative stable states in Australia include; Lake Mokoan (Victoria), Little White Lake (Western Australia), Crescent (Tasmania), Sorrell (Tasmania) and in New Zealand; Ellesmere (South Island), Horowhenua (North Island), Omapere (North Island) and Waikare (North Island). All of these lakes are characterized by switches to the turbid state except Little White Lake, which appears to be transitional, in which the transitional state appears to be regulated by changes in salinity.

The Australian saline lakes typically have fairly low species richness but can have high productivity.

Resistant propagules of invertebrates, plants and phytoplankton can rapidly germinate upon flooding. The salt-tolerant charophyte *Lamprothamnium papulosum* and the aquatic angiosperms *Ruppia* sp. and *Leilaena* tend to dominate the macrophyte community in the saline lakes. In Australia, most of the taxa found in saline lakes are endemic and typically species richness decreases with increasing salinity. Lake Corangamite (Victoria, Australia) has suffered from diversion of the major freshwater tributary from the lake and subsequently has experienced steadily increasing salinity, to the point where it hosts very few aquatic species and supports a fraction of the bird life that it did historically.

## Changes Associated with Colonization and Development

European colonization of New Zealand and Australia drastically altered the aquatic landscape. Development of catchments, damming of rivers, control of water levels, and the construction of reservoirs have led to a general deterioration of water quality. The water quality issues are similar to other countries, though they tend to be have arisen more recently, and organic chemical contaminants are not generally a major concern. Nutrients, cyanobacteria and pathogens are the principal water quality concerns. The first documented recording of toxicity by algae was made by George Francis in an article published in *Nature* in 1878. Francis noted a bloom of the toxic cyanobacterium *Nodularia spumigena* in Lake Alexandrina in South Australia (**Figure 6**) and the same species of alga still forms blooms in the lake today.

### Eutrophication

Both New Zealand and Australia are characterized by large proportions of human populations focused into coastal towns and cities, with most industrial and domestic wastewater discharges into the coastal environment. Eutrophication of lakes and reservoirs in both countries has resulted mostly from agricultural development. In New Zealand and south-east Australia, around 90% of wetlands have been cleared and drained, mostly for agricultural development. In New Zealand in particular, pastoral farmers traditionally used phosphatic fertilizers often supplemented with micronutrients (e.g., cobalt) to increase pastoral productivity, while relying on nitrogen fixation by clover to meet grassland nitrogen demands. With the advent of relatively cheap nitrogenous fertilizers, predominantly urea and ammonium, there has

**Figure 6** Holiday houses on the shore of Lake Alexandrina, the site of the first recorded toxic cyanobacterial bloom. (Photo: J. Brookes.)

been a rapid increase in their use in pastoral farming. The NZ Parliamentary Commissioner for the Environment provides statistics for nitrogenous fertilizers that show an increase in application rates of 160% in dairy farms and 680% in sheep and beef farms between 1996 and 2002. Nitrate loads to Lake Rotorua are projected to increase nearly 10-fold between the relatively unimpacted period of the 1950s and 2050, in response increased intensity of pastoral farming, notably nitrogen fertilizer applications, in the lake catchment. The duration of this response reflects the long lag time as water and nutrients are transported through the porous volcanic soils into the underlying aquifer and ultimately into the lake. Productivity in many New Zealand lakes, particularly the large volcanic lakes of the central North Island (e.g., Lake Taupo and Lake Rotorua), is limited by availability of nitrogen. Under these circumstances the rate of eutrophication is expected to increase rapidly and productivity may be increasingly controlled by availability of phosphorus rather than nitrogen. Specific case studies are given below of large New Zealand lakes that have undergone rapid deterioration in water quality.

### Lake Rotoiti – a Large Volcanic Lake in Central North Island

Lake Rotoiti is a deep (max. depth 124 m) volcanic lake in central North Island which receives most of inflowing water from Lake Rotorua. Despite rapid eutrophication

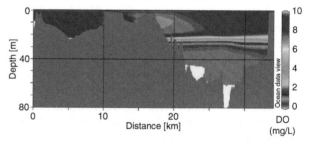

**Figure 7** A transect taken in summer 2005 from west to east across Lake Rotorua (0–10 km) through the Ohau Channel (10–12 km) and into Lake Rotoiti (12–34 km), with colors representing levels of dissolved oxygen and shaded area representing the lake bed.

of Lake Rotorua in the 1960s, attributed to wastewater discharges from the city of Rotorua, water quality in Lake Rotoiti initially remained relatively unimpacted. However, oxygen levels in the bottom waters of this lake during summer have declined over several decades from an annual minimum around three parts per million in 1956 to being devoid of oxygen (anoxic) for more than four months in 2003. A transect across Lake Rotoiti and Lake Rotorua shows well oxygenated waters of shallow Lake Rotorua and anoxic bottom waters of deep, stratified Lake Rotoiti (**Figure 7**). Lake Rotoiti, once New Zealand's second-most frequented lake by trout anglers, has in recent years been affected by recurrent blue-green algae (cyanobacteria) blooms. A diversion wall has recently been constructed in Lake Rotoiti to direct inflows from Rotorua via the Ohau Channel, directly towards the lake outflow.

### Lake Ellesmere – a New Permanently Turbid State

Lake Ellesmere is a large (area $182\,km^2$) but relatively shallow (2 m) lake on the east coast of South Island, New Zealand. Its waters were once clear and the lake had abundant aquatic weed beds that supported around 70 000 black swans (*Cygnus atratus*). In April 1968, a powerful cyclone (the 'Wahine' storm) produced New Zealand's highest ever recorded wind speed ($267\,km\,h^{-1}$), and resulted in the drowning of 51 people when a boat sank between the North and South Islands and also decimated the weed beds of Lake Ellesmere. The swan population was quickly reduced to below 10 000; weed beds have not re-established and bottom sediments in the lake are now readily disturbed by frequent coastal winds. Lake Ellesmere it still highly turbid (muddy) waters and visibility nowadays seldom exceeds 10 cm (**Figure 8**).

### Lake Alexandrina and Lake Albert

There are two lakes at the terminal of the Murray Darling Basin, Lake Alexandrina and Lake Albert. Historically these lakes were estuarine and had complete connection to the Coorong and Murray Mouth. Barrages were built in 1940 to stop salt water incursion into the lakes and upstream into the River Murray during periods of low flow. The barrages drastically altered the ecology of the lakes which typically have been maintained at full pool level. As a result of a drought in 2006, and over-allocation of water to upstream irrigators, the lakes are now facing significant pressure including no environmental water allocation, significant lake drawdown and salinization. Sea water leaking through the barrages is threatening industries that rely on water from the lake including stock watering and irrigation, and salinity is at concentrations high enough to compromise the health of the freshwater macrophytes. Plans to construct a temporary weir at the entrance to the lake are well advanced to ensure the supply to Adelaide city is secure and halt salt water incursion into the river channel. The lower River Murray lakes have experienced water quality problems, such as toxic cyanobacteria in the past, but the latest drought may irreversibly change the character of the lakes and there are genuine concerns for native fisheries, commercial fisheries and bird habitat.

### Contaminants of Concern

The contaminants of most concern are nutrients and pathogens. The pathogens of most concern are *Cryptosporidium*, *Giardia*, *Campylobacter*, and viruses. The expansion of dairying and beef production in

**Figure 8**   Lake Ellesmere, east coast of South Island, New Zealand. Note the Selwyn River inflow (left), the sand barrier separating the lake from the ocean to the east (top) and the encroachment of farmland to the lake edge (bottom), as well as the highly turbid nature of the lake. (Photo: Alex Hamilton.)

reservoir catchments has been a major concern because *Cryptosporidium parvum* is cross infectious between cattle and humans. In Australia, Cryptosporidiosis attributable to contamination of urban water supplies has not been documented. The number of notified Campylobacteriosis cases has increased steadily in New Zealand from 100 in 1980 to 12 000 by 2002. New Zealand also has higher rates of notified Giardiasis and Cryptosporidiosis than other developed countries.

## The Future of Australian and New Zealand Lakes

The climate change predictions for Australia are bleak but New Zealand lakes will be less impacted. The general consensus is that rainfall will decrease over the southern part of the Australian continent, storms will be more intense but less frequent and runoff from catchments to rivers will decrease markedly. Planning is underway to adapt water supply systems to be less impacted by climate variability and ensure greater continuity of supply. Desalination, reuse of wastewater, and aquifer storage and recovery are being considered and implemented by water utilities to decrease the reliance on surface water ecosystems. Issues of water quantity are only just starting to be addressed in New Zealand as irrigated areas expand in the more arid eastern areas of both islands. Expansion of irrigated dairy farming in these regions will affect both the quantity and quality of available water.

The catchments of both Australia and New Zealand have been heavily modified and it is likely that water quality will remain a challenge in the future. In Australia there are concerns that reforestation of catchments will decrease water yields whereas in New Zealand there is a current trend to replace existing pine forest plantations with dairy pasture. In both countries there is increasing awareness of the importance of lakes and rivers as valuable ecosystems and water supplies but prudent planning and management will be required to ensure their long-term water quality.

*See also:* Origins of Types of Lake Basins; Reservoirs; Saline Inland Waters.

## Further Reading

Bowler JM (1982) Aridity in the Late Tertiary and Quaternary of Australia. In: Barker WR and Greenslade PJM (eds.) *Evolution of the Flora and Fauna of Arid Australia*, pp. 35–45. Adelaide, Australia: Peacock Press.

Bowler JM, Jones R, Allen H, and Thorne AG (1970) Pleistocene human remains from Australia: A living site and human cremation from Lake Mungo, western New South Wales. *World Archaeology* 2: 39–60.

Chiew FHS, Piechota TC, Dracup JA, and McMahon TA (1998) El Nino/Southern Oscillation and Australian rainfall, streamflow and droughtl: Links and potential for forecasting. *Journal of Hydrology* 204: 138–149.

Davis JR and Koop K (2006) Eutrophication in Australian rivers, reservoirs and estuaries – A southern hemisphere perspective on the science and its implications. *Hydrobiologia* 559: 23–76.

Frakes LA (1999) Evolution in Australian Environments. In: Orchard AE (ed.) *Flora of Australia*, 2nd edn. vol. 1, pp. 163–203. Canberra: ABRS/CSIRO.

Harding J, Mosley P, Pearson C, and Sorrell B (2004) *Freshwaters of New Zealand*. NZ Hydrological Society/NZ Freshwater Sciences Society Publ.

Herczeg AL and Lyons WB (1991) A chemical model for the evolution of Australian sodium chloride lake brines. *Paleogeography, Paleoclimatology, Paleoecology* 84: 43–45.

McMahon TA, Finlayson BL, Haines AT, and Srinkanthan R (1992) Global runoff – Continental comparisons of annual flows and peak discharges. Cremlingen Destedt: Catena. 166 pp.

Philander SGH (1983) El Niño Southern Oscillation phenomena. *Nature* 302: 295–301.

Thorne A, Grun R, Mortimer G, Spooner NA, Simpson JJ, McCulloch M, Taylor L, and Curnoe D (1999) Australia's oldest human remains: Age of the Lake Mungo 3 skeleton. *Journal of Human Evolution* 36: 591–692.

White ME (2000) *Running Down: Water in a Changing Land*. Roseville, NSW: Kangaroo Press.

Vant WN (1987) *Lake Managers Handbook*, Water and Soil Miscellaneous Publ. No. 103. Ministry of Works and Development. 230 pp.

Viner AB (1987) *Inland Waters of New Zealand*, Wellington: Department of Scientific and Industrial Research. 494 pp. Department of Scientific and Industrial Research Bulletin 241.

# Europe

**G A Weyhenmeyer,** Swedish University of Agricultural Sciences, Uppsala, Sweden
**R Psenner,** University of Innsbruck, Innsbruck, Austria
**L J Tranvik,** Uppsala University, Uppsala, Sweden

## Introduction

Europe, covering about $10\,400\,000\ km^2$ or 2.0% of the Earth's surface, is a continent with a large variety of inland waters. Water resources are abundant compared with other continents, but the resources are unevenly distributed. Water resources in Europe generally decrease towards densely populated warmer regions. These regions are mainly located in the geographical region defined as Southern Europe by the United Nations. The three additional corresponding European geographical regions are Northern, Eastern, and Western Europe (**Figure 1**).

Lakes are most abundant in Northern Europe. Finland is the country in Europe that counts most lakes, i.e., almost 190 000 lakes that are larger than $500\ m^2$. Despite the high abundance, these northern lakes have captured little international attention. Northern Europe is relatively sparsely populated and water of sufficient quality is abundant. Therefore, most of these lakes are of limited economic, social, and political concern. Often lakes become of general interest only when acute water quality problems arise. A typical water quality problem of Europe has been the human-induced increase of nutrient concentrations, i.e., eutrophication. Especially the strong increase in phosphorus concentrations during the middle of the twentieth century, due to untreated sewage waters, resulted in heavy algal blooms. The algal blooms were visible for everyone, awakening strong pressure from the public to find solutions. In the 1970s and the 1980s, many European countries finally began to use and to improve sewage water treatment plants. At first phosphorus concentrations remained on a high level because of internal phosphorus loading from the lake sediments, but subsequently a decrease in phosphorus concentrations could be observed in many systems, a process known as reoligotrophication. As a consequence of this process, the occurrence of heavy algal blooms decreased in a variety of lakes but the problem became apparent again during recent exceptional warm summers, and debates are ongoing in how far climate change can cause heavy algal blooms.

Climate change is only one of multiple stressors affecting lakes of Europe. Well-known other stressors are intensive land use, increasing industrial activities, overfishing, invasive species, and changes in the atmospheric deposition of acids, metals, and organic pollutants. Consequences of stressed lake ecosystems can be, for example, water shortage, flooding, water pollution by various substances, harmful algal blooms, changes in biodiversity, extinction of fish species, and smell and taste problems of drinking waters. The challenge is to face these water problems in time by adequate water management.

The responses of lakes to stressors can vary widely, depending on the lake characteristics. The most important factors characterizing lakes are climate, geological history, land use and to some extent atmospheric deposition. In the following an overview is given about the main factors characterizing lakes of Europe. A description of common European lake characteristics follows and finally examples of lakes of special interest are given.

## Climate

Climate influences physical, chemical, and biological processes in and around lakes. Although much of Europe lies in the northern latitudes, the relatively warm seas that border the continent give most of Central and Western Europe a moderate climate, with cool winters and mild summers. From approximately central Poland eastward, the maritime effects are reduced, and cooler, drier, and more continental conditions prevail (**Figure 2**). Temperature differences between summer and winter are low, about 15 °C, in the coastal northwest regions of Europe but they are high in the northeast regions, especially north of the Caspian Sea. Here the temperature difference between summer and winter can be as high as 70 °C. Because of the pronounced seasonal cycles across Europe, many lakes are dimictic, i.e., they undergo two periods of mixing, one in spring and one in autumn, and are thermally stratified in summer and winter. However, large lakes at the northern and southern slopes of the Alps, such as Lake Garda, Lake Leman, and Lake Constance, are usually warm monomictic, i.e., they mix only once a year during winter. During very cold winters these large lakes can switch to a dimictic regime, e.g., Lake Constance became dimictic in February 1963 during a very cold North Atlantic Oscillation period – the only occasion in the twentieth century.

**Figure 1** Regions of Europe as delineated by the United Nations. The map has been taken from http://en.wikipedia.org/wiki/Image: Location-Europe-UNsubregions.png.

During winter, lakes especially in the north and the northeast and in the alpine regions are covered by ice. Ice cover and frozen ground in the drainage area diminishes the renewal of water from the catchment, decreases the underwater light conditions, and hampers water mixing. This influences the nutrient recycling and all biological processes in lakes. Toward the south of Europe an ice cover is less common, and lakes usually mix during winter. As soon as surface waters heat more rapidly than the heat is distributed by mixing, lakes become thermally stratified. Thermal stratification is one of the most important processes affecting the chemistry and biology of lake waters. Deep waters stay cool during summer, and are warmer than surface water in winter, with important implications for temperature-dependent biological processes. Even more importantly, the thermal stratification results in limited exchange of substances between deep (hypolimnion) and surface (epilimnion) waters. This frequently results in anoxic water in the

hypolimnion, and depletion of nutrients in the epilimnion, both strongly influencing biological processes. One of the consequences of thermal stratification in eutrophied waters is the occurrence of harmful algal blooms. The period of thermal stratification ends when water movements, usually induced by wind, cause a complete water column mixing.

The strong seasonal pattern of temperature does usually not apply to precipitation. Especially in the western part of Europe the prevailing westerly winds, warmed in part by passing over the North Atlantic Drift ocean current, bring precipitation throughout most of the year. Large parts of Europe receive between 500 and 1500 mm of precipitation per year (**Figure 2**) and most areas of Europe are classified as either mild and humid or cold and humid. In the Mediterranean region, almost all rainfall occurs in winter and the summer months are usually hot and dry. This is in contrast with the situation in the northern part of the Alps, where most of the precipitation

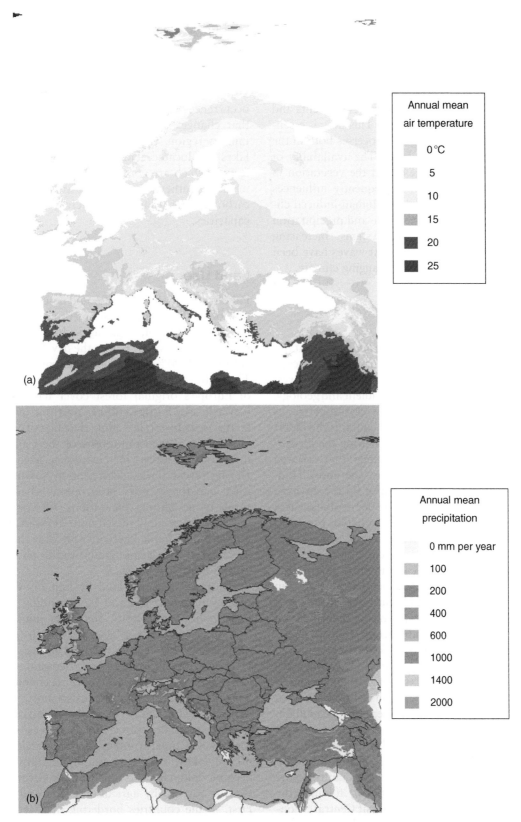

**Figure 2** Present annual mean air temperature (a) and annual mean precipitation (b) in Europe. The map has been created using the European Geo-Portal at http://eu-geoportal.jrc.it.

falls in the summer months while winters are usually dry. The difference in precipitation, evaporation, and most importantly, discharge is very pronounced. In the Alps, the average discharge is 980 mm m$^{-2}$, while it is only 270 mm m$^{-2}$ in Europe without the Alps.

The amount and intensity of precipitation influences the flow of water that transports nutrients and other substances into lake waters. This flow is essential for chemical and biological processes both in the catchment area and in the lakes. The availability of water determines to a large extent the vegetation in the catchment area, which subsequently influences the nutrient flow into the lakes. Human-induced climate change will alter temperature and precipitation patterns, and thereby lake ecosystems. Increasing risks for floods, droughts, and heat waves have been identified as major threats of a changing climate.

## Geological History

Lakes can be seen as a mirror of the landscape that has been formed over thousands of years. The landscape of Europe is diverse, and major controls on this diversity are the north–south climatic gradient and geology. In the north, an ancient mass of crystalline rocks created a stable shield. This shield contains the oldest rocks of the European continent. During the Pleistocene epoch, large ice sheets formed that scoured and depressed the shield's surface. The entire landscape of the north has been shaped by the advance and the retreat of glaciers, leaving depressions where thousands of lakes and streams originated. Mineralization in these cold northern regions has always been quite slow. Together with moist conditions and bedrock relatively resistant to weathering, this results in surface soil layers that are generally quite poor in calcium carbonate but rich in humic substances. These types of surface soil layers explain why many lakes in Northern Europe are naturally acidic.

Much of the landscape in Western and Eastern Europe is characterized by a belt of sedimentary rocks that are covered by a layer of glacially deposited debris. Some of Europe's best soils can be found in this region, which includes the Great European Plain. Agricultural activities are intensive and the quality of inland waters is much influenced by these activities. Bordered at the belt of sedimentary rocks follows a mixed geological structure created by faulting (the Vosges and Black Forest mountains), folding (the Jura range), and volcanism (the Massif Central). Consequently, mountains, plateaus, and valleys alternate, with lakes located at various altitudes.

Southern Europe is characterized by comparatively recent mountain-building activity. In mid-Tertiary time, about 40 Ma, the Afro-Arabian plate collided with the Eurasian plate, triggering the Alpine Orogeny and resulting in ranges such as the Pyrenees, Alps, Apennines, Carpathians, and Caucasus. These ranges are not only the highest mountains of Europe but also exhibit the steepest topography. The frequent occurrence of earthquakes in this region illustrate that changes are still taking place. In this mountainous region, Europe's most famous high altitude lakes are located. Further south lowland lakes are found, which are commonly nutrient-rich. The soils in these southern regions are usually rich in calcium carbonate, resulting in relatively good pH buffering capacities.

## Land Use

Lake water quality can often be described as a function of land use in the catchment area, expressed as the percentage of, e.g., agriculture, forest, mire, buildings, and water. Before the development of agriculture, 80–90% of Europe was covered by forest. Through the centuries of deforestation over half of Europe's original forest cover has disappeared. Nevertheless, Europe still has over one-quarter of its land area as forest, including the taiga of Scandinavia and Russia, mixed rainforests of the Caucasus, and the cork oak forests in the western Mediterranean. During recent times, deforestation has been diminished and forests have been reestablished. In many cases monoculture, plantations of conifers have replaced the original mixed natural forest. Relatively unaffected by human activities are only the forests in the most northerly mountains and in parts of north central European Russia.

The present use of land by humans varies from region to region (**Figure 3**), depending on economics, politics, traditions, and other human factors. The intensive human activities result in a clear north–south gradient in the land use of Europe. In Northern Europe in the Arctic coastal region and above the tree-line, the land cover is characterized by a tundra vegetation, which consists mostly of lichens, mosses, shrubs, and herbs. In the inland of Northern Europe, especially spruce and pine trees occur. Here forestry is frequently intensive in the catchment areas of the lakes. Further south agricultural activities become dominant and much of the Great European Plain is covered by agricultural land and grassland. First, in the countries bordering the Mediterranean fruit plantings, especially olives, citrus fruit, figs, apricots, and grapes, become dominant.

Land cover itself strongly influences the quality of lake waters but activities connected to the land cover,

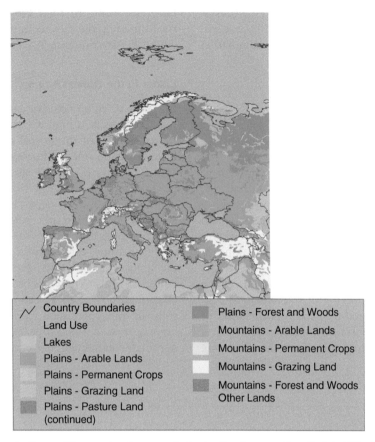

**Figure 3** Present land use in Europe. The map has been created using the European Geo-Portal at http://eu-geoportal.jrc.it/.

such as the distribution of fertilizers and pesticides, are sometimes even more important. The choice of the amount and type of fertilizers and pesticides used to be country specific, but new legislation within the European Union, such as the European Water Framework Directive, decreases the variability among countries. In general, land use practices within the European Union tend to become more homogenous with new legislations, and are influenced by the EU-wide aim to get waters of good ecological status. Many European lakes are still very nutrient-rich because of intensive agricultural activities, giving raise to harmful algal blooms.

## Atmospheric Deposition

Atmospheric deposition implies the input of dust, metals, acids, nutrients, and pollutants into terrestrial and aquatic ecosystems. Much attention has been given to the deposition of sulfur and nitrogen compounds released to the atmosphere by the burning of fossil fuels and a wide range of other industrial activities. These compounds affect lake waters to a large extent, since nitrogen is an important nutrient and sulfur can lead to acid conditions. Sulfur concentrations in lake waters began to increase after the industrial revolution in the nineteenth century, leading to acidified waters, especially in the northern parts of Europe where soils have a low pH buffering capacity, and lake waters became increasingly acidified. The consequences of a very low pH were far-reaching for the chemistry and the biology of the lakes, the leaching of aluminum being one example. To reduce the damaging effects of acid deposition on lake ecosystems, intensive liming activities were initiated in the northern part of Europe. In parallel, strong measures were taken to reduce the emission rates across Europe. In the 1990s, effects were seen and since then the deposition of sulfur and nitrogen compounds are decreasing in most regions of Europe.

The lakes of Europe have also been largely affected by the deposition of radioactive substances after the major accident at the Chernobyl Nuclear Power Plant on April 26, 1986. Radioactive $^{137}$Cs accumulated in soils, sediments, and flora and fauna. Especially fish showed high $^{137}$Cs concentrations, making the consumption in some northern European regions impossible. $^{137}$Cs can still be detected in lake ecosystems, mainly in lake sediments. Here the $^{137}$Cs signal is often used to determine sediment accumulation rates.

## Common European Lake Characteristics

To account for the most common characteristics of European lakes, a division of lakes according to their biogeographical region (**Figure 4**) has been implemented. This division might not be the most common one, but very appropriate for the entire Europe, since differences in climate, geological history, and land use are accounted for. Traditionally, lakes are classified according to their nutrient status. This kind of classification, where lakes are termed as oligotrohic, mesotrophic, and eutrophic, was developed in Europe by Naumann at the beginning of the twentieth century. Naumann also noted that the concentration of humic matter is an important characteristic of many lakes, which is accounted for when characterizing lakes according to their biogeographical region. In the following, the most common European lake characteristics are presented.

### Lakes in the Boreal Region

The boreal region covers a large part of Europe, and the lake density is high. Consequently, the lake types representative of the boreal region may also be the most common ones of Europe as a whole. The lakes of the region are usually small and shallow, and of glacial origin. They range from low to high contents of humic substances, depending on catchment characteristics and hydraulic retention time. The most common lake type in the boreal region in the northeast of Europe, especially in the two countries with the highest lake density, Finland and Sweden, is a humic lake, also known as brown water or dystrophic lake (**Figure 5**). Lakes become humic as a result of soils in the catchment containing high levels of organic matter. Upland peaty catchments and forest dominated catchments give a perfect prerequisite for

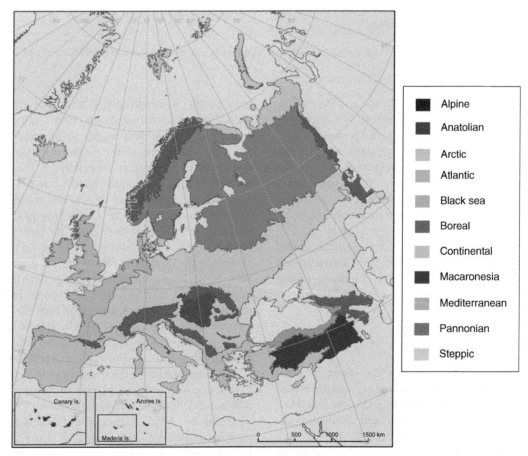

**Figure 4** Biogeographical regions of Europe. The map has been taken from the European Environmental Agency at http://dataservice.eea.europa.eu/atlas/.

**Figure 5** Example of a humic lake in the Boreal Region (Oppsveten, Sweden). Photograph: Cristian Gudasz.

high humic contents in lake waters. Humic lakes are usually characterized by brownish water with a low pH, low nutrient concentrations, and poor underwater light conditions. The flora and fauna is adapted to these conditions with dominance of flagellates that are able to survive under poor light conditions.

Humic substances result in high levels of dissolved organic carbon in lake waters, influencing the pathways of energy and nutrient flow within a lake ecosystem. Dissolved organic carbon in the water absorbs ultraviolet radiation, which has far-reaching consequences for aquatic organisms. Recently, an increase in dissolved organic carbon has been observed across northern Europe. This increase has been coupled to changes in the climate and atmospheric deposition. It has been identified as problematic for society since drinking water rich in dissolved organic carbon can produce carcinogens when chlorinated.

### Lakes in the Continental Region

The land use in the continental region in central Europe is dominated by agriculture and grassland, resulting in a high nutrient input into lakes. Consequently, many lakes are eutrophic (**Figure 6**). The continental region is relatively flat, giving raise to a variety of nutrient-rich shallow lakes. Shallow lakes have a special ecology because thermal stratification, influencing sediment–water interactions, is often lacking, making the lakes polymictic. Shallow lakes are frequently associated with murky water, troubled by heavy algal blooms and a high amount of suspended sediment particles. In traditional agricultural practices, the shorelines were frequently kept open by grazing cattle, but presently often overgrown by macrophytes surrounded by shrubs, with effects on nutrient retention and bird fauna. Extensive water level changes are common. These water level changes have a great impact on the biota. Intensive water management is needed to fulfill the requirements of the European Water Framework Directive for this region.

### Lakes in the Alpine Region

Europe's most famous alpine lakes are located in the Alps. Lakes in the Alps are often associated with alpine lakes but when the term alpine is based on a biogeographical region division, true alpine lakes are here restricted to mountainous regions above the treeline. Such alpine lakes are located at various altitudes in Europe, varying from the coastline in northern Norway to above 2000 m in the Alps. Lakes in the alpine region are usually headwater lakes (**Figure 7**), many of them without human settlements in the catchment and thus normally ultraoligotrophic and pristine in the sense that there is little local anthropogenic disturbance, with two exceptions especially in the Alps: one is the presence of sheep, a tradition that goes back 7000–9000 years, and the other is the atmospheric deposition. Atmospheric deposition includes not only natural dust, for instance of Saharan origin, but also acid rain and snow, and

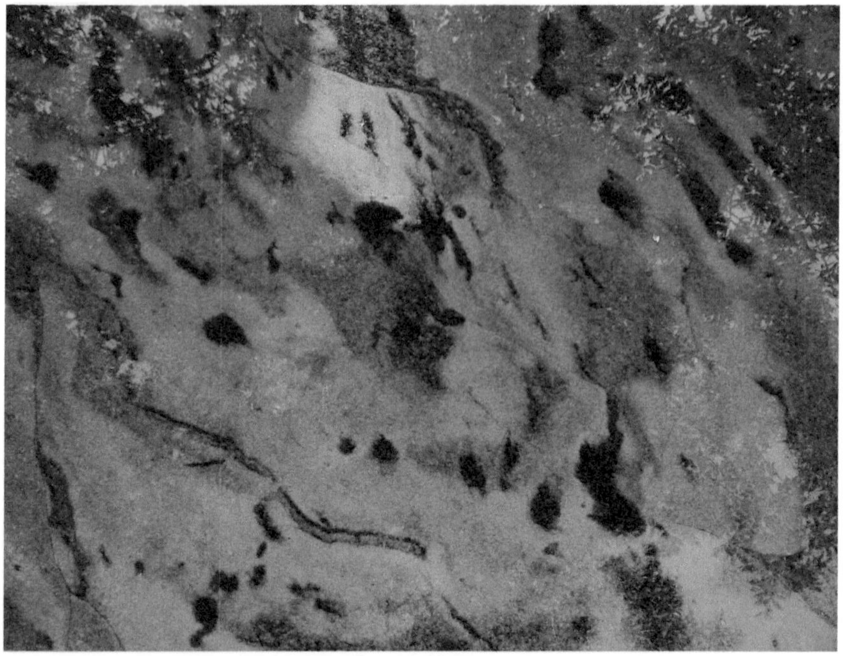

**Figure 6**  Example of an eutrophic lake with blooming algae. Photograph: Gesa Weyhenmeyer.

**Figure 7**  Example of a lake in the Alpine Region (Schwarzee ob Sölden in Oetztal, Tyrol, Austria, at 2796 m above sea level). Photograph: Roland Psenner.

organic pollutants, such as polychlorinated biphenyls (PCBs), which undergo global distillation and cold trapping, i.e., they accumulate in cold areas such as the alpine regions and the poles. Saharan dust has been identified as an important source of limiting nutrients (phosphorus) in lakes of the southernmost alpine lakes of Europe.

The numerous lakes in the Scandinavian alpine region are less known than the alpine lakes in the Alps, probably because they are very remote from larger human settlements. The lakes in the Scandinavian alpine region have much in common with the alpine lakes in the Alps as they are also mainly ultraoligotrohic and less influenced by local anthropogenic disturbance. However, there are also considerable differences in the biogeochemical recycling, resulting from a colder climate and another geology in the Scandinavian alpine region. In addition, lakes in the Scandinavian alpine region are not affected by sheep but by reindeer herding that has a great influence on the vegetation of many watersheds, with unclear effects on lakes.

Most of the alpine lakes are ice covered during winter. The lakes usually show a well-defined thermal winter stratification and clear thermal summer stratification, making them dimictic. Additionally, the lakes are often deep and nutrient-poor, giving them a romantic crystal clear appearance. Since the lakes in the alpine region are located above the tree-line, where organic soils are thin and export only small amounts of dissolved organic substances, they mostly show low concentrations of humic substances. This special feature makes those lakes very transparent to UV radiation that increases by 10–15% with each kilometer altitude. Major threats to lakes in the alpine region are changes in the climate and in atmospheric deposition. Especially, lakes in crystalline rock basins are often very sensitive to acidification, with natural (preindustrial) alkalinities of much less than $100\ \mu\mathrm{equiv}\ l^{-1}$.

### Lakes in the Atlantic and Mediterranean Region

Lakes that are close to the sea are influenced by a maritime climate with mild winters. Consequently, they are seldom covered with ice. These characteristics are similar for lakes in the Atlantic and lakes in the Mediterranean region. However, a variety of characteristics differ between lakes in the two regions, mainly due to differences in the wind and temperature pattern and the land use. Lakes in the Atlantic region are subjected to strong westerly winds, bringing high amounts of precipitation and sea-salt deposits. Frequent water turnover is common for these lakes, making them less exposed to harmful algal blooms. In contrast, lakes in the Mediterranean region experience hot and dry summers, resulting in a strong thermal summer stratification. Oxygen depletion occurs frequently and harmful algal blooms are observed in a variety of lakes. Water shortage during summer is of concern, and some lakes experience exceptional low water levels that affect the entire food web.

### European Lakes of Special Interest

Much of European's limnological understanding is traditionally based on research in lakes in Western Europe, and the oldest limnological stations were founded in Western Europe, i.e., in Plön, Germany in 1891 and in Lunz, Austria in 1905. In particular, research results from the largest lakes in Western Europe such as Lake Leman and Lake Constance reached international recognition as large lakes are worldwide regarded as fascinating. However, focusing limnological research on lakes in Western Europe gives a misleading picture of lakes in Europe. Most lakes, including the 18 largest lakes of Europe are located in Northern Europe in the boreal region (**Table 1**). These largest lakes are internationally not well known, probably because the lakes are located in sparsely populated regions and consequently of little general interest. Lakes in the boreal region are generally underrepresented when reading limnological publications. Considering that global biogeochemical cycles are affected by exchange processes between the lake water surface area and the atmosphere, more focus should be on limnological research of lakes in the boreal region since here the total lake water surface area exceeds by far the total lake water surface area of other European regions.

Apart from the large natural lakes, Europe has a variety of large artificial lakes, the largest being Lake Yssel ($1250\ \mathrm{km}^2$) in the Central Netherlands, where intensive research activities are ongoing. Artificial lakes, also known as reservoirs, are of

**Table 1** Europe's largest natural lakes by area

| Lake name | Country | Lake area (km²) | Mean depth (m) | Max depth (m) |
|---|---|---|---|---|
| Ladoga | Russia | 17 670 | 51 | 258 |
| Onega | Russia | 9670 | 30 | 120 |
| Vänern | Sweden | 5670 | 27 | 106 |
| Peipsi | Russia, Estonia | 3570 | 23 | 47 |
| Vättern | Sweden | 1912 | 39 | 128 |
| Vygozero | Russia | 1285 | 7 | 20 |
| Saimaa | Finland | 1147 | 12 | 82 |
| Mälaren | Sweden | 1140 | 13 | 61 |
| Il'men' | Russia | 1124 | 3 | 10 |
| Beloye | Russia | 1120 | 4 | 20 |
| Inari | Finland | 1102 | 14 | 96 |
| Päijänne | Finland | 1054 | 17 | 98 |
| Topozero | Russia | 1025 | 15 | 56 |
| Oulujärvi | Finland | 893 | 8 | 35 |
| Pielinen | Finland | 867 | 10 | 60 |
| Segozero | Russia | 781–910 | 23 | 97 |
| Imandra | Russia | 845 | 16 | 67 |
| Pyaozero | Russia | 660–754 | 15 | 49 |
| Balaton | Hungary | 596 | 3 | 11 |
| Leman | Switzerland, France | 584 | 153 | 310 |
| Constance | Germany, Switzerland, Austria | 540 | 90 | 252 |

Source: EEA 1995.

high interest, since they are used for the supply of drinking water. The proportion of reservoirs compared to natural lakes increases toward Southern Europe where water needs to be stored to supply densely populated areas.

The largest lakes are usually not the deepest lakes. Well-known deep lakes are the lakes in the Alps. Europe's deepest lake, however, is located in Northern Europe in Norway, named Hornindalsvatnet. It is a fjord lake, i.e., a land-locked lake in a narrow, glacially deepened valley adjacent to the sea. Its maximum depth is 514 m.

## Concluding Remarks: Lakes of Europe in Comparison to Other Lakes

Lakes of Europe are characterized by the climate, geological history, land use, and atmospheric deposition. The climate of Europe differs from other continents due to the strong influence of the Gulf Stream, bringing warm air masses to northern latitudes. At European northern latitudes, lakes show similar characteristics to lakes in Canada and the Asian part of Russia, all being classified as boreal lakes. In some areas, these boreal lakes at northern latitudes are subjected to acidification processes.

Lakes in the central continental region of Europe are similar to lakes in other regions that are heavily influenced by agriculture and grassland. These lakes face problems of eutrophication. Further south in the Mediterranean region, the problem of eutrophication becomes even more severe due to a dry and hot summer climate. Here harmful algal blooms are frequently observed, similar to those in other warm, densely populated regions of the world.

Rather specific for Europe are the lakes in the Alps. Lakes in steep mountainous regions occur also in other parts of the world but these lakes are usually located far away from densely populated areas. In contrast, the population density around the lakes in the Alps is generally high, resulting in intensive recreational use that affects the lakes.

Europe is presently striving to achieve a good ecological status of its waters by implementing the European Water Framework Directive. Data are collected from all member states of the European Union and waters are classified according to their ecological status. The most efficient ways to achieve a good ecological status of European inland waters depend on further research.

*See also:* Abundance and Size Distribution of Lakes, Ponds and Impoundments; Geomorphology of Lake Basins; Mixing Dynamics in Lakes Across Climatic Zones; Origins of Types of Lake Basins; Shallow Lakes and Ponds.

## Further Reading

Eloranta P (ed.) (2004) *Inland and Coastal Waters of Finland.* Finland: University of Helsinki, Palmenia Publishing.

Fott J (1994) *Limnology of Mountain Lakes.* The Netherlands: Kluwer Academic.

George GD (ed.) (2008) *The Impact of Climate Change on European Lakes. Aquatic Ecology Series.* Berlin: Springer.

Henrikson L and Brodin YW (eds.) (1995) *Liming of Acidified Surface Waters.* Germany: Springer.

Nõges T, Eckamann R, Kangur K, et al. (2008) *European Large Lakes – Ecosystem Changes and their Ecological and Socioeconomic Impacts.* Developments in Hydrobiology 199. The Netherlands: Springer.

Scheffer M (2004) *Ecology of Shallow Lakes.* The Netherlands: Kluwer Academic.

Steinberg CEW (2003) *Ecology of Humic Substances in Freshwaters.* Germany: Springer.

O'Sullivan PE and Reynolds CS (2004) *The Lakes Handbook.* Malden, MA: Blackwell Science.

# North America

**W H Renwick,** Miami University, Oxford, OH, USA

## Lakes and Reservoirs of North America

Lakes and reservoirs are found throughout North America. The combination of extensive glaciations and widespread dam building has resulted in a continent that is rich in lakes of all sizes, from the smallest features tens of meters across to the Great Lakes, which might be considered inland seas. In this article the terms 'lake' and 'reservoir' will be used to signify water bodies of natural and human origins, respectively, and this distinction is the primary framework for organizing the discussion. The term 'water body' is used to indicate a lake or reservoir without regard to its origin.

The distribution of large lakes and reservoirs in North America is shown in **Figure 1**. This map includes 1835 water bodies averaging about $39 \, km^2$ in area. As is evident, both the largest water bodies and their highest concentrations are in eastern Canada and in adjacent portions of the United States. This is the area that was covered by continental ice sheets during the Pleistocene Epoch (~2 million to 12 thousand years ago). Although most of the features in this area are natural, many are reservoirs, especially in Quebec. The map also shows many large features outside the glaciated region. Some of these, particularly in the mountain system known as the North American Cordillera and including the Rocky Mountains, are of natural origin. Most of the water bodies outside the Cordillera and the glaciated area are reservoirs.

## Natural Lakes and Their Spatial Distribution

For organizing the information about lakes and their distribution at the continental scale, we can divide the continent into several regions characterized by their landscapes with respect to lakes and their origins (**Figure 2**). These regions are essentially defined by their geologic histories, especially occurrence of glaciation and tectonic activity. Processes of lake formation vary from region to region; for more complete discussion of lake origins refer to 'See also' section. Because these regions are so large they inevitably mask great internal variability. The discussion that follows is organized around the regions shown in **Figure 2**.

### Lakes of the Canadian Shield

The Canadian Shield is the ancient core of the North American Continent. It is composed mainly of highly metamorphosed granite, with smaller areas of metamorphosed sedimentary and igneous rocks and some areas of relatively horizontal but still quite ancient sedimentary rocks. These rocks are generally quite resistant to weathering and erosion, but have been subjected to intense and repeated glaciation. The topography of the landscape reflects the interplay between rock resistance and glacial action, with areas of relatively weaker rocks having been eroded to lower elevations while more resistant rocks form topographic highs. The distribution of lakes strongly reflects patterns of rock resistance. Faults and bedding layers in rocks tend to cause many lakes in the shield to be long and narrow. Nowhere is this more evident in maps of the Shield than in Lake Manicouagan, a reservoir constructed in the 1970s and forms a circle roughly 70 km in diameter. The circular structure is the result of a meteor collision 214 million years ago. The deepest lake in Quebec, Pingualuit Crater (252 m deep) is also the result of a meteor impact. Lake Mistassini is another example of a structurally-controlled lake. It is more than $2000 \, km^2$ in area, consisting of a series of northwest-southeast trending arcs paralleling the underlying bedrock structures.

The number of lakes in the Canadian Shield is very large. Ontario and Quebec alone list over 12 000 lakes greater than $3 \, km^2$ in area; the number of smaller lakes is much larger (Refer to 'see also' section). They occur throughout the Shield region, but are especially numerous in the highlands of Quebec and Labrador, the Ungava peninsula, northern Manitoba and Saskatchewan, Nunavut, and the eastern portions of the Northwest Territories. They are less abundant in the Hudson Bay lowlands of northern Ontario and eastern Saskatchewan.

The age and mineralogical composition of Shield rocks means that they are relatively deficient in calcium. In addition, the cool temperatures of the region and widespread coniferous forest lead to development of highly acidic soils. Thus, the lakes of the region tend to be acidic and highly oligotrophic. In addition, of course, they are ice-covered for much of the year.

### The Great Lakes Region

Although it is common to speak of five Great lakes (Superior, Michigan, Huron, Erie, and Ontario) several more are part of the same group based on physiographic setting and origins. Among these are Great

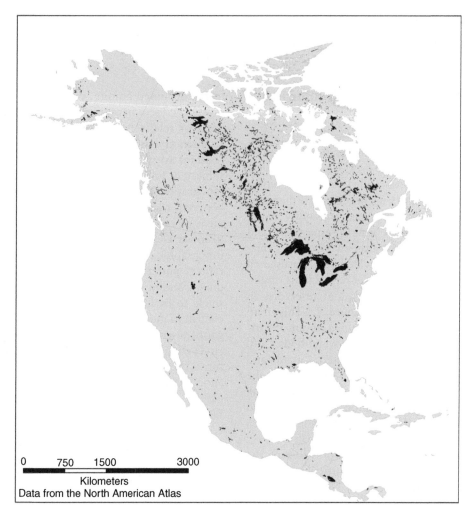

**Figure 1** Major lakes and reservoirs of North America.

Bear Lake, Lake Winnipeg, Lake Athabaska, Great Slave Lake, and Reindeer Lake. These ten total about 340 000 km$^2$ in area, and all are among the 25 largest lakes in the world (**Table 1**). They lie at or near the southern and western edge of the Canadian Shield, where it is overlapped by sedimentary rocks of the interior plateaus and Great Plains. Their location also coincides with the distal portions of the Laurentide ice sheets that formed repeatedly during the Pleistocene, spreading from the general region of Hudson's Bay. Most were formed by a combination of glacial scour and damming by terminal moraines.

Large as these lakes are, most were much larger in the past. Because the Canadian Shield lay under a great thickness of glacial ice during much of the Pleistocene, it was isostatically depressed below the elevation of the surrounding unglaciated areas. As the ice melted, huge lakes formed at the edge of the melting ice mass. Some of these overflowed to the mid-continent, draining either to the ancestral McKenzie

River or to the Mississippi, while others followed ice-marginal paths to the Arctic or the Atlantic via the Gulf of St. Lawrence. The Shield region rose with isostatic rebound, tilting these lake basins to the south and west. This tilting, combined with downward erosion of outflow channels gradually drained the lakes, reducing them to their current extent. Thus, many of the Great Lakes are bordered by large areas of former lake bed, exposing lake-bottom sediments and former shorelines.

The system of Great Lakes along the US–Canada border occupy a very high proportion of the watersheds within which they lie. Superior's land drainage area is only 1.55 times the area of the lake itself, and for the entire system from Lake Ontario upstream the ratio of land area to lake area is 2.13. They thus dominate the hydrology of the St. Lawrence River watershed. Superior is the highest (elevation 184 m above sea level) and largest, with an average depth of 147 m and a volume that exceeds that of the other

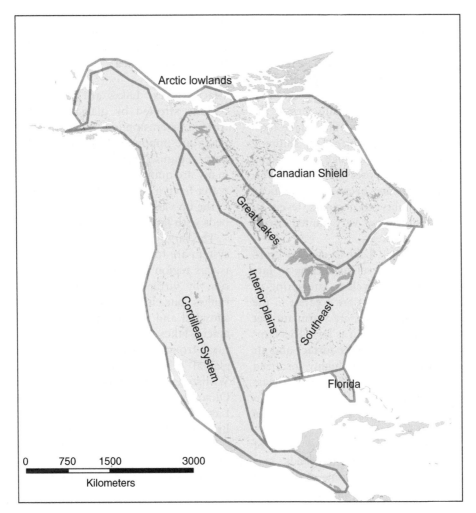

**Figure 2** Lake regions of North America.

**Table 1** Large Lakes of North America

| Name | Area (km²) | Volume (km³) | Rank in world (by area) |
|------|-----------|-------------|------------------------|
| Superior | 82 100 | 12 100 | 2 |
| Huron | 59 600 | 3540 | 4 |
| Michigan | 57 800 | 4920 | 5 |
| Great Bear | 31 000 | 2292 | 8 |
| Great Slave | 27 000 | 2088 | 10 |
| Erie | 25 700 | 484 | 11 |
| Winnipeg | 24 500 | 371 | 12 |
| Ontario | 18 960 | 1640 | 13 |
| Nicaragua | 8150 | 108 | 23 |
| Athabasca | 7935 | 110 | 24 |
| Reindeer | 5597 | 5 | 25 |

four combined. It drains into Lake Huron via the St. Mary's River. In 1921, a dam was built to regulate the lake's level and generate electricity, and locks were built at Sault Ste. Marie to facilitate shipping.

Huron and Michigan are actually one lake with a common surface elevation of 177 m, separated into two basins by the Mackinac Strait. Huron averages 59 m deep and is the larger of the two in area, while Michigan has a greater volume with an average depth of 85 m. They drain into the St. Clair River, which flows to Lake St. Clair and feeds the Detroit River, which flows to Lake Erie. The total elevation drop between Huron and Erie is 3 m, and flow through the St. Clair-Detroit River system is naturally regulated. Lake Erie (elevation 174 m) drains into Lake Ontario (elevation 75 m) via the Niagara River, with most of the elevation change accounted for by Niagara Falls. The level of Lake Ontario is regulated by an outlet structure on the St. Lawrence River. Erie and Ontario are each less than half the area of Huron or Michigan. Erie is considerably shallower (averaging 19 m) while Ontario has an average depth of 86 m. The total volume of these five lakes is roughly 20% of the total volume of freshwater lakes in the world.

Because of their large surface areas relative to drainage area the lakes exhibit long lags times in their response to variations in climate. Water levels vary up to 2 m at the decadal-scale, driven by interannual variations in precipitation and evaporation. In recent decades the lowest lake levels occurred in the mid-1960s, while high levels characterized the 1950s, mid-1970s, and mid-1980s. These lake level variations cause a variety of problems for shipping and shoreline management, with high levels being preferred for navigation purposes but lower levels desired by coastal residents because of reduced coastal erosion. The ability to regulate levels in Lakes Superior and Ontario has very limited effects on lake levels in the other lakes.

At the seasonal scale, Great Lake levels vary up to 0.5 m, with the lowest lake levels occurring in late summer or autumn and high levels in late spring or early summer. Evaporation losses are greatest in the autumn, when water temperatures are warm and yet the temperature of the overlying air is low. This temperature differential between the lake surface and air also contributes to the creation of highly localized zones of heavy 'lake-effect' snowfall on the southeast and east shores of the lakes. This effect is greatest in early winter, and smaller in late winter when lake surface temperatures are lower.

In addition to the several very large lakes of the Great Lakes Region are many thousands, and possibly millions, of smaller lakes. Some of these are simply depressions in an undulating surface of ground moraine; others called kettle lakes were formed by burial and later melting of large blocks of glacial ice, while still others such as the Finger Lakes of New York were formed by tongues of ice that scoured preexisting valleys. The northern portions of Minnesota, Wisconsin, and Michigan have tens of thousands of lakes, many of which are kettle lakes. One of the highest concentrations of these is found in Vilas County, Wisconsin, which has over 400 named lakes in an area of 2260 km². Southern New England is another area with a high concentration of lakes in a glacial depositional setting.

## Lakes of the Cordillera

The cordilleran system of North America, extending from the Brooks Ranges and Aleutian peninsula in Alaska to the isthmus of Panama, is home to a great many natural lakes. Two processes are responsible for the formation of most of these: glaciation and tectonics. The glacial lakes are mainly in the northern half of the cordillera, where Pleistocene ice was most extensive. Lakes of tectonic origin are found throughout the region, although in the north most of these

are also modified by glacial action. In the more arid portions of the region the existence of lakes is closely tied to precipitation and runoff, and many lakes have no outlets to the sea.

**Glacial lakes**    Lakes are common in mountainous regions affected by glaciation. Most of them are formed by erosion by glaciers flowing in valleys, and many are impounded by terminal moraines forming natural dams. They tend to have steep shorelines and many are quite deep. At one time they received meltwater from glaciers and some still do, forming deltas at the up-valley ends of the lakes and blanketing lake beds with layers of sediment.

The largest of the glacial lakes in the cordilleran system are found in the Canadian Rockies. Not only is this region geologically conducive to lake formation by virtue of numerous sub-parallel mountain ranges, but climatically it is an area of considerable snowfall and thus prone to glaciation. Most of the glacial lakes in the Canadian Rockies occupy structural troughs formed by faulting and enlarged by glacial action. Among the largest of these are Williston Lake, Ootsa lake, Kootenay Lake and Okanagan Lake in British Columbia, and Tagish Lake, Atlin Lake, and Lake Bennett spanning the British Columbia-Yukon border.

Thousands of smaller glacial lakes are found throughout the Rockies, generally sharing the characteristics of being deep and often long and narrow. The smallest of the lakes formed by glacial erosion occupy small bowl-shaped hollows called cirques. These are found high in the mountains and thus these lakes are typically ice-covered most of the year.

**Tectonic basins**    The Cordilleran system is a region is active mountain-building, and as such there are many areas with considerable recent tectonic activity. This inevitably results in the formation of topographic depressions that, if climate permits, become filled with water. These depressions can be very large, both in terms of area and depth. For example, at the southern end of the Cordillera Lake Nicaragua and Lake Managua lie adjacent to each other at the bottom of a tectonic depression between two mountain ranges. They are relatively low in elevation (32 and 37 m, respectively). Lake Nicaragua, with an area of over 8000 km², is the 9th largest in North America. Lake Nicaragua was once connected to the ocean and thus contains many marine species not found in inland lakes. Lake Chapala, the largest lake in Mexico, lies in a tectonic basin south of Guadalajara. Increasing demands for fresh water in the region have had significant impacts on the lake. Satellite imagery shows that the lake surface area decreased from 1048 to $812 \, km^2$ between 1986 and 2001, corresponding to a water level decrease of 2–4 m.

Active volcanism also contributes to formation of lake basins, such as Crater Lake in Oregon (a volcanic caldera) and Lake Catemaco in Mexico, (a basin dammed by a lava flow).

**Endorheic lakes** In the drier parts of the Cordillera the existence of lakes is dependent on climatic conditions as well as topography. If evaporation rates are high enough to equal the amount of runoff and precipitation entering the lake, then there may be no surface outflow. Such lakes are called endorheic. They tend to have high concentrations of dissolved solids as a result of evaporation, and concentrations fluctuate with water level, as higher water volumes dilute dissolved solids (refer to 'see also' section). Great Salt Lake is the best-known example in North America. Other prominent internally-draining lakes in the western United States include Pyramid Lake in Nevada, and Mono Lake in California. The lakes around which the Aztec culture flourished in the Valley of Mexico were also endorheic. They have since been drained and replaced by Mexico City.

Because there is no overflow control, the levels of endorheic lakes fluctuate substantially in response to climatic variations, rising in wet and/or cool periods and falling in dry and/or warm periods. During the Pleistocene, reduced temperatures cause large reductions in evaporation rates so that the level of water in what is now Great Salt Lake rose by more than 300 m, and the lake grew in area to more than $50\,000 \, km^2$, to form Lake Bonneville. This was the largest of several lakes known as pluvial lakes, associated with wetter periods of the Pleistocene in areas that were not glaciated. Lake Bonneville's maximum area was about ten times the present area of Great Salt Lake. At that maximum extent it overflowed to the north, into the headwaters of the Snake River. Another great pluvial lake called Lake Lahontan occupied several valleys in western Nevada and Eastern California, but did not overflow to the sea. Variations in levels of these prehistoric lakes caused massive floods whenever lake level rose to the elevation of a key overflow point, and their marks on the landscape are prominent today.

The shallowest endorheic lakes are ephemeral: they only contain water following brief periods of high runoff and disappear by evaporation between such periods. These are called playas, and are common in the intermountain western United States and northern Mexico. Many of these never disappear completely and at their smallest extent sustain populations of salt-tolerant organisms such as brine shrimp.

## The Arctic Lowlands

The coastal plain of northwest Canada and adjacent Alaska is a unique lake region. This area is one of extensive permafrost in a low-relief environment. Parts of this region were glaciated during the Pleistocene Epoch, but much was not. The Arctic lowlands are home to tens and perhaps hundreds of thousands of shallow lakes. Some of these are underlain by peat, while others lie on glacial deposits or marine sediments derived from prior high sea levels. In the present era of rising temperatures many lakes are being formed and subsequently drained by thawing of permanently frozen ground. When ice-rich sediments begin to thaw they may settle as water melts and is expelled from the sediments. Areas of shallow standing water form, and because in the brief warm season this water absorbs and stores more heat than surrounding vegetated areas the thawing and settling of sediments beneath is accelerated. This process is called thermokarst, and results in the formation of lakes. At the same time many lakes are being drained as the thawing and settling of sediments continues.

## Natural Lakes of the Interior Plains and Southeastern U.S.

**Glaciated interior plains** In the northern, glaciated, interior plains, the Pleistocene glaciers originating in the Canadian Shield flowed westward to the base of the Rocky Mountains. There they met ice flowing eastward out of the Rockies. Thus, the entire width of the plains was glaciated for most of the extent of the Cordillera in Canada. Lakes are much less abundant on the glaciated interior plains than in the Shield. Lake Claire ($1415 \, km^2$) and Lesser Slave Lake ($1167 \, km^2$), both in northern Alberta, are the largest natural lakes of the region. In northeastern North and South Dakota as well as in much of Manitoba and Saskatchewan a great many small lakes, known as prairie potholes, exist. These are small lakes formed in glacial depressions; some are merely low spots in an undulating surface while others are kettle lakes. They are particularly significant as habitat for migratory waterfowl. In semiarid portions of the plains numerous endorheic saline lakes exist, particularly in Saskatchewan and Alberta (refer to 'see also' section).

**Unglaciated interior plains** The southern, unglaciated portion of the interior plains is essentially without natural lakes. The absence of lakes is a result of the lack of a lake-forming process such as glaciation or tectonic activity. Some of the major rivers have substantial oxbow lakes, formed by abandonment of a portion of channel by an actively meandering river. There is a high density of natural playa lakes in the Texas panhandle and adjacent New Mexico. In this area over 20 000 small endorheic ephemeral lakes dot the prairie. The depressions were formed by wind erosion, and standing water is maintained by clay-rich sediments that limit percolation losses. They grow and shrink seasonally, and form important islands of biodiversity in an intensively agricultural landscape.

**Southeastern U.S.** The portion of the United States south of the Pleistocene glacial limit and east of the Great Plains has very few natural lakes, although as will be discussed below there are millions of reservoirs. Nearly all of the natural lakes that are present are of one of two types: oxbow lakes and karst sinkholes that intercept groundwater. The largest oxbow lakes are found in the lower Mississippi valley. In this area the channel of the Mississippi is generally 1–2 km wide, and meanders actively although artificial controls on the channel have reduced this activity in the last century. The oxbow lakes of the lower Mississippi are typically 1–2 km wide several km long, and generally not more than a few meters deep. Reelfoot Lake, lying in the Mississippi floodplain in northwest Tennessee, is thought to have been formed at least in part by the New Madrid earthquakes of 1811–1812. Smaller rivers draining the coastal plain also have scattered oxbow lakes on their floodplains, of smaller dimensions appropriate to the rivers that created them.

The Florida Peninsula contains the highest concentration of natural lakes outside the glaciated portion of North America. Lake Okeechobee is the largest ($1890 \, km^2$). These lakes are generally small, circular depressions formed by underground solution weathering of limestone bedrock and subsequent collapse, forming sinkholes. Landscapes with such features are called karst, and are common in the southeastern United States. In most U.S. karst areas the sinkholes do not reach the ground water table, and so do not form lakes. In Florida, however, a great many sinkholes were formed during periods of the Pleistocene when sea level was substantially lower than at present. As sea level rose so did water tables and these sinkholes became lakes. Lakes so formed are also known as cenotes. Some are relatively shallow, but many are quite deep and connected to underwater cave systems.

## Reservoirs and Their Spatial Distribution

The distribution of large reservoirs in North America is shown in **Figure 3**. This map shows only relatively large

features. It includes 654 reservoirs ranging in size from 11 to $8770 \, km^2$ in area. Millions of smaller reservoirs exist but cannot be shown at this scale. Reservoirs as a class of lakes are discussed in chapter 39.

### Brief History of Reservoir Construction

Reservoirs – artificial water bodies created normally by construction of a dam across a stream, sometimes in combination with excavation of a depression – have been built in North America for hundreds and perhaps thousands of years. The Aztecs who built their city in the Valley of Mexico constructed dams to regulate water flows to the endorheic lakes of the valley. When Europeans occupied the continent in the 17th century they quickly began to build small dams that created ponds supplying power to mills for various purposes. Tens of thousands of such ponds were built in the eastern United States in the 17th–19th centuries. In addition, larger reservoirs were built in many areas in the 19th century to provide water to navigational canal systems.

In the early 20th century the scale of reservoir construction increased dramatically (**Figure 4**). In this period reservoirs were built principally for public water supply, flood control, irrigation, and hydroelectric power generation. Later in the 20th century recreation became a major justification for reservoir construction. By the 1970s, large-scale reservoir construction in North America began to meet obstacles. Foremost among these was public concern about loss of land and free-flowing aquatic habitat. This opposition arose wherever dams were being built: on land used by Native Americans in central Quebec or within easy reach of metropolitan regions in the United States. Other factors contributing to the decline of reservoir construction include increasing costs and the development of alternative approaches to problems of flooding or water supply.

### Major Reservoir Development Programs

**Hydroelectric power in Canada** Hydroelectric power generation in Canada began in the last decades of the 19th century, with construction of facilities near waterfalls in Ontario, and grew steadily in the 20th century. Dam construction slowed in Canada during the Depression of the 1930s, but increased rapidly in the years following World War II. Several large reservoirs were constructed in the Ottawa River system bordering Ontario and Quebec. In the 1960s, a few very large hydroelectric dams were built in British Columbia, including a 244-m high dam that impounds Kinbasket Lake, and dams forming the Upper and Lower Arrow Lakes. The most extensive

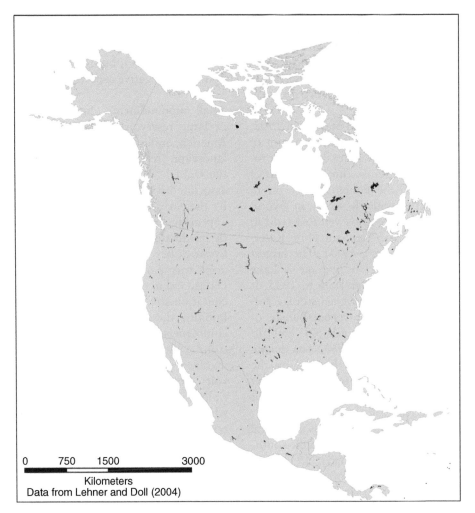

**Figure 3** Large reservoirs of North America.

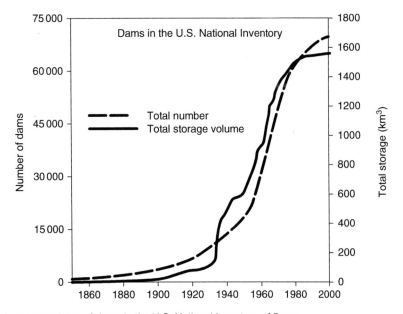

**Figure 4** Number and storage volume of dams in the U.S. National Inventory of Dams.

Canadian hydroelectric projects have been built by Quebec Hydro, which has constructed several very large reservoirs. Among these are Lake Manicouagan (mentioned above), and the La Grande 2 and 3 reservoirs of the James Bay Project. Smallwood Reservoir in Labrador is also linked to the Quebec Hydro network.

**Large dam programs in the United States** In 1902, the U.S. Reclamation Act established a program of government-sponsored irrigation development projects in the western United States. Among the first of these were the 32-km$^2$ Belle Fourche Reservoir in South Dakota (1907), Pathfinder Reservoir in Wyoming (91 km$^2$, 1909), Theodore Roosevelt Reservoir in Arizona (72 km$^2$, 1909), and Elephant Butte Reservoir in New Mexico (162 km$^2$, 1915). The largest Bureau of Reclamation Project by volume is Lake Mead, impounded by the Hoover Dam in 1935. It has an area of 660 km$^2$, and an original storage volume of 40 km$^3$. The most controversial Bureau of Reclamation reservoir is Lake Powell (687 km$^2$) in Arizona and Utah, completed in 1963. The debate over construction of this reservoir focused on the loss of a spectacular canyon wilderness upstream of the Grand Canyon, and raised public awareness of the negative aspects of reservoir development.

The Tennessee Valley Authority (TVA) is one of the largest reservoir development programs in the southeastern United States. It was established in 1933, with a mission to spur economic development through constructing reservoirs for electricity generation and flood control. Today its reservoirs are also major recreational facilities. They include Norris Lake (137 km$^2$) and Kentucky Lake (66 km$^2$).

The Bureau of Reclamation and the TVA are only two of many major governmental reservoir-construction programs in the United States. The U.S. Army Corps of Engineers, which has also built a great many reservoirs as part of its flood control mission, has compiled an inventory of dams in the United States. It lists over 75 000 features that are more than 2 m in height and impound more than 61 700 m$^3$, or are more than 8 m high and impound more than 18 500 m$^3$. Not all of these create permanent lakes, but this gives an indication of the numbers of medium-to-large artificial impoundments in the United States.

**Large reservoirs in Mexico and Central America** Several large reservoirs have been constructed in Mexico, primarily in the latter part of the 20th century. Among these are Lake Angostura (760 km$^2$), Lake Inhernillo (665 km$^2$), Lake Miguel Aleman (400 km$^2$), Lake Malpaso (381 km$^2$), and the Falcon and Amistad Reservoirs on the U.S.–Mexico Border (329 and 238 km$^2$, respectively). The Chixoy

Reservoir in Guatemala gained international notoriety because of violent evictions of people living in the area that was to be inundated. The dam went into operation in 1983, but the episode increased awareness of the negative aspects of reservoir construction.

**Small impoundments** In addition to the thousands of large reservoirs that have been built in North America, millions of small impoundments dot the landscape (refer to 'see also' section), especially in the United States. The earliest of these were those associated with the water power, mentioned earlier. By the early 20th century a new form of pond was appearing, primarily in agricultural areas where cattle production required additional water sources for livestock beyond what was available in streams. The need was particularly great in the eastern Great Plains, and in the southeast where despite ample precipitation streamflow can be very limited seasonally. Initially, these ponds were constructed by private landowners, but beginning with the 1936 Flood Control Act the Federal government became increasingly involved in subsidizing construction of small reservoirs. Construction of such impoundments was especially rapid in the decades following World War II, when machinery was widely available as were government subsidies for this work. The construction of ponds to support agricultural needs diminished in the 1980s and 1990s, but this was replaced by ponds built primarily for recreational or aesthetic purposes, largely in suburban areas, and for stormwater control in urban and suburban developments. In the United States these small impoundments now number in the millions (**Figure 5**).

### Transformation of the North American Hydrologic Landscape

The cumulative impact of the placement of millions of artificial water bodies on the landscape of North America has not been documented. Nonetheless, it is reasonable to assume that their impacts are both profound and pervasive. There are no major watersheds in the United States without reservoirs, and relatively few minor ones. In Canada the number of relatively undisturbed watersheds is much greater, especially in the north, and Mexico too has several large unregulated watersheds remaining although this is changing. One indication of the effect of these impoundments is that despite very large increases in soil erosion associated with deforestation and agricultural activities, the total flux of sediment from the continent to the seas is apparently little different today from what it was prior to European settlement. Other cumulative impacts are less well known. In arid

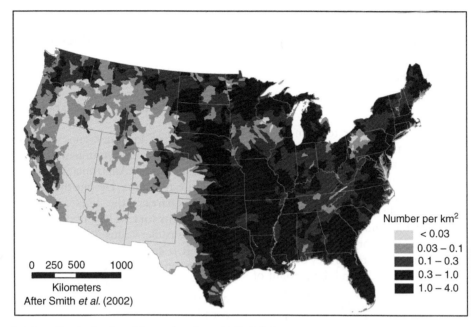

**Figure 5** Density of small water bodies of the conterminous United States.

and semiarid regions total evaporation from reservoirs can be a significant portion of total stream flow, but the effect is probably minor in humid regions. The greatest impacts are likely in the distribution of biota, both in terms of creation of new habitat for aquatic species where none existed before, and in creation of barriers to fish migration. Another as yet unquantified but potentially large impact of reservoirs is on the carbon cycle. Many reservoirs appear to be net sources of carbon dioxide to the atmosphere as well as significant generators of methane, but at the same time they are major sinks of particulate carbon.

*See also:* Abundance and Size Distribution of Lakes, Ponds and Impoundments; Geomorphology of Lake Basins; Origins of Types of Lake Basins.

## Further Reading

Canadian Dam Association (2003) *Dams in Canada.* Edmonton: Canadian Dam Association.

Dahl TE (2006) *Status and Trends of Wetlands in the Conterminous United States 1998 to 2004.* Washington D.C: U.S. Department of the Interior; Fish and Wildlife Service.

Graf WL (1999) Dam Nation: A geographic census of American dams and their large-scale hydrologic impacts. *Water Resources Research* 35: 1305–1311.

Hunt CB (1967) *Natural Regions of the United States and Canada.* San Francisco: WH Freeman and Co. 725 pp.

Lehner B and Döll P (2004) Development and validation of a global database of lakes, reservoirs and wetlands. *Journal of Hydrology* 296: 1–22.

Petts GE and Gurnell AM (2005) Dams and geomorphology: Research progress and future directions. *Geomorphology* 71: 27–47.

Smith SV, Renwick WH, Bartley JD, and Buddemeier RW (2002) Distribution and significance of small, artificial water bodies across the United States landscape. *Science of the Total Environment* 299: 2–36.

## Relevant Websites

http://www.iaglr.org/ – International Association for Great Lakes Research.

http://www.icold-cigb.net/ – International Commission on Large Dams.

http://www.nalms.org/ – North American Lakes Management Society.

http://www.worldlakes.org/ – World Lakes Network.

# South America

**M E Llames and H E Zagarese,** Instituto Tecnológico de Chascomús (CONICET – UNSAM), Buenos Aires, Argentina

## Introduction

Two obvious elements must concur to form a lake: a source of water and a land depression capable of storing water. Basin size matters, as deep lakes often stratify for long periods, while shallow lakes tend to mix more often or even continuously. The magnitude of water supply matters, as it affects the residence time and the tendency for diluting or concentrating solutes (i.e., salts, organic matter). Geographic location and climate matter, as they determine rain patterns, hydrologic periods, and the amount of solar and wind energy impinging upon the surface. Elevation and position within the landscape matter, as they affect daily and seasonal thermal amplitude, as well as the likelihood of receiving allochthonous inputs. South America is a diverse continent in terms of geomorphology and climates, which in turn translates into a wide diversity of lake types. In addition, South America is located in a predominantly maritime hemisphere. All in all, South America offers the opportunity to contrast empirical relationships developed for Northern Hemisphere lakes.

In this chapter, we summarize the salient geomorphologic features of the continent, its climate, and the main forces that have created its lakes. Then, we introduce the main types of lake exemplified by a few selected lakes.

## Geography and Geomorphology

South America covers an area of $17\,870\,218\,km^2$ and spans a broad latitudinal range, extending from 12° 28'N (Punta Gallinas, Colombia) to 55°59'S (Cabo de Hornos, Chile). Politically, the territory is divided into thirteen countries: Colombia, Venezuela, Guyana, Suriname, French Guiana, Brazil, Ecuador, Peru, Bolivia, Paraguay, Uruguay, Argentina, and Chile (**Figure 1(a)**). At a regional scale, four major geomorphologic regions may be identified:

1. The Guaiana and Brazilian Highlands
2. The interior lowlands
3. The Cordillera
4. The Patagonian steppe

The Guiana and the Brazilian Highlands form the continental shield of north-eastern South America. This region contains the continent's oldest rocks, which are mostly covered by ancient sediments. The Guiana and the Brazilian Highlands are separated by the Amazon geosyncline.

A series of lowlands runs southward, along the middle part of the continent, from the Llanos del Norte, across the Amazonia and Gran Chaco, to the Argentinean Pampa. Despite the absence of major geographical barriers, the climate, hydrology, and vegetation patterns vary widely among these lowlands.

The Cordillera runs North-South along the Pacific shore. It is composed of the Andes ranges and high intermountain valleys and plateaus. The region is seismically active and frequently affected by earthquakes. Several volcanoes spread along the Andean cordillera, but most of them are inactive.

The Patagonian steppe lies east of the Andes and extends south from the Colorado River (35° S 69'W). Tablelands are typical features of the Patagonian landscape. They are made up of tectonically uplifted basaltic plateaus and coarse grain fans originated in the Andes (**Figure 1**).

As regards to the origin of lakes, at least seven different processes have played significant roles:

1. Tectonic activity
2. Vulcanism
3. Glacial activity
4. Fuvial action
5. Aeolic action
6. Marine action
7. Human activity

These processes have acted either alone or in combination with others, and their importance differs among regions.

## Climate

Owing to the wide latitudinal coverage of the continent, a large variety of climates occurs in South America. Although much of the continental mass is located within the intertropical belt, large regions of Chile, Argentina, and Uruguay lay in the Southern Temperate Zone, while the southern tip of the continent actually extends into sub-Antarctic latitudes. The near-surface wind circulation, which determines the general water vapor transport, is dominated by the following:

1. The low pressure belt at the Intertropical Convergence Zone (ITCZ).

2. The semi-permanent anticyclones of the South Atlantic and South Pacific Oceans (approximately at 30°S).
3. The sub polar low pressure belt at approximately 60°S.
4. The low pressure center in the central-western part of Argentina and southern Bolivia.

As the position of the ITCZ oscillates seasonally, there is a north-south displacement of the South Atlantic and Pacific highs, which conditions the annual cycle of precipitations over the whole region.

In the Equatorial belt (roughly between 10N and 10S), the trade winds from both hemispheres (NE-SW or SE-NW, in the North and South Hemisphere, respectively) converge in the ITCZ and generate an upward motion of the air. This process becomes locally intensified in tropical storms resulting in heavy rains. The rainfall regime is driven by convection with a bimodal distribution and maximum values coincident with the equinoxes (i.e., March–April; September–October). Constant high temperatures and humidity, and the absence of a dry season are also characteristic of this region.

Below about 10°S appears a wide belt of arid land, the so-called arid diagonal of South America. This arid region runs in a roughly NW-SE direction from the Peruvian and northern Chilean Pacific coast, down to the Patagonian Atlantic coast in southern Argentina (yellow area in the precipitation maps of **Figure 1(b)**). Lack of moisture, the most salient feature of this area, results from several factors, including the cold Humboldt Current in the Pacific coast and mountain rain shadows.

Above the arid diagonal, the sources of water vapor originate in the Atlantic Ocean and the tropical rainforest. In general, the amount of precipitation decreases along an east-west gradient, although precipitation associated with mountain ranges is responsible for local rain maxima in the inner continent (i.e., parts of Bolivia and northern Argentina). The seasonal precipitation pattern has typical monsoon features, with maximum precipitation during the austral summer.

Below the arid diagonal, the predominant winds blow from the Pacific Ocean. The westerly winds discharge most of their moisture on the western slope of the Andes (Chile), causing a sharp precipitation gradient from about 4000 mm year$^{-1}$ on the western slope of the Patagonian Andes to less than 250 mm year$^{-1}$ in the Atlantic Patagonian coast. Precipitation peaks during winter due to the northward movement of the Pacific high, the intensification of the subpolar low, and the thermal difference between oceanic and land masses. This seasonal pattern is opposite to the one described for the region above the arid diagonal.

## Lakes and Reservoirs

Taken as a whole, South America is predominantly a rivers continent. In fact, many floodplain lakes, lagoons and reservoirs originate from rivers; particularly, from the Amazon, Orinoco and Paraná-Plata rivers, which rank among the largest river systems of the world. On the other hand, the proportion of 'archetypical' lakes is relatively small, and they could hardly compete for the size record with lakes in other continents. Nevertheless, South America does have a great number and diversity of lakes. In addition, its location in a predominantly maritime hemisphere offers a unique opportunity for comparing empirical patterns developed originally for Northern Hemisphere lakes.

The previous section described three factors that determine the diversity of standing waters in South America: geography, geomorphology and climate. Important morphological features, such as lake size and shape are largely a consequence of the processes that created the basins (for a detailed description of

**Figure 1a** Detailed scheme of main geomorphologic features of South America.

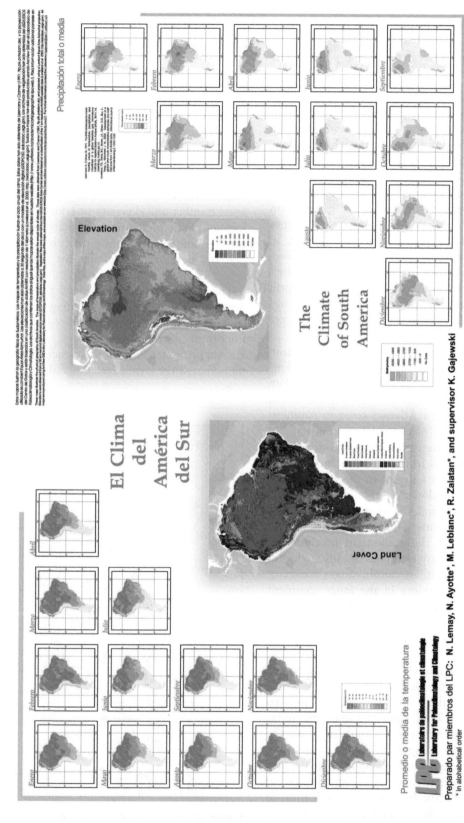

**Figure 1b**   Summary of main geomorphologic and climatic features of South America. Reprinted from http://www.pc.uottawa.ca/data/americas/samerica/sam_large.jpg.

the origin of lake basins and abundance of lakes refer to 'See also' section). As most of these processes act on regional scales, morphologically similar lakes tend to be grouped in clusters, or lake districts. In order to provide a simplified account of the main limnological characteristics of lakes and reservoirs of South America, we have grouped lakes according to a hybrid arrangement that mixes geomorphologic and climatic features as follows:

1. Natural basins:
   a. Andean lakes:
      i. Tropical mountain lakes:
      ii. Temperate mountain lakes:
   b. Inland lowland lakes:
      iii. Tropical lowland lakes
      iv. Temperate lowland lakes
   c. Coastal lagoons
2. Human-made basins
   a. Reservoirs
   b. Various types of excavations

## Natural Basins

### Andean Lakes

Most Andean lakes were formed by tectonic activity, vulcanism or glaciers. Tectonic and volcanic events played important roles in lake formation throughout the Andean ranges. However, given that these processes have been more active in arid regions, there are fewer tectonic lakes than glacial lakes. Lake Titicaca is perhaps the best known example of an Andean tectonic lake. Tectonic forces are also responsible for the creation of salt lakes in endorreic areas within the Altiplano (shared by Argentina, Bolivia, Chile and Perú). On the other hand, glaciation has widened, deepened, and dammed many valleys, and created an important lake district in the Andean region of southern South America. In fact, the latter region (between 39° S and 55° S) includes many of the largest and deepest lakes of South America.

**Tropical mountain lakes** In general, two kinds of systems can be distinguished among tropical mountain lakes:

1. High-elevation lakes from the Andes: these lakes occur in orographic deserts, such as the Puna, at high altitudes (ca. 3500–4000 m.a.s.l.). Because of their occurrence at high elevations, sometimes above the timberline, they are characterized by low water temperatures and rather particular chemo-optical characteristics. High salt and pH values, as well as low dissolved organic matter

(DOC) concentrations and high transparency, characterize these lakes. Moreover, some of the smaller lakes freeze during part of the year. The best known of these lakes is Lake Titicaca (**Table 1**) in the Andes on the border of Perú and Bolivia. Within high-altitude lakes, Titicaca is the largest and deepest lake in the world. It is one of the few cases in which the lake basin was formed by the uplifting of virtually undisturbed sections of ancient peneplains during the process of the mountain building. In addition, a large number of smaller salt lakes are spread all over the region.

2. Lakes from humid equatorial or seaward facing ranges: in contrast to tropical high altitudinal lakes, the high rainfall in these zones results in low salt concentrations. Overall, the limnological characteristics of these lakes are similar to those described for tropical lowland lakes (see later text). The best studied of these is Lake Valencia (**Table 1**). It is a graben lake located in the Aragua Valley of Venezuela. It was formed due to faulting and subsequent damming of the Valencia River. The level of the lake has fluctuated widely during the last 10 000 years in response to climatic variations. In recent years, the water level has decreased due to a shift towards drier conditions and to human use of water in the watershed.

**Temperate mountain lakes** Temperate lakes in the Southern Hemisphere are restricted to the southern tip of South America and New Zealand. Compared to North America, Europe and Asia, the continental southern temperate land masses are relatively small. As a result southern temperate lakes experience strong winds and maritime climates. Thus, large temperate lakes in the Southern Hemisphere are monomictic and do not freeze during winter. Interestingly, thermoclines develop deeper in the Southern Hemisphere than in similar lakes in the Northern Hemisphere. In turn, deeper epilimnia translates into lower chlorophyll concentrations at comparable levels of phosphorus in Southern Hemisphere temperate lakes than their Northern Hemisphere counterparts.

At least three lake types may be recognized in the temperate Andean region of South America:

1. Large, deep, piedmont lakes. These vary in size and depth, but can be as large as 1892 km$^2$ (Buenos Aires lake) and as deep as 836 m (O'Higgins-San Martín). These lakes are monomictic, oligotrophic and highly transparent. They tend to occur in transversal valleys running east-west and are therefore well exposed to the dominant winds from the Pacific. These lakes mostly occur in the forested areas close to the

**Table 1**　Location and morphological features of selected lakes

| Name | Location | Lake type | Origin | Surface area (km²) | Max depth (m) | Mean depth (m) | Altitude (m.a.s.l.) | Source |
|---|---|---|---|---|---|---|---|---|
| Valencia | 10°12'N 67°44'W | Tropical mountain | Tectonic | 350 | 39 | 19 | 420 | 8 |
| Titicaca | 16°S 69°W | Tropical mountain | Tectonic | 8400 | 284 | 100 | 3810 | 2, 13, 14 |
| Chungará | 18°14'S 69°09'W | Tropical mountain | Tectonic-volcanic | 260 | 34 | | 4520 | 7, 9 |
| Poopo | 18°47'S 67°7'30"W | Tropical mountain | Tectonic | 2600 | | 3–5 | 3695 | 16 |
| Iberá | 28°S 57°W | Temperate lowland | Fluviatile | 54 | 4.5 | 3 | | 3 |
| Mar Chiquita | 30°40'S 62°40'W | Temperate lowland | Tectonic | 1984 | | 7.3 | 69 | 4 |
| Patos | 31°9'S 51°05'W | Coastal lagoon | Marine | 9800 | 14 | <2 | 0 | 10, 12 |
| Mirim | 31°9'S 51°05'W | Coastal lagoon | Marine | 560 | | <2 | 0 | 10, 12 |
| Chascomús | 35°36'S 58°00'W | Temperate lowland | Eolic | 30.1 | 1.9 | 1.5 | 8 | 4 |
| Caviahue | 37°53'S 71°02'W | Temperate mountain | Volcanic | 9.22 | 95 | 51.4 | 1600 | 11 |
| Puyehue | 40°40'S 72°35'W | Temperate mountain | Glacial | 165 | 123 | 76 | 230 | * |
| Nahuel Huapi | 40°50'S 71°30'W | Temperate mountain | Glacial | 557 | 464 | 157 | 764 | 4 |
| Morenito | 41°03'S 71°31'W | Temperate mountain | Glacial | 0.83 | ~10 | | 770 | ** |
| Escondido | 41°04'S 71°35'W | Temperate mountain | Glacial | 7 | 38 | 73 | 790 | ** |
| Todos los santos | 41°04'S 72°16'W | Temperate mountain | Glacial | 178.5 | 335 | | 189 | *** |
| Schmol | 41°11'S 71°18'W | Temperate mountain | Glacial | | 5 | | 1950 | 15 |
| Jacob | 41°11'S 71°34'W | Temperate mountain | Glacial | | 25 | | 1550 | 15 |
| Musters | 42°22'S 69°11'W | Temperate mountain | Tectonic | 414 | 38.5 | 20 | 260 | 1, 4 |
| Colhué Huapi | 45°30'S 68°45'W | Temperate lowland | Tectonic-eolic | 810 | 5.5 | 2 | 258 | 1, 4 |
| Buenos Aires | 46°39'S 72°03'W | Temperate mountain | Glacial | 1892 | 463 | | 230 | 6, 16 |

| Lake | Coordinates | Origin | Type | | | | | Sources |
|---|---|---|---|---|---|---|---|---|
| O'Higgins–San Martín | 48°50'S 72°36'W | Glacial | Temperate mountain | 1058 | 836 | | 285 | 16 |
| Cardiel | 48°57'S 71°13'W | Tectonic | Temperate mountain | 460 | 80 | 49.1 | 300 | 1, 4 |
| La Tota | 5°34'N 72°56'W | Tectonic | Tropical mountain | 60 | 60 | 30 | 3015 | 5 |
| Argentino | 50°20'S 72°45'W | Glacial | Temperate mountain | 1466 | 500 | 150 | 187 | 4 |
| Ofhidro | 53°57'S 69°38'W | Glacial | Temperate mountain | 54 | | | 210 | 16 |
| Blanco Lake | 54°04'S 69°03'W | Glacial | Temperate mountain | 155 | | | 180 | 16 |
| Fagnano | 54°32'S 67°50'W | Tectonic | Temperate mountain | 590 | 200 | | 140 | 16 |

Sources
1. Baigún C and Marinone C (1995). Cold-temperate lakes of South America: Do they fit northern hemisphere models? *Archive für Hydrobiologie* 135(1): 23–51.
2. Baker PA, Fritz SC, Garland J, and Ekdhal E (2005) Holocene hydrologic variation at Lake Titicaca, Bolivia/Peru, and its relationship to North Atlantic climate variation. *Journal of Quaternary Science* 20(7–8): 655–662.
3. Brachini L, Cózar A, Dattilo AM, Picchi MP, Arena C, Mazzuoli S, and Loiselle SA (2005) Modelling the components of the vertical attenuation of ultraviolet radiation in a wetland lake ecosystem. *Ecological Modelling* 186: 43–45.
4. Calcagno A, Fioriti MJ, Lopez H, Pedrozo F, Razquin ME, Rey C, Quiros R, and Vigiliano P (1995) Catálogo de Lagos y Embalses de la Argentina, Direccion Nacional de Recursos Hidricos, Buenos Aires, Argentina.
5. Cordero RD, Ruiz JE, and Vargas EF (2005) Determinación espacio-temporal de la concentración de fósforo en el Lago de Tota. *Revista Colombiana de Química* 34(2): 211–218.
6. Díaz M, Pedrozo F, and Baccala N (2000) Summer classification of Southern Hemisphere temperate lakes (Patagonia, Argentina). *Lakes and Reservoirs: Research and Management* 5: 213–229.
7. Dorador C, Pardo R, and Vila I (2003) Variaciones temporales de parámetros físicos, químicos y biológicos de un lago de altura: el caso del lago Chungará. *Revista Chilena de Historia Natural* 76: 15–22.
8. Lewis W Jr (1983) Temperature, heat and mixing in lake Valencia, Venezuela. *Limnology and Oceanography* 28(2): 273–286.
9. Mülhauser H, Hrepic N, Mladinic P, Montecino V, and Cabrera S (1995) Water quality and limnological features of a high altitude Andean lake, Chungará, in northern Chile. *Revista Chilena de Historia Natural* 68: 341–349.
10. Niencheski LFH, Baraj B, Windom HL, and França RG (2004). Natural background assessment and its anthropogenic contamination of Cd, Pb, Cu, Cr, Zn, Al and Fe in the sediments of the southern area of Patos Lagoon. *Journal of Coastal Research.* 39: 1040–1043. (Special issue).
11. Pedrozo F, Kelly L, Díaz M, Temporetti P, Baffico G, Kringel R, Friese K, Mages M, Geller W, and Woelfl S (2001) First results on the water chemistry, algae and trophic status of an Andean acidic lake system of volcanic origin in Patagonia (Lake Caviahue). *Hydrobiologia* 452: 129–137.
12. Spivak E (1997) Cangrejos estuariales del Atlántico sudoccidental (25°–41° S) (Crustacea: Decapada: Brachyura). *Ivest. Mar. Valparaíso* 25: 105–120.
13. Taborga J and Campos J (1995) Les resources Hydriques de L'Altiplano. Bull. Inst. Fr. E'tudes Andines 24(3): 441–448.
14. Vincent W, Wurtsbaugh W, Vincent C, and Richerson P (1984) Seasonal dynamics of nutrient limitation in a tropical high-altitude lake (Lake Titicaca, Peru-Bolivia): Application of physiological bioassays. *Limnology and Oceanography* 29(3): 540–552.
15. Zagarese HE, Díaz M, Pedrozo F, and Úbeda, C. (2000) Mountain lakes in orthwestwen Patagonia. Verhandlungen Internationale Vereinigung fu"r theoretische und angewandte. *Limnologie.* 27: 533–538.
16. Ziesler R and Ardizzone GD (1979) COPESCAL, Doc. Téc./COPESCAL Tech. Pap., (1): Las Aguas Continentales de América Latina. http://www.fao.org/docrep/008/ad770b/AD770B00.HTM
Refer to * Soto (2002), ** Morris et al. (1995), *** Marinone et al. (2006) in Further readings section.

Andes, but some of them also occur in (or extend into) the Patagonian steppe.

2. Small to medium size, piedmont lakes. They share many characteristics with the previous type, the main difference being their relatively shallowness (less than 15–20 m) and as a consequence, polymictic thermal regimes. Least exposed lakes freeze during winter. They tend to be more productive and less transparent, partly due to the lack of permanent stratification (i.e., epilimnetic water is periodically in contact with the sediments), and partly because of higher levels of colored dissolved organic matter (CDOM) from the surrounding forests.

3. Small to medium size mountain lakes. These lakes are similar in size and depth to the previous group of lakes, but they occur at higher elevations, some even above the timberline. Given that the soils in their watersheds are poorly developed and the dominant rock, granite, is highly refractory, these lakes are typically oligotrophic or ultraoligotrophic. Conductance may be as low as $10–15\,\mu S\,cm^{-1}$, CDOM concentrations are close to the lower limit reported for lakes, and transparency to PAR and ultraviolet radiation is only slightly lower than that of the most oligotrophic oceanic waters (e.g., Lake Schmol). Most of these lakes freeze from late fall until late spring or early summer. Summer stratification is also unstable, since these lakes frequently mix completely due to their relative shallowness and high exposure to strong winds. Most, if not all, of these lakes lack fish and have extremely simple food webs.

### Inland Lowland Lakes

**Tropical lowland lakes**   As already mentioned, tropical South America is dominated by river systems. Thus, shallow fluvial lakes and lagoons associated with floodplains are the rule. In contrast to more populated regions of the world, the majority of South American floodplains retain most of their natural hydrological characteristics. These extensive floodplains are associated with the large rivers of the region. The running water of the rivers possesses considerable erosive power, which creates lake basins along their courses as a result of a variety of depositional and erosive processes, while on their deltas; sedimentation can also produce large shallow basins. Unlike temperate regions, the alternation of wet and dry seasons is biologically more important than temperature and photoperiod fluctuations. Because floodplain wetlands of large river systems are analyzed in later chapters, only their general characteristics of these environments are described here.

South American tropical lowland lakes experience considerable water level fluctuation that frequently results in substantial areas of exposed sediments every year. Such fluctuations are due to nonsynchronous water gains and losses occurring at different times of the year. This seasonal pattern of volume changes depends upon the interaction of water exchange with vegetation, rainy seasons and catchment morphology.

Tropical lowland lakes stratify and mix more readily than temperate ones in response to changes in wind strength or reversal in heat flux. These characteristics have great significance for the efficiency of nutrient recycling and for the setting of the successional clock for biological communities. These shallow lakes are polymictic since they are permanently affected by wind, even frequently undergoing a diurnal pattern of daytime stratification and nightly thermocline breakdown. On the other hand, deeper lakes (>10 m) are predominantly warm monomictic and show great regularity in seasonal mixing, which coincides with the corresponding hemispheric winter. Stratification is seasonally persistent, but less stable than at higher latitudes, and the amount of heat exchange required to cause important changes in stability is smaller than in temperate regions.

Tropical waters appear to have a lower ratio of dissolved inorganic nitrogen to soluble reactive phosphorus than typical lakes in other regions. This difference in the N:P ratio is related to (i) a more efficient chemical weathering of phosphorus due to higher temperatures, (ii) higher rates of evapoconcentration of phosphorus at low latitudes, and (iii) higher nitrogen losses from tropical environments than at temperate latitudes because of the high water temperatures of tropical lakes.

As regards to natural communities, these lakes tend to be: (i) more efficient in producing phytoplankton biomass on a given nutrient base, (ii) inefficient in transferring primary production to higher trophic levels, (iii) similar to temperate lakes as regards to phytoplankton and zooplankton composition, and (iv) affected by strong sporadic variability superimposed to the seasonal cycle.

**Temperate lowland lakes**   Under this section we have grouped a rather heterogeneous set of permanent, semipermanent, and temporary lakes located in the Pampa plains and the Patagonian steppe. These lakes range from a few ha to several hundred $km^2$, and with a few exceptions (e.g., lakes Cardiel and Musters), they are typically shallow. With the exception of a few tectonic examples, aeolic lakes are common in the closed depressions of the Patagonian plateau. Vulcanism has also had a strong influence, and lakes located on basins filled with volcanic ashes or basaltic rocks are common in the ecotone between the Patagonian steppe and the Andes. On the other hand, fluvial-aeolic lakes dominate in the Pampa region. In general, most lakes have rounded contours, pan-shaped profiles, and are relatively shallow relative to their surface area.

Shallow lakes often alternate between contrasting states: a turbid state, in which production is dominated by phytoplankton, and a vegetated, relatively clear state, in which production is dominated by rooted vegetation. These switches are particularly evident in Pampean lakes. The chronicles of ninetieth century travelers suggest that, before the settlement of Europeans and their descendants, most Pampean lakes were dominated by macrophytes. However, because of their occurrence in fertile lands, these already eutrophic lakes have experienced increased levels of eutrophication in recent times, and many of them have switched to a turbid state. Presently, they are shallow, polimyctic, eutrophic or hypertrophic lakes, with highly variable water renewal time and salinity. Pampean lakes have rich and diverse fish communities. In contrast, many Patagonian lakes lack autochthonous fish communities, while others have been stocked with indigenous (i.e., *Percichthys*, *Odonthestes*) or introduced (salmonids) fish.

### Coastal Lagoons

Lagoon environments comprise approximately 12% of the coastal areas of South America. The marine transgressions that occurred during the Quaternary favored the development of coastal lagoons, particularly in the Caribbean Sea and Atlantic Ocean shorelines. In general terms, littoral sand deposition promoted the building of bars across irregularities or indentations of the coastline (i.e., bays). As no significant erosive actions took place, such spits finally cut off the bays as coastal lakes or lagoons. Patos Lagoon and Mirim Lagoon (southern Brazil) form the largest lagoon system of South America and the watershed discharging into these lagoons is about $202\,000\,km^2$. The only contact with the sea is through an inlet at the southern end of Patos Lagoon. Geomorphologically the southern region of the lagoon has the characteristics of a bar-built estuary with a 30 km wide upper limnic part which gradually, over 50 km, narrows into a 700 m wide access channel.

In general terms, coastal lagoon ecosystems are directly related to the physical and chemical environment; viz, coastal lagoons are dynamic, open systems where functions are dominated and controlled by physical processes. The driving forces of these systems are characterized by: (i) flux vectors (currents, tide, solar energy, rain), (ii) marine inputs (sediment, coastal waters and associated elements such as nutrients, plankton), and (iii) continental inputs (rivers, groundwater, nutrients, sediment, organic matter).

## Human-Made Basins

Human activity has also played an important role in the development of lake basins all over the continent. Strip mining and road construction has produced a large number of artificial excavations. Once abandoned and flooded, these depressions form lakes. One example is Topibo Meer in Suriname, which is a large red-mud lake complex at an old mine site near Paranam. Nevertheless, the damming of streams and rivers has been, and continues to be, the most important human action responsible for the origin of lake basins throughout South America. Many artificial lakes have been constructed to serve as water supplies, to provide power, to aid in navigation, to increase the supply of fish and as other aquatic products or for defense.

### Reservoirs

South American reservoirs occur at a wide range of latitudes, from the tropics to the Temperate Zone, but are particularly numerous in Brazil. **Table 2** summarizes some morphometric parameters and theoretical residence times of 55 South American reservoirs. Two periods can be distinguished in South American reservoir construction: (i) end of 19th century-middle 20th century, and (ii) final decades of 20th century-present. In the early period the main objectives were local water storage for domestic and agricultural purposes, fish production and small-scale hydroelectricity production. Thus, reservoirs were small with relatively simple morphometry and volumes less than 100 million $m^3$. In the second period of construction, size, volume and morphometric complexity increased and uses extended to large-scale hydroelectric power generation, irrigation and fishery. Particularly, hydroelectric power generation is of paramount importance in South America and will further increase in future years. For instance, Brazil generates >93%, Paraguay nearly 100%, Peru 74%, Venezuela 73%, Ecuador 68%, Colombia 68%, and Chile 57% of their electrical energy from hydropower (**Figure 2**).

The need for reservoir management dramatically increased after the second phase of large-scale reservoir construction. Nevertheless, this management is complicated due to the morphometric, functional and operational characteristics of large reservoirs. As a consequence, many limnological studies were developed in the rivers prior to the reservoir

**Table 2** Morphometric parameters and theoretical residence times of some South American reservoirs

| Reservoir | Latitude | Longitude | Elevation (m.a.s.l.) | Area (km²) | Mean depth (m) | Max depth (m) | Theoretical residence time (days) |
|---|---|---|---|---|---|---|---|
| Gurí | 6°59′N | 62°40′W | 270 | 4250 | 30 | 170 | |
| Aguada Blanca | 2°16′S | 72°16′W | 3666 | 3980 | | | |
| Tucuri | 3°4′S | 49°12′W | 72 | 2430 | 17.3 | 75 | 51 |
| Yacyretá | 27°30′S | 56°48′W | 82 | 1600 | 13.1 | 26 | |
| Itaipú | 25°33′S | 54°37′W | 223 | 1460 | 21.5 | 140 | 40 |
| Três Marias | 18°15′S | 44°18′W | 585 | 1120 | 6.8 | ~30 | 29 |
| Chongon | 2°13′S | 80°06′W | 15 | 1100 | | | |
| El Fraile | 16°08′S | 71°08′W | 4000 | 1087 | | 72 | |
| Bonete | 32°49′S | 56°25′W | | 1070 | | 32 | 150 |
| Ezequiel Ramos Mexia | 39°30′S | 69°00′W | 381 | 816 | 24.7 | 60 | 422 |
| Salto Grande | 31°00′S | 57°50′W | 35 | 783 | 6.4 | 33 | 12 |
| Jurumirim | 23°29′S | 49°52′W | 568 | 446 | 12.9 | 40 | 322 |
| Los Barreales | 38°35′S | 68°50′W | 421 | 413.1 | 67 | 120 | 864 |
| Casa de Piedra | 38°15′S | 67°30′W | 285 | 360 | 11.1 | 39 | 375 |
| Petit Saut | 5°05′N | 53°15′E | | 360 | | 35 | 150 |
| Palmar | 33°03′S | 57°27′W | | 320 | | 15 | 16 |
| Piedra del Águila | 40°20′S | 70°10′W | 590 | 305 | 41.3 | 120 | 202 |
| Boa Esperança | 6°45′S | 43°34′W | 304 | 300 | ~35 | 196 | |
| Río Hondo | 27°30′S | 65°00′W | 275 | 296.7 | 5.3 | 24.5 | 173 |
| Volta Grande | 20°10′S | 48°25′W | | 222 | 10.2 | | 25 |
| El Pane | 15°18′S | 71°00′W | 4500 | 185 | | | |
| Mari Menuco | 38°36′S | 68°37′W | 414 | 173.9 | 79.3 | 140 | 432 |
| Cabra Corral | 25°18′S | 65°25′W | 1037 | 115 | 27 | 67 | 936 |
| Baygorria | 32°53′S | 56°25′W | | 100 | | 20 | 3 |
| El Nihuil | 35°04′S | 68°45′W | 1325 | 96 | 4 | 20 | 141 |
| Urre Lauquen | 38°05′S | 65°49′W | 219 | 95 | 1.6 | | |
| Urugua-i | 25°55′S | 54°22′W | 197 | 88.4 | 13.5 | 69 | |
| Amutui Quimey | 43°03′S | 71°42′W | 485 | 86.7 | 64.7 | | 368 |
| Betania | Colombia | | 564 | 74 | 28 | | |
| Tucupido | 9°31′N | 70°18′W | 260.5 | 68.6 | 27 | 72 | |
| Alicurá | 40°40′S | 71°00′W | 705 | 67.5 | 48.4 | 110 | 137 |
| Florentino Ameghino | 43°42′S | 67°27′W | 169 | 65 | 24.6 | 61.5 | |
| Guatapé | Colombia | | 1887 | 63.4 | | | |
| Río Tercero I | 32°12′S | 64°27′W | 661 | 54.3 | 13.5 | 46.5 | 303 |
| Paranoá | 15°48′S | 47°75′W | 1000 | 40 | 14.3 | 38 | 300 |
| Arroyito | 39°14′S | 68°40′W | 315 | 38.6 | 7.7 | 15 | 5 |
| Boconó | 9°31′N | 70°18′W | 260.5 | 38.4 | 26 | 52 | |
| Quebrada de Ullum | 31°30′S | 68°39′W | 768 | 32 | 15 | 40 | |
| Los Molinos | 31°50′S | 64°32′W | 765 | 21.1 | 16.3 | 53 | 368 |
| San Roque | 31°22′S | 64°27′W | 643 | 15 | 14.1 | 35.3 | 231 |
| El Cadillal | 26°35′S | 65°14′W | 611 | 13.5 | 17.8 | 72 | 184 |
| Cerro Pelado | 32°12′S | 64°40′W | 876 | 12.4 | 29.8 | 46 | 368 |
| Guavio | Colombia | | 1640 | 11.6 | | | |
| Cruz del Eje | 30°45′S | 64°45′W | 567 | 10.9 | 9.5 | 37.2 | 476 |
| Agua del Toro | 34°35′S | 69°05′W | 1240 | 10.5 | 36.2 | 119 | 112 |
| Neusa | Colombia | | 2997 | 9.5 | | | |
| Jacaré-Pepira | 22°26′S | 48°01′W | 800 | 3.7 | 3.0 | 12 | |
| Pampulha | 19°55′S | 43°56′W | | 2.4 | 5.0 | 16 | 120 |
| La Fe | 6°06′N | 75°30′W | 2156 | 1.39 | | 20 | |
| Agua Fría | 10°24N | 67°10′W | 1700 | 0.44 | 13.2 | | 38 |
| Das Garças | 23°39′S | 46°37′W | 798 | 0.09 | 2.1 | 46 | 69 |
| Dourada | 22°11′S | 47°55′W | 715 | 0.08 | 2.6 | ~6.3 | |
| Monjolinho | 22°01′S | 47°53′W | 812 | 0.05 | 1.5 | 30 | −10 |
| Jacaré | 22°18′S | 47°13′W | 600 | 0.003 | 0.9 | ~2.2 | 11 |
| Poechos | 4°03′S | 80°02′W | 98 | | | 46 | |

Sources: Tundisi and Matsumura Tundisi (2003) and Calcagno *et al*. (1995). Refer to **Table 1** and further readings section.

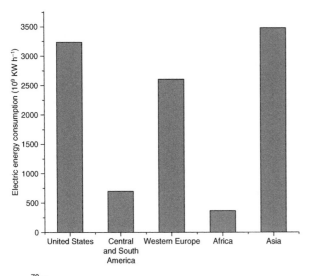

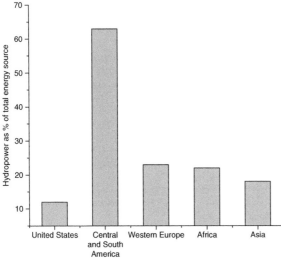

**Figure 2** Comparison of electric energy consumption and hydropower contribution to total electric generation.

construction so that environmental changes could be anticipated and subsequently documented. Such is the case of those studies developed on Itaipú reservoir. Predicted effects of reservoir aging and watershed uses have become important tools in reservoir management and operation in South America.

Thus, to maintain a sustainable use of reservoirs, the need of tightly coupled interaction between fundamental research and reservoir management becomes evident. As an example, the management of drinking water reservoirs in the Metropolitan Region of São Paulo (Brazil) was improved by studies on phytoplankton ecology and biogeochemical cycles.

## A Note on Natural Communities

In South America, there are little changes in the composition of phytoplankton communities at the level of genus or species from temperate to tropical latitudes. Obviously, seasonal succession patterns that are the rule in temperate lakes are absent in tropical lakes. In fact, the phytoplankton communities of shallow tropical lakes appear to remain arrested in a single successional stage, which may explain the simplicity of these communities. As regards zooplankton, several groups, which are well represented in temperate lakes, become scarce as one approaches the equator. Notably, the genus *Daphnia* (Cladocera) is seldom reported in the tropics, where only small to medium-size species (e.g., *D. gessneri* and *D. leavis)* occur at low densities. Similarly the calanoid copepods, which are among the largest zooplankton in temperate lakes, are absent or rare in many tropical lakes.

Endemisms are frequent in South America. For example, recent evidence suggests that 14 of 19 species of daphniids reported for South America are endemic. Particularly, three of them (*Daphnia inca, Daphniopsis chilensis* and *D. marcahuasensis*) are considered to be Andean endemisms. Many copepod species belong to the family Centropagidae, which is almost entirely restricted to the Southern Hemisphere. As for rotifers, roughly 40% of the species of the genus *Brachionus* and 40% species of the genus *Keratella* reported for South America are endemic. Examples include *Brachionus ahlstromi, B. gillardi, B. incertus, B. insuetus, B. kultrum, B. mirus, B. trahea, B. voigti, Keratella caudata, K. kostei, K. thomassoni, K. ona* and *K. yaman.*

Freshwater South American fishes belong to the Neotropical ichthyofauna, which contains 5000–8000 different species. The native populations are mostly dominated by five major groups: siluriforms, characiforms, gymnotiforms, cyprinodontiforms and cichlids. The Amazon Basin is home of the largest fish diversity on Earth, with over a thousand different species. The Orinoco, Parana, and other large, tropical river basins are also rich in fish species. In contrast, fish diversity drops sharply in the watersheds that drain into the Caribbean Sea and the Pacific Ocean. In the temperate region, fish diversity is also lower and the taxonomic composition differs markedly from subtropical and tropical areas.

## Highlights

- Tropical South America is dominated by floodplain lakes and lagoons which derive from the main river systems, while 'archetypical' lakes are scarce and restricted to the mountain ranges. These tropical lakes share most of their limnological characteristics with tropical lakes elsewhere, even those at high elevations. Nevertheless, in contrast to the more populated regions of the world; most tropical floodplain lakes and lagoons maintain most of their natural hydrological characteristics.

- Temperate South America is restricted to the southern part of the continent. As a consequence of the geomorphologic forces that have acted in the region, an important lake district has developed which limnologically differs from similar systems in the Northern Hemisphere: these systems develop deeper thermoclines which translate into lower chlorophyll concentrations at comparable levels of phosphorus. Moreover, the strength and persistence of the westerlies preclude the occurrence of cold dimictic lakes.

- Reservoirs for diverse uses are numerous throughout the continent. The experience of reservoir research and management in the region is part of the worldwide tendency on reservoir development, which is aimed at balancing multiple uses based on modeling development and intensive use of data-bases from long-term studies.

## Acknowledgments

We thank Rosmery Ayala, Ricardo Coelho, Fernando Coronato, Konrad Gajewski, Rolando Quirós, Doris Soto, and José Galizia Tundisi for making available valuable sources of information, and Claudio Baigún, Cristina Marinone, and Peter C. Schulze for critical reading of the manuscript. This work was supported by Consejo Nacional de Investigaciones Científicas y Técnicas and Agencia Nacional de Promoción Científica y Tecnológica (PICT 13550 and 25325).

*See also:* Abundance and Size Distribution of Lakes, Ponds and Impoundments; Africa: North of Sahara; Origins of Types of Lake Basins; Reservoirs.

## Further Reading

Adamowicz SJ, Herbert PDN, and Marinone MC (2004) Species diversity and endemism in the *Daphnia* of Argentina: A genetic investigation. *Zoological Journal of the Linnean Society* 140: 171–205.

Baigún C and Marinone MC (1995) Cold-temperate lakes of South America: Do they fit northern hemisphere models? *Archive für Hydrobiologie* 135(1): 23–51.

Carvalho P, Bini LM, Thomaz SM, Gonçalves de Oliveira L, Robertson B, Gomez Tavechio WL, and Darwisch AJ (2001) Comparative limnology of South American floodplain lakes and lagoons. *Acta Scientiarum* 23(2): 265–273.

Cressa C (1993) Aspectos generales de la limnología en Venezuela. *Interciencia* 8: 237–248.

Diaz M, Pedrozo F, Reynolds C, and Temporetti Po (2007) Chemical composition and the nitrogen-regulated trophic state of Patagonian lakes. *Limnologica* 37(1): 17–27.

Dumont HJ (1983) Biogeography of rotifers. *Hydrobiologia* 104: 19–30.

Lewis WM Jr. (1996) Tropical lakes: How latitude makes a difference. In: Schiemer F and Boland KT (eds.) *Perspectives in tropical limnology.* Amsterdam: SPB Academic Publishing.

Marinone MC, Menu Marque S, Añón Suárez D, Diéguez MC, Pérez AP, De Los Ríos P, Soto D, and Zagarese HE (2006) UVR Radiation as a Potential Driving Force for Zooplankton Community Structure in Patagonian Lakes. *Photochemistry and Photobiology* 82: 962–971.

Morris DP, Zagarese HE, Williamson CE, Balseiro EG, Hargreaves BR, Modenutti B, Moeller R, and Queimaliños C (1995) The attenuation of solar UV radiation in lakes and the role of dissolved organic carbon. *Limnology and Oceanography* 40(8): 1381–1391.

Paruelo JM, Beltran A, Jobbagy E, Sala OE, and Golluscio RA (1998) The climate of Patagonia: General patterns and controls on biotic processes. *Ecologia Austral* 8(2): 85–101.

Payne AI (1986) *The Ecology of Tropical Lakes and Rivers.* Chichester, UK: Wiley.

Quirós R and Drago E (1999) The environmental state of Argentinean lakes: An overview. *Lakes and Reservoirs: Research and Management* 4: 55–64.

Ringuelet RA (1975) Zoogeografía y Ecología de los Peces. *Ecosur (Argentina)* 2(3): 1–122.

Soto D (2002) Oligotrophic patterns in southern Chilean lakes: The relevance of nutrients and mixing depth. *Revista Chilena de Historia Natural* 75(2): 377–393.

Suárez-Morales E, Reid JW, and Elías Gutiérrez M (2005) Diversity and Distributional patterns of Neotropical Freshwater Copepods (Calanoida: Diaptomidae). *International Review of Hydrobiologia* 1: 71–83.

Tundisi JG (1994) Tropical South America: Present and perspectives. *Limnology Now* 353–424.

Tundisi JG and Matsumura-Tundisi T (2003) Integration of research and management in optimizing multiple uses of reservoirs: The experience in South America and Brazilian case studies. *Hydrobiologia* 500: 231–242.

# LAKES AND RESERVOIRS: POLLUTION, MANAGEMENT AND SERVICES

## Contents

## Shallow Lakes and Ponds

**M Meerhoff,** University of Aarhus, Denmark; Universidad de la República, Uruguay
**E Jeppesen,** University of Aarhus, Denmark

### Definition of Shallow Lakes and Ponds and World Distribution

Shallow lakes and ponds are the most abundant lake types in the global landscape. They provide a myriad of ecosystemic and social services as well as goods and materials, while having great conservation value (for example for migratory birds). Such shallow systems usually occur in lowland areas, many in association to seasonal changes in the flood regimes of rivers. The origin of many of these systems can also be related to geological disturbances, such as glacial movements or other processes that generate depressions in the landscape. Also, many shallow lakes and ponds have been created by humans after millennia of landscape modification, such as stream and river impoundment or diversion, or when digging for metals, sand, gravel and peat (particularly common in Medieval times in Europe) or by establishing fish ponds.

The perception that despite constituting the most abundant freshwater systems, shallow lakes and ponds still do not represent a large percentage of the world's total freshwater area has been recently modified. Millions of water bodies smaller than $1 \, km^2$ occur, and small lakes and ponds ($0.001–0.1 \, km^2$) represent most of the world's lacustrine area. The historical undercounting of small lakes has led to a significant underestimation of the world's lake and pond area.

Shallow lakes can be defined as more or less permanent standing water bodies that are shallow enough as to, potentially, allow light penetration to the bottom.

They can thus be colonized by higher aquatic plants over the entire basin or at least in large sections (**Figure 1**), although this is not necessarily the case in all lakes. Ponds are particularly small shallow lakes, typically less than 1 hectare in area. Shallow systems are also defined as polymictic, implying that the entire water column circulates continuously or for long periods and that long-lasting thermal stratification in summer is lacking. The maximum depth for these processes to occur is around 5–7 m in the temperate region, but most typically shallow lakes are <3 m deep. In contrast to deep water bodies, shallow systems are often characterized by a greater littoral zone and a closer contact with the surrounding lands, entailing a stronger aquatic-terrestrial exchange of matter and organisms. The water volume is relatively low. Together with polymixis, this results in enhanced benthic-pelagic coupling and greater impact of sediment processes in the water column. Typical shallow lakes and ponds are likely more productive than deep lakes, owing to the greater recycling of nutrients that become available for phytoplankton and the contribution to productivity by the aquatic plants and their attached epiphyte communities. This is reinforced by the fact that these systems are often distributed in the open cultivated landscape and therefore receive higher nutrient input.

Shallow lakes and ponds often show higher species richness per unit of area than large lakes owing to the strong interaction between the littoral and the pelagic habitats (**Figure 1**). Furthermore, lack of fish

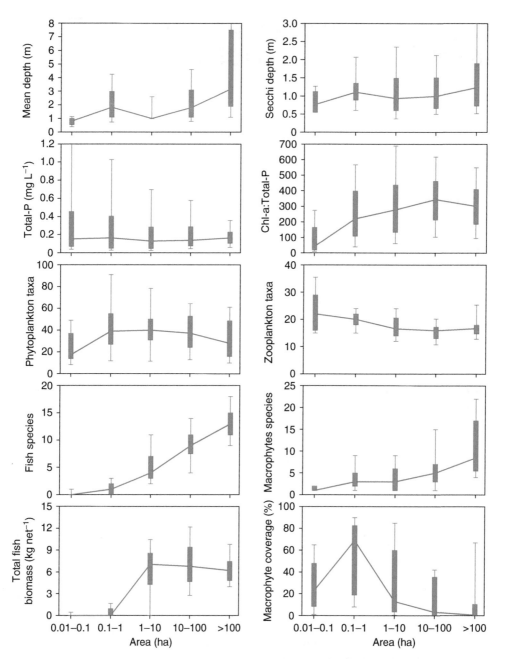

**Figure 1** Main physico-chemical and biological variables of temperate shallow Danish lakes with different lake surface area, shown as box-plots ($n = 796$ lakes). Each box represents the 25% and 75% percentiles, the horizontal line the mean value, and the top and bottom of the thin line depict the 90% and 10% percentiles. Redrawn from Søndergaard et al. (2005) Archiv für Hydrobiologie 162: 143–165.

sometimes occurring in small, isolated water bodies (owing to high risk of extinction and low colonization opportunities or owing to strong changes in water depth) may lead to enhanced abundance of waterfowl, amphibians, invertebrates, and plants.

## Alternative States Hypothesis

Despite their prevalence in the world landscape, shallow lakes and ponds had historically received little scientific attention than deep lakes until the early 1990s, when several seminal works proposed ideas to explain the functioning of these unique and complex ecosystems.

Shallow lakes do not always change smoothly from one condition to another in response to external pressures. Instead, changes are often sudden and may trigger important modifications in the complex network of stabilizing mechanisms that previously kept the lake in a stable condition. The 'alternative stable state hypothesis' states that different environmental states can occur under similar conditions with relative stability against external perturbations. Particularly, over an intermediate range of nutrient concentrations, temperate shallow lakes may exhibit extensive coverage of submerged (or floating-leaved) plants, clear water, and high biodiversity or, conversely, have few or no plants and a turbid-water, often phytoplankton-dominated state (**Figure 2**). Even different basins of the same lake may show such alternative states. The range of nutrient concentrations (total phosphorus) is typically considered to be between 0.025 and 0.15 mg TP L$^{-1}$, although the upper limit may be much higher (several mg l$^{-1}$), particularly in small lakes and ponds, notably if fish abundance is low or fish are absent.

At low nutrient concentrations (usually below 0.025 mg TP L$^{-1}$), the plant-dominated state is the most likely to occur, because phytoplankton is nutrient-limited while plants can use nutrients from the sediments. Lakes dominated by plants have a higher biodiversity of invertebrates, amphibians, fish, and waterfowl. By contrast, at high nutrient levels plants are rare and lakes are typically turbid. The shift to the turbid state does frequently not happen gradually together with the increase in nutrient level, but abruptly when a particular lake-specific nutrient threshold is reached. With nutrient enrichment, several changes in the structure of fish communities occur and small specimens (such as cyprinids in temperate European lakes) become dominant. This has a consequent strong predation pressure on zooplankton and grazer macroinvertebrates, such as snails. The increased biomass of bottom-feeding fish also facilitates the resuspension of sediments and higher inorganic turbidity. The biomass ratio of zooplankton to phytoplankton typically decreases

from 0.5–0.8 in mesotrophic lakes to less than 0.2 with phosphorus concentrations above 0.10–0.15 mg TPl$^{-1}$, implying that the zooplankton is no longer able to control phytoplankton biomass. The reduced grazing pressure on phytoplankton and epiphytes, leads, in turn, to adverse growing conditions for submerged plants. The transitions between states also result in changes in the habitat of dominant algae (from benthic to pelagic), but not necessarily in increased abundance of primary producers. Owing to the new conditions in the lake, the reduction of external nutrient loading does not always easily result in the return of the plants.

Many large shallow lakes lack vegetation despite having low or moderate nutrient concentrations. This likely happens because the effect of the winds on sediment destabilization and resuspension, particularly strong in large systems, prevents the successful establishment of rooted plants. These often turbid lakes present generally low productivity (as also phytoplankton is limited, in this case for light), and lower biodiversity.

Another alternative state is represented by the dominance of free-floating plants (from small duckweeds to large-bodied water hyacinth). This may occur at high nutrient concentrations in the water and usually high water column stability (**Figure 2**). Since very low air temperatures can strongly damage these plants, their dominance is more likely to occur in warm regions of the globe, such as the subtropics and tropics. In contrast to rooted plants, free-floating plants do not have access to the nutrient pool in the sediments. Their leaves are in larger contact with the atmosphere than with the water, thus reducing the possibility of taking up nutrients other than carbon through their leaves. Free-floating plants are

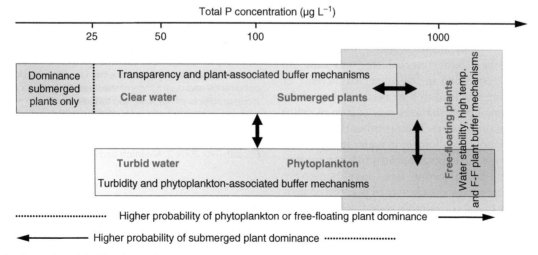

**Figure 2** General model of the alternative states in shallow lakes, over the gradient of nutrients (phosphorus) where the three main alternative states: phytoplankton-dominated, submerged plant-dominated and free-floating plant-dominated, may occur. Modified (adding free-floating plants) with permission from the original model for temperate lakes published in Moss *et al.* (1996).

superior competitors for light, whereas submerged plants can grow at lower nutrient concentrations in the water and reduce these concentrations further. Nutrient enrichment therefore reduces the resilience of lakes against a shift to dominance of free-floating plants and phytoplankton. The probability for these three main states to occur changes along gradients of water transparency (and therefore light availability) and nutrients and is likely affected by morphological features of the lakes, such as size or shoreline development (**Figure 3**).

Each of these environmental states is stabilized by several physical, chemical, and biological processes that act as positive feedbacks. For instance, the presence of submerged plants may promote a local increase in water clarity of up to 90% compared to control conditions, and under this light-enhanced environment, plants, algae on the plant surfaces (periphyton) and benthic algae grow even better while reducing nutrient and the light availability for phytoplankton competitors. The following higher physical stability of the sediments (and consequent reduction of nonbiotic turbidity) helps maintain and expand a clear-water, submerged plant-dominated state. Submerged plants also reduce available nutrients for phytoplankton, epiphyton, and floating plants, either by direct luxurious uptake or enhanced denitrification. Besides, submerged plants can indirectly promote higher grazing pressure on the algae competitors by offering habitat and refuge to large-bodied zooplankton and several grazing macroinvertebrates (e.g., snails, mussels) against visual predators. According to laboratory studies, some plant species (e.g., *Myriophyllum spicatum*, *Chara* spp, *Ceratophyllum demersum*, *Stratiotes aloides*) may excrete allelopathic substances that can inhibit algal growth, although this has not yet been fully substantiated by field experiments.

Perturbations of different nature (e.g., climate-related or most likely anthropogenic) must occur to lose some of these buffer mechanisms and shift the lake from one condition to the other. An increase in nutrient loading will lead to a weaker stability of the system, and small perturbations may suffice to cause a switch to the turbid state. Factors other than turbidity, such as pronounced changes in water level, waterfowl herbivory, wind exposure, and a fish community dominated by benthivorous fish, may negatively affect the development of submerged plants and promote the shift to a less desirable state (turbid waters, or phytoplankton or free-floating plant dominance). The upper limit in nutrients to promote the loss of submerged macrophytes and a shift to these alternative states is not absolute and seems to depend on e.g., lake morphometry and size.

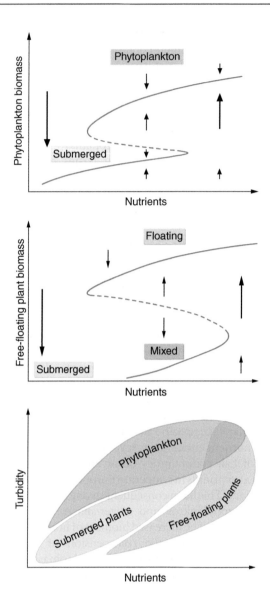

**Figure 3** Alternative states in shallow lakes. Effect of nutrient loading on the equilibrium biomass of phytoplankton (above) and free-floating plants (medium) with respect to the biomass of submerged plants. The arrows indicate the direction of change if the system is out of equilibrium (i.e., the dashed equilibrium is unstable). Catastrophic shifts to an alternative equilibrium occur as vertical transitions in the scheme. Lower panel: probability of occurrence and dominance of the three alternative communities along gradients of nutrients and turbidity. Redrawn with permission after Scheffer *et al.* (1993) *TREE* 8: 275–279, (Copyright Elsevier); Scheffer *et al.* (2003) *PNAS* 100: 4040–4045, (Copyright (2003) National Academy of Sciences, USA, and *Ecosistemas* 2004/2, Meerhoff and Mazzeo (2004).

## Functioning of Shallow Lakes and Ponds: Nutrient Dynamics and Retention

Most shallow lakes and ponds tend to be located in the more fertile lowland regions and are therefore more sensitive to human activities in the catchments.

Nutrient loading is proportionally higher in shallow than in deep systems. In deep stratified lakes, nutrients are typically lost through sedimentation from the surface layer (epilimnion) to the bottom layers (hypolimnion) during the summer stratification. These nutrients only return to the epilimnion when the water column becomes mixed again during the autumn turnover (in the temperate region). Shallow systems, in contrast, present a strong interaction between the sediments and the water column, which ensures a fast recirculation of the settled material and less net losses of nutrients to the sediments during the critical growing season. The activities in the catchments therefore acquire greater importance for lake functioning with decreasing lake size and depth.

In pristine waters, total phosphorus (TP) concentrations are around a few micrograms to a few tens of micrograms per liter, while total nitrogen (TN) concentrations will usually be ca. 10–20 times as high. Since P compounds are less soluble than nitrogen, P is usually scarce in aquatic systems and frequently limits algal growth. However, the sediments in shallow systems experience relatively high temperatures in summer. This leads to an increase in mineralization rates and higher release of nutrients from the sediment when the supply of organic matter is very high. These processes may enhance the eutrophication effects in shallow lakes. Much of the P that has been trapped in the sediments during periods of high loading or, in the temperate zone, during cold winters can be easily released to the water column, a process called 'internal loading.' This phenomenon is frequent in shallow eutrophic lakes even after the external loading has been reduced (due to improved wastewater treatment, for example), and may take place during most of the summer in temperate lakes. In lakes with a high phosphorus accumulation in the sediments, the recovery process may last several years and even decades, regardless of the residence time of the lakes. During the recovery, a progressive decline in the P release rate and duration takes place, particularly in winter, followed by spring and autumn. Many north temperate lakes often reach a new equilibrium with respect to TP < 10–15 years after the external loading has been reduced, whereas a new equilibrium regarding TN is typically reached after <5–10 years. However, there are examples of more than 40 years response time. Besides the consequent nutrient control of phytoplankton biomass, increased top-down control due to fast response of the fish community seems often to accompany the reduced external loading. A decline in fish biomass and an increase in the percentage of piscivores and the ratio between zooplankton to phytoplankton biomass occur in many lakes. Lakes with reduced fish populations, such as those subjected to biomanipulation measures, exhibit a shorter period of P release during summer, while the TP load may be reduced to 50%.

Contrarily, N will eventually be lost or removed from the system through denitrification. However, N availability and concentrations in many ground and surface waters have increased remarkably since the 1950s, mostly as a result of agricultural activities. Typically, primary production in lakes is thought to be P-limited, although the nutrient controls seem different in lakes dominated by plants from those turbid, often dominated by phytoplankton. Low N concentrations seem to entail higher aquatic plant species richness and even increase the potential for submerged plants to occur. A diverse plant community assemblage is likely more stable against external perturbations. In contrast, TN concentrations higher than 1–2 mg $L^{-1}$, under moderate to high TP concentrations, seem to trigger the loss of submerged plants or at least a strong impoverishment of plant diversity. These findings impose a reconsideration of the TP and TN thresholds to be achieved after nutrient control during a restoration strategy.

## Functioning of Shallow Lakes and Ponds: Biological Structure

The functioning of shallow water bodies is, as described above, strongly affected by the presence of aquatic plants and the fish structure. Aquatic plants exert multiple effects on the structure and functioning of shallow lakes and ponds. They affect the communities that live permanently or temporarily in the littoral area, through a series of physical, chemical and behaviorally-mediated processes. The structure of the fish community can, in turn, affect the lake functioning in contrasting ways than the plants. Fish, due to their mobility and flexible feeding behavior, link the littoral, benthic, and pelagic habitats very significantly and thereby affect the nutrient transport and predator–prey interactions in the littoral and the pelagial.

### Trophic Interactions in the Pelagial

The biomass and production of fish per unit of area at a given nutrient level do not depend on lake depth. Therefore, shallow lakes have a substantially higher fish biomass per unit of volume than deep lakes. This relationship may reflect the higher nutrient recycling and availability from settled material and probably as well the availability of feeding and spawning sites for fish offered by the aquatic plants. The biomass and production of benthic invertebrates are also generally higher in shallow lakes (**Figure 4**), probably due to the

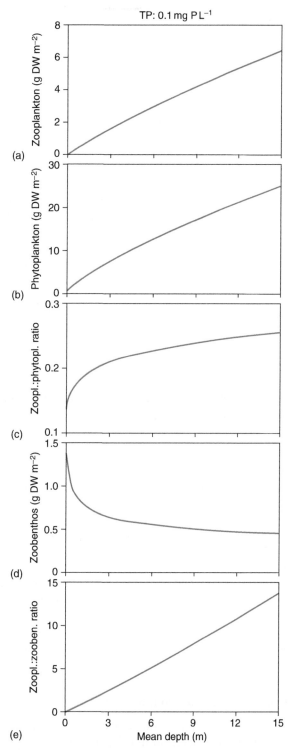

**Figure 4** Changes in some trophic variables in temperate lakes along a mean depth gradient derived from published empirical equations. Biomass per unit area of zooplankton (a) and phytoplankton (b), zooplankton:phytoplankton biomass ratio that typically declines with increasing fish predation (c), zoobenthos biomass (d), and the zooplankton:zoobenthos ratio (e) in lakes with an epilimnion concentration of 0.1 mg TP L$^{-1}$. Reproduced with kind permission from Jeppesen *et al.* (1997) *Hydrobiologia* 342/343: 151–164.

faster settlement of organic matter from the water column (more fresh material) and higher oxygen concentrations due to lack of stratification. The easier access to benthic invertebrates in shallow lakes also helps maintain a relatively higher fish biomass and a consequent high predation pressure on zooplankton (a process called 'benthic facilitation'), as most fish can shift between benthic and pelagic feeding. Moreover, the biomass ratio of zoobenthos:zooplankton increases with decreasing depth allowing stronger facilitated control of zooplankton in shallow lakes (**Figure 4**). In recent years, analysis of the stable isotopic signals of freshwater temperate communities has shown that most fish species are actually omnivores, to a much greater extent than historically perceived. The relatively higher impact of fish likely leads to a higher predation pressure on macroinvertebrates and zooplankton with decreasing mean depth, with consequent impacts on periphyton and phytoplankton communities (**Figure 4**).

However, in small lakes and ponds, fish may be scarce or even absent, which increases the likelihood of a clear-water, plant-dominated state. As a consequence of the lack of fish predation and higher habitat heterogeneity caused by the macrophytes, other groups may flourish. Although lakes and ponds behave similarly in many aspects, ponds do not necessarily respond with high phytoplankton biomass to an increase in nutrient concentrations as do larger lakes, due to other controlling factors. Also, the phytoplankton community responds differently in shallow than in deep lakes. Green algae (Chlorophyceae) frequently dominate in shallow hypertrophic lakes, whereas in deep lakes cyanobacteria are usually dominant under such conditions. The high nutrient release from the sediments and continuous mixing conditions may favor green algae over nutrient-storing, slow-growing cyanobacteria.

## Littoral–Pelagic Coupling: Horizontal Migration of Zooplankton

The spatial distribution of predators, such as fish, can have important consequences for the spatial distribution of their preys, which can modify the expected outcome of direct predator–prey interactions. The phenomenon is described as 'behavioral cascade' in opposition to the classic 'trophic cascade' concept that implies direct effects of fish on the abundance of lower trophic levels (e.g., phytoplankton) due to predation on the intermediate levels (e.g., zooplankton and macroinvertebrates). In the temperate region, submerged plants help stabilize trophic interactions between fish predators and cladoceran and macroinvertebrate preys by offering a physical refuge. In deep lakes vulnerable zooplankton migrate to the bottom layers during the day, thus minimizing the predation risk by fish, and migrate upwards during the night to graze with lower chances of being eaten ('diel vertical migration'). Shallow lakes typically lack a hypolimnetic refuge that would favor vertical migration. Instead, the perception of predation risk through chemical cues from potential predators (fish and pelagic macroinvertebrates, such as the phantom midge *Chaoborus*) may lead the large-bodied zooplankton to perform 'diel horizontal migration' (DHM) into the littoral zone, particularly into the submerged and floating-leaved plants. These zooplankters move back to the pelagic zone at night to graze on phytoplankton when the predation risk decreases. Also, emergent plants may occasionally be important for the migration pattern of vulnerable zooplankton. In temperate lakes, submerged plants would offer little refuge for zooplankton at both extremes of the nutrient (and turbidity) gradient. In low-nutrient lakes, clear water and the low density of plants enhance fish predation on zooplankton, whereas the refuge effect is weak under hypertrophic conditions because of the scarcity of submerged plants and often high densities of benthi-zooplanktivorous fish. Accordingly, macrophyte refuges seem more efficient under eutrophic clear-water conditions and in small and dense plant beds, zooplankton being more concentrated in the edges of the plant patches during the day.

The consequent plant-mediated maintenance of a herbivorous zooplankton population that can effectively graze on algae has consequences for the pelagic habitat and even ecosystem impacts in the temperate region, such as the enhancement of water transparency.

## Benthic–Pelagic Coupling

The periphyton developed on sediments (or epipelon) may contribute substantially to the whole-lake primary production in shallow water bodies. Turbid lakes seem more productive than clear lakes if phytoplankton is the only contributor to total annual production. However, the proportional habitat distribution of the primary producers changes under contrasting turbidity and nutrient states, benthic production being highest in clear lakes. Under nutrient-rich conditions, high phytoplankton production in the water column may suppress the epipelic production throughout the year. Lakes with low nutrient supply, in contrast, tend to have low phytoplankton production, clear water, and therefore a high proportion of benthic production throughout the year, with a slight increase of pelagic production in spring or autumn. At intermediate nutrient levels, lakes may shift from epipelon dominance in winter to phytoplankton dominance in summer (**Figure 5**).

On the other hand, fish are considered key determinants in shallow lake functioning, not least because of their positive cascading effects on periphyton via feeding on macroinvertebrate grazers (particularly snails), and the consequent deterioration of the light climate for the submerged plants (**Figure 6**). Direct or indirect positive effects of fish on periphyton biomass have

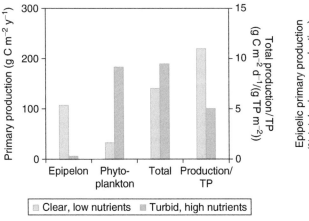
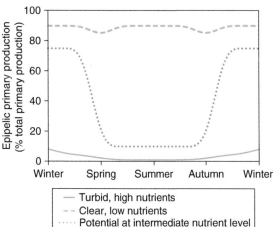

**Figure 5** Benthic-pelagic coupling in shallow lakes. Left panel: Comparison of annual estimates of epipelon, phytoplankton, total microalgal primary production and the ratio of total primary production per pelagic TP in a clear and a turbid shallow temperate lake. Right panel: Tentative conceptual model showing the seasonal variation in the relative contribution of epipelic primary production to total primary production in temperate shallow lakes without submerged macrophytes and with different nutrient levels. Reproduced from Liboriussen and Jeppesen (2003) *Freshwater Biology* 48: 418–431, with permission from Blackwell.

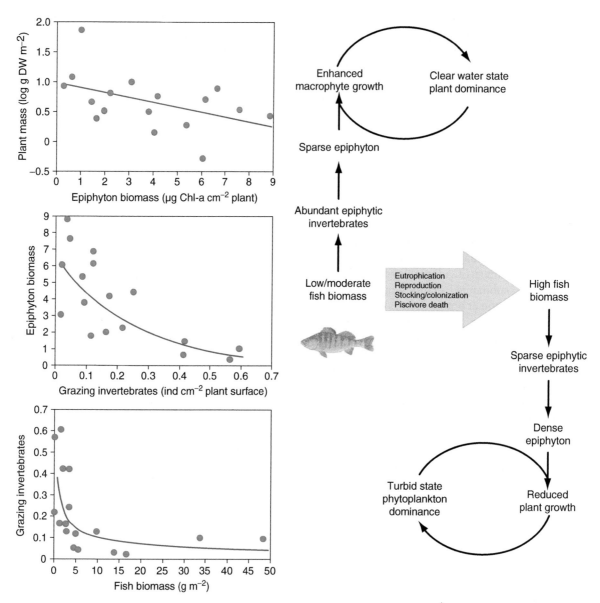

**Figure 6**   Littoral trophic cascade, linking submerged plants to fish via epiphyton and grazing macroinvertebrates, which could mediate the shift from turbid to clear conditions in shallow systems. A high abundance of fish (caused by several processes) would eliminate epiphyton grazers, allowing epiphyton to accumulate on plant surfaces. Consequently, macrophyte loss would occur owing to light limitation, and the absence of macrophytes would allow a turbid state to dominate. Modified with kind permission of Springer Science and Business Media from Burks *et al.* (2006), **Figure 3** (graphs originally published by Jones and Sayer (2003) *Ecology* 84: 2155–2167).

been registered under different environmental conditions. The strength of trophic cascades in the littoral and benthic zones may therefore determine the fate (the maintenance or loss) of the submerged plants.

## Functioning of Shallow Waters in Different Climatic Regions

Most of the previously described processes hold true for shallow lakes and ponds in temperate climates, whereas information on shallow lakes and ponds in

other climates is unfortunately limited. The climate regime imposes, however, important differences in the structure and functioning of these ecosystems, not least because of different patterns in the trophic interactions (**Figure 7**). In many (sub)tropical and warm temperate lakes, the climate is characterized by dry and wet seasons, with relatively high temperatures and long day lengths. In the wet season, nutrients enter from the catchment. Usually, the soils of warm regions have long been leached of nutrients due to their old age. By contrast, the soils in the catchments of most cold temperate lakes are

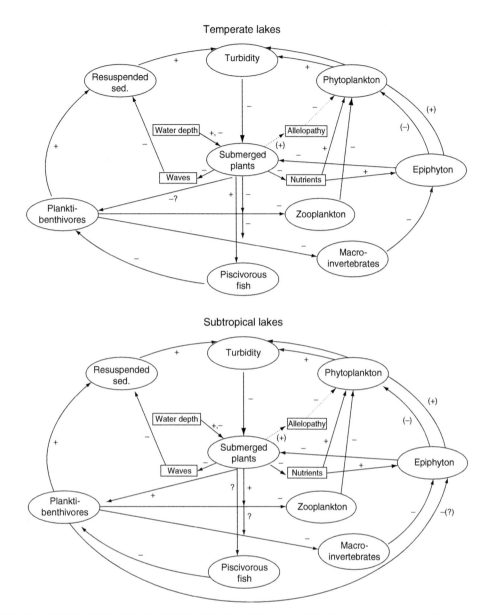

**Figure 7** Role of submerged plants in the functioning of shallow lakes and ponds, showing the effects on different physical, chemical and biological processes that ultimately affect turbidity, in temperate (above) and subtropical (below) systems. Note that the benthic-pelagic coupling is included in the scheme in a partial manner and requires further consideration. The qualitative effect of each route can be determined by multiplying the signs along the way. Symbols: +positive, –negative, ? scarce data, () occasionally. Modified and updated after original model proposed for temperate lakes in Scheffer *et al.* (1993) *TREE* 8: 275–279, (Copyright Elsevier).

relatively young and may provide abundant nutrients. However, warm lakes tend to be more productive than similar cold lakes. For a 50-fold range of light energy available from the poles to the equator, a 1000-fold range in algal production is found. Therefore, the 'latitude effect' may thus be related to a greater rate of nutrient recycling and mineralization with the higher temperatures at decreasing latitudes. Internal nutrient cycling in warm lakes is thus relatively more important and external loading relatively less important than in cold temperate lakes. On the other hand, patterns regarding nutrient additions and

the impacts of fish are more unpredictable with increasing latitude, probably due to more variable weather.

### Cold Lakes

Shallow lakes and ponds are extremely frequent in the Arctic landscape. These systems are tightly linked to climate, both directly and via catchment inputs, with comparatively small or highly localized anthropogenic disturbances. The shallow lakes in Antarctica are simple, fishless ecosystems ranging in productivity from extremely low productive melt-water lakes to highly productive guano lakes. Shallow systems located in cold and very cold regions are very prone to anoxic conditions and fish kills or are fishless because they freeze solid in winter, while typically have a very poorly developed plant littoral zone. In systems where fish are present, the zooplankton community is usually very depressed and dominated by small taxa. The thresholds of fish density required to lose *Daphnia* and other large-bodied cladocerans are much lower in these systems due to several co-occurring factors. The usual oligotrophic conditions and high water transparency make visual-feeding fish extremely efficient. Large-bodied cladocerans are thus easier to spot, not least because they often present dark pigmentation as a protection against the enhanced UV radiation. The resulting predation pressure is higher than in

comparative temperate systems, also because fish biomass is sustained by the usually high benthic production (benthic facilitation) in these nutrient poor systems. This also results in a much lower biomass of benthic invertebrates than in lakes without fish.

The cascading effect of fish in phytoplankton is, however, much weaker than in other lakes as nutrient limitation of algae growth in the water is often strong. High water clarity allows dominance of benthic production, enhanced by the nutrients in the sediments.

### Mediterranean Lakes

The Mediterranean climate is usually characterized by dry arid summers and wet winters. Climate and hydrological regimes are likely to affect lake communities differently in warmer regions. Mediterranean

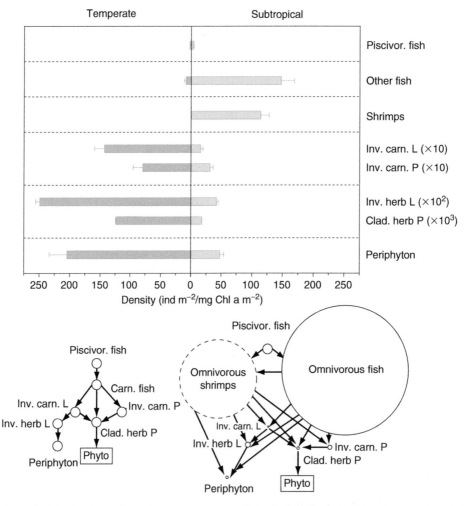

**Figure 8** Comparison of the structure of the main communities associated to aquatic plants in temperate (left) and subtropical (right) shallow lakes. Above and from top to bottom: density of potentially piscivorous fish, all other fish, shrimps, littoral (L) carnivorous (carn) invertebrates (inv), pelagic (P) carnivorous invertebrates, littoral herbivorous (herb) macroinvertebrates, pelagic herbivorous cladocerans (clad), and biomass of periphyton. Below: simplified scheme of trophic interactions among the same trophic groups. The densities in the subtropics are expressed in relation to those in the temperate lakes, and except fish the same taxa share the same trophic classification in both climate regions. Shrimps were absent in the temperate lakes. Data are sample means (±1 SE) of five lakes varying in size and nutrient state, but paired between climate regions in terms of phytoplankton biomass and physico-chemical and morphometric characteristics. Modified from Meerhoff *et al.* (2007). *Global Change Biology* 13: 1888–1897.

shallow water bodies have significantly higher annual water temperatures and seasonal fluctuations in water level, but smaller seasonal changes in light and temperature than temperate lakes. Many Mediterranean small lakes and ponds are temporary and often dry out in summer, with consequent fish kills. The thresholds of nutrient loading and in-lake TP required to avoid a turbid state in the permanent lakes seem lower than those described from temperate shallow lakes, and it is suggested to be $\leq 0.05$ mg TP L$^{-1}$. The high annual temperatures influence fish growth and enhance predation rates on zooplankton and macroinvertebrates. Owing to mild winter temperatures and lower density of large crustacean grazers, Mediterranean shallow lakes seem suitable habitats for cyanobacteria over a wide range of nutrient concentrations and water turbidity. On the other hand, the lower water levels in summer may improve the light conditions for submerged plant development, which can afterwards persist all the year round owing to the higher temperatures.

### (Sub)Tropical Lakes

(Sub)tropical shallow systems share some characteristics with Mediterranean lakes. Some of the most important differences from temperate lakes involve the fish community: the species richness and proportion of omnivores are frequently high, while many species (both juveniles and adults) are strongly associated to the submerged plant habitats. Also density, but not necessarily biomass, is substantially higher (1–2 orders of magnitude) than in comparative temperate lakes, independently of trophic state (**Figure 8**). Owing to high predation by fish, the zooplankton communities in tropical and subtropical lakes are frequently dominated by small cladocerans (e.g., *Diaphanosoma, Ceriodaphnia* and *Bosmina*) and rotifers, and by copepodites among the copepods. The littoral macroinvertebrate communities also seem impoverished in terms of taxon richness and densities compared with similar temperate lakes. Although submerged and free-floating plants (e.g., *Eichhornia crassipes, Salvinia* spp., *Pistia stratiotes*) may occur all year-round in these systems, their effects on water transparency seem much weaker than in temperate lakes. In particular, submerged plants offer scarce refuge towards large-bodied zooplankton against fish predation. Consequently, the predator-avoidance behavior of zooplankton seems to differ, and diel vertical migration seems more frequent than in similar temperate shallow lakes, where diel horizontal migration prevails. The trophic web appear more truncated in the (sub)tropics, likely due to the omnivorous nature of most fish species and the structure of the predatory assemblages.

Shallow lakes and ponds located in warm climate regions therefore seem much more sensitive than cold lakes to external impacts, such as temperature increases (due to climate warming), water level changes (natural or anthropogenic), and nutrient loading increases (eutrophication).

*See also:* Abundance and Size Distribution of Lakes, Ponds and Impoundments; Aquatic Ecosystem Services; The Benthic Boundary Layer (in Rivers, Lakes, and Reservoirs); Benthic Invertebrate Fauna, Lakes and Reservoirs; Biomanipulation of Aquatic Ecosystems; Ecological Zonation in Lakes; Effects of Climate Change on Lakes; Lakes as Ecosystems; Mixing Dynamics in Lakes Across Climatic Zones; Origins of Types of Lake Basins; Trophic Dynamics in Aquatic Ecosystems.

## Further Reading

Beklioğlu M, Romo S, Kagalou I, Quintana X, and Bécares E (2007) State of the art in the functioning of shallow Mediterranean lakes: Workshop conclusions. *Hydrobiologia* 584: 317–326.

Brönmark C and Hansson LA (1999) *The Biology of Lakes and Ponds*. Oxford University Press, Avon, UK.

Burks RL, Mulderij G, Gross E, *et al.* (2006) Center stage: The crucial role of macrophytes in regulating trophic interactions in shallow lake wetlands. In: Bobbink R, Beltman B, Verhoeven JTA, and Whigham DF (eds.) *Wetlands: Functioning, Biodiversity, Conservation, and Restoration*, pp. 37–59. Berlin: Springer.

Hosper H (1997) *Clearing Lakes. An Ecosystem Approach to the Restoration and Management of Lakes in the Netherlands*. Lelystad: Ministry of Transport, Public Works and Water Management (RIZA).

Jeppesen E, Søndergaard M, Søndergaard M, and Christoffersen K (eds.) (1997) *The Structuring Role of Submerged Macrophytes in Lakes*. New York: Springer.

Jeppesen E, Jensen JP, Jensen C, *et al.* (2003) The impact of nutrient state and lake depth on top-down control in the pelagic zone of lakes: Study of 466 lakes from the temperate zone to the Arctic. *Ecosystems* 6: 313–325.

Meerhoff M, Clemente JM, Teixeira de Mello F, Iglesias C, Pedersen AR, and Jeppesen E (2007) Can warm climate-related structure of littoral predator assemblies weaken the clear water state in shallow lakes? *Global Change Biology* 13: 1888–1897.

Moss B (1998) Ecology of freshwaters. Man and Medium, Past to Future. London: Blackwell.

Moss B, Madgwick J, and Phillips GL (1996) *A Guide to the Restoration of Nutrient-Enriched Shallow Lakes*. Norwich, UK: Broads Authority and Environment Agency (CE).

Scheffer M (1998) *Ecology of Shallow Lakes*. London: Chapman & Hall.

Scheffer M, Hosper SH, Meijer ML, Moss B, and Jeppesen E (1993) Alternative equilibria in shallow lakes. *Trends in Ecology and Evolution* 8: 275–279.

Scheffer M, Szabó S, Gragnani A, *et al.* (2003) Floating plant dominance as an alternative stable state. *Proceedings of the National Academy of Sciences of USA* 100: 4040–4045.

Søndergaard M, Jeppesen E, and Jensen JP (2005) Pond or lake: Does it make any difference? *Archiv für Hydrobiologie* 162: 143–165.

Vadeboncoeur Y, Vander Zanden MJ, and Lodge DM (2002) Putting the lake back together: Reintegrating benthic pathways into food web models. *Bioscience* 52: 44–55.

# Lake Management, Criteria

**A C Cardoso,** Institute for Environment and Sustainability, Ispra, VA, Italy
**G Free,** Institute for Environment and Sustainability, Ispra, VA, Italy
**P Nõges,** Institute for Environment and Sustainability, Ispra, VA, Italy
**Ø Kaste,** Norwegian Institute for Water Research, Grimstad, Norway
**S Poikane,** Institute for Environment and Sustainability, Ispra, VA, Italy
**A Lyche Solheim,** Norwegian Institute for Water Research, Oslo, Norway

## Introduction

Lakes are of importance for their heritage, ecological, and aesthetical value; however, current management strategies regard lakes primarily as a resource. They have been and are at present mainly used for drinking water supply, fishing, irrigation, waste disposal, industrial processes and cooling, transportation, hydroelectric power generation, and for recreation. These uses and further anthropogenic activities in the catchments of lakes have caused an enormous impact on these ecosystems. Consequently, many lakes worldwide are currently threatened by many types of pollution and suffer from several ecological disturbances.

Effective management and protection of lakes are therefore of critical importance for their preservation as both a heritage asset and also as a beneficial resource for humans. Management of water quality in lakes has been an important issue worldwide during the past decades and important improvements have been reached. Yet, the goal of sustainable management is still not being achieved for many lakes. In part, this is explained by the complex nature of lakes requiring consideration of multiple interacting factors by managers, such as climate, catchment characteristics, lake morphometry, stratification patterns as well as the biota present. Currently, it is recognized that sustainable management is possible only as part of a comprehensive ecosystem river basin management strategy.

### Management Principles

Management as prescribed by legislation was in the past founded on the regulation of emissions at their source through the establishment of emission limit values or by the setting of chemical quality standards for particular uses of water (**Table 1**). In both cases the aim was to ensure the continued use of lakes as a resource. However, these approaches have proved not to be sufficient for the protection of ecological integrity.

More recently, attention has turned to the establishment of ecological objectives for these ecosystems. There are several approaches to ecosystem management, but generally the common goal is to restore and sustain ecosystem functions based on the assumption that provided the ecosystem is protected most uses are also protected. The core concepts of ecosystem management are as follows:

1. assessment of status expressed as quality of composition, structure and functioning of the ecosystem components;
2. definition of reference baseline conditions for ecosystem functioning against which a measured deviation can be evaluated.

The European Water Framework Directive is an example of a recent legislation on water management that is based on these two core concepts.

### Pressures and Drivers

Industrial emissions, sewage, agricultural run-off, hydrological modifications, and human induced climate changes are the major drivers of pressure on lakes. These are responsible for the six major pressures affecting lakes today:

1. excessive inputs of nutrients leading to eutrophication
2. hydromorphological modifications
3. acidification
4. alien species
5. hazardous substances
6. climate changes.

Effective management requires a focus on criteria indicative of the ecosystem response to such pressures. Having a comprehensive suite of criteria at a lake manager's disposal will allow the evaluation of the extent of pressure affecting a lake and its influence on the ecosystem's structure and functioning. Such criteria form an important input to the river basin management plan.

## Lake Management Criteria for Main Pressures

### Eutrophication

**General introduction** The word eutrophication is of Greek origin and it means food/ nutrient (='Trophi')

**Table 1** Common water uses, related quality problems, and monitoring requirements in a sample of legislative texts

| Water use | Pollution problem | Monitoring requirements | Source |
|---|---|---|---|
| Drinking water supply | Toxic pollution | Health related inorganic and organic compounds | 1, 2, 3, 4 |
| | Microbial pollution | Microbiological parameters | |
| | Toxins from algal blooms | Microcystin | |
| | | Turbidity | |
| | | Total dissolved solids | |
| Bathing, recreation | Microbial pollution | Microbiological parameters | 1, 5, 6 |
| | Cyanobacteria toxins and products | Turbidity, transparency | |
| | Turbidity | Cyanobacteria abundance and toxins | |
| Fish resources | Oxygen depletion | Dissolved oxygen | 7, 8 |
| Aquaculture | Toxic compounds | Suspended solids | |
| | Temperature | Toxic compounds | |
| Agricultural water supply (irrigation and livestock) | Chemical contamination | Salinity | 9, 10, 11 |
| | Salinization | Sodium content | |
| | Bacteriological contamination | Toxic compounds | |
| Industrial water supply | Chemical/bacteriological contamination | Nutrients | 12 |
| | Excessive alga/plants | Suspended sediment pH, salinity (industry dependent) | |
| Aquatic ecosystem | Eutrophication | Biological, physical, chemical hydromorphologcial variables | 5, 8, 12, 13, 14, 15 |
| | Acidification | | |
| | Habitat degradation | | |
| | Toxic pollution | | |

Sources

1. WHO (2006) *Guidelines for drinking water quality* (3rd ed.), *Vol. 1: Recommendations.* Geneva: World Health Organization.

2. US EPA (2006) *Drinking Water Standards and Health Advisories.* Washington, DC: Office of Water, U.S. Environmental Protection Agency.

3. Council Directive 98/83/EC (1998). Quality of water intended for human consumption. *Official Journal of the European Communities*, 5.12.2998, L330 41: 32–54.

4. Health Canada (2007) *Guidelines for Canadian Drinking Water Quality.* Ottawa: Federal-Provincial-Territorial Committee on Drinking Water (CDW), Health Canada.

5. WHO (2003) *Guidelines for Safe Recreational Water Environments, Vol. 1: Coastal and fresh waters.* Geneva: World Health Organization.

6. Council Directive 2006/7/EC (2006) Directive 2006/7/EC of the European Parliament and of the Council of 15 February 2006 concerning the management of bathing water quality and repealing Directive 76/160/EC. *Official Journal of the European Communities*, 4.3.2006, L64: 37–51.

7. Council Directive 98/83/EC of 18 July 1978 on the quality of fresh waters needing protection or improvement in order to support fish life (78/659/EEC).

8. CCME (Canadian Council of Ministers of the Environment) (1999) *Canadian Environmental Quality Guidelines (Water: Aquatic Life).* Ottawa: Canadian Council of Ministers of the Environment.

9. CCME (Canadian Council of Ministers of the Environment) (1999) *Canadian Environmental Quality Guidelines (Water: agriculture).* Ottawa: Canadian Council of Ministers of the Environment.

10. FAO (Food and Agriculture Organization of the United Nations) (1994) Water *Quality for Agriculture. Irrigation and Drainage.* Paper 29, Rev. 1. Rome: Food and Agriculture Organization of the United Nations.

11. FAO (Food and Agriculture Organization of the United Nations) (1986) *Water for Animals.* Report AGL/MISC/4/85. Rome: Food and Agriculture Organization of the United Nations.

12. Japan: Basic Environmental Law (Law No. 91) (1993) Environment Quality Guidelines. Available at http://www.env.go.jp/en/water/wq/wp.html.

13. European Commission (2000) Directive 2000/60/EC of the European Parliament and of the Council of 23rd October 2000 establishing a framework for Community action in the field of water policy. *Official Journal of the European Communities*, 22.12.2000, L 327/1.

14. Australian and New Zealand Environment and Conservation Council (ANZECC) and Agriculture and Resource Management Council of Australia and New Zealand (ARMCANZ) (2000) Auckland, New Zealand: Australian and New Zealand Guidelines for Fresh and Marine Water Quality.

15. European Community (1992) Directive 92/43/EEC of the Council of 21 May 1992 on the conservation of natural habitats and wild fauna and flora. *Official Journal of the European Communities* LZ06: 7–50.

in abundance (= 'eu'). Currently, several definitions can be found in generalist, scientific, and policy texts. Most agree in defining eutrophication as a process wherein enrichment of aquatic systems by nutrients, usually phosphorus and nitrogen compounds, causes excessive growth of plants and algae, leading to an imbalance between the processes of algal production and consumption. This is often reflected in enhanced sedimentation of algal-derived organic matter, stimulation of microbial decomposition and oxygen consumption, with depletion of oxygen in the hypolimnion.

Eutrophication becomes a serious problem when one or more of the following symptoms are observed:

1. aquatic plants hinder uses of the waterbody
2. toxic algal species proliferate in large numbers
3. noxious odors are common
4. water becomes highly turbid
5. depletion of dissolved oxygen and fish kills.

A lake's natural productivity reflects the biogeochemical characteristics and hydrological processes of the watershed in which the lake is located, as well as the physicochemical and biological dynamics within the lake's waters and sediments. These characteristics and processes are responsible for a differential vulnerability of lakes to alterations in nutrient loading. Specifically, the vulnerability of lakes to enhanced nutrient loading is mediated by lake size and depth, flushing rate, patterns of stratification and mixing, and alkalinity. The establishment of criteria for eutrophication management needs to consider these conditions, for example by defining lake types and setting type specific reference conditions and objectives.

The issue of lake eutrophication started to receive public and scientific attention in the late 1960s and early 1970s in the industrialized countries, where it was linked to a growing population and to the intensification of industrial and agricultural activities. Since then, for many lakes in industrialized countries, wastewater treatment to remove phosphorus and/or nitrogen has reduced water quality degradation. However, the impact of human activities on lakes is today several times higher than 40–50 years ago, given the growth in population, in industrial and agricultural production, and in transport.

**Criteria for assessment** In 1967, the Organization for Economic Cooperation and Development (OECD) initiated a Cooperative Program on eutrophication, designed to quantify the relationship between nutrient loading in inland aquatic systems (lakes and reservoirs) and their trophic response. This program covered a wide variety of geographic and limnological situations and aimed at developing a sound database for evaluating the response of inland aquatic systems to various rates of nutrient supply over a large geographical area and provided the foundation for the development of scientific lake management principles for the control of excessive nutrient loading to lake water. The main results of this program have since then constituted the basis for lake eutrophication control worldwide and are as follows:

1. recognition that phosphorus is the factor determining eutrophication in most cases
2. definition of the relationship between nutrient loading to lake waters and their trophic response for a high number of lakes
3. development of models allowing the establishment of nutrient load limits compatible with the water use objectives.

The OECD proposed quantitative classification criteria for the previously vague, qualitative, description of trophic classes, i.e., oligotrophic, mesotrophic, eutrophic, and their border categories ultra-oligotrophic and hypereutrophic. Two classification systems, fixed boundary and open boundary, were developed, integrating the criteria for in-lake total phosphorus concentration, yearly average chlorophyll-*a* concentration, maximum chlorophyll-*a* concentration, yearly average Secchi depth transparency, and minimum Secchi depth transparency into a trophic scale from ultra-oligotrophic to hypertrophic. **Table 2** shows the boundaries for a selection of published classification systems that typically have a 4–5 trophic scale based on average chlorophyll-*a* concentrations as found in studies published from 1966 to 1999.

However, the success of management strategies based on OECD and analogous approaches proved to have some limitations. Improvement of water quality after the implementation of restoration measures in some lakes has been relatively slow, despite the implementation/introduction of nutrient abatement measures being tailored to achieve the established water quality criteria. Several cases exist where the trophic status could not be reversed owing to several buffering mechanisms, for instance phosphorus release from sediments, interactions in the food web, and interactions with other pressures (e.g., climate changes). For example, if piscivores are substantially reduced by changes induced by eutrophication, planktivores often increase and exert strong predation on the larger zooplankton. Hence, the grazing pressure on phytoplankton declines and algal blooms may increase in severity. Furthermore, the size distribution of the phytoplankton may shift to larger species, which sink faster and may decompose at different rates than smaller algae.

Since the OECD studies and following improvement of the scientific knowledge on the eutrophication process in lakes, including knowledge on biotic interactions, there has been a shift in the management approach from a use-orientated management to an ecosystem approach. This has stimulated the development of ecological classifications incorporating biological criteria (**Table 3**).

**Table 2**  Eutrophication class boundaries based on average phytoplankton chlorophyll-*a*

*(a) Trophic class boundaries based on average chlorophyll-a concentration (before 2000)*

| Ultra-oligotrophic | Oligotrophic | Mesotrophic | Eutrophic | Politrophic | Hyper-eutrophic | Source |
|---|---|---|---|---|---|---|
| | 0.3–2.5 | 1–5 | 5–140 | | | 1 |
| | 0–4 | 4–10 | >10 | | | 2 |
| | 0–4.3 | 4.3–8.8 | >8.8 | | | 3 |
| | <7 | 7–12 | >12 | | | 4 |
| | <2.6 | 2.6–6.4 | 6.4–56 | | >56 | 5 |
| | 0–2 | 2–6 | >6 | | | 6 |
| | <3 | 3–7 | 7–40 | | >40 | 7 |
| | <3 | 3–10 | 10–40 | 40–60 | >60 | 8 |
| <1.0 | <2.5 | 2.5–8 | 8–25 | | >25 | 9 |
| | 0.8–3.4[a] | 3.0–7.4[a] | 6.7–31[a] | | | 9 |
| | 0.4–7.1[b] | 1.0–11.6[b] | 3.1–66[b] | | | 9 |
| | | <10 (dimictic) | <30 (dimictic) | | >30 (dimictic) | 10 |
| | | <30 (polymictic) | 30–100 (polymictic) | | >100 (polymictic) | |
| <0.5 | 0.5–4.2 | 1.7–21 | 9.6–97.4 | | >45.4 | 11 |
| | <8 | ≤8 | ≤15 | | ≤25 | 12 |
| | | 8–25 | 25–35 (moderate) 35–55 (strong) 55–75 (high) | | >75 | 13 |
| | <3 | 3–9.7 | 9.7–31 | 31–100 | >100 | 14 |

*(b) New class boundaries according to the WFD, numbers given as chlorophyll-a (µg l⁻¹)*

| Type | Lake type characterization | Reference conditions | High/good boundary | Good/moderate boundary |
|---|---|---|---|---|
| L-AL3 | Lowland or mid-altitude, deep, moderate to high alkalinity, large | 1.5–1.9 | 2.1–2.7 | 3.8–4.7 |
| L-AL4 | Mid-altitude, shallow, moderate to high alkalinity, large | 2.7–3.3 | 3.6–4.4 | 6.6–8.0 |
| L-A1/2 | Lowland, shallow, calcareous | 2.6–3.8 | 4.6–7.0 | 8–12 |
| L-CB1 | Lowland, shallow, calcareous | 2.6–3.8 | 4.6–7.0 | 8–12 |
| L-CB2 | Lowland, very shallow, calcareous | 6.2–7.4 | 9.9–11.7 | 21–25 |

Continued

**Table 2** Continued

(a) Trophic class boundaries based on average chlorophyll-a concentration (before 2000)

| Ultra-oligotrophic | Oligotrophic | Mesotrophic | Eutrophic | Politrophic | Hyper-eutrophic | Source |
|---|---|---|---|---|---|---|
| L-CB3 | Lowland, shallow, siliceous | 2.5–3.7 | 4.3–6.5 | 8–12 | | |
| L-M5/7 | Reservoirs, deep, large siliceous, 'wet areas' | 1.4–2.0 | – | 6.7–9.5 | | |
| L-M8 | Reservoirs, deep, large, calcareous | 1.8–2.6 | – | 4.2–6.0 | | |
| L-N1 | Lowland, shallow, siliceous (moderate alkalinity) clear, large | 2.5–3.5 | 5.0–7.0 | 7.5–10.5 | | |
| L-N2a | Lowland, shallow, siliceous (low alkalinity) clear, large | 1.5–2.5 | 3.0–5.0 | 5.0–8.5 | | |
| L-N2b | Lowland, deep, siliceous (low alkalinity) clear, large | 1.5–2.5 | 3.0–5.0 | 4.5–7.5 | | |
| L-N3 | Lowland, shallow, siliceous (low alkalinit), humic, large | 2.5–3.5 | 5.0–7.0 | 8.0–12.0 | | |
| L-N5 | Mid-altitude, shallow, siliceous (low alkalinity) clear, large | 1.5 | 2.0–4.0 | 3.0–6.0 | | |
| L-N6 | Mid-altitude, shallow siliceous (low alkalinity), humic, large | 2.5 | 4.0–6.0 | 6.0–9.0 | | |
| L-N8 | Lowland, shallow, siliceous (moderate alkalinity), humic, large | 3.5–5 | 7.0–10.0 | 10.5–15.0 | | |

All chlorophyll values are in $\mu g\, l^{-1}$.

[a] $x \pm 1$ SD.

[b] $x \pm 2$ SD.

Sources

1. Sakamoto M (1966) Primary production by the Phytoplankton community in some Japanese lakes and its dependence on lake depth. *Archives of Hydrobiology* 62: 1–28.
2. National Academy of Sciences (NAS) (1973) Guidelines and FDA action levels for toxic chemicals in shellfish. Water Quality Criteria 1972. *EPA Ecological Research Series.* EPA-R3-73-033. U.S. Environmental Protection Agency, Washington, D.C.
3. Dobson II F, Gilbeitson M, and Sly PG (1974) A summary and comparison of nutrients and related water quality in Lakes Eric, Ontario, Iluron, and Superior. *Journal the Fisheries Research Board of Canada* 31: 731–738.
4. US Environmental Protection Agency (US EPA) (1974) *The relationships of phosphorus and nitrogen to the trophic state of northeast and north central lakes and reservoirs.* National Eutrophication survey Working Paper No. 23. Corvallis: US EPA Pacific Northwest Environmental Protection Agency Laboratory.
5. Carlson RE (1977) A trophic state index for lakes. *Limnology and Oceanography* 22: 361–369.
6. Rast W and Lee GF (1978) *Summary analysis of the North American (US portion) OECD eutrophication project: Nutrient loading-lake response relationships and trophic state indices.* Corvallis: US EPA Environmental Research Laboratory.
7. Forsberg C and Ryding S-O (1980) Eutrophication parameters and trophic state indices in 30 Swedish waste-receiving lakes. *Archives of Hydrobiologia* 89: 189–207.
8. Technical Standard (1982) *Utilization and protection of waterbodies. Standing inland waterbodies. Classification.* Berlin: Technical Standard 27885/01.
9. Organisation of Economic Co-operation and Development (OECD) (1982) *Eutrophication of waters: monitoring, assessment and control.* Paris: OECD.
10. Hillbricht-Ilkowska A (1993) Phosphorus loading to lakes of Suwalki Landscape Park (north-eastern Poland) and its relation to lake trophy indices. *Ekol. Pol./Polish Journal of Ecology* 41: 221–235.
11. Yoshimi H (1987) Simultaneous construction of single-parameter and multiparameter trophic state indices. *Water Research* WATRAG 21(12): 1505–1511.
12. Soszka H and Kudelska D (1994) An approach to lake monitoring in Poland. In Adriaanse M, der Kraats J, Stoks PG et al. (eds.) *Proceedings of the International Workshop on Monitoring and Assessment in Water Management*, pp. 232–240. Lelystad: RIZA.
13. Bowman JJ, Clabby KJ, Lucey J, et al. (1996) *Water quality in Ireland 1991–1994.* Wexford: Environmental Protection Agency.
14. Länderarbeitsgemeinschaft Wasser (LAWA) (1999) Gewässerbewertung – stehende gewässer – Vorläufige richtlinie für eine erstbewertung von naturlich entstandenen seen nach trophischen kriterien 1998. Berlin: Kulturbuch-Verlag GmbH.

**Table 3** Models or classification schemes forming the basis for development/or outlining the eutrophication criteria based on biological elements for lakes

| Region | Notes | Source |
|---|---|---|
| *Phytoplankton* | | |
| USA | EPA lake classification scheme, which uses phytoplankton abundance, chlorophyll concentrations, and transparency (Secchi disc depth). | 1, 2, 3 |
| UK European Alps | Quantitative direct relationships between phytoplankton composition and nutrient conditions for planktonic diatoms, based on relative abundances within surface sediment sub-fossil assemblages. Individual species TP optima are dataset specific and cannot necessarily be applied outside the region of development or for different lake types. | 4, 5 |
| Europe, Denmark, UK, Norway | Preliminary phytoplankton classifications (composition and abundance) developed specifically for the WFD. | 6, 7, 8, 9 |
| *Macrophyte* | | |
| UK | The Trophic Ranking Scheme (TRS) used the macrophyte composition recorded in 1224 standing waters to classify UK lakes into 10 vegetation groups, which were related to lake alkalinity, pH, and conductivity. Each of these 10 groups were allocated site types based on lake trophy. Individual macrophyte species were also allocated a TRS based on the range of site types within which they were found. The average site TRS can be used to infer whether eutrophication has occurred. | 10 |
| Sweden | The Swedish Environmental Quality Criteria (SEQC) scheme assesses the state of lakes using a variety of factors, including nutrients and species richness of macrophytes. Comparisons of the current condition with reference values are used in appraisals. The SEQC scheme defines conditions for both nutrient loadings and macrophytes at high to low ecological status as well as deviations from the high reference state. The macrophyte scheme is based on the UK TRS (Palmer *et al.*, 1992). | 11 |
| USA | The US EPA lakes and reservoir bioassessment and biocriteria uses submerged macrophytes as one of 7 biological monitoring elements for assessing the condition of US lakes. Lake condition is assessed using additive indices that integrate both habitat and biological scores. The LRBB scheme provides reference values for macrophyte metrics and nutrients but does not directly relate them together. This multimetric index indicates the overall biological condition of a lake; however, it cannot quantify the actual cause of degradation, although it does suggest where eutrophication may be the cause. | 12 |
| Northern Ireland | The Northern Ireland Lake Survey quantified macrophyte species – environmental relationships from over 500 lakes using Generalized Additive Models (GAM) and Canonical Correspondence Analysis (CCA). Nutrient concentrations appeared influential in 'explaining' species distribution, but were highly correlated with alkalinity and altitude. | 13 |
| Romania | A classification of Danube Delta lakes based on aquatic macrophytes and turbidity. | 14 |
| Europe, Denmark, Finland, Germany, Northern Ireland | Preliminary macrophyte classification schemes developed specifically for evaluation of ecological status in relation to the European WFD. | 6, 7, 15, 16, 17 |
| *Benthic invertebrate* | | |
| Palearctic | Benthic Quality Index (BQI) used to assess trophic status of Palearctic lakes. In this system in eutrophic lakes the BQI value is considered to be 1 and *Chironomus plumosus* the dominant taxon; a BQI value of 5 is characteristic of oligotrophic lakes. If no indicator species are present then a value of 0 is scored indicating a hypereutrophic lake. | 18 |
| Lake Geneva, Switzerland | Indices of trophy were developed based on the structure of the *tubificid* and *lumbriculid* (Oligochaeta) communities. | 19 |
| Sweden | Studies linking chironomid communities with lake trophic state (indicated by variables such as oxygen concentration, total phosphorus, chlorophyll-*a*, and algal biovolume). | 20 |
| The Netherlands | | 21 |
| Norway | | 22 |
| Finland | | 23 |
| Denmark | | 24 |
| Alberta, Canada | Empirical model linking profundal macroinvertebrate biomass (PMB) and water chemistry and morphometric variables from 26 lakes located within the Boreal mixed wood and Boreal Subarctic ecoregions. | 25 |
| Florida, USA | Five macroinvertebrate core metrics were linked to surface-water TP from an observed P gradient and a P-dosing experiment in coastal wetlands of south Florida to estimate numerical water quality criteria for TP. | 26 |
| Lake Ladoga, NW Russia | Model of the annual dynamics and distribution of zoobenthos biomass with respect of lake phosphorus dynamics. | 27 |

Continued

**Table 3**    Continued

| Region | Notes | Source |
|---|---|---|
| *Fish* | | |
| Denmark | Fish species richness, biodiversity, and trophic structure are linked to a trophic gradient of TP. | 28 |
| Europe | Classification system developed for 48 European lake ecotypes using 28 variables for water quality status with 3 fish parameters: fish community, fish biomass, and piscivorus-zooplanctivorus biomass ratio. | 6 |
| Europe, Danube Delta lakes | Classification of fish communities in relation with a lake ecology guild, such as riverine/white fish, euritopic/grey fish, and limnophilic/black fish. | 29, 30, 31 |
| The Netherlands | Occurrence of three fish communities (perch-type, pike-perch-type, and pike – perch-bream-type) and the size of total fish stock, perch stock, pike stock, and cyprinid stock linked with summer average TP concentration. | 32 |
| Austria | A relationship between TP (mg m$^{-3}$) and total fish biomass in kg ha$^{-1}$ were calculated for prealpine oligo- and oligo-mesotrophic lakes using: $BM = 3.8148 \times TP^{1.0940}$, $r^2 = 0.59$, $n = 10$. | 33 |
| UK, Europe, USA | Classification criteria between threshold requirement for oxygen concentrations and fish communities (e.g., in UK: dissolved oxygen $>8$ mg l$^{-1}$ for salmonids, $>6$ for non-salmonids, and $>1$ for tolerant species of cyprinids). Similar or slightly different thresholds have been forwarded by the European Inland Fishery Advisory Commission (EIFAC), National Research Council of Canada (NRCC), and US Environmental Protection Agency (US EPA). | 34, 35, 36, 37 |
| Sweden | Method for classifying fish communities in rivers and lakes using parameters from fish surveys and physical characteristics of the water bodies. The classifications are not directly related to specific physico-chemical parameters, but aim to define the extent to which these communities may deviate from 'undisturbed' waters. | 38 |
| Romania | Classification systems based on fish ecology and oxygen sensitivity. | 39, 40 |
| Romania | Direct nutrient stress on fish. Toxic effects of nitrite and nitrate, depend on species, salinity, concentration, and time of exposure. Maximum level of Romanian STAS range of 10–45 mg l$^{-1}$ for nitrate ($-NO_3$) and 0–1 mg.l$^{-1}$ for nitrite ($-NO_2$), depending of water category use. | 41 |

The table integrates and modifies information contained in tables 1.2, 1.3, 1.4, and 1.5 of the REBECCA report edited by Solimini *et al.* (2006).

Sources

1. Gibson G, Carlson R, Simpson J, *et al.* (2000) *Nutrient criteria technical guidance manual: Lakes and reservoirs*. Washington DC: US EPA.
2. Carlson RE (1977) A trophic state index for lakes. *Limnology and Oceanography* 22: 361–369.
3. Carlson RE (1980) More complications in the chlorophyll-Secchi disk relationship. *Limnology and Oceanography* 25: 379–382.
4. Bennion H (1994) A diatom-phosphorus transfer-function for shallow, eutrophic ponds in southeast England. *Hydrobiologia* 276: 391–410.
5. Wunsam S and Schmidt R (1995) A diatom-phosphorus transfer function for alpine and pre-alpine lakes. *Memorie del Istituto Italiano di Idrobiologia* 53: 85–99.
6. Moss B, Johnes P, and Phillips G (1996) The monitoring of ecological quality and classification of standing water in temperate regions: A review and proposal based on a worked scheme for British waters. *Biological Review* 71: 301–339.
7. Sondergaard M, Jeppesen E, Jensen JP, *et al.* (2003) *Vandrammedirektivet og danske søer* Undertitel: Del 1: Søtyper, referencetilstand og økologiske kvalitetsklasser. Copenhagen: Danmarks Miljøministeriet.
8. Carvalho L, Maberly, S, May L, *et al.* (2004) *Risk assessment methodology for determining nutrient impacts in surface freshwater bodies*. Environment Agency of England and Wales and Scottish Environment Protection Agency.
9. Solheim A Lyche, Andersen T, Brettum P, *et al.* (2004) *BIOKLASS – Klassifisering av økologiske status i norske vannforekomster: Forslag til aktuelle kriterier og foreløpige grenseverdier mellom god og moderat økologisk status for utvalgte elementer og påvirkninger*. Oslo: Norsk institutt for vannforskning (NIVA).
10. Palmer MA. and Bell SL, *et al.* (1992) A botanical classification scheme of standing waters in Britain, applications for conservation and monitoring. *Aquatic Conservation: Marine and Freshwater Ecosystems* 2: 125–143.
11. Swedish Environmental Protection Agency (2000) *Lakes and watercourses: Environmental quality criteria*. Report 5050.
12. US EPA (1998) *Lake and reservoirs bioassessment and biocriteria*. Washington DC: Environmental Protection Agency.
13. Heegaard E, Birks HH, *et al.* (2001) Species environmental relationships of aquatic macrophytes in Northern Ireland. *Aquatic Botany* 70: 175–223.
14. Coops H, Hanganu J, Tudor M, *et al.* (1999) Classification of Danube Delta lakes based on aquatic vegetation and turbidity. *Hydrobiologia* 415: 187–191.
15. Kanninen A, Leka J, Vallinkosk V-M, *et al.* (2003) Aerial photograph interpretation and field surveys of aquatic macrophytes in assessing the ecological status of small boreal lakes – preliminary results. In: Ruoppa M, Heinonen P, Pilke A, *et al.* (eds.) *How to assess and monitor ecological quality in freshwaters*. pp. 123–126. TEMANORD Report No. 547.
16. Schaumberg J, Scranz C, *et al.* (2004) Macrophytes and phytobenthos as indicators for ecological status in German lakes – Contributions to the implementation of the WFD. *Limnologica* 34(4): 302–314.
17. Dodkins I, Rippey B, and Hale P (2003) The advantage of metrics for aquatic macrophyte assessment in Northern Ireland. In Ruoppa M, Heinonen P, Pilke A, *et al.* (eds.) *How to assess and monitor ecological quality in freshwaters*. pp. 29–34. TEMANORD Report No. 547.
18. Wiederholm T (1980) Using benthos for lake monitoring. *Journal of the Water Pollution Control Federation* 52: 537–547.

19. Lang C (1998) Using oligochaetes to monitor the decrease of eutrophication: the 1982–1996 trend in Lake Geneva. *Archiv für Hydrobiologie* 141: 447–458.

20. Brodin Y (1982) Paleoecological studies of the recent developments of the lake Vaxjosjon: Interpretetation of eutrophication process through analysis of subfossil chironomids. *Archiv für Hydrobiologie* 93: 313–326.

21. Heinis F and Crommentuijn T (1992) Behavioural response to changing oxygen concentration of deposit feeding chironomid larvae (Diptera) of littoral and profundal habitats. *Archiv für Hydrobiologie* 124: 173–185.

22. Saether OA (1979) Chironomid communities as water quality indicators. *Holarctic Ecology* 2: 65–74.

23. Kansanen PH, Aho J, and Paasivirta L (1984) Testing the benthic lake type concept based on chironomid associations in some Finish lakes using multivariate statistical method. *Annales Zoologici Fenici* 21: 55–76.

24. Brodersen KP and Lindegaard C (1999) Classification, assessment and trophic reconstruction of Danish lakes using chironomids. *Freshwater Biology* 42: 143–157.

25. Dinsmore PW, Scrimgeour GJ, and Prepas EE (1999) Empirical relationships between profundal macroinvertebrate biomass and environmental variables in boreal lakes of Alberta, Canada. *Freshwater Biology* 41: 91–100.

26. King RS and Richardson CJ (2003) Integrating bioassessment and ecological risk assessment: An approach to developing numerical water-quality criteria. *Environmental Management* 31: 795–809.

27. Astrakhantsev GP, Minina, TR, Petrova NA, *et al.* (2003) Modeling Lake Ladoga zoobenthos and studying its role in phosphorus exchange at the water–bed interface. *Water Resources* 30: 209–220.

28. Jeppesen E, Jensen JP, Sondergaard M, *et al.* (2000) Trophic structure, species richness and biodiversity in Danish lakes: changes along a phosphorous gradient. *Freshwater biology* 45: 201–218.

29. Regier HA, Welcomme RL, and Steedman RJ (1998) Rehabilitation of degraded river ecosystems. *Proceedings of the International Large River Symposium (LARS). Canadian Special Publication of Fisheries and Aquatic Sciences* 106: 86–97.

30. Navodaru I, Buijse AD, and Staras M (2002) Effects of Hydrology and Water Quality on the Fish Community in Danube Delta Lakes. *International Review of Hydrobiology* 87: 329–348.

31. Oosterberg W, Sataras M, Buijse AD, *et al.* (2000) *Ecological gradients in the Danube Delta lakes. Present state and man-induced changes.* Lelystad: RIZA.

32. Ligtvoet W and Grimm MP (1992) Fish in clear water. Fish-stock development and management in Lake Volkerak/Zoom. How an estuary changed into freshwater lake: The water management of Lake Volkerak/ Zoom. Hooghart JC and Posthumus CWS (eds.) Proceedings and Information, CHO-TNO, 46: 9–84.

33. Gassner H, Tischler G, and Wanzenböck J (2003) Ecological integrity assessment of lakes using fish communities – Suggestions of new matrix development in two Austrian prealpine lakes. *International Review of Hydrobiology* 88: 635–652.

34. Carvalho L, Benninon H, Darwell A, *et al.* (2002) Physico-chemical conditions for supporting different levels of biological quality for the Water Framework Directive for freshwater. Edinburgh: Centre for Ecology and Hydrology.

35. EIFAC 1973 *Water Quality Criteria for European Freshwater Fish. Report on Dissolved Oxygen and Inland Fisheries.* EIFAC Tech. Paper 86, pp. 19.

36. US EPA 1986 Ambient Water Quality Criteria for Dissolved Oxygen (Freshwater). Washington, DC: Environmental Protection Agency.

37. US EPA 1997 Ambient Water Quality Criteria for Dissolved. Washington, DC: Environmental Protection Agency.

38. Swedish Environmental Protection Agency (2000) *Lakes and Watercourses: Environmental Quality Criteria.* Report 5050, 102 pp.

39. Busnita T and Brezeanu G (1967) Ihtiofauna. In: *Limnologia sectorului romanesc al Dunarii*, pp. 325–370. Romania, Bucuresti.

40. Nikolski GV (1962) *Ecologia pestilor.* Bucuresti. 362 pp.

41. Diudea M, Tudor S, and Igna A (1986) Toxicologie acvatica. Cluj Napoca. 183 pp.

## Acidification

**General introduction** Acidification of soils and surface waters as a result of elevated sulphur (S) and nitrogen (N) deposition has been widely documented from many sites in Europe and North America, and more recently in Southeaste Asia and parts of southern Africa. The acidification process is related to leaching of atmospheric-derived sulphate ($SO_4^{2-}$) and excess nitrate ($NO_3^-$) from soils to surface waters. In acid-sensitive ecosystems with slow-weathering bedrock and limited or depleted pools of base cations, $SO_4^{2-}$ and $NO_3^-$ in runoff will to a large extent be accompanied by acidifying hydrogen ions ($H^+$) and inorganic aluminium ($Al_i$) that are toxic for many aquatic organisms.

In Europe and eastern North America emissions of S and N oxides increased steadily from the second half of the 19th century due to accelerating industrialization with extensive burning of fossil fuels. The highest S emission levels were reached in the late 1970s, whereas N emissions peaked about ten years/ a decade later. The reduction of S and N emissions over the last few decades has largely been a result of international efforts and legislation aiming at reducing the problem of acidification of soil and surface waters.

Rapidly growing economies in parts of southern and eastern Asia have led to large increases in S and N emissions during the past 20 years, and further large increases are expected. Acidification problems have been reported in several areas and these can be expected to grow unless emission reduction measures are introduced.

In response to the reduced deposition of acidifying compounds in Europe and eastern North America, many acid-sensitive freshwater ecosystems have started to recover from acidification damage. The regional differences are large, however, and the response of water chemistry to changes in acid deposition often includes time lags of years to decades, depending on site history and its physical-chemical characteristics. Biological recovery, often requiring several generation times for recolonization and readaptation after several keystone species might have been lost, usually requires longer time.

**Criteria for assessment**  There is no global and harmonized approach on lake management criteria related to acidification. In Europe, the Critical Loads concept has been widely accepted as a basis for negotiating control strategies for transboundary air pollution. Critical loads have been defined as: 'the highest load that will not cause chemical changes leading to long-term harmful effects in the most sensitive ecological systems.' Critical loads are the maximum amount of pollutants that ecosystems can tolerate without being damaged. In freshwater ecosystems critical limits are often defined for brown trout and invertebrates (**Table 4**).

Key chemical parameters related to biological effects are pH and concentrations of inorganic labile aluminium ($Al_i$), calcium and Acid Neutralising Capacity (ANC; defined as the equivalent sum of base cations minus the sum of strong acid anions). The most widely used criterion for modeling and assessment has been ANC, which shows good empirical relationships to, for instance, brown trout. The critical limits for brown trout can be modified in water with extremely low solute concentrations and high concentrations of total organic carbon (TOC).

In many surface waters, especially rivers, there can be large differences between 'average water chemistry' and the chemistry during acidification episodes. The potential damage to aquatic organisms during episodes is often a function of 'intensity' and 'duration.' Many organisms can recover and survive after short acid episodes, but the probability of serious damage or death increase with acidity level and the duration of the acidification event. The water quality requirements vary not only between species, but also between different life stages of the same species. An example here is the smolt stage of Atlantic salmon, which is extremely sensitive to pH and $Al_i$ before and during migration from the river to the sea.

The concept of Critical Loads is a good example of how water quality criteria can form the basis for international negotiations on the reduction of long-range transported air pollution, as evidenced by the UN-ECE Convention on Long-range Transboundary Air Pollution with the Second Sulphur Protocol in 1994, and the Multi-pollutant, Multi-effect Protocol (the Gothenburg protocol) in 1999. In addition,

Critical Loads and more simple water quality criteria are often used when defining restoration targets for, for example, liming of acidified lakes and streams.

### Hazardous Substances

**General introduction**  Hazardous substances are substances or groups of substances that are toxic, persistent and liable to bio-accumulate, and other substances or groups of substances which give rise to an equivalent level of concern.

Priority substances in terms of water management are those which present a significant risk to or via the aquatic environment, including such risks to waters used for the abstraction of drinking water. For those pollutants, measures should be aimed at their progressive reduction. Among these substances the European Community (EC) legislation distinguishes priority hazardous substances, which means substances of concern undertaken in the relevant Community legislation or relevant international agreements. For those pollutants, measures are to be aimed at the cessation or phasing-out of discharges, emissions and losses.

Derivation of quality standards (QS) for hazardous substances is intended concomitantly to protect human beings from all impacts on health by drinking water uptake or ingestion of fishery products as well as ecosystems in inland, transitional, coastal, and territorial waters from adverse effects. Further, all relevant modes of toxicity both for ecosystems and man must be considered, such as direct toxicity, carcinogenicity, mutagenicity, and adverse effects on reproduction for humans or effects on endocrine regulation in other animals.

**Criteria for assessment**  Quality criteria for priority substances in the context of the EU WFD.

The WFD (2000/60/EC) set out a 'Strategy against pollution of water,' the first step of which was the establishment of a list of priority substances. The preparation of the priority list of substances included a combined monitoring-based and modeling-based priority setting (COMMPS) procedure in which about 820 000 monitoring data from waters and sediments from all Member States were evaluated and data for more than 310 substances on production, use and distribution in the environment were used for modeling where the available monitoring information was insufficient.

The Directive setting environmental QS for the priority substances adopted by the European Commission on 17 July 2006 (COM(2006)397) sets environmental QS for surface waters for 41 dangerous chemical substances including 33 priority substances and 8 other pollutants (**Table 5**). Within this list,

**Table 4** Some example studies on effects of acidification on aquatic biota

| Species/functional group | Region | Note | Source |
|---|---|---|---|
| Phytoplankton | North America | Phytoplankton community responses to acidification | 1 |
| | Europe and North America | Diatom-based pH reconstruction studies of acid lakes in Europe and North America | 2 |
| | Sweden | Acidification and phytoplankton development in West-Swedish lakes | 3 |
| | North America | Phytoplankton succession during acidification with and without increasing aluminium levels | 4 |
| | North America | Responses of phytoplankton and epilithon during acidification and early recovery of a lake | 5 |
| Periphyton | North America | Effect of stream acidification on periphyton composition, chlorophyll, and productivity | 6 |
| | North America | Early responses of periphyton to experimental lake acidification | 7 |
| | North America | Changes in epilithon and epiphyton associated with experimental acidification of a lake to pH 5 | 8 |
| Macrophytes | North America | Patterns of species composition in relation to environment | 9 |
| | | The effects of lake acidification on aquatic macrophytes (review) | 10 |
| | Europe | The effect of acidification, liming, and reacidification on macrophyte development, water quality, and sediment characteristics of soft-water lakes | 11 |
| Zooplankton | North America | Effects of experimental acidification on zooplankton population and community dynamics | 12 |
| | North America | Lake acidification: Effects on crustacean zooplankton populations | 13 |
| | Czech Republic and Slovakia | Acidification of lakes in Šumava (Bohemia) and in the High Tatra Mountains (Slovakia) | 14 |
| Benthic invertebrates | Norway | Monitoring of acidification by the use of aquatic organisms | 15 |
| | Sweden | Acid-stress effects on stream biology | 16 |
| | Europe | Critical limits of acidification to invertebrates in different regions of Europe | 17 |
| Fish | | | |
| Brown trout | Norway | Brown trout (*Salmo trutta*) status and chemistry from the Norwegian Thousand Lake survey | 18 |
| | Norway | A critical limit for acid neutralizing capacity in Norwegian surface waters, based on new analyses of fish and invertebrate responses | 19 |
| | North America | Episodic acidification of small streams in the northeastern United States: Effects on fish populations | 20 |
| | Norway | Assessment of damage to fish populations in Norwegian lakes due to acidification | 21 |
| | Nordic countries | Fish status survey of Nordic lakes: Effects of acidification, eutrophication, and stocking activity on present fish species composition | 22 |
| Atlantic salmon | Canada | A summary of the impact of acid rain on atlantic salmon (*Salmo salar*) in Canada | 23 |
| | Norway | Water quality requirement of Atlantic salmon (*Salmo salar*) in water undergoing acidification or liming in Norway | 24 |
| | Norway | Low concentrations of inorganic monomeric aluminum impair physiological status and marine survival of Atlantic salmon | 25 |

Sources

1. Findlay DL and Kasian SEM (1986) Phytoplankton community responses to acidification of lake 223, experimental lakes area, northwestern Ontario. *Water Air Soil Pollution* 30: 719–726.
2. Battarbee RW and Charles DF (1986) Diatom-based pH reconstruction studies of acid lakes in Europe and North America: A synthesis. *Water Air Soil Pollution* 30: 347–354.
3. Morling G and Willen T (1990) Acidification and phytoplankton development in some West-Swedish lakes 1966–1983. *Limnologica* 20: 291–306.
4. Havens KE and Heath RT (1990) Phytoplankton succession during acidification with and without increasing aluminium levels. *Environmental Pollution* 68: 129–145.
5. Findlay DL, Kasian SEM, Turner MT, *et al.* (1999) Responses of phytoplankton and epilithon during acidification and early recovery of a lake. *Freshwater Biology* 42: 159–175.
6. Mulholland PJ, Elwood JW, Palumbo AV, *et al.* (1986) Effect of stream acidification on periphyton composition, chlorophyll, and productivity. *Canadian Journal of Fisheries and Aquatic Sciences* 43: 1846–1858.
7. Turner MA, Jackson MB, Findlay DL, *et al.* (1987) Early responses of periphyton to experimental lake acidification. *Canadian Journal of Fisheries and Aquatic Sciences* 44: 135–149.
8. Turner MA, Howell ET, Summerby M, *et al.* (1991) Changes in epilithon and epiphyton associated with experimental acidification of a lake to pH 5. *Limnology and Oceanography* 36: 1390–1405.
9. Jackson ST and Charles DF (1988) Aquatic macrophytes in Adirondack (New York) lakes: Patterns of species composition in relation to environment. *Canadian Journal of Botany* 66: 1449–1460.
10. Farmer AM (1990) The effects of lake acidification on aquatic macrophytes – A review. *Environmental Pollution* 65: 219–240.
11. Roelofs JGM, Smolders AJP, Brandrud T-E, *et al.* (1995) The effect of acidification, liming and reacidification on macrophyte development, water quality and sediment characteristics of soft-water lakes. *Water Air Soil Pollution* 85: 967–972.
12. Locke A and Sprules WG (1993) Effects of experimental acidification on zooplankton population and community dynamics. *Canadian Journal of Fisheries and Aquatic Sciences* 50:1238–1247.
13. Havens KE, Yan ND, and Keller W (1993) Lake acidification: Effects on crustacean zooplankton populations. *Environmental Science Technology* 27: 1621–1624.
14. Fott J, Pražáková M, Stuchlík E, *et al.* (1994) Acidification of lakes in Šumava (Bohemia) and in the High Tatra Mountains (Slovakia). *Hydrobiologia* 274: 37–47.
15. Raddum GG, Fjellheim A, and Hesthagen T (1988) Monitoring of acidification by the use of aquatic organisms. *Internationale Vereinigung fur Theoretische und Angewandte Limnologie, Verhandlungen* 23: 2291–2297.
16. Herrmann J, Degerman E, Gerhardt A, *et al.* (1993) Acid-stress effects on stream biology. *Ambio* 22: 298–307.
17. Raddum GG and Skjelkvåle BL (1995). Critical limits of acidification to invertebrates in different regions of Europe. *Water Air Soil Pollution* 85: 475–480.
18. Bulger AJ, Lier L, Cosby BJ, *et al.* (1993) Brown trout (*Salmo trutta*) status and chemistry from the Norwegian Thousand Lake Survey: statistical analysis. *Canadian Journal of Fisheries and Aquatic Sciences* 50: 575–585.
19. Lien L Raddum GG Fjellheim A and Henriksen A (1996) A critical limit for acid neutralizing capacity in Norwegian surface waters, based on new analyses of fish and invertebrate responses. *Science of Total Environment* 177: 173–193.
20. Baker JP, Van Sickle J, Gagen CJ, *et al.* (1996) Episodic acidification of small streams in the northeastern United States: Effects on fish populations. *Ecological Applications* 6: 422–437.
21. Hesthagen T, Sevaldrud IH, and Berger HM (1999) Assessment of damage to fish populations in Norwegian lakes due to acidification. *Ambio* 28: 112–117.
22. Tammi J, Appelberg M, Beier U, *et al.* (2003) Fish status survey of Nordic lakes: Effects of acidification, eutrophication, and stocking activity on present fish species composition. *Ambio* 32: 98–105.
23. Watt WD (1987) A summary of the impact of acid rain on Atlantic salmon (*Salmo salar*) in Canada. *Water Air Soil Pollution* 35: 27–35.
24. Staurnes M, Kroglund F, and Rosseland BO (1995) Water quality requirement of Atlantic salmon (*Salmo salar*) in water undergoing acidification or liming in Norway. *Water Air Soil Pollution* 85: 347–352.
25. Kroglund F and Finstad B (2003) Low concentrations of inorganic monomeric aluminum impair physiological status and marine survival of Atlantic salmon. *Aquaculture* 222: 119–133.

**Table 5** Water quality criteria for priority polluting substances and other pollutants. Combined list based on EPA National Recommended Water Quality Criteria for Priority Pollutants and Nonpriority Pollutants[1] (http://www.epa.gov/waterscience/criteria/wqcriteria.html), and EU Environmental Quality Standards for Priority Substances and Other Pollutants.[2] Values in the table should be taken with precaution as a number of explanatory remarks were removed. Hence, it is always recommended to consult the original data sources

| CAS number[a] | Name of pollutant | EU priority hazardous substance | EU priority substance | EU other pollutant | EPA priority pollutant | EPA Non-priority pollutant | AA-EQS (µg l⁻¹)[b] | MAC-EQS (µg l⁻¹)[c] | 2005 CERCLA PRIORITY LIST score[d] | CMC (acute, µg l⁻¹)[e] | CCC (chronic, µg l⁻¹)[f] | Human health for the consumption of water + organism (µg l⁻¹) | Human health for the consumption of organism only (µg l⁻¹) |
|---|---|---|---|---|---|---|---|---|---|---|---|---|---|
| 110758 | 2-Chloroethylvinyl Ether | | | | + | | | | — | | | | |
| 91587 | 2-Chloronapthalene | | | | + | | | | — | | | 1000 | 1600 |
| 95578 | 2-Chlorophenol | | | | + | | | | — | | | 81 | 150 |
| 534521 | 2-Methyl-4,6-Dinitrophenol | | | | + | | | | 822.35 | | | 13 | 280 |
| 88755 | 2-Nitrophenol | | | | + | | | | — | | | | |
| 59507 | 3-Methyl-4-Chlorophenol | | | | + | | | | — | | | | |
| 101553 | 4-Bromophenyl Phenyl Ether | | | | + | | | | — | | | | |
| 7005723 | 4-Chlorophenyl Phenyl Ether | | | | + | | | | — | | | | |
| 100027 | 4-Nitrophenol | | | | + | | | | 566.05 | | | 670 | 990 |
| 83329 | Acenaphthene | | | | + | | | | 729.63 | | | | |
| 208968 | Acenaphthylene | | | | + | | | | 1057.7 | | | 190 | 290 |
| 107028 | Acrolein | | | | + | | | | 528.09 | | | | |
| 107131 | Acrylonitrile | | | | + | | | | | | | 0.051[g] | 0.25[g] |
| n.a. | Aesthetic Qualities | | | | | + | | | | Narrative statement | Narrative statement | Narrative statement | Narrative statement |
| 15972608 | Alachlor | | + | | | | 0.3 | 0.7 | | | | | |
| 309002 | Aldrin | | | + | + | | Σ = 0.010 | | 1116.9 | | | 0.000049[g] | 0.000050[g] |
| 60571 | Dieldrin | | | + | + | | | n.a. | 1153.2 | 0.24 | 0.056 | 0.000052[g] | 0.000054[g] |
| 72208 | Endrin | | | + | + | | | | 1040.9 | 0.086 | 0.036 | 0.059 | 0.06 |
| 465736 | Isodrin | | | + | | | | | — | | | | |
| n.a. | Alkalinity | | | | | + | | | | | 20000 | | |
| 319846 | alpha-BHC | | | | + | | | | 807.72 | | | 0.0026[g] | 0.0049[g] |
| 959988 | alpha-Endosulfan | | | | + | | | | 1000.8 | 0.22 | 0.056 | 62 | 89 |
| 7429905 | Aluminum pH 6.5–9.0 | | | | | + | | | 688.21 | 750 | 87 | | |
| 764417 | Ammonia | | | | | + | | | 744.67 | pH, temperature and life-stage dependent | pH, temperature and life-stage dependent | | |
| 120127 | Anthracene | + | | | | + | 0.1 | 0.4 | — | | | 8,300 | 40,000 |
| 7440360 | Antimony | | | | + | | | | 606.3 | | | 5.6 | 640 |
| 7440382 | Arsenic | | | | + | | | | 1668.6 | 340 | 150 | 0.018[g] | 0.14[g] |
| 1332214 | Asbestos | | | | + | | | | 842.16 | | | 7 million fibers l⁻¹ | |
| 1912249 | Atrazine | | + | | | | 0.6 | 2 | — | | | | |
| n.a. | Bacteria | | | | | + | | | — | For primary recreation and shellfish uses | For primary recreation and shellfish uses | For primary recreation and shellfish uses | For primary recreation and shellfish uses |
| 7440393 | Barium | | | | | + | | | 812.12 | | | 1000 | |
| 71432 | Benzene | | + | | + | | 10 | 50 | 1353.5 | | | 2.2[g] | 51[g] |
| 92875 | Benzidine | | | | + | | | | 1114.1 | | | 0.000086[g] | 0.00020[g] |

Continued

**Table 5**  Continued

| CAS number[a] | Name of pollutant | EU priority hazardous substance | EU priority substance | EU other pollutant | EPA priority pollutant | EPA Non-priority pollutant | AA-EQS ($\mu g\ l^{-1}$)[b] | MAC-EQS ($\mu g\ l^{-1}$)[c] | 2005 CERCLA PRIORITY LIST score[d] | CMC (acute, $\mu g\ l^{-1}$)[e] | CCC (chronic, $\mu g\ l^{-1}$)[f] | Human health for the consumption of water + organism ($\mu g\ l^{-1}$) | Human health for the consumption of organism only ($\mu g\ l^{-1}$) |
|---|---|---|---|---|---|---|---|---|---|---|---|---|---|
| 56553 | Benzo(a)anthracene | | | | + | | | | 1055.6 | | | 0.0038[g] | 0.018[g] |
| 50328 | Benzo(a)pyrene | + | + | | + | | 0.05 | 0.1 | 1307.8 | | | 0.0038[g] | 0.018[g] |
| 205992 | Benzo(b)fluoranthene | + | + | | + | | Σ = 0.03 | n.a. | 1263.1 | | | 0.0038[g] | 0.018[g] |
| 207089 | Benzo(k)fluoranthene | + | + | | + | | | | 980.61 | | | 0.0038[g] | 0.018[g] |
| 191242 | Benzo(g,h,i)perylene | + | + | | + | | Σ = 0.002 | n.a. | 698.45 | | | | |
| 193395 | Indeno(1,2,3-cd)pyrene | + | + | | + | | | | 1046.5 | | | 0.0038[g] | 0.018[g] |
| 7440417 | Beryllium | | | | | | | | 1056.5 | | | j | |
| 319857 | beta-BHC | | | | + | | | | 976.59 | | | 0.0091[g] | 0.017[g] |
| 33213659 | beta-Endosulfan | | + | | + | | | | 858.19 | 0.22 | 0.056 | 62 | 89 |
| 111911 | Bis(2-Chloroethoxy) Methane | | | | + | | | | | | | | |
| 111444 | Bis(2-Chloroethyl) Ether | | | | + | | | | | | | 0.030[g] | 0.53[g] |
| 108601 | Bis(2-Chloroisopropyl) Ether | | | | + | | | | | | | 1400 | 65000 |
| n.a. | Boron | | | | | + | | | — | Narrative statement | Narrative statement | Narrative statement | Narrative statement |
| 75252 | Bromoform | | | | + | | | | 644.75 | | | 4.3[g] | 140[g] |
| 85687 | Butylbenzyl PhthalateW | | | | + | | | | 657.37 | | | 1500 | 1900 |
| 85535848 | C10-13-chloroalkanes | + | + | | | | 0.4 | 1.4 | — | | | | |
| 7440439 | Cadmium and its compounds | + | + | | + | | 0.15[h] | 0.9[h] | 1321.5 | | 0.25[h] | j | |
| 56235 | Carbontetrachloride | | | + | + | | 12 | n.a. | 1022.7 | | | 0.23[g] | 1.6[g] |
| 57749 | Chlordane | | | | + | | | | 1133.3 | 2.4 | 0.0043 | 0.00080[g] | 0.00081[g] |
| 470906 | Chlorfenvinphos | | + | | + | | 0.1 | 0.3 | — | | | | |
| 16887006 | Chloride | | | | | + | | | | 860000 G | 230000 G | | |
| 7782505 | Chlorine | | | | | + | | | | 19 | 11 | | |
| 108907 | Chlorobenzene | | | | + | | | | 840.04 | | | 130[j] | 1600 |
| 124481 | Chlorodibromomethane | | | | + | | | | 817.28 | | | 0.40[g] | 13[g] |
| 75003 | Chloroethane | | | | + | | | | 580.41 | | | | |
| 93721 | Chlorophenoxy Herbicide (2,4,5,-TP) | | | | | + | | | 692.77 | | | 10 | |
| 94757 | Chlorophenoxy Herbicide (2,4-D) | | | | | + | | | 584.13 | | | 100[j] | |
| 2921882 | Chlorpyrifos | | + | | + | + | 0.03 | 0.1 | 805.19 | | 0.041 | | |
| 16065831 | Chromium (III) | | | | + | + | | | 1149.7 | 570[h] | 74[h] | j | |
| 18540299 | Chromium (VI) | | | | + | | | | 799.59 | 16 | 11 | 0.40[g] | 13[g] |
| 218019 | Chrysene | | | + | + | | | | — | | | 0.0038[g] | 0.018[g] |
| n.a. | Color | | | | | + | | | — | Narrative statement | Narrative statement | Narrative statement | Narrative statement |
| 7440508 | Copper | | | | + | + | | | 802.6 | 13[h] | 9.0[h] | 1,300 | |
| 57125 | Cyanide | | | | + | | | | 1098.8 | 22 | 5.2 | 140 | 140 |
| 72548 | DDD 4,4'- | | | | + | | | | 1121.4 | | | 0.00031[g] | 0.00031[g] |
| 72559 | DDE 4,4'- | | | | + | | | | 1135.8 | | | 0.00022[g] | 0.00022[g] |
| n.a. | DDT total | | | | + | | 0.025 | n.a. | 1038.1 | | | | |
| 319868 | delta-BHC | | | | + | | | | | | | | |
| 8065483 | Demeton | | | | | + | | | | | 0.1 | | |
| 117817 | Di(2-ethylhexyl)phthalate (DEHP) | | + | | + | | 1.3 | n.a. | 918.6 | | | 1.2[g] | 2.2[g] |

(table continued)

| CAS No. | Constituent | + | Guideline | Aquatic A | Value | Aquatic B | Aquatic C | HH (water + org) | HH (org only) |
|---|---|---|---|---|---|---|---|---|---|
| 333415 | Diazinon | + | | 0.17 | 1001.9 | 0.17 | 0.17 | | |
| 53703 | Dibenzo[a,h]Anthracene | + | | | 1165.5 | | | 0.0038[g] | 0.018[g] |
| 95501 | Dichlorobenzene 1,2- | + | | | 703.53 | | | 420 | 1,300 |
| 541731 | Dichlorobenzene 1,3- | + | | | 630.23 | | | 320 | 960 |
| 106467 | Dichlorobenzene 1,4- | + | | | 735.51 | | | 63 | 190 |
| 91941 | Dichlorobenzidine 3,3'- | + | | | 1051.3 | | | 0.021[g] | 0.028[g] |
| 75274 | Dichlorobromomethane | + | | | 737.82 | | | 0.55[g] | 17[g] |
| 75343 | Dichloroethane 1,1- | + | | | 865.6 | | | 0.38[g] | 37[g] |
| 107062 | Dichloroethane 1,2- | + | 10 / n.a. | | 894.91 | | | 330 | 7100 |
| 75354 | Dichloroethylene 1,1- | + | | | 700.56 | | | 140[j] | 10000 |
| 156605 | Dichloroethylene Trans-1,2- | + | | | 886.69 | | | 4.6[g] | 590[g] |
| 75092 | Dichloromethane | + | 20 / n.a. | | 615.86 | | | 77 | 290 |
| 120832 | Dichlorophenol 2,4- | + | | | — | | | 0.50[g] | 15[g] |
| 78875 | Dichloropropane 1,2- | + | | | — | | | 0.34[g] | 21[g] |
| 542756 | Dichloropropene 1,3- | + | | | | | | 17000 | 44000 |
| 84662 | Diethyl PhthalateW | + | | | — | | | 270 000 | 1 100 000 |
| 131113 | Dimethyl PhthalateW | + | | | — | | | 380 | 850 |
| 105679 | Dimethylphenol 2,4- | + | | | 684.1 | | | 2000 | 4500 |
| 84742 | Di-n-Butyl PhthalateW | + | | | 1005.4 | | | 69 | 5300 |
| 51285 | Dinitrophenol 2,4- | + | | | 860.43 | | | 69 | 5300 |
| 25550587 | Dinitrophenols | + | | | 834.71 | | | 0.11[g] | 3.4[g] |
| 121142 | Dinitrotoluene 2,4- | + | | | 554.5 | | | 0.036[g] | 0.20[g] |
| 606202 | Dinitrotoluene 2,6- | + | | | 713.7 | | | 62 | 89 |
| 117840 | Di-n-Octyl Phthalate | + | | | 578.36 | | | 0.29 | 0.3 |
| 122667 | Diphenylhydrazine 1,2- | + | | | 1004.3 | | | 0.00010[g] | 0.00029[g] |
| 330541 | Diuron | + | 0.2 / 1.8 | | 1004.9 | | | 530 | 2100 |
| 115297 | Endosulfan | + | 0.005 / 0.01 | | 954.5 | | | 130 | 140 |
| 1031078 | Endosulfan Sulfate | + | | | 830.62 | | | 1,100 | 5,300 |
| 7421934 | Endrin Aldehyde | + | | | 810.29 | | | 0.98 | 1.8 |
| 542881 | Ether, Bis(Chloromethyl) | + | | | | | | 0.00010[g] | 0.00029[g] |
| 100414 | Ethylbenzene | + | 0.1 / 1 | | 1080.4 | | | 530 | 2100 |
| 206440 | Fluoranthene | + | | | 802.32 | | | 130 | 140 |
| 86737 | Fluorene | + | | | | | | 1,100 | 5,300 |
| 58899 | gamma-BHC (Lindane) | + | | 0.95 | | 0.95 | Narrative statement | 0.98 | 1.8 |
| n.a. | Gases, Total Dissolved | + | | | | | | Narrative statement | Narrative statement |
| 86500 | Guthion | + | | Narrative statement | | Narrative statement | 0.01 | | |
| n.a. | Hardness | + | | Narrative statement | | Narrative statement | Narrative statement | Narrative statement | Narrative statement |
| 76448 | Heptachlor | + | 0.01 | 0.52 | 1070.8 | 0.52 | 0.0038 | 0.000079[g] | 0.000079[g] |
| 1024573 | Heptachlor Epoxide | + | 0.1 | 0.52 | 1028.3 | 0.52 | 0.0038 | 0.000039[g] | 0.000039[g] |
| 118741 | Hexachlorobenzene | + | 0.02 / 0.05 | | 838.26 | | | 0.00028[g] | 0.00029[g] |
| 87683 | Hexachlorobutadiene | + | 0.1 / 0.6 | | 1130.7 | | | 0.44[g] | 18[g] |
| 608731 | Hexachlorocyclohexane | + | 0.02 / 0.04 | | 774.6 | | | 0.0123 | 0.0414 |
| 319868 | Hexachlorocyclohexane-Technical | + | | | 1038.1 | | | | |
| 77474 | Hexachlorocyclopentadiene | + | | | 718.71 | | | 40 | 1100 |
| 67721 | Hexachloroethane | + | | | 652.73 | | | 1.4[g] | 3.3[g] |
| 7439896 | Iron | | 1000 | | | | 1000 | 300 | |
| 78591 | Isophorone | + | 0.3 | | | | | 35[g] | 960[g] |
| 34123596 | Isoproturon | + | 7.2 / 1 | | | | | | |
| 7439921 | Lead and its compounds | | n.a. | | | | | | |
| 121755 | Malathion | + | 0.1 | | 1534.5 | | 2.5[h] | | |
| 7439965 | Manganese | + | | | 808.16 | | 0.1 | 50 | 100 |

Continued

**Table 5**   Continued

| CAS number[a] | Name of pollutant | EU priority hazardous substance | EU priority substance | EU other pollutant | EPA priority pollutant | EPA Non-priority pollutant | AA-EQS ($\mu g\,l^{-1}$)[b] | MAC-EQS ($\mu g\,l^{-1}$)[c] | 2005 CERCLA PRIORITY LIST score[d] | CMC (acute, $\mu g\,l^{-1}$)[e] | CCC (chronic, $\mu g\,l^{-1}$)[f] | Human health for the consumption of water + organism ($\mu g\,l^{-1}$) | Human health for the consumption of organism only ($\mu g\,l^{-1}$) |
|---|---|---|---|---|---|---|---|---|---|---|---|---|---|
| 7439976 | Mercury and its compounds | + | | | + | | 0.05 | 0.07 | 1507.3 | | 0.77 | | |
| 72435 | Methoxychlor | | + | | | + | | | 993.32 | | 0.03 | 100[j] | |
| 74839 | Methyl Bromide | | | | + | | | | | | | 47 | 1500 |
| 74873 | Methyl Chloride | | | | + | | | | 674.44 | | | | |
| 22967926 | Methylmercury | | | | | + | | | 806.47 | 1.4 | 0.77 | | 0.3 |
| 2385855 | Mirex | | | | | + | | | | | 0.001 | | |
| 91203 | Naphthalene | | + | | + | | 2.4 | n.a. | 895.49 | | | | |
| 7440020 | Nickel and its compounds | | + | | + | | 20 | n.a. | 1004 | | 52[h] | 610 | 4600 |
| 14797558 | Nitrates | | | | | + | | | | | | 10000 | |
| 98953 | Nitrobenzene | | | | + | | | | 609.85 | | | 17 | 690 |
| n.a. | Nitrosamines | | | | | + | | | | | | | |
| 924163 | Nitrosodibutylamine, N | | | | | + | | | | | | 0.0008 | 1.24 |
| 55185 | Nitrosodiethylamine, N | | | | | + | | | | | | 0.0063[g] | 0.29[g] |
| 930552 | Nitrosopyrrolidine, N | | | | | + | | | | | | 0.0008[g] | 1.24[g] |
| 62759 | N-Nitrosodimethylamine | | | | + | | | | 650.41 | | | 0.016[g] | 34[g] |
| 621647 | N-Nitrosodi-n-Propylamine | | | | + | | | | 811.01 | | | 0.00069[g] | 3.0[g] |
| 86306 | N-Nitrosodiphenylamine | | | | + | | | | 623.8 | | | 0.0050[g] | 0.51[g] |
| 1044051 | Nonylphenol | + | | | | | 0.3 | 2 | | | | 3.3[g] | 6.0[g] |
| 25154523 | Nonylphenols | | | | | + | | | | 28 | 6.6 | | |
| n.a. | Nutrients | + | | | | | | | | EPA's Ecoregional criteria for total phosphorus total nitrogen chlorophyll a and water clarity | EPA's Ecoregional criteria for total phosphorus total nitrogen chlorophyll a and water clarity | EPA's Ecoregional criteria for total phosphorus total nitrogen chlorophyll a and water clarity | EPA's Ecoregional criteria for total phosphorus total nitrogen chlorophyll a and water clarity |
| 1806264 | Octylphenols | | + | | | + | 0.1 | n.a. | | | | | |
| n.a. | Oil and grease | | | | | + | | | | Narrative statement | Narrative statement | Narrative statement | Narrative statement |
| 7782447 | Oxygen, dissolved | | | | | + | | | | Warm water and coldwater matrix | Warm water and coldwater matrix | Warm water and coldwater matrix | Warm water and coldwater matrix |
| 50293 | para-para-DDT | | | | + | | 0.01 | n.a. | 1195 | 1.1 | 0.001 | 0.00022[g] | 0.00022[g] |
| 56382 | Parathion | | | | | + | | | 784.02 | 0.065 | 0.013 | | |
| n.a. | Pentabromodiphenylether | | | + | | | 0.0005 | n.a. | | | | | |
| 608935 | Pentachlorobenzene | | + | | | + | 0.007 | n.a. | 753.47 | | | 1.4 | 1.5 |
| 87865 | Pentachlorophenol | | | | + | | 0.4 | 1 | 1024.4 | 19[i] | 15[i] | 0.27[g] | 3.0[g] |
| n.a. | pH | | | | | + | | | | | 6.5–9 | 5–9 | 5–9 |
| 85018 | Phenanthrene | | | | + | | | | 595.25 | | | | |
| 108952 | Phenol | | | | + | | | | 692.66 | | | 21 000 | 1 700 000 |
| 7723140 | Phosphorus Elemental | | | | | | | | 1144.7 | | | | |
| n.a. | Polyaromatic hydrocarbons | | + | | | + | n.a. | n.a. | | | | | |

| CAS[a] | Parameter | | | | EQS-AA[b] | Score[d] | EQS-MAC[c] | CMC[e] | CCC[f] | HH[g] | HH[g] |
|---|---|---|---|---|---|---|---|---|---|---|---|
| n.a. | Polychlorinated Biphenyls (PCBs) | | + | | | — | | | | 0.014 | 0.000064g | 0.000064g |
| 129000 | Pyrene | + | | | | 575.34 | | | | | 830 | 4000 |
| 7782492 | Selenium | + | + | | | 777.65 | | | | 5 | 170i | 4200 |
| 7440224 | Silver | + | + | | | 613.48 | | | 3.2h | | | |
| 122349 | Simazine | + | | | 1 | | | | | | | |
| n.a. | Solids Dissolved and Salinity | | | + | 4 | — | | | | | 250 000 | |
| n.a. | Solids Suspended and Turbidity | | + | | | | | | | Narrative statement | Narrative statement | Narrative statement |
| 7783064 | Sulfide-Hydrogen Sulfide | + | + | | | 672.87 | | | | 2 | | |
| n.a. | Tainting Substances | + | + | | | | | | | Narrative statement | Narrative statement | Narrative statement |
| 1746016 | TCDD (Dioxin) 2,3,7,8- | | + | | | 937.04 | | | | | 5.0E-9g | 5.1E-9g |
| n.a. | Temperature | | | + | | | | | | Species dependent criteria | Species dependent criteria | Species dependent criteria |
| 95943 | Tetrachlorobenzene 1,2,4,5- | | | + | | | | | | | 0.97 | 1.1 |
| 79345 | Tetrachloroethane 1,1,2,2- | + | + | + | | 778.29 | | | | | 0.17g | 4.0g |
| 127184 | Tetrachloroethylene | + | + | + | 10 | 1084.9 | | | | | 0.69g | 3.3g |
| 7440280 | Thallium | | + | | n.a. | | | | | | 0.24 | 0.47 |
| 108883 | Toluene | + | + | | | 946.04 | | | | | 1 300i | 15,000 |
| 8001352 | Toxaphene | | + | | | 1086.2 | | 0.73 | | 0.46 | 0.00028g | 0.00028g |
| n.a. | Tributyltin (TBT) | | | + | 0.0002 | | 0.002 | | | | | |
| 688733 | Tributyltin compounds | + | | | | 803.07 | | | 0.46 | 0.072 | 35 | 70 |
| 120821 | Trichlorobenzene 1,2,4- | | + | + | 0.4 | 644.17 | | | | | | |
| 12002481 | Trichlorobenzenes | | + | + | n.a. | 558.13 | | | | | | |
| 71556 | Trichloroethane 1,1,1- | + | | | | 835.27 | | | | | 0.59g | |
| 79005 | Trichloroethane 1,1,2- | | | + | | 722.98 | | | | | 2.5g | 16g |
| 79016 | Trichloroethylene | + | + | | 10 | 1158.2 | | | | | 5.7g | 30g |
| 67663 | Trichloromethane (Chloroform) | | + | | 2.5 | 1224.2 | | | | | | 470g |
| 95954 | Trichlorophenol 2,4,5- | | + | | | 604.44 | | | | | 1800j | 3600j |
| 88062 | Trichlorophenol 2,4,6- | | + | | | 863.1 | | | | | 1.4g | 2.4g |
| 1582098 | Trifluralin | | + | | 0.03 | 770.06 | | | | | | |
| 75014 | Vinyl chloride | + | + | | n.a. | 1389 | | | | | 0.025g | 2.4g |
| 7440666 | Zinc | + | + | | | 930.42 | | | 120h | 120h | 7400 | 26 000 |

[1] (Printed brochure EA-822-F-04-010;

[2] Commission Proposal (COM(2006)397)

[a] Chemical Abstracts Service registry number.

[b] This parameter is the Environmental Quality Standard expressed as an annual average value (EQS-AA).

[c] This parameter is the Environmental Quality Standard expressed as a maximum allowable concentration (EQS-MAC). Where the MAC-EQS are marked as 'not applicable,' the AA-EQS values are also protective against short-term pollution peaks since they are significantly lower than the values derived on the basis of acute toxicity.

[d] The ranking of hazardous substances on the priority list is based on three criteria (frequency of occurrence, toxicity, and potential for human exposure), which are combined to get the total score for each substance (details in text).

[e] The Criteria Maximum Concentration (CMC) is an estimate of the highest concentration of a material in surface water to which an aquatic community can be exposed briefly without resulting in an unacceptable effect.

[f] The Criterion Continuous Concentration (CCC) is an estimate of the highest concentration of a material in surface water without resulting in an unacceptable effect.

[g] This criterion is based on carcinogenicity of $10^{-6}$ risk. Alternate risk levels may be obtained by moving the decimal point (e.g., for a risk level of $10^{-5}$, move the decimal point in the recommended criterion one place to the right).

[h] The freshwater criterion for this metal is expressed as a function of hardness (mg l$^{-1}$) in the water column. The value given here corresponds to a hardness of 100 mg l$^{-1}$.

[i] Values displayed in table correspond to a pH of 7.8.

[j] The organoleptic effect criterion is more stringent than the value presented or a more stringent Maximum Contaminant Level (MCL) has been issued by EPA under the Safe Drinking Water.

11 substances have been identified as priority hazardous substances which are of particular concern for the inland, transitional, coastal, and territorial waters. These substances will be subject to cessation or phasing out of discharges, emissions and losses within a period that should not exceed 20 years. A further 14 substances were identified as being subject to review for identification as possible 'priority hazardous substances.'

QS ensuring a good chemical status of the Communities' surface waters are based on the parallel assessment of three compartments: water, sediment and biota. QS were assessed by means of pre-defined trigger criteria (**Table 6**) that determine whether a substance may place a certain objective at risk. For those objectives for which a possible risk was identified (trigger value exceeded), specific quality standards were derived. In a subsequent step, the lowest of the standards derived for the individual protection objectives was selected as the overall QS.

The Environmental Quality Standards are expressed as maximum allowable concentrations (MAC-EQS) or as annual average values (AA-EQS). Where the MAC-EQS are marked as 'not applicable,' the AA-EQS values are also protective against short-term pollution peaks since they are significantly lower than the values derived on the basis of acute toxicity.

**Quality criteria for pollutants recommended by US EPA**  In the United States water quality criteria for the protection of aquatic life and human health in surface waters have been derived for approximately 150 pollutants. The compilation in **Table 5** includes all priority toxic pollutants and some nonpriority toxic pollutants, and both human health effect and organoleptic effect criteria issued pursuant to the Clean Water Act (CWA) §304(a). Blank spaces indicate that the Environment Protection Agency (EPA) has no CWA §304(a) criteria recommendations.

Analogously to the MAC-EQS and AA-EQS used in the EC, the US EPA uses the Criteria Maximum Concentration (CMC) and Criterion Continuous Concentration (CCC). The CMC is an estimate of the highest concentration of a material in surface water to which an aquatic community can be exposed briefly without experiencing an unacceptable effect. The CCC is an estimate of the highest concentration of a material in surface water to which an aquatic community can be exposed indefinitely without experiencing an unacceptable effect. The CMC and CCC are just two of the six parts of an aquatic life criterion; the other four parts are the acute averaging period, chronic averaging period, acute frequency of allowed exceedence, and chronic frequency of allowed exceedence.

For a number of nonpriority toxic pollutants not listed, CWA §304(a) 'water + organism' human health criteria are not available, but the EPA has published Maximum Contaminant Levels (MCLs) under the Safe Drinking Water Act (SDWA) that may be used in establishing water quality standards to protect water supply designated uses. Drinking water and health advisory summary tables are prepared periodically. They contain drinking water standards in the form of nonenforceable concentrations of drinking water contaminants, Maximum Contaminant Level Goals (MCLGs), or enforceable Maximum Contaminant Levels (MCLs). Maximum Contaminant Levels are the maximum permissible level of a contaminant in water delivered to users of a public water system. Health Advisories provide information on contaminants that can cause human health effects and are known or anticipated to occur in drinking water. Health Advisories are guidance values based on noncancer health effects for different durations of exposure (e.g., one day, ten days, and life time). **Table 5** includes footnotes for pollutants with MCLs that are more stringent than the recommended water quality criteria (CMCs or CCCs). MCLs for these pollutants are not included in the compilation, but can be found in the appropriate drinking water regulations (40 CFR 141.11–16 and 141.60–63).

**Table 5** includes also 23 nontoxic pollutants with organoleptic effects for which criteria have been recommended by the US EPA based on the Clean Water Act. Organoleptic effects (e.g., taste and odor) would make water and edible aquatic life unpalatable but not toxic to humans. Pollutants with organoleptic

**Table 6**  Trigger criteria to derive quality standards that refer to specific protection objectives

| Water | Sediment | Biota (secondary poisoning) |
|---|---|---|
| No trigger criterion applies, QS derived for all substances | QS derived for all substances if experimental sediment–water partition coefficient $K_{s\text{-}w} \geq 1000$. If a reliable $K_{s\text{-}w}$ is not available, the octanol–water partition coefficient criterion $\log K_{o\text{-}w} \geq 3$ is used as the trigger value. | Derivation of QS for substances with an experimental bioconcentration factor (BCF) $\geq 100$ or biomagnification factor (BMF) $>1$. If a reliable BCF or BMF is not available, the trigger is $\log K_{o\text{-}w} \geq 3$ |

effect criteria more stringent than the criteria based on toxicity are footnoted in **Table 5**.

The Comprehensive Environmental Response, Compensation, and Liability Act (CERCLA) created the Superfund Program to clean up uncontrolled or abandoned hazardous-waste sites (National Priority List sites, NPLs) and to respond to accidents, spills, and other emergency releases of pollutants and contaminants. Among others, hazardous substances identified in CERCLA include all chemicals on the Clean Water Act list of hazardous substances and priority pollutants. This priority list is revised and published on a 2-year basis. The list from the year 2005 contained 275 substances, which were ranked based on a combination of their frequency, toxicity, and potential for human exposure at NPL sites.

## Hydromorphology Modifications

**General introduction**  Lake hydromorphology concerns aspects of the physical lake environment: a lake's morphology and the quantity and dynamics of the water it contains. It requires specific attention in management strategies as it concerns the physical basis for both the water quality and the ecological status of lakes. This section deals with lakes close to natural condition, the more substantial alterations to hydromorphology found in reservoirs are dealt with in separate chapters.

Lake hydromorphological modifications may be separated into three subcategories:

1. *Hydrodynamic alteration* concerns changes in water quantity and seasonal dynamics and connectivity to rivers and groundwaters. The main pressures are water abstraction or augmentation, watershed land drainage and dams.
2. *Sediment alteration (quality, deposition, and erosion)* can occur through watershed activities such as changes in land use (e.g., afforestation/deforestation, agriculture) as well as through hydrodynamic alteration and eutrophication.
3. *Direct disturbance to a lake or riparian zone* includes building activities such as construction of piers, artificial banks, recreational activities and alteration of the near lake environment: the riparian zone.

**Criteria for assessment**  A list of criteria for the assessment of hydromorphological pressure has been developed in the United Kingdom (**Table 7**). The criteria were chosen to represent the effect of 12 main pressures: impounding and lowering that influence hydrodynamics, modification to the

**Table 7**  List of hydromorphological pressures and criteria (modified from Rowan and Soutar, 2005)

| Pressure | Criteria |
| --- | --- |
| Impounding | Impoundment height (m) |
| | Vertical level range (% or m) change |
| | Level seasonality change (fraction of month) |
| | Area exposed on drawdown (%) |
| | Increased residence time (%) |
| Lowering | Reduction in surface area (%) |
| | Reduction of depth (%) |
| | Change in littoral/profundal ratio (%) |
| | Reduction in residence time (%) |
| Bank construction/ reinforcement | Percent of shoreline constructed |
| | Percent of hard closed shoreline |
| | Percent of hard open shoreline |
| Construction | Percent in-lake area with construction |
| | Density of projections (# per km) |
| | Percent area isolated by causeways etc. |
| | Percent area divided by roads etc. |
| | Percent area fish cages/platforms etc. |
| Dredging/extraction | Percent of shoreline affected |
| | Percent spawning/nursery habitat lost |
| Dumping | Percent of lake area affected by dumping |
| Macrophyte management | Percent of lake area with macrophyte cutting/control |
| Intensive riparian land-use | Percent of shore forestry/arable/urban |
| | Percent of shore without buffer zone |
| | Percent of shore pasture/other grazing |
| | Percent of shore with roads/railways |
| | Percent of shore with grazed moorland |
| Channelization of inflows/outflows | Length of shoreline affected (%) |
| | Loss of in-lake spawning/nursery (%) |
| | Percent of feeder stream modified |
| | Physical/hydrobarrier |
| Modification to sediment regime | Area of unvegetated bars (%) |
| | Area of profundal accretion (%) |
| | Habitat smothering (# plots) |
| | Shoreline erosion extent (%) |
| Recreation | Percent shore trampled/eroded etc. |
| Watershed development | Extent of watershed with urban (%) or other non-natural land-cover % |

sediment regime and many criteria indicating direct disturbance to a lake or riparian zone: bank construction and reinforcement, construction, dredging and extraction, dumping, macrophyte management, intensive riparian land use, watershed development, recreation and channelization of inflows and outflows that affect connectivity. Many of these pressures are interrelated, such as the influence of impounding on the sediment regime. Gathering information on all these criteria requires data collation from maps, satellite images, hydrological databases as well as specific field surveys. A recent development, building on the work done by the Environmental Monitoring and Assessment Program (EMAP) of the US-EPA, is the lake habitat survey (LHS) used in the UK. The LHS involves a field survey of important habitat characteristics: riparian zone vegetation, shore and littoral zone composition and structure, hydrology as well as a visual assessment of pressures to collect information on some of the criteria listed in **Table 7**.

Lake macrophytes are an example of a biological group particularly sensitive to water level regulation and shoreline alteration because a substantial portion of the community is located at the land-water interface and in the shallow littoral area. They are an important biological group in their own right and also provide a habitat for macroinvertebrates, fish, waterfowl and other organisms, thereby forming an important functional component affecting system biodiversity. The regulation of lakes has a significant effect on macrophytes with a loss or decline of sensitive species. However, several studies have found biodiversity to increase as a result of fluctuating lake levels, reducing dominance and promoting species competition or through lower water levels leading to a greater colonizable habitat in shallow lakes. Impacts are lake specific; with the highering or lowering of lake level the zonation of macrophytes may shift with time, but if the depth profile or substrate prove unfavorable it can result in a lower habitable area for some species. In lakes with low light penetration caused by humic substances or phytoplankton abundance, fluctuating water levels may restrict macrophyte growth further. Impacts will also vary depending on the rate of change of the water level as well as its seasonal occurrence and whether it occurs at a sensitive time in the life cycle, such as at flowering, seed dispersal or when macrophytes are seedlings. Changes in sediment composition are one of the principal factors leading to changes in macrophyte communities in lakes and can occur either as a result of hydromorphological or eutrophication pressure and may be prohibitively costly to restore.

Another group of aquatic biota that is highly sensitive to hydromorpphological modifications are fishes. In the short term, alterations in flow and level regimes can modify the distribution, seasonal range and migratory behavior of fish species, with direct effects on fish catches. Restricted access to spawning grounds and sheltered areas in rivers by bank constructions, damming, and lowering the water levels can affect fish recruitment success. Hydroconstructions may partly or totally block the migration routes while low water levels reduce the surface area of spawning grounds and limit access to them. In the long term, variations in the hydrological regime affecting differently the reproductive processes of fish species can influence the fish community composition by changing the relative abundance of the affected species. Only the integrity of the natural flood cycle can ensure the maintenance of a diverse and resilient fish community.

Information gathered on lake hydromorphological criteria must be managed in an effective way to express attainment or failure of management goals. A decision support framework proposed for the UK suggests that each criterion is assigned threshold values depending on the sensitivity of each individual lake. Points for each criterion are then awarded depending on the threshold crossed. The points are then summed to give an 'alteration of lake morphology score,' which expresses the overall extent of lake alteration. This approach may be complemented by separate criteria for the lake water level by establishing an acceptable percentage of deviation away from the natural water level. Such acceptable standards should also be set according to factors affecting a lake's sensitivity, such as light climate, season, altitude, size, basin form and depth.

At the present time the ecological responses to hydromorphological alterations are not fully understood. Therefore, current management approaches rely on information gathered on the observable physical alterations of lake hydromorphology, which are then interpreted in terms of perceived or known relevance to lake use and flora and fauna. Therefore, in the future the criteria selected, their threshold values, and information regarding a lake's sensitivity as well as incorporation into an overall index will need to be analyzed alongside ecological data and be re-evaluated.

## Alien Species

**General introduction** Invasive alien species represent a significant threat to the biodiversity and functioning of lake ecosystems. Alien species have been defined by the conference of the parties to the convention on biological diversity (Decision VI/23) as referring to a species, subspecies or lower taxon introduced outside its natural past or present distribution; it includes any part, gametes, seeds, eggs, or propagules of such species that might survive and subsequently reproduce; whereas an invasive alien species (IAS) means an alien species whose introduction and/or spread threaten biological diversity.

Lakes are vulnerable to colonization as they are a focal point for human activities such as water supply or recreation, thereby increasing the incidence of deliberate or accidental introductions. The connectivity of lakes with rivers and canals greatly facilitates the spread of IAS throughout watersheds with downstream transfer occurring at a faster rate than upstream. The main pathways of introduction of freshwater IAS are through activities associated with aquaculture and recreational fisheries, aquaria and ornamental ponds, organisms used for bait and through the relocation of equipment such as boats and fishing gear. An introduced species may also act as a carrier for accompanying microscopic organisms including pathogens, parasites, and algae.

**Criteria for assessment** There is a current lack of criteria adequately defining the effect of IAS in lakes. **Table 8** suggests potential criteria. These should focus on the early detection and location of IAS to facilitate their eradication or containment. Many assessment criteria such as the extent of displacement of native species, alteration of natural community composition, loss of natural biodiversity as well as alteration of genetic characteristics of populations require quantitative information on a lake's prior ecological status or

possibly from an adjacent uncolonized lake of similar type as a basis for assessment.

The introduction of the spiny-cheek crayfish Orconectes limosus from Pennsylvania (USA) into Europe is an example of an IAS that can have a substantial impact on lake ecosystems. A study on Lake Großer Vätersee in Germany found that the introduced crayfish was playing a substantial role in the lake's food-web and accounted for 49% of the biomass of macroinvertebrates and formed 15 and 48% of the diet of pike and predatory perch. Owing to the lack of historical information, including colonization by native crayfish, it was not possible to estimate the change to the food-web, but the biomass of the IAS relative to other biological groups served as an indicative criterion of their importance.

IAS can have an affect on lake ecosystems of similar magnitude to other pressures such as nutrient enrichment or hydromorphological alteration. Key factors in the management of IAS are risk assessment, preventative measures and early detection followed by

**Table 8** List of criteria that could be indicative of alien species pressure in lakes (many criteria require data from pre-colonization or from an adjacent uncolonized lake of similar type)

| Assessment criteria | Sources |
|---|---|
| Presence in lake (determined using early detection methods: risk assessment or specific targeted monitoring such as traps) | 1 |
| Spatial dispersal of alien species in lake | 1 |
| Extent of displacement of native species | 1 |
| Loss or threat to rare species | 7 |
| Alteration of natural community composition, abundance, function, and interaction | 1, 6 |
| Loss of natural biodiversity | 4, 5 |
| Alteration of natural habitat (including physical and chemical conditions) | 1, 3 |
| Alteration of genetic characteristics of existent species | 2 |
| Disruption to anthropogenic lake use such as abstraction, navigation, recreation, and commercial fisheries | 9 |
| Degree of synergistic action with other pressures such as eutrophication | 3 |
| Risk to upstream and downstream waters | 10 |
| Introduction of transmissible diseases | 8 |

Sources
1. Olenin S, Minchin D, and Daunys D (2007) Assessment of biopollution in aquatic ecosystems. *Marine Pollution Bulletin* 55: 379–394.
2. Perry WL, Feder JL, and Lodge DM (2001) Implications of hybridization between introduced and resident *Orconectes* crayfishes. *Conservation Biology* 15: 656–1666.
3. Urban RA, Titus JE, and Zhu W-X (2006) An invasive macrophyte alters sediment chemistry due to suppression of a native isoetid. *Oecologia* 148: 455–463.
4. European Community (2006) Halting the loss of biodiversity by 2010 – and beyond. Sustaining ecosystem services for human well-being. COM: 216.
5. Sala OE, Chapin FS, Armesto JJ, Berlow E, Bloomfield J, Dirzo R, Huber-Sanwald E, Huenneke LF, Jackson RB, Kinzig A, Leemans R, Lodge DM, Mooney HA, Oesterheld M, LeRoy Poff N, Sykes MT, Walker BH, Walker M, and Wall DH (2000) Global biodiversity scenarios for the year 2100. *Science* 287: 1770–1774.
6. Gherardi F and Acquistapace P (2007) Invasive crayfish in Europe: the impact of *Procambarus clarkii* on the littoral community of a Mediterranean lake. *Freshwater Biology* 52: 1249–1259.
7. Zettlera ML and Daunys D (2007) Long-term macrozoobenthos changes in a shallow boreal lagoon: comparison of a recent biodiversity inventory with historical data. *Limnologica* 37: 170–185.
8. Matthews MA and Reynolds JD (1990) Laboratory investigations of the pathogenicity of *Aphanomyces astaci* for Irish freshwater crayfish. *Hydrobiologia* 203: 121–126.
9. Pimentel D (2005) Aquatic nuisance species in the New York State Canal and Hudson River systems and the great lakes basin: An economic and environmental assessment. *Environmental Management* 35: 692–701.
10. Kolar CS and Lodge DM (2002) Ecological predictions and risk assessment for alien fishes in North America. *Science* 298: 1233–1236.

a rapid response (eradication or containment) following a set of action plans prepared in advance following information from the initial risk assessment.

### Climate Change

**General introduction** Aquatic ecosystems are very vulnerable to climate change. Alterations in climate pose a serious risk to inland freshwater ecosystems and coastal wetlands, and adversely affect numerous services provided to human populations. Impacts on lakes include the increase in nuisance algae and the reduction of fish habitat with the warming of lakes. These impacts will be especially severe in shallow lakes. In addition, there will be changes in runoff-increases and decreases-which will in turn affect lake levels. Changes in land cover and land use in the catchment and stronger thermal stratification of lakes will change nutrient mobility and their availability for plant growth. Finally, distribution areas of aquatic species may expand or contract. In most cases the negative impacts of climate change to aquatic ecosystems cannot be mitigated by management measures in the river basin. As a result of climate change, water bodies, especially those located near the type boundaries may change their type. Compared to typology characteristics, water quality parameters are even more labile and may be easily affected by climate change. The concept of reference conditions is the anchor point of most methodologies for water quality assessment and any adjustment of a reference condition to account for a change climate can be justified only if it

1. does not counteract the main goal of protecting and improving the quality of waters, and
2. prevents the setting of unachievable goals and making unreasonable investments into measures that cannot be effective under the changed climatic boundary conditions.

**Criteria for assessment** So far, no specific climate change criteria have been developed for inland water management due to the high uncertainty of climate change projections and, especially, the poor predictability of ecological responses to climate change. This is due to their high complexity and the novel situations caused by interactions of native and non-native species. Most of the adaptive management measures aimed to protect the inland waters from the adverse effects of climate change represent complimentary approaches as 'prudent and responsible' actions for citizens and policymakers to minimize human pressures on the global and local environment in order to reduce the vulnerability of ecosystems. Prudent actions include: reducing nutrient loading

to lakes and rivers, protecting the quality of water supplies and aquatic habitats, reducing habitat destruction and fragmentation, protecting healthy wetlands and restoring degraded wetlands to enhance nutrient uptake and maintain biodiversity, preventing the spread of invasive non-native species. Adaptations may include: shifts in fisheries management and farming activities, changes in hydraulic regulation schemes to prepare for extreme events, and planning for the impacts of climate change to reduce future damage. New reservoirs should be located off-channel so as not to disrupt the natural downstream flow of water and sediments to critical riverine ecosystems. Groundwater pumping for irrigation, human consumption, etc., removing water from aquatic and wetland ecosystems, should be minimized. In addition to preventing or minimizing environmental impacts, these actions will also result in collateral benefits that include cost savings, cleaner water, improved habitat and recreation, and enhanced quality of life.

### Conclusion

So far lake management is found on criteria derived from the knowledge of how lake ecosystems respond to the impact of single pressures. However, today many lakes suffer from the impact of several pressures, which act simultaneously and interact in a complex manner influencing all levels of the ecosystem. In such cases, the ecological status is varying according to the sensitivity of lake ecosystems and the combination of pressures.

Lake management is further complicated because the reduction/increase of one pressure may alter the fate of the other. Thus, it is necessary to understand the interactive mechanisms between the pressures and the internal ecosystem factors that may influence these mechanisms, e.g., in the case of eutrophication and chemical contamination the mechanisms that may alter bioavailability and the fate of contaminants need to be considered. In this example, the implementation of measures to reduce nutrients would potentially improve the trophic status but could also lead to an increase of contaminant concentrations in the system and higher bioaccumulation of toxicants in the pelagic biota. Improved oxygen conditions at the bottom may lead to recolonization of sediments by benthic fauna but as a result of bioturbation and geochemical processes, contaminants that were previously trapped in the sediments may be mobilized. The level of contaminants may, vice versa, influence the primary production and indirectly the status of eutrophication.

The impact from combined pressures is generally assessed through the response of biological groups such as fish or macroinvertebrates that are sensitive to several pressures, e.g., benthic invertebrates respond sensitively to a number of human impacts (hydrological, climatological, morphological, navigational, recreational, and others) and could potentially be used for a holistic indication of lake ecosystem health. In general, single biological group multimetric indices composed of several metrics sensitive to different pressures are combined in a unique index integrating the response to a wide range of alterations are but of limited use in management.

In summary, the response of freshwater organism groups to single stressors has been described in numerous studies but there are only a limited number of studies on the response of organism groups to multiple-stress situations. Thus, there is a general need to quantify these relationships for the improvement of the current lake ecological models and to support lake management.

In the absence of such tools and to resolve the problem, managers may need to define a hierarchy among the pressures to identify priority actions in the appropriate geographical scale.

The appropriate unit of management of water quality is at watershed level. It is within this context that indicator's criteria relating to single pressures and general pressure indicator's criteria in lakes should be incorporated into a decision support system to allow interpretation regarding their influence on the structure and functioning of lake ecosystems. The efficacy of any restorative or conservative management measures to achieve the goal of sustainable catchment and lake use should be reviewed periodically. The usefulness of criteria should be reevaluated both within context of their ability to reflect pressure and also their relevance to lake biota.

*See also:* Effects of Climate Change on Lakes; Mixing Dynamics in Lakes Across Climatic Zones.

## Further Reading

Ciruna KA, Meyerson LA, and Gutierrez A (2004) *The Ecological and Socio-Economic Impacts of Invasive Alien Species on Inland Water Ecosystems.* Report to the Convention on Biological Diversity on behalf of the Global Invasive Species Programme, Washington, D.C., 34 pp.

COM (2006) 398. *Proposal for a Directive of the European Parliament and of the Council on Environmental Quality Standards in the Field of Water Policy and Amending Directive 2000/60/EC,* 17.7.2006, 25 pp. Brussels: Commission of the European Communities.

Eisenreich SJ (ed.) (2005) *Climate Change and the European Water Dimension.* A Report to the European Water Directors, EU Report No. 21553. Ispara, Italy: European Commission – DG Joint Research Centre. Available at http://ies.jrc.eu.int.

Gibson GR, Carlson J, Simpson E, *et al.* (2000) *Nutrient Criteria Technical Guidance Manual: Lakes and Reservoirs.* Washington, DC: US EPA.

Haertel-Borer SS, Zak D, Eckmann R, *et al.* (2005) Population Density of the Crayfish, Orconectes limosus, in Relation to Fish and Macroinvertebrate Densities in a Small Mesotrophic Lake – Implications for the Lake's Food Web. *International Review of Hydrobiology* 90(5–6): 523–533.

Henriksen A, Posch M, Hultberg H, *et al.* (1995) Critical loads of acidity for surface waters – Can the ANC-limit be considered variable? *Water Air Soil Pollution* 85: 2419–2424.

Jenkins A, Camarero L, Cosby BJ, *et al.* (2003) A modelling assessment of acidification and recovery of European surface waters. *Hydrology and Earth System Sciences* 7: 447–455.

Kling G, Hayhoe K, Johnson LB, *et al.* (2003) Confronting climate change in the Great Lakes region: Impacts on our communities and ecosystems. Union of Concerned Scientists/The Ecological Society of America. Cambridge, MA. Available at. ucsusa.org/greatlakes.

Kuylenstierna JCI, Rodhe H, Cinderby S, *et al.* (2001) Acidification in developing countries: Ecosystem sensitivity and the critical load approach on a global scale. *AMBIO* 30: 20–28.

Nilsson J and Grennfelt P (eds.) (1988) Critical loads for sulphur and nitrogen. Report from a workshop held at Skokloster (NORD 1988:15), Sweden 19–24 March, 1988. Copenhagen, Denmark: Nordic Council of Ministers.

Poff NL, Brinson MM, and Day JW Jr. (2002) Aquatic ecosystems and global climate change: Potential impacts on inland freshwater and coastal wetland ecosystems in the United States. Arlington, VA: Pew Center on Global Climate Change.

Rowan J and Soutar I (2005) *Development of Decision Making Frameworks for Managing Alterations to the Morphology of Lakes.* Edinburgh: Sniffer. www.sniffer.org.uk.

Rowan JS, Duck RW, Carwardine J, *et al.* (2004) *Development of a Technique for Lake Habitat Survey (LHS): Phase 1.* Edinburgh: Sniffer. www.sniffer.org.uk.

Skei J, Larsson P, Rosenberg R, Jonsson P, Olsson M, and Broman D (2000) Eutrophication and contaminants in aquatic ecosystems. *AMBIO* 29: 184–194.

US EPA (1998) *Lake and Reservoir Bioassessment and Biocriteria.* Washington DC: US Environmental Protection Agency.

US EPA (1998) *National Strategy for the Development of Regional Nutrient Criteria.* Washington DC: US Environmental Protection Agency.

Vollenweider RA (1976) Advances in defining critical loading levels for phosphorus in lake eutrophication. *Memorie dell'Istituto Italiano di Idrobiologia* 33: 53–83.

Wiederholm T (1989) Bedömningsgrunder för sjöar och vattendrag. Bakgrundsdokument 1. Näringsämnen, syre, ljus, försurning. *Naturvaralsverket Rappost* 3627.

## Relevant Websites

http://www.epa.gov/.
http://www.atsdr.cdc.gov/.
http://ime.fraunhofer.de/.
http://ec.europa.eu/.

# Lake and Reservoir Management

**E Jeppesen,** University of Aarhus, Denmark
**M Søndergaard,** University of Aarhus, Denmark
**H S Jensen,** University of Southern Denmark, Odense, Denmark
**A-M Ventäla,** Pyhäjärvi Institute, Kauttua, Finland

## Introduction

Freshwater lakes and reservoirs (lakes) constitute a vitally important resource for humans. They supply water for consumption and irrigation, are used for harvesting fish and other food resources, and for recreational activities such as angling, boating, and swimming. Moreover, lakes and to some extent also reservoirs add significantly to biodiversity on Earth and act as important foraging areas for many terrestrial animals and waterfowl.

For fifty years or longer eutrophication has represented the most serious environmental threat to lakes worldwide. High loading of nutrients of lakes has resulted in turbid water, excessive blooms of often toxic cyanobacteria, and loss of biodiversity. The shift towards blooming reflects in part the increasing nutrient loading to and ensuing enhanced nutrient concentration in lakes (reduced resource control also called 'bottom-up control' of phytoplankton). However, eutrophication is also a result of higher predator control (also called 'top-down' control) of invertebrates by fish. Increasing nutrient concentrations leads to a major increase in fish density, a shift in dominance from piscivorous fish to plankti-benthivorous fish and, consequently, a higher predation on the vital large-bodied zooplankton, such as *Daphnia*, and with it less grazing by zooplankton on phytoplankton. Moreover, benthivorous fish stir up sediment, which results in enhanced turbidity in the water column above and less light for growth of submerged plants. An increase in fish predation also reduces snail abundance and thereby also the snail grazing of epiphytes on plants, which further impoverishes the growth conditions for the submerged vegetation. The plants disappear and the food source and feeding habitats of aquatic birds become reduced.

Although many countries in the developing world at present face an alarming increase in lake eutrophication as a result of the fast economic development, major efforts have been made in the Europe and North America to combat eutrophication. Substantial investments have been made to improve wastewater treatment and other pollution-combating measures. However, despite reductions in nutrient loading, eutrophication remains a major problem. Today, the highest pollution input to lakes in the developed world is often derived from diffuse sources in the lakes' catchment areas, mainly from cultivated land, while sewage water and pollution from in-lake fish or crab/shrimp farming often play a more central role in the developing countries.

Restoration of eutrophicated waterbodies will always require reductions of external nutrient loading and often supplementary in-lake measures to speed up the recovery process.

## Similarities and Differences among Shallow Lakes, Deep Lakes, and Reservoirs

Although the total volume of freshwater water in the world is dominated by a few large and deep lakes, most lakes are small and shallow. Shallow and deep lakes exhibit significant differences in trophic structure and dynamics as well as in sensitivity to threats such as that posed by increasing nutrient loading. An essential difference is that deep lakes often show thermal stratification in summer, which largely cuts off the upper water layers (epilimnion) from the colder deep water (hypolimnion), thereby preventing interaction with the sediment.

Reservoirs are human-made lakes, created for storage of water mainly for drinking water supply and hydropower generation. Except for small reservoirs in farmland areas, their catchments are typically large compared with ordinary lake catchments. Accordingly, nutrient and sediment loading to and storage in reservoirs are often larger than in natural lakes. Major differences exist in the biological community structure of lakes and reservoirs, which to a large extent reflect the greater temporal water level changes in reservoirs. Many reservoirs lack a littoral zone because they are typically constructed in places in narrow steep-sided gorges (e.g., the world's largest reservoir, 'Three Gorges Dam,' in China). This implies a limitation of shallow water areas. In addition, the fluctuating water levels of reservoirs impede the development of a plant-rich littoral zone. In natural lakes, in contrast, the littoral zone can have a huge impact on the entire lake ecosystem owing to its use as a refuge by large-bodied zooplankton and small fish and as spawning grounds for fish. Often reservoirs have lower fish recruitment success, which may lead to reduced density of planktivorous fish, enhanced zooplankton grazing on phytoplankton, and clearer water than suggested by the nutrient level (for instance as seen in the London Reservoirs). Fish production may also be lower unless fish stocking is regularly repeated.

Additional measures to sustain recovery after a loading reduction vary between deep and shallow lakes and between lakes and reservoirs. Here, we will treat lakes and reservoirs together and only highlight differences in approaches for the specific water types when relevant. We will focus on measures to alleviate eutrophication and not include restoration measures applied to combat, for instance, acidification.

## Measures Taken to Reduce the External Loading

The key restoration target for all lakes is to reduce the external nutrient loading. Unless the loading has been reduced sufficiently to create a shift to a good ecological state in the long term, additional in-lake measures can only be considered as symptom treatments. Several simple empirical models have been developed to predict total phosphorus (TP) and total nitrogen (TN) concentrations in lakes on the basis of information on external TP and TN loading and between ecological state variables and lake TP and TN concentrations. Such models are useful tools for setting critical loading targets for the different types of lakes. Phosphorus (P) is invariably considered the most critical nutrient for the ecological state of lakes, though recent studies suggest nitrogen (N) to be of higher importance than previously assumed.

A multi-approach is often needed to reach a sufficiently low external nutrient loading to lakes to achieve good ecological state in populated areas. Approaches include P stripping and occasionally N removal at sewage works, sewage diversion, increased use of phosphate-free detergents, establishment of regulations concerning animal fertilizer storage capacity, fertilizer application practices, fertilization plans, and green cover in winter. In addition, various measures can be implemented to enhance the nutrient retention and N loss capacity in lake catchments by re-establishing wetlands, stabilizing river banks to reduce erosion, re-establishment of a natural riparian zone, and by allowing flooding of riverine areas.

## Response to Nutrient Loading Reduction

The lakes do not always respond immediately to a nutrient loading reduction; persistence of eutrophication may be attributed to continuously high external loading preventing a shift to a clearer state. However, even when the P loading has been sufficiently reduced, resistance against improvements occurs. This resistance may be 'chemical': the P concentrations remain high due to release of P from nutrient pool accumulated in the sediment at high loading (**Figure 1**). It typically takes 10–15 years, but sometimes longer, before this surplus pool of P in the lake sediment is released or permanently buried, the duration depending both on the thickness of the nutrient enriched sediment layer, the nature of P binding sites in the sediment, and on the flushing rate in summer when P accumulates in the lake water.

Biological resistance also affects internal P loading and the physicochemical environment. Particularly planktivorous and benthivorous fish contribute to biological resistance. Continuously high fish predation prevents both the appearance of large herbivorous zooplankton, and thus a higher grazing on phytoplankton, and diminishes the number of benthic animals that stabilize and oxidize the sediment. However, recent studies have shown surprisingly fast (within less than 5–10 years) reductions in fish biomass and increases in the proportion of piscivores when both external and internal nutrient loading is efficiently controlled. Grazing by herbivorous waterfowl as coot (*Fulica atra*) and mute swan (*Cygnus olor*) or lack of plant seeds or turions may also create resistance by delaying the recolonization of submerged macrophytes.

## Reinforcing Recovery

To reinforce recovery, numerous physicochemical and biological restoration methods have been developed. Here we present the most frequently applied methods and an overview of some of the syntheses and notable case studies on each method (**Table 1**). Informative books or overview papers on the response of lakes to nutrient loading reduction and restoration are listed in 'Further reading' section.

### Physico-Chemical Methods

Various physicochemical methods have been used to reduce internal P loading. These include sediment removal in shallow lakes and chemical treatment of the sediment with alum, calcium, or iron salts in both deep and shallow lakes. In deep lakes, injections with oxygen or nitrate to the bottom layer or continuous destabilization of the thermocline by effective circulation of the water column have been employed. Also withdrawal of hypolimnion water or flushing has been used in several case studies.

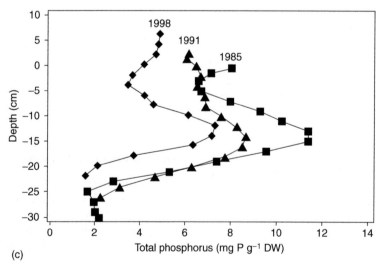

**Figure 1** Ratio of annual mean total phosphorus (TP) concentration measured and predicted in the surface water of lakes at maximum nutrient loading and 5, 10, 15, and 20 years after loading reduction (a); hydraulic retention time ($t_w$) and mean depth ($Z_{mean}$) of the lakes (b); and slope (mean ± SE) of linear regressions (forces through the origin) of observed versus predicted annual mean TP concentration in different years following loading reduction in shallow (c) and deep lakes (d). The slopes for 5 and 10 years in shallow

**Sediment removal by dredging** Dredging is an efficient, but a relatively costly technique to reduce an internal loading problem when the right equipment is used (cutterhead suction dredging is often efficient: **Figure 2**). It may also serve to deepen a lake that is gradually filling in. Due to the potential content of toxic substances in the sediment a key problem of dredging is how to dispose of the sediment. Other concerns are to obtain sufficient storage capacity during dredging, disturbance of wildlife during the process and release of toxic substances. Also, redistribution of sediment during the dredging period may take several years depending on lake size. Typically, the upper 20–60 cm is removed, sometimes modified so that more sediment is removed from TP 'hot spot' areas.

Sediment removal may also be used following an entire or partial water-level drawdown. Removal of sediment has in many cases reduced the internal loading immediately and substantially – a pioneering and notable example is that of Lake Trummen in Sweden. However, long-term success has frequently been hindered by a continuously too high external loading. Sediment removal is probably the most reliable method to reduce internal loading, as it permanently removes the source of nutrient loading. Other methods designed to eliminate internal loading have often proved less stable in the long term. Dredging is most useful in shallow lakes and reservoirs, in the latter preferably combined with water level drawdown. However disposal of dredged sediment may be a problem at places as the concentration of various toxic substances may exceed critical levels.

**Hypolimnetic withdrawal and flushing** Hypolimnetic withdrawal aims at removing nutrient-enriched hypolimnion water to maximize nutrient concentrations exported in the lakes outlet. Coincidentally, the retention time in the hypolimnion is shortened, reducing the risk of oxygen depletion. Typically, a pipe is installed in the deepest part of the lake with an outlet downstream below the lake level enabling the pipe to act as a siphon. Destratification and lake level reduction should be avoided by using a low outflow rate from the pipe. The method has generally generated good results. In most cases, hypolimnion TP decreases and so does the depth of anoxia layer, leading to reduced internal loading. After some years also the epilimnion TP will typically decline, the effect being stronger the longer the pipe has been in action. The drawback of the method is potential pollution of downstream systems (high loading of TP, ammonia, low oxygen concentration and occasionally an obnoxious smell of $H_2S$). The most notable case is Lake Kortowo, Poland, where hypolimnion withdrawal has been conducted since 1956 (it was the first example) and is still ongoing due to still too high external loading. The method is particularly easy to apply in reservoirs where withdrawal can be established at the dam, but is also useful in summer stratified lakes. An alternative method is flushing of lake water with water, low in nutrient or rich in binding substances like iron or calcium, which are added at the time when the lake water concentration is high. Examples are Green Lake, Seattle, USA and Lake Veluwe, The Netherlands.

**Aluminum, iron, and calcium treatment** This method aims at supplying new sorption sites for phosphate onto the surface sediment. Phosphate adsorbs readily to calcite ($CaCO_3$) and hydroxides of oxidized iron ($Fe^{3+}$) and aluminum ($Al^{3+}$). Phosphate precipitation with calcite has been used in hardwater bodies. An example is Frisken lake, British Columbia, Canada. However, the method appears unpredictable because pH often drops to below 7.5 in the sediment at which level calcite is dissolved. Phosphate adsorption onto $Fe^{3+}$ and $Al^{3+}$ is widely used to precipitate P in waste water treatment plants. Here, a molar (e.g., Al:P) precipitation ratio of 1:1 can be obtained due to high concentrations of phosphate when the ions are added, but in lakes the metal ions will normally form hydroxides with a lesser binding capacity for

lakes and after 5 years in deep lakes were significantly different from the slopes for years 0 and 15 and 10 years in deep lakes (A). From Jeppesen et al., 2005 (full reference in Further Reading). Conceptual scheme describing the changes in retention and internal loading of phosphorus in a shallow lake with changing external loading. Different thickness of arrows indicates different loading and release rates (B). Sediment profiles of total phosphorus ($TP_{sed}$) in Lake Søbygaard in 1985 (square), 1991 (triangle), and 1998 (star) (C). The sediment phosphorus profile of Lake Søbygaard changed markedly during 13 years. In the upper 25–30 cm of the sediment, $TP_{sed}$ has decreased at all depths. During the first 6 years, phosphorus was primarily released from the very high concentrations found at 15–20 cm depth, but during the latter 7 years $TP_{sed}$ has decreased at all depths. At most depths down to 25–30 cm, $TP_{sed}$ has been reduced by 3–4 mg P $g^{-1}$ DW. Calculations based on comparisons of the 1985 and 1998 profiles show that a total of 57 g P $m^{-2}$ has been released from the upper 20 cm sediment. In the same period, mass balance measurements show a total release of approximately 40 g P $m^{-2}$. At the present release rate this means that another 15–20 years will pass before the lake will be in equilibrium, implying that the transient phase after reduced external loading in total will last for more than 30 years. Sediment cores were sampled from a central location and sectioned into 2-cm slices. Sediment from three different cores was pooled into one sample before analysis. Sediment profiles were adjusted to the 1985 level using a sedimentation rate of 0.6 cm years$^{-1}$. From Søndergaard M, Jensen JP, and Jeppesen E (1999) Internal phosphorus loading in shallow Danish lakes. Hydrobiologia 408/409, 145–152.

**Table 1** Some key references to papers on the different restoration methods discussed in the paper

| Methodology | Reference |
|---|---|
| *Physical methods* | |
| Water level drawdown | Beard TD (1973) *Owerwinter drawdown. Impact on the aquatic vegetation in Murphy Flowage, Wisconsin*. Technical Bulletin No. 61. Wisconsin Department of Natural Resources, Madison. |
| | Ploskey GR, Aggus LR, and Nestler JM (1984) *Effects of water levels and hydrology on fisheries in hydropower storage, hydropower mainstream and flood control reservoirs*. Technical Report E-84-8, U.S. Army Engineer Waterways Experiment Station, Vicksburg, MS., NTIS No. AD A146 239. |
| | Randtke SJ, de Noyelles F, Young DP, Heck PE, and Tedlock RR (1985) *A critical assessment of the influence of management practices on water quality, water treatment and sport fishing in multi-purpose reservoirs in Kansas*. Kansas Water Resources Research Institute, Lawrence. |
| Hypolimnion withdrawal or flushing | Dunalska JA, Wiśniewski G, and Mientki C (2007) Assessment of multi-year (1956–2003) hypolimnetic withdrawal from Lake Kortowskie, Poland. *Lake and Reservoir Management* 23, 377–387. |
| | Hosper H (1985) Restoration of Lake Veluwe, The Netherlands by reduction of phosphorus loading and flushing. *Water Science and Technology,* 17, 757–768 |
| | Lathrop RC, Astfalk TJ, Panuska JC, and Marshall DW (2004) Restoring Devil's Lake form the bottom up. *Wisconsin Natural Resources* 28, 4–9. |
| | McDonald RH, Lawrence GA, and Murphy TP (2004) Operation and evaluation of hypolimnetic withdrawal in a shallow eutrophic lake. *Lake and Reservoir Management* 20, 39–53. |
| | Nürnberg GK (1987) Hypolimnetic withdrawal as lake restoration technique. *Journal of Environmental Engineering* 113, 1006–1016. |
| Destratification | Fast AW (1973) Effects of artificial destratification on primary production and zoobenthos of El Capitan Reservoir, California. *Water Resources Research* 9, 607–623. |
| | Gâchter R (1987) Lake restoration. Why oxygenation and artificial mixing cannot substitute for a decrease in the external phosphorus loading. *Schweizerische Zeitung für Hydrologie* 49, 170–185. |
| | Pastorak RA, Lorenzen MW, and Ginn TC (1982) *Environmental aspects of artificial aeration and oxygenation of reservoirs: a review of theory, techniques and experiences*. Technical Report No. E-82-3. U.S. Army Corps of Engineers, Vicksburg, MS. |
| Sediment removal | Björk S (1974) *European Lake Rehabilitation Activities*. Institute of Limnology Report. Sweden: University of Lund. |
| | Hinsman WJ and Skelly TM (1987) *Clean lakes Program Phase 1 Diagnostic/Feasibility Study for the Lake Springfield Restoration* Plan. Springfield City Water, Light and Power, Springfield, IL. |
| *Oxygenation* | |
| Air/oxygen | Fast AW, Moss B and Wetzel RG (1973) Effects of artificial aeration on the chemistry and algae of two Michigan lakes. *Water Resources Research* 9, 624–647. |
| | Gächter R and Wehrli B (1998) Ten years of artificial mixing and oxygenation: No effect on the internal P loading of two eutrophic lakes. *Environmental Science and Technology* 32, 3659–3665. |
| | McQueen DJ and Lean DRS (1984) Aeration of anoxic hypolimnetic water: effect on nitrogen and phosphorus concentrations. *Verhandlungen der Internationale Vereinigung der Limnologie* 22, 268. |
| | Pastorak RA (1981) Prey vulnerability and size selection by *Chaoborus* larvae. *Ecology* 62, 1311–1324. |
| Nitrate | Noon TA (1986). Water quality in Long Lake, Minnesota, following Riplox sediment treatment. *Lake and Reservoir Management* 2, 131. |
| | Ripl W (1976) Biochemical oxidation of polluted lake sediment with nitrate – A new restoration method. *Ambio* 5, 132–135. |
| | Ripl W (1986) Internal phosphorus recycling mechanisms in shallow lakes. *Lake and Reservoir Management* 2, 138. |
| | Søndergaard M, Jeppesen E, and Jensen JP (2000) Hypolimnetic nitrate treatment to reduce internal phosphorus loading in a stratified lake. *Lake and Reservoir Management* 16, 195–204. |
| *Chemical methods* | |
| Iron | Boers PJ, Vand der Does J, Quaak M, and Van der Vlugt J (1994) Phosphorus fixation with iron(III) chloride: A new method to combat internal phosphorus loading in shallow lakes? *Archiv für Hydrobiologie* 129, 339–351. |
| | Hayes CR, Clark RG, Stent RF, and Redshaw CJ (1984) The control of algae by chemical treatment in a eutrophic water supply reservoir. *Journal of the Institute of Water and Engineering Science* 38, 149–162. |
| | Young SN, Clough WT, Thomas AJ, and Siddall R (1988) Changes in plant community at Foxcote Reservoir following use of ferric sulphate to control nutrient levels. *Journal of the Institute of Water and Engineering Science* 2, 5–12. |

Continued

**Table 1** Continued

| Methodology | Reference |
|---|---|
| Aluminum | Barko JW, James WF, Taylor WD, and McFarland DG (1990) Effects of alum treatment on phosphorus and phytoplankton dynamics in Eau Galle Reservoir: A synopsis. *Lake and Reservoir Management* 4, 63–72. |
| | Rydin E, Huser B, and Welch EB (2000) Amount of phosphorus inactivated by alum treatments in Washington lakes. *Limnology & Oceanography* 1, 226–230. |
| | Jernelov A (Ed.) (1971) Phosphate reduction in lakes by precipitation with aluminum sulphate. In: *5th International Water Pollution Research Conference*, San Francisco, CA, 26 July–1 May 1970. New York: Pergamon Press. |
| | Reitzel K, Hansen J, Andersen FØ, Hansen KS, and Jensen HS (2005) Lake restoration by dosing aluminum relative to mobile phosphorus in the sediment. *Environmental Science and Technology* 39, 4134–4140. |
| | Smeltzer E, Kirn RA, and Fiske S (1999) Long-term water quality and biological effects of alum treatment of Lake Morey, Vermont. *Lake and Reservoir Management* 15, 173–184. |
| | Welch EB and Cooke GD (1999) Effectiveness and longevity of phosphorus inactivation with alum. *Lake and Reservoir Management* 15, 5–27. |
| *Biological methods* | |
| Fish removal/stocking | Benndorf J (1995) Possibilities and limits for controlling eutrophication by biomanipulation. *Internationale Revue der gesamten Hydrobiologie* 80, 519–534. |
| | Drenner RW and Hambright KD (1999) Review: Biomanipulation of fish assemblages as a lake restoration technique. *Archiv für Hydrobiologie* 146, 129–165. |
| | Gulati RD, Lammens EHHR, Meijer M-L, and van Donk E (1990) Biomanipulation, tool for water management. *Hydrobiologia* 200/201, 1–628. |
| | Hansson L-A, Annadotter H, Bergman E, Hamrin SF, Jeppesen E, Kairesalo T, Luokkanen E, Nilsson P-Å, Søndergaard M and Strand J (1998) Biomanipulation as an application of food-chain theory: constraints, synthesis, and recommendations for temperate lakes. *Ecosystems* 1, 558–574. |
| | Lazzaro, X. (1997). Do the trophic cascade hypothesis and classical biomanipulation approaches apply to tropical lakes and reservoirs? *Verhandlungen Internationale Vereinigung der Limnologie* 26, 719–730. |
| | McQueen DJ (1998) Freshwater food web biomanipulation: A powerful tool for water quality improvement, but maintenance is required. *Lake and Reservoir Management* 3, 83–94. |
| | Mehner T, Benndorf J, Kasprzak P, and Koschel R (2002) Biomanipulation of lake ecosystems: Successful applications and expanding complexity in the underlying science. *Freshwater Biology* 47, 2453–2465. |
| | Meijer ML, De Boois I, Scheffer M, Portielje R, and Hosper H (1999) Biomanipulation in Shallow Lakes in the Netherlands: An Evaluation of 18 Case Studies. *Hydrobiologia* 409, 13–30. |
| | Perrow MR, Meijer M-L, Dawidowicz P, and Coops H (1997) Biomanipulation in shallow lakes: State of the art. *Hydrobiologia* 342/343, 355–365. |
| | Shapiro J (1979) The need for more biology in lake restoration. In: *U.S. Environmental Protection Agency National Conference on Lake Restoration*. USEPA 440/5–79–001. pp. 161–167. |
| Macrophyte protection/ transplantation | Cooke GD, Welch EB, Peterson SA and Nichols SA (2005) *Restoration and management of lakes and Reservoirs*, 3rd edn. Boca Raton, FL: Taylor & Francis. |
| | Lauridsen TL, Sandsten H, and Møller PH (2003) The restoration of a shallow lake by introducing *Potamogeton* spp. The impact of waterfowl grazing. *Lakes & Reservoirs: Research and Management* 8, 177–187. |
| | Smart RM, Dick GO, and Doyle RD (1998) Techniques for establishing native aquatic plants. *Journal of Aquatic Plant Management* 36, 44–49. |
| | Søndergaard M, Bruun L, Lauridsen TL, Jeppesen E, and Vindbæk Madsen T (1996) The impact of grazing waterfowl on submerged macrophytes. *In situ* experiments in a shallow eutrophic lake. *Aquatic Botany* 53, 73–84. |
| Macrophyte removal | Engel S (1990) Ecological impacts of harvesting macrophytes in Halverson Lake, Wisconsin. *Journal of Aquatic Plant Management* 28, 41–45. |
| | Pieterse A and Murphy K (Eds.) (1990) *Aquatic Weeds. The Ecology and Management of Nuisance Aquatic Vegetation*. Oxford, UK: Oxford University Press. |

phosphate. The formation of tri-valent hydroxides from reaction with water results in a marked drop in lake water pH, depending on the alkalinity level. Indeed, lake water alkalinity limits the amount of acid salts that can be added since pH has to be in the range of 5.5–9 to obtain hydroxide floc formation. Thus, in most cases of aluminum treatment of lakes the dose has been determined by the lake water alkalinity to avoid a pH drop to below 6 (for safety reasons).

**Figure 2** A cutterhead suction dredging machine 'Mudcat' used for dredging 500 000 m$^3$ sediment from shallow Lake Brabrand, Denmark.

Of the two hydroxides, iron has the higher affinity for phosphate, and P release from oxic sediment surfaces is often controlled by $Fe^{3+}$ when present in a molar ratio higher than 8:1 relative to potentially mobile phosphate. Meanwhile $Fe^{3+}$ is reduced to $Fe^{2+}$ in oxygen-depleted sediments and, so far, all documented examples of iron addition have shown a time-limited effect. Also, $Fe^{3+}$ is a high quality electron acceptor for bacterial respiration (as $NO_3^-$ and $O_2$) which may stimulate mineralization, releasing organic P that would otherwise be buried in the reducing sediment.

Contrary to $Fe^{3+}$, $Al^{3+}$ is stable independent of redox conditions. Only, the 'pH window' is narrower (6–9) due to formation of soluble hydroxides at lower or higher levels. Therefore, aluminum should normally not be applied to sediments that may be resuspended in epilimnetic water where pH is high. A fresh $Al(OH)_3$ floc strips lake water for phosphate while sinking to the bottom, but with ageing the floc loses some 75% of its binding capacity and affinity for phosphate due to crystal (gibbsite) formation. Still, it seems that permanent immobilization of P can be reached with a 10:1 ratio of Al and P. Al addition has been used for restoration in some 120 lakes in USA and Europe and the longevity of positive effects varies from a few to 10–15 years. Factors that may influence P adsorption onto $Al(OH)_3$ are high concentrations of humic acids and silicate (e.g., in pore-water). Still, distinct layers of Al and aluminum-bound P can be found in the sediment decades after treatment and the longevity of improved conditions is, therefore,

most likely limited by continuously high external P-loading, for example from diffuse run-off.

Of the three chemicals listed here, treatment with aluminum (**Figure 3.**) seems most promising and at the same time it provides a cost-efficient solution compared with sediment dredging. Toxic effects of the $Al^{3+}$ ion in treated lakes have only been reported in cases where the application technique failed and pH dropped below 6 during the treatment. If chemical treatment is to be successful it generally requires that the main problem is a very limited wash out of excess P during internal loading events and that the overall annual external P-loading is low; otherwise, the new sorption sites added to the sediment will be rapidly saturated.

**Hypolimnetic oxidation** This method employs addition of oxidizers to the hypolimnion to improve the redox sensitive sorption of phosphate to iron in the sediment and thereby reduce internal P loading. Most often oxygen is used (an example in **Figures 4 and 5**), but alternatively electron acceptors such as nitrate can be applied. The presence of both oxygen and nitrate will maintain a redox potential above approximately 200 mV where iron is in its oxidized form ($Fe^{3+}$). Both oxygen and nitrate can oxidize reduced iron species present in the sediment by microbial metabolism. Oxygenation may also serve to improve the living conditions for fish and invertebrates.

Oxidation has been conducted using different techniques: oxygen is added either as pure oxygen or as atmospheric air. Oxygen or atmospeheric air may be

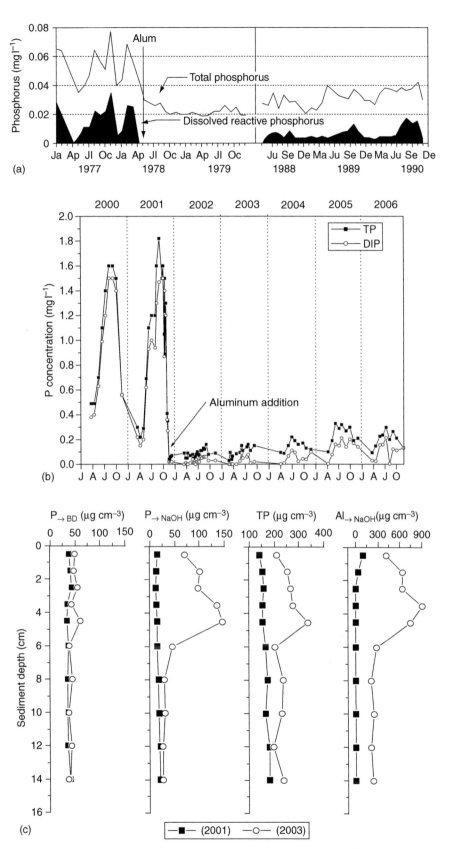

**Figure 3** Examples of aluminium treatment. In Shadow Lake (WI, USA) aluminium (5.7 g Al m$^{-3}$) was applied to hypolimnetic water in 1978. The dose was calculated from lake water alkalinity and resembled approx. 11 g Al m$^{-2}$. Concomitantly external P loading was reduced by 58%. TP declined instantly from 0.06 to 0.02 mg l$^{-1}$ and phosphate from 0.02 to 0 mg l$^{-1}$. In 1988 concentrations had

injected via a number of diffusers creating fine bubbles at the deepest part of the lake or hypolimnetic water may be pumped on land where it is oxidized before being returned to the deep waters. Also, an 'oxidation tower' that circulates hypolimnetic water while oxidizing it has been used in German lakes. When using nitrate, a liquid solution of nitrate is added by stirring it into the upper sediment layer or

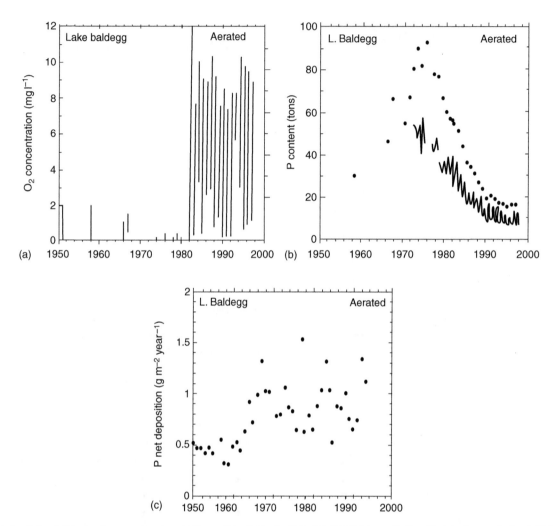

**Figure 4** Annual changes in oxygen concentrations close to the bottom of Lake Baldegg, Switzerland, before and after initiation of oxygenation in 1982 (upper). Below are the changes in P content at overturn (dots) and seasonal changes of hypolimnetic ($z > 20$ m) P content (lines) and annual P deposition rates derived from freeze core analyses. From Gachter R and Wehrli B ( 1998 ) Ten years of artificial mixing and oxygenation: No effect on the internal phosphorus loading of two eutrophic lakes. Environmental Science & Technology 32: 3659–3665.

increased to around 50% of the pre-treatment values, likely reflecting the reduced external loading (a). Reproduced from Cooke *et al*. (2005) (full reference in Further Reading). In Lake Sønderby (Denmark), the aluminium dose was calculated from potentially mobile P (P in lake water and loosely adsorbed P, iron-bound P, and NaOH-extractable organic P in the surface 10 cm of the sediment) by assuming that a 4:1 Al:P ratio would immobilize the pool. The dose was 31 g Al m$^{-2}$. TP declined from 1.60 to 0.08 mg l$^{-1}$ and phosphate from 1.50 mg l$^{-1}$ to 0 in the year after treatment. TP and seasonal internal loading increased slowly during the following 5 years and reached a TP value of 0.3 mg l$^{-1}$ in summer 2006. Lake Sønderby had 15 times higher lake water P concentrations than Shadow Lake and was probably under-dosed since the ratio of Al:P should have been 10:1. Shadow Lake probably received a higher dose than needed to immobilize the P pool in water and sediment, but burial of the reactive aluminium may limit the effectiveness for trapping new P to a few years. Slowly, a new surface sediment with a reactive P pool built up in both lakes as new P is supplied from external sources and (at least in Lake Sønderby) from deeper sediment layers (b). Sequential chemical extraction is used to quantify P pools and reactive Al and Fe in sediments. Depth distributions of iron-bound P (P$_{BD}$), aluminium-bound P (P$_{NaOH}$), total sediment P (TP), and eactive aluminium (Al$_{NaOH}$) in Lake Sønderby demonstrate accumulation of aluminium-bound P in the surface 4 cm following the aluminium treatment in 2001. No change in iron-bound P has been observed (c). Reproduced from Reitzel *et al*. ( 2005 , full reference in **Table 1**).

**Figure 5** Oxygenation equipment used in various lakes in Denmark.

by injecting it into the water just above the sediment. The advantage of using nitrate instead of oxygen is that much higher oxidation equivalents can be added because being a salt nitrate is more soluble than oxygen (a gas). Thus, nitrate will penetrate the sediment to a greater depth than oxygen. In contrast to oxidation of bottom water by oxygen, the objective of using nitrate is not only to oxidize the surface sediment, but also to increase the turnover of organic matter via denitrification and, with it, rapidly reduce the sediment oxygen consumption and increase its P binding potential.

The results obtained for oxidation are variable. While some oxidation experiments have clearly led to lower accumulation of P and reduced elements (e.g., ammonium) in the hypolimnion, others showed neither reduced release of P from the sediment nor its higher retention due to increased hypolimnetic dissolved oxygen concentrations. The explanation is that oxygenation enhances P retention only if the sulphide production is lowered and more ferrous phosphate (e.g., vivianite) and less FeS are deposited in the anoxic sediment. Also, oxygenation needs to be conducted for many years. Basically, hypolimnetic oxygenation treats the symptoms rather than the underlying and fundamental problems, namely the sediment's large consumption of oxygen and the release of

P from a mobile pool. Only when the redox conditions are permanently improved or the P pool is buried deep down the sediment, can the sediment oxygen consumption be expected to decline, leading to increasing and more permanent P retention. The method is potentially applicable to eutrophic deep and stratified lakes.

**Water level alterations** Water level management has been used extensively as a tool to improve the habitats for waterfowl and promote game fishing and water quality. The ultimate regulation is a complete draw-down, which has been used to control nuisance plant growth. It may also facilitate a shift to clear-water conditions in nutrient-rich turbid lakes, at least in the short term, as drying out may consolidate the sediment. Moreover, fish kills mediated by the draw-down enhance zooplankton grazing on phytoplankton, which in turn improves water clarity and thus growth conditions for submerged macrophytes and food resources for waterfowl. Changes in the water level may also influence lakes indirectly by affecting fish recruitment. Lack of flooding of marginal meadows in spring has been suggested as an important factor for poor recruitment of pike (*Esox lucius*) in regulated lakes. Short-term partial draw-down has been used to improve game fishing, which enhances the biomass and size of predatory fish at the expense of planktivorous and benthivorous fish. This may be because the lower water table augments the predation risk or that it dries the fertilized eggs in their early ontogeny. Water level regulation as a restoration tool may be particularly useful in shallow lakes and in reservoirs.

## Biological Methods

In recent years a number of biological methods (often termed *biomanipulation* in the literature) have been developed of which especially fish manipulation has been widely used.

**Fish manipulation**   Various methods have been used to overcome biological resistance. The typical measure is removal of plankti-benthivorous fish. This method has been extensively used in north temperate lakes in Europe, such as The Netherlands and Denmark, during the past 20 years. Removal of approx. 75% of the planktivorous and benthivorous fish stock during a 1–2 year period has been recommended to avoid regrowth and to stimulate the growth of potentially piscivorous fish (**Figure 6**). A simple, feasible strategy of fish removal is to *catch passive fish with active gear and active fish with passive gear* using information on the seasonal behavior of fish, such as spawning or foraging migration and shoaling of the target species. An alternative or supplementary method to fish removal is stocking of piscivores. The basic tools are stocking with nursery or pond-raised fingerlings, catch regulations and habitat management. Stocking with pelagic species (zander, *Stizostedion lucioperca*, walleye, *Sander vitreum*) should be accompanied by littoral species (pike, largemouth bass, *Micropterus salmoides*). Otherwise, the nursery areas of the target species are not affected. Stocking of zander or walleye should be accompanied by catch and mesh size limits for fishing.

Dramatic, cascading short-term effects are generally achieved in eutrophic lakes and reservoirs with efficient fish reduction in the form of reduced phytoplankton biomass, dominance by large sized zooplankton and improved transparency (**Figures 6** and **7**). The effects of fish manipulation may cascade to the nutrient level as well. From 30% to 50% reduction in lake concentrations of TP has been recorded in the relatively more successful fish manipulation experiments in shallow and stratified eutrophic lakes. So far, the long-term perspectives are less promising. A gradual return to the turbid state and higher abundance of zooplanktivorous fish after 5–10 years have been reported in many case-studies; however, also long term or permanently lower abundance of benthivores, such as bream (*Abramis brama*), has occurred, resulting in improved water clarity in the long term. It

is therefore recommended to repeat the fish removal at intervals to maintain the clear-water state, expectedly involving a gradual diminishing of efforts. A drawback of the method is that the lakes retain more P in the clear-water state after manipulation and therefore have more P available for internal loading if the system returns to the turbid state than if fish manipulation had not been conducted. Stocking of piscivorous fish has often been less successful than fish removal.

Fish manipulation is potentially useful for all lakes and is probably particularly effective in the temperate zone when TP has reached concentrations below $0.05\,\mathrm{mg\,P\,l^{-1}}$ in shallow lakes and below, say, $0.02\,\mathrm{mg\,P\,l^{-1}}$ in deeper lakes and reservoirs. However, the exact values may vary along a gradient in climate and may also depend on the external N loading.

**Protection of submerged macrophytes and transplantation**   Construction of exclosures to protect macrophytes against waterfowl grazing has been employed as an alternative or supplementary restoration tool to fish manipulation. The exclosures enable the macrophytes to grow in a grazer-free environment from where they may spread seeds, turions or plant fragments augmenting colonization. Moreover, the exclosures also serve as a daytime refuge from fish predation for the zooplankton. The usefulness of plant refuges as a restoration tool is probably particularly high in small lakes although 11 ha enclosures have been established in the large Lake Wuli in China. Transplantation of plants or seeds is an alternative method. The methods described are particularly useful in shallow lakes.

**Combating nuisance plant growth**   Although re-establishment of submerged macrophytes is the goal of many a lake restoration project, dense plant beds appearing in nutrient-enriched lakes may occasionally be considered a nuisance since they impede navigation and reduce the recreational value for anglers. Moreover, excessive growth of invading species, like the Eurasian milfoil, *Myriophyllum spicatum*, *Pistia stratiotes*, or *Eichhornia crassipes* in many lakes in the US, South America, and Africa, or the North American *Elodea canadensis* in Europe, may substantially alter lake ecosystems and constitute a serious threat to the native flora and fauna. Methods to combat such nuisance plant growth are manual harvesting, introduction of specialist phytophagous insects such as weevils or herbivorous grass carps (*Ctenopharyngon idella*), water level draw-down, coverage of the sediment with sheets, or chemical treatment with herbicides. Often, harvesting and water level draw-down have only a transitory effect because of fast regrowth of the plant community and

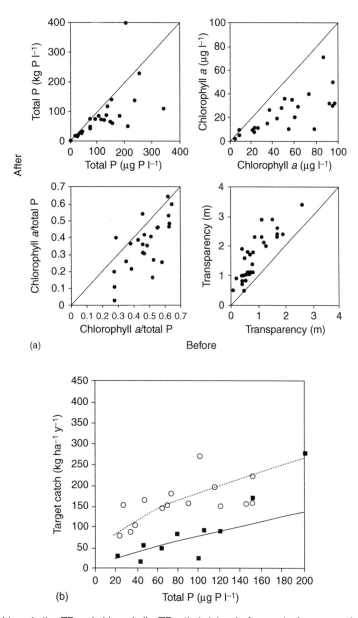

**Figure 6** Transparency, chlorophyll *a*, TP and chlorophyll *a*/TP ratio in lakes before and a few years after an effective fish removal (upper). Below the amount of fish removed during biomanipulation of eutrophic European lakes dominated by planktivorous and benthivorous fish. White circles denote the annual catch in cases with effective fish removal leading to an improvement in water quality (increased transparency and decline in biomass of cyanobacteria) at least in the short term or an increase in piscivorous perch. Black dots indicate lakes in which fish removal was too limited to have an effect on water quality or fish density. Data compiled by Jeppesen and Sammalkorpi ( 2002 ). In: Perrow MR and Davy AJ *Handbook of Ecological Restoration, Vol. 2: Restoration in practice*, pp. 297–324.

high external loading. Grass carp may have a strong effect on plant growth and is currently used in many parts of the world to reduce macrophyte abundance, but a shift to a turbid state is a typical side effect (**Figure 8**). The method should therefore be used with caution. Moreover, before planning plant removal one has to bear in mind that these plants generally have a positive effect on the water clarity and biodiversity of the lakes.

**Mussels and lake restoration** Mussels are efficient filter feeders in lakes. Large unioids like *Anodonta*, *Unio* and *Hyridella* are sometimes abundant in well-mixed macrophyte-dominated lakes and can filter the entire water volume in a few days. They often disappear, however, in turbid lakes probably due to predation by fish larvae. Re-introduction of these species may therefore be a useful tool but it has so far received little attention. Also, the zebra mussel,

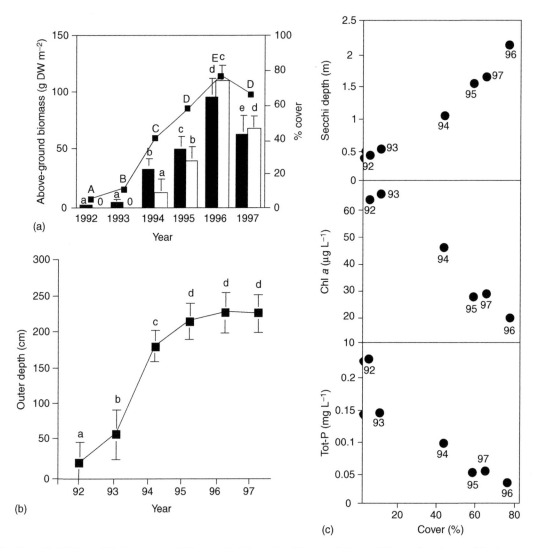

**Figure 7** Mean (± 1 SD, $n = 10$) above-ground biomass (<1.7 m = black bars, >1.7 m = white bars) and cover (black squares) of the total submerged vegetation in areas above 3 m water depth at 10 sites in Lake Finjasjoen (12 km² surface area), Sweden during the period 1992–97. Fish removal (393 kg ha⁻¹) was conducted during 1992– 1994 (a). Bars or squares that share a common letter are not significantly different. Mean (± 1 SD, $n = 10$) outer depth of the submerged vegetation at 10 sites during the period 1992–97 (b). Squares that share a common letter are not significantly different. Relation between two submerged macrophyte variables (above-ground biomass and cover), and Secchi depth, chlorophyll a, TP and PO₄-P for the years 1992–97. TP, Po₄-P, Secchi depth and chlorophyll a are summer means (c). From Strand JA and Weisner SEB (2001) Dynamics of submerged macrophyte populations in response to biomanipulation. *Freshwater Biology* 46: 1397– 1408.

*Dreissena polymorpha*, which colonized Europe in the 19th century and more recently also in Great lakes in the United States, may have a major impact on water clarity when abundant. Significant effects on water clarity have also been shown in enclosure experiments. However, spreading *Dreissena* to other water systems is problematic as demonstrated in North America where lack of natural enemies in the rapid colonization phase allows *Dreissena* to reach enormous densities. This has resulted in significantly reduced chlorophyll of the Great lakes, but also in fouling of water intakes in reservoirs and uncontrolled impacts on the entire lake ecosystem. The

method is potentially most useful in shallow lakes, but the long term perspective is not clear.

## Differences in Lake Restoration Strategies in Cold and Warm Climate Regions

Most lake restoration methods have been developed for temperate lakes and cannot be readily transferred to subtropical lakes and particularly not to tropical water bodies because they differ in many aspects from lakes in the temperate zone. For instance, in the

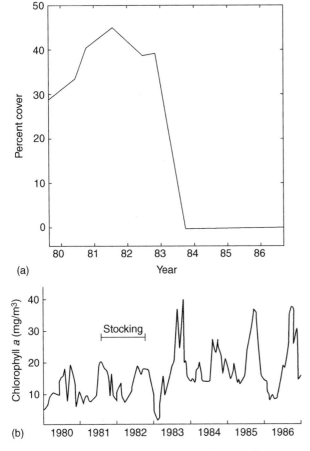

**Figure 8** Percent cover of aquatic macrophytes and chlorophyll *a* in Lake Conroe, Texas, before and after diploid grass carps were stocked in 1981–82 (33 fish ha⁻¹). Modified from Cook GD *et al.* (2005), originally from Maceina MJ *et al.* (1992). *Journal of Freshwater Ecology* 7: 81–95.

tropics, the faster nutrient cycling combined with higher predation pressure on zooplankton, as a consequence of a lower percentage of pelagic piscivores and repeated fish recruitment over the season, result in reduced grazer control of phytoplankton. There is also a higher risk of dominance by cyanobacteria owing to a higher probability of N limitation in the warmer climate and also to greater dominance of nuisance floating plants. Little is, however, known about the long-term effects of restorations in (sub)tropical lakes.

## Gaps in Knowledge

There are clear gaps in our knowledge on lake restoration. We need to:

- define critical nutrient concentrations in different parts of the world (climate zones) for shifting lakes to a good and high ecological state and to set target concentrations for initiating in-lake measures in addition to external loading reduction.
- adapt existing methods and develop new viable ecological restoration methods for lakes in the tropics.
- determine the long-term stability of the different restoration measures and evaluate the need for and how to undertake follow-up measures in the years after restoration.
- conduct controlled experiments and obtain experience on the effects of using combined restoration methods.

## Acknowledgments

We thank Anne Mette Poulsen, Ramesh Gulati and Tinna Christensen for editorial and layout assistance. The study was supported by The Danish project CLEAR (a Villum Kann Rasmussen Centre of Excellence project), the EU EUROLIMPACS project (www. eurolimpacs.ucl.ac.uk), and the Finnish project CARE.

*See also:* Aquatic Ecosystem Services; Benthic Invertebrate Fauna, Lakes and Reservoirs; Biological-Physical Interactions; Biomanipulation of Aquatic Ecosystems; Lakes as Ecosystems; Trophic Dynamics in Aquatic Ecosystems.

## Further Reading

Björk S (1985) Scandinavian lake restoration activities. In: Proceedings of European Water Pollution Control Association, International Congress on Lake Pollution and Recovery, pp. 293–301. Rome, April 15–18.

Cooke GD, Welch EB, Peterson SA, and Nicholas SA (2005) *Restoration and Management of Lakes and Reservoirs*. Boca Raton, FL: CRC Press.

Duncan A (1990) A review: Limnological management and biomanipulation in the London reservoirs. *Hydrobiologia* 200/201: 541–548.

Gulati RD and Van Donk E (2005) Restoration of Freshwater Lakes. In: van Andel J and Aronson J (eds.) *Restoration Ecology*. Oxford, UK: Blackwell.

Gulati RD, Dionisio Pires LM, and Van Donk E (2008) Lake restoration studies: Failures, bottlenecks and prospects of new ecotechnological measures. *Limnologica* 38: 233–247.

Jeppesen E, Søndergaard M, Jensen JP, *et al.* (2005) Lake responses to reduced nutrient loading – An analysis of contemporary long-term data from 35 case studies. *Freshwater Biology* 50: 1747–1771.

Lathrop D (ed.) (2007) *Long-term Perspectives in Lake Management. Lake and Reservoir Management* 23(4): 321–477.

Moss B, Madgwick J, and Phillips G (1996) *A Guide to the Restoration of Nutrient Enriched Shallow Lakes*. Norwich, UK: Environment Agency and Broads Authority.

Perrow M and Davy T (eds.) (2002) *Handbook of Restoration Ecology*. Cambridge, UK: Cambridge University Press.

Reddy MV (ed.) (2005) *Tropical Eutrophic Lakes: Their Restoration and Management.* New Delhi: Science Publishers.

Sas H (ed.) (1989) *Lake Restoration by Reduction of Nutrient Loading. Expectation, Experiences, Extrapolation.* Sankt Augustin: Academia Verlag Richarz.

Søndergaard M, Jeppesen E, Lauridsen TL, *et al.* (2007) Lake restoration in Denmark and The Netherlands: Successes, failures and long-term effects. *Journal of Applied Ecology* 44: 1095–1105.

Søndergaard M, Liboriussen L, Pedersen AR, and Jeppesen E (2008) Lake restoration by fish removal: Short- and long-term effects in 36 Danish lakes. *Ecosystems* (in press).

Straskraba M (1994) Ecotechnical models for reservoir water quality management. *Ecological Modelling* 74: 1–38.

Zalewski M (1999) Minimising the risk and amplifying the opportunities for restoration of shallow reservoirs. *Hydrobiologia* 395/396: 107–114.

# Effects of Recreation and Commercial Shipping

**J E Vermaat,** Institute for Environmental Studies, Vrije Universiteit Amsterdam, Amsterdam, the Netherlands

## Introduction

Mankind has used inland waters for transport, hunting and fisheries since the paleolithic times, as evident from archeological finds. Quarried neolithic stone masonry has probably been transported over water (Stonehenge megaliths, for example, came from 200 km westward), as it certainly has been during the last two millennia. For example, Roman occupation forces of the present Netherlands extensively used the river Rhine for transport and to patrol their boundaries. The Romans are also known to have dug out canals to facilitate navigation between different naval bases in the Netherlands (after 47 AD at Forum Hadrianum, presently Voorburg). Commercial shipping developed in medieval Europe in parallel with the expansion of marine and coastal traffic, and the Vikings, for example, expanded their vast network of destinations across Russian rivers into the Black Sea. Similar developments must have unfolded elsewhere in the world, and the spread of the steam engine during the nineteenth century has only enhanced the use of waterways with a parallel expansion of canal networks during the industrial revolution across the Western World. Only the affluence of the twentieth century has witnessed the spread of recreative use of inland waters among wider strata of the population than the aristocracy. Until then, inland waters certainly have been used extensively, but to make a living, not for leisure.

This article reviews the impacts of the present-day variety of human uses on a range of inland waters. For the purpose of structure, a matrix is laid out combining types of use with types of water bodies. This involves categories, which can only be artificial where water bodies form a continuum of size (area, volume), flushing (or residence time), and position in the catchment. The author has tried to adhere to conventional categories of the textbooks and employed the following: (a) lakes and lake districts, (b) temporary wetlands, marshes and ponds, (c) rivers and streams, and (d) artificial waterways, such as canals and ditches. A range of human use categories is listed as rows in the matrix (**Table 1**). Also, the types of human use differ in the intensity and scale of their impact, ranging from virtually zero in the case of an incidental skating tour over a frozen lake to the trivial, massive effects of land fill on wetlands and excavation of new canals and irrigation networks. The latter may stop rivers to flow and totally alter the hydrology of a region, as in the case of the Aral Sea. The major patterns emerging from the table are discussed first, and a few important issues are then highlighted. Here a basic attempt is carried out to evaluate the effect of a use category by asking the question whether a factor would affect the provision of ecosystems goods and services, positively, negatively or neutrally. This crude three-point scale is entered in each cell of **Table 1**.

## Major Emergent Patterns in the Matrix

The first emergent point is that the manifold of human interactions with freshwater habitats can have positive and negative effects. Second, clines in extent and severity of these impacts span across the table, from incidental recreative visits to a lake in sparsely inhabited boreal zones, via subsistence fisheries and littoral reed harvesting to depletive fisheries and wholesale reclamation. Thus, a simple question on the effect of recreation and navigation in inland waters will require further specification. The presented matrix is only a broad-brushed effort to do so.

The prime service of inland waters to human society worldwide is the provision of clean fresh water. This too has been recognized early in human history, as witnessed by the impressive waterworks still remaining from ancient cultures. High density lakeside development and intensive navigation are both a consequence of increased population density, economic development and affluence. Both may severely affect this ecosystem service in the competitive struggle for water as a resource and a geographic feature. A range of other services (i.e., flood control and storm buffering, water supply, water quality improvement, habitat and nursery for commercial interesting species, recreational hunting and fishing, raw materials, amenity and biodiversity) has been identified for wetlands and inland waters alike. The overall economic value connected to these services is still the subject of methodological and conceptual study and debate. Generally, these values were found to be a positive function of national GDP and population density, and covary negatively with total wetland area. Hence in more affluent societies, these services are valued higher, although the remaining area of unaffected habitat is declining and remaining fragments have received a protection status.

**Table 1** Breakdown of the effects of human use in four different categories of inland water bodies

| Type of human use | Larger lakes and lake districts | Wetlands, marshes, and ponds | Rivers and streams | Artificial waterways | Explanatory remarks |
|---|---|---|---|---|---|
| 1. Recreative trekking, canoeing and swimming | 0 | 0 | 0 | 0 | Low intensity recreation is generally not considered to have negative effects (1) |
| 2. Recreative hunting and fishing | –/0 | –/0 | –/0 | –/0 | Recreative fishing may alter the composition of fish stocks and can be accompanied by artificial stocking with target species |
| 3. Recreative sailing and boating | –/0 | –/0 | –/0 | –/0 | Littoral fringes, reed beds and submerged vegetation may be affected by intense boat traffic (2–5) |
| 4. Commercial and subsistence fisheries, trapping, and hunting | –/0 | –/0 | –/0 | –/0 | Commercial fisheries as well as trapping and hunting have greatly altered densities of target vertebrate species (6, 7) |
| 5. Planned or unplanned food-web manipulation; introduction of new species, biological invasions | –/0/+ | –/0 | –/0 | –/0/+ | Fish stock manipulation may be considered positive and negative. Released large herbivores (e.g., muskrat) may affect wetland vegetation. Altered balances of planktivorous and piscivorous fish will affect zoo- and phytoplankton. Classical examples: the nile perch in Lake Victoria for introduction of nonnative fish and filtering zebra mussels into the North American Great Lakes as unplanned biological invasion. (6, 8–10) |
| 6. Exploitation of other natural resources | 0 | –/0/+ | –/0/+ | 0/+ | Managed exploitation may have beneficial effects on biodiversity of semi-natural wetlands, such as reed-cutting to maintain reed bed flora and fauna. Exploitation of sand and gravel in unregulated river beds is an alteration of habitat and may affect spawning grounds of fish. Regular dredging and clearing of ditches maintains the aquatic habitat (11) |
| 7. Commercial shipping | –/0 | – | – | –/0 | Enhanced shipping intensity may increase biological invasions; the North American Great Lakes and the Danube-Rhine canal are examples. Artificial waterways may serve in connecting previously isolated lake systems. Otherwise see 3 above (12–14) |
| 8. Shoreline development with cottages, mooring piers, and boat houses | –/0 | – | –/0 | 0/+ | With growing affluence, recreative settlements develop along lake shores. These generally affect littoral wetland vegetation first, but all littoral habitats follow suit and fish communities change (15, 16) |
| 9. Permanent settlement: trading posts, villages, and towns | – | – | – | –/0/+ | The consequence of economic development of society along rivers and other navigation networks. Effects generally aggravate those mentioned under 8. Deposition of human solid waste and disposal of domestic and industrial sewage are a consequence of the development of human settlements. Since these water bodies are created by man's activities, urban settlement is a primary cause of their existence, and hence should be judged positively when judged in isolation |
| 10. Irrigation and drainage, flow diversion, flow regulation, damming for hydropower or navigation | – | – | – | + | Upstream irrigation schemes may cause massive degradation in downstream lakes, wetlands and rivers. Classical examples are the drying-up of the Aral Sea and the reduction of freshwater flows and sediments to coastal wetlands and mangroves (14, 17–21). Flood regulation along river corridors will affect habitat prevalence and cause reduced biodiversity (22–24). Otherwise, like the previous entry, human activity is the root cause of the mere existence of these water bodies, and hence positive |

| | | | |
|---|---|---|---|
| 11. Reclamation | − | − | + | Reclamation of wetlands generally involves the creation of artificial waterways such as drainage ditches as new aquatic habitat, but overall the balance is negative (20, 21, 25) |

Key references are given in parentheses. Effects are judged negative, neutral, or positive (−, 0, +). Often, two or even three options are possible; the most probable is given in bold. The perspective is the provision of services to human society, including their mere existence.

Sources

1. Vermaat JE, Goosen H, and Omtzigt N (2007) A multivariate analysis of biodiversity patterns in Dutch wetland marsh areas: Urbanisation, eutrophication or fragmentation? *Biodiversity Conservation* 16: 3585–3595.

2. Mosisch TD and Arthington AH (1998) The impacts of power boating and water skiing on lakes and reservoirs. *Lakes Reservoir Research Management* 3: 1–17.

3. Hilton J and Phillips GL (1982) The effect of boat activity on turbidity in a shallow Broadland river. *Journal of Applied Ecology* 19: 143–150.

4. Vermaat JE and De Bruyne RJ (1993) Factors limiting the distribution of submerged waterplants in the lowland river Vecht (The Netherlands). *Freshwater Biology* 30: 147–157.

5. Murphy KJ and Eaton JW (1983) Effects of pleasure boat traffic on macrophyte growth in canals. *Journal of Applied Ecology* 20: 713–729.

6. Balirwa JS, Chapman CA, Chapman LJ, *et al.* (2003) Biodiversity and fishery sustainability in the Lake Victoria Basin: An unexpected marriage? *Bioscience* 53: 703–715.

7. Coble DW, Bruesewitz RE, Fratt TW, *et al.* (1990) Lake trout, sea lampreys, and overfishing in the upper Great Lakes: a review and reanalysis. *Transactions American Fisheries Society* 119: 985–995.

8. Andersson G, Berggren H, Cronberg G, *et al.* (1978) Effects of planktivorous and benthivorous fish on organisms and water chemistry in eutrophic lakes. *Hydrobiologia* 59: 9–15.

9. Danell K (1979) Reduction of aquatic vegetation following the colonization of a Northern Swedish lake by the muskrat, *Ondatra zibethica*. *Oecologia* 38: 101–106.

10. Witte F, Msuku BS, Wanink JH, *et al.* (2000) Recovery of cichlid species in Lake Victoria: An examination of factors leading to differential extinction. *Reviews in Fish Biology and Fisheries* 10: 233–241.

11. Guesewell S and Le Nedic C (2004) Effects of winter mowing on vegetation succession in a lakeshore fen. *Applied Vegetation Science* 7: 41–48.

12. Gutreuter S, Vallazza JM, and Knights BC (2006) Persistent disturbance by commercial navigation alters the relative abundance of channel-dwelling fishes in a large river. *Canadian Journal of Fisheries and Aquatic Sciences* 63: 2418–2433.

13. Lodge DM, Williams S, Macisaac HJ, *et al.* (2006) Biological invasions: recommendations for U.S. policy and management. *Ecological Applications* 16: 2035–2054.

14. Nentwig W (ed.) (2007) *Biological Invasions. Ecological Studies,* vol. 193. Springer.

15. Scheuerell MD and Schindler DE (2004) Changes in the spatial distribution of fishes in lakes along a residential development gradient. *Ecosystems* 7: 98–106.

16. Brazner JC (1997) Regional, habitat, and human development influences on coastal wetland and beach fish assemblages in Green Bay, Lake Michigan. *Journal of Great Lakes Research* 23: 36–51.

17. Aladin NV and Potts TW (1992) Changes in the Aral Sea ecosystems during the period 1960–1990. *Hydrobiologia* 237: 67–79.

18. De Groot SJ (2002) A review of the past and present status of anadromous fish species in the Netherlands: Is restocking the Rhine feasible? *Hydrobiologia* 478: 205–218.

19. Thampanya U, Vermaat JE, Sinsakul S, *et al.* (2006) Coastal erosion and mangrove progradation of Southern Thailand. *Estuarine and Coastal Shelf Science* 68: 75–85.

20. Brinson MM and Malvarez AI (2002) Temperate freshwater wetlands: types, status, and threats. *Environmental Conservation* 29: 115–133.

21. Tamisier A and Grillas P (2003) A review of habitat changes in the camargue: An assessment of the effects of the loss of biological diversity on the wintering waterfowl community. *Biological Conservation* 70: 39–47.

22. Pollock MM, Naiman RJ, and Hanley TA (1998) Plant species richness in riparian wetlands—A test of biodiversity theory. *Ecology* 79: 94–105.

23. Girel J and Manneville O (1998) Present species richness of plant communities in alpine stream corridors in relation to historical river management. *Biological Conservation* 85: 21–33.

24. Sabo JL, Sponseller R, Dixon M, *et al.* (2005) Riparian zones increase regional species richness by harboring different, not more, species. *Ecology* 86: 56–62.

25. Van Turnhout CAM, Foppen RPB, Leuven RSEW, *et al.* (2007) Scale-dependent homogenization: Changes in breeding bird diversity in the Netherlands over a 25-year period. *Biological Conservation* 134: 505–516.

## Alterations of the Littoral Landscape

With reclamation and settlement at lake shores and along river corridors, the landscape is greatly altered by man. This leads to the decline of specific habitats and the typical species that inhabit these. For example, open marshland declines with increasing population density and recreative visitor frequency across the Netherlands (**Figure 1**), leading to declines in characteristic and red-list angiosperm plants and changes in bird communities. Lake fish communities adjacent to human development had fewer species and lower abundance than pristine sites, and were also characterized by a more disturbance-tolerant species assemblage. Along North American lake shores, the density of woody debris fragments in the littoral zone

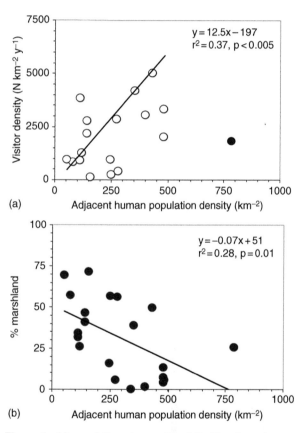

(a)

(b)

**Figure 1** (a) covariation of recreative visitor density and population density in 21 wetland complexes across the Netherlands; (b) percentage of open marshland (fens, reedbeds) remaining in the wetland complex versus population density. In a multivariate analysis, angiosperm plant diversity declined in parallel with the percentage marshland, and characteristic reed bed bird species declined overall during the period 1965–1995. The filled symbol in (a) is excluded from the regression because in this wetland complex visitor entry is strictly regulated (the Naardermeer). Data re-analysed from Vermaat JE, Omtzigt N, and Goosen H (2007) A multivariate analysis of biodiversity patterns in Dutch wetland marsh areas: urbanisation, eutrophication or fragmentation? *Biodiversity Conservation* 16: 3585–3595.

declined precipitously with increasing cottage density, thus altering the inshore habitat and affecting benthic macroinvertebrate communities. In the Great Lakes area of North America, the high biodiversity and large numbers of ecologically and economically important fish associated with wetlands are a key reason to conserve and restore these habitats. Equally, river floodplains will be affected, when the main channel is straightened, parts of the floodplain will be cut off by dams and side arms and backwaters will be drained and reclaimed. In a comparative study across the United States, it was found that the protection of sites sought after by real-estate developing agencies would greatly help biodiversity conservation because of the large number of characteristic, but endangered species associated notably with shorelines. In short, large changes in habitat prevalence will affect presence and abundance of species with potentially unforeseen changes in ecosystem services provided.

During the last decade of the twentieth century, however, a trend reversal in awareness and management practices can be observed. Increasing attention is given to the amelioration and restoration of riparian habitat, including the protection of river banks from boat traffic-induced erosion, the (re-)creation of shallow wetland fringes and fish spawning habitat as well as passageways, and the (re-)opening of formerly straightened meanders.

## Food-Web Manipulation

For fisheries enhancement purposes, nonnative species have been introduced into waters across the world. Subsequent consequences for food web composition were often unforeseen, as exemplified by the case of Lake Victoria. Here, introduction of the nile tilapia and nile perch brought species rich communities of native cichlid fish to extinction, but subsequent overfishing of these target species may well have led to recovery of some of these native species. In highly eutrophic lakes purposeful alterations of this community involve increases in piscivorous predators and decreases in benthivores and planktivores. The reduced predation on zooplankton and reduced sediment resuspension have led to drastic reductions in turbidity, and clear water (Secchi disk transparency >1 m) is an important target for lake management.

## Biological Invasions

Increased regulation, connection of river networks, and enhanced shipping transport during the twentieth century have augmented the transport of aquatic organisms across the world. Successful colonization of habitat by invasive species has had substantial consequences for food-web structure and ecosystem functioning. In North America's Great Lakes, the settlement of zebra mussels (*Dreissena polymorpha*) has greatly enhanced the benthic filtering capacity and thus probably assisted in countering the eutrophication symptom of high algal densities. The completion of the Main-Donau canal has opened up the Rhine network for colonization by a Ponto-Caspian fauna from the Black Sea region, and vice versa. Several species are now greatly expanding their ranges and alter benthic communities. The arrival of the water hyacinth (*Eichhornia crassipes*) in Lake Victoria (East Africa) has precipitated major changes in the littoral and impeded access to the lake for riparian human communities. This was brought to a standstill by a combination of biological control (introduction of the weevil Neochetina) and an unprecedented severe El Nino event.

*See also:* Aquatic Ecosystem Services; Biomanipulation of Aquatic Ecosystems; Eutrophication of Lakes and Reservoirs; Lake and Reservoir Management; Lake Management, Criteria; Littoral Zone.

## Further Reading

Aladin NV and Potts TW (1992) Changes in the Aral Sea ecosystems during the period 1960–1990. *Hydrobiologia* 237: 67–79.

Andelman SJ and Fagan WF (2000) Umbrellas and flagships: Efficient conservation surrogates or expensive mistakes? *Proceedings of the National Academy of Science* 97: 5954–5959.

Balirwa JS, Chapman CA, Chapman LJ, *et al.* (2003) Biodiversity and fishery sustainability in the Lake Victoria Basin: An unexpected marriage? *Bioscience* 53: 703–715.

Balmford A, Bruner A, Cooper P, *et al.* (2002) Economic reasons for conserving wild nature. *Science* 297: 950–953.

Brander L, Vermaat JE, and Florax RJGM (2006) The empirics of wetland valuation: A meta-analysis. *Environmental Resource Economics* 33: 223–250.

Dobson M and Frid CLJ (1998) *Ecology of Aquatic Systems.* Longman.

Kalff J (2002) *Limnology, Inland Water Ecosystems.* Prentice Hall.

Lodge DM, Williams S, Macisaac HJ, *et al.* (2006) Biological invasions: Recommendations for U.S. policy and management. *Ecological Applications* 16: 2035–2054.

McGowan S, Leavitt PR, Hall RI, *et al.* (2005) Controls of algal abundance and community composition during ecosystem state change. *Ecology* 86: 2200–2211.

Murphy KJ, Willby NJ, and Eaton JW (1995) Ecological impacts and management of boat traffic on navigable inland waterways. In: Harper DM and Ferguson AJD (eds.) *The Ecological Basis for River Management*, pp. 427–442. Chichester: Wiley.

Naeem S and Wright JP (2003) Disentangling biodiversity effects on ecosystem functioning: deriving solutions to a seemingly unsurmountable problem. *Ecology Letters* 6: 567–579.

Nentwig W (ed.) (2007) *Biological Invasions. Ecological Studies* vol. 193. Springer.

Perrow MR and Davy AJ (2002) *Handbook of Ecological Restoration.* Cambridge: Cambridge University Press.

Vitousek PM, Mooney HA, Lubchenco J, *et al.* (1997) Human domination of Earth's ecosystems. *Science* 277: 494–499.

Williams AE, Hecky RE, and Duthie HC (2007) Water hyacinth decline across Lake Victoria – Was it caused by climatic perturbation or biological control? A reply. *Aquatic Botany* 87: 94–96.

# Biomanipulation of Aquatic Ecosystems

**L-A Hansson and C Brönmark,** Lund University, Lund, Sweden

## Introduction and Problem Identification

Because of industrialization and intensified agriculture many lakes in urban areas have received high external loading of antropogenically derived nutrients, especially phosphorus (P) and nitrogen (N). These nutrients sooner or later transform pristine lakes into water bodies with dense algal blooms, bad odor, and mucky bottoms; a process called eutrophication (from Greek *eutrophus*: highly nutritious). In most lakes, algal growth is limited by phosphorus availability, and thus, eutrophication results in a massive increase in phytoplankton growth. The increased biomass of phytoplankton leads to a reduction in water transparency and, further, when the algae die they sink to the bottom, resulting in an increased amount of dead organic matter accumulating as sediment. Bacteria mineralizing this organic material consume large amounts of oxygen, which they derive from the water, leading to reduced oxygen concentrations. Because of wind-induced mixing, the oxygen concentration in the entire column is reduced to very low levels often leading to massive fish kills. Moreover, because visually hunting predatory fish become less efficient when turbidity (caused by massive algal growth) increases, a common feature of eutrophic lakes is an increasing proportion of planktivorous prey fish (**Figure 1**). These fish feed efficiently on large-bodied zooplankton, resulting in reduced abundances and size of grazing zooplankton, which in turn leads to an increase in the abundance of planktonic algae (phytoplankton), further deteriorating the underwater light climate. Hence, the eutrophication process leads to considerable food-web changes in the lake ecosystem, and also reduces the potential of using these lakes for recreation, fishing, and as a source of drinking water.

The logical solution to prevent the aforementioned water quality changes is to divert the runoff nutrient-rich waste water away from the polluted lake or to treat it by removing nutrients before they enter the lake, i.e., to prevent the factors causing eutrophication and pollution of lakes. Such preventive measures may, in some cases, lead to improvement of the water quality. One important process preventing lakes from recovering from eutrophication is that large amounts of nutrients are released from the mineralization of dead organic matter accumulated at the lake bottom. This release of nutrients from the sediment into the water column is called internal nutrient loading, a process that may proceed at high rates for decades. Hence, there is often a need for restoring eutrophicated lakes by, in addition to nutrient reduction, bringing about changes in food-web interactions by using biological methods. However, as for most restoration attempts, such measures rather rehabilitate than completely restore lakes to their pristine state.

## Lake Rehabilitation

If the problem with eutrophication is not solved prophylactically, i.e., by reducing the external nutrient input, it can be therapeutically approached by different rehabilitation methods, sometimes with good, but sometimes with less satisfactory results. Such rehabilitations involve physical removal of nutrient-rich sediment by dredging or chemically binding with e.g., ferric chloride or aluminium sulphate, or inactivating the phosphorus at the sediment and thereby considerably reducing the release rate in order to also reduce algal growth.

Another therapeutic measure is to use biomanipulation, which has its theoretical roots in food-web theory. The term biomanipulation refers to the manipulation of biota in order to make a water body more desirable for humans. In general, the goal of biomanipulation is to reduce the algal biomass in eutrophic lakes. Biomanipulations are generally performed by reducing the abundance of planktivorous fish (i.e., fish eating zooplankton), either by addition of piscivorous fish (fish eating other fish), or by manually removing or reducing the biomass of undesired fish (e.g., by trawling). If the amount of planktivorous fish is sufficiently reduced, the predation pressure on zooplankton should decrease and the grazing rate on algae should increase. In this way, the likelihood of algal blooms will decrease and water transparency increase.

## Simple Theory Versus Complex Real Ecosystems

Biomanipulation of lake ecosystems is based on the 'food chain theory,' which, for inland waters, states that phytoplankton are eaten by zooplankton, which are eaten by planktivorous fish, which, in turn, are eaten by piscivorous fish (**Figure 1**). If a lake is undergoing eutrophication, excess supply of nutrients enters the food chain and promotes rapid and high

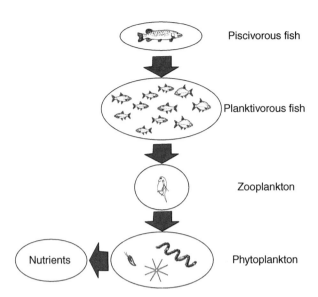

**Figure 1** A simplified food chain, constituting the theoretical basis for biomanipulation. The food chain starts with an increased nutrient input ('eutrophication'), which makes the phytoplankton (algae) abundant. Zooplankton feed on algae and planktivorous fish on zooplankton. Finally, piscivorous fish feed on other fish.

phytoplankton growth so that they become very abundant. In theory, the increasing amount of phytoplankton will provide abundant food supply for the zooplankton community, which, in turn, will provide excess food for planktivorous fish, and which will finally cause an increase in the next level of the food chain, piscivorous fish (**Figure 1**). Thus, if nutrients are provided at the bottom of the food chain, we would, logically, get a bottom up increase in abundances at all trophic levels in the food chain. In aquatic systems this is, however, not as simple since top predators (piscivorous fish) are visual feeders, i.e., they need to see their prey in order to catch it. As the water becomes green and turbid, their hunting success decreases and eventually the piscivorous fish cannot regulate the trophic level below the planktivorous fish. These planktivores then have excess food and are not efficiently preyed upon by the piscivorous fish, and therefore they become very abundant. Hence, a common and complex problem in lakes undergoing eutrophication is that the water becomes green and the planktivorous fish, such as cyprinids, become abundant, whereas zooplankton and piscivorous fish become much less conspicuous (**Figure 1**). The logic behind this chain of events is appealing and easy to understand. Especially appealing is the prediction that if planktivorous fish are removed, or piscivorous fish added, zooplankton will be released from the predation pressure allowing them to increase their grazing pressure on the phytoplankton, thereby causing lake water to become clear.

The relative simplicity of the theory, and the clear predictions derived from it, have attracted considerable interest from researchers and engineers. The road from basic science to its application is, however, not always straight. Although predictions from theory may be correct, the response of natural ecosystems subject to biomanipulation has often been more complex than expected or desired, and thus not always successful. Hence, some biomanipulations have failed to reach their predicted goals as many unexpected problems have arisen. On the other hand, there have also been several unexpected gains. Obviously, therefore, the original basis for performing biomanipulations – the theory of food chain manipulation – may not suffice to explain the complex responses of natural ecosystems. In the following we attempt to unravel some of this complexity.

## Resolving the Complexity

The application of food chain theory to lake biomanipulation has been surprisingly fast and many studies using this approach have been performed to restore eutrophicated lakes and to prevent algal bloom nuisance that often accompanies eutrophication of lakes. A major lesson learnt from the first biomanipulations is that the predictions from the food chain theory are not the only processes involved and that the theory may be too simple for a complex reality. Removal of planktivorous fish from a lake or addition of piscivorous fish to a lake, or both, clearly have effects on lower trophic levels, but not necessarily to the extent as predicted by the simple and appealing food chain theory. The importance of processes other than pure food chain dynamics has been demonstrated mainly in shallow lakes where the theoretical developments have progressed in closer cooperation with lake managers than in deep lakes. In many of the biomanipulation experiments performed, strong reductions in phytoplankton biomass have been observed due to increased grazing pressure from zooplankton. However, as zooplankton abundance often decreases, algal blooms often return, i.e., the effects are sometimes transitory and difficult to sustain over longer periods.

**The risk of increased recruitment of young fish** A major cause for these, sometimes short-lived, positive effects of biomanipulations is the increased recruitment of young fish. When the abundance of cyprinid fish is reduced through biomanipulation, competition for zooplankton food decreases and the recruitment of strong year classes of young fish is enhanced. Even though they are very small, the young fish grow fast, thereby mobilizing and excreting high amounts of nutrients. Hence, high recruitment of young fish and their predation on zooplankton may strongly counteract or offset the effect of the biomanipulation (**Figure 2**). This 'baby boom' of fish is commonly observed 1–4 years after a biomanipulation and the young fish strongly reduce the abundance of large, efficient herbivorous zooplankton. Hence, the very high reproductive potential of fish is, indeed, a problem that has to be considered when planning a biomanipulation. Possible solutions are additional fish reductions, prevention of recruitment by egg destruction, or addition of piscivorous fish to the lake.

**The feeding of benthic fishes**  As already stated, the theory behind biomanipulation rests mainly on pelagic processes, such as fish predation on zooplankton and zooplankton grazing on phytoplankton, whereas the importance of bottom feeding fish species is not always considered. Specialized benthic feeding fish, such as bream (*Abramis brama*) and gizzard shad (*Dorosoma cepedianum*), are 'vacuum cleaning' the sediment surface in their search for prey, such as chironomid larvae. This feeding behavior causes resuspension of sediment particles, which in shallow lakes reduces the light penetration in

the entire water column decreasing the foraging efficiency of piscivorous fish even more. Although most planktivorous fish mainly feed on zooplankton, they are forced to switch to feeding on benthic animals when zooplankton become scarce. Thus, such a shift from planktivorous to benthic feeding causes bioturbation and resuspension of sedimented material and also results in increased leakage of nutrients from the sediment and their diffusion to the overlying water layer, providing phytoplankton with additional nutrient resources. Hence, there is a strong argument for reducing the abundance not only of planktivorous, pelagic fish, but also of benthic feeding fish.

**Expansion of submerged macrophytes**  A general attribute of most lakes is that large amounts of phytoplankton and submerged macrophytes only rarely occur simultaneously. Moreover, after a more successful biomanipulation, the cover of submerged macrophytes generally increases at the expense of phytoplankton (**Figure 2**). There are several factors that explain the macrophyte expansion: the main ones are (1) increased light availability due to zooplankton grazing on the phytoplankton and (2) reduction of direct, physical disturbance of the sediment and macrophytes by the benthic feeding fish.

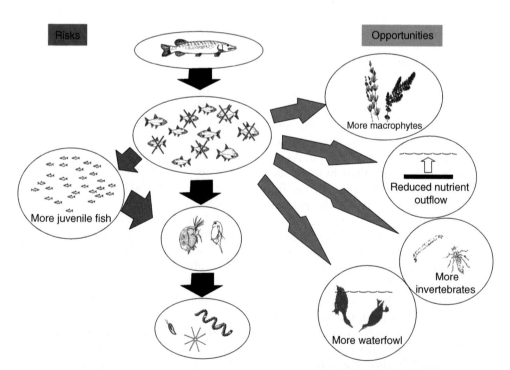

**Figure 2** During a biomanipulation, processes other than the feeding links within the food chain are active (the removal of fish is illustrated by crossed-over fish). For example, reduced competition between planktivorous fish is a risk leading to strong survival of young fish which may shortcut the expected decrease in predation pressure on zooplankton. Positive processes and opportunities may be reduced bioturbation by benthic feeding fish, which leads to improved conditions for submersed macrophytes and reduced internal loading of nutrients; processes leading to increased water clarity. Moreover, reduced predation by fish also results in higher amounts of benthic invertebrates, which, in turn, improves the conditions for waterfowl.

Moreover, submerged macrophytes and the algae attached to them are able to absorb large quantities of nutrients during the summer season, thereby outcompeting the phytoplankton. Macrophytes also stabilize the sediment surface, reducing resuspension of sediment particles, thereby improving the underwater light climate. Macrophytes may also function as a refuge for zooplankton from fish predation, and they are reported to favor predatory fish such as pike (*Esox esox*), more than cyprinids, such as roach (*Rutilus rutilus*) and bream (*A. brama*).

**Reduced nutrient concentrations** At low nutrient concentrations, lake water is generally clear, macrophytes are abundant and piscivores such as pike are common in the fish community. As nutrient concentrations increase, phytoplankton and planktivorous fish become abundant, macrophytes become rare or even disappear if the nutrient concentrations further increase. Such lakes are generally dominated by cyanobacteria or green algae and planktivorous fish. Several studies show that total phosphorus concentration should be less than 100 μg phosphorus l$^{-1}$ to obtain a

long-term, i.e., sustainable effect of biomanipulation in shallow lakes. However, a biomanipulation may be performed at higher phosphorus concentrations and still be successful as the manipulation often leads to a reduction in total phosphorus, at least temporarily. Such reductions have been reported in the literature and are most probably a combined result of reduced internal loading of phosphorus from the sediments, lower abundances of bottom feeding fish and phosphorus absorption by macrophytes. In addition, improved light climate at the sediment surface stimulates periphytic algal growth (algae growing at surfaces such as macrophyte leaves), and thereby oxygen production through photosynthesis, chemical sorption, and biological uptake of phosphorus at the sediment surface.

**Benthic invertebrates and waterfowl** Biomanipulation generally also leads to a strong increase in the amount of benthic invertebrates at the sediment surface (**Figure 3**), mainly due to reduced activities of benthic fish. Moreover, an expansion of submersed macrophytes (explained earlier) stimulates the growth

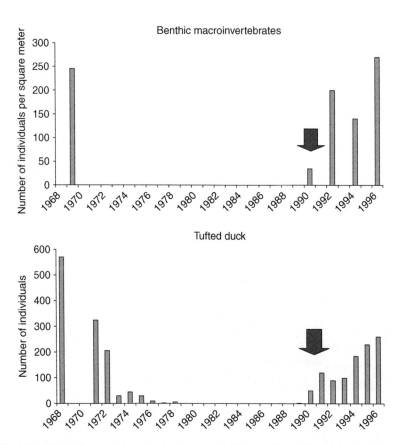

**Figure 3** The figure illustrates the positive effects on both macroinvertebrates and waterfowl following biomanipulation. Number of benthic invertebrates (no. per square meter) and the tufted duck (*Aythia fuligula*) in the western basin of Lake Ringsjön, Sweden, before eutrophication (until about 1970), during the severe eutrophication period (from about 1971 onwards), to the mid-1990s. The arrows indicate the period of biomanipulation efforts.

of benthic invertebrates even further. As benthic invertebrates and submersed macrophytes become more abundant following a biomanipulation, new food resources become available for waterfowl, such as coot (*Fulica atra*), swans (*Cygnus* spp.), and diving ducks (**Figure 3**). This causes such lakes to shift from a state characterized by high abundances of planktivorous fish and algal blooms into lakes with well-structured meadows of submerged macrophytes, clear water, and numerous waterfowl. Hence, the recreational value may increase considerably.

## Practicalities

In practice, biomanipulation can be performed either as an addition of piscivorous fish or as a removal of planktivorous fish, or by applying both measures simultaneously. Adding piscivores will, if successful, increase the predation pressure on planktivores, thereby allowing zooplankton to escape predation from planktivores and thus flourish and feed on phytoplankton, thereby increasing the clarity of the water. This method alone has, however, not proved to be very successful. The more common way of performing a biomanipulation is, instead, to remove planktivorous fish. Although labor-intensive, this is a simple method, most efficiently performed by using two boats with a trawl in between (**Figure 4**). Trawling may result in catches of several tons of fish per day but one problem is that the catch has to be sorted in order to return any piscivorous fish caught. Moreover, the disposal of the fish removed is generally a major problem.

## What is a Successful Biomanipulation?

Although the decision whether a rehabilitation effort is successful or not is subjective, some more or less general criteria may be put forward. One criterion for a successful biomanipulation is that the algal turbidity decreases and the water becomes clearer. Another criterion is that the amount of cyanobacteria decreases. The problems with cyanobacteria are that they tend to form nuisance 'blooms' that may be toxic. A third criterion lies in the stability of the improvements, that is, if the effects are only temporary or if they last for a longer period. On the basis of these criteria and of diagnostic variables such as (a) increased Secchi depth (water clarity), (b) reduced algal chlorophyll, (c) increased biomass of zooplankton, and (d) increased coverage of submerged vegetation, the success of specific biomanipulation measures undertaken may be roughly quantified. Biomanipulations may be ranked from those where all (100%) of the chosen diagnostic variables (a–d) were improved, to those where only a few, or even none, of the variables improved. If biomanipulation through fish reduction really has a positive effect on lake status, the portion of the diagnostic variables that are still improved after, for example 5 years, might be expected to be positively related to the intensity of the measure; that is, to the portion of fish removed from the lake. This would be a critical test of whether the fish removal is the probable cause of the observed

**Figure 4**  Biomanipulation in practice showing the trawl catch and the table for separating planktivorous fish from piscivorous fish, which are thrown back to the lake. When the catch is sorted, the sorting table is opened and the planktivorous fish put in a container under the table and later frozen (Photo: Richard Nilsson; www.richardnilsson.com).

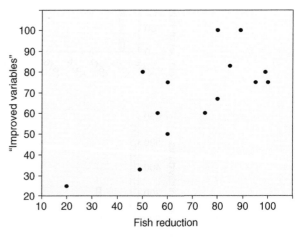

**Figure 5**  The portion of planktivorous fish removed in 14 biomanipulation cases plotted against the portion (%) of the diagnostic variables (water clarity, chlorophyll, submersed macrophytes, zooplankton (*Daphnia*), total phosphorus, and cyanobacteria) that were improved. The figure shows that there is a positive relation between the effort (i.e., how much fish that is removed) and the gain; here expressed as the portion of 'diagnostic variables' that were still improved after 5 years.

improvements. As seen in **Figure 5**, which is based on 14 biomanipulation studies, there is indeed a relationship between the proportion of improved diagnostic variables and the proportion (%) of the planktivorous fish community that was removed from the lake. In other words, the intensity of the measure is proportional to how 'successful' the biomanipulation is likely to become (**Figure 5**). A major conclusion is that if the planktivorous fish proportion is reduced by less than 50%, for example, because of a restricted budget, biomanipulation as a rehabilitation method should not be recommended since only a limited number of the diagnostic variables are expected to be found to have improved 5 years after the biomanipulation (**Figure 5**). However, if the budget allows a reduction of 75% of the cyprinid fish assemblage, the probability is high that the biomanipulation will be viewed as a success even after 5 years (**Figure 5**). Moreover, also the intensity of the fish reduction is important; that is, a fish reduction with 10 tons during one year is not the same as 1 ton each year during 10 years, since the fish community is able to compensate mortality with reproduction if the intensity is too low.

## Synthesis – What Happens after a Biomanipulation?

As stated in the introduction, the main theory behind biomanipulation is still focused on pelagic processes. However, both littoral and benthic factors, such as submerged macrophytes and benthic feeding fish, strongly affect the success of a biomanipulation, suggesting that these processes should receive equal consideration. Instead of being the only mechanism involved, alterations in the food chain may be viewed as triggers that initiate other processes.

A biomanipulation has several 'primary effects' such as an increase in zooplankton abundance within 1–2 years (**Figure 6**). This results in a reduced algal biomass, which in turn leads to improved light conditions and improved possibilities for submerged macrophytes to establish. The macrophytes absorb nutrients, leading to a further reduction in algal biomass and a 'positive spiral' is created (**Figure 6**). Furthermore, the reduction in benthic feeding fish results in reduced resuspension of sediment particles and thereby to a reduced turbidity, which improves light penetration, and also reduces the damage to macrophytes. Less feeding by fish at the sediment surface also reduces the nutrient transport from sediment to water, thereby reducing the algal growth potential. Moreover, once the macrophytes have established they also stabilize the sediment surface,

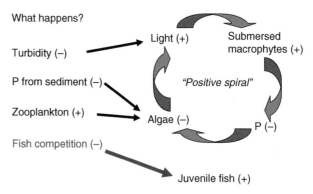

**Figure 6** An overview of important processes triggered by a reduction in cyprinid fish abundance (biomanipulation). Primary positive effects are increased zooplankton abundances and reduced turbidity and internal nutrient (P) loading from the sediment. These in turn have effects on other variables in the system by creating a 'positive spiral' of reduced algal growth, improved light penetration and growth of submerged macrophtes. The removal of fish also causes reduced competition among planktivorous fish which leads to a higher survival of juvenile fish; a process that may counteract the positive effects.

thereby further reducing the resuspension. These processes strengthen the 'positive spiral.' However, another 'primary effect' of a fish reduction is reduced food competition among fish (**Figure 6**). If the biomanipulation is intense enough and the density of piscivorous fish is high enough, the recruitment of young fish may not become a problem. However, if an insufficient amount of fish is removed, the young fish, recruited years following the biomanipulation, will have access to an almost unlimited food resource, since competition with larger fish is negligible. However, this 'negative spiral' may, at least partly, be avoided by addition of piscivorous fish.

Some lakes present better conditions than others for using biomanipulation as a therapeutic tool. Low external and internal phosphorus loading and large areas suitable for colonization of submerged macrophytes increase the probability of a successful biomanipulation. Another important conclusion is that the former completely dominant view that the only mechanism involved in a biomanipulation is the pelagic food chain (fish–zooplankton–algae) should be less pronounced. Instead this process may be viewed as the trigger for secondary, mainly benthic and littoral, processes, such as establishment of submerged macrophytes and reduction in benthic feeding by fish. A final conclusion may be that a well-planned biomanipulation in combination with a reduction in the external nutrient input is an attractive lake rehabilitation method likely to be successful in improving the water quality in most types of eutrophic lakes.

## Further Reading

Benndorf J (2005) Ecotechnology: Basis for a new immission concept in water pollution control. *Water Science and Technology* 52: 17–24.

Carpenter SR, Kitchell JF, and Hodgson JR (1985) Cascading trophic interactions and lake productivity: Fish predation and herbivory can regulate lake ecosystem. *Bioscience* 35: 634–639.

Gulati R and van Donk E (2002) Lakes in the Netherlands, their origin, eutropication and restoration: State-of-the-art review. *Hydrobiologia* 478: 73–106.

Hairston NG, Smith FE, and Slobodkin LB (1960) Community structure, population control, and competition. *American Naturalist* 94: 421–425.

Hansson L-A and Bergman E (eds.) (1998) *The Restoration of Lake Ringsjön. Developments in Hydrobiology (Hydrobiologia)* vol. 140). Dordrecht, The Netherlands: Kluwer Academic.

Hansson L-A, Annadotter H, Bergman E, *et al.* (1998) Biomanipulation as an application of food chain theory: Constraints, synthesis and recommendations for temperate lakes. *Ecosystems* 1: 558–574.

Horppila JH, Peltonen T, Malinen EL, and Kairesalo T (1998) Top-down or bottom-up effects by fish – Issues of concern in biomanipulation of lakes. *Restoration Ecology* 6: 1–10.

Hosper SH and Meijer M-L (1993) Biomanipulation, will it work for your lake? A simple test for the assessment of chances for clear water, following drastic fish-stock reduction in shallow eutrophic lakes. *Ecological Engineering* 2: 63–72.

Jeppesen E, Søndergaard M, Meerhoff M, Lauridsen T, and Jensen JP (2007) Shallow lake restoration by nutrient loading reduction– Some recent findings and challenges ahead. *Hydrobiologia* 584: 239–252.

Meijer M-L, Jeppesen E, van Donk E, Moss B, Scheffer M, Lammens E, and van Nes E (1994) Long-term responses to fish-stock reduction in small shallow lakes: Interpretation of five-year results of four biomanipulation cases in The Netherlands and Denmark. *Hydrobiologia* 275/276: 457–466.

Moss B, Stephen D, Balayla D, *et al.* (2004) Continental-scale patterns of nutrient and fish effects on shallow lakes: Synthesis of a pan-European mesocosm experiment. *Freshwater Biology* 49(12): 1633–1650.

Perrow MR, Meijer M-L, Dawidowicz P, and Coops H (1997) Biomanipulation in shallow lakes: State of the art. *Hydrobiologia* 342/343: 355–365.

Reynolds CS (1994) The ecological basis for the successful biomanipulation of a aquatic communities. *Archiv für Hydrobiologie* 130: 1–33.

Scheffer M and van Nes E (2007) Shallow lakes theory revisited: Various alternative regimes driven by climate, nutrients, depth and lake size. *Hydrobiologia* 584: 455–466.

Shapiro J and Wright DI (1984) Lake restoration by biomanipulation: Round lake, Minnesota, the first two years. *Freshwater Biology* 14: 371–383.

Søndergaard M, Jeppesen E, Lauridsen T, Skov C, Van Nes E, Roijackers R, Lammens E, and Portielje R (2007) Lake restoration: Successes, failures and long-term effects. *Journal of Applied Ecology* 44: 1095–1105.

# Vector-Borne Diseases of Freshwater Habitats

**V H Resh,** University of California, Berkeley, CA, USA

## Introduction

Diseases transmitted by organisms living in fresh water have been a major force in shaping both the size of the human population and the development of our civilization. Certainly, far more people have been killed by diseases transmitted by freshwater vectors than those caused by all the wars in the history of the world. This is true even in recent times; during the Vietnam War, many times more casualties resulted from malaria than from combat.

Water-related human diseases currently kill more than 5 million people a year, and some 2.3 billion people suffer from diseases linked to water. Water-borne diseases can be classified into four major categories: (1) water-borne diseases that are spread through contaminated drinking water, such as cholera; (2) water-washed diseases arising from poor sanitation, such as typhoid; (3) diseases that are directly transmitted by vectors (usually biting flies) that depend on fresh water for one or more of the stages of their life cycle, such as malaria; and (4) diseases that are transmitted to humans that involve an aquatic animal that serves as an intermediate host, usually snails, such as schistosomiasis. Other articles in the Encyclopedia concentrate on the first two categories (e.g., coliforms); this article concentrates on the latter two categories.

## Human Diseases and Freshwater Vectors

A variety of human diseases are transmitted by vectors that have life cycles associated with various types of water bodies, including lakes, ponds, rivers, streams, reservoirs, and irrigated fields. Of these diseases, malaria, schistosomiasis, lymphatic filariasis, onchocerciasis, Japanese encephalitis, and dengue are generally considered now to be the most important in terms of their infection rate, morbidity (i.e., either the incidence or to prevalence of a disease), or mortality (i.e., the number of deaths resulting from a particular disease).

These diseases have had and continue to have a devastating toll on human life. For example, malaria is endemic in more than 100 developing countries and more than 2 billion humans are at risk. 80–90% of the cases occur among the poorest countries of the world in Africa, but malaria is also a major problem in parts of South America, India, and Southeast Asia.

Between 100 and 200 million people are estimated to be infected worldwide, and ~2 million deaths occur per year (even higher during epidemic years). An estimated cost of $1.7 billion in lost productivity and treatment costs applies in sub-Saharan Africa alone.

Malaria is caused by several species of *Plasmodium* protozoans: *P. vivax, P. falciparum, P. malariae,* and *P. ovale.* It is vectored by various species of *Anopheles* mosquitoes that often occur in high numbers in irrigation ditches and canals, lakes and ponds, riverine floodplains, wetland-rice cultivation areas, and human settlements. Current control measures involve personal protection (e.g., repellents, pesticide-impregnated bed nets, anti-malarial drugs), vector reduction through insecticide applications, and parasite elimination in infected humans through drug therapy.

Schistosomiasis occurs in most subtropical and tropical areas of the world, and 200 million people being affected with this debilitating disease; 20 million suffer consequent renal failure, bladder cancer, and liver fibrosis. Like malaria, the majority of the schistosomiasis cases occur in sub-saharan Africa. Children are particularly at risk; 88 million children under 15 are infected each year. The passing of blood in the urine of adolescent boys, a consequence of the disease, is so common in some parts of West Africa that it is considered normal and analogous to menarche in adolescent girls. Certain species of snails serve as the intermediate host of the blood fluke parasites in the genus *Schistosoma*; *S. mansoni, S. japonicum,* and *S. haematobium* are the most important agents of this disease. Control involves snail elimination through molluscicide applications or providing drug therapy to infected individuals. The latter is complicated by the frequency of reoccurrence of infections that occurs among individuals that have already been treated.

Lymphatic filariasis (or elephantiasis), which is transmitted by a few genera of culicine mosquitoes (culicine is a subfamily of the mosquito family Culicidae), is a global problem with high morbidity but low mortality. It is estimated that this disease affects some 120 million people living in 80 countries, and high infection rates were noted in Egypt, sub-Saharan Africa, India, the western Pacific Islands, and parts of the Caribbean and South America. The disease is caused by two species of filarial nematodes *Wucheria bancrofti* and *Brugia malayi.* It is especially prevalent in areas with irrigation systems and in weed-infested reservoirs. Control involves mosquito reduction and drug therapy, the latter either as a prophylactic or a treatment.

Japanese encephalitis (or brain fever) is also transmitted by some genera of culicine mosquitoes and is the result of a flavivirus infection. It occurs in South, Southeast, and East Asia, and in some Pacific Islands. During epidemic outbreaks, high mortality rates occur, especially among children. Control involves mosquito reduction through habitat modification or insecticidal applications, and through vaccination.

Dengue fever (or hemorrhagic fever) occurs mostly during and shortly after the rainy season in most tropical and subtropical areas of world. More than 100 million people suffer from it each year and the number of infections is rising annually. Likewise, its distribution has spread greatly over the past three decades. It is caused by four related forms of flavovirus and is vectored by *Aedes* mosquitoes, most commonly *Aedes aegypti*. This species of mosquito is also the infamous vector of yellow fever, a disease that has been eliminated in much of the world but which is now gaining renewed importance because of its transmission of dengue.

Adult females of many *Aedes* mosquitoes lay their eggs on moist soil that will later be flooded and from which the mosquitoes emerge as adults. Control is directed at mosquito reduction and biting prevention; no vaccine is available.

In contrast to diseases mentioned so far, in which the vector occurs in still or at least very slow-moving water, onchocerciasis is vectored by black flies whose immature stages occur attached to vegetation in fast-flowing streams. The parasite is the filarial nematode *Onchocerca volvulus*. Although most cases occur in Africa, it is also found in parts of South America and the Arabian Peninsula. It has been successfully controlled in West Africa through insecticide applications to reduce aquatic stages of the black fly vector and through the distribution of antihelminthic drugs (see later text) but it is still a problem in central and east Africa, and in parts of South America.

## Insect Vectors of Human Disease

Biting flies (insects of the arthropod order Diptera) are the most important of the aquatic vectors of human disease. Of the flies, mosquitoes (the family Culicidae) are the most important vectors because of the high mortality and morbidity of the many diseases they transmit and because of the range of diseases for which they can serve as vectors (**Table 1**). For example, certain species of mosquitoes can transmit many

**Table 1** Examples of human diseases vectored by mosquitoes

| Disease vectored | Parasite transmitted |
|---|---|
| Lymphatic filariasis | *Brugia* and *Wucheria* nematodes |
| Malaria | *Plasmodium* protozoans |
| Yellow fever | Flavovirus |
| Dengue | Flavovirus |
| St. Louis encephalitis | Flavovirus |
| Japanese encephalitis | Flavovirus |
| Murray Valley encephalitis | Flavovirus |
| Russia spring-summer encephalitis | Flavovirus |
| Omsk Hemorrhagic fever | Flavovirus |
| West Nile fever | Flavovirus |
| Kyasanur forest disease | Flavovirus |
| Louping III | Flavovirus |
| California encephalitis | Bunyavirus |
| La Crosse encephalitis | Bunyavirus |
| Rift Valley fever | Bunyavirus |
| Eastern equine encephalitis | Togavirus |
| Western equine encephalitis | Togavirus |
| Venezuelan encephalitis | Togavirus |
| Ross River fever | Togavirus |

types of nematodes, protozoans, or viruses. However, it should be remembered that while ~3000 described species of mosquitoes take blood for nourishment, most are nuisance biters (and some not even of humans) and do not transmit human diseases.

As with terrestrial insect vectors, disease transmission involves two different blood meals. Only adult females take blood meals, using the blood of their victims to provide protein for their eggs to develop and pass through their larval and pupal stages in the oftentimes nutrient-poor aquatic habitats that they occupy. Using mosquitoes as an example, during the first blood meal a female injects saliva into the feeding wound; the saliva contains substances to reduce blood clotting. Using muscle pumps in her head, she then ingests blood from an infected human. Adult mosquitoes generally cannot undergo ovarian development until they have taken a blood meal (although some, referred to as being autogenous, can do this). Females tend to be attracted to their hosts by the heat, moisture, and carbon dioxide that they emit but certainly other factors (e.g., volatile chemicals emitted through the skin) are involved. Species can be selective in terms of hosts; some choose only birds, while some others chose only mammals. In species where both are selected, disease transmission may be enhanced when one of the hosts can serve as a disease reservoir (e.g., birds in West Nile Fever; dogs, cats, and monkeys in *Brugia* lymphatic filariasis).

Depending on the disease agent resulting in disease, there are three processes that can occur within the mosquito as the ingested pathogen migrates from her gut to her salivary glands. The pathogen can multiply but not undergo developmental changes (e.g., as in yellow fever), the pathogen can undergo developmental changes but not multiply (e.g., onchocerciasis), or the pathogen can do both (e.g., malaria). In all cases, when a mosquito takes a blood meal from another host, the disease agent may be transmitted through the injection of the anticoagulant saliva.

Other Diptera are important as well in disease transmission to humans and their animals. Black flies (the family Simuliidae) transmit onchocerciasis (river blindness), which is caused by a nematode (roundworm). This filarial worm lives in the human body for over a decade, and adult females produce millions of larvae that migrate throughout the body causing a series of progressive symptoms as the number of parasitic worms increases, including rashes and lesions, intense itching, loss of skin pigmentation, general debilitation, and eventually blindness.

Many species of biting flies that occur in moist or semi-aquatic habitats as larvae or that are common riparian-dwelling species as adults also transmit diseases. For example, some sand flies (Ceratopogonidae) can transmit the protozoan causing leishmaniasis, tsetse (Glossinidae) transmit the protozoan causing African sleeping sickness, and some deer flies (Tabanidae) can transmit *Loa loa*, the eye worm of humans.

Even biting flies that do not transmit disease can be major nuisances. In North America, horse flies and deer flies (both in the family Tabanidae) and black flies are irritating pests, oftentimes with painful bites that produce allergic reactions and infections that result from scratching the bites. There is also an economic cost associated with them in terms of lost tourist revenue and lower population densities in areas where they are a problem.

Other insects that are not important as direct transmitters of diseases (e.g., aquatic bugs, dragonfly larvae, beetle larvae, along with snails, small fish, and the even biofilm of aquatic plants) may serve as reservoirs of a water-associated disease, Buruli ulcer. Some naucorid bugs have even been shown experimentally to have the ability to transmit this disease.

## Snails and Crustacenas as Intermediate Hosts of Human Disease

Freshwater snails are the intermediate hosts of a variety of trematodes (flukes) and some nematodes (roundworms) that cause many human diseases. The most important of these diseases is schistosomiasis (sometimes referred to as Bilharzia). This disease is caused by parasitic trematodes (blood flukes) that must find and invade a particular species of snail to

continue their life cycle. The ecological requirements of these snails are a key determinant in the distribution and prevalence of this disease.

Schistosomiasis is a major public health problem in the world and blood flukes in general are important parasites of cattle and other large animals that humans depend on for survival. In humans, the blood flukes causing schistosomiasis lie in small veins in the lower abdomen; here males and females copulate and the females lay millions of eggs for 10 years or more. The eggs are passed into water supplies from feces and urine of infected people. After hatching the parasite actively swims and invades the body of a snail, usually by penetrating through the snail's foot. The parasite requires the presence of certain species of snails (e.g., *Oncomelania*, *Biomphalaria*, *Bulinus*) to transform (and multiply) itself into an infective stage, called cercaria. This life stage actively swims to the human host and penetrates the skin, which leads to infection.

A human's susceptibility to schistosomiasis depends on the species and strain of the invading blood fluke, the person's age, number of other parasites carried, nutritional status, and previous exposure to the parasite. Infection may cause incapacitation but the main effect is long-term damage to the intestine, bladder, and liver. Consequently, death may result from other infective agents attacking an already debilitated body.

Although schistosomiasis commonly occurs in developing countries, an infection related to this disease that occurs in developed countries is cercarial dermatitis or 'swimmer's itch'. This disease occurs worldwide and affects people who are swimming, wading, or working in littoral areas of both marine and freshwater habitats. Infection results when cercaria meant for birds or small mammals such as muskrats penetrate human skin, eliciting an immune response. The resulting skin rash goes away in about a week, the parasites degenerate, and the disease does not progress because humans are not the appropriate host for these animal schistsomes. Prevention involves brisk toweling after leaving the water (because penetration increases with drying and exposure length) and snail control.

Snails are vectors to several other human diseases (although much less common than schistosomiasis) as well, such as clonorchiasis and fasciolopsiasis. Like with schistosomiasis, the most medically important members of the Phylum Platyhelminthes (the flatworms) have definitive hosts that are vertebrates and intermediate hosts that are mollusks (mostly gastropods).

Another human disease that has been important in Africa and parts of the developing world is Guinea worm or dracunculiasis. The large nematode

(~1 m long) causing this disease releases larvae from an adult female worm (usually embedded in a human leg) through a skin lesion, when an infected human comes in contact with a pond or well. A larva is then ingested by a water flea (*Cyclops*), in which it develops and becomes infective. When a person drinks this infected water, the *Cyclops* is dissolved by the acidity in their stomach. The nematode larva is then activated, migrates through the subcutaneous tissue, and stays within the new host for about a year. It then emerges and starts the life cycle again. The treatment for guinea worm that has been used for thousands of years is to slowly wind the emerging worm around a stick over a several day period. This treatment is immortalized in the staff of Asclepius wound with a serpent, which is also the symbol of the modern-day healer or physician. An extensive Guinea worm eradication program has been highly successful and has eliminated the disease in most parts of the world and reducing human cases from 3.5 million to ~32 000.

## Habitats of Human-Disease Vectors

Three important categories can be distinguished that provide habitat for vectors and intermediate stages of the causative agents of human diseases: (1) natural water bodies; (2) human-made water bodies, and (3) water bodies that form in human settlements and household environments. Freshwater vectors of human disease can occur in all of these habitat categories.

In terms of natural water bodies, streams and rivers are sources of the black flies that serve as vectors of onchocerciasis, and lakes and ponds provide habitats for snails that are intermediate hosts of schistosomiasis. Although mosquito vectors do occur in natural systems as eggs, larvae, and pupae, their densities tend to be far higher when they occur in human-made or settlement habitats, probably because of reductions in predation and competition compared to natural systems. Likewise, standing water in barrels and pots in human settlements, habitats where predation and competition are also lacking, are important habitats for mosquitoes.

The creation of human-made water bodies such as impoundments and irrigation ditches often result in hydrological changes that favor intensified vector breeding. This can result in increased *Anopheles* mosquito populations that may transmit malaria or in snail populations that may increase prevalence of schistosomiasis. Likewise, shifts in species composition may occur that reduce numbers of predators or competitors, and allow vectors to increase in number. Recent studies have also suggested that habitat disturbance may increase the prevalence of Buruli ulcer in ponds.

The creation of the Aswan High Dam on the Nile River in Egypt is an example of a project that resulted in an expansion of schistosomiasis. This disease was long prevalent in Lower Egypt, which is in the northern part of the country, such as in the Nile delta. Schistosomes have been found in mummies that are thousands of years old, and in the biblical story about the pharaoh's daughter finding Moses' floating cradle in the weeds, she certainly would have been attacked by cercariae when she went in to get him! In contrast, schistosomiasis was far less common in Upper Egypt (which is located in the south). However, the construction of the dam moved infected people from Lower Egypt to Upper Egypt and created favorable snail habitat because snails thrived in the slower-moving water conditions provided by the dam. Consequently, the disease became established there as well.

Potential increases in disease vectors are now generally a design consideration in development projects and this has been especially important when irrigation schemes are planned because their introduction may create new or more favorable habitats for disease vectors. A variety of hydraulic engineering approaches such as improvements in drainage, concrete lining or covering of canals, land leveling, filling in land depressions, controlling seepages, diking, and dewatering are useful additions. For example, it has been estimated that water improvements may result in disease reductions of 80% of Gambian sleeping sickness, 20% for onchocerciasis, and 10% for yellow fever. In some cases, however, pesticide applications to control insect vector and snail problems may be inevitable. Many of these habitat modifications and pesticide applications, however, may have a deleterious effect on the nontarget fauna that keeps vectors under control or is of economic or conservation value.

Tires are not a typical freshwater habitat but they are extremely important breeding sources for some mosquitoes. Tires are perfectly designed to hold water (there is no orientation from which water can drain from it), and many tree-hole occupying species use tires as breeding sites. The shipping of tires from Asia to North America for recapping and reselling (they do not recap tires in Asia) introduced the Asian tiger mosquito *Aedes albopictus* to North America. This mosquito is an effective transmitter of 23 human viruses, including those that cause St. Louis encephalitis, dengue, and yellow fever. Likewise, it has been speculated that reintroduction of yellow fever to Mexico was the result of shipments of

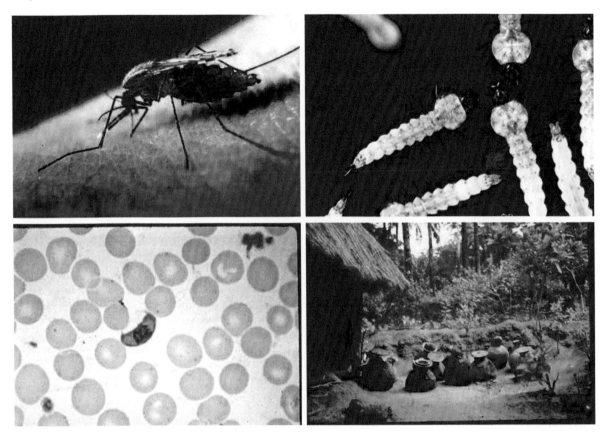

**Plate 1** Top left. *Anopheles gambiae*: an adult female mosquito taking a blood meal and possibly vectoring malaria (WHO/TDR/ Stammers). Top right. *Anophelese gambiae*: fourth-instar mosquito larva; all of the immature stages of mosquitoes occupy aquatic habitats (WHO/TDR/Stammers). Bottom left. *Plasmodium falciparum*: Mature female gametocyte stage of the malaria pathogen (WHO/ MAP/TDR). Bottom right. Water-storage pots that *Anopheles gambiae* and other mosquitoes can use as breeding sites (WHO/TDR/ Service).

discarded tires from Texas to Mexico to make huarache sandals because of a shortage of tires used in making them there!

## Vector Control Strategies

In addition to advances in chemotherapy and vaccination strategies to control the disease agents, a great deal of attention has been applied to control strategies that can reduce populations of freshwater vectors of human diseases. The most commonly used approach is the application of toxicants to control the vectors. These can be done over large spatial scales (e.g., aerial applications of insecticides in 11 West African counties to control black fly vectors of onchocerciasis) or locally (e.g., the distribution of pesticide impregnated bed nets to control malaria vectors at the village level).

A second approach is the introduction of environmental modifications. These include draining standing or stagnant water, creating ditches for recirculation of water, and diverting flow so streams go dry for part of the year.

A third approach, the introduction of biological control agents such as predatory fish (e.g., *Gambusia*) and invertebrates, along with a variety of bacteria, fungi, and other organisms, has been widely used with mixed success and, like the above control approaches, sometimes with major undesired ecological consequences. In general, predators do not restrict themselves to mosquitoes, black flies, and other vectors as prey. However, the more specific the control agent (such as the bacterial toxins from *Bacillus thuringiensis israelensis*), the less likely the chance of such unwanted consequences.

Finally, the addition of competitive organisms that are not vectors (e.g., closely related snails to compete with schistosomiasis vectors) has worked in isolated areas. Swamping populations with sterilized males (as was effective in the screw worm control program) and the development of genetic modifications may have future value but have not thus far been found to be effective.

**Plate 2** Top left. *Simulium damnosum:* An adult black fly female taking a blood meal and possibly vectoring river blindness or onchocerciasis (WHO/TDR/Stammers). Top right. *Simulium damnosum:* Larva of the black fly; all of the immature stages of black flies occupy aquatic habitats (WHO/TDR/Stammers). Bottom left. *Onchocerca volvulus:* The adult worms that causes onchocerciasis (WHO/TDR/OCP). Bottom right. A young girl leads her father, blind from onchocerciasis, through their village, a common scene in West Africa before this disease was controlled (WHO/TDR/Crump).

The large-scale control of aquatic vectors of human disease has had mixed success. Malaria, for which eradication through DDT and other pesticide applications was discussed as a possibility in the 1950s, soon re-emerged as a major source of human mortality. Malaria and other mosquito-vectored diseases continue to be an area of research in which both 'high' and 'low tech' solutions are being tested. Many of the former involve molecular biology and genetic engineering approaches. The latter involve widespread distribution of pesticide-impregnated bed nets to protect people during sleep.

Control of schistosomiasis has achieved somewhat better success than malaria and has involved breaking the life cycle of the parasite or interrupting the chain of infection. Clearly, if all snails are eliminated or all humans treated and remain free of infection, the chain of infection would be interrupted. Snail control is achieved through chemical control and, although biological control (e.g., with predatory snails or birds) has not been generally effective, habitat control

has been very effective in China. There, removing snails from canals with chopsticks, sealing latrines, and preventing human wastes from reaching water have been very successful measures. Habitat modification for snail control, such as modifying habitats by reducing vegetation, and altering stream bed and channel conditions to increase water velocity are also effective but these approaches require continuous long-term effort.

Guinea worm control has revolved around providing clean water through filtration, treating contaminated water, and preventing persons who have active infections from contact with water supplies. This approach has been very successful in Africa. However, when political instability and migration of humans from infected to disease-free areas occurs, the threat of re-invasion of this and other diseases with freshwater vectors increases.

The true success story about the control of aquatic vectors of human disease, and really of any human disease, is the control of river blindness in

**Plate 3** Top left. *Bulinus globosus*: The snail that is the intermediate host for *Schistosoma haematobium* (WHO/TDR/Stammers). Top right. *Schistosoma haematobium*: The adult worm that causes schistosomiasis (WHO/TDR/Stammers). Bottom left. *Schistosoma haematobium*: The cercaria that penetrates the skin of humans and leads to schistosomiasis (WHO/TDR/Stammers). Bottom right. A sample of normal urine and blood-containing urine from a child suffering from schistosomiasis (WHO/TDR/Lengeler).

West Africa. Although river blindness is known from 35 different countries, 99% of the cases occur in 26 countries of Africa. Prior to intensive control efforts, as many as 30% of adults living in streamside villages were blind, and blindness was viewed as an inevitable part of a person's life cycle if they lived near the fertile flood plains of West African rivers.

The Onchocerciasis Control Programme in West Africa (OCP) was begun in 1974 and continued through 2002. Covering an operational area of over 1.2 km², it included 30 million people who were vulnerable to the disease. OCP initially was based entirely on aerial application of insecticides to rivers to control black fly larvae. The extent of the program was enormous; over 50 000 km of streams received applications of insecticides at the peak of activity, and some rivers for as long as 20 years.

Beginning in the 1980s, ivermectin, a drug used in veterinary practice such as for treatment of dog heartworm, was distributed to humans to control the larval worms in their bodies and to complement the aerial application of insecticides in breaking the cycle of transmission of this disease.

OCP has been heralded as both a public health (the elimination of river blindness as a public health concern) and economic development (the opening of the river valleys free of onchocerciasis to settlement has enabled food to be grown for an estimated 17 million Africans) success. It has been expanded to 19 other African countries, primarily based on ivermectin distribution. Likewise, the success of the control program in West Africa prompted efforts to eliminate onchocerciasis from affected areas of Latin America.

A remarkable feature of OCP was the establishment of the first large-scale, long-term monitoring of the fish and benthic macroinvertebrate populations in the areas treated with insecticides. This, coupled with earlier experimental and taxonomic studies, has been the basis for much of what is known today about tropical African streams and rivers.

## Human Culture, Freshwater Vectors and Disease

Cultural practices may make humans more susceptible to certain diseases with freshwater vectors.

For example, fish and other organisms can be hosts to parasites, some of which can be transmitted to humans through eating raw or partially cooked fish such as salmonids, pike, perch, and burbot, carp and other cyprinids. Jewish women have had higher rates of infection by the broad fish tapeworm (*Diphyllobothrium latum*) than women of other cultures. This is because they often ingested infective stages of this tapeworm while tasting raw fish during the preparation of a traditional dish, gefilte fish. The tendency of restaurants to cook fish 'rare' and the consumption of more raw fish may lead to an increase in the rate of other such parasites in humans.

Many Asian cultures use aquatic insects as part of their diet. Ingestion of uncooked caddis fly larvae while working in rice paddies is known to pass parasites on to humans. Likewise, eating uncooked freshwater prawns (*Macrobrachium*) may pass on parasites as well.

The creation of cities has certainly affected the prevalence of freshwater vectors and human diseases. For example, when early Africans were hunters and gatherers, malaria was likely to have been far less of a problem to the human population than it is today because of their continual movements and migrations to other areas to find food. However, when people settled in villages and cities with the onset of agriculture and animal domestication, the transmission of the protozoan and its cycling within the population increased greatly. Likewise, the introduction of root crops in agriculture created ideal habitat in which mosquitoes can propagate.

The modification of cultural practices may have disease consequences as well. For example, river blindness was likely less of a problem in West Africa before European colonization because people tended to live some distance (1 km) from the rivers and avoided going there during hours of peak black fly biting. However, when the European colonists and missionaries encouraged settlement close to the river, these long-standing cultural taboos disappeared and the infection rate greatly increased.

Finally, the enhancement of global biodiversity has become a cultural goal of many developed countries. The issue of biodiversity in human disease has been a major topic in the conservation biology literature; however, it is invariably about the loss of biodiversity from large-scale disease control programs (loss of fish in OCP and nontarget terrestrial and riparian fauna in tsetse control programs). However, biodiversity is also an important component in the control of freshwater vectors of human diseases as well. For example, all major malaria vectors consist of species complexes, and differences are evident among *Anopheles* strains in their vector competency,

resting-site choice, biting behavior, and development of insecticide resistance. In cost-effective control of schistosomiasis, knowledge of snail diversity, distribution, and the epidemiology of parasite susceptibility by different snail species have been essential.

Onchocerciasis may provide the best example of how biodiversity relates to freshwater vectors and the diseases that they transmit. For example, cryptic species or strains, with varying competence in transmitting the disease or influencing its severity, are characteristic of both the black fly vector and the parasitic roundworms. Likewise, much of the success of OCP was also based on biodiversity. For example, the choice of insecticides used was based on maintenance of fish and nontarget invertebrate biodiversity. But even more unusual, the drug used to control onchocerciasis was also based on biodiversity. It was derived from a natural strain of *Streptomyces avermitilis*, a fungus discovered while screening a soil core from a golf course in Japan for potential nematocidal toxicity.

## Glossary

**Bilharzia** – A human disease caused by various species of trematode worms that use snails as an intermediate host; also called schistosomiasis.

**Black flies** – Insects in the family Simuliidae of the order Diptera; some species are the vectors of the nematode worms that cause onchocerciasis.

**Buruli ulcer** – A human disease caused by infection with the bacteria *Mycobacterium ulcerans*.

**Dengue fever** – A human disease caused by a flavovirus that is transmitted by infected mosquitoes.

**Dranunculiasis** – A human disease caused by nematode worms, with the crustacean *Cyclops* as an intermediate host; also known as guinea worm.

**Elephantiasis** – A human disease caused by nematode worms that are transmitted by infected mosquitoes; also called lymphatic filariasis.

**Filariasis** – A group of human diseases caused by the presence of nematode worms.

**Guinea worm** – A human disease caused by nematode worms, with the crustacean *Cyclops* as an intermediate host; also known as dranunculiasis.

**Intermediate host** – A host that a human parasite uses as part of its life cycle.

**Japanese encephalitis** – A human disease caused by a flavovirus that is transmitted by infected mosquitoes.

**Lymphatic filariasis** – A human disease caused by nematode worms and transmitted by infected mosquitoes.

**Malaria** – A human disease caused by a protozoan that is transmitted by infected mosquitoes.

**Morbidity** – The incidence of a disease in a population.

**Mortality** – The death rate caused by a disease in a population.

**Mosquitoes** – Insects in the family Culicidae of the order Diptera that transmit many human and animal diseases.

**Onchocerciasis** – A human disease caused by nematode worms and transmitted by infected black flies; also known as river blindness.

**River blindness** – A human disease caused by nematode worms and transmitted by infected black flies; also known as onchocerciasis.

**Schistosomiasis** – A human disease caused by various species of trematode worms that use snails as an intermediate host; also called bilharzia.

**Swimmer's itch** – A mild form of schistosomiasis, which results in a skin rash, and is caused by trematode worms, that have snails as the intermediate host.

**Vector of disease** – An organism, such as a biting fly, that transmits an infectious disease.

**Yellow fever** – A human disease caused by a flavovirus that is transmitted by infected mosquitoes.

## Further Reading

Desowitz RS (1991) *The Malaria Capers.* NY: W.W. Norton & Co.

Gratz NG (1999) Emerging and resurging vector-borne diseases. *Annual Review of Entomology* 44: 51–75.

Lane RP and Crossky RW (1993) *Medical Insects and Arachnids.* London: Chapman & Hall.

Mahmoud AAF (ed.) (2001) *Schistosomiasis.* London: Imperial College Press.

Marquardt WC, Demaree RS, and Grieve RB (2000) *Parasitology & Vector Biology,* 2nd edition, p. 702. San Diego: Academic Press.

Merritt RW, Benbow ME, and Small PLC (2005) Unravelling an emerging disease associated with disturbed aquatic environments: the case of Buruli ulcer. *Frontiers in Ecology and Environment* 3: 323–331.

Mullen G and Durden L (2002) *Medical and Veterinary Entomology.* San Diego: Academic Press.

Patz JA, Graczyk TK, Geller N, and Vittor AY (2000) Effects of environmental change on emerging parasitic diseases. *Australia Society for Parasitology* 30: 1395–1405.

Remme JHF (2004) Research for control: The onchocerciasis experience. *Tropical Medicine and International Health* 9: 243–254.

Resh VH, Lévêque C, and Statzner B (2004) Long-term, large-scale biomonitoring of the unknown: Assessing the effects of insecticides to control river blindness (onchocerciasis) in West Africa. *Annual Review of Entomology* 49: 115–139.

Ross AGP, Li YS, Sleigh AC, and McManus DP (1997) Schistosomiasis control in the People's Republic of China. *Parasitology Today* 13: 152–155.

Sakanari J, Moser M, and Deardoff TL (1995) *Fish Parasites and Human Health* (Rept. T-CSGP034). LaJolla: California Sea Grant.

Secor WE and Colley DG (eds.) (2005) *Schistosomiasis.* New York: Springer.

Spielman A and D'Antonio M (2001) *Mosquito, A Natural History of our Most Persistent and Deadly Foe.* NY: Hyperion.

Yaméogo L, Resh VH, and Molyneux DH (2004) Control of river blindness in West Africa: Case history of biodiversity in a disease control program. *EcoHealth* 1: 172–183.

# Conservation of Aquatic Ecosystems

**R Abell,** WWF-United States, Washington, DC, USA
**S Blanch,** WWF-Australia, Darwin, NT, Australia
**C Revenga,** The Nature Conservancy, Arlington, VA, USA
**M Thieme,** WWF-United States, Washington, DC, USA

## Introduction

Freshwater species and their habitats are on average among the most imperiled worldwide. Because they drain surface runoff from the landscape, freshwater ecosystems – also called inland aquatic systems or wetlands – are subject to impacts from land-based activities in addition to threats like direct habitat alteration and invasive species. Although limnology and related scientific disciplines are arguably well-developed, the field of freshwater biodiversity conservation lags behind that of the terrestrial and marine realms. This article details the state of freshwater biodiversity and habitats, summarizes major threats to freshwater systems, discusses conservation challenges for freshwaters, and provides an overview of more common conservation tools and strategies.

Recent studies show that freshwater species are on average more threatened than those in the terrestrial and marine realms. This is not surprising, as proximity to water bodies has been a preference for the establishment of human settlements for millennia. Society has used rivers for transport and navigation, water supply, waste disposal, and as a source of food. As a consequence we have heavily altered waterways to fit our needs by building dams, levies, canals, and water transfers and by heavily polluting our rivers, lakes, and streams with fertilizers and pesticides, industrial discharges, and municipal waste. And while freshwater ecosystems are very resilient, with examples of species refugia found in highly altered river systems, this resiliency is finite. We know there are thresholds that, once crossed, can put entire ecosystems at risk, with severe consequences for human well-being and biodiversity.

Given the importance of freshwater ecosystems in sustaining human well-being, it is surprising how little we know about their changing condition, their dependent species, or the roles that these species play in sustaining ecological functions. Knowledge is particularly poor for lower taxonomic groups (freshwater plants and invertebrates), especially in tropical regions. Here is a summary of the current status of freshwater biodiversity, given these gaps in our knowledge.

## Status of Freshwater Biodiversity

Data on the condition and trends of freshwater species are for the most part poor at the global level, although some countries (e.g., Australia, Canada, New Zealand, South Africa, and the United States) have better inventories and indicators of change of freshwater species. Much of the problem originates from the fact that large numbers of species have never been catalogued and baselines on population status rarely exist, with the exception of a few highly threatened species (e.g., river dolphins) or species of commercial value (e.g., Pacific salmon in the United States).

The leading global effort to monitor the conservation status of species, the World Conservation Union (IUCN) Red List, has limited coverage of freshwater species, although a large effort is ongoing to fill this gap. Because of its harmonized category and criteria classification (i.e., all contributing experts follow the same methodology and guidelines), the IUCN Red List is the best source of information, at the global level, on the conservation status of plants and animals. This system is designed to determine the relative risk of extinction, with the main purpose of cataloguing and highlighting those taxa that are facing a higher risk of extinction globally (i.e., those listed as Critically Endangered, Endangered, and Vulnerable).

The 2006 Red List highlighted that freshwater species have suffered some of the most marked declines. For instance, of the 252 endemic freshwater Mediterranean fish species, 56% are threatened with extinction, and seven species are now extinct. This represents the highest proportion of imperiled species of any regional freshwater fish assessment that IUCN has conducted so far. Similarly, in East Africa, one in four freshwater fish is threatened with extinction. Odonates, another taxonomic group assessed by IUCN, also show high levels of imperilment, with almost one-third of the 564 species assessed being listed as threatened.

In 2004, IUCN completed the first global assessment of more than 5500 amphibian species, which was updated in 2006 to include 5918 species. This assessment considerably improved our overall knowledge of the condition of freshwater species, though its scope and representativeness are limited by lack of information, with 107 species still listed as Data Deficient and therefore unassigned a threat category. The Global Amphibian Assessment serves to reinforce the reality of the imperiled status of freshwater species, with close to a quarter of all assessed species listed as threatened, 34 as extinct, and as many as 165 species described as probably extinct. Overall, 43% of all amphibian species are declining in population, indicating that the number of threatened species can be expected to rise in the future.

Even large freshwater mammals are at increasing risk. For instance, the common hippopotamus, which until recently was not thought to be endangered, was listed in 2006 as threatened because of drastic and rapid declines in its population figures, with recorded reductions of up to 95% in the populations of the Democratic Republic of Congo, because of illegal hunting for meat and ivory. Overall, 41 species of freshwater mammals, including many otter species, freshwater dolphins, two freshwater feline species, as well as freshwater ungulates and rodents are threatened with extinction.

Data on freshwater reptiles, namely freshwater turtles and crocodilians (i.e., crocodiles, caimans, and gharials) also show declining trends. According to the IUCN/SSC (Species Survival Commission) Tortoise and Freshwater Turtle Specialist Group and the Asian Turtle Trade Working Group, of the 90 species of Asian freshwater turtles and tortoises, 74% are considered threatened, including 18 critically endangered species, and 1 that is already extinct: the Yunnan box turtle. The number of critically endangered freshwater turtles has more than doubled since the late 1990s. Much of the threat has come from overexploitation and illegal trade in Asia. The status of crocodilians presents a similar pattern, particularly in Asia. Of the 17 freshwater-restricted crocodilian species, as of 2007, 4 are listed by IUCN as critically endangered (3 of which are in Asia), 3 as endangered, and 3 as vulnerable. The most critically endangered is the Chinese alligator. The major threats to crocodilians worldwide are habitat loss and degradation caused by pollution, drainage and conversion of wetlands, deforestation, and overexploitation.

While information on freshwater plants and invertebrates are not readily available to portray population trends, available data give insight into the condition of freshwater ecosystems and species. In terms of freshwater plants, while many macrophytic species are probably not threatened at a global or continental scale, many bryophytes with restricted distributions are rare and threatened. In the United States, one of the few countries to assess more comprehensively the conservation status of freshwater molluscs and crustaceans, The Nature Conservancy has assessed that one-half of the known crayfish species and two-thirds of freshwater molluscs are at risk of extinction, with severe declines in their populations in recent years. Furthermore, of the freshwater molluscs, at least 1 in 10 is likely to have already become extinct (Master et al., 1998).

While the Red List focuses only on threatened species and therefore does not look at population trends of nonthreatened species, it does provide a good measure of progress in attenuating species loss. Other measures of the change in vertebrate species populations, such as WWF's Living Planet Index (LPI), show a similar downward trend.

As these indices and examples show, freshwater species are in serious decline all over the world. However, available data and information are predominantly from temperate and developed regions. Some progress is being made to collect and compile information elsewhere, but progress is slow and resources needed are high, particularly in developing countries where capacity is limited.

## Major Threats

Threats to freshwater systems and species are numerous, overlapping, and operate over a range of scales. The embeddedness of freshwaters within the larger landscape, coupled with the fact that human communities require freshwater resources to survive, means that few freshwaters around the world remain pristine. Most freshwaters are subject to multiple anthropogenic stresses, and this multiplicity can complicate the identification of threat pathways and appropriate conservation levers. Threats can be variously classified, but here we recognize habitat degradation, water pollution, flow modification, species invasion, overexploitation, and climate change as major, often overlapping categories. These threats can be further described in terms of their origins (**Table 1**).

Habitat degradation encompasses habitat alteration, outright habitat destruction, and loss of access due to fragmentation, all of which are described briefly here. Virtually any modification to natural land cover within a catchment has the potential to alter downstream freshwater habitats, including floodplains. Land cover conversion for agriculture, urbanization, forestry, road-building, or other activities can result in changes in flow, sediment regimes, riparian and aquatic vegetation, water chemistry, and other parameters that together define freshwater habitats (**Figure 1**). Direct modifications like streambank mining may also make freshwaters inhospitable for some native species

**Table 1** Major threats to freshwater species and habitats

| Major threats to freshwater ecosystems | Description | Origin | | |
|---|---|---|---|---|
| | | Local | Catchment | Extra-catchment |
| Habitat degradation | Degradation and loss | X | X | |
| | Fragmentation by dams and inhospitable habitat segments | X | | |
| Flow modification | Alteration by dams | X | X | |
| | Alteration by land-use change | | X | |
| | Alteration by water abstraction | X | X | |
| Overexploitation | Commercial, subsistence, recreational, poaching | X | X | |
| Water pollution | Agricultural runoff (nutrients, sediments, pesticides) | | X | |
| | Toxic chemicals including metals, organic compounds, endocrine disruptors | X | X | |
| | Acidification due to atmospheric deposition and mining | | | X |
| Species invasion | Altered species interactions and habitat conditions resulting from accidental and purposeful introductions | X | X | |
| Climate change | Results in changes to hydrologic cycle and adjacent vegetation, affects species ranges and system productivity | | | X |

Note that, in nearly all cases where both local and catchment origins are listed, local stresses are transferred downstream to become catchment impacts elsewhere. Introduced species originate outside a catchment but introductions occur at individual locations and can spread both up- and downstream. Modified with permission from Abell R, Allan JD, and Lehner B (2007) Unlocking the potential of protected areas for freshwaters. *Biological Conservation* 134: 48–63, with permission from Elsevier; major categories from Dudgeon D, Arthington AH, Gessner MO, Kawabata Z, Knowler DJ, Lévêque C, Naiman RJ, Prieur-Richard A, Soto D, Stiassny MLJ, and Sullivan CA (2006) Freshwater biodiversity: Importance, threats, status and conservation challenges. *Biological Reviews* 81: 163–182.

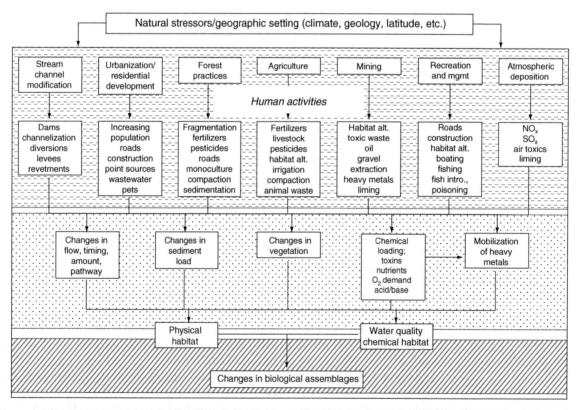

**Figure 1** Threats and threat pathways in freshwater ecosystems. Modified from Bryce SA, Larsen DP, Hughes RM, and Kaufmann PR (1999) Assessing relative risks to aquatic ecosystems: A mid-Appalachian case study. *Journal of the American Water Resources Association* 35: 23–36, with permission from Blackwell.

without destroying habitats entirely. Habitat loss can take a variety of forms, such as through wetland draining, dewatering of a river system, disconnecting a river from its floodplain, or conversion of lotic to lentic habitat through reservoir construction. Freshwater species may lose access to habitat when their dispersal or migratory routes are impeded, either by constructed barriers like dams or virtual barriers like highly degraded, and therefore impassable, river reaches. Assessments of the extent of habitat alteration, loss, and fragmentation are notoriously difficult to undertake at broad scales (**Table 2**).

Water pollution is related to habitat degradation and is typically classified as either point or nonpoint source. Point source pollution can be traced to an identifiable, single source like a pipe draining directly into a freshwater. Nonpoint source pollution, like runoff containing fertilizers from agricultural activities or oil from urban centers, comes from multiple diffuse sources and can be far more difficult to mitigate. Acid deposition and other toxic substances transported by air from outside a drainage basin are a special kind of nonpoint source pollution. Many pollutants are chemicals, such as pesticides and endocrine disruptors, but sediments, nutrients, and other 'natural' materials can also act as pollutants when present at abnormal levels. Even temperature can serve as a pollutant, such as when discharge from a power plant is hotter than normal river water or that from a deep reservoir is colder.

Like habitat degradation, flow modification can also result from either landscape activities or direct modifications to freshwaters, and often both simultaneously. Any landscape activity that alters infiltration and associated runoff, or even precipitation in the case of broad-scale climatic impacts, can change a freshwater system's hydrograph or hydroperiod, in the case of flowing and still waters, respectively. The effect of urbanization on reducing infiltration opportunities is well-documented, and soil compaction from activities like forestry can have similar consequences. River impoundments designed for hydropower generation, irrigation, flood control, navigation, or other uses generally alter the timing and volume of flows, as well as sediment and thermal regimes. Water withdrawals as well as returns and interbasin transfers alter flow regimes as well, even if the total volume of water over time may be relatively unchanged. In general, any modification to the natural flow regime, defined by flow magnitude, timing, duration, frequency, and rate of change, has the potential to affect native species adapted to it.

Overexploitation and species invasion can both affect species populations and communities directly. Overexploitation, or the unsustainable removal of individual animals or plants for commercial or other purposes has primarily affected some species of larger fish, some reptiles, as well as mussels and other large macroinvertebrates. Overexploitation has only rarely been implicated as the single cause in the extinction of individual species, but it has likely been a contributing factor in the decline of many. Species invasion, through accidental or intentional introductions of nonnative species, including through the opening up of previously inaccessible habitats, has had severe consequences for freshwater species in some instances. Impacts can include direct competition with or predation on native species, hybridization, habitat modification, and the introduction of disease and parasites. Species living in closed habitats like lakes appear to be particularly vulnerable to impacts from species invasion.

**Table 2** Alteration of freshwater systems worldwide

| Alteration | Pre-1900 | 1900 | 1950–1960 | 1985 | 1996–1998 |
|---|---|---|---|---|---|
| Waterways altered for navigation (km) | 3125 | 8750 | – | >500 000 | – |
| Canals (km) | 8750 | 21 250 | – | 63 125 | – |
| Large reservoirs[a] | | | | | |
| Number | 41 | 581 | 1105 | 2768 | 2836 |
| Volume (km³) | 14 | 533 | 1686 | 5879 | 6385 |
| Large dams (>15 m high) | – | – | 5749 | – | 41 413 |
| Installed hydrocapacity (MW) | – | – | <290 000 | 542 000 | ~660 000 |
| Hydrocapacity under construction (MW) | – | – | – | – | ~126 000 |
| Water withdrawals (km³/year) | – | 578 | 1984 | ~3200 | ~3800 |
| Wetlands drainage (km²)[b] | – | – | – | 160 000 | – |

[a]Large reservoirs are those with a total volume of 0.1 km³ or more. This is only a subset of the world's reservoirs.

[b]Includes available information for drainage of natural bogs and low-lying grasslands as well as disposal of excess water from irrigated fields. There is no comprehensive data for wetland loss for the world.

Reproduced with permission from Revenga C, Brunner J, Henninger N, Kassem K, and Payne R (2000) *Pilot analysis of global ecosystems: Freshwater systems.* Washington, DC: World Resources Institute.

Climate change is a final major category of threats to freshwaters, overlapping with habitat degradation, flow modification, and species invasion. Changes in global surface temperature and precipitation patterns will translate to changes in water temperature, water quantity, and water quality in the world's rivers, lakes, and other wetlands. Freshwater biodiversity will be affected indirectly through habitat alteration, and directly where species' life histories are tightly adapted to particular temperature or flow regimes. Dispersal opportunities to more hospitable habitats may be highly limited, especially in systems already fragmented or otherwise modified.

## Conservation Challenges

Although recognition of the looming freshwater crisis is growing, freshwater systems and their inhabitants are often still forgotten in local, national, regional, and international processes and plans. In part, this is due to the hidden nature of many freshwater species – they are literally 'out of sight and out of mind' underneath the water's surface. Additionally, many freshwater species are indistinct and small and thus do not engender the same emotional response as the large, colorful, charismatic species found in terrestrial and marine environments. Knowledge about freshwater species and habitats also lags behind that of their terrestrial counterparts. For example, about 3000 freshwater fish species are currently known in the Amazon Basin, but experts estimate that up to 5000 species will be discovered once the basin has been fully explored. A low profile and a lack of knowledge about freshwater systems' biology and ecology make the need for increased awareness from local to international levels even more critical for their conservation.

An even greater challenge to the sustainability of freshwater ecosystems is the direct competition that they are under with human societies for water resources. As the global human population increases, societal needs for water for agriculture, industry, energy generation, and human consumption will continue to grow and put freshwater ecosystems under mounting pressure. Demand is expected to grow fastest in developing countries and agriculture is expected to continue to be the largest consumer of water withdrawals (**Table 3**).

Conservation of freshwater ecosystems requires a paradigm different from that which guides terrestrial conservation activities. Traditional terrestrial approaches to biodiversity conservation center on setting aside areas of high conservation value as networks of protected areas. The inherent connectivity of freshwater systems limits the effectiveness of this approach within the freshwater realm. For example, water withdrawals or a dam upstream of a protected wetland can significantly alter that wetland's hydrology, thus undercutting the conservation effort. Basin-wide processes and interconnectivity must be a central objective in any effective freshwater conservation plan.

## Conservation Strategies

Effective freshwater conservation often requires the use of multiple complementary strategies. The most appropriate mix of strategies may depend on the scale of conservation significance of the ecosystem, as shown in **Figure 2**. Here we detail a subset of possible strategies, focusing on several with direct and more frequent applications to conserving freshwater biodiversity. Many additional strategies found in **Figure 2** are addressed elsewhere.

**Table 3**   Water withdrawals for world regions, by sector

| Region | Total (million $m^3$) 2000 | Per capita ($m^3$ per person) 2000 | Sector Withdrawals (%), 2000[a] | | |
|---|---|---|---|---|---|
| | | | Agriculture | Industry | Domestic |
| Asia (excluding Middle East) | 2 147 506 | 631 | 81 | 12 | 7 |
| Europe | 400 266 | 581 | 33 | 52 | 15 |
| Middle East and North Africa | 324 646 | 807 | 86 | 6 | 8 |
| Sub-Saharan Africa | 113 361 | 173 | 88 | 4 | 9 |
| North America | 525 267 | 1663 | 38 | 48 | 14 |
| Central America and Caribbean | 100 657 | 603 | 75 | 6 | 18 |
| South America | 164 429 | 474 | 68 | 12 | 19 |
| Oceania | 26 187 | 900 | 72 | 10 | 18 |
| Developed | 1 221 192.0 | 956 | 46 | 40 | 14 |
| Developing | 2 583 916.4 | 545 | 81 | 11 | 8 |
| Global | 3 802 320 | 633 | 70 | 20 | 10 |

[a]Sectoral withdrawal data may not sum to one hundred because of rounding.

Source: World Resources Institute, EarthTrends Freshwater Resources 2005.

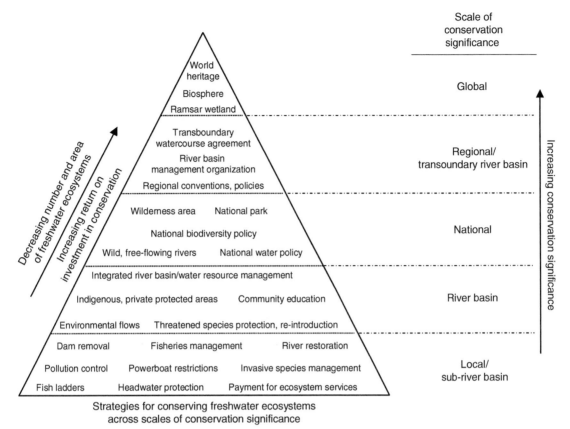

Figure 2 Strategies for conserving freshwater ecosystems. Modified with permission from Blanch SJ (2006) Securing Australia's natural water infrastructure assets. Solutions for protecting high conservation value aquatic ecosystems. A proposal. Sydney: WWF-Australia.

## Integrated Water Resources Management and Integrated River Basin Management

The concept of integrated water resources management (IWRM) is based on the interconnected nature of water bodies across landscapes, as well as along the river corridor from headwaters to the coast. IWRM promotes the need for participatory planning and implementation processes that bring stakeholders together to determine how to meet society's long-term needs for water while maintaining essential ecological services and economic benefits. A particular adaptation of the principles of IWRM to a river or lake basin is known as integrated river basin management (IRBM). The IRBM approach advocates managing a river and its entire catchment as a single system, and coordinating all the user group activities that take place within this geographic unit.

One of the key elements of IRBM is that it follows the principles of the ecosystem approach. The ecosystem approach framework is based on the central concept of managing water resources as integral parts of

the ecosystem, rather than just as a resource to be exploited without regard to the system that nurtures it. Under this approach, water managers must do more than just satisfy one or two key users, but instead accommodate the wide array of economic and social benefits that people derive from aquatic environments, such as recreation, transportation, local livelihoods, cultural identity, and so on. The practical effect of this is that it widens the group of users who have a legitimate say in how the resource is managed.

Applying the ecosystem approach to managing water would ensure, at least in theory, that all goods and services derived from ecosystems, including inherent ecological functions, are taken into account when assessing development plans for a given river or lake. But despite the commitment by many countries to implement IWRM and IRBM approaches, such plans are still in their infancy. In most river basins around the world, allocation of water for irrigation and hydropower continues to take precedence over other water uses, as countries prioritize food and electricity production. Part of the problem is that implementing an IRBM approach requires different legal and institutional frameworks that go beyond our existing national government agencies. It requires cross-sectoral collaboration, and at times, new institutions, such as river basin organizations (RBOs), which link adjacent states along the river corridor in a legal framework that allows for the cooperative management of water resources within a single basin. RBOs can provide a forum for dialogue where the wide array of stakeholders can participate. As a result, development plans and water-use strategies can become more balanced, minimizing environmental and social impacts. For RBOs to be effective, however, they need to be given the authority, funding, and legal mandate to implement long-term water management policies – something that to date has been the exception rather than the norm. They must also have the wide participation of riparian states. The success of such approaches, however, ultimately depends heavily on cooperative governance and political commitment, which unfortunately are still lacking in many parts of the world.

### Environmental Flows

Water management laws share water among users, such as towns, agriculture, industry, and the environment. Changing the shares among different user groups is often contentious, particularly in arid areas or where existing water rights are infringed upon, but presents a key conservation opportunity by reserving water for ecosystems. Termed 'environmental flows,' water allocated for ecosystems protects or reinstates key aspects of a river's flow. For example, pumping water may be restricted or prevented when river flows are low to allow aquatic organisms to move along a stream and allow wildlife to drink. Environmental flows do not just benefit the river channel but also floodplains, wetlands, estuaries, and coastal environments. These environments rely upon freshwater, sediments, nutrients, and carbon to be delivered from river channels. Small- and medium-sized floods may be protected from overextraction to ensure lateral hydrological connectivity between a river and its floodplain to allow wetlands to be refilled and fish species to migrate.

Groundwater-dependent ecosystems may be particularly dependent upon environmental flows, even if they occur at long distances from the river channel. Laws exist in some countries to legally protect water from being extracted, or to protect water that is specifically released from dams to maintain or recreate pulses and small floods. For example, environmental water reserves are guaranteed in water statutes in South Africa and the Australian state of Victoria to sustain river ecological functions.

### Dams: Operation, Design, Removal

Building, operating, and removing dams, weirs, and barrages (collectively 'dams' for short) can affect river ecosystem function more than nearly any other set of activities. There are an estimated 45 000 large dams worldwide, and millions of small ones. In many parts of the world, new large dams are under construction or planned, especially for hydropower generation, whereas in a few countries like the United States, select dams are now being removed.

River ecology considerations should be fully incorporated into decision making about whether or not to dam a river, as recommended by the World Commission on Dams. For example, leaving the mainstem or large tributary of a river undammed will retain significant ecosystem benefits for river communities. Where dams are in place or under development, trying to mimic natural flow patterns as much as possible can recreate aspects of the flow regime that have been lost. Mitigating unnaturally warm or cold water discharges from dams is often needed to reestablish temperature regimes that trigger fish spawning and allow growth. For example, average water temperatures immediately below deep bottom-release dams in Australia's Murray-Darling Basin may be 5–10 °C cooler than natural. Building a multiple-level off-take at the dam wall allows dam managers to selectively draw surface water from the warmer epilimnion as dam levels rise and fall, thus warming river water below the dam, and hence increasing growth and survival of native warm-water fish.

Old, unsafe, or unnecessary dams can be removed to reinstate more natural flow patterns and permit freshwater organisms to move up and down a stream. Dam removal also reestablishes sediment transport regimes that continually erode and deposit sediments, thus continually creating new habitats. If removal is not feasible, a fish ladder or fish lift may be built to enable some fish and other freshwater animals to move upstream past dams, particularly during spawning migrations. Fish ladders and fish lifts are often poor alternatives to removal, however. They may not allow all fish species to use them, their effectiveness can be reduced by becoming blocked, and they require ongoing expensive maintenance.

### Protected Areas

Protected areas – defined as 'areas of land and/or sea especially dedicated to the protection and maintenance of biological diversity, and of natural and associated cultural resources, and managed through legal or other effective means' – have received far less attention as tools for conserving freshwater species and habitats than they have for terrestrial and more recently marine features. Where protected areas have been used for freshwaters it has most frequently been through the establishment of Ramsar sites, identified under the Convention on Wetlands. Traditional protected areas have often been dismissed as ineffective for conserving freshwaters because of the connected and often linear nature of the systems. However, nontraditional protected areas, embedded within basin-wide integrated management efforts, are receiving increased attention. For example, riparian buffer zones can protect critical stream- or lake-side vegetation that filters pollutants, contributes organic material, moderates water temperatures, and provides woody debris for instream habitat. Floodplain reserves are in effect a special kind of riparian buffer zone, typically much wider and designed to protect the highly productive transitional areas that provide habitat to large numbers of both freshwater and terrestrial species. Fishery, or harvest, reserves are designed to provide spatial or temporal refuges for exploited freshwater species, so that populations can be sustainably fished over the long term; broader biodiversity conservation may be a secondary benefit. And protecting rivers as free-flowing, as described later, is a potentially powerful conservation tool gaining traction in certain countries.

### Retaining Wild or Free-Flowing Rivers

Maintaining rivers without dams is arguably the single most effective river conservation strategy. Free-flowing rivers have been defined elsewhere as any river that flows undisturbed from its source to its mouth, either at the coast, an inland sea, or at the confluence with a larger river, without encountering any dams, weirs, or barrages and without being hemmed in by dykes or levees. Wild or free-flowing rivers allow water, sediment, nutrients, and biota to move longitudinally from the headwaters to the sea, terminal wetland or lake. Options for designating wild rivers are significantly reduced in most regions (**Table 4**), as only one-third of the world's large rivers remain free-flowing. But, some of the world's greatest rivers remain free-flowing along their mainstem, and sometimes on major tributaries too, including the Amazon, Okavango, Irawaddy, Sepik, and Mackenzie Rivers. No global free-flowing river conservation framework currently exists, although various countries have laws and programs to legally recognize and protect unimpounded rivers. The United States' Wild and Scenic Rivers program protects reaches of over 150 rivers covering 11 000 miles from new dams and some other types of development, but other damaging catchment uses may not be regulated. Wild and Heritage rivers programs and laws also exist in Canada and in the Australian States of Queensland and Victoria.

### Restoration

Freshwater restoration involves recreating key features of a stream's ecological processes that have been impaired or lost, and potentially reintroducing species that have become locally extinct. The profession of stream restoration has developed significantly in recent decades, with investments in restoration by communities, industries, and governments totaling billions of dollars annually. Restoration techniques are many and varied. For example, fencing out stock and revegetating riparian zones with indigenous plants helps filter nutrients, reduce erosion, improve habitat, and shade the water. Reintroducing rare and endangered species to streams from which they have disappeared may achieve high community support and serve to raise the profile of stream restoration more broadly, but will only be effective if the threatening processes that drove the species extinct in the first place have been mitigated or stopped. Relatively drastic restoration techniques are sometimes the only realistic option for highly degraded streams. Bulldozers may be required to remove contaminated sediments that pose an ongoing risk to human health or ecosystems. Formerly channelized streams with low habitat and biodiversity value may have meander bends and rock bars added back. Similarly, logs may be put back into streams after decades of desnagging for navigation to reproduce the eddies and submerged habitats that fish and invertebrates need.

**Table 4** Regional distribution of rivers longer than 1000 km and percentage of rivers remaining free-flowing

| Region | Number of large rivers | Percent free-flowing | Example of free-flowing rivers |
| --- | --- | --- | --- |
| Australia/Pacific | 7 | 43 | Cooper Creek, Sepik, Fly |
| Europe (west of Urals) | 18 | 28 | Oka, Pechora, Vychegda |
| Africa | 23 | 35 | Chari, Rufiji, Okavango |
| North America | 33 | 18 | Fraser, Mackenzie, Liard |
| South America | 37 | 54 | Amazon, Orinoco, Beni |
| Asia | 59 | 37 | Lena, Amur, Brahmaputra |

Modified with permission from WWF (2006) *Free-Flowing Rivers. Economic Luxury or Ecological Necessity?* Zeist: WWF.

### Convention Programs of Work

No single comprehensive international convention currently exists for the conservation and sustainable use of freshwater ecosystems. Rather, global efforts are primarily underpinned by two global conventions, namely the Convention on Wetlands (Ramsar) and the Convention on Biological Diversity, plus a host of transboundary watercourse agreements and widely espoused best management principles. The Convention on Wetlands commits signatory nations to the wise use of all wetlands, the designation of wetlands of international importance, and international cooperation. The Convention's definition of a wetland is very broad and provides a basis for providing some form of conservation for all forms of freshwater and coastal ecosystems. Ramsar sites arguably form the world's largest network of conserved aquatic ecosystems with 1650 sites covering 150 million ha of freshwater and some coastal systems globally as of April 2007, although effective legal protection and on-ground management is lacking for many.

The Convention on Biological Diversity's Programme of Works on Inland Waters and Programme of Work on Protected Areas have goals and actions regarding protecting representative types of freshwater ecosystems within IRBM. Formal cooperation exists between these two Conventions to harmonize global efforts on the conservation of freshwater ecosystems. The United Nations Convention on the Law of the Non-Navigational Uses of International Watercourses contains commitments relevant ·to freshwater conservation, such as protecting ecosystems, but has not been ratified by enough countries to come into force legally. Notwithstanding, transboundary watercourse agreements have been negotiated for many river basins and provide a framework, at least on paper, for freshwater ecosystem conservation.

### Conclusions

The imperilment of freshwater species and habitats around the world is of urgent concern. Not only is a large fraction of the Earth's biodiversity threatened, but essential ecosystem services upon which human communities depend are at risk as well. Conservation strategies for addressing degraded ecosystems will have to be developed and applied within the context of integrated basin management to ensure that threats are mitigated and critical ecosystem processes, often linked to hydrology, can function within natural ranges of variation. Many of the world's freshwaters are already irreparably damaged, but there is time to secure protection for those that remain relatively intact if the political will for such protection can be generated and sustained.

*See also:* Aquatic Ecosystem Services; Biomanipulation of Aquatic Ecosystems; Effects of Climate Change on Lakes; Effects of Recreation and Commercial Shipping; Eutrophication of Lakes and Reservoirs; International Water Convention and Treaties; Lake and Reservoir Management; Lake Management, Criteria; Lakes as Ecosystems.

### Further Reading

Abell R, Allan JD, and Lehner B (2007) Unlocking the potential of protected areas for freshwaters. *Biological Conservation* 134: 48–63.

Allan JD (2004) Landscapes and riverscapes: The influence of land use on stream ecosystems. *Annual Review of Ecology Evolution and Systematics* 35: 257–284.

Allan JD, Abell RA, Hogan Z, Revenga C, Taylor BW, Welcomme RL, and Winemiller K (2005) Overfishing of inland waters. *BioScience* 55: 1041–1051.

Allan JD and Flecker AS (1993) Biodiversity conservation in running waters. *Bioscience* 43: 32–43.

Dudgeon D, Arthington AH, Gessner MO, *et al.* (2006) Freshwater biodiversity: Importance, threats, status and conservation challenges. *Biological Reviews* 81: 163–182.

Harrison IJ and Stiassny MLJ (1999) The quiet crisis: A preliminary listing of the freshwater fishes of the world that are extinct or 'missing in action.' In: MacPhee RDE and Sues HD (eds.) *Extinctions in Near Time: Causes, Contexts, and Consequences*, pp. 271–332. New York: Kluwer Academic/Plenum.

Master LL, Flack SR, and Flack BA (eds.) (1998) Rivers of Life: Critical Watersheds for Protecting Freshwater Biodiversity. Arlington: The Nature Conservancy.

Poff NL, Allan JD, Bain MB, *et al.* (1997) The natural flow regime: A paradigm for river conservation and restoration. *Bioscience* 47: 769–784.

Revenga C and Kura Y (2003) Status and trends of biodiversity of inland water ecosystems. Technical Series No. 11. Montreal: Secretary of the Convention on Biological Diversity.

Revenga C, Brunner J, Henninger N, Kassem K, and Payne R (2000) Pilot Analysis of Global Ecosystems: Freshwater Systems. Washington, DC: World Resources Institute.

Silk N and Ciruna K (eds.) (2004) A Practitioner's Guide to Freshwater Biodiversity Conservation. Boulder: The Nature Conservancy.

WWF (2006) Free-Flowing Rivers. Economic Luxury or Ecological Necessity? Zeist: WWF.

### Relevant Websites

http://www.dams.org – World Commission on Dams.

http://www.biodiv.org – Convention on Biodiversity (CBD).

http://www.earthtrends.org – Watersheds of the World, and Freshwater Resources 2005.

http://www.fishbase.org – FishBase.

http://www.gemswater.org – UNEP Global Environment Monitoring System (GEMS), Water.

http://www.giwa.net/ – Global International Waters Assessment (GIWA).

http://www.globalamphibians.org – Global Amphibian Assessment.

http://www.issg.org – Global Invasive Species Database.

http://www.iucn.org – Red List of threatened species.

http://www.millenniumassessment.org – Millennium Ecosystem Assessment.

http://www.natureserve.org – NatureServe.

http://www.panda.org – Living Planet Report.

http://www.ramsar.org – Ramsar.

# Aquatic Ecosystem Services

**K E Limburg,** State University of New York, Syracuse, NY, USA

## Introduction

What's a sunset on a still lake worth? Or a sip from a spring on a summer bike ride? What inspires novels, poems, and essays about water? These questions raise the idea of the value, in terms that might relate directly to many people's experience. Other kinds of value are not as readily as perceived. For example, there is value in a clean river, or in ground water that is available for plant uptake and evapotranspiration. There is value in the cooling and heating properties of lakes, rivers, and the ocean, and there is value in the existence of healthy fish communities that may provide both protein and recreational pleasure.

Collectively, ecologists and others have begun to refer to these different kinds of values as *ecosystem services*. The concept gained attention during the 1990s as scientists began to realize that natural ecosystems were being damaged and destroyed by humans at unprecedented rates, due to human population growth and the resulting increased exploitation of natural resources. Further, there was concern that this environmental damage might not only be irreparable in many cases, but might also fundamentally alter global cycles, such as the hydrological cycle or the global climate. Scientists feared that many ecosystem properties would be altered or lost before understanding their importance, for example, biodiversity in many systems. Ecologists coined the term 'ecosystem services' as a way of conveying the idea that ecological systems provide services, in addition to goods, that underpin human well-being. These goods and services provide value to humanity in both direct and indirect ways.

### Notions of Value

No one knows precisely how long humans have held formalized concepts of value, but the philosophical roots go back at least several thousand years. The Greek philosopher Aristotle struggled with the value of using things, versus the value of exchanging things, and how there could be parity between them. An often-cited example of the difference between use and exchange value is the "diamonds and water paradox." Water, which is essential for life, has been extremely cheap, historically; on the other hand, diamonds, which are exchangeable luxury items, are extremely expensive. The history of economic thought embodies the search for measures of value, wealth, and exchange. This chapter will not go into this history, but references are included in Further Reading.

Essentially, value is the difference that something makes to someone; it may be tangible and apparent, like a durable good, or it may be something that people overlook in their day-to-day activities. The value of many ecosystem services falls into the latter category. It is also true that value is subjective and contextual. The value of some ecosystem services in a generally deteriorating environment may be higher than in a 'pristine' world.

## Classification of Ecosystem Services and Natural Capital

Ecosystem services are broken down into a number of categories. The Millennium Ecosystem Assessment listed four broad categories:

- *Provisioning services* are those that provide goods such as food and water;
- *Regulating services* are those that control various processes, such as flood control or suppression of disease outbreaks;
- *Supporting services*, such as nutrient recycling, maintain material and energy balances; and
- *Cultural services* are those that provide spiritual, moral, and aesthetic benefits.

To this, we may add other types of services, such as provision of habitat, or information flow (science and education).

Selective examples of ecosystem services from aquatic systems are given in **Table 1**. Some are obvious, such as the production of food and other exploitable products, or the provision of recreational opportunities. Some are less apparent, such as the storage and purification of water in aquifers, the ameliorating influence of large lakes on local climate, the role of the world's oceans in regulating global climate, or the assimilation of wastes by biogeochemical processing.

Another distinction that is often made is between 'ecosystem services' and 'natural capital.' Following economic terminology, capital is the standing stock of a good or information (represented by money); hence, natural capital is the standing stock of environmental goods. Natural capital generates flows of ecosystem services, either on their own or together

**Table 1** Examples of aquatic ecosystem services and the underlying ecosystem processes, components, and functions that generate the services

| Type of service | Function | Ecosystem process and components | Specific service |
|---|---|---|---|
| Provisioning | Biogeochemical conversion | Energy and matter flow through food webs | Production of fish, shellfish, algae, and other consumables |
| Provisioning | Genetic resources | Evolution and natural selection | Production of novel compounds used in medicine, industry, engineering, etc. |
| Habitat | Nursery | Habitat suitable for reproduction and early life stages | Promote survival and maintenance of species, e.g., estuarine-dependent fish species |
| Habitat | Refugia | Habitat necessary for some other life stage | Promote survival and maintenance of species |
| Habitat (ecosystem structure) | Flood and erosion control | Specific structures, e.g., mangrove swamps, salt marshes, and riparian vegetation that break the force of storm surges and floods | Mitigating the force of natural disasters |
| Supporting | Nutrient cycling | Biotic and abiotic storage, transformation, and uptake of nutrients | Control of eutrophication; waste assimilation |
| Supporting | Primary production | Transformation of solar energy into biochemical energy through photosynthesis | Provision of food and other products that are directly or indirectly consumed |
| Supporting | Wetland soils formation | Production and partial decomposition of organic matter; mixing with inorganic sediment | Provision of peat or other fertile soils |
| Regulating | Gas regulation | Biogeochemical processes involved in $O_2$ and $CO_2$ exchanges between air and water | Maintenance of gas balances in water and air |
| Regulating | Water supply and regulation | Filtering, retention, and storage of fresh water | Provision of water for consumptive use |
| Regulating | Trophic feed-back effects | Top-down predatory control in food webs; trophic cascades | Maintenance of low populations of nuisance algae; control of algally-derived turbidity |
| Regulating | Climate regulation | Temperature regulation, hydrologic cycle, biotically mediated processes (e.g., production of dimethyl sulfide aerosols) | Maintenance of favorable climatic conditions for humans and their production systems |
| Cultural | Aesthetics and art | Attractive features (lakes, rivers, marshes, shorelines) | Enjoyment of scenery, inspiration for art, music, and literature |
| Cultural | Recreation | Variety in land- and waterscapes that promote recreation | Enjoyment of scenery, activities, light exploitation |
| Cultural | Spiritual | Whole or partial ecosystem functioning or features | Use of water or aquatic resources for religious purposes |
| Informational | Scientific knowledge creation | Whole or partial ecosystem functioning or features that promote inquiry and learning | Use of aquatic systems in research and education |

Note that services are couched in anthropocentric terms.

with capital flows from other resources. An example of autonomous production of a service from natural capital might be the provisioning of protein from an animal community, such as oysters in an estuary. An example of natural capital flows combining from different systems might be the movement of detritus through a river ecosystem into a recipient estuary; the detritus is transformed on its journey and fuels different food webs, some of which may ultimately provide food or other services.

The concept of ecosystem services is inherently biased toward the anthropocentric perspective. The reason is twofold. First, a major point of discussing ecosystem services is to highlight their utility and essentiality for humans in an economic world, which increasingly marginalizes the value of undeveloped, natural ecosystems. Second, although ecosystems do not 'care' about whether *Homo sapiens* exists among all other species, *we do*, and thus we recognize that ecosystem goods and services are the natural components and processes that cannot be compromised, if our species is to persist. Moreover, functionality of ecosystem depends on access to these goods and services by nonhuman organisms as well. In contrast to the commonly held, layperson's view that the natural world is a subsystem of the human world, the concept of ecosystem services helps us to see how human societies are, in fact, embedded within the natural world.

## Valuing Ecosystem Goods and Services

Although the importance of ecosystem goods and services is recognized, quantifying these has been at times challenging and controversial. Some quantification has taken the form of economic valuation, using accepted methods. Where markets exist, valuation in monetary terms is fairly straightforward, even if it does not necessarily capture all the value. For example, fish markets capture the value of fish as food, but not their value in food webs per se.

Methods used by environmental and ecological economists include:

- *Avoided cost:* the cost that an individual or society avoids paying because of the natural service. For example, fringing mangroves and other coastal wetlands can buffer coastlines from storm damage. In 2005, Hurricanes Katrina and Rita inflicted damage along the U.S. Gulf of Mexico coast that ran into tens of billions of dollars; much of that could have been avoided, had wetlands been left intact.
- *Replacement cost:* the cost that individuals or society would have to pay if the natural service did not exist. An example is the use of cypress wetlands in the Southeastern United States to 'polish' treated wastewater, stripping out nutrients that otherwise would have to be treated with expensive tertiary wastewater treatment.
- *Travel cost:* some ecosystem goods and services require travel to appreciate, implying the costs willingly borne by people to 'utilize' the good or service. This approach is a common method to assess the minimum value of parks, recreational areas and activities.
- *Factor income:* the degree to which a service enhances incomes. For example, improvements in water quality may enhance the tourism sector.
- *Hedonic pricing:* a de facto analysis of goods associated with ecosystem services. Real estate prices near a lake or a beach typically are much greater than a few miles inland, reflecting the appeal of the water or shoreline to humans.
- *Contingent valuation:* a survey-based method to evaluate individuals' willingness to pay for increased flow of a service, or willingness to accept the costs of maintaining a service, through the posing of hypothetical scenarios. An example would be the willingness not to develop near a sensitive wetland area.
- *Option and insurance values:* although these may be more difficult to compute, these are values of

ecosystem services that provide options in the face of uncertainty. An example is the potential for new uses to be found through 'prospecting' for pharmaceuticals in corals, or maintenance of intact riparian zones along a floodplain (rather than building thereon). A variant on option value is 'bequest value,' that is, the value that is to be left to future generations.

A landmark study of the total value of Earth's ecosystem goods and services (see Further Reading), conducted as a valuation exercise in the mid-1990s, arrived at an estimated worth of $33.3 trillion, nearly double the global Gross Domestic Product (GDP). Controversy ensued, as economists faulted the methods used, and ecologists argued that monetary valuation did not truly capture ecosystem value. Nevertheless, this and other case studies have brought to light that ecosystems, even disturbed ones, provide value to society.

## Examples of Ecosystem Services Generation in Aquatic Ecosystems

### Wetlands

Wetlands have received a great deal of study over the past 40 years, and have become recognized as ecosystems having many valuable properties. As a result, many wetlands are now protected the world over, or if they are destroyed, replacement wetlands are created. Wetland ecosystem services include provisioning of food, freshwater, and building materials; water filtration and purification; critical habitat for many species of plants, amphibians, fish, and birds; storm abatement, flood control, and erosion control; microclimate regulation; and at larger scales, wetlands are important sites for nutrient cycling and carbon sequestration.

Because of their recognized ecological importance and also their particular vulnerability to development pressures, more attempts at economic valuation of wetlands have been undertaken than at perhaps any other ecosystem. One study from the early 1970s estimated the value of tidal marshes at $2500–$4000 per acre per year (approximately $6200–$9900 per hectare per year), summing up all nonoverlapping services. In India, hurricane damages cost nearly five times as much to villages not protected by mangroves compared to a village in the wind shadow of a mangrove protected area. More recent estimates are lower, but still run high enough to place the value generation of the world's wetlands in the tens of billions of dollars per year.

### Rivers and Estuaries

Although composing a small fraction of the world's surface water, rivers and estuaries are important producers of ecosystem services. Historically, waterways have been critical for transportation, and human settlements often sprung up around the intersection of a river or estuary with another geographic feature (e.g., a natural harbor). Rivers have also been used for drinking water and to remove or dilute waste from populated areas, although often their assimilation capacities, such as to break down organic matter and still provide sufficient dissolved oxygen to support aquatic life, have been exceeded.

Rivers and estuaries are important habitat for fish and shellfish; estuaries, on an areal basis, are some of the most productive ecosystems on Earth. As systems that link continents to the oceans, rivers and estuaries play key roles for many species that use them for all or part of their life cycles. Many commercially important species of fish, for example, use estuaries and rivers as 'nursery' habitat wherein reproduction and early life stages play out. Oysters and other bivalves filter enormous quantities of water as they feed, reducing turbidity, and translocating nutrients to the benthos.

Among the less obvious services of these systems is the connection to land through floodplains and their riparian zones. During flooding, these areas receive silts that increase their fertility, and debris is removed as floodwaters recede. Flooding also connects aquatic and terrestrial food webs, so that fish may literally, as in the Amazon, forage in the trees during flooding. When dry, floodplains are important habitat to many species that are exploited, including large mammals and birds.

Rivers also play a role in the spiritual life of civilizations. For example, the Hindu religion particularly reveres large rivers, which symbolize the washing away of pollution and sin. In the Shinto religion of Japan, springs are thought to be inhabited by deities called Kami. Rivers have been harnessed for a thousand years for irrigation on the island of Bali, through a system of water temples that are managed by priests. Many religions also worshipped water spirits of various forms.

## Lakes

Lakes hold a special place of importance for many people and societies, as they provide freshwater, food fish, and many opportunities for recreation. Lakeshores hold strong attraction for many people, such that real estate values of lakeside are multiplicatively higher than that of adjacent areas without lakes. In a study for the State of Maine, it was found that good water quality increased the collective value of lakeshore homeowners' property by $6 billion over their purchase costs.

Lake productivity can provide recreational fishing opportunities in the billions of dollars. A recent study of sport fishing in New York State found that inland recreational fishing generated $1.2 billion annually in direct expenditures and an additional billion dollars' worth of indirect expenditures (e.g., dining out at local restaurants, motel stays, and so on).

### Coral Reefs

Although not typically occurring inland, coral reefs rank among the most biodiverse ecosystems on Earth, and there is great interest in evaluating their ecosystem goods and services contributions. Reefs form in clear, tropical waters, concentrating biomass and building structure in what are often otherwise nutrient-poor systems. Corals themselves are formed of symbiotic associations of coral polyps and zooxanthellae, a type of dinoflagellate that provides the corals with photosynthate while receiving protection from the coral. Coral reefs attract many fish and invertebrates, and can also support algae.

Services that coral reefs provide include direct support of fisheries and recreation. Reefs also generate the fine white sands that attract tourists to tropical resorts, and many provide other commercial products such as shells or fish for the aquarium trade. Where reefs are located close to shore, they also play an important role in coastal protection. Studies have shown that this service alone is worth $1–$12 million per kilometer of shoreline. Animals that live on coral reefs but that make excursions to nearby seagrass beds export nutrients to these outlying areas, further enhancing productivity.

## Issues in Ecosystem Service Assessment and Valuation

Ecosystem services are not always easy to quantify for a number of reasons. One reason is that information about a particular service, or the natural capital that generates it, may be imperfect or even completely lacking. For example, the benefits of aquatic ecosystems in developing countries may be unaccounted for because no research has been conducted on the scope and status of these systems. In developed nations, markets may distort the value of ecosystem goods and services, inflating or deflating them due to factors such as the influence of media attention or politics. An example is the recent, rapid colonization of inland North American waters by the zebra mussel (*Dreissena polymorpha*). This invasive species is generally regarded as a pest that devalues water bodies through removal of plankton that fuels the food web and hence affects fisheries. Although true, increased water clarity from zebra mussel filtering is highly valued as well and can quite likely be seen by examining changes in waterfront property values over time.

Another problem is that ecosystem goods and services are often 'multifunctional,' and involved in more than one process. Thus, it has been difficult to parse the multiple values generated by the same component (e.g., marsh vegetation is important as structure, as habitat, and as food or building material). One approach has been to calculate the 'total ecosystem value,' which sums all the values; another is to estimate the major value, or the values of the most clearly distinguishable values. Clearly, these different approaches will yield different estimates with different ability to capture the full value.

Other issues are those of scale and uncertainty, even when ecosystems are fairly well researched. Different ecosystems may contribute one type of service at local scales (or short-time horizons), but contribute, either collectively or individually, different services at larger (or longer) scales. Uncertainty arises not only from spatiotemporal variability, but also from the system's degree of resilience, that is, its responses to stochastic events (e.g., oil spills or hurricanes). Some aquatic ecosystems, such as lakes, can be 'pushed' from one stable state (e.g., oligotrophic) to another (e.g., eutrophic) through pollution with excess nutrients that ultimately build up in sediments, and are released for many years after the pollution load has diminished or ceased. The specifics of lake morphometry, residence time, climate, and trophic structure all play a role in lake resiliency and maintenance of trophic state.

## Threats to Ecosystem Services

Aquatic ecosystems and the services they generate are threatened by direct and indirect anthropogenic insults. These all-too-familiar threats include pollution, habitat loss and fragmentation, and overuse. The force of humanity over the past century has overwhelmed many aquatic ecosystems, or altered their functioning in ways that compromises delivery of ecosystem services.

Some human alterations of aquatic systems have apparently opposite impacts. Dams and dikes obstruct the connectivity of systems, impairing the movement of organisms. Conversely, canals increase connectivity, and promote the movement of organisms, including exotic or invasive species, often with disastrous consequences. Reservoirs, important for drinking water supplies, hydropower generation, agricultural production, or other uses, sometimes enhance ecosystem services (e.g., wildlife habitat value), but also 'starve'

rivers, estuaries, and coastal zones of sediments and nutrients. Moreover, reservoirs often change in productivity and species richness over time, many becoming warm-water havens undesirable for fish, for example.

Species richness and diversity is threatened in many aquatic ecosystems. The causes include habitat loss or alteration, overharvesting, and pollution. In North America alone, over 350 species or subspecies of fish are considered endangered, threatened, or of special concern. This list includes species that only a few decades ago were so abundant that it was difficult to imagine they would become scarce. We are only beginning to understand the implications of depleted food webs, and their impaired flows of services.

Climate change is a specter that threatens aquatic ecosystems and their services in multiple ways. Calculations have been done to estimate the loss of cold-water habitat under different scenarios of global warming; many temperate and boreal aquatic systems will become unsuitable for coldwater species, such as trout, salmon, and whitefish. Warming will also alter hydrologic cycles, causing more loss of small and ephemeral water bodies and streams. Climate change will also involve more extremes of weather: for example, although the northeastern United States is predicted to become wetter on average, this increase in precipitation will be delivered through more and larger storms. Already, the shallower Laurentian Great Lakes of North America are generating more winter snow storms, as they store more heat and therefore interact more (because of less and shorter duration of ice cover) with passing cold fronts and increased winds. It may be that intact, functioning ecosystems will become increasingly important and valuable as buffers against increasing numbers of catastrophic weather events.

## Conclusions

Although still in its infancy, the study of ecosystem services has brought out the tangible importance of aquatic systems to humanity. Much work remains, in terms of identifying the services, developing criteria for their measurement, and quantifying them. Nevertheless, it is clear that aquatic ecosystem services and goods provide billions and perhaps trillions of dollars' worth of benefit to societies. Historically, people often viewed Nature not only as a wily adversary, fraught with danger and disease, but also with opportunity. Today, the scale of human impacts on ecosystems is such that it is overwhelming or destroying many at an unprecedented rate, and humans will ultimately pay a price through such effects as decreased productivity and lower quality of life. The identification and quantification of natural capital and ecosystem services is one means of revealing hidden subsidies of Nature to societal functioning and well-being, and thus help humanity avoid the cost of ignoring these services.

## Glossary

**Benefit** – The amount of 'good' received in consuming a physical good or a service; related to utility.

**Cost** – The amount of money or other resource required to commensurate for a given good or service.

**Good** – A good is a physical item to which value can be attributed. Goods may be durable (e.g., wood, metal, stone) or nondurable (e.g., food). Goods possess economic utility.

**Service** – A service is a nonmaterial analog to a good, in the sense that it can also be valued and possesses utility. An example of an ecological service is remineralization of nitrogen by soil microbes.

**Utility** – In economics, utility is a measure of the satisfaction derived by humans from consuming goods and services.

## Further Reading

Badola R and Hussain SA (2005) Valuing ecosystem functions: an empirical study on the storm protection function of Bhitaranika mangrove ecosystem, India. *Environmental Conservation* 32: 85–92.

Boyle K and Bouchard R (2003) Water quality effects on property prices in northern New England. *LakeLine* 23: 24–27.

Costanza R (2004) Value theory and energy. In: Cleveland CJ (ed.) *Encyclopedia of Energy*, pp. 337–346. Amsterdam: Elsevier Science.

Daily G (ed.) (1997) *Nature's Services*. Washington, DC: Island Press.

Farber SC, Costanza R, and Wilson MA (2002) Economic and ecological concepts for valuing ecosystem services. *Ecological Economics* 41: 375–392.

Costanza R, D'Arge R, de Groot R, *et al.* (1997) The value of the world's ecosystem services and natural capital. *Nature* 387: 253–260.

Gosselink JG, Odum EP, and Pope RM (1974) *The Value of the Tidal Marsh*. Baton Rouge: Louisiana State University, Center for Wetland Resources. Publication no. LSU-SG-74–03.

Limburg KE and Folke C (eds.) (1999) The ecology of ecosystem services. *Ecological Economics* 29(2); special issue.

Millennium Ecosystem Assessment (2006) Millennium Ecosystem Assessment Synthesis Reports. Available at http://www.millennium assessment.org/.

New York Sea Grant (2001) The economic contributions of the sport fishing, commercial fishing, and seafood industries to New York State. Stony Brook, New York: New York Sea Grant.

Schumpeter JA (1978) *History of Economic Analysis*. New York: Oxford University Press.

# International Water Convention and Treaties

**A T Wolf,** Oregon State University, Corvallis, OR, USA

*"Fierce competition for fresh water may well become a source of conflict and wars in the future."*

Kofi Annan, March 2001

*"But the water problems of our world need not be only a cause of tension; they can also be a catalyst for cooperation...If we work together, a secure and sustainable water future can be ours."*

Kofi Annan, February 2002

## Water Conflict and Cooperation

Water management is, by definition, conflict management. Water, unlike other scarce, consumable resources, is used to fuel all facets of society, from biology and economics to aesthetics and spiritual practice. Moreover, it fluctuates wildly in space and time; its management is usually fragmented and is often subject to vague, arcane, and/or contradictory legal principles. Within a nation, the chances of finding mutually acceptable solutions to the conflicts among water users drop exponentially as more stakeholders are involved. Add international boundaries, and the chances decrease exponentially still further.[1]

Surface and groundwater crossing international boundaries presents increasing challenges to regional stability, because hydrologic needs can often be overwhelmed by political considerations. There are 263 rivers around the world that cross the boundaries of two or more nations, and an untold number of international groundwater aquifers. The basin areas that contribute to these rivers (**Figure 1**) comprise approximately 47% of the land surface of the Earth,

include 40% of the world's population, and contribute almost 60% of freshwater flow.

Within each international basin, demands from environmental, domestic, and economic users increase annually, while the amount of freshwater in the world remains roughly the same as it has been throughout history. Given the scope of the problems and limited resources available to address them, avoiding water conflict is vital because conflict is expensive, disruptive, and interferes with the efforts to relieve human suffering, reduce environmental degradation, and achieve economic growth.

A closer look at the world's international basins gives a greater sense of the magnitude of the issues: First, the problem is growing; there were 214 international basins listed in a 1978 United Nations study, the last time any official body attempted to delineate them, and there are 263 today. The growth is largely the result of the 'internationalization' of national basins through political changes, such as the break up of the Soviet Union and the Balkan states, as well as access to today's better mapping sources and technology. Even more striking than the total number of basins is a breakdown of each nation's land surface that falls within these watersheds. Twenty-one nations lie entirely within international basins; including these, a total of 33 countries have over 95% of their territory within these basins. These nations are not limited to smaller countries, such as Liechtenstein and Andorra, but include such sizable countries as Hungary, Bangladesh, Belarus, and Zambia. A final way to visualize the dilemmas posed by international water resources is to consider that 19 basins are shared by five or more riparian nations.

Disparities between riparian nations – whether in economic development, infrastructural capacity, or political orientation – add further complications to water resources development, institutions, and management. As a consequence, development, treaties, and institutions are regularly seen as, at best, inefficient, often ineffective, and, occasionally, as a new source of tensions themselves.

There is room for optimism, though, notably in the global community's record of resolving water-related disputes along international waterways. For example, the record of acute conflict over international water resources is overwhelmed by the record of cooperation. Despite the tensions inherent in the international setting, riparian countries have shown tremendous creativity in approaching regional development, often

---

[1] The Register of International River Basins of the World defines a 'river basin' as the area which contributes hydrologically (including both surface- and groundwater) to a first order stream, which, in turn, is defined by its outlet to the ocean or to a terminal (closed) lake or inland sea. Thus, 'river basin' is synonymous with what is referred to in the U.S. as a 'watershed' and in the UK as a 'catchment,' and includes lakes and shallow, unconfined groundwater units (confined or fossil groundwater is not included). We define such a basin as 'international' if any perennial tributary crosses the political boundaries of two or more nations.

Similarly, the 1997 UN Convention on Non-Navigational Uses of International Watercourses defines a 'watercourse' as 'a system of surface and underground waters constituting by virtue of their physical relationship a unitary whole and flowing into a common terminus.' An 'international watercourse' is a watercourse, parts of which are situated in different states [nations].

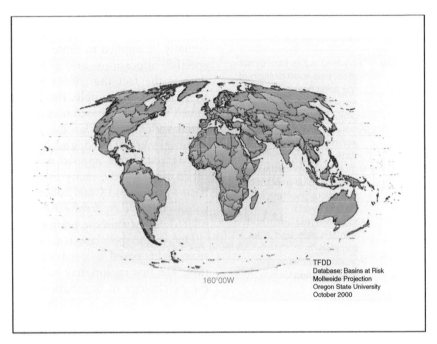

**Figure 1** International basins of the world.

through preventive diplomacy, and the creation of 'baskets of benefits' which allow for positive-sum, integrative allocations of joint gains. Vehement enemies around the world have negotiated water sharing agreements, and once cooperative water regimes are established through treaty, they turn out to be impressively resilient over time, even as conflict rages over other issues. Shared interests along a waterway seem to consistently outweigh the conflict-inducing characteristics of water.

Moreover, international organizations such as the UN International Law Commission, through its work on shared natural resources (see **Transboundary Aquifers Resources**), the International Law Association, and even the Government of Germany have hosted efforts to offer guidelines for the legal resolution of international water issues.

## Overcoming the Costs of Non-Cooperation: From Rights to Needs to Interests

International negotiations are often hamstrung because of entrenched and contradictory opening positions. Generally, parties base their initial positions in terms of rights – the sense that a riparian country is entitled to a certain allocation based on hydrography or chronology of use. Upstream riparian countries often invoke some variation of the Harmon Doctrine, claiming that water rights originate where the water falls. India claimed absolute *sovereignty* in the early

phases of negotiations over the Indus Waters Treaty, as did France in the Lac Lanoux case, and Palestine over the West Bank aquifer. Downstream riparian countries often claim absolute *integrity*, claiming rights to an undisturbed system or, if on an exotic stream, historic rights based on their history of use. Spain insisted on absolute *integrity* regarding the Lac Lanoux project, while Egypt claimed historic rights to the Nile, first against Sudan, and later against Ethiopia.

However, in almost all the disputes, particularly over arid or exotic streams which have been resolved, the paradigms used for negotiations were not 'rights-based' at all, but rather were 'needs-based.' 'Needs' are defined by irrigable land, population, or the requirements of a specific project.[2] [See **Table 1** – Examples of needs-based criteria.] In the agreements between Egypt and Sudan signed in 1929 and in 1959, for example, allocations were arrived at on the basis of local needs, primarily of agriculture. Egypt argued for a greater share of the Nile because of its larger population and extensive irrigation. In 1959, Sudan and Egypt then divided future water from development equally between them. Current allocations of 55.5 BCM/yr. for Egypt and 18.5 BCM/yr. for Sudan reflect these relative needs.

---

[2] Here we distinguish between 'rights' in terms of a sense of entitlement, and legal rights. Obviously, once negotiations lead to allocations, regardless of how they are determined, each riparian has legal 'rights' to that water, even if the allocations were determined by 'needs.' The point is that it is generally easier to come to a joint definition of 'needs' than it is of 'rights.'

**Table 1** Examples of needs-based criteria

| Treaty | Criteria for allocations |
| --- | --- |
| Egypt/Sudan (1929, 1959, Nile) | 'Acquired' rights from existing uses, plus even division of any additional water resulting from development projects |
| Johnston Accord (1956, Jordan) | Amount of irrigable land within the watershed in each state |
| India/Pakistan (1960, Indus) | Historic and planned use (for Pakistan) plus geographic allocations (western vs. eastern rivers) |
| South Africa (Southwest Africa)/Portugal (Angola) (1969, Kunene) | Allocations for human and animal needs, and initial irrigation |
| Israel-Palestinian Interim Agreement (1995, shared aquifers) | Population patterns and irrigation needs |

Likewise, the Johnston Accord emphasized the needs rather than the inherent rights of each of the riparian countries in the Jordan River basin and was the only water agreement ever negotiated (although not ratified) for the basin until very recently. Johnston's approach, based on a report performed under the direction of the Tennessee Valley Authority, was to estimate, without regard to political boundaries, the water needs of all the irrigable land within the Jordan Valley basin which could be irrigated by gravity flow. This was not only an acceptable formula to the parties at the time, but it also allowed for a breakthrough in negotiations when a land survey of Jordan concluded that its future water needs were lower than previously estimated. Years later, Israel and Palestine came back to the needs in the Interim Agreement of 1995, where Israel first recognized Palestinian water rights to the West Bank; a formula for agriculture and per capita consumption determined future Palestinian water needs at 70–80 MCM/yr and Israel agreed to provide 28.6 MCM/yr towards those needs.

Outside of the Middle East, needs are the prevalent criteria for allocations along arid and exotic streams as well. Allocations of the Rio Grande/Rio Bravo and the Colorado between Mexico and the USA are based on the Mexican irrigation requirements. Similarly, a 1975 Mekong River agreement among the four lower riparian states of Laos, Vietnam, Cambodia, and Vietnam defined 'equality of right' not as equal shares of water, but as equal rights to use water on the basis of each riparian country's economic and social needs.[3]

---

[3] In the context of navigation, the 1995 Mekong River agreement, which superceded the 1975 agreement, again referenced, but in this case did not define, the concept of 'equality of right.'

Interestingly, once the need-based allocations are determined, it is *not* generally required that water actually be applied to those needs, and furthermore, specific allocations are generally *not* readjusted, despite the fact that needs may change drastically over time. For example, the Johnston Accord determined allocations based on potential gravity-fed irrigated agriculture *within* the Jordan basin. Once the numbers were derived, and Jordan and Israel implicitly agreed, Israel applied most of its allocation to other uses entirely, many of them being outside the basin. Jordan and Israel adhere to the Johnston allocations to this day, despite dramatic changes to water-related uses within the basin over the last 50 years.

One might speculate as to why negotiations move from rights-based to needs-based criteria for allocation. The first reason may have something to do with the psychology of negotiations. Some workers point out that negotiation ideally moves through three stages: the adversarial stage, where each side defines its positions, or rights; the reflexive stage, where the needs of each side are addressed; and finally, the integrative stage, where negotiators brainstorm together to address each side's underlying interests. The negotiations here seem to follow this pattern from rights to needs, and occasionally, to interests. While the negotiators may initially identify the rights of the people in their own country as paramount, eventually one seems to empathize to some degree, noticing that even one's enemy requires the same amount of water for the same use with the same methods as oneself.

The second reason for the shift from rights to needs may simply be that rights are not quantifiable whereas needs are. If two nations insist on their respective rights to upstream versus down, for example, there is no spectrum along which to bargain, and no common frame of reference. One can determine a needs-based criterion – irrigable land or population, for example – and quantify the needs of each nation much more readily Even with differing interpretations, once both sides feel comfortable that their minimum quantitative needs are being met, talks eventually turn to straightforward bargaining over numbers within a common spectrum.

## From Rights and Needs to Interests: 'Baskets of Benefits'

Traditionally, co-riparian countries have focused on water as a commodity to be divided – a zero-sum, rights-based approach. Precedents now exist for determining formulas that equitably allocate the *benefits* derived from water, not the water itself – a positive-sum, integrative approach. For example, as part of the 1961 Columbia River Treaty, the United

States paid Canada for the benefits of flood control and Canada was granted rights to divert water between the Columbia and Kootenai for hydropower purposes. The result is an arrangement by which power may be exported out of the basin for gain, but the water itself may not be. Likewise, the relative nature of 'beneficial' uses is exhibited in a 1950 agreement on the Niagara, flowing between the USA and Canada, which provides for a greater flow of the famous falls during 'show times' of summer daylight hours, when tourist dollars are worth more per cubic meter than the alternative use in hydropower generation.

In many water-related treaties, water issues are dealt with alone, separate from any other political or resource issues between countries. By separating the two realms of 'high' (political) and 'low' (resource economical) politics, or by ignoring other resources which might be included in an agreement, some have argued that the process is either likely to fail, as in the case of the 1955 Johnston accords on the Jordan, or likely to more often achieve a suboptimum development arrangement, as is currently the case on the Indus agreement, signed in 1960. However, linkages are being made increasingly between water and politics, and between water and other resources. These multi-resource linkages may offer more opportunities for creative solutions to be generated, allowing for greater economic efficiency through a 'basket' of benefits.

## Institutional Development—Contributions from the International Community[4]

The international community has long advocated the development of cooperative water management institutions for the world's international waterways, and has focused considerable attention in the 20th century on developing and refining principles of shared management. In 1911, the Institute of International Law published the Madrid Declaration on the International Regulation regarding the Use of International Watercourses for Purposes other than Navigation. The Madrid Declaration outlined certain basic principles of shared water management, recommending that coriparian states establish permanent joint commissions and discouraging unilateral basin alterations and harmful modifications of international rivers. Expanding on these guidelines, the International Law Association developed the Helsinki Rules of 1966 on the Uses of Waters of International Rivers. Since then, international freshwater law has matured through the work of these two organizations, as well as the United Nations and other governmental and nongovernmental bodies.

However, the past decade has witnessed perhaps an unprecedented number of declarations as well as organizational and legal developments toward the international community's objective of promoting cooperative river basin management. The decade began with the International Conference on Water and the Environment in the lead-up to the 1992 UN Conference on Environment and Development (UNCED) in Rio.[5] Subsequent actions taken by the international community have included the pronouncement of nonbinding conventions and declarations, the creation of global water institutions, and the codification of international water principles. While clearly more work is required, these initiatives have not only raised awareness of the myriad issues related to international water resource management, but have also led to the creation of frameworks in which the issues can be addressed.

## Conventions, Declarations, and Organizational Developments

Although management of the world's water resources was only one of several topics addressed by the 1992 UNCED forum, it was, however, the primary focus of the International Conference on Water and the Environment (ICWE). The ICWE participants, representing governmental and nongovernmental organizations, developed a set of policy recommendations outlined in the Conference's Dublin Statement in 2002 on Water and Sustainable Development, which the drafters entrusted to the world leaders gathering in Rio for translation into a plan of action. While covering a range of water resource management issues, the Dublin Statement specifically highlights the growing importance of international transboundary water management and encourages greater attention to the creation and implementation of integrated water management institutions endorsed by all affected basin states. Moreover, the drafters of the Dublin Statement outlined certain essential functions of international water institutions including "reconciling and harmonizing the interests of riparian countries, monitoring water quantity and quality, development of concerted action programmes, exchange of information, and enforcing agreements."

At the Rio Conference, water resource management was specifically addressed in Chapter 18 of Agenda 21, a nonbinding action plan for improving the state of the globe's natural resources in the 21st century adopted by UNCED participants. The overall objective of Chapter 18 is to ensure that the supply

---

[4] This section draws from Giordano and Wolf 2003.

[5] The UN Conference on Environment and Development is often referred to as the Rio Earth Summit.

and quality of water is sufficient to meet both human and ecological needs worldwide, and measures to implement this objective are detailed in the Chapter's ambitious, seven-part action plan.

One result of the Rio Conference and Agenda 21 has been an expansion of international freshwater resource institutions and programs. The World Water Council, a self-described 'think tank' for world water resource issues, for example, was created in 1996. Since its inception, the World Water Council has hosted three World Water Forums – gatherings of government, nongovernment, and private agency representatives to discuss and determine collectively a vision for the management of water resources over the next quarter century.[6] These forums have led to the creation of the World Water Vision, a forward-looking declaration of philosophical and institutional water management needs, as well as the creation of coordinating and implementing agencies such as the World Commission on Water for the 21st Century and the Global Water Partnership. The Second World Water Forum also served as the venue for a Ministerial Conference in which the leaders of participating countries signed a declaration concerning water security in the 21st century. The recent World Summit on Sustainable Development (WWSD) has helped to sustain the momentum of these recent global water initiatives. In the Johannesburg Declaration on Sustainable Development, delegates at the WWSD reaffirmed a commitment to the principles contained in Agenda 21 and called upon the United Nations to review, evaluate, and promote further implementation of this global action plan (United Nations, 2002a).

While many of the strategies in Agenda 21 and subsequent statements are directed primarily at national water resources, their relevance extends to international transboundary waters. In fact, the Ministerial Declaration at the Second World Water Forum included 'sharing of water' (between different users and states) as one of its seven major challenges to achieving water security in the 21st century. The other six challenges, which include meeting basic needs, securing food supply, protecting ecosystems, managing risks, valuing water, and governing water wisely, can be translated to an international setting. Furthermore, policy measures prescribed by the international community to build greater institutional capacity, such as integrated water resource management, expanded stakeholder participation, and improved monitoring and evaluation schemes, are likewise important components of international watercourse management. Nonetheless, while many of the principles of national water management apply to international waters, the political, social, and economic dynamics associated with waters shared between sovereign states require special consideration.

## States can Require Special Consideration

Nevertheless, after decades of institutional risk-aversion and a general lack of leadership in international waters, the 1990s and 2000s are turning out to be a period of tremendous momentum on the ground as well: the World Bank and UNDP have collaborated to facilitate the Nile Basin Initiative, which looks close to establishing a treaty framework and development plan for the basin, and the Bank is taking the lead in bringing the riparian countries of the Guarani Aquifer in Latin America to dialogue. The US State Department, a number of UN agencies, and other parties have established a Global Alliance on Water Security, aimed at identifying the priority regions for assistance, which may help countries get ahead of the crisis curve. The EU has developed its Water Framework Directive, which includes specific guidelines for transboundary water management to the EU and its Accession States. The UNECE has programs on 10 European and Central Asian basins, and supports the International Water Assessment Center. The Global Environment Facility (GEF) is now active in 55 international basins. The Southern African Development Community and the Economic And Social Commision for Asia and the Pacific have been taking the lead in holding dialogues on transboundary issues within their respective regions. The International Network of Basin Organizations (INBO) has created a thriving network of those managing international and transboundary rivers, including a 'twinning' program that brings together diverse basin managers to share experiences and best practices. The International Water Academy has engaged researchers from around the world to address these difficult issues, as has the Universities Partnership for Transboundary Waters. And UNESCO and Green Cross International have teamed up for a broad-based, multiyear project called, "From Potential Conflict to Cooperation Potential," working also with the Organization for Security and Cooperation in Europe on their project on international waters (United Nations Educational, Scientific and Cultural Organization—UNESCO-PCCP, 2007). Moreover, UNESCO is taking the lead in helping to develop a global 'Water Cooperation Facility,' to help prevent and resolve the world's water disputes.

## Legal Principles

The UN Convention on the Law of the Non-navigational Uses of International Watercourses (UN Convention), adopted in 1997 by the UN General

---

[6] The Fourth World Water Forum will take place in March 2006 in Mexico City.

Assembly, is one post-Rio accomplishment that specifically focuses on international transboundary water resources.[7] The UN Convention codifies many of the principles deemed essential by the international community for the management of shared water resources, such as equitable and reasonable utilization of waters with specific attention to vital human needs; protection of the aquatic environment; and the promotion of cooperative management mechanisms. The document also incorporates provisions concerning data and information exchange and mechanisms for conflict resolution. Once ratified, the UN Convention will provide a legally binding framework to its signatories for managing international watercourses.

However, the UN's approval of the Convention does not entirely resolve many legal questions concerning the management of internationally shared waters. To date, five years after its adoption by the UN General Assembly, only 14 countries are party to the UN Convention, well below the requisite 35 instruments of ratification, acceptance, accession, or approval needed to bring the Convention into force (United Nations, 2002b).[8] Additionally, international law only guides conduct *between* sovereign nations. Thus, grievances of political units or ethnic groups *within* nations over the domestic management of international waterways would not be addressed. Another problem, in the words of Biswas in 1999, is that the "vague, broad, and general terms" incorporated in the UN Convention "can be defined, and in certain cases quantified, in a variety of different ways..." leading to potentially varied and conflictive interpretations of the principles contained therein. Moreover, there is no practical enforcement mechanism to back up the Convention's guidance. The International Court of Justice, for example, hears cases only with the consent of the parties involved and only on very specific legal points. In its 55-year history, the Court has decided only one case, apart from those related to boundary definitional disputes, pertinent to international waters – that of the Gabçikovo-Nagymaros Project on the Danube between Hungary and Slovakia in 1997.[9] Finally, the Convention only addresses those groundwater bodies that are connected to surface water systems – i.e., unconfined aquifers, yet several nations have already begun to tap into confined groundwater systems, many of which are

shared across international boundaries. Nevertheless, and despite the fact that the process of ratification is moving extremely slowly, the Convention's common acceptance, and the fact that the International Court of Justice referred to it in its decision on the 1997 case on the Gabçikovo Dam, gives the Convention increasing standing as an instrument of customary law.

## Institutional Developments in Basin-level Transboundary Water Management

The history of international water treaties dates as far back as 2500 BC, when the two Sumerian city-states of Lagash and Umma crafted an agreement ending a water dispute over the Tigris River, bringing an end to the first and only 'water war' in history. Since then, a large body of water treaties has emerged. The Food and Agricultural Organization of the United Nations has identified more than 3600 treaties dating from AD 805 to 1984. While the majority of these relate to some aspect of navigation, a growing number address non-navigational issues of water management, including flood control, hydropower projects, or allocations for consumptive or nonconsumptive uses in international basins.

At least 54 new bilateral and multilateral water agreements have been concluded since the Rio Conference, representing basins in Asia, Africa, Europe, North America, and South America. As in the past 50 years as a whole, European water accords continue to dominate; however, agreements from Asia in particular, have grown disproportionately.[10]

In addition to greater geographic representation, a number of improvements can be seen in this more recent set of treaties compared with the last half-century as a whole. First, a growing percentage of treaties address some aspect of water quality, a finding consistent with Rio's obective of both managing and protecting freshwater resources. Second, a number of agreements establish joint water commissions with decision-making and/or enforcement powers, a significant departure from the traditional advisory standing of basin commissions. Third, country participation in basin-level accords appears to be expanding. Although few of the agreements incorporate all basin states, a greater proportion of treaties are multilateral and many incorporate all major hydraulic contributors. Finally, although the exception, a 1998 agreement on the Syr Darya Basin, in which water management is exchanged for fossil fuels, provides a

---

[7] UN General Assembly document A/RES/51/229 of 8 July 1997.

[8] As of January 2006, Finland, Hungary, Iraq, Jordan, Lebanon, Libya, Namibia, The Netherlands, Norway, Portugal, Qatar, South Africa, Sweden, and Syria were party to the Convention.

[9] The ICJ was established in 1946 with the dissolution of its predecessor agency, the Permanent Court of International Justice. This earlier body did rule on four international water disputes during its existence from 1922–1946.

[10] The fact that agreements representing European basins dominate the treaty record is not surprising given that Europe has the largest number of international basins (69) followed by Africa (59), Asia (57), North America (40), and South America (38).

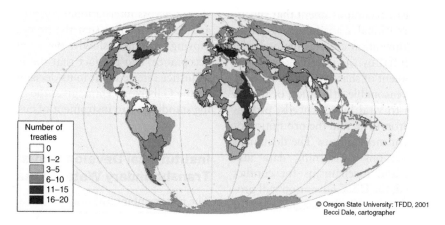

© Oregon State University: TFDD, 2001
Becci Dale, cartographer

**Figure 2** Number of agreements per international river basin.

post-Rio example of basin states broadly capitalizing on their shared resource interests.

While a review of the past century's water agreements highlights a number of positive developments, institutional vulnerabilities remain. Water allocations, for example, the most conflictive issue area between co-riparian states, are seldom clearly delineated in water accords. Moreover, in the treaties that do specify quantities, allocations are often in fixed amounts, thus ignoring hydrologic variation and changing values and needs.[11] Formal

management institutions have been established in only 106 of the 263 international basins (see **Figure 2**), and even within these, few include all nations riparian to the affected basins, which precludes the integrated basin management advocated by the international community. Moreover, treaties with substantive references to water quality management, monitoring and evaluation, and conflict resolution remain in the minority. Enforcement measures and public participation, two elements that can greatly enhance the resiliency of institutions, are also largely overlooked.[12] As a result of these circumstances, most existing international water agreements continue to lack the tools necessary to promote long-term, holistic water management.

Notably, the important hydrologic link between groundwater and surface water is recognized but understood only at a reconnaissance level even in the most studied basins in the world.[13] Several experts underscore that current international law does not adequately define groundwater, much less the spatial flow of groundwater. The transboundary movement or 'silent trade' of hazardous wastes into Lebanon

---

[11] The treaty record is replete with agreements which do not allow for the vagaries of nature and the scientific unknown, misunderstandings which often lead to tense political standoffs:

The waters of the Colorado were already overallocated between the upper and lower US states when a treaty with Mexico was signed in 1944, which also neglected the entire issue of water quality. After legal posturing on both sides as water quality continued to degrade, the US subsequently built a massive desalination plant at the border so that the water delivered would at least be usable. Currently, the fact that shared groundwater is likewise not covered in the treaty is leading to its share of tensions between the two nations.

In December, 1996, a treaty between India and Bangladesh was finally signed, allocating their shared Ganges waters after more than 35 years of dispute. In April 1997, however – the very first season following signing of the treaty – the two countries were involved in their first conflict over cross-boundary flow: water passing through the Farakka dam dropped below the minimum provided in the treaty, prompting Bangladesh to insist on a full review of the state of the watershed.

In 1994, Israel and Jordan signed one of the most creative water treaties on record. It has Jordan store winter runoff in the only major surface reservoir in the region – the Sea of Galilee – even though that lake happens to be in Israel; it has Israel lease from Jordan in 50 year increments wells and agricultural land on which it has come to rely; and it created a Joint Water Committee to manage the shared resources. But it did not adequately describe what would happen to the prescribed allocations in a drought. In early 1999, this excluded issue roared into prominence with a vengeance, as the worst drought on record caused Israel to threaten to renege on its delivery schedule, which in turn caused protests in the streets of Amman, personal outrage on the part of the King of Jordan, and, according to some, threatened the very stability of peace between the two nations before a resolution was found.

[12] A consensus is generally emerging that regional agreements, while proliferating, have *less* impact than bilateral agreements, precisely because they are unenforceable guidelines rather than detailed agreements. Likewise, bilateral agreements are, in general, *easier* to negotiate than multilateral agreements, simply because of the truism that, "the more people (or interests) in the room, the more difficult it is for them to agree (or the less the final document will say). Oftentimes, however, even multilateral basins are effectively managed through sets of bilateral agreements. The Jordan comes to mind, where agreements exist between Syria–Jordan, Jordan–Israel, and Israel–Palestine, and, while no multilateral agreement has regional oversight, the basin is managed relatively effectively.

[13] For a summary of international groundwater issues, see, Jarvis T, Giordano M, Puri S, Matsumoto K, and Wolf A. "International Borders, Ground Water Flow, and Hydroschizophrenia." Ground Water. Vol. 43 #5, Sept.-Oct. 2005., from where this paragraph is drawn.

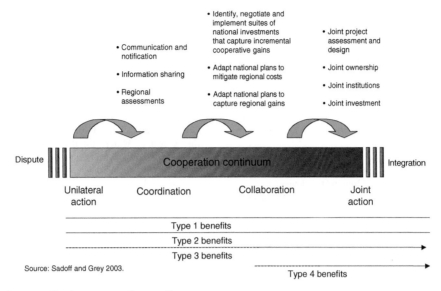

**Figure 3** Types of cooperation in a cooperation continuum.

provides an example of the need to increase 'global harmonization' of international water and waste treaties. Therefore, while the effects of groundwater use may be contained within national boundaries, the water laws of few states or provinces address groundwater management due to the 'invisible' nature of the resource, or the technical challenges in predicting spatial and temporal changes in the groundwater system with increased use. Of the nearly 400 treaties inventoried in the Transboundary Freshwater Dispute Database (TFDD) and UNEP, only 62 treaties have recognized groundwater; groundwater quality has been addressed only in the past few years.

## Types of Institutional Arrangements

An agreement or institution may be thought of as a sociopolitical analogue to a vibrant ecosystem, and thus vulnerable to the same categories of stressors that threaten ecosystem sustainability. In this regard, water management treaties and institutions must sustain resilience despite the following types of stressors:

- Biophysical stressors: Are there mechanisms to account for droughts and floods, or shifts in the climate or river course?
- Geopolitical stressors: Will the agreement or institution survive dramatic changes in government, both internal and international?
- Socioeconomic stressors: Is there public support? Is there a stable funding mechanism? Will the agreement or institution survive changing societal values and norms?

Similar to an ecosystem, the best management is *adaptive* management, i.e., the institution has mechanisms to adapt to changes and stresses, and to mitigate their impact on its sustainability.[14]

Crafting institutions requires a balance between the efficiency of integrated management and the sovereignty-protection of national interests. Along with greater integration of scope and authority may come greater efficiency, but also comes greater potential for disagreements, greater infringement on sovereignty, and greater transaction costs for more information. Some possible institutional models are offered in **Figure 3**. Nevertheless, for every set of political relations, there is some possible institutional arrangement which will be acceptable and, if its management is iterative and adaptive, responsibility can be regularly 're-crafted' to adapt or even lead political relations.

## Further Reading

Biswas AK (1999) Management of International Waters. *Water Resources Development* 15: 429–441.

Dublin Statement (1992) *International Conference on Water and Environment: Development Issues for the 21st Century, 26–31 January 1992, Dublin*. Geneva: World Meteorological Organization.

Feitelson E and Haddad M (1998) *Identification of Joint Management Structures for Shared Aquifers*. Washington, DC: World Bank, Technical Paper #415.

Giordano MA and Wolf AT (2003) Sharing Waters: Post-Rio International Water Management. *Natural Resources Forum* 27: 163–171.

---

[14] See Lee (1995) for the classic text on adaptive management.

Jarvis T, Giordano M, Puri S, Matsumoto K, and Wolf A (2005) International Borders, Ground Water Flow, and Hydroschizophrenia. *Ground Water.* 43(5): 764–770.

Jurdi M and Ibrahim S (2003) Differential Water Quality in Confined and Free-flowing Water Bodies, Lebanon. *International Journal of Environment and Pollution* 19(3): 271–291.

Lee KN (1995) *Compass and Gyroscope: Integrating Science and Politics for the Environment.* Washington, DC: Island Press.

Main, Chas T, Inc. (1953) *The Unified Development of the Water Resources of the Jordan Valley Region.* Knoxville: Tennessee Valley Authority.

Matsumoto K (2002) *Transboundary Groundwater and International Law: Past Practices and Current Implications.* Corvallis, OR: Unpublished Master's thesis, Oregon State University.

Puri S, Appelgren B, Arnold G, Aureli A, Burchi S, Burke J, Margat J, Pallas P, and von Igel W (2001) *Internationally Shared (transboundary) Aquifer Resources Management, their Significance and Sustainable Management: a framework document.* IHP-VI, International Hydrological Programme, Non Serial Publications in Hydrology SC-2001/WS/40. Paris, France: UNESCO.

Puri S (2003) Transboundary Aquifer Resources: International Water Law and Hydrogeological Uncertainty. *Water International* 28(2): 276–279.

Rothman J (1995) Pre-negotiation in Water Disputes: Where Culture is Core. *Cultural Survival Quarterly* 19(3): 19–22.

UN Environment Programme and Oregon State University (2002) *Atlas of International Freshwater Agreements.* Nairobi: UNEP Press.

United Nations (2002a) *Report of the World Summit on Sustainable Development.* Document A/CONF.199/20, Sales No. E.03.II.A.1. New York, United Nations.

United Nations (2002b) *Report of the World Summit on Sustainable Development.* E.03.II.A.1. United Nations.

Varady RG and Iles-Shih M (2005) Global Water Initiatives: What Do the Experts Think? Report on a Survey of Leading Figures in the 'World of Water.' In: Biswas AK (ed.) *Impacts of Mega-Conferences on Global Water Development and Management,* pp. 1–58, Springer Verlag.

Waterbury J (1979) *Hydropolitics of the Nile Valley.* New York: Syracuse University Press.

Wolf AT (1998) Conflict and Cooperation along International Waterways. *Water Policy* 1(2): 251–265.

Wolf AT (1999) Criteria for Equitable Allocations: The Heart of International Water Conflict. *Natural Resources Forum* 23(1): 3–30.

Wolf AT (1999) The Transboundary Freshwater Dispute Database Project. *Water International* 24(2): 160–163.

Wolf AT (2002) *Conflict and Cooperation: Survey of the Past and Reflections for the Future.* Geneva: Green Cross International.

Wolf AT (ed.) (2002) Conflict Prevention and Resolution in Water Systems. Cheltenham, UK: Elgar.

# Subject Index

Notes

**Cross-reference** terms in italics are general cross-references, or refer to subentry terms within the main entry (the main entry is not repeated to save space). Readers are also advised to refer to the end of each article for additional cross-references - not all of these cross-references have been included in the index cross-references.

The index is arranged in set-out style with a maximum of three levels of heading. Major discussion of a subject is indicated by bold page numbers. Page numbers suffixed by T and F refer to Tables and Figures respectively. vs. indicates a comparison.

This index is in **letter-by-letter** order, whereby hyphens and spaces within index headings are ignored in the alphabetization. For example, acid rain is alphabetized after acidity, not after acid(s). Prefixes and terms in parentheses are excluded from the initial alphabetization.

Where index subentries and sub-subentries pertaining to a subject have the same page number, they have been listed to indicate the comprehensiveness of the text.

**Abbreviations**

DCAA - dissolved combined amino acids

DFAA - dissolved free amino acids

DIN - dissolved inorganic nitrogen

DOC - dissolved organic carbon

DOM - dissolved organic matter

DON - dissolved organic nitrogen

TDS - total dissolved solids

TSS - total suspended solids